Principles of Molecular Regulation

Principles of Molecular Regulation

Edited by

P. Michael Conn

*Oregon Regional Primate Research Center,
Beaverton, and Oregon Health Sciences University,
Portland, OR*

and

Anthony R. Means

*Department of Pharmacology and Cancer Biology,
Duke University Medical Center, Durham, NC*

Foreword by
Bert W. O'Malley

© 2000 Humana Press Inc.
999 Riverview Drive, Suite 208
Totowa, New Jersey 07512

For additional copies, pricing for bulk purchases, and/or information about other Humana titles, contact Humana at the above address or at any of the following numbers: Tel.: 973-256-1699; Fax: 973-256-8341; E-mail:humana@humanapr.com; Website: http://humanapress.com

All rights reserved.

No part of this book may be reproduced, stored in a retrieval system, or transmitted in any form or by any means, electronic, mechanical, photocopying, microfilming, recording, or otherwise without written permission from the Publisher.

All articles, comments, opinions, conclusions, or recommendations are those of the author(s), and do not necessarily reflect the views of the publisher

Cover Illustration: The cover depicts a montage of tyrosine hydroxylase immunoreactive neurons labeled with fluorescent dye CY2 in the hypothalamic arcuate nucleus and their axon terminals in the median eminence.

Cover figure courtesy of Dr. Anna Lerant, Department of Anatomy, University of Mississippi Medical Center, Jackson, MS.

Cover design by Patricia F. Cleary.

This publication is printed on acid-free paper. ∞
ANSI Z39.48-1984 (American National Standards Institute) Permanence of Paper for Printed Library Materials.

Photocopy Authorization Policy:
Authorization to photocopy items for internal or personal use, or the internal or personal use of specific clients, is granted by Humana Press Inc., provided that the base fee of US $10.00 per copy, plus US $00.25 per page, is paid directly to the Copyright Clearance Center at 222 Rosewood Drive, Danvers, MA 01923. For those organizations that have been granted a photocopy license from the CCC, a separate system of payment has been arranged and is acceptable to Humana Press Inc. The fee code for users of the Transactional Reporting Service is: [0-89603-630-8/00 $10.00 + $00.25].

Printed in the United States of America. 10 9 8 7 6 5 4 3 2 1

Library of Congress Cataloging-in-Publication Data

Principles of molecular regulation / edited by P. Michael Conn and Anthony R. Means
 p.cm.
  Includes bibliographical references and index.
  ISBN 0-89603-630-8 (alk. paper)
  1. Molecualr endocrinology. 2. Cellular signal transduction. 3. Nuclear receptors
(Biochemistry I. Conn, P. Michael. II. Means, Anthony R.
  QP187.3M64 P75 2000
  573.4'4—dc21

00-027312

FOREWORD

THE STATE OF HORMONE ACTION AND MOLECULAR REGULATION

Molecular Endocrinology and Regulation is a burgeoning field, having experienced a remarkable period of growth since the late 1960s. At that time, there was no field of Hormone Action. The prevailing view of how hormones worked ranged from effects on membrane transport of nutrients and precursors for RNA and protein synthesis, to effects on the translation of mRNA at the level of ribosomes. There was, however, a small cadre of voices that predicted a possible nuclear action on mRNA synthesis. The first such paper was presented at the national endocrine meetings in 1967 and dealt with hormonal stimulation of oviductal protein synthesis. It was about this time that a small group of scientists interested in hormone effects in cells attended a Gordon Conference in New Hampshire organized by Jim Florini; this meeting represented one of the first conferences to focus an entire program on hormone action and mechanisms. The attendees were primarily involved in aspects of steroid hormone and thyroid hormone actions; peptide hormone action was yet to experience its birth and neonatal growth. Only a short time previously Elwood Jensen had discovered the estrogen binding protein that eventually became the "estrogen receptor"; it was logical that the conference dealt mainly with steroid receptors plus a few papers presenting data that steroid hormones could induce specific enzyme/protein synthesis in target cells. The mechanisms of these effects were the subject of great debate at this first conference on Steroid Hormone Action, a meeting that persists to this day in New England each summer.

Following the monumental discovery of peptide immunoassays, most of the peptide field was immersed in measuring their hormones, ranging from insulin to LH, FSH, TRH, GH, etc. For over a decade, little attention was being given to the difficulty in measuring receptors for these and other membrane acting ligands. Nevertheless, the advent of this assay methodology, including the ability to synthesize radiolabeled peptide hormones, eventually allowed the identification and quantitation of cell-surface receptors. The time of this application was about 1970. Soon researchers noted that cAMP levels were induced in concert with ligand occupation of certain membrane receptors and this second messenger (cAMP) was postulated to initiate intracellular phosphorylation of unknown targets. At this point, the field of Peptide Hormone Action also came into being.

These fields developed together for much of the next decade. Hormone Action Conferences invariably contained talks on both types of receptors and progress was rapid and in concert with the explosion molecular biology. In the steroid field, progress was more rapid initially, but by the mid 1980s, the peptide field attained equal mechanistic status and the source of specialized conferences of its own.

Steroid hormone action investigators concentrated on first understanding the "pathway" of action for their hormones. Scientists looked for model systems showing large responses to steroids. Changes in enzyme levels in cultured cells was one approach. The intact chicken oviduct was another of the more notable in that regard because of the ability of sex steroids (estrogen, progesterone) to induce both large increases in certain egg-white proteins (i.e., ovalbumin, ovomucoid) and more modest and specific responses in others (i.e., avidin). Soon the mRNA for ovalbumin was shown to be induced by steroid hormones and induction was postulated to occur at the nuclear level. Viral proteins were shown to be induced by glucocorticoids next, and the field was off and running. The demonstrations that steroid receptors could bind DNA directly and the purification of the progesterone receptor led a maturing field that felt it was "hot stuff" because the pathway of steroid to intracellular receptor to DNA to mRNA to protein to function was now understood. They were unsuspecting of the many complexities yet to be uncovered.

The peptide action researchers may have been a bit envious at that time because they were initially "stuck" at the point of cAMP induction and protein kinase A activation. But, where did the pathway lead from there? This was a slightly depressing thought considering there could be hundreds of target proteins for phosphorylation—only made worse by the early descriptions of other kinases under regulation—and the realization that calcium, diacylglycerol, and a number of other novel molecules also appeared to be second messengers for select membrane acting hormones. The complexity of signaling pathways emanating from membrane receptors did nothing but increase with the discovery of numerous protein kinases that phosphorylated serine, threonine, and tyrosine, kinase-kinases, phosphatases, calcium regulators, and regulators of all of the phosphorylation intermediates themselves. The types of receptors that proliferated ranged from seven membrane (PKA), growth factors (TK, PKC), cytokine, and eventually even chemokine in nature. The appreciation of G-proteins (1960s and early 1970s) as upstream targets of the cAMP pathway was key to eventual solutions of the signaling cascade. The discovery of the ras/raf pathway for mediating the effects of mitogens, and the JAK/Stat pathway as mediator for cytokines and certain peptide hormone effects were also important milestones in unraveling signal transduction in eukaryotic cells. The realization that CREB and Stat proteins were regulatable transcription factors that eventually acted on DNA united the steroid–peptide field, in part, at the level of the nucleus. That order was made out of this putative chaos is a striking tribute to the intellectual prowess and perseverance of the workers in this field. Most important, the signaling pathways emanating from the membrane have brought new insights to pathologies such as cancer, and have led to an explosive development of new pharmaceutical stimulators and inhibitors with good promise for therapies of neuropsychiatric disorders, cancer, and other human diseases.

Over in the steroid action field, a brief pause occurred at the 1970–1980 transition, and then the steroid action field heated up again with the cloning of the steroid (first glucocorticoid and then all others within one year), thyroid, vitamin D, and retinoic receptors. Then a strange thing happened, investigators began to clone (by cross screening) numerous molecules that were similar to steroid receptors, but which were not associated with any ligand. The term orphan receptors was born and the deduction was made that the steroid receptors were part of a giant superfamily of nuclear receptor transcription factors, numbering over 50 at least. The availability of cloned cDNAs and other reagents allowed mutational analyses followed by reintroduction of receptors into the cell to monitor their effects on synthetic reporter genes; structure–function relationships proved the existence of receptor domains for transcriptional activation, nuclear translocation, and DNA binding. At a now frenetic pace, information on dimeric DNA binding and heterodimeric partners (RXR), receptor crosstalk with peptide pathways, ligand independent receptor activation, and receptor phosphorylation was accumulated. More definitive appreciation of the biology of classical and orphan receptors was accomplished by the emerging transgenic technologies and gene knockout strategies. The ability to screen for new ligands and for orphan receptors extended the range of hormones to lipids (PPAR) and other previously unsuspected metabolic regulators. The yearly stream of publications on the physiology of orphan receptors and their novel ligands continues to bring excitement and more expansion to the field. The mouth of big pharma watered at the possibilities of new drugs acting at the nuclear level. Finally, the structure–function studies began to pay off and basic scientists convinced the pharmaceutical industry that "designer specific" ligands were possible. This was based on the premise that intermediate forms of receptors, other than "on" and "off," were possible, depending upon which specific ligand occupied the receptor. The tamoxifen paradox (it acts in one tissue as an agonist and in another as an antagonist) provided encouragement for the generation of a successful search for SERMS (Selective Estrogen Receptor Modulators) that contain tissue and function-specific profiles.

Still, the field clamored for a greater molecular understanding of how the nuclear receptors worked at the level of DNA. The discovery of receptor-associated regulatory proteins provided this mechanism and changed the field of hormone action further. We moved from a situation where many believed that intracellular receptors themselves carried out the transcriptional regulation inherent to the actions of steroid/thyroid hormones and vitamins to an understanding that the receptor-associated "coregulators" are the primary mediators of this genetic response. We now know that the receptors act as low-specific transducers of hormone action by serving as a foundation or scaffold for recruitment of a host of other regulatory proteins. The coregulators can be divided loosely into two camps: the coactivators and corepressors. Taken together, these molecules mediate the two main tasks of receptors—stimulation and the repression of gene expression. The coactivators amplify the inductive influence of receptors, whereas the corepressors provide repression of gene transcription. The first functional coregulator for the nuclear receptor family was the coactivator SRC-1, followed by the corepressors NCoR and SMRT. Since SRC-1 was predicted by squelching and uncovered by in vitro technologies, a level of comfort was provided by the phenotype (partial resistance to hormones) displayed by SRC-1 knockout mice.

Although key to understanding the mechanism bywhich nuclear receptors modulate gene transcription, the discovery of coregulators has complicated, temporarily, the picture of how receptors specifically function. The diversity of coactivators and the fact that many of them are themselves enzymes (acetyl transferases, ubiquitin ligases, methyl transferases, kinases, etc.) capable of modifying chromatin or other local transcription factors, comprise an exciting discovery. Even a novel RNA (SRA) has been shown recently to selectively coactivate steroid receptors. The diversity of coactivators can explain the developmental, cell, and promoter context-specific expression of regulated genes and function within a two-step model for coactivators to: (1) enzymatically remodel chromatin locally at the target gene and provide access of general transcription factors to the promotor; and (2) touch and stabilize these promoter-bound GTFs for efficient RNA polymerase recruitment. Most important, perhaps, the discovery of coregulators may have led us finally to a point where we will not only understand the molecular nature of the active receptor complex bound to the HRE sequences at all target genes, but will allow us to explain tissue and gene specific transcription, the kinetics of transcriptional regulation, inherited resistances to hormone, the effects of hormones on the cell cycle and in tumor progression, and even individual sensitivities to hormones in humans. Together with the exciting recent crystal structures of nuclear receptors, this new molecular understanding will allow the refinement of pharmaceutical assays and screens, and likely, lead to new drugs that can target a subset of hormone's effects for pharmaceutical application—something long doubted possible by the pharmaceutical industry.

In *Principles of Molecular Regulation*, one has the opportunity to survey the state of membrane receptor initiated cell signaling and nuclear receptor-initiated gene regulation from the viewpoints of a selected cadre of leading investigators. I hope the reader can savor the current progress and excitement in this vibrant field of molecular endocrinology.

Bert W. O'Malley

PREFACE

The striking observations that humans and worms share about half their genome and that fundamental metabolic pathways are common over eons of evolution, provokes the inescapable conclusion that nature has built living organisms as modifications on a theme. It is overly simplistic, of course, to conclude that "to understand one is to understand all"—but, certain fundamentals must be accepted.

Principles of Molecular Regulation is designed to fill a void in the discipline: a basic, yet up-to-date text, to cover certain fundamental aspects of the living state. This book was written to accommodate individuals new to the area and offer fresh insights to those familiar with certain aspects of regulation, yet looking for a source that contains a broad base of information for reference purposes.

We expect that this volume will prove valuable for beginning graduate and medical students, fellows, residents, as well as those in clinical practice who wish to understand basic phenomena. Seasoned researchers may find it particularly useful as a shelf reference text.

Textual material is supported by clear and comprehensive tables and figures, which should make *Principles of Molecular Regulation* a valuable learning and teaching tool.

The authors were selected as individuals who are recognized in their field and who are excellent writers. We thank the authors and publishers for adhering to a tight schedule needed to produce an up-to-date work.

P. Michael Conn
Anthony R. Means

CONTENTS

Foreword v
Preface ix
Contributors xiii

Part I. Signaling Mechanisms Initiated by Cell Surface Receptors

1. G Protein-Coupled Receptors and G Proteins 3
   Alfredo Ulloa-Aguirre and P. Michael Conn

2. G Proteins Large and Small Guanine Nucleotide Binding Proteins:
   Cellular Comunication Networks 27
   Thomas M. Wilkie and Michael A. White

3. The Insulin Receptor Kinase: *Regulation and Therapeutic Significance* 41
   Barry I. Posner

4. Growth Hormone Action: *Signaling via a JAK/STAT-Coupled Receptor* 55
   David J. Waxman and Stuart J. Frank

5. Regulation of *Drosophila* Visual Transduction Through a Supramolecular
   Signaling Complex 85
   Craig Montell

6. *Dictyostelium*: *A Model Experimental System for Elucidating
   the Pathways and Mechanisms Controlling Chemotaxis* 99
   Chang Y. Chung and Richard A. Firtel

7. gp130-Related Cytokines 115
   Christoph J. Auernhammer and Shlomo Melmed

Part II. Signaling Mechanisms Mediated by Ion Channels, Calcium, and Lipids

8. Ion Channels 135
   Deborah J. Nelson and Harry A. Fozzard

9. Calcium Ions as Intracellular Messengers: *Pathways and Actions* 149
   **Stanko S. Stojilkovic with Melanija Tomić, Taka-aki Koshimizu,
   and Fredrick Van Goor**

10. Calcium and Calmodulin-Mediated Regulatory Mechanisms 187
    Anthony R. Means

11. Protein Kinase C 205
    Alexandra C. Newton

12. Nitric Oxide 219
    Randy Krainock and Sean Murphy

13. Phospholipases 229
    J. H. Exton

Part III. Cyclic AMP, Protein Kinases and Protein Phosphatases

14 Adenylyl Cyclases 249
Lutz Birnbaumer

15 Cyclic Nucleotide Phosphodiesterases 261
Marco Conti

16 Cyclic Nucleotide-Dependent Protein Kinases 277
Sharron H. Francis and Jackie D. Corbin

17 Protein Serine/Threonine Kinases 297
*Jörg Heierhorst, Richard Pearson, James Horne,
Steven Bozinovski, Bostjan Kobe, and Bruce E. Kemp*

18 Protein Serine/Threonine Phosphatases 311
Shirish Shenolikar

19 Protein Tyrosine Phosphatases 323
Cynthia V. Stauffacher and Harry Charbonneau

Part IV. Signaling Mechanisms Initiated by Nuclear Receptors

20 The Mechanism of Action of Steroid Hormone Receptors 351
Donald P. McDonnell

21 Orphan Nuclear Receptors 363
Deepak S. Lala and Richard A. Heyman

22 Coactivators and Corepressors 385
Neil J. McKenna, Zafar Nawaz, Sophia Y. Tsai, and Ming-Jer Tsai

23 CREB Binding Protein–Coactivator Complexes 395
Riki Kurokawa and Christopher K. Glass

Part V. Molecular Regulation of Cell Proliferation and Death

24 Cell Cycle Checkpoints 407
Sally Kornbluth

25 Molecular Regulation of Apoptosis 415
Rosemary B. Evans-Storms and John A. Cidlowski

26 T Cells and Immunosuppression 429
Andrew W. Taylor

Part VI. Rational Drug Discovery

27 Rational Drug Discovery and the Impact of New, Advanced Technologies ... 439
Jonathan Greer

Index 461

CONTRIBUTORS

CHRISTOPH J. AUERNHAMMER, MD • *Department of Academic Affairs, Cedars-Sinai Research Institute, Los Angeles, CA*

LUTZ BIRNBAUMER, PHD • *Departments of Molecular, Cell, and Developmental Biology, Biological Chemistry, and Anesthesiology, University of California, Los Angeles, CA*

STEVEN BOZINOVSKI, PHD • *St. Vincent's Institute of Medical Research, Fitzroy, VIC, Australia*

HARRY CHARBONNEAU, PHD • *Department of Biochemistry, Purdue University, West Lafayette, IN*

CHANG Y. CHUNG, PHD • *Department of Biology, Center for Molecular Genetics, University of California at San Diego, La Jolla, CA*

JOHN A. CIDLOWSKI, PHD • *Laboratory of Signal Transduction, National Institute of Environmental Health Sciences, National Institutes of Health, Research Triangle Park, NC*

MARCO CONTI, MD • *Division of Reproductive Biology, Department of Obstetrics and Gynecology, Stanford University Medical School, Stanford, CA*

P. MICHAEL CONN, PHD • *Oregon Regional Primate Research Center, Beaverton, and Oregon Health Sciences University, Portland, OR*

JACKIE D. CORBIN, PHD • *Department of Molecular Physiology and Biophysics, Vanderbilt University School of Medicine, Nashville, TN*

ROSEMARY B. EVANS-STORMS, PHD • *Laboratory of Signal Transduction, National Institute of Environmental Health Sciences, National Institutes of Health, Research Triangle Park, NC*

J. H. EXTON, MD, PHD • *Department of Molecular Physiology and Biophysics and Howard Hughes Medical Institute, Vanderbilt University Medical Center, Nashville, TN*

RICHARD A. FIRTEL, PHD • *Department of Biology, Center for Molecular Genetics, University of California at San Diego, La Jolla, CA*

HARRY A. FOZZARD, MD • *Department of Neurobiology, Pharmacology, and Physiology, University of Chicago, Chicago, IL*

SHARRON H. FRANCIS, PHD • *Department of Molecular Physiology and Biophysics, Vanderbilt University School of Medicine, Nashville, TN*

STUART J. FRANK, MD • *Division of Endocrinology and Metabolism, Departments of Medicine, University of Alabama Medical Service, Birmingham VAMC, Birmingham, AL*

CHRISTOPHER K. GLASS, MD, PHD • *Divisions of Cellular and Molecular Medicine, School of Medicine, University of California at San Diego, La Jolla, CA*

JONATHAN GREER, PHD • *Department of Structural Biology, Pharmaceutical Products Division, Abbott Laboratories, Abbott Park, IL*

JÖRG HEIERHORST, MD • *St. Vincent's Institute of Medical Research, Fitzroy, VIC, Australia*

RICHARD A. HEYMAN, PHD • *Department of Retinoid Research, Ligand Pharmaceuticals, San Diego, CA*

JAMES HORNE, PHD • *St. Vincent's Institute of Medical Research, Fitzroy, VIC, Australia*

BRUCE E. KEMP, PHD • *St. Vincent's Institute of Medical Research, Fitzroy, VIC, Australia*

BOSTJAN KOBE, PHD • *St. Vincent's Institute of Medical Research, Fitzroy, VIC, Australia*

SALLY KORNBLUTH, PHD • *Department of Pharmacology and Cancer Biology, Duke University Medical Center, Durham, NC*

TAKA-AKI KOSHIMIZU, MD, PHD • *Endocrinology and Reproduction Research Branch, National Institute of Child Health and Human Development, National Institutes of Health, Bethesda, MD*

RANDY KRAINOCK, DVM • *Department of Pharmacology, College of Medicine, University of Iowa, Iowa City, IA*

RIKI KUROKAWA, PHD • *Divisions of Cellular and Molecular Medicine, Department of Medicine, University of California at San Diego, La Jolla, CA*

DEEPAK LALA, PHD • *Department of Retinoid Research, Ligand Pharmaceuticals, San Diego, CA*

DONALD P. MCDONNELL, PHD • *Department of Pharmacology and Cancer Biology, Duke University Medical Center, Durham, NC*

NEIL J. MCKENNA, PHD • *Department of Molecular and Cellular Biology, Baylor College of Medicine, Houston, TX*

ANTHONY R. MEANS, PHD • *Department of Pharmacology and Cancer Biology, Duke University Medical Center, Durham, NC*

SHLOMO MELMED, MD • *Department of Endocrinology, Cedars-Sinai Medical Center, Los Angeles, CA*

CRAIG MONTELL, PHD • *Departments of Biological Chemistry and Neuroscience, The Johns Hopkins University School of Medicine, Baltimore, MD*

SEAN MURPHY, PHD • *Department of Pharmacology, College of Medicine, University of Iowa, Iowa City, IA*

ZAFAR NAWAZ, PHD • *Department of Molecular and Cellular Biology, Baylor College of Medicine, Houston, TX*

DEBORAH J. NELSON, PHD • *Departments of Neurobiology, Pharmacology, and Physiology, University of Chicago, Chicago, IL*

ALEXANDRA C. NEWTON, PHD • *Department of Pharmacology, University of California at San Diego, La Jolla, CA*

RICHARD PEARSON, PHD • *St. Vincent's Institute of Medical Research, Fitzroy, VIC, Australia*

BARRY I. POSNER, PHD • *Department of Medicine, McGill University and the Sir Mortimer B. Davis Jewish General Hospital, Montreal, Quebec, Canada*

SHIRISH SHENOLIKAR, PHD • *Department of Pharmacology and Cancer Biology, Duke University Medical Center, Durham, NC*

CYNTHIA V. STAUFFACHER, PHD • *Department of Biological Sciences, Purdue University, West Lafayette, IN*

STANKO S. STOJILKOVIC, PHD • *Endocrinology and Reproduction Research Branch, National Institute of Child Health and Human Development, National Institutes of Health, Bethesda, MD*

ANDREW W. TAYLOR, PHD • *Schepens Eye Research Institute and The Department of Ophthalmology, Harvard Medical School, Boston, MA*

MELANIJA TOMIC´, PHD • *Endocrinology and Reproduction Research Branch, National Institute of Child Health and Human Development, National Institutes of Health, Bethesda, MD*

MING-JER TSAI, PHD • *Department of Molecular and Cellular Biology, Baylor College of Medicine, Houston, TX*

SOPHIA Y. TSAI, PHD • *Department of Molecular and Cellular Biology, Baylor College of Medicine, Houston, TX*

ALFREDO ULLOA-AGUIRRE, MD • *Research Unit in Reproductive Medicine, Hospital de Ginecoobstericia, Instituto Mexicano del Seguro Social, Col. Tizapán San Angel, Mexico*

FREDRICK VAN GOOR, PHD • *Endocrinology and Reproduction Research Branch, National Institute of Child Health and Human Development, National Institutes of Health, Bethesda, MD*

DAVID J. WAXMAN, PHD • *Division of Cell and Molecular Biology, Department of Biology, Boston University, Boston, MA*

MICHAEL A. WHITE, PHD • *Department of Cell Biology, University of Texas-Southwestern Medical Center, Dallas, TX*

THOMAS M. WILKIE, PHD • *Department of Pharmacology, University of Texas-Southwestern Medical Center, Dallas, TX*

PART I
SIGNALING MECHANISMS INITIATED BY CELL SURFACE RECEPTORS

1

G Protein-Coupled Receptors and G Proteins

Alfredo Ulloa-Aguirre and P. Michael Conn

CONTENTS
 INTRODUCTION
 STRUCTURE–FUNCTION RELATIONSHIP OF G PROTEIN-COUPLED RECEPTORS
 THE HETEROTRIMERIC G PROTEIN FAMILY
 SELECTED READINGS

1. INTRODUCTION

G protein-coupled receptors (GPCRs) form a large and functionally diverse superfamily of cell membrane receptors. Many signaling cascades use GPCRs to convert a large diversity of external and internal stimuli including photons, odorants, and ions as well as hormones and neurotransmitter agonists into intracellular responses. Following agonist binding and receptor activation, GPCRs interact with one or more members of the guanine nucleotide binding signal transducing proteins (G proteins); binding and activation of G proteins usually occur within the intracellular domains of the GPCRs. Activated G proteins carry the information received by the receptor to downstream specific cellular effectors such as enzymes and ion channels. Effector enzymes generate second messengers which in turn regulate a wide variety of cellular processes including cell growth and differentiation. The regulation of receptor–G protein signal selectivity and intensity is highly complex and multifactorial and involves the activation of a network of fine-tuning mechanisms that eventually lead to highly specific biological responses.

The general structure of the GPCRs consists of a single polypeptide chain of variable length that traverses the lipid bilayer seven times, forming characteristic transmembrane helices and alternating extracellular and intracellular sequences. Based on nucleotide and amino acid sequence similarity, the G protein receptor superfamily can be further separated into three collections or families of receptors whose protein sequences share significant similarity (i.e., ≥20% sequence identity over the predicted hydrophobic transmembrane segments). Three main mammalian families of GPCRs have been characterized. The receptors related to rhodopsin/β-adrenergic receptor (Family A, Table 1), to which the majority of G protein-linked receptors identified to date belong, are the best studied from the structural and functional aspects. This family comprises receptors that are activated by a large variety of stimuli including photons and odorants as well as hormones and neurotransmitter agonists of variable molecular structure, ranging from small biogenic amines (catecholamines and histamine) to peptides (substance P and gonadotropin-releasing hormone) and complex glycoproteins (such as the gonadotropic hormones, luteinizing hormone [LH] and follicle-stimulating hormone [FSH]) (Table 1). Because of the large variability in the structure of ligands that bind these receptors, both the NH_2 (N)- and the COOH (C)-termini, but not the transmembrane domains, may be highly variable in length. The secretin/vasointestinal

From: *Principles of Molecular Regulation* (P. M. Conn and A. R. Means, eds.), © Humana Press Inc., Totowa, NJ.

Table 1
G Protein-Coupled Receptor Superfamily[a]

Family	Group I	Group II	Group III	Group IV	Group V	Group VI
A: Receptors related to rhodopsin/ β-adrenergic receptor	Olfactory, adenosine, melanocortin, cannabinoid, and several orphan receptors	Serotonin, α- and β-adrenergic receptors, dopamine, histamine, muscarinic, octopamine, and orphan receptors	Bombesin/ neuromedin, cholecystokinin, endothelin, growth hormone secretagogue, neuropeptide Y, neurotensin, opsins, tachykinin, thyrotropin-releasing hormone and orphan receptors	Bradykinin, invertebrate opsins, and orphan receptors	Angiotensin, C3a and C5a, chemokine, conopressin, eicosanoid, fMLP, FSH, LH and TSH, galanin, GnRH, leukotriene, P2 (nucleotide), opioid, oxytocin, platelet activating factor, thrombin and protease activated, somatostatin, vasopressin, vasotocin, and orphan receptors	Melatonin and orphan receptors
B: Receptors related to the calcitonin and parathyroid hormone receptors	Calcitonin, calcitonin gene-related peptide, corticotropin-releasing factor, insect diuretic hormone, and orphan receptors	Parathyroid hormone/ parathyroid-related peptide receptors, orphan receptors	Glucagon-like peptide, glucagon, gastric inhibitory peptide, growth hormone releasing hormone, pituitary adenylyl cyclase activating peptide, vasoactive intestinal peptide, secretin, and orphan receptors	Latrotoxin and orphan receptors		
C: Receptors related to the metabotropic glutamate receptors	Metabotropic glutamate receptors	Extracellular calcium ion sensor receptor	GABA-B receptor	Putative pheromone receptors		
D: Receptors related to the STE2 pheromone receptor	STE2-α-factor-pheromone receptor					
E: Receptors related to the STE3 pheromone receptor	STE3-alpha-factor-pheromone receptor					
F: Receptors related to the cAMP receptor	*Dictyostelium discoideum* CAR2 to CAR4 receptors					

[a]GCRDb data base maintained by L.F. Kolakowski and J. Zhuang at http://www.gcrdb.uthscsa.edu/FA_intro.htm

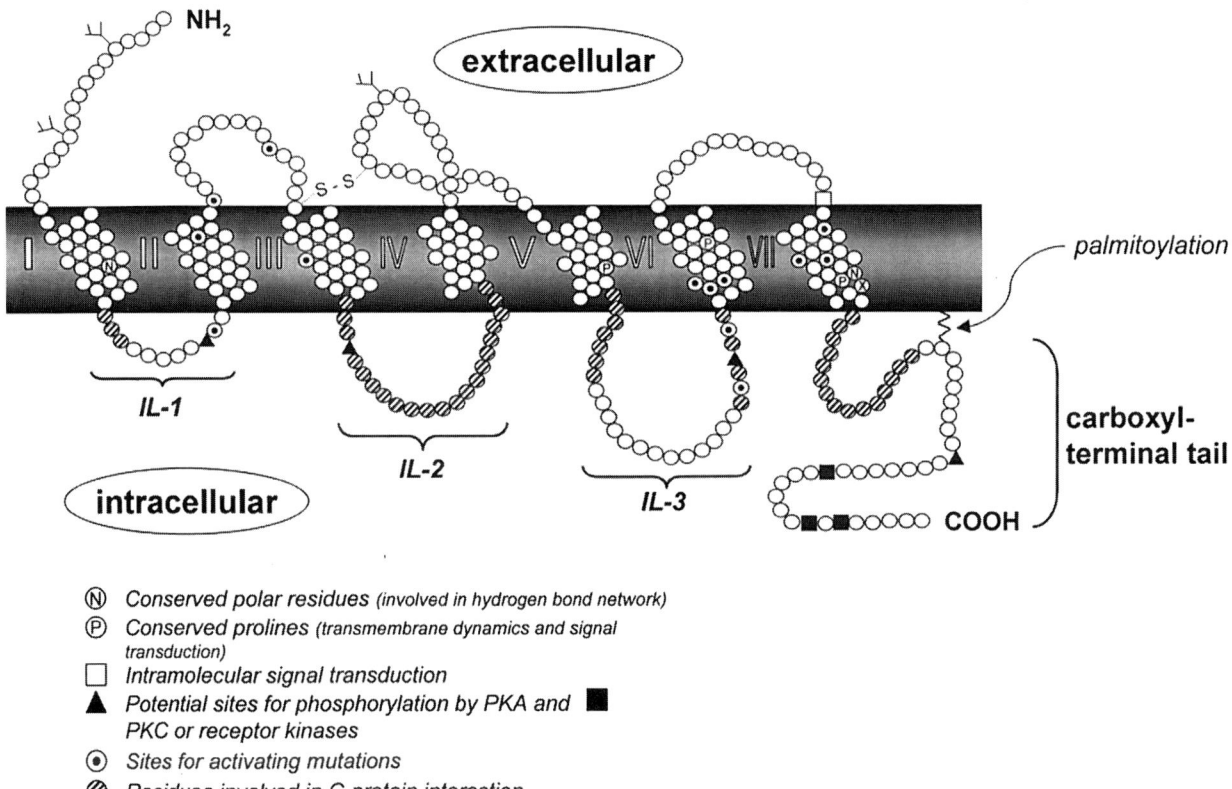

Fig. 1. Model of the proposed seven transmembrane spanning domains of a prototypic G protein-coupled receptor belonging to the rhodopsin/β-adrenergic family showing some structural characteristics including putative glycosylation sites as well as some amino acid residues involved in signal transduction, receptor phosphorylation, sequestration, and palmitoylation. The structure also shows the location of some spontaneously occurring mutations leading to constitutive activation of the receptor (e.g., thyrotropin, melanocyte-stimulating hormone, luteinizing hormone, rhodopsin, and adrenergic receptors). A region in the N-terminal end of the first intracellular loop (IL-1) involved in effector activation by the human calcitonin receptor and the murine GnRH receptor is also shown.

peptide (VIP) family (Family B), which binds several neuropeptides and peptide hormones (Table 1), exhibits none of the fingerprint residues characteristic of the rhodopsin/β-adrenergic receptor family with the exception of the putative disulfide bridge between the third transmembrane domain (or the C-terminus end of the first extracellular loop) and the second extracellular loop (Fig. 1). These receptors are characterized by large N-termini with at least six highly conserved Cys residues conceivably involved in ligand binding. Finally, the metabotropic glutamate receptor family (Family C), which comprises at least six closely related subtypes of receptors, binds glutamate, the major excitatory neurotransmitter in the central nervous system (CNS); all receptor subtypes exhibit long N-terminus domains but C-terminus domains of variable length.

Although the calcium-sensing receptor exhibits only modest identity in its amino acid sequence with the metabotropic glutamate receptor (18–24%), it shares striking topological similarity and thus is included as a member of this restricted GPCRs family. Three additional families that encompass nonmammalian receptors may be also included within the GPCRs superfamily (Families D–F; Table 1).

Even though the receptor–G protein system is a highly efficient means through which the cell responds to such a wide variety of extracellular stimuli, in some abnormal conditions "loss-of-function" or "gain-of-function" mutations in the receptor molecule or the G proteins may modify the activity of the signal transduction pathways and lead to altered cell function, including abnormal growth and tumorigenesis.

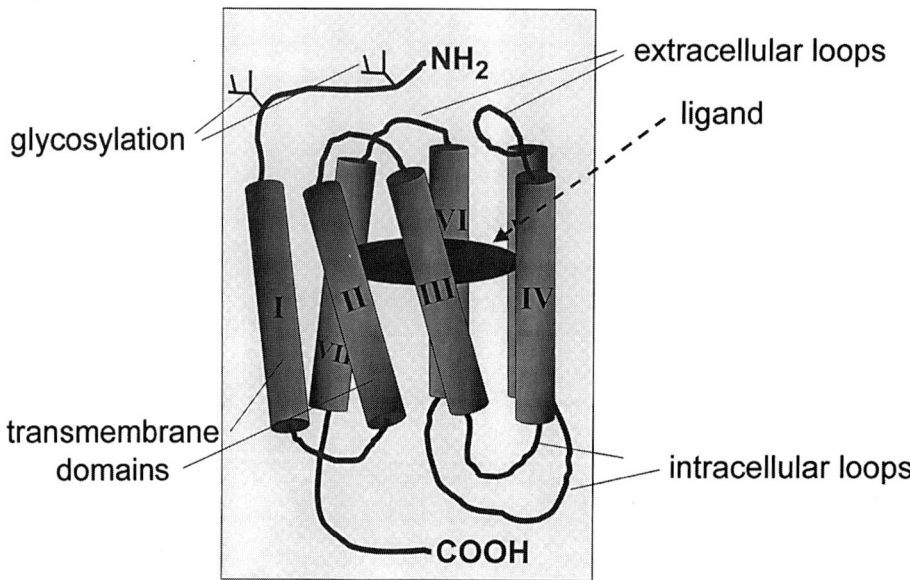

Fig. 2. Counterclockwise orientation of GPCRs from transmembrane domain I–VII. The closed loop structure is representative of receptors for small ligands, such as biogenic amines and nucleosides. In this arrangement, the core is mainly comprised by domains II, III, V, and VI, whereas domains I and IV are peripherally sequestered. Note the proximity between helix 2 and helix 7, which is characteristic of this family of GPCRs.

2. STRUCTURE–FUNCTION RELATIONSHIP OF G PROTEIN-COUPLED RECEPTORS

2.1. General Features

Considerable structural homology exists within each G protein-linked receptor family. Cloning and sequence determination as well as hydrophobic analysis of the primary sequences of at least 300 members of the main family (Family A) of receptor proteins show the characteristic existence of seven stretches of 20–25 predominantly hydrophobic amino acids (Fig. 1). By extrapolation from the structure of bacteriorhodopsin (a seven transmembrane [TM] domain protein from *Halobacterium halobium* that is not coupled to G proteins), these stretches are predicted to form α-helical membrane-spanning domains, connected by alternating extracellular and intracellular loops (which are predicted to be between 10 and 40 amino acids in length with the exception of the third intracellular loop, which may exhibit more than 150 residues), oriented to form specific binding domains. The seven transmembrane domains are thought to form a barrel shape, oriented roughly perpendicular to the plane of the membrane with an extracellular N- and an intracellular C-terminus (Fig. 1). Most of the primary sequence homology among the different groups of this type of protein receptors is contained within the hydrophobic transmembrane domain. Residues that are highly conserved among the members of this large receptor family (and probably among members of other families as well) apparently represent essential structural determinants of receptor structure and function. For example, in the majority of the receptors belonging to the rhodopsin/β-adrenergic family, two highly conserved residues are an Asp residue in TM-2 and an Asn in TM-7; the exact location of and interaction between these residues seems to be essential to keep helix 2 and helix 7 in close proximity and allow receptor activation and signal transduction (Fig. 2). On the other hand, binding specificity requires differences in residues between receptor subclasses but a high degree of conservation for subtypes of receptors with related ligands. This is the case of several receptors such as the muscarinic, adrenergic, dopaminergic, and serotonin receptors, which exhibit a highly conserved third transmembrane domain Asp residue that facilitates the formation of an ionic interaction with their corresponding ligands (Fig. 3). Alternately, the hydrophilic loops connecting the transmembrane segments as well as the N- and C-termini may vary substantially. Thus, the N-terminus may be formed by a relatively short peptide chain, such as in the photoreceptor, rhodopsin, and the $β_2$-adrenergic receptor, by medium length chains as in the members of the secretin/VIP receptor subfamily, to very long, such as the receptors for the gonadotropic hormones and the calcium-sensing

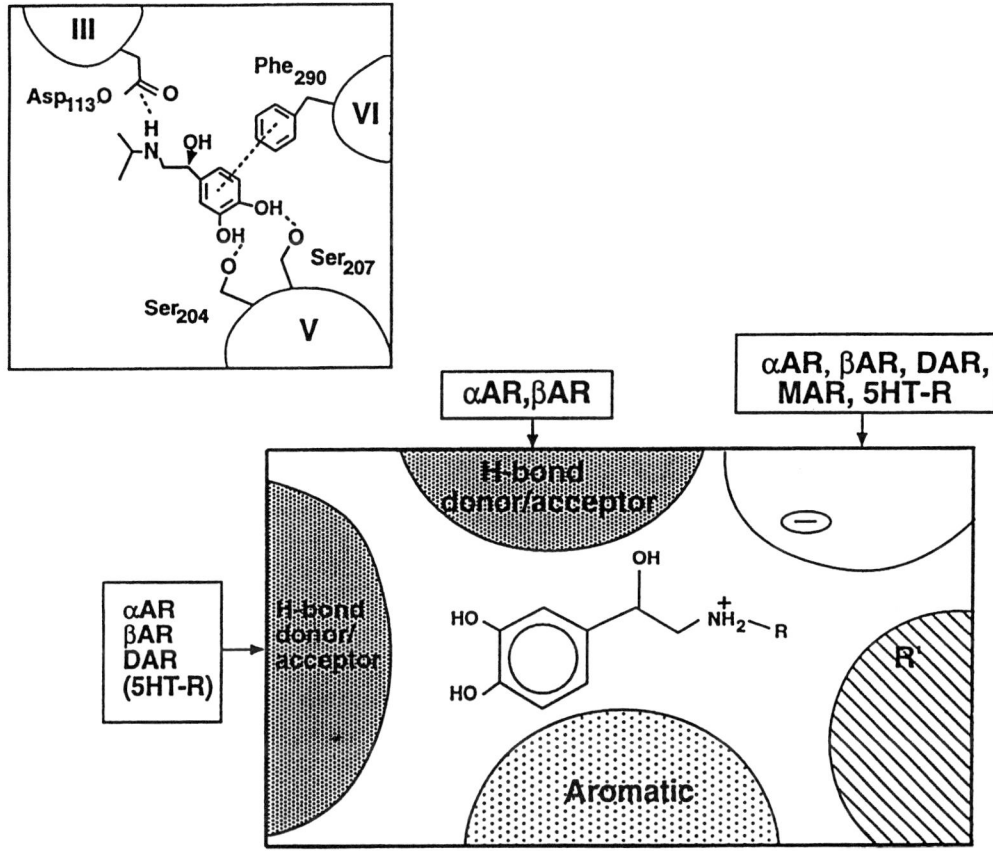

Fig. 3. Pharmacophore map of the catecholamine binding site of the β-adrenergic receptor. A catecholamine ligand is shown in a hypothetical binding site intercalated among the transmembrane helices of the receptor. Each of the *large semicircles* represents a transmembrane helix of the receptor, inscribed with the type of binding interaction expected. Other GPCRs that would be expected to have similar interactions with their specific ligands are designated in *boxes* next to each helix. αAR, α-adrenergic receptor; βAR, β-adrenergic receptor; DAR, dopaminergic receptor; MAR, muscarinic acetylcholine receptor; 5HT-R, 5-hydroxytryptamine (serotonin) receptor. **Inset:** Model for the ligand binding site of the β-adrenergic receptor showing key interactions with the agonist isoproterenol (*see* text for details). (Reproduced from Strader et al., Trends Pharmacol Sci 1989; [Suppl.]: 26 and FASEB J 1995;9:745, with permission).

receptor. Although most G protein-linked receptors exhibit a C-terminal domain that is rich in Ser and Thr residues (which are potential sites for phosphorylation by kinases that induce GPCRs desensitization), in some rare instances, such as in the case of the mammalian GnRH receptor, there is a complete absence of the intracellular C-terminal domain in several species examined.

In the large family of receptors belonging to the rhodopsin/β-adrenergic class, the general structural homology (primary and tertiary) shared by these receptor proteins reflect their common mechanism of action. This is the case, for example, of the photosensitive G protein-linked rhodopsin receptor molecule present in the photoreceptor cells and the widely distributed catecholamine receptors. In the former, its covalently linked ligand, all-*cis*-retinal, isomerizes to the all-*trans* isomer upon activation by photons; this structural change in retinal triggers the occurrence of a series of conformational modifications in rhodopsin which in turn binds the trimeric G protein, transducin (G_t). The α-subunit of rhodopsin-activated transducin activates the effector enzyme, cGMP phosphodiesterase, which hydrolyzes GMP, with a consequent decrease in intracellular cGMP levels, closure of cGMP-gated membrane Na^+-specific channels, hyperpolarization of the plasma membrane, and reduction of the rate of neurotrasmitter release from the synaptic region; this sequence of intracellular events allows light to free the neurons from neurotransmitter inhibition and excite them. A similar pathway of G protein-coupled, receptor-mediated signal transduction may be found in the adrenergic receptors, which bind the endogenous catecholamines, epinephrine and norepi-

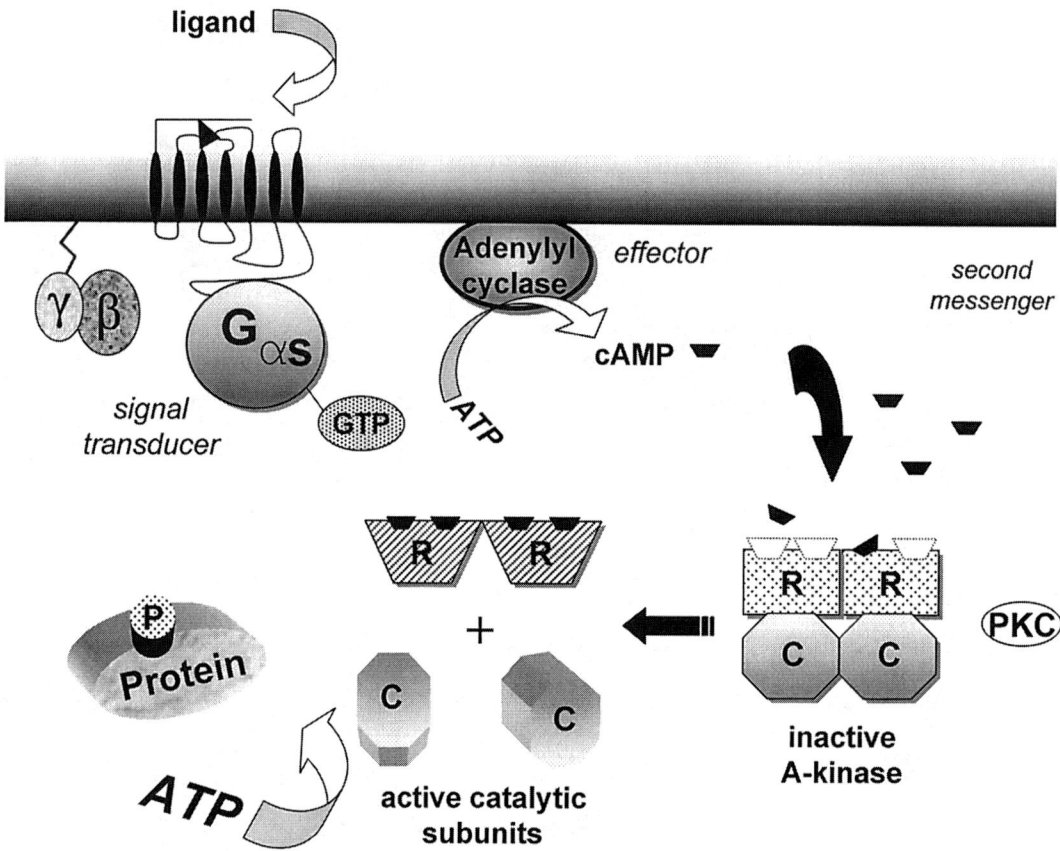

Fig. 4. Signal transduction mediated by the G_s pathway. After ligand-induced receptor activation, the $G\alpha_s$-subunit in its activated (GTP-bound) form stimulates the effector enzyme adenylyl cyclase and the production of a second messenger, cAMP, leading to activation of protein-kinase A (PKA) and the initiation of a potential phophorylation cascade. R, regulatory subunit; C, catalytic subunit.

nephrine, as well as a large number of synthetic agonists and antagonists. Upon specific agonist binding, β-adrenergic receptors (including its $β_1$, $β_2$, and $β_3$ subtypes) activate a stimulatory G protein (G_s) which in turn activates the effector, adenylyl cyclase; this enzyme is responsible for adenosine 3′,5′-monophosphate (cAMP) formation, one of the main intracellular mediators of extracellular signals (Fig. 4) (see also Section 3.1.).

GPCRs contain a variable number of Cys residues. Conserved Cys residues in the N-terminus and the extracellular loops (particularly in the first and second) may potentially form disulfide bonds that stabilize the structure of the functional protein (Fig. 1). In fact, it has been shown that substitution of either of the Cys residues present in the first and the second extracellular loops of the $β_2$-adrenergic receptor (Cys^{106} and Cys^{184}) or rhodopsin (Cys^{110} and Cys^{187}) results in destabilization of the tertiary structure of the protein and alterations in the binding properties of the receptors. For some receptors in which the principal determinants of ligand binding reside within large extracellular N-terminus domains (see Section 2.2), such as the calcium-sensing receptor, the metabotropic glutamate receptors, and the glycoprotein hormone receptors, the numerous conserved Cys residues located within the extracellular domain may contribute to organize this region of the receptor into a binding pocket appropriate for interacting with small charged ligands such as glutamate or Ca^{2+} or for allowing recognition and interaction of several discontinuous extracellular domains of the receptor with complex ligands such as FSH or LH.

Most GPCRs exhibit one or two Cys residues in the membrane-proximal portion of the C-terminus; these residues are particularly important because they are susceptible to undergo palmitoylation, that is, reversible thioesterification of the C_{16} fatty acid palmitate, allowing formation of an anchor to create a fourth intracellular loop (Fig. 1). Palmitoylation of a number of GPCRs has been documented. These include rhodopsin (in fact, the first G protein-linked receptor shown to be modified by palmitoylation), the β_2- and α_{2A}-adrenergic, D_1-dopaminergic and $5HT_{1B}$-serotonergic receptors, and the LH/choriogonadotropin (CG) receptors. All these receptors have in common the presence of one or two Cys residues in the membrane-proximal domain of their C-terminus. Although a strict consensus sequence that determines the occurrence of covalent *S*-palmitoylation of the Cys residues present in the proximal segment of the C-terminus tail of these receptor has not yet been identified, the consensus sequence $F/Y(X)n_1B(X)n_2C^P$ ... (where B is a hydrophobic residue, n_1 and n_2 are residues that may vary between between 0 and 4, and C^P is the palmitoylated Cys) has been proposed based on the primary sequence of the palmitoylated receptors mentioned earlier. Modulation of receptor palmitoylation by agonist has been documented for the β_2- and α_{2A}-adrenergic receptors and for the dopamine-D_1 receptor; these studies have suggested that receptor stimulation by its specific signaling molecule affects the turnover rate of the palmitate moiety linked to the receptor by increasing its relative rate of depalmitoylation. The physiological significance of receptor palmitoylation still remains to be fully understood; it has been shown, however, that mutations that prevent palmitoylation of the β_2-adrenergic receptor promote the functional uncoupling of the receptor and consequently a loss of G protein activation. Palmitoylation increases the hydrophobicity of a protein and may favor redistribution to the membrane compartment. Because prevention of palmitoylation of the β_2-adrenergic receptor is accompanied by an increased protein kinase A (PKA)-dependent phosphorylation of the receptor, it has been proposed that for those receptors harboring phosphorylation sites near potentially palmitoylated Cys residues, there is a close relationship between palmitoylation, phosphorylation, and receptor desensitization (*see* Section 2.4.2.). Although uncoupling of G proteins after mutation of a palmitoylated Cys residue has also been observed for the µ opiate receptor, for other receptors such as rhodopsin, α_{2A}-adrenergic, M2-muscarinic, and the LH/CG receptor, abolition of palmitoylation has no effect on G protein and effector activation. Nevertheless, the fact that the presence of Cys residues susceptible to palmitoylation is highly conserved among this G protein-linked receptor subfamily strongly suggests that this post-translational modification may be important for the function of these receptors.

Another posttranslational modification found in most GPCRs is the presence of one or more consensus sequences (Asn-Xaa-Ser/Thr) for N-linked glycosylation. The N-linked oligosaccharides are usually located near the N-terminus of the receptor molecule, although for some receptors there may be predicted glycosylation sites in the first and second extracellular loops as well. For some receptors (e.g., the β-adrenergic receptor and rhodopsin), prevention of receptor glycosylation either by inhibiting posttranslational glycosylation or by site-directed mutagenesis leads to decreased cell-surface receptor expression without any significant change in binding or functional activity. In others (e.g., the GnRH receptor), incorporation of an additional glycosylation may result in enhancement of the level of cell surface expression of the receptor. On the other hand, whereas glycosylation of the FSH receptor is not involved directly in hormone binding, the receptor strictly requires at least one N-linked glycosylated site for proper folding, membrane expression, and function. This is in contrast to the LH/CG receptor in which prevention of receptor glycosylation does not significantly affect its cell-surface expression or binding capacity. Thus, it seems that interactions between *N*-glycosyl carbohydrate side chains of receptors and their ligands are independent events and that for some types of GPCRs glycosylation is functionally important for correct trafficking of the receptor. Some GPCRs may exhibit a certain degree of heterogeneity or variability in the type of oligosaccharide residues attached to their N-terminus domain. The impact of such a type of microheterogeneity on receptor function and/or expression is still unkown.

Diversity of receptors among same receptor subtypes may be provided by alternate splicing. Even though not all mRNA spliced variants of a given receptor may be expressed as receptor protein, when this occurs the receptor isoform either may or may not exhibit distinct functional characteristics. For example, alternative splicing of the human α_{1C}-adrenergic receptor produced three isoforms differing by their C-terminus domain that exhibit no differences in functional parameters when expressed in Chinese hamster ovary (CHO) cells. In other instances, divergent structures of the receptor may profoundly alter ligand binding or G protein activation. For instance, the prostaglandin E type EP_3 receptor gene may produce at least four receptor isoforms by alternate mRNA splicing that differ only in the sequence of their C-terminus. The EP_{3A} couples to the G protein class $G\alpha_{i/o}$; the EP_{3B} and EP_{3C} variants to $G\alpha_s$; and EP_{3D} more permissively to $G\alpha_{i/o}$, $G\alpha_s$, and $G\alpha_{q/11}$ G proteins. These receptor variants also exhibit different abilities to undergo agonist-induced desensitization.

Thus, diversity in receptor isoform expression may influence the specificity, sensitivity, and magnitude of response of a given receptor–G protein system.

2.2. Ligand Binding Domain

The specific receptor region at which the ligand binds to the receptor varies depending on the particular subfamily of G protein-coupled receptor as well as on the size and chemical structure of the ligand. Receptors for small ligand molecules, such as the biogenic amines and small peptides, characteristically bind the ligand through a "pocket" involving highly conserved residues located in the middle and extracellular thirds of hydrophobic transmembrane helices (Fig. 2). On the other hand, in the calcium-sensing and the metabotropic glutamate receptors the principal determinants of their corresponding binding pockets reside in the N-terminus domain. For larger ligands, such as moderate-sized and large peptides and complex glycoproteins, the binding site usually resides in both the extracellular and the transmembrane domains or within the extracellular N-terminus alone. In fact, receptors that bind large ligands usually exhibit a long N-terminus. For example, in the parathyroid hormone/calcitonin receptor subfamily, an approx 100-residue extracellular N-terminus contains regions shown to be critical for ligand binding; the LH/CG and FSH receptors have a long N-terminus region, constituted by 338 and 349 amino acid residues respectively, which exhibit high affinity and specificity for their ligands even when the receptor is expressed in the abscence of the transmembrane domains or when it is expressed in a chimeric form with transmembrane and C-terminus domains from other G protein receptors.

The receptor ligand-binding determinants for small molecules that bind to the large rhodopsin/β-adrenergic family of receptors (such as retinal and the biogenic amines) have been extensively studied. Experiments involving construction of chimeric receptors, specific amino acid residue substitutions (site-directed mutagenesis), and affinity labeling of ligands have indicated that no single transmembrane domain is the dominant contact site for ligand binding to these receptors. Similar to rhodopsin, whose binding site for retinal is buried within the core of the receptor protein, in the α_2- and β_2-adrenergic receptors various sites in several of the hydrophobic transmembrane domains (TM-2 to -7) contribute to forming the binding pocket for catecholamines (Fig. 2). It as been shown that two particular residues (Ser^{204} and Ser^{207}) in TM-5 of the β_2-adrenergic receptor (both of which are highly conserved among the receptors for catecholamines) are required for normal binding of agonists, but not of antagonists (noncatechol), to the catecholamine receptors, thus indicating that the catechol hydroxyl groups of the agonist form hydrogen bonds with the side chains of the two Ser residues. Another important binding determinant in these receptors is the Phe^{290} residue in TM-6; whereas the catechol hydroxyl groups of the agonist form hydrogen bonds with the side chains of the two Ser residues at TM-5, the catechol-containing aromatic ring interacts with the side chain of Phe^{290} in TM-6 (Fig. 3). Analogous to the Glu^{113} residue in TM-3 of rhodopsin, systematic mutagenesis of the negatively charged residues in the transmembrane domain of the β-adrenergic receptor has disclosed that Asp^{113} in TM-3 acts as the counterion for the basic amines in both agonists and antagonists. In contrast to the interaction of the ligand with residues in TM-5 and TM-6 which is important for both binding and receptor activation, the ionic interaction in TM-3, although important as a source of binding energy between the ligand and the receptor protein, is not essential for receptor activation. Mutagenesis analysis of other receptors for biogenic amines (dopamine, α_2-adrenergic, muscarinic, serotonin, and histamine receptors) has clearly demonstrated the importance of these key amino acid positions in ligand binding and receptor activation, albeit with the expected variations in specific residues according to the chemical structure of the ligand. Therefore, binding of small ligands (such as the biogenic amines) to their corresponding receptors is characterized by a complex network of interactions involving several transmembrane domains but in which key residues in TM-5 and TM-6 are esssential to form a binding pocket that is critical for ligand binding, with the specificity for agonist recognition determined by the particular chemical nature of the side chains of these residues. Based on the combination of mutagenesis, amino acid replacements, medicinal chemistry, and molecular modeling, it has been possible to generate a pharmacophore map from which specific interactions between biogenic amines and the transmembrane domains of the receptor may be inferred based on specific amino acid residues present within the transmembrane receptor domains and the particular chemical structure of the ligand (Fig. 3).

In contrast to cationic neurotransmitter receptors, the ligand binding site has not been clearly defined for G protein-linked receptors that bind small peptide ligands. Mutagenesis on residues located in the transmembrane domains at positions corresponding to the monoamine binding pocket (*see* above) do not always affect the binding of peptide ligands. On the other hand, mutations in extracellular domains often lead to significant decreases in peptide binding. Thus, the binding site for peptide ligands may involve a larger surface area comprising either the extracellular or the transmembrane domains of the receptor or both. For example, the binding pocket for the tripeptide thyrotropin-releasing hormone is localized within the transmembrane domains of the receptor and involves specific residues in TM-3 (Tyr^{106}, Asn^{110}), TM-6 (Arg^{283}), and TM-7 (Arg^{306}). On the other hand, the binding of the decapeptide GnRH requires, in addition to the proximity of helix 2 and helix 7 (a general feature of the rhodopsin/β-adrenergic family), an ionic interaction between the Arg^8 of the decapeptide and the carboxylate side chain of Glu^{301} in the third extracellular loop. Tachykinin receptors are another example. Site-directed mutagenesis on the NK_1 receptor, which binds substance P, has shown that three residues in the first extracellular segment (Asn^{23}, Gln^{24}, and Phe^{25}) and two in the second (Asp^{96} and His^{108}) are particularly required for ligand binding; several residues in the second and the seventh transmembrane domains of this receptor (Asn^{85}, Asn^{89}, Tyr^{92}, and Asn^{96} in TM-2 and Tyr^{287} in TM-7) are, however, also important in determining the affinity of the receptor for its ligand. In fact, these residues in TM-2 would be predicted to be positioned on the same face of the α-helical region of TM-2, thus indicating that their side chains may form a hydrogen-bonding surface projecting into the ligand binding site. In other peptide-binding receptors, such as the angiotensin, vasopressin, oxytocin, and neuropeptide Y receptors, the ligands also bind to determinants located in both the first extracellular loop and several transmembrane domains (TM-2 to -7), emphasizing the importance of the transmembrane domains in the binding of peptide ligands to GPCRs. In many cases, the sites required for high-affinity binding of antagonists of these peptide ligands may be distinct from those of the naturally occurring ligands.

The N-terminus region of receptors for other relatively larger peptide ligands and glycoprotein molecules plays a major role in determining high-affinity binding for their corresponding ligands. In the secretin/VIP receptor subfamily, there is a sequence conservation in their N-terminus region that strongly suggests a role of this region in mediating receptor–ligand interactions. For example, the N-terminal extracellular regions of the parathormone and secretin receptors contain domains that largely determine the binding affinity of these hormones. Nevertheless, some of these ligands may also interact with other extracellular domains; although the first 10 residues of the N-terminus of the secretin receptor are critical for ligand binding, other extracellular domains located at the C-end of the first extracellular loop as well as at the N-terminal half of the second loop also provide critical determinants for ligand binding. Other receptors in which the N-terminal region appears to be critical in determining ligand binding include the interleukin-8, the C5a, the glycoprotein hormones (LH/CG, FSH, and thyroid-stimulating hormone [TSH]), and the thrombin receptors. Glycoprotein receptors have large extracellular domains (300–400 amino acids in length) and bind complex ligands with molecular masses ranging from 28 to 38 kDa. These receptors exhibit an extraordinary amino acid homology (30–50%) in the extracellular domains in which multiple discontinuous segments spanning the entire domain contribute to the binding site and ligand specificity. Interestingly, the extracellular domain of the gonadotropin receptors exhibits multiple ($n=14$) Leu-rich repeats that are involved in high-affinity specific binding. Additional sites located in the first extracellular loop, the second transmembrane domain, and the third extracellular connecting loop are also important for the binding affinity of these complex ligands. Apparently, these binding sites are critical to activate or deactivate specific transmembrane conductors that may eventually lead to the generation of distinct intracellular signals. Furthermore, the receptors for these particular ligands exhibit a high degree of plasticity which allows the induction of a variety of conformational changes in response to variations in the molecular structure of the ligand; depending on the conformation change elicited, distinct G protein coupling may occur. Finally, the thrombin receptor exhibits in its N-terminus a unique mechanism for ligand binding and receptor activation. The extracellular N-terminus of the receptor contains both a ligand-specific anion binding site as well as the specific proteolytic cleavage substrate site; the receptor is irreversibly activated by binding its tethered ligand, which is unmasked upon receptor cleavage by thrombin.

2.3. G Protein Coupling Domain

G protein-linked receptors characteristically bind G proteins which in turn act as mediators of receptor-induced effector activation. The nature of the second-messenger pathways activated in response to agonist binding to a G protein-linked receptor is essentially determined by the type of G protein or proteins coupled to each particular receptor. As can be expected from both the general topography of these receptors and the fact that G protein-linked receptors couple and interact with a number of distinct G proteins (at least 17 different species of Gα genes coding for four main classes and their numerous corresponding subtypes have been identified [see Section 3.1.]), the receptor G protein coupling domains are expected to lie within the divergent sequences of its intracytoplasmic domains. In fact, when a particular ligand binds to its receptor, the molecular perturbation is relayed from the membrane-embedded helices to the cytoplasmic face of the protein molecule. These intracellular domains, particularly the regions closest to the plasma membrane, as well as some specific regions located in the intracellular ends of the transmembrane helices and in the membrane-proximal portion of the C-terminus, are important for receptor–G protein interaction (Fig. 1).

Within members of the various G protein-coupled receptor families there are some classes of receptors with their corresponding subtypes that may show particular preferences or specificity toward a particular G protein class (G_s, G_i/G_o, G_q or G_{12}). This is the case, for example, of the muscarinic receptor class, in which the M2 and M4 receptors preferentially couple to G proteins of the G_i/G_o class, which are sensitive to pertussis toxin, whereas the M1, M3 and M5 receptors predominantly stimulate pertussis toxin-insensitive G proteins of the G_q/G_{11} class. The different subclasses of the serotoninergic (5-hydroxytryptamine [5-HT]) receptors ($5-HT_1$ to $5-HT_7$) with their corresponding subtypes (e.g., $5-HT_{1A}$, $5-HT_{1B}$, etc) also show particular preferences; $5-HT_1$ receptors bind preferentially to G_i/G_o proteins whereas the $5-HT_2$ and the $5-HT_4$, and $5-HT_6$ and $5-HT_7$ subtypes are coupled to members of the G_q/G_{11} and to G_s protein classes respectively. In the metabotropic glutamate receptor (mGluR) family, the $mGluR_1$ (including its splice variants $mGluR_{1\alpha}$, $mGluR_{1\beta}$, and $mGluR_{1C}$) and $mGluR_5$ types are associated with effects most likely mediated by the G_q/G_{11} class whereas the $mGluR_2$, $mGluR_3$, $mGluR_4$, and $mGluR_6$ classes exhibit some preference toward the G_i/G_o class. A further level of complexity in receptor–G protein coupling is added by the fact that certain receptors in a given family may show a strong preference for a particular member of a G protein class; for example, the C5A receptor prefers G_{16} to other members of the G_q class. Finally, there are receptors normally coupled to different G protein classes. The receptors for glycoprotein hormones (LH/CG, TSH, and FSH) stimulate (albeit not to the same extent) both cAMP and inositol phosphate production throughout the pathways defined by G_s and G_q proteins respectively; further, the FSH receptor apparently couples to three different G proteins (G_s, $G_{q/11}$, and G_i). Thus, there is the potential for a high degree of promiscuity in signaling of these receptors.

It is well known that the second and third intracellular loops as well as the membrane-proximal region of the C-terminus of several G protein-linked receptors are tightly involved in G protein coupling and specificity determination. In rhodopsin, the second and third intracellular loops as well as the fourth loop created by palmitoylation of Cys residues in the membrane-proximal portion of the the C-terminus are involved in the interaction between the receptor and G_t, which indicates that determinants present in several intracellular domains are important for receptor–G protein interaction. In the muscarinic and catecholamine receptors [and also presumably in other receptors for biogenic amines and peptide hormones], both the N- and C-terminal portions of the third intracelullar loop appear to be critical determinants of G protein binding and activation. Specifically, in the β_2-AR (which is coupled to the G_s protein) deletions in limited regions of the N-terminal (residues 222–229) and the C-terminal (residues 258–270) portions of its third intracellular loop as well as construction of chimeric α_1/β_2-adrenergic receptors and several β_2-adrenergic receptor mutants, have indicated that these particular regions are necesary for the β_2-adrenergic receptor-mediated activation of adenylyl cyclase. Apparently, in the β-adrenergic receptor the C-terminal region of this loop is more important for G protein coupling specificity than the opposite N-terminal region of the same loop, as substitution of residues 216–237 in the latter with the corresponding residues from the α_2-AR (which is coupled to the G_q/G_{11} class) fails to affect the ability of the β_2-AR to couple G_s whereas substitution of residues 263–274 corresponding to the C-terminal region of the loop results in a severe impairment in coupling this G protein. On the other hand, it is known that a restricted region composed of 27 amino acid residues of the N-terminal region from the third intracellular loop of the α_{1B} receptor plays a major role in determining the selectivity for receptor–G_q/G_{11} protein coupling. Thus, in all these receptors, both specific amino acid residues from the N- and the C-terminal portions of the third intracellular loop and the secondary structure of these regions, which form amphipathic helical extensions of the adjacent transmembrane domains, play a crucial role in determining coupling specificity, selectivity, and efficiency to activate a particular class of G protein.

For several G protein-linked receptors, the importance of the second intracellular loop in G protein coupling and specificity has been well documented. It must be emphasized that the amino acid sequence of this loop is among the most highly conserved in the G protein-linked receptor superfamily and that substitutions in some of its highly conserved residues may severely impair G protein coupling. It has been shown that the second intracellular loop is essential for maintaining the configuration of the G protein binding site and thus for the normal interaction between the β_2-AR and G_s. On the other hand, construction of a chimeric α_{1B}/β_2-AR in which the third intracellular loop of the α_1-adrenergic receptor replaced the corresponding loop in the β_2-receptor resulted in a chimeric receptor capable of activating both the G_q/G_{11} and the G_s pathways; in the M2-muscarinic receptor, which normally couples to the G_i protein class, replacement of the third intracellular loop of this receptor by the corresponding loop of the β_2-AR results in a chimeric receptor species coupled to both G_i and G_s. Thus, specificity determinants for G protein coupling in these receptors may lie in regions other than the third intracellular loop, probably in the second loop. Other lines of evidence strongly suggest that the second intracellular loop is involved in G protein coupling and selectivity. Mutational analysis of the M1-muscarinic receptor has shown that a highly conserved Asp residue located at the begining of the second intracellular loop is important for efficient G protein coupling. In addition, it has been shown that specific regions located in the C-terminal region of this loop in the M2- and M4-muscarinic receptors are involved in selection and activation of G_i. These receptors as well as some adrenergic receptors (e.g., the β_2-adrenergic receptor) exhibit distinct structural determinants (either a B-B-X-X-B or B-B-X-B motif, where B is a basic amino acid residue and X is a nonbasic residue) for G protein activation. Interestingly, in other GPCRs, such as the FSH receptor, these motifs may be present in other intracellular loops but in a reversed fashion (B-X-X-B-B), notwithstanding determining coupling to the G_s protein class. The importance of the second intracellular loop in receptor interaction with G proteins is also exemplified in rhodopsin, the angiotensin II receptor, and more recently the GnRH receptor and the interleukin-8 receptor.

There are several studies indicating that regions in both the N- and C-ends of the C-terminal domain from GPCRs may be involved in G protein coupling and activation. For example, in rhodopsin, it has been shown that a synthetic polypeptide from the fourth intracellular loop is capable of interacting with G_t. In the β_2-adrenergic receptor, substitution of residues forming its fourth intracellular loop with the corresponding residues from the α_2-receptor has resulted in a reduced efficiency of coupling to its corresponding G protein. In this particular receptor, substitution of the conserved Cys residue (Cys^{341}) within the N-terminal segment of the cytoplasmic tail produces a significant reduction in its ability to stimulate adenylyl cyclase; as mentioned earlier, palmitoylation of this residue may induce anchoring of this portion of the C-terminus to the cell membrane, leading to the formation of a short fourth intracellular loop that in turn may facilitate the receptor–G protein coupling. There are, on the other hand, other receptors of the same family (e.g., the α_2-adrenergic receptors as well as in the LH/CG and M1- and M2-muscarinic receptors) for which this structural feature might not be important for G protein coupling. Thus, this shared structural motif may play differing roles in different receptor–G protein interactions. Finally, the FSH receptor contains a region closest to the C-end of the C-terminal domain exhibiting the B-B-X-B motif, which allows coupling of this region to the G_s protein.

There is some evidence indicating that the first intracellular loop may also be involved in G protein binding. This is observed, for example, in the case of the calcitonin receptor (G_s and $G_{q/11}$ coupled), in which substitution with the first intracellular loop of a particular isoform of the human calcitonin receptor containing a unique insertion of 16 amino acids in this loop may completely abolish production of inositol phosphates while allowing stimulation of the cAMP pathway by the host porcine receptor. Conversely, in the murine GnRH receptor cAMP signaling is highly dependent on particular residues located in the first intracellular loop, which apparently are not essential for activation of the phosphoinositide signaling pathway.

In conclusion, many intracellular domains are involved in receptor–G protein coupling. The most important domains lie within the membrane-proximal regions of the second and third intracellular loops. Although these domains are crucial for coupling specificity and activation of G proteins, other regions or particular residues located in the transmembrane helices or even in some extracellular loops may play major roles in signal transduction and G protein activation. In this vein, recent studies have shown that certain receptors related to the rhodopsin/β-adrenergic family carrying the AsnProXXTyr motif in their seventh transmembrane domain may associate with small G proteins, such as ARG and RhoA, leading to activation of the effector enzyme phospholipase D.

2.4. Mechanisms that Regulate Receptor Function

Several feedback mechanisms that control the functional status of GPCRs have been extensively analyzed. Among these, the following two are worthy of particular consideration.

2.4.1. Receptor Conformation and Signaling

It has been observed that GPCRs may exist in several dynamic states as disclosed by marked shifts in agonist affinity. Receptors are assumed to exist in an equilibrium between an inactive (R) and an active (R*) conformation. Binding of agonists exhibiting a preferentially higher affinity for R* promotes an isomerization step that stabilizes the receptor in a "relaxed" state and shifts the equilibrium toward the active R* conformation, promoting its interaction with G proteins and allowing the formation of a highly efficient ternary complex (agonist–R*–G protein). The requirement of such an isomerization step in the receptor molecule to switch into the R* conformation was unmasked by the particular behavior exhibited by a series of mutant adrenergic receptors, disclosing that certain intracellular regions of the receptor are critical to maintain the receptor in a relatively constrained, "unrelaxed" state in the absence of the agonist. Mutations in specific amino acid residues in the C-terminus of the third intracellular loop of the α_{1B}- (coupled to G_q-phospholipase Cβ) and α_{2A}- adrenergic receptors (coupled to the G_i-adenylyl-cyclase pathway), as well as in the same loop of the G_s-coupled β_2-adrenoreceptor, results in elevated basal, agonist-independent signaling activity as well as in a supersensitivity to agonists. Apparently, these specific mutations allow for conformational changes that eventually lead to a high level of constitutive receptor activation and interaction with the G proteins in the unliganded state, recreating the R* state promoted in intact wild-type receptors only by agonist binding. Analysis of several spontaneously occurring gain-of-function mutations that induce constitutive receptor hyperactivity have revelaed that multiple regions of the receptor molecule, including the extracellular, intracellular, and transmembrane domains, may be involved in the molecular mechanisms that constrain the G protein coupling of the receptor, a constrain normally relieved upon agonist occupancy. Several physiological implications arise from this mechanism of receptor activation. The equilibrium between R and R* implies that at any time a certain proportion of receptors resides in a constitutively active state even in the absence of an activating ligand. In this regard, some endogenous GPCRs may exhibit minimal, albeit detectable activity even in the absence of agonist. Thus, different receptor isoforms that bind the same ligand and activate the same effector may vary in their ability to spontaneously isomerize into the active state and therefore differ in their affinity for endogenous agonists. In this setting, it may be expected that agonists acting through receptor variants with an elevated constitutive isomerization rate will have a higher potency, whereas those acting through variants with a low isomerization rate will provoke a more graded range of responses. Such differences in their intrinsic capacity to isomerize could be accounted for by alterations of one or more critically placed amino acid residues and could be a reasonable explanation for the existence of distinct receptor subtypes for a single ligand. The existence of naturally occurring constitutively active GPCRs also may allow certain antagonistic drugs and endogenous ligands to elicit negative activity or inverse agonism (i.e., lowering the basal activity); as opposed to agonists, inverse agonists exhibit a high affinity for R and therefore decrease the R*/R ratio. On the other hand, abnormal constitutive activation of GPCRs provoked by particular mutations may lead to a variety of pathological conditions characterized by agonist-independent activation of the receptor and indirectly the G protein pathway.

2.4.2. Receptor Desensitization

Continuous or prolonged stimulation of a cell generally results in progressively attenuated responses to subsequent stimulation by the same agonist. This decrease in cellular sensitivity to further stimulation, or desensitization, may be considered as one of the most important mechanisms to protect the cell from excessive stimulation, particularly in conditions of high or prolonged agonist exposure. Phosphorylation-mediated deactivation of rhodopsin exemplifies the paramount importance of this protective mechanism; in this system, signaling is rapidly (high millisecond–second time scale) inactivated following exposure to light, thereby preventing a brief flash of light to be perceived as continuous illumination. In fact, transgenic mice bearing a mutant rhodopsin variant that is incapable of rapid desensitization, exhibit a marked prolongation of light-evoked electrical responses from single rods, an abnormality that eventually leads to retinal degeneration. Thus, for many cells bearing GPCRs the functional status of such receptors is determined by their phosphorylation-induced early desensitization state.

Early receptor desensitization occurs rapidly, within seconds to minutes of agonist exposure, and involves phosphorylation of intracellular domains of the stimulated receptors by second messenger dependent activated kinases (PKA or PKC) and/or by a special class of Ser/Thr-specific kinases called G protein-coupled receptor kinases (GRK1–6) with the subsequent uncoupling of the receptors from their respective G proteins and loss of downstream-signaling events. Homologous short-term desensitization occurs very rapidly, with a half-life from milliseconds to a few seconds and is observed when the cell's response decreases after high receptor occupancy by a specific agonist. In contrast, heterologous desensitization (nonagonist specific) is observed after several minutes of low amounts of agonist exposure and usually occurs when a cell's response to various agonists acting through different receptors decreases after receptor activation by a given agonist. Whereas heterologous desensitization generally involves phosphorylation by second messenger dependent kinases, homologous desensitization may be mediated through receptor phosphorylation triggered by both second messenger activated protein kinases and GRKs. Once the receptor is phosphorylated near sites that are required for G protein interaction, the subsequent binding of a group of soluble inhibitory proteins, the arrestins, amplifies the desensitization process initiated by phosphorylation and turns "off" the receptor by impeding its coupling to G proteins. Although the available information does not yet make it possible to unambiguously clarify the in vivo specificity of the several GRKs and arrestin isoforms, it is possible that, with few exceptions, short-term desensitization of GPCRs may be regulated by more than one GRK or arrestin isoform. Short-term desensitized receptors are resensitized by a complex process that involves arrestin-mediated receptor sequestration into endosomes, where it becomes dephosphorylated by a particular group of membrane-associated receptor Ser/Thr phosphatases.

A more profound deactivation process, called long-term desensitization, that involves a decrease in the net complement of receptors specific for a particular agonist, without detectable changes in receptor affinity, may be observed after hours or days of high levels of agonist exposure. Several biochemically distinct mechanisms, including receptor down-regulation (i.e., loss in total cellular content of functional receptors) and internalization (loss of surface receptors), subserve this type of cellular desensitization. Although long-term desensitization does not require previous phosphorylation of the receptor, in some systems it amplifies the loss of functional receptors triggered by short-term deactivation. For certain GPCRs, the net loss in receptor number and/or function involves the occurrence of various receptor and postreceptor events, including internalization, sequestration, and degradation of the receptor as well as changes in the rate of receptor synthesis by alterations in receptor mRNA levels and translation. In others, down-regulation does not require immediate changes in protein synthesis, and decreased receptor mRNA levels are observed only after prolonged exposure to the agonist. The overall process leads to a significant reduction in the number of functionally available receptors on the cell surface, where they could interact with agonists and activate signaling. Finally, in certain cells bearing GPCRs, the loss of capability to respond to prolonged agonist stimulation may also be achieved by mechanisms unrelated to the receptor itself but to downstream and negative feedback mechanisms involving inactivation of Gα-subunits by RGS proteins (regulators of G protein signaling; Watson et al., 1996) or loss of functional activity of effectors. An example of these postreceptor mechanisms is the GnRH responsive cell in which loss of responsiveness is due, on one hand, to receptor loss and on the other to the action of GAPs (GTPase-activating proteins) or RGS proteins, decrease in the functional activity of the Ca^{2+} ion channel, and reduced efficiency of mobilization by inositol triphosphate (a second messenger generated by the action of phospholipase Cβ) of Ca^{2+} from the intracellular pool.

3. THE HETEROTRIMERIC G PROTEIN FAMILY

3.1. General Features

Heterotrimeric G proteins are signal-transducing molecules belonging to a superfamily of proteins regulated by guanine nucleotides. These signal-carrying proteins are heterotrimers individually termed α- (molecular mass 39–52 kDa), β- (35–37 kDa), and γ- (6–10 kDa) subunits that are encoded by distinct genes. The most diverse genes are those encoding for the α-subunits; in fact, molecular cloning has revealed

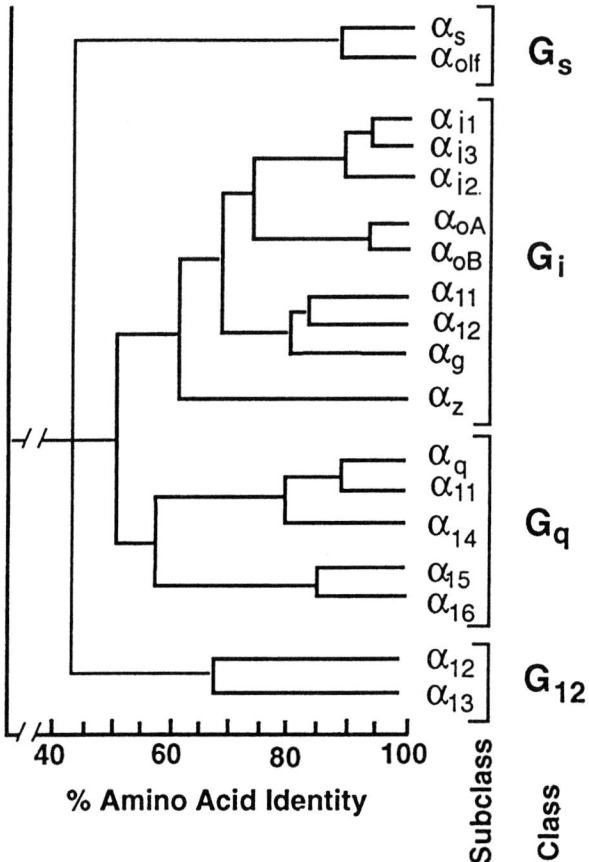

Fig. 5. Sequence relationships between mammalian Gα-subunits and family groupings. (Reproduced from Hepler and Gilman, Trends Biol Sci 1992; 17:383, with permission.)

in the regulation of Ca^{2+} and Na^+ channels. The $G\alpha_i$ class includes $G\alpha_{gust}$, $G\alpha_t$ ($G\alpha_{t1}$ and $G\alpha_{t2}$), $G\alpha_{i1}$–$G\alpha_{i3}$, $G\alpha_o$ ($G\alpha2_{o1}$ and $G\alpha_{o2}$), and $G\alpha_z$; they mediate different effects such as stimulation of cGMP phosphodiesterase ($G\alpha_t$), inhibition of adenylyl cyclase and activation of K^+ channels ($G\alpha_i$ class), and Ca^+ channel closure and inhibition of inositol phosphate turnover ($G\alpha_o$). The $G\alpha_q$ class includes $G\alpha_q$, $G\alpha_{11}$, and $G\alpha_{14}$–α_{16}; they are predominantly associated with activation of the enzyme phospholipase $C\beta_{1-4}$ which catalyzes the hydrolysis of the lipid phosphatidylinositol 4,5-biphosphate to form two second messengers, inositol 1,4,5-triphosphate (IP_3) and diacylglycerol, and activate PKC (Fig. 6). Finally, the $G\alpha_{12}$ class, defined by the α-subunits of G_{12} and G_{13}, constitutes the fourth and the most recently identified family; although the downstream effector systems for these G proteins still remain to be identified, recent studies have indicated that the signaling pathways regulated by these G proteins include modulation of the ubiquitous sodium-proton exchanger NHE_1 as well as regulation of cell growth and differentiation.

Gβ- and γ-subunits are also diverse, and at least five different β-subunits and eight γ-subunits have been identified (Table 3). Although this diversity in β- and γ-subunits may theoretically yield 30 or more different βγ complexes, not all the possible pairs can combine and some β-subunits are more able to interact and form stable combinations with particular γ-subunits than others. This specific dimerization raises the possibility of differential effector regulation by unique Gβγ complexes; in fact, regions have been localized on both subunits that are important in defining dimerization specificity. For example, construction of chimeras of $G\gamma_1/G\gamma_2$ has shown Phe[40] of $G\gamma_t$ to be particularly important for discriminating between dimer combinations; $G\gamma_t$ will not bind Gβ-subunits other than $G\beta_t$ whereas different Gγ-subunits with smaller side chains in this position will bind $G\beta_t$. Thus, diversity of α, β-, and γ-subunits would theoretically allow a limit of nearly 1000 oligomeric combinations between the three components comprising G proteins; such a high number of possible combinations not only facilitates the transduction of signals with a high degree of specificity but also helps to define receptor and effector signal sorting and specificity.

As discussed earlier, G proteins interact with the intracellular elements (loops and C-terminal domain) of the receptor, which implies that they must be associated in some way with the plasma membrane. In spite of this, G proteins do not bear strongly hydro-

the existence of at least 17 Gα genes that encode different members of the four main classes of Gα-subunits (G_s, G_i, G_q, and G_{12}) grouped on the basis of amino acid identity and effector regulation. Figure 5 shows the structural relationship between the different G protein α-subunits and Table 2 lists the different classes of Gα-subunits and some of their specific effects on their corresponding effectors. The tissue distribution of Gα-subunits is highly variable; it may be ubiquitous (or nearly ubiquitous as in the case of $G\alpha_s$ and some members of the $G\alpha_i$ and $G\alpha_q$ classes) or expressed in selected tissues ($G\alpha_{t1}$ in photoreceptor rod outer segments, $G\alpha_{olf}$ in the olfactory neuroepithelium, $G\alpha_g$ [gusducin] in the taste buds, $G\alpha_{15}$ and $G\alpha_{16}$ in some hematopoietic cells). The $G\alpha_s$ class includes $G\alpha_s$ and $G\alpha_{olf}$ and is involved in the activation of the various types (I–VIII) of the enzyme adenylyl cyclase to enhance the synthesis of the second messenger cAMP, which in turn activates PKA (Fig. 4); in some tissues this Gα-subunit also participates

Table 2
Characteristics of Mammalian Gα-Subunits

Gα-Subunit	Tissue Distribution	Size (kDa)	Effect(s)	Sensitivity to Toxins
G_s class				
$G\alpha_s 1-4$	Ubiquitous	45–52	↑ Adenylyl cyclase; regulation of Ca^{2+} channels	CTX
$G\alpha_{olf}$	Olfactory neurons			
G_i class				
$G\alpha_i 1$ to 3	Ubiquitous	40–41	↓ Adenylyl cyclase	PTX
$G\alpha_o 1$ and 2	Neural, endocrine	39	↓ Ca^{2+} channels	PTX
$G\alpha_t 1$ and 2	Retina	39–40	↑ cGMP-PDE	PTX/CTX
$G\alpha_{gust}$	Taste buds	41	↑ cGMP-PDE	PTX
$G\alpha_z$	Neural, platelets	41	↓ Adenylyl cyclase; ↓ K^+ channel; cell growth	—
G_q class				
G_q	Ubiquitous	41–43	↑ Phospholipase Cβ	—
$G\alpha_{11}$	Ubiquitous			—
$G\alpha_{14}$	Liver, lung,			—
$G\alpha_{15}$	kidney			—
$G\alpha_{16}$	Blood cells			—
G_{12} class				
$G\alpha_{12}$	Ubiquitous	44	Na^+, H^+ antiporter?; cell growth	—
$G\alpha_{13}$				—

cGMP-PDE, cyclic GMP-dependent phosphodiesterase; CTX, cholera toxin; PTX, pertussis toxin.

phobic membrane-spanning domains; they are usually found tightly associated with the cytoplasmic surface of the plasma membrane through anchoring mediated by myristoylation and/or palmitoylation of their α-subunit as well as by isoprenylation of their γ-subunit. Observations of the pheromone-responsive Gα protein of yeast and of $G\alpha_o$ strongly suggest that myristoylation of some particular α-subunits is also essential for high-affinity interaction with βγ dimers. Myristoylation and palmitoylation of Gα-subunits not only allow for subunit anchoring to the plasma membrane and subunit–subunit interaction but also influence the activity of the G protein. In contrast to myristoylation, palmitoylation is reversible and can be dynamically regulated. Further, this modification is closely related to the functional state and localization of some Gα monomers; for example, agonist or cholera toxin activation of the β-adrenergic receptor–G_s protein complex leads to dramatic changes in the palmitoylation status of $G\alpha_s$, causing its release from the plasma membrane and either its reversible redistribution along the inner surface of the plasma membrane and/or into the cytoplasm or its degradation. Apparently, a similar mechanism also operates in receptors coupled to the G_q/G_{11} class of proteins. Thus, control of Gα palmitoylation by receptor activation might be an additional mechanism to regulate the rate and intensity of signal transduction and effector activation.

Although no specific posttranslational lipidation has been described for the Gβ-subunits, γ-subunits have at their C-terminus a CAAX consensus polyisoprenylation signal. In fact, Gγ-subunits may be either farnesylated (e.g., the rod photoreceptor γ-subunit [γ_1]) or geranyl geranylated (e.g., the brain γ-subunits [$G\gamma_{2-4}$]) at Cys residues, depending on the nature of the amino acid in the X position. Although this lipophilic modification in the γ-subunit is not important for β/γ association, it is essential for colocalization and orientation of the subunits on the inner membrane surface as well for high-affinity interactions of the dimer with both Gα subunits and effectors.

Some types of Gα-subunits and Gβγ complexes can be phosphorylated in vivo and in vitro. Even though Gα-subunit phosphorylation on Ser, Thr, or Tyr residues does not change the activity of the protein, it may modify its association with other proteins. In some systems (e.g., human leukemia cells), the

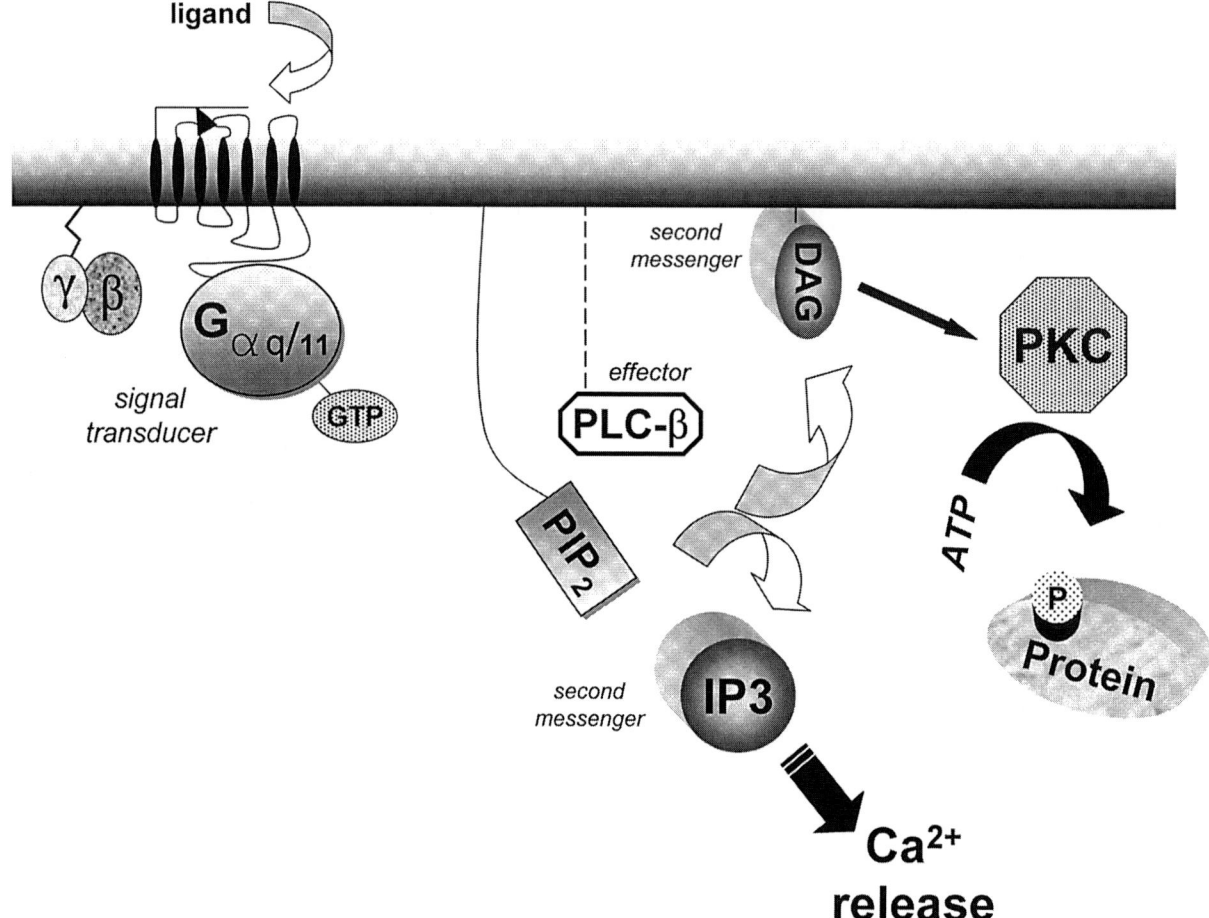

Fig. 6. Activation of the $G_{q/11}$-mediated signal transduction pathway. After ligand-induced receptor activation, the activated (GTP-bound) Gα-subunit of the heterotrimer dissociates from the Gβγ complex and stimulates the effector enzyme phospholipase Cβ (PLCβ) which in turn induces phosphatidylinositol 4,5-bisphosphate (PIP_2) hydrolysis leading to formation of the second messengers inositol 1,4,5-triphosphate (IP_3), which diffuses through the cytoplasm and releases Ca^{2+} from the endoplasmic reticulum, and diacylglycerol (DAG), which helps activate the enzyme protein kinase C (PKC), initiating a cascade of potential protein phosphorylation events.

Gβγ-subunits may be phosphorylated on His residues; the functional significance of this modification still remains to be elucidated.

3.2. The G Protein Cycle

Association and interaction of a receptor with a G protein promotes the development of a high-affinity binding state of the receptor for its corresponding agonist. Agonist binding and activation of a receptor exhibiting a high-affinity binding state is followed by a conformational change in the receptor and by activation of the trimeric Gαβγ protein complex by guanine nucleotide. The Gα-subunits have a single, high-affinity binding site for the guanine nucleotide that is comprised within a highly conserved amino acid sequence that forms a guanine nucleotide binding pocket. The GDP–Gα complex binds tightly to the βγ dimer and as such is inactive (i.e., in an "off" state). Receptor-promoted and Mg^{2+}-dependent GDP→GTP exchange within the Gα guanine nucleotide binding site leads to a conformational change of the subunit that causes G protein activation and dissociation of the trimeric complex into Gα–GTP and Gβγ ("on" state), allowing their interaction with and activation of different effector enzymes or ion channels (Fig. 7). Detailed structural analysis of G_t strongly suggests that Gα activation occurs through its nucleotide-dependent structural reorganization, whereas Gβγ activation results solely as a consequence of its release from the Gαβγ complex. This raises the interesting possibility that Gα may act as a negative regulator

Table 3
Characteristics of Mammalian Gβ- and Gα-Subunits

Subunit	Size (kDa)	Tissue Distribution	Gβγ Effectors
β$_1$	36	Ubiquitous	K$^+$ channel
β$_2$	35	Ubiquitous	(I_{KACh});
β$_3$	36	Retina	phospholipase A2;
β$_4$	37	Ubiquitous	Adenylyl cyclase
β$_5$	39	Neural	I, II and IV; phospholipase Cβ1–3; GRK (β-adrenergic); c-Src kinase
γ$_1$	8	Retina	
γ$_2$	6	Neural, endocrine	
γ$_3$	7	Neural, endocrine	
γ$_4$	7	Kidney; retina?	
γ$_5$	7	Ubiquitous	
γ$_6$	7	Neural	
γ$_7$	7	Ubiquitous	
γ$_{10}$	7	Ubiquitous	
γ$_{11}$	7	Ubiquitous	

of Gβγ by restricting its degrees of freedom and/or masking sites on the surface of Gβ that interact with downstream signaling components (*see* Section 3.3.). GTP-mediated G protein activation and subunit dissociation are accompanied by its separation from the receptor whose ligand binding state is then switched to the low-affinity state. Intrinsic GTPase activity of the α-subunit leads to hydrolysis of the terminal phosphate of bound GTP to yield GDP, release of inorganic phosphate, dissociation of the Gα-subunit from effector, reassociation with the βγ dimer, and switching of the G protein complex to the "off" membrane-associated state (Fig. 7). Hydrolysis of GTP is a relatively slow process that contrasts in some manner with the rapid termination of signaling that is observed in many tissues. Despite the slow in vitro rates of GTP hydrolysis by the Gα-subunits, some effectors, such as the myocardial Gα$_i$-activated K$^+$ channel and the Gα$_t$-cyclic GMP phosphodiesterase, are deactivated with extraordinary short $t_{1/2}$ values, which implies that at least some cells possess alternative mechanisms to accelerate Gα-subunit deactivation by GTP hydrolysis. In this regard, it has been observed that the effector enzyme phospholipase Cβ$_1$, which is activated by the G$_q$/G$_{11}$ protein class, can stimulate the steady-state GTPase activity of G$_q$ up to 50-fold when GTP/GDP exchange is catalyzed by the activated M$_1$-muscarinic receptor, thus acting as a GTPase-activating protein (GAP) specific for the members of this particular heterotrimeric G protein class. There is also evidence for a GAP effect of the retinal cyclic GMP phosphodiesterase on Gα$_t$. RGS proteins (regulators of G protein signaling) also act as GAPs that selectively and potently deactivate members of the Gα$_{i/o}$ and Gα$_q$ classes by accelerating the rate of intrinsic GTPase activity of the subunits almost 100-fold. RGS proteins, initially identified in screens for negative regulators of the yeast *Saccharomyces cerevisiae* mating pheromone responses and *Caenorhabditis elegans* egg-laying behavior, comprise more than 19 mammalian genes encoding for a conserved ~130 amino acid RGS core domain. Several studies suggest that RGS proteins act as GAPs by stabilizing the transition state conformation of the switch regions (those regions whose conformations are sensitive to the identity of the bound nucleotide), particularly the switch I and/or II, of the Gα-subunit (Fig. 8), thus favoring the transition state of the reactants. Structural studies on RGS4 suggest that RGS proteins may additionally act by competitively inhibiting effector binding to Gα switch regions. It has been recently shown that the palmitoylation of certain Gα-subunits may inhibit the GAP activity of RGS proteins, which raises the interesting possibility that nonconserved parts of RGS proteins may interact with Gα outside the switch regions.

3.3. Structural and Functional Relationships of G Protein Subunits

3.3.1. Gα-Subunits

Gα-subunits, the main effector activators, are at least 45–80% similar in amino acid sequence (Fig. 5); the regions of higher identity lie mainly in the domains that form the guanine nucleotide binding pocket. Whereas the conserved regions apparently perform similar functions in the various Gα-subunits, the variable domains reflect unique properties such as specific receptor, effector, and Gβγ interactions. Although the three-dimensional structure in the majority of the Gα proteins is still unknown, the resolved structures of Gα$_t$ and Gα$_{i1}$ and more recently Gα$_s$ have suggested mechanisms for GTP hydrolysis and revealed the conformational changes occurring during the transition from the inactive to the active state (Fig. 8). In general, the architecture of these Gα-subunits is likely to be shared by all members of the Gα-subunit protein family albeit with some potential differences when comparing the conforma-

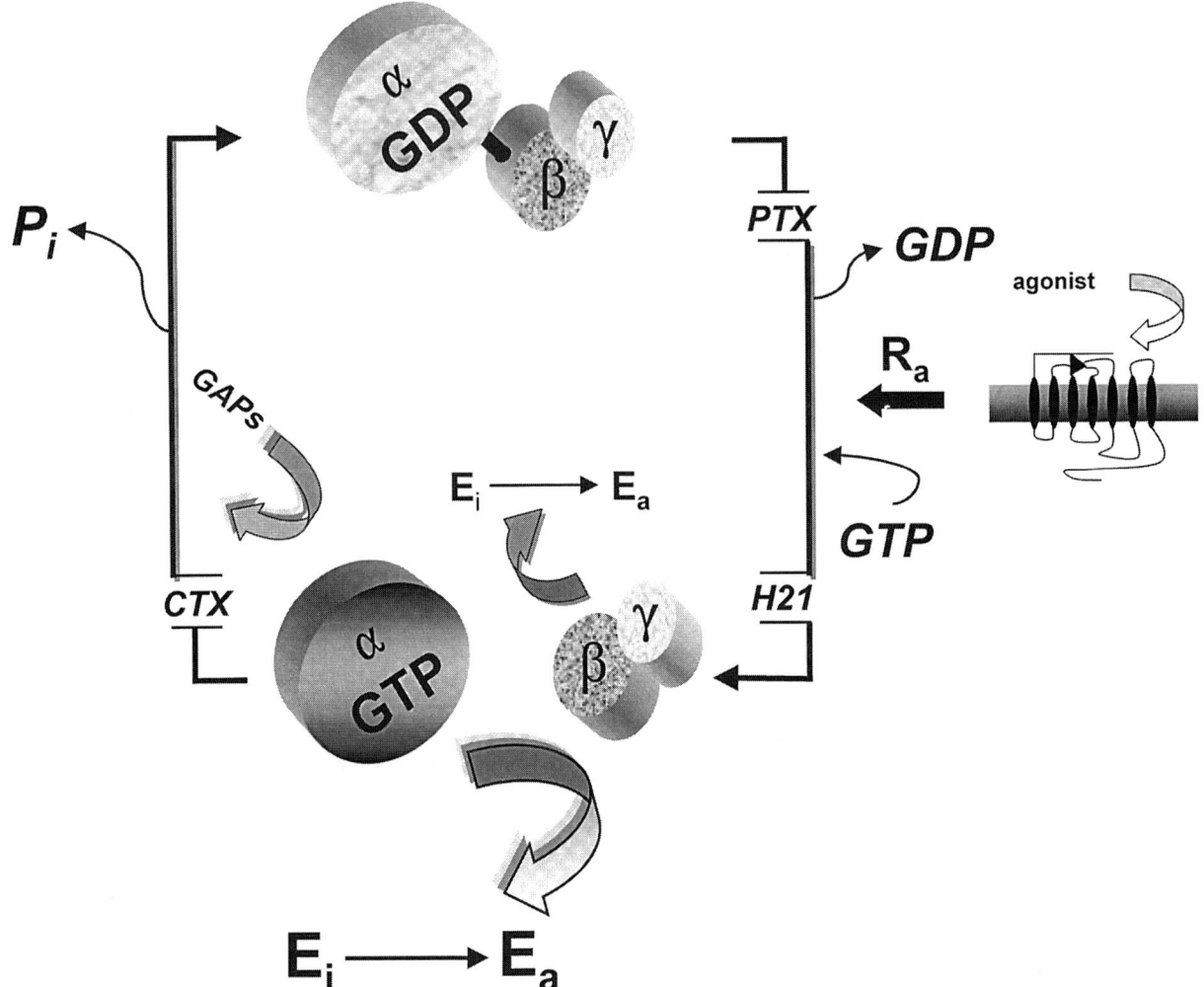

Fig. 7. The basic regulatory cycle of a G protein involving both GTP-induced activation and subunit dissociation and GTPase-dependent inactivation and subunit reassociation. The unoccupied receptor interacts with a specific agonist leading to the activation of the receptor (R_a). R_a interacts with the trimeric G$\alpha\beta\gamma$ protein complex promoting Mg^{2+}-dependent GDP→GTP exchange and subunit dissociation, allowing their interaction with effectors (E) (*see* text for details). The sites of action of pertussis toxin (PTX), which blocks signal transduction by uncoupling several G proteins from receptors as well as that of cholera toxin (CTX), which constitutively activates its substrate on α_s causing agonist-independent cAMP formation, are also shown. The *H21* mutation on α_s blocks signal transduction by preventing GTP activation. GAPs, GTPase-activating proteins; E_i, inactive effector; E_a, activated effector.

tional change between the GDP (inactive) and GTP (active) forms of the subunits. Basically, the Gα-subunit consists of two domains separated by a deep guanine nucleotide binding cleft (Fig. 8). One domain (the GTPase or Ras-like domain) contains the GTP binding motif, and is joined through two linker polypeptides to an α-helical domain that is unique to heterotrimeric G proteins. The GTPase domain contains the consensus sequences involved in GTP binding of all the GTPasas, the phosphate-binding loop, the Mg^{2+}-binding domain, and the guanine ring binding region. Interestingly, comparison of the recently resolved crystal structure of activated Gα_s with that of Gα_i suggests that their corresponding effector specificity is dictated primarily by the shape of the binding surface formed by homologous elements (switch II helix and the α_3–α_5 loop), whereas sequence divergence between these subunits may explain the inability of RGS to stimulate the GTPase activity of Gα_s. The α-helical domain consists of a long central helix surrounded by several shorter helices. Possible roles for the helical domain include increasing the affinity of GTP binding, acting as a tethered intrinsic GAP, and participating in effector recognition.

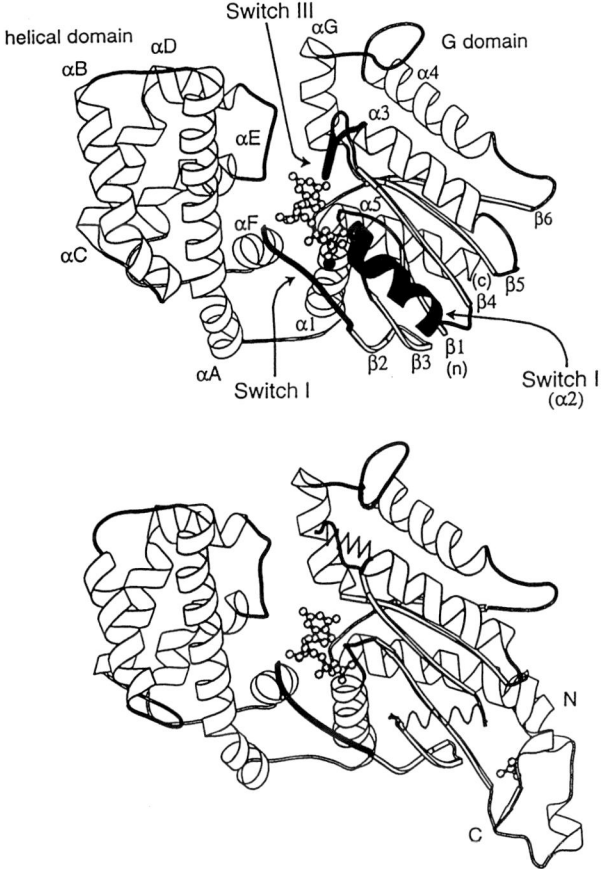

Fig. 8. (**A**) $G\alpha_{i1}$ in the GTPγS–Mg^{2+} complex with switch segments darkened and secondary structure elements labeled. (n) and (c) mark the positions at which the N- and C-termini become ordered in the crystal structure (residues 33 and 343, respectively). In the GDP complex (**B**), switches II and III are disordered, whereas the N-terminus, from residue 8, and the C-terminus, residue 354, are ordered. The GTP analog is depicted by a ball-and-stick model. The single solid sphere represents Mg^{2+} (Reproduced from Sprang, Annu Rev Biochem 1997; 66:639, with permission.)

The C-terminus region of Gα proteins is involved in the interaction and specificity of the protein with its cognate receptor and effectors. However, specificity may involve additional sites, as several Gα-subunits (e.g., $G\alpha_{oA}$ and $G\alpha_{oB}$) interact with different receptors despite being identical at their extreme C-termini. On the other hand, the N-terminus of the Gα-subunit, particularly its first 60 amino acid residues, appears to be one of the sites of interaction with the Gβγ dimer. As this region also associates with effectors, it is unlikely that the effector and the βγ-complex may bind the α-subunit at the same time. In this regard, when an effector is activated by both the α-subunit and the βγ-complex, they probably bind to distinct sites on the effector. Another possible site of Gα-subunit interaction with the Gαβ dimer lies within the switch regions (i.e., regions that undergo nucleotide-dependent conformational changes of the subunit [Fig. 8]), specifically in the switch II region. Apparently, the Gβγ dimer binds to a hydrophobic pocket that is present in Gα–GDP but that is closed by the GTP-dependent conformational change; this pocket is made up of the switch II region and the β-sheet hydrophobic core. In fact, a Cys in this switch region (Cys^{215} in $G\alpha_o$, cognate to Cys^{210} in $G\alpha_t$) can be chemically crosslinked to βγ, and the Gly^{226} to Ala mutation in $G\alpha_s$, which impairs the GTP-induced conformational changes in the switch II region, also prevents dissociation of the Gα-subunit from the βγ dimer. The Gα-subunit is apparently able to interact with both the β- and the γ-subunits.

Certain subclasses of Gα-subunits are ADP-ribosylated by the bacterial toxins from *Bordetella pertussis* (pertussis toxin or PTX) and *Vibrio cholerae* (cholera toxin or CTX). Cholera toxin catalyzes ADP-ribosylation of an Arg residue present within the GTP-binding domain of $G\alpha_s$ and $G\alpha_{olf}$ in a ligand-independent fashion and of $G\alpha_t$ and $G\alpha_i$ in a ligand-dependent manner. This CTX-induced modification of the Gα-subunit considerably decreases its intrinsic GTPase activity, causing constitutive activation of the protein. Mutation of the same Arg residue also inhibits GTPase activity, leading to a comparable constitutive activation. $G\alpha_i$, $G\alpha_t$, $G\alpha_{gust}$, and $G\alpha_o$ can be ADP-ribosylated by PTX on the fourth Cys residue from the C-terminus; this modification, which is enhanced by the presence of the Gβγ dimer (i.e., the presence of the G protein heterotrimeric state), leads to uncoupling of the G protein from the receptor without modifying other functions such as the guanine nucleotide exchange or GTPase activity (Fig. 7).

3.3.2. Gβγ-Subunits

Gβγ-subunits are tightly bound to each other through noncovalent hydrophobic interactions; the complex is resistant to tryptic cleavage and can be dissociated only by treatment with denaturing agents. Separate expression of these subunits leads to unstable Gβ-subunits and unfolded Gγ subunits. The Gβ-subunits display 58–90% identity, whereas the Gγ-subunits are more diverse. This probably accounts for a high degree of specificity for subunit association; $Gβ_1$ can associate with $Gγ_1$ or $Gγ_2$, whereas $Gβ_2$ can associate with $Gγ_2$ but not with $Gγ_1$, and $Gβ_3$ does not associate with either $Gγ_1$ or $Gγ_2$. Primary structures of the cloned Gβ-subunits as well as the crystal structure of the heterotrimeric $Gα_iβ_1γ_2$

protein and the transducin βγ dimer, indicate the existence of two distinct Gβ domains. The relatively conserved N-terminal domain forms an amphipathic α-helical coil that lies closely parallel to and interacts with the N-terminal helix of Gγ; this proximity is supported by the fact that Cys residues located in both subunits can be chemically crosslinked. The second domain of Gβ consists of seven 40- to 43-residue repetitive segments that contain a characteristic tryptophan–aspartic acid pair, termed WD40, which may be found in a variety of functionally unrelated proteins engaged in a variety of functions including signal transduction, cell division, transcription, cytoskeletal asembly, and vesicle function. The WD repeat regions in Gβ are probably involved in the regulation of the selectivity of the β-subunit to dimerize with a particular γ-subunit as well as in the capacity of the β-subunits to adopt multiple conformations, which in turn could greatly expand their number of protein–protein interactions and signaling pathways. On the other hand, Gγ-subunits are largely α-helical and their selectivity for the various β-subunits is determined by a stretch of 14 amino acids located in the middle of the γ-subunit; this region contains the Cys residue that is susceptible to crosslinking with the β-subunit (*see above*). In some systems, this subunit appears to be critical for determining the function of the Gβγ dimer. For example, the function of the $Gβ_1γ_1$ (localized in the retina) is quite different from that of $Gβ_1γ_2$ (found in the brain and other tissues); because the Gβ-subunits are the same, the functional differences between the two dimers are probably due to the different γ-subunits. In contrast, in other systems the Gβ-subunits appear to be the determinant factor for effector selectivity; this effect of Gβ is exemplified by the fact that although both $Gβ_1$ and $Gβ_5$ are able to activate phospholipase-$Cβ_2$ in a $γ_2$-dependent manner, $Gβ_5$, unlike $β_1$, does not stimulate the mitogen-activated protein (MAP) kinase or the c-Jun N-terminal kinase (JNK) pathways.

The association of the Gβγ dimer with Gα is the most well defined protein–protein interaction of the dimer. The Gβγ dimer is associated with the α-subunit through direct interactions between the prenylated form of the Gγ-subunit and Gα; the Gβ- and Gα-subunits are in close proximity but apparently they do not interact independently of the Gγ-subunit. Gβγ association with Gα increases the affinity of the latter subunit for GDP by about 100-fold, stabilizing the inactive state of the protein. On the other hand, the Gβγ complex associates with its cognate receptor through the C-terminus of the γ-subunit, which becomes physically available for such interaction during the course of receptor–Gβγ complex coupling. Thus, it seems that the Gγ-subunit plays a key role in conferring the specificity of the G protein interaction with receptors.

The Gβγ dimer play a significant role in signal transduction. It is required for efficient association of the Gα-subunit with the plasma membrane and for an effective interaction of the Gα-subunit with the receptor. In addition to these well-known effects, the Gβγ complex regulates the activity of several effectors such as the adenylyl cyclase types I, II, and IV. The extent of activation of these isoforms may vary depending on the specific βγ dimer involved; the type II isoform is effectively activated by $Gβ_1γ_2$, $Gβ_2γ_2$, and $Gβ_2γ_3$, whereas the activation induced by transducin $Gβ_1γ_1$ is about one-tenth as effective as that mediated by the other dimers. The activity of the other types of adenylyl cyclase is not apparently altered by the Gβγ complex. The Gβγ dimer also activates isoforms 1–3 of phospholipase Cβ with a rank order of effect of phospholipase $Cβ_3 \geq β_2 >> β_1$, different from that of $Gα_{q/11}$. Another important role of the Gβγ complex is the regulation of the muscarinic-gated K^+ channels, in which the $Gα_i$-subunit is also involved. However, the Gβγ dimer activates this particular channel more effectively than $Gα_i$. Gβγ-subunits activate also phospholipase A_2, the enzyme responsible for arachidonic acid formation; although arachidonic acid and its metabolites have been implicated in the activation of the muscarinic-dependent cardiac K^+ channels, their effect is relatively weaker as compared to that elicited by the Gβγ complex. In contrast to the members of the $Gα_i$ family, the dimer does not significantly influence the activity of the cardiac ATP-sensitive K^+ channel. The Gβγ dimer has been found to produce a 10-fold increase in the agonist-dependent phosphorylation of purified $β_2$-adrenergic receptor; this increase in phosphorylation, which eventually leads to receptor desensitization, occurs through a mechanism that involves interaction of the dimer with a specific region of the pleckstrin homology domain of the β-adrenergic receptor kinase and translocation of the kinase from the cytosol to the plasma membrane by the prenylated Gβγ-subunit, allowing the enzyme to recognize and interact with those intracellular elements of the receptor susceptible to modification by phosphorylation.

All classes of G proteins may also be potential mediators of long-term effects on gene expression and cell growth induced by mitogenic hormones that bind to G protein-linked receptors. The best established pathway by which heterotrimeric G proteins transduce signals destined to phosphorylate nuclear transcription factors promoting the transcription of genes controlling cellular growth is the G protein-activated/Ras-regulated MAP kinase cascade, a series of protein–protein interactions and phosphorylations. Ras proteins are low molecular weight GTPases involved in mitogenesis and differentiation triggered by activated enzyme (tyrosine kinase)-linked receptors. Activated (GTP-bound) Ras binds to and activates the Raf-1 Ser/Thr kinase; assembly of an activated Ras–Raf complex on the membrane promotes the phosphorylation of a second kinase, the cytosolic MEK (mitogen-activated/extracellular signal-regulated kinase). MEK specifically activates, through dual phosphorylation of Tyr and Thr residues, the MAP kinases (MAPKs) which are then translocated from the cytoplasm to the nucleus, where they phosphorylate a variety of transcription factors (Fig. 9). MAPKs activity can be regulated by PTX-sensitive and insensitive heterotrimeric G proteins in a Ras-dependent manner. Certain G_i-linked receptors (e.g., the M2-muscarinic and α_2-adrenergic receptors) triggers the Ras–MAPK cascade through a complex mechanism that may involve stimulation of Src tyrosine kinase activity by their Gβγ dimer. Src kinase, in turn, phosphorylates several substrates including tyrosine kinase receptors (e.g., the epidermal growth factor or the platelet-derived growth factor receptors) and/or the Shc adapter molecule, which binds to a second adaptor, the Grb2 protein, through Src-homology 2 (SH2) domains; since Grb2 and the guanosine nucleotide exchange factor Sos1 are constitutively associated, the Gβγ-mediated recruitment of Grb2 to activated (tyrosine-phosphorylated) Shc augments the Shc-associated guanine nucleotide exchange factor activity leading to GDP→GTP exchange and activation of Ras. G_q protein-linked receptors may activate MAPK through their corresponding Gα-subunits; this may occur through either Ras-dependent or -independent pathways (Fig. 9).

Ras-independent activation of MAPKs involves $G\alpha_q$ activation of PKC, which in turn phosphorylates and activates Raf-1, whereas the Ras-dependent mechanism involves $G\alpha_q$-promoted activation of downstream signals that lead to the formation of the Shc–Grb2–Sos signaling complex (Fig. 9), and the subsequent activation of Ras. Apparently, MAPK activation by $G\alpha_q$ proteins may be further modulated by the contribution of their Gβγ-subunits. Interesting, it has been recently shown that activation of the cytosolic MEK–MAPK cascade involves GPCR sequestration, as inhibitors of receptor internalization specifically block Raf-mediated activation of MEK. Another effector pathway whereby G proteins may transduce signals to the nucleus involves activation of JNK, a kinase related to MAPK but differentially regulated. Activation of JNK also proceeds through a cascade of protein–protein interactions, triggered by the low molecular weight cdc42 and Rac GTPases, that involve the activation of several kinases including the p21-activated kinase (PAK), MEK kinase (MEKK-1), and JNK kinase; JNK, like MAPK, phosphorylates and increases the activity of several transcription factors. Activation of M1-muscarinic receptors as well as expression of constitutively active $G\alpha_q$, $G\alpha_{12}$, or $G\alpha_{13}$ results in increased JNK activity. Although cAMP is not mitogenic in most cell types, G_s activation may also induce growth responses. In fact, constitutively active mutants of the G_s-linked thyroid-stimulating hormone and growth hormone receptor have been isolated from thyroid and pituitary tumors.

3.4. G Protein-Mediated Regulatory Mechanisms

In addition to the particular characteristics of the individual α- and β-subunits that allow for some degree of selectivity for interaction with each other, with the receptor, and with their corresponding effectors there are other mechanisms that tightly regulate the selectivity and intensity of the signal transduced by these molecules. In some systems the activation of particular effectors by the Gβγ complex is conditioned by Gα priming (e.g., stimulation of adenylyl cyclase types II and IV) whereas in others it seems to be independent (e.g., activation of phospholipase $C\beta_{1-3}$). On the other hand, whereas in certain systems the effects mediated by the Gβγ-dimer are the opposite of those triggered by $G\alpha_s$ (e.g., activation of type I adenylyl cyclase by the latter and inhibition by the former) in others, stimulation of the same effector (e.g., activation of the acetylcholine-regulated cardiac K^+ channels) occurs as a result of the activation of convergent pathways triggered independently by both elements, the Gα-subunit and the Gβγ complex. Activation of G protein-dependent divergent pathways leading to stimulation of different effectors may also occur as exemplified by phospholipase Cβ activation by PTX-sensitive G proteins (G_i and G_o) through their corresponding Gβγ complexes. Thus, the existence of convergent, divergent, and opposite pathways for effector activation as well as of differences in selectivity for protein–protein interactions among the G pro-

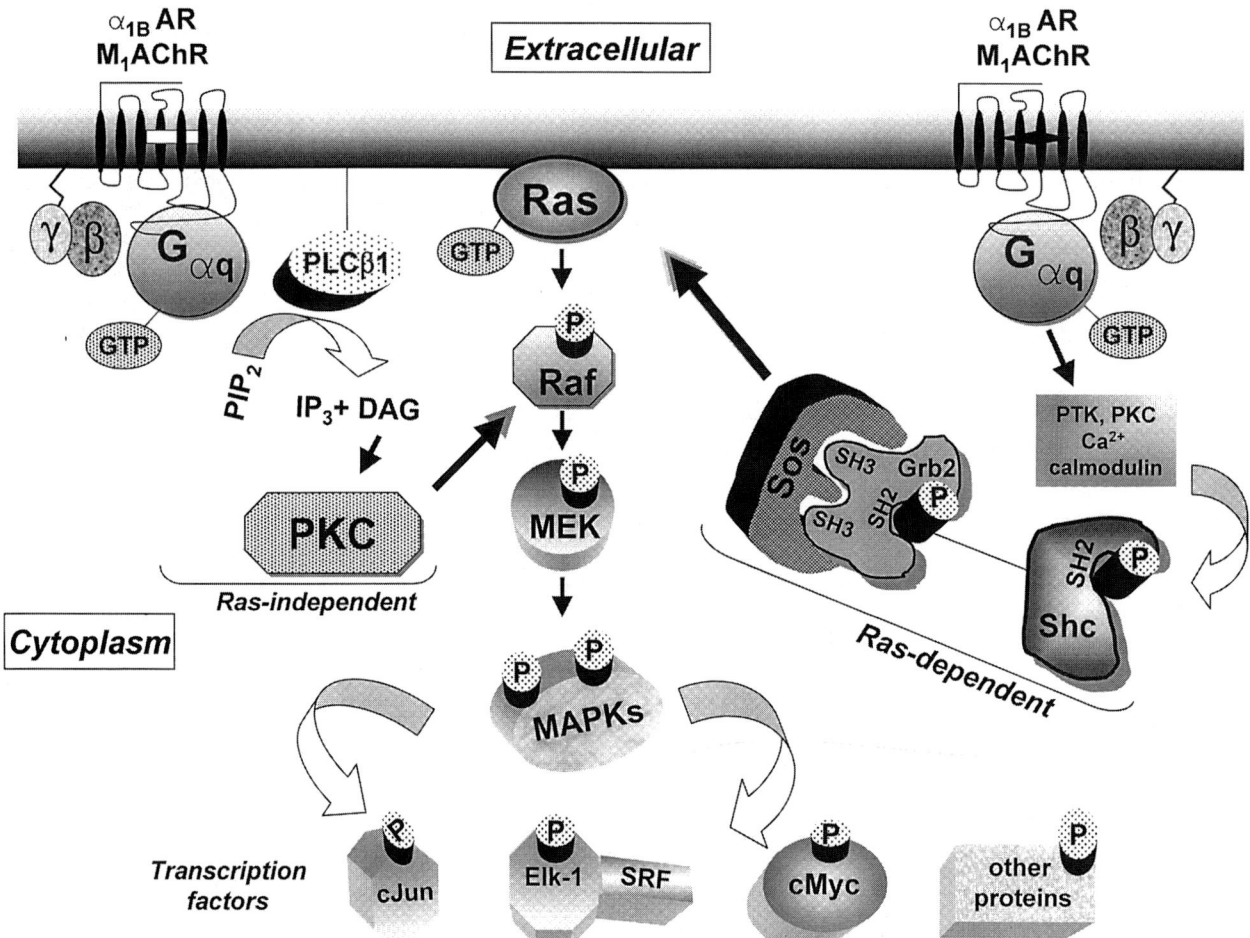

Fig. 9. Ras-dependent (**right**) and independent (**left**) pathways of MAPKs activation by GPCRs [e.g., α_{1B}-adrenergic receptor (α_{1B} AR) and muscarinic M1-acetylcholine receptor (M_1AChR)] coupled to PTX-insensitive ($G\alpha_q$) G proteins. $G\alpha_q$ activates phospholipase $C\beta_1$ resulting in the conversion of phosphatidylinositol 4,5-biphosphate to inositol 1,4,5-triphosphate (IP_3) and diacylglycerol (DAG), which in turn contributes to the activation of protein kinase C (PKC). PKC then activates Raf kinase by a still poorly understood mechanism. Activated $G\alpha_q$ also interacts with other intracellular signaling molecules (PKC, protein tyrosine kinases [PTK], etc.) to activate the Ras-regulated MAP kinase cascade (see text for details).

teins, allow for the existence of distinctly different mechanisms through which the cells may regulate the specificity of the signal conveyed by these effector-activating protein molecules.

Cells also limit their extent of response to external stimuli by decreasing the levels of specific G proteins involved in the signal transduction triggered by agonist-activated GPCRs. G protein down-regulation or desensitization has been demonstrated to occur in several in vitro systems in which a specific receptor is expressed at high levels. For example, high level expression and activation of the β_2-adrenergic receptor in NG108-15 cells (a neuroblastoma–glioma hybrid cell line) results in a significant decrease in the intracellular levels of $G\alpha_s$ (but not in $G\alpha_{i/o}$ or $G\alpha_{q/11}$) as compared to clones with a low expression level of the same receptor. Down-regulation of the $G\alpha_i$ and $G\alpha_{q/11}$ proteins has also been documented. Further, this desensitization is not apparently mediated by changes at the transcriptional or translational levels but rather by the enhancement in turnover of the specific G proteins that occurs upon receptor activation.

Another mechanism by which the activity of G proteins may be regulated is that mediated through phosducin. This phosphoprotein was originally purified from retina as a complex with the $G\beta\gamma$-subunit ($\beta_1\gamma_1$) of transducin as well as from the pineal gland. Phosducin inhibits retinal cGMP phosphodiesterase in vitro probably by preventing the association of the G protein subunits with the activated rhodopsin. In addition, phosducin mRNA has been detected in a variety of tissues, including the brain, heart, kidney, liver, lung, spleen, and skeletal muscle. In vitro, recombinant phosducin inhibits the GTPase activity of $G\alpha_s$, and $G\alpha_{i/o}$ as well as $G\alpha_s$-mediated adenylyl cyclase activation in a PKA-regulated manner. Apparently, regulation on effector activity by phosducin is exerted through the formation of a complex between phosducin and the heterotrimeric G protein, thus inhibiting G protein-mediated signaling either by preventing the formation of an activated state after GTP binding, or by sterically interfering with the G protein–effector interaction.

In addition to the receptor and G protein-dependent mechanisms described in the preceding sections, the selectivity and intensity of the signal triggered upon agonist-induced receptor activation may also be influenced by other factors. These include the functional expression of receptor splice variants (or isoforms) exhibiting different sensitivities and specificities to ligands and G proteins, receptor coupling to more than one G protein class, crosstalking between different G protein-transduced signals, different GPCRs or even different receptor systems, and compartmentation of signaling proteins. This latter control point for transmembrane signals may contribute not only to the specificity but also to the integrity of signal transduction pathways. Thus, the mechanisms controlling the activation of a particular GPCR by a given extracellular signal, the specificity of receptor–G protein–effector interaction, and the efficiency of the signal transduced are clearly multifactorial.

SELECTED READINGS

Berman DM, Giloman AG. Mammalian RGS proteins: barbarians at the gate. J Biol Chem 1998; 1269–1272.

Fields TA, Casey PJ. Signalling functions and biochemical properties of pertussis toxin-resistant G- proteins. Biochem J 1997; 321:561.

Freedman NJ, Lefkowitz RJ. Desensitization of G protein-coupled receptors. Rec Prog Horm Res 1996; 51:319.

Guderman T, Kalkbrenner F, Schultz G. Diversity and selectivity of receptor-G protein interaction. Annu Rev Pharmacol Toxicol 1996; 36:429.

Iñiguez-Lluhi J, Kleuss C, Gilman AG. The importance of G-protein βγ subunits. Trends Cell Biol 1993; 3:230.

Kjeldgaard M, Nyborg J, Clark FC. The GTP binding motif: variations on a theme. FASEB J 1996; 10:1347.

Spiegel AM. Defects in G protein-coupled signal transduction in human disease. Annu Rev Physiol 1995; 58:143.

Stern-Marr R, Benovic JL. Regulation of G protein-coupled receptors by receptor kinases and arrestins. Vitam Horm 1995; 51:193.

Strader CD, Fung TM, Tota MR, Underwood D. Structure and function of G protein-coupled receptors. Annu Rev Biochem 1994; 63:101.

Van Biesen T, Luttrell LM, Hawes BE, Lefkowitz RJ. Mitogenic signaling via G protein-coupled receptors. Endocrinol Rev 1996; 17:698.

2
G Proteins Large and Small Guanine Nucleotide Binding Proteins
Cellular Communication Networks

Thomas M. Wilkie and Michael A. White

CONTENTS
INTRODUCTION
SMALL GTPASES: RAS-LIKE G PROTEINS
LARGE GTPASES: HETEROTRIMERIC G PROTEINS
CONCLUSIONS
SELECTED READINGS

1. INTRODUCTION

All life on earth depends on the GTP hydrolyzing proteins EF-Tu (in bacteria) and EF-2 (in eukaryotes) for protein translation. The cycle of GTP binding and hydrolysis helps regulate peptide elongation as the ribosome processes along the mRNA. You can imagine that regulation of protein translation must be responsive to a cell's environment. The on/off switch of GTP binding and hydrolysis provides an effective regulatory mechanism. EF-Tu appears to be the primordial GTPase that gave rise to more than 100 distinct genes that encode large and small GTPases in eukaryotes. The roles of these later GTPases have diversified during the evolution of more complex eukaryotic organisms. The large GTPases, encoded by α-subunits of heterotrimeric G proteins, regulate numerous functions, including gametogenesis and fertilization, cell motility, feeding behavior, and responses to light and other environmental and hormonal cues. The small GTPases, such as Ras and its relatives, act as information nodes that process intracellular signaling cascades in response to growth

From: *Principles of Molecular Regulation* (P. M. Conn and A. R. Means, eds.), © Humana Press Inc., Totowa, NJ.

and cell survival factors. Other small GTPases regulate the transport of proteins into and out of the cell nucleus and the transport of proteins between cell organelles and the cell surface, including vesicle fusion during synaptic transmission. Thus, proteins of the GTPase superfamily serve essential functions in intercellular and intracellular communication. In each case, a similar mechanism of GTP binding and hydrolysis regulates protein activity.

2. SMALL GTPases: RAS-LIKE G PROTEINS

2.1. The Ras Superfamily

More than 100 monomeric G proteins have been identified in mammals. The diversity of this family of proteins in mammalian cells most likely reflects an evolutionary conservation of the use of GTPase switches to modulate dynamic cellular processes, such as growth, differentiation, and responsiveness to the external environment. The first monomeric or small GTPase identified, and by far the most intensely studied, is the Ras protooncogene. Within the last decade, a large family of Ras-like proteins has been identified whose members control diverse regulatory processes within eukaryotic cells. The superfamily of

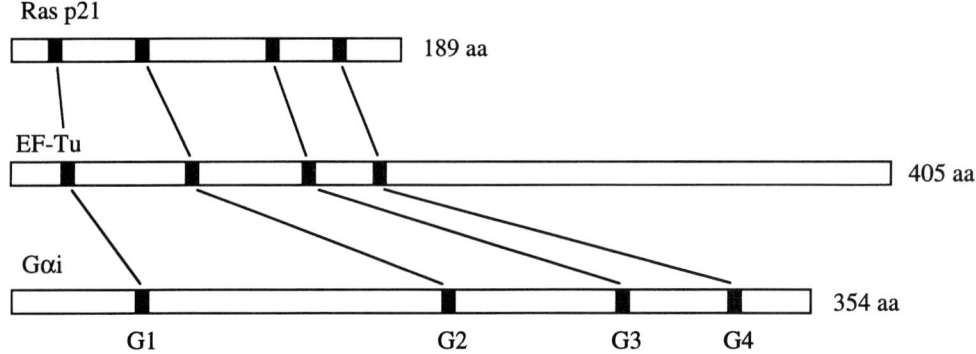

Fig. 1. Conserved amino acid sequence motifs in large and small GTPases. The regions indicated as G1–G4 contain amino acids that contribute to GTP binding and hydrolysis and are present in the same relative order in Ras-like GTPases, EF-Tu, and heterotrimeric Gα-subunits. The consensus sequence in Gα-subunits are G1, KLLLGAGESGKSTIV; G2, DXLRXRVTT GIVE, G3, MFDVGGQR; and G4, ILFLNKKD.

Ras-like GTP binding proteins are small, 20–25 kDa proteins, which share a common regulatory mechanism that cycles between an inactive GDP-bound state and an active GTP-bound state.

The Ras superfamily GTPases are classified into five major groups on the basis of structural and functional relationships, termed the Ras, Rho/Rac, Ran, Rab, and Arf subfamilies. There are at least 10 members of the Rac/Rho family of GTPases which regulate cytoskeletal organization and gene expression. More than 30 Rab GTPases, together with the ARF family of GTPases, are essential for regulation of vesicle budding, transport, and fusion. The Ran GTPases regulate the trafficking of proteins through the nuclear pore complex. Finally, there are at least 11 Ras family GTPases that play a major role in regulating mitogenic responses, and in coordinating the regulation of other Ras superfamily GTPases. Sequence divergence between the different classes of Ras-like proteins contributes to their distinct cellular functions. The characteristic sequence features of Ras superfamily GTPases and heterotrimeric Gα proteins are the residues required for binding and hydrolysis of the guanine nucleotide (Fig. 1). The Ras superfamily exhibits additional shared sequence features, including the switch regions, which change conformation and affinity for regulatory proteins depending on GDP or GTP binding and the C-terminal domain required for posttranslational isoprenylation.

2.2. Three-Dimensional Crystal Structures of GTP Binding Proteins

Regulatory features of the GTP-binding and hydrolysis cycle are shared by most of the small and large GTPase proteins (Fig. 2). Each step in the cycle is controlled by proteins that are usually dedicated to the regulation of closely related set of GTP binding proteins. Inside cells this is a dynamic process but key steps in the cycle of GTP binding and hydrolysis have been captured, like photographic snapshots, by X-ray crystallography. Three-dimensional molecular structures reveal similarities in form and function between Ras, EF-Tu, and Gα. Structures have now been solved for many GTP binding proteins in the GDP or GTP bound state, either alone or as complexes with regulatory and/or effector proteins (Table 1).

2.3. GTPase Binding Proteins Regulate GTP Binding and Hydrolysis

Table 2 provides examples of the various regulatory and effector proteins for each subfamily in the GTPase superfamily. Guanine nucleotide dissociation inhibitors (GDIs) stabilize the GDP-bound (inactive) state. GTP binding proteins are activated by guanine nucleotide exchange factors (GEFs) which promote the release of GDP and binding of GTP (and are sometimes referred to as GRFs-guanine nucleotide releasing factors). Intracellular signals are mediated by effector protein interaction with active GTP-bound proteins. GTPase activating proteins (GAPs) accelerate GTP hydrolysis to return GTP binding proteins to the inactive state. Specificity and duration of signaling is provide by these regulatory and effector proteins.

2.4. Regulation of Ras

Among the small GTPases, the regulation and function of the Ras subfamily is the best understood. This is primarily because of the intensive study of the role of Ras as an oncogene. H-Ras was originally identified as the transforming agent of Rous sarcoma virus, and

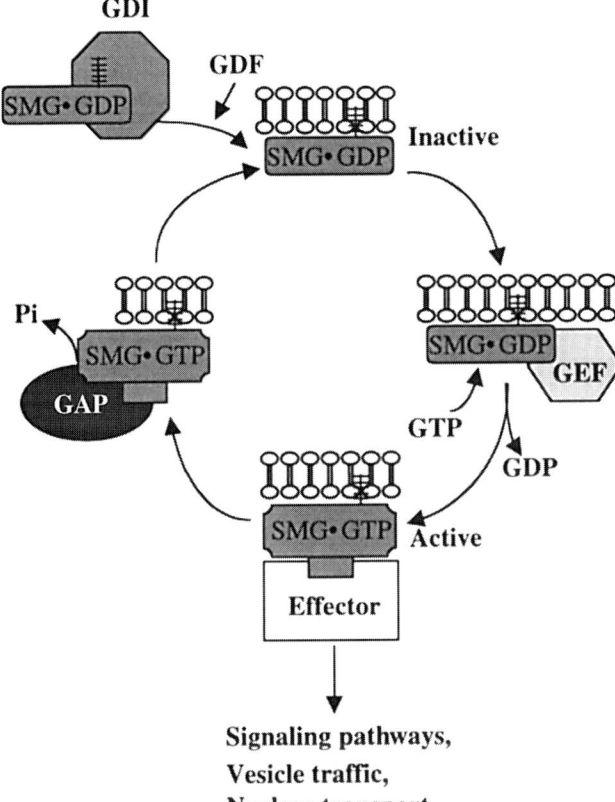

Fig. 2. Regulatory proteins interact with small molecular weight GTP binding proteins (smg) to control the cycle of GTP binding and hydrolysis. Smg proteins are inactive in the GDP-bound state and active in the GTP-bound state. Guanine nucleotide dissociation inhibitors (GDI) are either transport proteins that bind isoprenylated smg proteins (or, in the case of Gβγ-subunits, can stabilize the inactive GDP-bound state), GDI dissociation factors (GDF) promote membrane association of isoprenylated smg proteins, guanine nucleotide exchange factors (GEF) stimulate GDP dissociation from smg proteins, effector proteins are regulated by GTP-bound smg proteins, and GTPase activating proteins (GAP) accelerate GTP hydrolysis to attenuate smg protein acitvity.

independently, as a transforming element present in the genome of human bladder carcinoma cells. Two additional Ras isoforms were later found to be expressed in mammalian cells, K- and N-Ras (named for Kirsten, and neuronal Ras, respectively). Ras family proteins play a central role in the transmission of positive growth regulatory signals from the cell surface to the nucleus. This role is highlighted by the observation that close to 30% of all human tumors contain an activating mutation in one of the Ras isoforms. Ras activating mutations disrupt the cycle of GTP binding and hydrolysis. Mutations or posttranslational modifications of other G proteins have also

been identified that cause constitutive activation. In addition G proteins can be uncoupled from exchange factor catalyzed activation by posttranslational modification or mutation (Table 3). In the following sections we focus on the molecular events mediating Ras regulation of cell function as an example of the integral use of GTPases for the control of dynamic cellular processes.

Ras protein binds guanine nucleotide with high affinity, and has a relatively slow rate of GTP hydrolysis and exchange in vitro. Ras is localized to the inner leaflet of the plasma membrane of cells by the addition of lipid moieties to the C-terminus of the protein. This places Ras in close proximity to the transmembrane receptors responsible for sensing the external cellular environment. Ras can be activated upon agonist binding to tyrosine kinase associated receptors or heterotrimeric G protein-coupled receptors. G protein activation of Ras is mediated by Gβγ recruitment of RasGEFs through unknown mechanisms. The molecular events responsible for tyrosine kinase receptor mediated Ras activation are relatively well understood. Agonist engagement by the receptors leads to receptor clustering and activation of the cytoplasmic kinase domain and autophosphorylation of tyrosine residues. The phosphorylated tyrosines are recognized by adaptor molecules containing phosphotyrosine binding modules (SH2 domains) and polyproline binding modules (SH3 domains). These adaptor proteins, such as GRB2 in the case of Ras, recruit Ras GEFs (mSOS, mCDC25, RasGRF) to the activated receptor complex by associating with polyproline SH3 binding sites on the exchange factor. The formation of this multiprotein complex places the Ras exchange factors in the appropriate spatial context to act on neighboring Ras molecules, resulting in association of Ras with GTP.

Receptors can also have SH2 binding sites that are recognized by SH2 domain containing Ras GAPs (p120 GAP, NF-1). Recruitment of GAPs to activated receptors may result in negative feedback regulation of Ras proteins. GAPs selectively associate with the GTP-bound form of Ras to catalyze the GTP hydrolysis reaction. Mutations in the Ras protein that inactivate Ras GTPase activity render Ras insensitive to inactivation by GAPs. Mutations in the GEF or GAP proteins can also disregulate Ras activity and lead to oncogenic transformation.

2.5. Ras-Mediated Signal Transduction

GTP-bound Ras adopts a conformation that is competent to bind downstream effector molecules. Ras

Table 1
G Protein Crystal Stuctures[a]

G Protein Family	Structure		PDB No.[a]	Author
Ras-like	H-Ras P21-GppNp		5P21	E.F. Pai
	H-Ras P21(G12v)-GTP		521P	U. Krengel
	H-Ras/Sos	Complex	1BKD	P.A. Boriack
	Ras/RalGDS	Complex	1LFD	L. Huang
	Ras/RasGAP	Complex	1WQ1	K. Scheffzek
	Rap1A/Raf	Complex	1GUA	N. Nassar
	Rac1-GppNp		1MH1	M. Hirshberg
	Rho-GDP-AlF$_4^-$/RhoGAP	Complex	1TX4	K. Rittinger
	Rab3A-GTPγS/rabphilin	Complex		C. Ostermeier
	ARF1-GDP		1HUR	J.C. Amor
	ARF1-GDP/ARFGAP	Complex		J. Goldberg
	Ran-GDP		1QMA	R. Bayliss
EF-Tu	Ef-Tu-GDP		1EFM	F. Jurnak
	Ef-Tu/Ef-Ts	Complex	1AIP	Y. Wang
	Cysteinyl-tRNA/EF-Tu-GTP		1B23	P. Nissen
Heterotrimeric G proteins	Gα_{i1}-GDP		1GDD	M.B. Mixon
	Gα_{i1}-GDP-AlF$_4^-$		1GFI	D.E. Coleman
	Gα_t-GDP-AlF$_4^-$		1TAD	J. Sondek
	Gαt-GTPγS		1TND	J.P. Noel
	Gβγ		1TBG	J. Sondek
	Gβγ/Gα_{i1}-GDP	Complex	1GP2	M.A. Wall
	Gβγ-G$\alpha_{t/i1}$-GDP	Complex	1GOT	D.G. Lambright
	Gβγ/Phosducin	Complex	2TRC	R. Gaudet
	Gα_s-GTPγS/adenylyl cyclase	Complex	1AZS	J.J. Tesmer
	Gα_{i1}-GDP-AlF$_4^-$/RGS4	Complex	1AGR	J.J. Tesmer

[a]Protein Data Bank (PDB) at the Brookhaven National Laboratory
Web site address - http://www.rcsb.org

effectors have been operationally defined as molecules that mediate Ras-induced signal transduction events through direct association with GTP-bound "active" Ras. Work from a number of laboratories within the past 5 years has led to the notion that Ras functions as a key regulatory node in a complex signal transduction network. Ras appears to be able to integrate multiple upstream signals with multiple downstream events. Signaling specificity may be regulated by Ras interaction with different GEFs that are responsive to distinct inputs. Ras regulates distinct outputs through association with multiple effector molecules.

Ras can regulate at least three distinct signal transduction cascades. Each of these routes of information flow can function independently or coordinately to regulate distinct biological responses to Ras activation. The first of the three pathways to be discovered was Ras activation of a mitogen-activated protein (MAP) kinase cascade. Ras-GTP binds Raf, a serine/threonine kinase, to activate the MAP kinase cascade and complete a signaling circuit from the outside of the cell to the nucleus. The second Ras-activated pathway stimulates a lipid second messenger mediated cascade. In this pathway, Ras activation of phosphoinositol-3-kinase (PI$_3$K) results in the generation of phosphatidylinositol-3,4,5-trisphosphate [PI(3,4,5)P$_3$] second messengers that complete signaling circuits from the cell surface to the actin cytoskeleton. The third Ras activated pathway is a GTPase cascade, in which the Ras effector protein Ral guanine nucleotide dissociation stimulator (RalGDS) potentially links Ras activation to vesicle transport. While each of these three effector pathways can mediate distinct cellular responses to Ras activation, there are multiple integration points that result in coordinated regulation of cell proliferation. For example, activation of Raf, PI$_3$K, and RalGDS all lead to induction

Table 2
GTPase Superfamily Signaling Components[a]

GTPase	GEF	GDI	GAP	Effector	Second Messenger	Targets	Cell Response
Ras	SOS	?	RasGAP	Raf-1		MEK1	Proliferation, differentiation
Rho	p115	RhoGDI	RhoGAP	PIP-5K	PIP_3	Cytoskeletal proteins	Stress fiber formation
EF-Tu	EF-Ts		mRNA-ribosome	aa-tRNA		Ribosome-elongating peptide	Protein translation
$G\alpha_s$	GPCR β-adrenergic	Gβγ	?	↑ AC	↑ cAMP	↑ PKA ↑ Na channels	Various
$G\alpha_i$	GPCR M2 muscarinic	Gβγ	RGS	↓ AC	↓ cAMP	↓ PKA ↑ K^+ channels	Various
$G\alpha_t$	GPCR rhodopsin	$G\beta_2\gamma_1$	RGS9	↑ cG-PDE	↓ cGMP	↓ cGMP-gated Na^+ channels	Hyperpolarization
$G\alpha_q$	GPCR M1 muscarinic	Gβγ	RGS	↑ PLCβ	↑ IP_3 + DAG	↑ PKC, ↑ Ca^{2+} release	Various
$G\alpha_{13}$	GPCR thromboxane A_2	Gβγ	p115/RhoGEF	↑ p115/RhoGEF	?	RhoA	Contraction

[a] Only one of several possible regulatory or effector proteins is shown with each GTPase.

Table 3
Examples of Diseases and Mutations in the GTPase Superfamily

Mutation/Disease Organism	Mutation	Activity	Signal	Symptom
Ras protooncogene (various aa substitutions)	Somatic	↑	↑	Neoplasia
RasGAP protooncogene (various mutations)	Somatic	↓	↑	Neoplasia
NF1/GAP protooncogene (various mutations)	Somatic	↓	↑	Neoplasia
Clostridium botulinum C3 toxin Rho ADP-ribsoylation	NA	↓	↓	Inhibition of Rho-mediated cell shape changes and transcriptional activation
$G\alpha_s$ (R201H/C or Q227L) gsp	Somatic	↑	↑	Pituitary and thyroid adenomas
$G\alpha_s$ (R201H) McCune–Albright syndrome (MAS)	Germline	↑	↑	Acromegaly
Vibrio cholerae: Cholera toxin $G\alpha_s$ ADP-ribosylation	NA	↑	↑	Intestinal secretion of salt/water causing severe dehydration
$G\alpha i_2$ (R179) gip	Somatic	↑	↑	Adrenal and ovary adenomas
$G\alpha_s$ (various aa substitutions) pseudohypoparathyroidism type Ia Albright's Hereditary Osteodystrophy (AHO)	Germline	↓	↓	Obesity and short stature
$G\alpha_{t1}$ (G38D)	Germline	↓	↓	Night blindness
Bordetella pertussis $G\alpha_i$ ADP-ribosylation	NA	↓	↓	Whooping cough
Axin-RGS protein (deletion/insertion)	Germline	?	?	Developmental defect causing anterior duplciation

of active cyclin–cdk complexes that are required for cell cycle progression (*see* Part III).

2.6. Crosstalk on Small G Protein Signaling Pathways

The observation that Ras regulates multiple effector pathways explains how activated Ras can induce diverse cellular responses. However, it also begs the question of how cells determine which Ras effector pathway(s) are activated in response to various stimuli. In the context of expression of oncogenic Ras, it appears that convergence of several coactivated Ras effector pathways contribute to an oncogenic phenotype. The transient activation of wild-type Ras in response to upstream stimuli may result in activation of distinct effector pathways depending on the cellular context of the signaling events.

Recent investigations suggest at least three mechanisms controlling specificity of Ras signal transduction events. These include: 1) cellular compartmentalization of signal transduction molecules through scaffolding proteins that organize functional multiprotein complexes, 2) posttranslational modification of Ras effectors in coordination with other signaling events, and 3) Ras-isoform specific association with Ras effectors. Specificity in signaling requires appropriate regulation through each of these mechanisms.

Crosstalk between small G proteins fine tunes the regulation of their signaling pathways. Ras activates other small G proteins by stimulating their guanine nucleotide exchange factors. In one case, Ras indirectly activates Rac family GEFs, which are stimulated by association with PIP_3 produced in response to PI3K activation by Ras. The Rac subfamily of GTPases also interact with multiple effectors to regulate the actin cytoskeleton, induction of immediate early gene expression, and cell cycle progression. Rac is likely to be integrated with the regulation of Raf kinases because the Rac effector protein called p21-activated kinase (PAK) activates Raf kinase. Recall that Raf kinase is also activated via direct interaction with Ras. Another arm of the Ras effector pathway is through Ras binding to the guanine nucleotide exchange factor RalGDS, which activates Ral GTPases. Interestingly, it appears that Ral provides Ras with a negative feedback mechanism because a Ral effector protein is RacGAP, which attenuates signaling through Rac. These observations highlight the complexities of crosstalk in the small G protein signaling pathways which are required to produce coordinated cellular responses to distinct stimuli.

3. LARGE GTPases: HETEROTRIMERIC G PROTEINS

3.1. G Proteins Transduce Extracellular Signals to Regulate Intracellular Responses

The family of large GTPases are encoded by α-subunits of heterotrimeric G proteins. All higher eukaryotes, including yeast, *Dictyostelium*, plants, and animals utilize G protein signaling to mediate intracellular responses to extracellular stimuli. G protein signaling systems have several components, each encoded by distinct multigene families. Extracellular signals received by seven transmembrane domain receptors are coupled by heterotrimeric G proteins to the regulation of effector proteins that generate intracellular second messengers. Thus, G proteins act as signal transducers. Cells must also recover from stimulation to maintain homeostasis. Recovery can be mediated by feedback regulation. Receptor desensitization is one method of regulation, discussed in Chapter 1. A new family of proteins was recently identified, termed Regulators of G protein Signaling (RGS), which can rapidly attenuate G protein signaling through feedback regulatory mechanisms. Activation and inactivation may occur within a G protein signaling complex, composed of perhaps six distinct proteins, which acts like a molecular machine to convey extracellular signals and coordinate intracellular responses. All of the molecular components of G protein signaling cascades, from ligands to effector proteins and their targets, are required to relay information from outside to inside the cell.

3.2. Regulation of GTP Binding and Hydrolysis

The essential regulatory feature of G protein signaling is the cycle of GTP binding and hydrolysis on the Gα-subunit. Many of the proteins that interact with the Gα-subunit regulate its transit through this cycle. This is similar to the mechanisms that regulate the activity of Ras and other small GTP binding proteins (Table 2). In heterotrimeric G protein pathways, signaling is initiated upon receptor binding of extracellular agonists, such as nucleotides, small peptides, or the activation of a prebound chromophore by light. These receptor-dependent steps in the signaling pathway are discussed in Chapters 1 and 2.

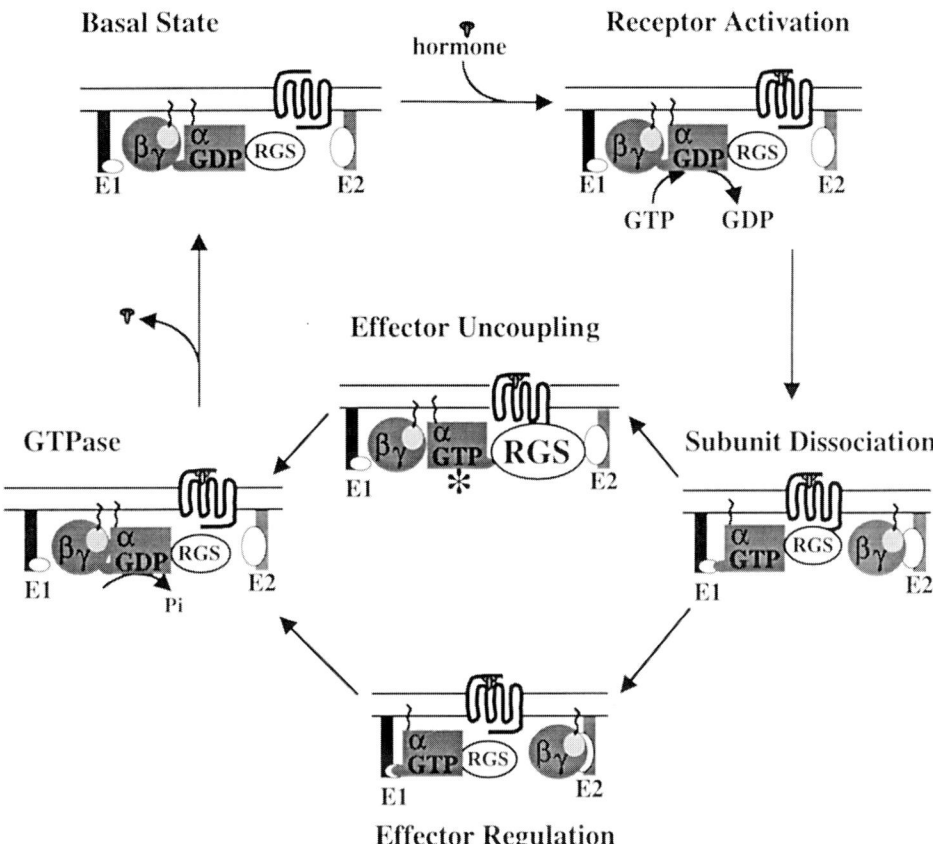

Fig. 3. Heterotrimeric G protein activation cycle. In the basal state, inactive Gα–GDPβγ form a stable heterotrimer. Hormone binding to receptor catalyzes the exchange of GDP for GTP on the Gα-subunit. Crystal structures of Gα complexed with Gβγ, effectors, and RGS proteins reveal that the switch regions of Gα (represented by the tongue on Gα) contribute intermolecular contacts to each of these Gα binding proteins. GTP binding promotes the dissociation of Gβγ from Gα-GTP, which can then independently regulate their respective effector proteins. The intrinsic GTPase of the Gα-subunit can hydrolyze GTP to terminate signaling. Alternatively, regulators of G protein signaling (RGS) proteins may initially associate as an inactive component of the receptor complex. Activation of RGS protein GAP activity can uncouple GTP binding on the Gα-subunit from effector protein regulation. Thus, signaling can be regulated both by the binding of extracellular hormones and the activation of intracellular RGS proteins.

The heterotrimeric G protein complex, composed of α-, β-, and γ-subunits, is required for receptor to stimulate intracellular signaling (Fig. 3). Gα is the largest subunit in the complex, ranging from 41 to 45 kDa, and binds the guanine nucleotide. Gβγ forms a stable heterodimer of 38 and 8 kDa, respectively. In the inactive state, Gα binds GDP to assume a conformation with high affinity for Gβγ. Gα–GDPβγ can interact with the intracellular loops and tail of the receptor. Agonist binding to receptor on the outside of the cell conveys conformational changes to the intracellular surfaces of the receptor that catalyze the dissociation of GDP from the Gα-subunit. As true of Ras-like GTPases, the Gα-subunit is most stable when guanine nucleotide sits in the binding pocket. In the open state, the Gα-subunit will bind GDP or GTP with equal affinity but the cytosolic concentration of GTP is 100-fold higher that GDP, which promotes GTP binding and progression of the activity cycle. GTP binding induces a conformational change in the switch regions of the Gα protein. The switch regions bind at least three different proteins, Gβγ, effector proteins, and RGS proteins, at different times during the cycle of GTP binding and hydrolysis. Conformational changes in the switch regions favor dissociation of the previous binding partner and association of the next partner in the cycle. GTP binding to the Gα-subunit weakens its interaction with Gβγ and receptor but induces a conformational change that favors interaction with effector proteins. Both Gα-GTP and Gβγ can regulate effector proteins. Thus information is transduced from extracellular ligand binding to effector proteins which regulate the production of second messengers, such as cAMP and Ca^{2+}, inside the cell. Gα remains active until GTP is hydrolyzed. Conformational changes in Gα-GDP drives dissociation of the effector protein, thus terminating signaling, and favors reassociation of Gβγ. The cycle of GTP binding and hydrolysis can be repeated in the presence of persistent agonist.

The tertiary and quaternary structures of Gα, Gβγ, and the heterotrimer have been solved by X-ray diffraction (Table 1). Gα and the small GTPases share structural similarities in the guanine nucleotide binding pocket and regulation of the switch regions that change conformation in response to GDP or GTP binding. In addition, Gα-subunits have a large helical domain which provides an arginine residue that catalyzes GTP hydrolysis. Protein–protein interactions with GEFs, GAPs, and effectors are mediated by amino acids that compose unique features on the surface of the various GTPases. The crystal structures of G proteins complexed with their regulatory proteins provide molecular rationales for the functional similarities and differences in large and small GTPases.

The RGS proteins are important regulators of G protein activity. They were first identified in genetic studies in yeast and nematode worms to act in opposition to G proteins. This lead to the discovery that RGS proteins are GTPase-activating proteins (GAPs) for Gα-subunits, thereby inactivating or attenuating signaling. Mammals express more than 20 different proteins that contain a common sequence motif of about 130 amino acids known as the RGS domain. The RGS domain from several different RGS proteins can stimulate GTP hydrolysis about 2000-fold above the basal activity of the Gα-subunit. An X-ray crystal structure of RGS4 bound to $Gα_{i1}$–GDP–AlF_4^- (Table 1) indicates that RGS proteins stabilize the transition state and thereby accelerate GTP hydrolysis on the Gα-subunit. All of the catalytic residues are supplied by the Gα-subunit. This is in contrast to Ras, which depends on Ras-GAP to provide the catalytic arginine finger that stabilizes the gamma phosphate leaving group (see Table 1). Recent studies have shown that some RGS proteins also serve as scaffolds to help assemble active receptor complexes that are tightly coupled to effector proteins (Fig. 4). Thus, RGS proteins regulate the kinetics of the G protein activity cycle, and some apparently can accelerate both the activation and inactivation of signaling. This leads to

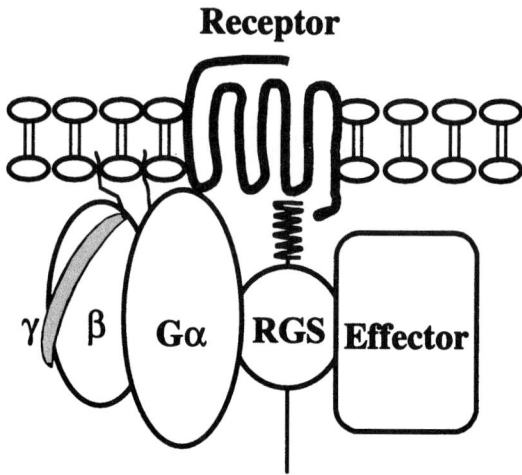

Fig. 4. A model of the G protein-coupled receptor signaling complex. RGS4 has been modeled to reside within a complex of M1 receptor, $G_oαβγ$, and GIRK1/2 channel (effector) in Xenopus oocytes injected with recombinant RNA and also M3 receptor, $G_qαβγ$, and PLCβ (effector) in pancreatic acinar cells dialyzed with recombinant RGS4 protein.

the idea that G protein signaling initiates within a multiprotein complex including receptor, G protein, effector, RGS protein, and perhaps membrane lipids and other proteins. In addition, some RGS proteins inhibit signal transduction from certain receptors better than other receptors coupled to the same G protein. This allows the cell to regulate signaling specificity through a combination of receptor-dependent activation by agonists and attenuation by RGS proteins.

3.3. Signaling Specificity of Heterotrimeric G Proteins

Hormones are molecular messengers that instruct or alert the body to physiologic changes. Epinephrine is released into the bloodstream from the adrenal medulla in an abrupt response to immediate danger, commonly known as the fight-or-flight response. Air intake is increased to power the body's response to danger. This is aided by epinephrine causing relaxation of bronchial smooth muscle cells. Epinephrine also vasodilates arterioles in skeletal muscle. At the same time, blood flow to the skin is restricted by epinephrine causing constriction of smooth muscle cells that line blood vesicles. How does the same hormone cause similar cell types in different locations to undertake opposite reactions? The answer is that homologous receptors of distinct functions are expressed in these different cells. Both types of receptor bind epinephrine but they couple to different G proteins inside the cell, which activate different effector proteins to generate distinct second messengers. Thus, similar cell types respond in opposite manner to the same hormone.

Receptor diversity and mechanism of action are described in Chapters xxx. The specificity of G protein interaction with receptors partly depends on amino acids in the third cytoplasmic loop and carboxy tail of heptahelical receptors convey specificity but other regions of the receptor are also required for interaction. A series of genetic experiments indicates that the identity of Gβ and Gγ also influences receptor selectivity and G protein coupling. Mammals express 16 Gα-, 5 Gβ-, and 12 Gγ-subunits, many of which are coexpressed in the same cell and can interchangeably form heterotrimeric proteins. The combinatorial matrix could provide a repertoire of distinct heterotrimeric G proteins to selectively couple all of the receptors expressed in the body. Although not the whole story, complexes of distinct heterotrimeric G proteins contribute to signaling specificity.

G protein α-subunits in fungi, plants, *Dictyostelium,* and animals (which compose the crown of eukaryotic organisms) are recognized as descendants of a common progenitor gene. The Gα-subunits have sequence similarities within the GTP binding motifs and switch regions, which are clearly distinct from the small GTP binding proteins. This is reflected by at least 35% amino acid identity among all Gα-subunits. The regions of sequence diversity convey specificity in the various signal transduction pathways.

In animals, sequence diversity between different Gα classes contributes to specific effector protein interactions. Mammals express 16 Gα genes that are grouped into four classes, G_s, G_i, G_q, and G_{12}, based on sequence, gene structure, effector protein targets, and regulation by RGS proteins (Fig. 5). The G_s class activate adenylyl cyclase, G_i are ADP ribosylated by pertussis toxin (with the exception of Gα$_z$) and either inhibit adenylyl cyclase, activate cGMP phosphodiesterase, or regulate ion channels, G_q activate phospholipase Cβ (in a pertussis toxin resistant manner), and one member of the G_{12} class was recently shown to activate p115Rho-GEF. Mammals may express other Gα genes but they have not been found after extensive searches using degenerate polymerase chain reaction (PCR) or sequence searches of the expressed sequence tag (est) databases. By contrast, all G protein genes in the nematode *Caenorhabditis elegans* are known because the genomic sequence is completely determined. The *C. elegans* genome encodes 2 Gβ-, 2 Gγ-, and 20 Gα-subunit genes. *C. elegans* (and the fruit fly *Drosophila melanogaster*) express one obvious ortholog within each Gα class. Surprisingly, the other 16 Gα genes encoded in the worm genome are not clearly related to any Gα class, or any other Gα gene expressed in yeast, *Dictyostelium*, or plants. These so-called "new" Gα genes were presumably acquired after the nematode lineage diverged from arthropod and hemicordate progenitors some 700 million years ago. They are predominantly expressed in sensory neurons that allow *C. elegans* to negotiate its environment. Perhaps these new Gα-subunits regulate unique effector proteins or couple novel chemosensory receptors to Gβγ activation. The human genome is scheduled to be 99% sequenced by 2005. All the genes will be there to find in the database. The challenge will be to determine their function, to identify the proteins they interact with, and to understand the mechanisms that drive diversity, and thereby the acquisition of new functions within multigene families.

3.3.1. THE G_s CLASS: ADENYLYL CYCLASE ACTIVATION

The G_s class is composed of two genes, Gα$_s$ and Gα$_{olf}$ (Fig. 5). The Gα$_s$ gene in various species are termed orthologs whereas Gα$_s$ and Gα$_{olf}$ are termed paralogs within the same species. All mammals that have been tested express these two proteins, but whereas Gα$_s$ is ubiquitously expressed, Gα$_{olf}$ is predominantly expressed in the olfactory epithelium and is required for odor recognition. Both proteins, and the othologous G_s class proteins in invertebrates, activate the effector protein adenylyl cyclase. Gα$_s$ can also potentiate voltage-gated Ca^{2+} channels. These activities are unique to G_s class α-subunits. Thus, hormone binding to receptors that are coupled to Gα$_s$ specifically activate adenylyl cyclase to generate the intracellular second messenger cAMP. Cells have at least three targets for cAMP regulation, protein kinase A (*see* Chapter 16), cAMP gated channels (*see* Chapter 8), and a recently identified guanine nucleotide exchange factor for the small GTPase Rap1, called Rap1-GEF. Gα$_s$ stimulation of adenylyl cyclase causes an amplification or broadcast release of a stimulus (cAMP) to the cytoplasm and all of its targets which are tethered at specific sites throughout the cell. cAMP-dependent phosphodiesterases can rapidly dampen stimulation by hydrolyzing cAMP to AMP. In the fight-or-flight response discussed previously, vascular and pulmonary dilation is stimulated by β$_1$-adrenergic receptor activation of Gα$_s$. The first

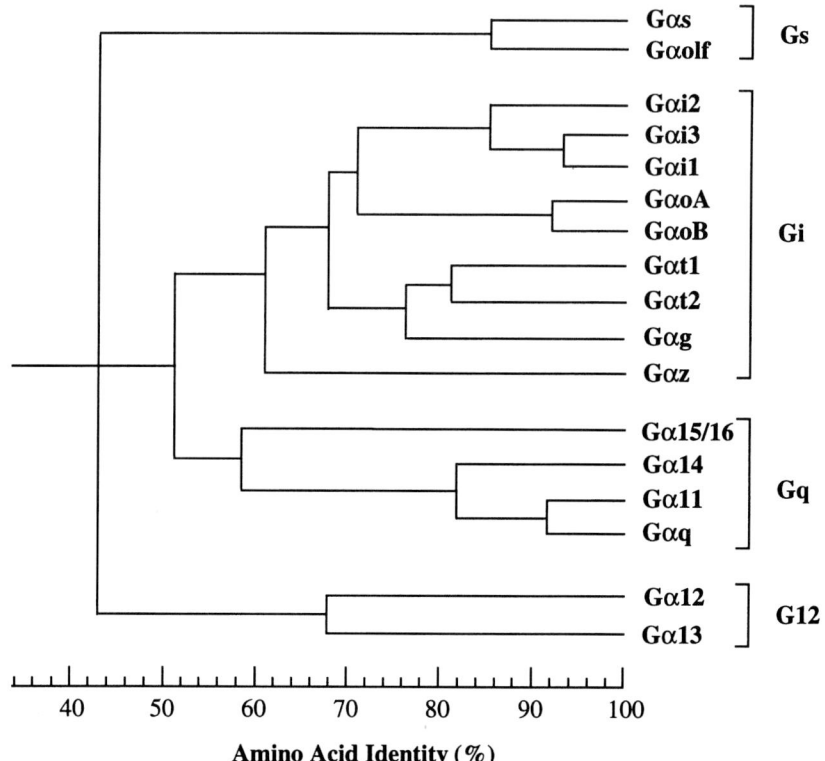

Fig. 5. Mammalian Gα multigene family. Four classes of Gα-subunits, G_s, G_i, G_q, and G_{12}, are defined based on amino acid identity and gene structure comparisons, and effector and RGS protein interactions. All known mammalian Gα orthologs, except for mouse and human $Gα_{15/16}$, are at least 97% identical.

cAMP response to be discovered was the release of glucose from glycogen (glycogenolysis) in muscle and liver, which also occurs in response to epinephrine.

3.3.2. THE G_i CLASS: DIVERSE PTX-SENSITIVE FUNCTIONS

G protein α-subunits in other classes have distinct functions. The G_i class is the most diverse, including nine genes subdivided according to function and sequence similarity. Three members of the G_i subclass inhibit adenylyl cyclase. This provides a hormone responsive mechanism to counteract the effects of $Gα_s$ stimulation of adenylyl cyclase. In addition, Gβγ-subunits released from $Gα_i$ subtypes also regulate some isoforms of adenylyl cyclase, phospholipase Cβ, and K^+ channels.

The transducins are closely related to $Gα_i$ but they specifically activate cGMP phosphodiesterase. Transducins are also expressed in specific cell types; $Gα_{t1}$ in rod photoreceptors, $Gα_{t2}$ in cones, and gustducin in taste bud cells. Their functions are extremely important to an organism in a variety of ways, not the least for identification of food and avoidance of predators. Despite regulating distinct effector proteins, the $Gα_i$ and transducin genes are clearly related. They are expressed from three sets of tandemly duplicated genes that presumably arose from a common ancestor by gene duplication followed by chromosomal duplications.

Another featured shared by $Gα_i$, transducin, and another member of the G_i class, $Gα_o$, is that pertussis toxin (expressed by the bacterium *Bordetella pertussis*) ADP-ribosylates a conserved cysteine residue located four residues from the C-terminus of each protein. Modification of this residue inhibits activation by disrupting the interaction of G_i class α-subunits with their receptors. $Gα_z$ is an interesting outlier of the G_i class. $Gα_z$ is the only G_i class gene that is not a substrate for pertussis toxin and it also possesses the lowest intrinsic GTPase activity. Nevertheless, $Gα_z$ is clearly a member of the G_i class as it inhibits adenylyl cyclase and is regulated by many of the same RGS proteins as other G_i class α-subunits.

3.3.3. THE G_q CLASS: PHOSPHOLIPASE Cβ ACTIVATION

All G_q class genes in mammals and invertebrates activate phospholipase Cβ isotypes in response to a wide variety of stimuli. In the mammalian fight-or-flight response, epinephrine stimulates G_q-coupled receptors in vascular smooth muscle cells to restricted blood flow to the skin. Another G_q-coupled signal transduction pathway is used in invertebrate photoreceptor cells to convert light into chemical signals that enable vision. In all cell types, $G\alpha_q$-GTP activates phospholipase $C\beta$ to hydrolyze the membrane-bound lipid phosphoinositde 4,5-bisphosphate (PIP_2) to produce diacylglycerol (DAG) and inositol trisphosphate (IP_3). DAG is a membrane-bound activator of many protein kinase C (PKC) isotypes. IP_3, released into the cytoplasm, activates IP_3-regulated channels that control the release of Ca^{2+} from intracellular compartments within the endoplasmic reticulum (see Chapter 9). Ca^{2+} is a small molecule that acts as a second messenger, like cAMP, to regulate the activity of numerous proteins, either directly or when bound by calmodulin. Unlike cAMP, Ca^{2+} cannot be broken down but cells can rapidly dissipate Ca^{2+} signaling with the buffering capacity of Ca^{2+} binding proteins or membrane pumps that extrude Ca^{2+} from the cell or return it to cytoplasmic stores.

Intracellular Ca^{2+} levels can oscillate, up and down, from several times a minute. These oscillations are thought to be regulated by the Ca^{2+} channels and pumps. It appears that G_q-coupled receptors can initiate Ca^{2+} oscillations inside cells. An oscillator requires an on switch and an off switch. The on switch is provided by agonist binding to the receptor, which activates G_q, and then phospholipase C (PLC), resulting in IP_3-dependent Ca^{2+} release from intracellular stores. RGS proteins that accelerate GTP hydrolysis on $G\alpha_q$ can periodically uncouple receptor-catalyzed GTP binding on $G\alpha_q$ from PLCβ activation. The combination of persistent hormone-stimulated GTP loading and periodic $G\alpha_q$ inactivation could initiate an oscillation in Ca^{2+} concentration near the G_q-coupled receptor. Ca^{2+} oscillations control progression through the cell cycle, cell differentiation, release of neurotransmitters, and activation of transcription factors.

Mammals express four G_q class α-subunits, called $G\alpha_q$, $G\alpha_{11}$, $G\alpha_{14}$, and $G\alpha_{15}$. Null mutations in each gene have been obtained by homologous recombination in mouse embryonic stem cells. Surprisingly, mice with a deficiency of the $G\alpha_{11}$, $G\alpha_{14}$, or $G\alpha_{15}$ gene appear fairly normal. Only the $G\alpha_q$ gene knockout has obvious phenotypic effects, including ataxia and a bleeding disorder caused by defects in platelet aggregation. These particular defects may be explained by the observation that $G\alpha_q$ is usually more abundant than $G\alpha_{11}$, especially in the tissues that are most severely affected in the $G\alpha_q$ knockout mice (i.e., the cerebellum, a region of the brain that coordinates movement, and platelets). In other tissues, $G\alpha_q$ and $G\alpha_{11}$ proteins probably compensate for each other's absence in the homozygous null mutants because they are both widely expressed and couple similar sets of receptors to PLCβ activation. By contrast $G\alpha_q$ and $G\alpha_{11}$ double homozygous null fetuses die during mid gestation due to malformations of the heart and fetuses with one active copy of either $G\alpha_q$ or $G\alpha_{11}$ die shortly after birth. These gene dosage effects are consistent with conclusions from biochemistry and pharmacology that $G\alpha_q$ and $G\alpha_{11}$ have very similar roles in regulating PLCβ. Functional redundancy within multigene families is common in mammals and is also observed among the G_i and G_{12} class knockout mice.

3.3.4. THE G_{12} CLASS: A HETEROTRIMERIC TO SMALL GTPASE SIGNALING PATHWAY IN GASTRULATION

Gastrulation is the first developmental event in the life of metazoan organisms wherein groups of cells coordinate invagination and migration within the embryo. The three primary lineages of somatic cells—the endoderm, mesoderm, and ectoderm—are established during gastrulation. These three primary cell lineages give rise to all of the somatic organs of the larva and adult organism. Mechanisms that efficiently regulate cell shape changes and coordinate the movement of groups of cells are critical to the development of complex multicellular organisms.

Proteins in the GTPase superfamily regulate morphogenesis during gastrulation in the fruit fly *Drosophila melanogaster*. Gastrulation initiates with flattening of the cells along the ventral midline of the *Drosophila* embryo. These cell shape changes require reorganization of the actin cytoskeleton. At the beginning of ventral furrow formation individual cells begin apical constriction in a random pattern along the midline. When about half the cells destined to begin migration have flattened, a transition occurs that allows all of the remaining cells to rapidly and coordinately constrict and complete the formation of the ventral furrow. Invagination of the ventral furrow allows cells to migrate into the embryonic cavity during gastrulation. Genetic analysis revealed that *fog*, a putative G protein-coupled agonist, and *concertina*, a member of the G_{12} class of heterotrimeric G protein α-subunits, were required for the proper formation of the ventral furrow. This G protein signaling pathway coordinates the simultaneous activation

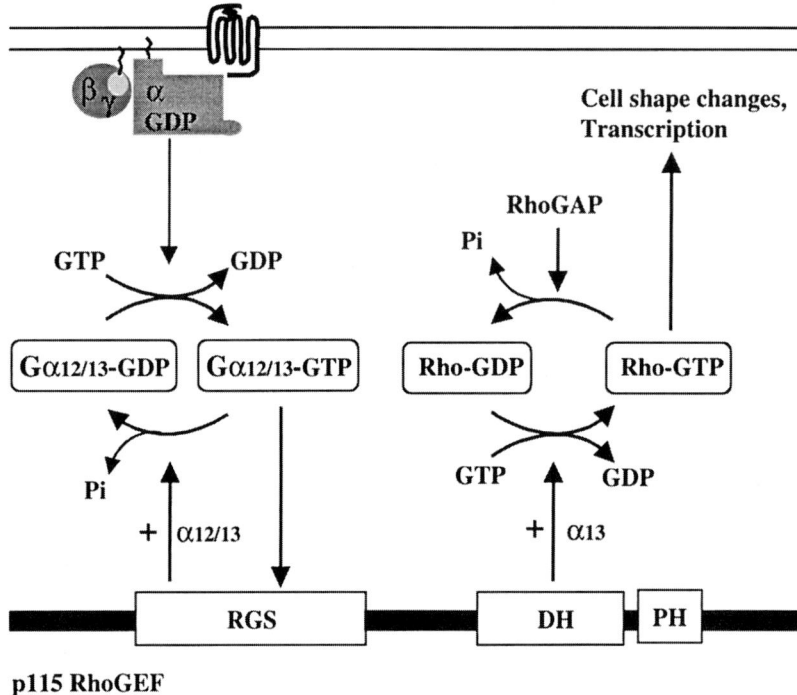

Fig. 6. p115RhoGEF is both a G_{12} class GAP and effector protein. A conserved signaling pathway that couples large ($G\alpha_{12/13}$) and small (Rho) GTP binding proteins regulates cell shape changes and transcription in *Drosophila* and mammals. $G\alpha_{13}$ binds to the RGS-like domain of p115RhoGEF to activate the guanine nucleotide exchange factor (GEF) mediated by the Dbl homology (DH) and plextrin homology (PH) domain. The RGS-like domain also accelerates GTP hydrolysis on both $G\alpha_{12}$ and $G\alpha_{13}$ subunits.

of *concertina* effector protein(s) in each cell along the ventral furrow.

Cell shape changes in the ventral furrow are regulated by the small GTPase Rho, which controls the assembly of actin stress fibers. Rho is activated by DRhoGEF, a Rho-specific guanine nucleotide exchange factor. Gastrulation defects were similar in *Drosophila* mutants lacking Rho, DRhoGEF, *concertina*, or *fog*, suggesting these genes may act in a common signaling pathway. Epistasis analysis and other genetic tests established a signaling pathway in which *fog*, the presumed agonist of an unidentified seven transmembrane domain receptor, was coupled by *concertina* to DRhoGEF-dependent activation of Rho. Thus, a signaling pathway that is essential for gastrulation is regulated by the sequential activation of a large GTP binding protein, in response to an extracellular ligand, and then a small GTP binding protein, which regulates the reorganization of the intracellular cytoskeleton.

The Rho-mediated signaling pathway that regulates cell shape changes is conserved through metazoan evolution. Mammalian tissue culture cells also undergo RhoA-mediated cell shape changes in response to activation of G_{12} class heterotrimeric G proteins. For example, the bacterial toxin C3, which blocks Rho-dependent signaling (Table 3), also inhibited actin cytoskeletal reorganization in response to the G_{12}-coupled agonist lysophosphatidic acid (LPA). Vertebrates and invertebrates share features of coordinated cell movements and their biochemical regulatory mechanisms.

The link in the signaling pathway between heterotrimeric G proteins and small GTP binding proteins was made by the demonstration that $G\alpha_{13}$ activated a guanine nucleotide exchange factor for the small GTPase RhoA, called p115Rho-GEF (Fig. 6). This protein has two domains with sequence homology to two different types of proteins. The homologous regions were identified by computer searches of expressed sequence tags (est). The first domain had weak homology to RGS proteins. The vague sequence similarity between RGS proteins and the p115Rho-GEF domain suggested these proteins may bind different classes of heterotrimeric G proteins. The second recognizable domain in p115Rho-GEF was similar to proteins such as DRhoGEF which catalyzed nucleotide exchange on RhoA and other small GTP binding proteins. To test the idea that p115Rho-GEF coupled large and small G proteins in a signaling pathway, the nucleotide sequence of the newly identified p115Rho-GEF was used to make oligonucleotide primers, the cDNA was cloned into an *E. coli* expression vector, and the catalytic activity of recombinant protein was assayed. p115Rho-GEF was found to accelerate GTP hydrolysis by $G\alpha_{12}$ and $G\alpha_{13}$, and to catalyze GTP exchange on RhoA. Thus, a direct interaction between G_{12} class α-subunits and RhoA activation was established for p115Rho-GEF. A powerful conclusion of many biochemical and genetic studies is that invertebrate and vertebrate signaling pathways use homologous components to regulate similar cellular activities.

4. CONCLUSIONS

GTP binding and hydrolysis provides a mechanism used by the small and large GTPases to drive cycles of protein–protein interactions that mediate signaling inside cells. Heterotrimeric G protein α- and/or $\beta\gamma$-subunits from each class can regulate the activity of small GTPases, and they are subject to feedback regulation through their attendant RGS proteins. Many of the same proteins that regulate cell shape changes and migration during gastrulation, such as G_{12} and RhoA, are also critical later in development for dorsal neural tube closure and other morphogenic processes. In each case, we see that a series of molecular switches, through both large and small G proteins, control the signal transduction pathways that link membrane receptors to the cytoskeleton. Analogous signaling pathways between G_s and Rap1, and G_i and G_q regulation of Ras have been discovered. Signaling proteins appear to be organized into complexes assembled around scaffolding proteins. Although scaffolding proteins often lack obvious enzymatic activity they provide an essential function by providing a platform for signaling proteins to rapidly activate specific target proteins. Genetic analysis of signaling in the unicellular eukaryot *Saccharomyces cerevisiae* (baker's yeast) suggests that G protein $\beta\gamma$-subunits, small GTP binding proteins, their GEFs and GAPs, kinases, transcription factors, and actin all assemble on a single scaffold complex to regulate psuedohyphal extension. Considering the complexity of building and maintaining a multicellular animal, multimeric protein assemblies are probably a common feature of signal transduction in metazoan organisms.

5. SELECTED READINGS

Avery L. *Caenorhabditis elegans* Genetics and Genomics web page: www.elegans.swmed.edu/genome.

Barrett K, Leptin M, Settleman J. The Rho GTPase and a putative RhoGEF mediate a signaling pathway for the cell shape changes in *Drosophila* gastrulation. Cell 1997; 91:905.

Berman DM, Gilman AG. Mammalian RGS proteins: barbarians at the gate. J Biol Chem 1998; 273:1269.

Boguski MS, McCormick F. Proteins regulating Ras and its relatives. Nature 1993; 366:643.

Bourne HR, Sanders DA, McCormick F. The GTPase superfamily: conserved structure and molecular mechanism. Nature 1991; 349: 117.

Farfel Z, Bourne HR, Iiri T. The expanding spectrum of G protein diseases. N Engl J Med 1999; 340:1012.

Hart MJ, Jiang X, Kozasa T, Roscoe W, Singer WD, Gilman AG, Sternweis PC, Bollag G. Direct stimulation of the guanine nucleotide exchange activity of p115 RhoGEF by $G\alpha 13$. Science 1999; 280:2112.

Geyer M, Wittinghofer A. GEFs, GAPs, GDIs and effectors: taking a closer (3D) look at the regulation of Ras-related GTP-binding proteins. Curr Opin Struct Biol 1997; 7:786.

Gilman AG. G proteins: transducers of receptor-generated signals. Annu Rev Biochem 1987; 56:615.

Offermanns S, Simon MI. Genetic analysis of mammalian G-protein signalling. Oncogene 1998; 17:1375.

Protein Data Bank (PDB) at the Brookhaven National Laboratory: www.rcsb.org

Ross E, Wilkie TM. Mammalian RGS proteins. Annu Rev Biochem 2000; 69; in press.

Simon MI, Strathmann MP, Gautam N. Diversity of G proteins in signal transduction. Science 1991; 252:802.

Wilkie TM, Gilbert DJ, Olsen AS, et al. Evolution of the mammalian G protein α subunit multigene family. Nat Genet 1992; 1:85.

3

The Insulin Receptor Kinase
Regulation and Therapeutic Significance

Barry I. Posner

CONTENTS

INTRODUCTION
THE INSULIN RECEPTOR KINASE (IRK): STRUCTURE AND ACTIVITY
INTERNALIZATION OF ACTIVATED IRKs
INSULIN DEGRADATION: THE ENDOSOMAL ACIDIC INSULINASE
THE IRK-ASSOCIATED PHOSPHOTYROSINE PHOSPHATASE (PTP)
ENDOSOMAL ACIDIFICATION AND IRK DEACTIVATION
REGULATION OF IRK ACTIVITY: SUMMARY
REFERENCES

1. INTRODUCTION

Polypeptide hormone receptors were first identified using direct binding techniques in which radiolabeled ligands of high purity were incubated with tissue preparations, cells, or subcellular membrane fractions in the presence and absence of unlabeled hormone (Cuatrecasas et al., 1971; Freychet et al., 1971). In this manner the key receptor properties of specificity and high affinity were characterized (Roth, 1973; Posner et al., 1974). Using this relatively simple approach peptide hormone receptors were found in tissues not previously regarded as targets for the hormones (Gavin et al., 1972; Posner et al., 1973; Kelly et al., 1974). Thus insulin receptors were observed in brain and placenta in addition to being present in the classical targets of muscle, liver, and adipose tissue (Posner et al., 1973, 1974; Kelly et al., 1974). It was also soon shown that tissue peptide hormone receptor levels were modulated, and that the hormone itself played a role in both down-regulating (Gavin et al., 1974) and up-regulating (Posner et al., 1975) receptors, in part depending on ambient ligand concentration (Barash et al., 1988). Using receptor preparations it was possible to measure hormone levels in analogous fashion to radioimmunoassay except that a higher degree of hormonal integrity was assessed in radioreceptor assays (Posner et al., 1980).

2. THE INSULIN RECEPTOR KINASE (IRK): STRUCTURE AND ACTIVITY

The application of molecular technology led to the cloning of intrinsic membrane proteins including peptide hormone receptors. Initial studies established the subunit structure of the IRK (Czedh and Massague, 1982, Yip et al., 1982). Subsequently the insulin receptor was cloned and was shown to be a heterotetramer consisting of two α- and two β-subunits linked by disulfide bonds (Ebina et al., 1985; Ullrich et al., 1985). The affinity labeling and cloning studies indicated that the α-subunits were entirely exterior to the plasma membrane and contain the insulin binding site whereas the β-subunits were intrinsic membrane proteins that traverse the membrane once and contain the canonical sequence of a tyrosine kinase. Following insulin binding the IRK was shown to

From: *Principles of Molecular Regulation* (P. M. Conn and A. R. Means, eds.), © Humana Press Inc., Totowa, NJ.

undergo autophosphorylation on tyrosine residues in the cytosolic domain of the β-subunit (Avruch et al., 1982; Kasuga et al., 1982) in which circumstance it manifested augmented tyrosine kinase activity against exogenous substrates (Kasuga et al., 1983; Stadtmauer and Rosen, 1983). Of significance was the recognition that once activated the level of tyrosine kinase activity of the IRK was sustained in the absence of insulin binding to the α-subunit as long as the β-subunit remained tyrosine phosphorylated (Yu et al., 1985a; Khan et al., 1986; Kadota et al., 1987a). It was shown more recently that autophosphorylation occurs by transphosphorylation in which each β-subunit tyrosine phosphorylated its opposite member in the heterotetrameric receptor (Ballotti et al., 1989).

The tyrosine kinase activity of the IRK was shown to be both necessary and sufficient for the realization of insulin action consequent to insulin binding to the IRK. Thus patients with severe insulin resistance were identified in whom insulin binding remained intact but whose IRK was either inactive or substantially impaired (Grigorescu et al., 1984; Grunberger et al., 1984). Antibodies to the kinase domain of the β-subunit could inhibit insulin action after accumulating in cells (Morgan et al., 1986). Most importantly it was found that mutating the ATP binding site of the IRK completely inactivated kinase activity and the insulin response without impairing insulin binding (Chou et al., 1987). In addition it was shown that the IRK can be fully activated in the absence of insulin and that this will entrain the full insulin response (Kadota et al., 1987). Thus activation of the IRK is both necessary and sufficient for realization of the insulin response.

Non-insulin-dependent diabetes mellitus (NIDDM) is characterized by resistance to the action of insulin (DeFronzo and Prato, 1996). This fact has given impetus to the study of the mechanism of insulin action. Considerable work has focused on defining the events that follow IRK activation. The activated IRK tyrosine phosphorylates substrates that act as docking proteins in that phosphotyrosine motifs on these proteins bind enzymes and transducers containing SH2 binding sites (White, 1994; Kahn, 1985). These molecules transfer the signal downstream largely through the activation of Ser/Thr protein kinases (Avruch et al., 1990). The molecular defect responsible for the insulin resistance of NIDDM has not yet been identified in the vast majority of cases. However, there is good reason to think that it lies at or very close to IRK activity itself. It is thus of considerable interest to identify the mechanisms that control IRK function.

3. INTERNALIZATION OF ACTIVATED IRKS

Following insulin binding to the IRK there rapidly occurred internalization of the insulin–IRK complex into the cell (Bergeron et al., 1979; Posner et al., 1980, 1986; Khan et al., 1985). Internalized insulin–IRK complexes were shown to concentrate in a novel cellular organelle (Khan et al., 1981) that became known as the endosomal apparatus or endosomes (ENs). It was appreciated early that endosomal IRK remained activated for a significant period of time subsequent to internalization and even appeared to undergo augmentation of its tyrosine kinase activity within ENs (Khan et al., 1986, 1989; Burgess et al., 1992). More recently it was shown that an activated IRK in ENs is capable of effecting signaling (Fig. 1) (Bevan et al., 1995a). Indeed the internalization of peptide–receptor complexes has been shown to play a significant role in cell signaling of many systems (Bevan et al., 1996).

These considerations indicate that IRK function depends on activation through autophosphorylation. In addition evidence has been adduced that the topological location of the IRK may influence the character and/or the intensity of insulin signaling. A third factor that has been implicated is Ser/Thr phosphorylation of the IRK. Several studies have shown that phosphorylation of the IRK on Ser/Thr residues significantly dampens the level of IRK activity (Takayama et al., 1984; Yu et al., 1985a; Stadtmauer and Rosen, 1986). It is still uncertain the extent to which this functions physiologically or pathologically nor is it clear which enzyme(s) act to effect such phosphorylation of the IRK.

4. INSULIN DEGRADATION: THE ENDOSOMAL ACIDIC INSULINASE

In the course of studying the internalization of the IRK it was shown that insulin accumulating in endosomal fractions underwent degradation in a time-dependent manner (Table 1) (Posner et al., 1980). Studies demonstrated that the acidotropic agent chloroquine concentrated in ENs and that there was a corresponding retention and augmentation of insulin within these structures (Posner et al., 1982). These studies thus implicated ENs as an important site for insulin clearance by a mechanism that was operative

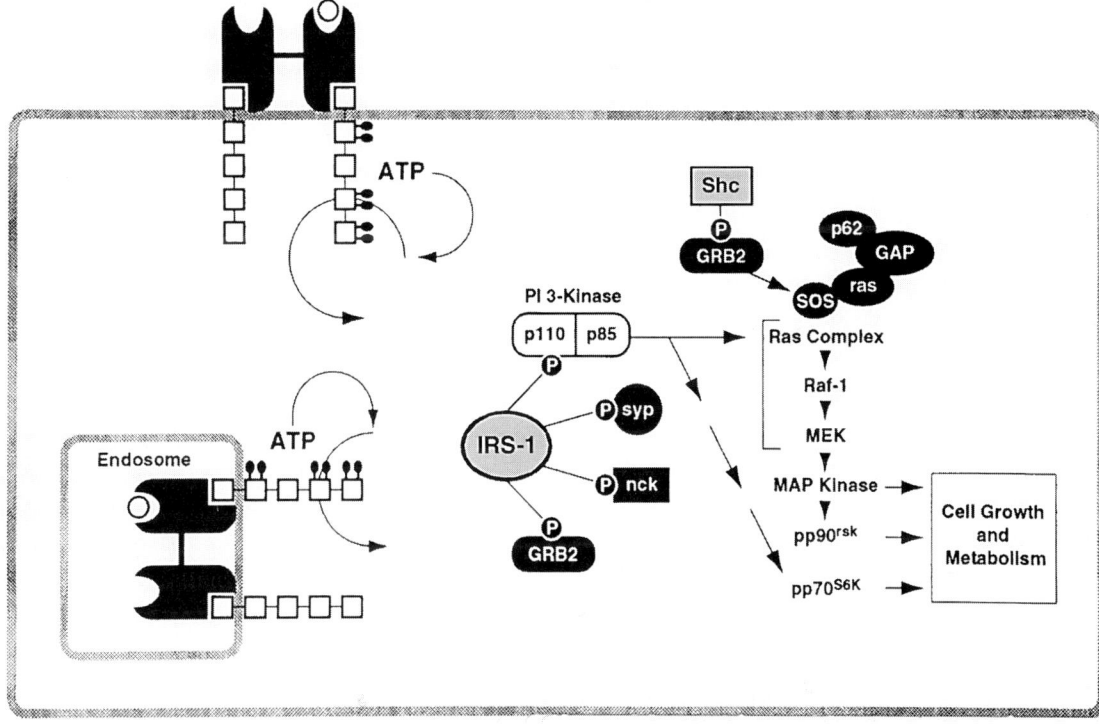

Fig. 1. Scheme of insulin action emanating from both cell surface and endosomal activated IRKs. This scheme is modified from Cheatham et al. (1994).

Table 1
Integrity of ^{125}I-Insulin Extracted from ENs

Insulin Preparation	Time Post-Injection (min)	Radioactivity in peak (% applied)	Rebinding (% spec.)	Integrity of ^{125}I-Insulin
Control		99.4	25.4 ± 3.0	100
ENs	2	89.5	21.0 ± 2.2	75.0 ± 3.2
	10	84.1	16.6 ± 2.5	53.6 ± 5.6
	20	81.5	15.9 ± 2.7	50.5 ± 3.0

at an acid pH. Subsequent studies, employing high-performance liquid chromatography (HPLC), showed that during internalization of ^{125}I-insulin, the peptide fragments generated in endosomal fractions were the same as those generated in the course of clearance of ^{125}I-insulin by intact liver (Hamel et al., 1988). Thus ENs appeared to be the physiologically relevant location at which insulin degradation and hence tissue clearance is initiated.

When isolated ENs, containing internalized ^{125}I-insulin, were incubated in a suitable buffer it was possible to show ATP-dependent degradation of the labeled insulin whereas comparably internalized ^{125}I-prolactin and epidermal growth factor (EGF) remained essentially intact (Fig. 2) (Doherty et al., 1990). This pointed to the existence of an endosomal insulinase with relative specificity for insulin. Because of its acidic pH optimum the enzyme was referred to as endosomal acidic insulinase (EAI). Studies documented that EAI was not the classical insulinase (IDE) described to that point in that IDE was absent from ENs and had a neutral pH optimum (Table 2) (Authier et al., 1994). Indeed IDE was shown to be located in both peroxisomes and cytosol (Authier et al., 1995).

Recent work has shown that inhibiting the clearance of endosomal insulin with chloroquine is associated with the prolongation and augmentation of IRK activation (Bevan et al., 1997). Thus EAI plays a role in regulating IRK activation and function.

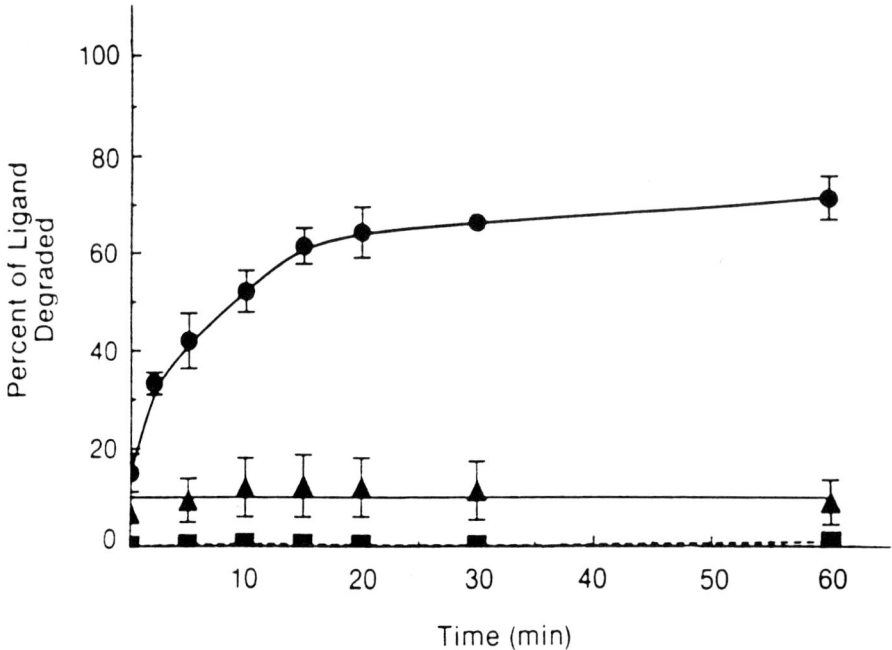

Fig. 2. Selective degradation of endosomal ^{125}I-insulin. Endosomes containing internalized ^{125}I-insulin (●), ^{125}I-EGF (▲), or ^{125}I-prolactin (■) were incubated in a cell-free system for the indicated times. The integrity of each labeled peptide was determined by precipitability in cold 10% trichloroacetic acid (TCA) as described in detail elsewhere (Doherty et al., 1990). Each point is the mean ± SD of three determinations.

Table 2
Insulin Degradation in Rat Liver

Physiologic process that occurs in ENs at pH less than 7
Observed to occur first within the endosomal system
IDE–Nonendosomal neutral protease located in peroxisomes and cytosol
EAI–Unique endosomal protease with pH optimum of 5–6

The above statements represent a summary of observations made by Authier et al. (1994, 1995) and Posner et al. (1980).

5. THE IRK-ASSOCIATED PHOSPHOTYROSINE PHOSPHATASE (PTP)

5.1. The Peroxovanadium Compounds as Insulin Mimetic Agents

In 1985 it was observed that the inclusion of vanadate in the drinking water of streptozotocin-diabetic rats resulted in the reduction of their elevated blood glucose levels into the normal range (Heyliger et al., 1985). It had previously been observed that vanadate mimicked insulin action when incubated with adipocytes and other tissues (Dubyak and Kleinzeller, 1980; Hori and Oka, 1980; Shechter and Karlish, 1980). In seeking to correlate the biological effect of vanadate and other insulin mimetic substances with their effect on activation of the IRK we observed that the combination of vanadate and H_2O_2 resulted in striking insulin mimesis and correlatively powerful activation of the IRK in rat adipocytes (Kadota et al., 1987a,b). We subsequently showed that the combination of vanadate and H_2O_2 resulted in the generation of peroxovanadium complexes or pervanadate (pV) which were responsible for the observed effects (Kadota et al., 1987). Soon thereafter synthetic pVs, each containing an oxo ligand, one or two peroxo anions, and an ancillary ligand in the inner coordination sphere of vanadium, were synthesized, crystallized, and characterized by ^{51}V nuclear magnetic resonance (NMR) as >95% pure. These compounds activated the IRK of cultured cells and that of rat liver following in vivo administration (Posner et al., 1994) and were shown to be potent insulin mimetics in adipocytes, liver, skeletal muscle, and cultured cells (Table 3) (Fantus et al., 1989; Foot et al., 1991; Bevan et al., 1995b; McIntyre et al., 1998). It was further observed that these compounds were potent hypoglycemic agents when administered intravenously to rats (Bevan et al., 1995b; Yale et al., 1995). The recognition that pVs activate the IRK and insulin responses in the complete absence of insulin strongly indicated that activation of the IRK was not only necessary but also sufficient for entraining insulin responses in tissues.

Table 3
Peroxovanadium Compounds Mimic Insulin Action

Activation of IRK in:
  (a) Cultured cells in absence of insulin
  (b) Tissues in vivo

Decrease blood glucose in N & BB diabetic rats

Adipocytes:
  (a) ↑ Lipogenesis, protein synthesis, glucose transport, and IGFR II translocation
  (b) ↓ Lipolysis

Skeletal muscle:
  ↑ Glucose transport, oxidation, and glycogen synthesis

A summary of the work on which this list is based may be found in the review by Bevan et al. (1995c).

5.2. Mechanism of pV Action

In studies involving the ^{32}P labeling of adipocytes it was shown that both insulin and pV stimulated tyrosine phosphorylation of the IRK whereas vanadate had no such effect (Fantus et al., 1989). Adding pV to IRK, partially purified by lectin chromatography, had no effect on IRK autophosphorylation or kinase activity (Kadota et al., 1987). The puzzle was resolved by studying the activated IRK in rat liver endosomal preparations. In these studies ENs prepared 2 min after in vivo insulin administration were incubated with [^{32}P]ATP in the presence and absence of pV (Fig. 3). In the absence of pV [^{32}P]tyrosine labeling of the IRK reached a maximum after 5 min of incubation and decreased to low levels thereafter. This decline in labeling paralleled the decline in ATP levels in the incubation medium and indicated the presence of a PTP able to dephosphorylate the IRK. In contrast, the addition of pV resulted in an augmented level of tyrosine phosphorylation and abrogated the decline in IRK labeling even though medium ATP levels fell as in the absence of pV (Faure et al., 1992).

Using ENs, prepared after in vivo insulin administration, it was possible to devise an assay for the IRK-associated PTP and demonstrate that pVs inhibit this activity (Fig. 4). It was suggested that under usual conditions there is a dynamic balance between the two processes; however, when autophosphorylation fails to proceed (as in the absence of ambient ATP) there occurs a rapid net removal of phosphotyrosine from the IRK and the restoration of the basal state. The demonstration of a similar process for the EGF receptor indicated that this is a general phenomenon

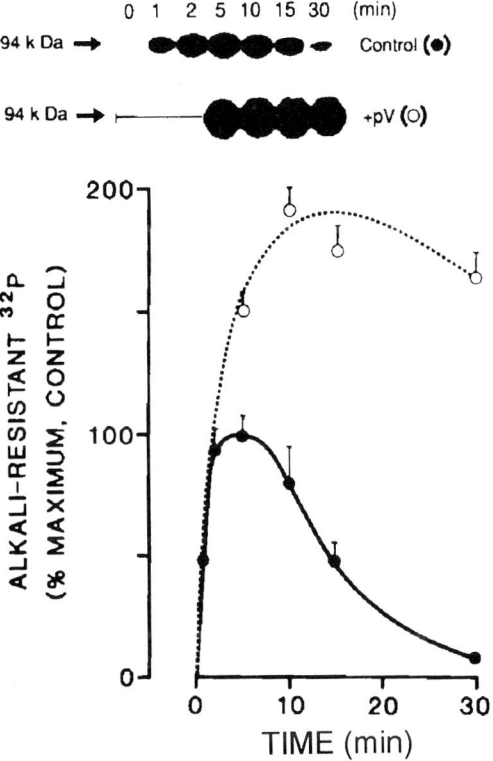

Fig. 3. Time course of IRK ^{32}P phosphorylation in ENs. Rats were injected via the portal vein with insulin (1.5 µg/100 g, body weight). After 2 min livers were excised and ENs were isolated, suspended in assay buffer, and subjected to phosphorylation in the presence of ^{32}P]ATP at 37°C for the indicated times. At the noted times ENs were solubilized, receptors were immunoprecipitated, and the immunoprecipitates were subjected to sodium dodecyl sulfate-polyacrylanide gel electrophoresis (SDS-PAGE) and radioautography to determine the extent of labeling of the 94-kDa β-subunit of the IRK (control, ●; pV [100 µM], ○). The data are expressed as a percentage of the maximal autophosphorylation observed. Each point is the mean (± SD) of three experiments. This figure is adapted from Faure et al. (1992).

that may well apply to all receptor tyrosine kinases (Faure et al., 1992).

These data support the view that tyrosine phosphorylation of the IRK is regulated by autophosphorylation on the one hand and by PTP activity on the other and that the balance between these two determines the IRK phosphorylation and hence activation state. It was thus inferred that pVs mimic insulin by inhibiting the IRK-associated PTP permitting autoactivation of the IRK in the absence of insulin. In support of this was the good correlation between the ability of 12 different pV compounds to activate hepatoma IRK on the one hand, and inhibit endosomal IRK dephosphorylation on the other (Posner et al., 1994). This

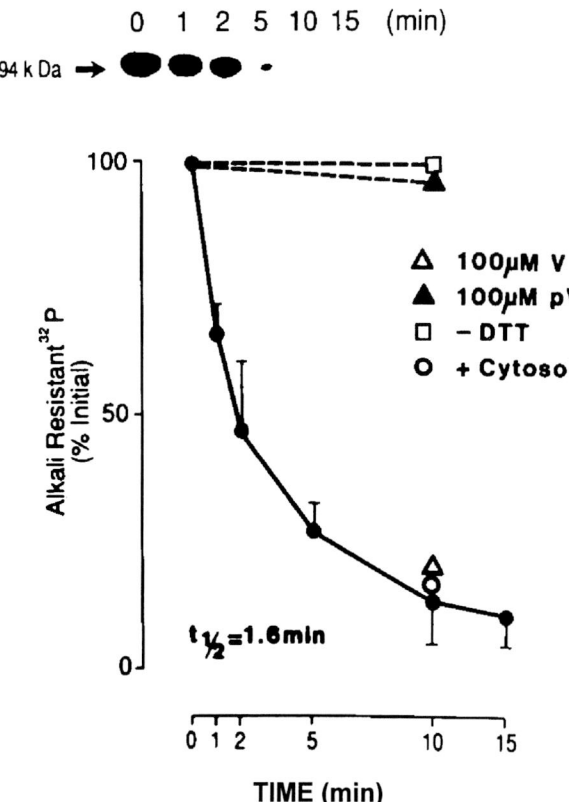

Fig. 4. Time course of dephosphorylation of ^{32}P-labeled IRK. Isolated ENs were subjected to autophosphorylation with [^{32}P]ATP for 5 min before addition of EDTA/unlabeled ATP (zero time) to terminate ^{32}P-labeling and initiation of the dephosphorylation assay. The **top panel** depicts labeling of the 94-kDa β-subunit of the IRK. The **bottom panel** represents quantitation following densitometric analysis of the extent of β-subunit labeling. Each point is the mean (± SD) of three experiments. Insulin receptor dephosphorylation was assessed in the absence of dithiothreitol (DTT, □) and in the presence of pV (▲) (100 μM), vanadate (△) (100 μM), and rat liver cytosol (○) as described in detail by Faure et al. (1992).

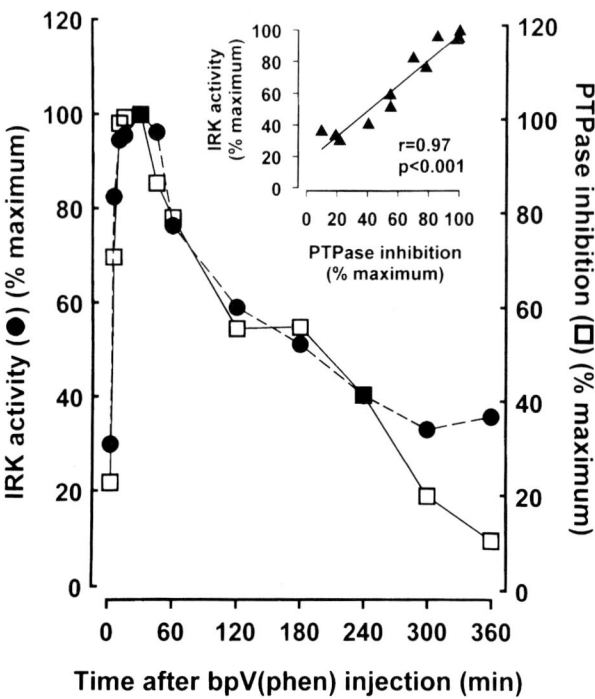

Fig. 5. Correlation between IRK activation and PTP inhibition in ENs after treatment with bpV(phen) and insulin. Rats were fasted overnight, injected with bpV (phen) (0.6 μmol/100 g body wt) at zero time, and insulin (1.5 μg/100 g body wt) at 2 min prior to sacrifice at the indicated times. Hepatic ENs were prepared and IRK and PTP activities were measured as described by Bevan et al. (1995a). **Inset**: Linear correlation between IRK activation and PTP inhibition. **Main panel**: Time course of IRK activation (●) and the corresponding inhibition of PTP activity (□). Each is expressed as a percentage of maximum activity observed at 15 min post-injection of bpV (phen). (Adapted from Bevan et al., 1995a.)

was followed by the demonstration in rat liver ENs that, following pV administration, there was a marked correlation between endosomal IRK activation and the inhibition of IRK dephosphorylation (Fig. 5) (Bevan et al., 1995a).

This still left open the question as to how IRK tyrosine phosphorylation was effected in the presence of pV. It was possible that pV inhibited a PTP leading to the activation of another tyrosine kinase which phosphorylated the IRK. Alternatively IRK tyrosine phosphorylation in the presence of pV could be effected by the intrinsic activity of the IRK itself. Studies on hepatoma cells bearing normal and kinase dead mutant IRKs were carried out to assess which model was correct. These studies showed that pVs could not effect tyrosine phosphorylation of the kinase dead IRK but only of the IRK with normal kinase activity (Fig. 6) (Band et al., 1997).

5.3. IRK-Associated PTP and Insulin Action

Of great interest was the recognition that following the in vivo administration of pV the IRK in ENs was activated before and to a greater extent than that in PM fractions (Fig. 7, lower left). This was consistent with a primary effect of pVs on endosomal IRK through the inhibition of its associated PTP, and with the recycling of activated IRKs to the cell surface. Colchicine was then employed to inhibit recycling of receptors from ENs to PM. In this circumstance it

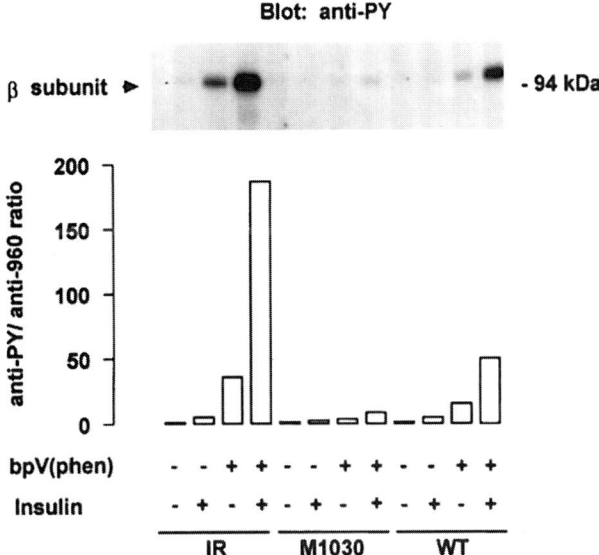

Fig. 6. Effect of bpV (phen) on tyrosine phosphorylation of overexpressed normal and kinase dead IRK. HTC control cells (WT), and those overexpressing normal IRK (IR) or kinase dead IRK (M1030) were incubated with insulin (100 nM) for 5 min, bpV (phen) (0.1 mM) for 20 min, or the latter for 15 min before addition of insulin for 5 min. The cells were then washed, solubilized, and IRK was partially purified by wheat germ agglutinin (WGA) chromatography, immunoprecipitated with a specific antibody to the β-subunit of the IRK, subjected to SDS–PAGE, and immunoblotted with antibodies to the β-subunit of IRK or phosphotyrosine as described by Band et al (1997). In this manner the phosphotyrosine content per β-subunit was determined. **Top**: Autoradiography of phosphotyrosine immunoblots. **Bottom**: Quantitation of the immunoblots was obtained by densitometry and expressed as the ratio of phosphotyrosine per β-subunit. It is clear that whereas bpV (phen) treatment effects considerable tyrosine phosphorylation of the β-subunit of normal IRKs there was virtually no such phosphorylation observed in M1030 (i.e., kinase-deficient IRKs). Thus an intrinsic kinase is required to effect tyrosine phosphorylation of the IRK when the IRK-associated PTP is inhibited by a pV compound.

was observed that, in the presence of colchicine, pV-induced endosomal IRK activation was unaffected whereas that of the PM was totally abrogated (Fig. 7, lower right). This in essence defined an in vivo model in which only the endosomal IRK was activated. In this circumstance substantial tyrosine phosphorylation of insulin receptor substrate protein-1 (IRS-1) was observed consistent with significant signaling occurring from the endosomal compartment (Fig. 7, top panel) (Bevan et al., 1995a). Indeed a large number of studies have demonstrated that the internalization of peptide–receptor complexes does not simply mediate degradation of these complexes but contributes significantly to biological responses (Bevan et al., 1996).

5.4. The Nature of IRK-Associated PTP(s)

The design of more specific PTP inhibitors for the management of diabetes should be facilitated by identification of the physiologically relevant PTP(s) that dephosphorylate(s) the IRK in vivo. PTP(s) have been classified into three main groups: (1) receptor-like PTP(s)—composed of a variable extracellular domain, a single hydrophobic transmembrane domain and one or two intracellular catalytic domains; (2) intracellular PTP(s)—which contain a single catalytic domain and possess C- and N-terminal extensions thought to be responsible for subcellular localization and/or enzymatic regulation and; (3) dual-specificity PTP(s)—a more recently identified class of PTP able to dephosphorylate both phosphotyrosine and phosphoserine/threonine residues (reviewed in Tonks and Neel, 1996).

Studies with purified enzymes and recombinant catalytic domains have revealed several PTPs that are active against the autophosphorylated IRK in vitro including the receptor-like PTPs, CD45, leukocyte antigen-related PTP (LAR), and leukocyte common antigen-related PTP (LRP, R-PTPα, or PTPα), and the cytosolic PTPs, PTP1B and TC-PTP. Using phosphopeptides modeled on the kinase regulatory domain of the IRK, differences have been observed in the in vitro specificity and rates by which particular PTPs dephosphorylate the three phosphotyrosines in the IRK kinase domain (Hashimoto et al., 1992). However, the lack of a defined three-dimensional structure may represent a limitation in using phosphopeptides to assess PTP specificity. In addition, PTP specificity appears to be determined, in part, by subcellular localization where associated proteins, targeted to specific domains of the cell, could influence enzyme activity. As a consequence, in recent years, work has focused on assessing PTP activity toward cellular substrates in intact cells.

The use of insulin-responsive cells transfected with and overexpressing selected PTPs has been one approach for assessing the role of a particular PTP in vivo. With this technique it has been shown that overexpression of LAR in CHO cells reduces insulin-stimulated IRK autophosphorylation (Zhang et al., 1996). As overexpression of a cytosolic, truncated form of LAR was without effect on insulin-stimulated IRK autophosphorylation this suggested that LAR required a transmembrane localization to directly

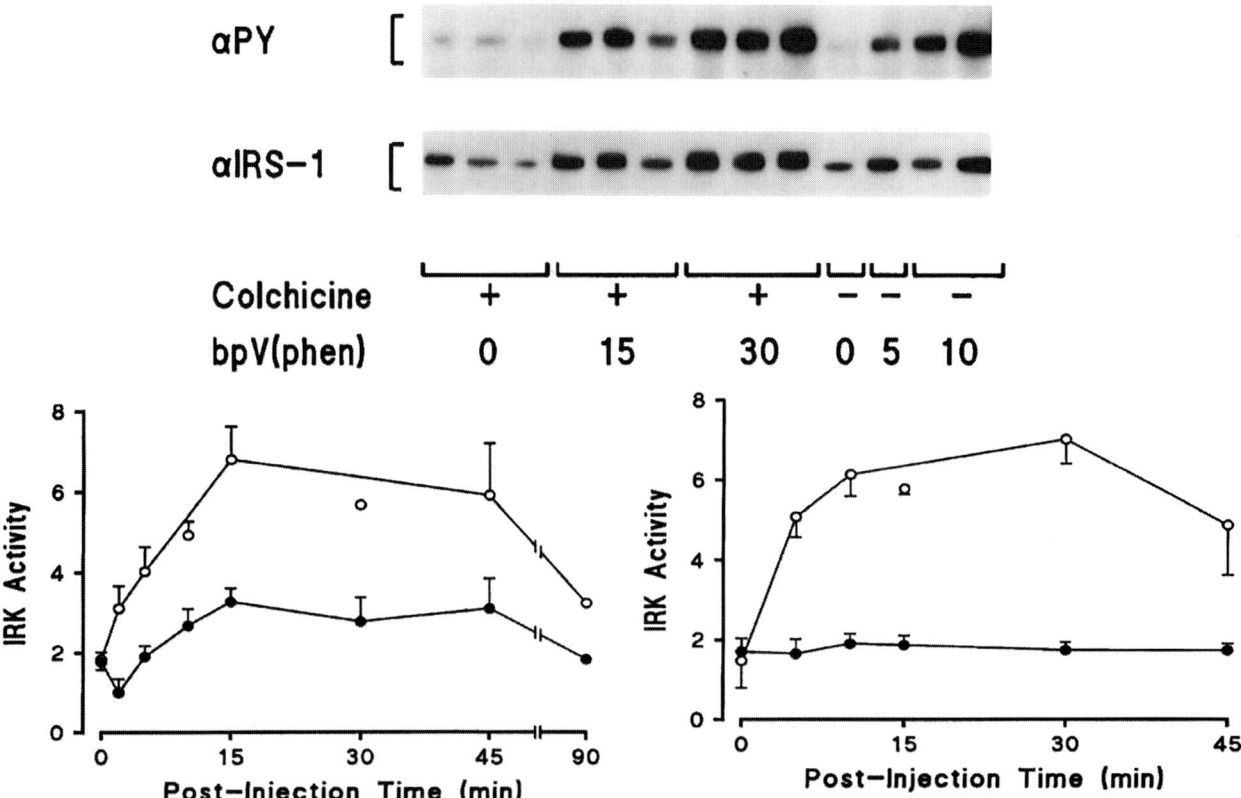

Fig. 7. Time course of IRK activation in PM (●) and ENs (○) and of IRS-1 tyrosine phosphorylation following bpV (phen) administration. After an overnight fast rats were pretreated with colchicine or buffer alone and then received bpV (phen) (0.6 µmol/100 g body wt). The animals were sacrificed at the noted times after bpV (phen), and hepatic ENs, PM, and cytosolic fractions were prepared. IRKs were partially purified from ENs and PM fractions, and IRK activity per unit receptor was assayed as described previously (Bevan et al., 1995a). IRS-1 was immunoprecipitated from cytosol followed by resolution on SDS-PAGE. The phosphotyrosine and IRS-1 content of the immunoprecipitates was determined by immunoblotting with specific antibodies, followed by autoradiography and laser densitometry of the immunoblots to quantitate yields. **Top**: Immunoblots of IRS-1 (αIRS-1) and IRS-1 phosphotyrosine (αPY) content. **Left**: IRK activity in PM and ENs from rats receiving no pretreatment with colchicine. **Right**: IRK activity in PM and ENs from rats pretreated with colchicine. *Note:* bpV (phen) treatment induced substantial IRS-1 tyrosine phosphorylation following colchicine pretreatment under which conditions there was no activation of IRKs in PM but only in ENs. Thus signaling can occur exclusively from activated IRKs in the endosomal compartment. (Adapted from Bevan et al., 1995a.)

interact with the IRK *in situ* (Zhang et al., 1996). In contrast overexpression of LAR in hamster kidney (BHK-IRK) cells had no effect on insulin-stimulated IRK autophosphorylation whereas overexpression of the receptor-like PTPs, PTPα (R-PTPα or LRP) and PTPε significantly inhibited IRK autophosphorylation and prevented insulin-stimulated growth inhibition of adherent cells (Moller et al., 1995). Overexpression of PTP1B was without effect in BHK-IRK cells (Moller et al., 1995), but inhibited insulin-stimulated IRK autophosphorylation in kidney 293 cells (Lammers et al., 1993) or Rat-1 fibroblasts (Kenner et al., 1996). The reasons for these conflicting observations are unclear but may reflect the use of different cells lines in which the relative levels of overexpressed PTP and IRK vary.

Using a "knockout" approach to evaluate an in vivo role for a particular PTP could circumvent possible nonphysiological interactions between highly overexpressed PTPs and IRKs, perhaps created through errant subcellular localization. Use of antisense RNA, one strategy for selectively reducing PTP levels in vivo, demonstrated that suppression of LAR levels by ~60% in rat hepatoma cells increased both insulin-stimulated IRK autophosphorylation and exogenous kinase activity (Kulas et al., 1995). Reducing PTP1B levels by osmotically loading rat hepatoma cells with PTP1B antibodies was also observed to augment insulin-stimulated IRK tyrosine phosphorylation and activity (Ahmad et al., 1995). Although some studies suggested a role for SH-PTP2, the observation that IRK phosphorylation was unaffected in hemizygous knockout mice suggested no role for this phosphatase in IRK dephosphorylation in vivo (Arrandale et al., 1996).

As significant IRK dephosphorylating activity has been described in hepatic ENs (Faure et al., 1992; Drake et al., 1996) this subcellular compartment represents a key site in which to identify the physiologically relevant PTP(s). Studies of LAR have revealed an insulin-stimulated recruitment of this enzyme to rat hepatic ENs that was associated with a decrease in levels at the cell surface (Ahmad and Goldstein, 1997). However, it is important to note that activated IRK is rapidly internalized into ENs following insulin administration, reaching maximal levels at 2–5 min post-injection before returning to basal levels by ~15 min (Khan et al., 1989; Burgess et al., 1992) whereas LAR internalization was maximal at 30 min following insulin administration (Ahmad and Goldstein, 1997). Incubation of endosomal fractions with neutralizing antibodies to LAR was reported to reduce in situ IRK dephosphorylating activity in this intracellular compartment by ~25% (Ahmad and Goldstein, 1997). However, the fact that endosomal IRK dephosphorylation was reduced but not prevented by LAR inhibition raises the possibility that multiple PTPs may be involved in effecting the dephosphorylation of activated IRKs.

Although certain interesting candidates have been identified the identity of the physiologically relevant PTP(s) effecting IRK dephosphorylation remains uncertain. The preceding observations raise the possibility that different PTPs may dephosphorylate specific tyrosine residues on the IRK. Thus IRK dephosphorylation may involve several PTPs acting in tandem or at different stages of the IRK activation "lifespan."

5.5. IRK-associated PTP(s): A Key Therapeutic Target

The recognition that the in vivo administration of pVs to rats induced hypoglycemia led to studies in diabetic animals. It was shown that in insulinoprivic BB Wistar diabetic rats the parenteral administration of pV but not vanadate returned blood glucose levels to normal and fully abrogated the associated ketoacidosis (Fig. 8) (Yale et al., 1995). This established the concept that inhibition of the IRK-associated PTP

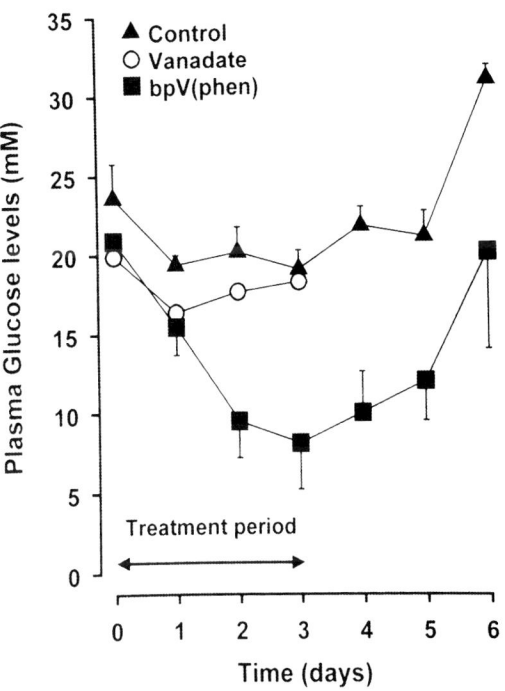

Fig. 8. Effect of bpV (phen) and vanadate on fasting glucose levels in insulin-deprived BB rats. Diabetic BB rats were withdrawn from insulin treatment and 16 h later were given subcutaneous injections of bpV (phen) (36 µmol/kg body wt), vanadate (36 µmol/kg body wt), or saline at 8:00 A.M. and 4:00 P.M. for 3 d (Treatment period). Throughout the experiment the rats were fed *ad libitum* from 4:00 P.M. to 8:00 A.M. and fasted from 8:00 A.M. to 4:00 P.M. Each data point is the mean (± SEM) of values from six animals. (This figure was adapted from Yale et al., 1995, which should be consulted for experimental details.)

could lead to insulin effects in vivo not only in normal but also in diabetic animals. Hence the IRK-associated PTP is an excellent target for the development of a drug designed to replace insulin.

6. ENDOSOMAL ACIDIFICATION AND IRK DEACTIVATION

On incubating ENs with ATP there occurred a concentration-dependent augmentation and then inhibition of IRK activity (Fig. 9). Thus at 0.1 and 1 mM ATP an increase in IRK activation level occurred whereas at ambient ATP levels of 5 or 10 mM a marked decrease of IRK activity occurred. These changes were accompanied by corresponding changes in the level of IRK tyrosine phosphorylation. This phenomenon was not seen when the ENs were incubated with nonhydrolyzable ATP analogs (viz AMP-PNP) nor was the effect observed when PM fractions

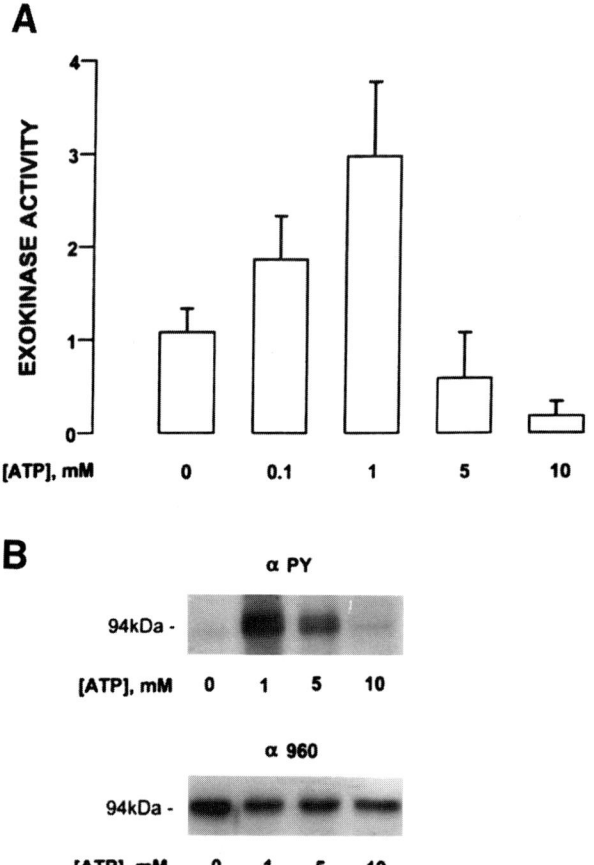

Fig. 9. Rats were injected with insulin (1.5 µg/100 g body wt) and sacrificed after 2 min. Fresh hepatic ENs were incubated for 15 min at 37°C with ATP at the indicated concentration. IRK was partially purified from solubilized ENs and assayed for tyrosine kinase activity and β-subunit phosphotyrosine content. (A) Effect of ATP on IRK activity. Equal amounts of WGA-purified IRK were assayed for exogenous tyrosine kinase activity using Poly (Glu/Tyr) (4:1) as substrate. Exokinase activity was expressed as pmol/10 min/10 fmol insulin binding. Each point is the mean ± SEM of determination from four to six separate experiments. (B) Effect of ATP on β-subunit phosphotyrosine content. Wheat germ agglutinin (WGA) eluates were subjected to SDS-PAGE and then immunoblotted with αPY or α960 antibodies (with ^{125}I-GAR as second antibody) and exposed for autoradiography to measure phosphotyrosine and IRK content respectively. (This figure is from Contreres et al., 1998, in which experimental details may be found.)

were incubated with ATP. The effect thus would appear to require intact ENs and the hydrolysis of ATP. Not only was this specific to ENs but it was also relatively specific to the IRK, as this phenomenon was not observed with endosomal EGF receptors (Contreres et al., 1998).

The role of endosomal acidification was implicated when it was observed that bafilomycin, an ATP-dependent proton pump inhibitor, largely prevented ATP-induced inhibition of IRK tyrosine phosphorylation and activation (Fig. 10). This acidification effect was not due to augmented serine phosphorylation of the IRK or to activation of the IRK-associated PTP. Furthermore, it was shown that even highly tyrosine phosphorylated IRKs could be inactivated when ENs were incubated with ATP implying some kind of conformational change in the IRK responsible for this phenomenon. An acidification-dependent conformational change in the IRK was shown when it was found that dithiothreitol (DTT)-dependent reduction of the IRK was significantly impeded by prior endosomal acidification (Fig. 11). This led to the postulate that endosomal acidification may effect a conformational change in the IRK during its intracellular itinerary, and hence reduce its kinase activity leading to a shortening of the signaling interval within the endosomal compartment. It is of interest that several studies have employed the acidophilic agent chloroquine, previously shown to accumulate in ENs (Posner et al., 1982), to essentially abolish insulin resistance in type II diabetics (Smith et al., 1987; Powrie et al., 1991). This is consistent with the view that endosomal acidification might contribute to the insulin resistance of type II diabetes by reducing the duration of the activated state of internalized IRKs.

7. REGULATION OF IRK ACTIVITY: SUMMARY

The foregoing has identified a number of factors involved in the regulation of IRK activity. Those recognized to date are summarized in Table 4 where it can be seen that four out of the six known mechanisms are processes that occur in the endosomal system. This highlights the significance of the endosomal system in mediating and regulating the insulin response. As alluded to previously it is possible that disturbances in endosomal regulatory processes might contribute to the pathogenesis of NIDDM.

ACKNOWLEDGMENTS

The author wishes to express his appreciation to Ms. Sheryl Jackson for help with the manuscript. I am grateful to the MRC of Canada, the National Cancer Institute of Canada, the Fonds de la Recherche en Santé du Quebec, the Pattie Cleghorn Fund in Diabetes Research at McGill University, and the Morris Pollack Foundation for their support of the work that formed the basis of this chapter.

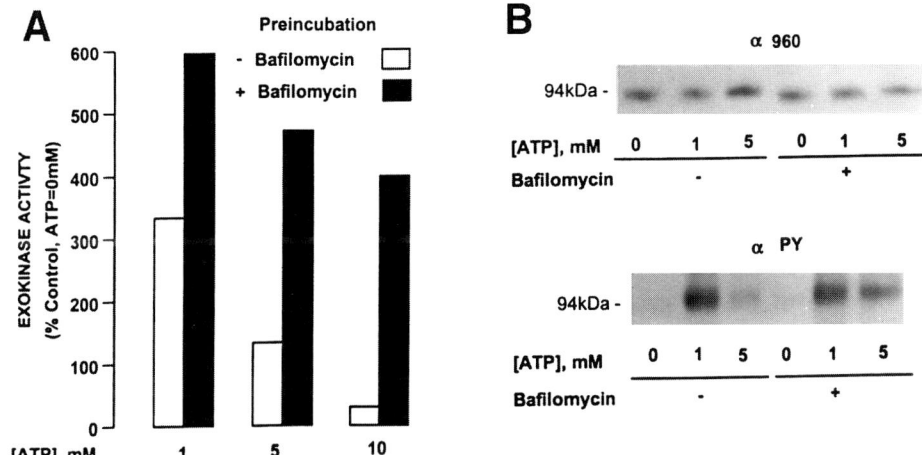

Fig. 10. Effect of the ATPase inhibitor, Bafilomycin, on the ATP-dependent deactivation of endosomal IRK activity. Hepatic ENs, prepared 2 min post-insulin (1.5 µg/100 g body wt), were preincubated in the absence or presence of 1 µM Bafilomycin A_1. After 30 min at 37°C the incubation was continued for 15 min with ATP at the indicated concentrations. **(A)** Effect of Bafilomycin A_1 on ATP-dependent inhibition of IRK activity. Wheat germ agglutinin (WGA)-purified IRK tyrosine kinase activity was assayed using Poly (Glu/Tyr) (4:1) as substrate and expressed as a percent of that obtained from incubations conducted without ATP. **(B)** Effect of Bafilomycin on ATP-dependent regulation of IRK β-subunit autophosphorylation. WGA eluates (5 µg of protein) were subjected to SDS-PAGE followed immunoblotting with αPY or α960 antibodies (^{125}I-GAR was second antibody) and autoradiography. (This figure was taken from Contreres et al., 1998, which can be consulted for further details.)

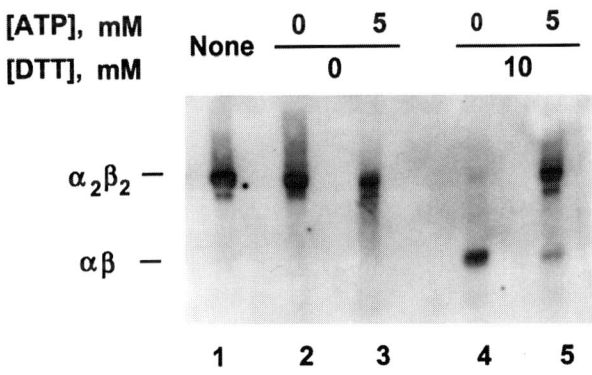

Fig. 11. Effect of ATP on dithiothreitol (DTT)-dependent reduction of endosomal IRK species. Hepatic ENs, prepared 2 min after insulin (1.5 µg/100 g body wt), were incubated for 15 min at 37°C in the absence or presence of 10 mM DTT and 5 mM ATP. ENs were centrifuged, resuspended without heating, and subjected to SDS-PAGE in the absence of reductant. Proteins were transferred to Immobilon-P membranes and probed with α960 and ^{125}I-GAR antibodies prior to autoradiography. (This figure is from Contreres et al., 1998, where further experimental detail may be found.) *Note:* Under these experimental circumstances the heterotetrameric ($\alpha_2\beta_2$) form of the receptor is seen. Reduction to the heterodimer (αβ) form occurs in the presence DTT but this is substantially inhibited by incubating with 5 mM ATP. This indicates that ATP-dependent acidification of ENs effected a conformational change in the IRK rendering the disulfide bond between the α-subunits in the heterotetramer inaccessible to DTT, presumably due to a conformational change in the molecule.

Table 4
Regulation of IR Kinase Activity

Autophosphorylation of tyrosine residues
Ser/Thr phosphorylation
Endosomal processes:
  1. Accumulation of activated IRK
  2. Insulin degradation by EAI
  3. Dephosphorylation by associated PTP
  4. Acidification-dependent IRK inactivation

REFERENCES

Ahmad F, Goldstein BJ. Functional association between the insulin receptor and the transmembrane protein-tyrosine phosphatase LAR in intact cells. J Biol Chem 1997; 272:448–457.

Ahmad F, Li PM, et al. Osmotic loading of neutralizing antibodies demonstrates a role for protein-tyrosine phosphatase 1B in negative regulation of the insulin action pathway. J Biol Chem 1995; 270:20503–20508.

Arrandale JM, Gore-Willse A, et al. Insulin signaling in mice expressing reduced levels of Syp. J Biol Chem 1996; 271:21353–21358.

Authier F, Rachubinski RA, et al. Endosomal proteolysis of insulin by an acidic thiol metalloprotease unrelated to insulin degrading enzyme. J Biol Chem 1994; 269:3010–3016.

Authier F, Bergeron JJ, et al. Degradation of the cleaved leader peptide of thiolase by a peroxisomal proteinase. Proc Natl Acad Sci USA 1995; 92:3859–3863.

Avruch J, Nemenoff RA, et al. Insulin-stimulated tyrosine phosphorylation of the insulin receptor in detergent extracts of human placental membranes. Comparison to epidermal growth factor-stimulated phosphorylation. J Biol Chem 1982; 257:15162–15166.

Avruch J, Tornqvist HE, et al. The role of tyrosine-and serine-threonine-protein phosphorylation in insulin action. Adv Sec Messen Phosphoprot Res 1990; 24:295–300.

Ballotti R, Lammers R, et al. Intermolecular transphosphorylation between insulin receptors and EGF-insulin receptor chimerae. EMBO J 1989; 8:3303–3309.

Band CJ, Posner BI, et al. Early signaling events triggered by peroxovanadium [bpV(phen)] are Insulin Receptor Kinase (IRK)-dependent—specificity of inhibition of IRK-associated Protein Tyrosine Phosphatase(s) by bpV(phen). Mol Endocrinol 1997; 11:1899–1910.

Barash I, Cromlish W, et al. Prolactin (PRL) receptor induction in cultured rat hepatocytes: dual regulation by PRL and growth hormone. Endocrinology 1988; 122:1151–1158.

Bergeron JJ, Sikstrom R, et al. Binding and uptake of ^{125}I-insulin into rat liver hepatocytes and endothelium. An in vivo radioautographic study. J Cell Biol 1979; 80:427–443.

Bevan AP, Burgess JW, et al. Selective activation of the rat hepatic endosomal insulin receptor kinase. Role for the endosome in insulin signaling. J Biol Chem 1995a; 270:10784–10796.

Bevan AP, Burgess JW, et al. In vivo insulin mimetic effects of pV compounds: role for tissue targeting in determining potency [published erratum appears in Am J Physiol 1996 Jan; 20(1 Pt 1): section E following table of contents]. Am J Physiol 1995b; 268(1 Pt 1):E60–66.

Bevan AP, Drake PG, et al. Peroxovanadium compounds: biological actions and mechanism of insulin-mimesis. Mol Cell Biochem 1995c; 153:49–58.

Bevan AP, Drake PG, et al. Intracellular signal transduction: the role of endosomes. Trends Endocrinol Metab 1996; 7:13–21.

Bevan AP, Krook A, et al. Chloroquine extends the lifetime of the activated insulin receptor complex in endosomes. J Biol Chem 1997; 272:26833–26840.

Burgess JW, Wada I, et al. Decrease in β-subunit phosphotyrosine correlates with internalization of the endosomal insulin receptor kinase. J Biol Chem 1992; 267:10077–10786.

Cheatham B, Vlahos CJ, et al. Phosphatidylinositol 3-kinase activation is required for insulin stimulation of pp70 S6 kinase, DNA synthesis, and glucose transporter translocation. Mol Cell Biol 1994; 14:4902–4916.

Chou CK, Dull TJ, et al. Human insulin receptors mutated at the ATP-binding site lack protein tyrosine kinase activity and fail to mediate postreceptor effects of insulin. J Biol Chem 1987; 262:1842–1847.

Contreres JO, Faure R, et al. ATP-dependent desensitization of insulin binding and tyrosine kinase activity of the insulin receptor kinase. The role of endosomal acidification. J Biol Chem 1998; 273:22007–22013.

Cuatrecasas P, Desbuquois B, et al. Insulin-receptor interactions in liver cell membranes. Biochem Biophys Res Commun 1971; 44:333–339.

Czech MP, Massague J. Subunit structure and dynamics of the insulin receptor. Fed Proc 1982; 41:2719–2723.

DeFronzo RA, Prato SD. Insulin resistance and diabetes mellitus. J Diabetes Complicat 1996; 10(5):243–245.

Doherty JJD, Kay DG, et al. Selective degradation of insulin within rat liver endosomes. J Cell Biol 1990; 110:35–42.

Drake PG, Bevan AP, et al. A role for tyrosine phosphorylation in both activation and inhibition of the insulin receptor tyrosine kinase in vivo. Endocrinology 1996; 137:4960–4968.

Dubyak GR, Kleinzeller A. The insulin-mimetic effects of vanadate in isolated rat adipocytes. Dissociation from effects of vanadate as a (Na^+-K^+)ATPase inhibitor. J Biol Chem 1980; 255:5306–5312.

Ebina Y, Ellis L, et al. The human insulin receptor cDNA: the structural basis for hormone-activated transmembrane signalling. Cell 1985; 40:747–758.

Fantus IG, Kadota S, et al. Pervanadate [peroxide(s) of vanadate] mimics insulin action in rat adipocytes via activation of the insulin receptor tyrosine kinase. Biochemistry 1989; 28:8864–8871.

Faure R, Baquiran G, et al. The dephosphorylation of insulin and epidermal growth factor receptors. Role of endosome-associated phosphotyrosine phosphatase(s). J Biol Chem 1992; 267:11215–11221.

Foot EA, Bliss T, et al. Dose related stimulation of glucose metabolism by peroxovanadate in rat skeletal muscle preparations in vitro. Biochem Soc Transasc 1991; 19(2):133S.

Freychet P, Roth J, et al. Monoiodoinsulin: demonstration of its biological activity and binding to fat cells and liver membranes. Biochem Biophys Res Commun 1971; 43:400–408.

Gavin JRD, Roth J, et al. Insulin receptors in human circulating cells and fibroblasts. Proc Natl Acad Sci USA 1972; 69:747–751.

Gavin JRD, Roth J, et al. Insulin-dependent regulation of insulin receptor concentrations: a direct demonstration in cell culture. Proc Natl Acad Sci USA 1974; 71:84–88.

Grigorescu F, Flier JS, et al. Defect in insulin receptor phosphorylation in erythrocytes and fibroblasts associated with severe insulin resistance. J Biol Chem 1984; 259:15003–15006.

Gunberger G, Comi RJ, et al. Tyrosine kinase activity of the insulin receptor of patients with type A extreme insulin resistance: studies with circulating mononuclear cells and cultured lymphocytes. J Clin Endocrinol Metab 1984; 59:1152–1158.

Hamel FG, Posner BI, et al. Isolation of insulin degradation products from endosomes derived from intact rat liver. J Biol Chem 1988; 263:6703–6708.

Hashimoto N, Feener EP, et al. Insulin receptor protein-tyrosine phosphatases. Leukocyte common antigen-related phosphatase rapidly deactivates the insulin receptor kinase by preferential dephosphorylation of the receptor regulatory domain. J Biol Chem 1992; 267:13811–13814.

Heyliger CE, Tahiliani AG, et al. Effect of vanadate on elevated blood glucose and depressed cardiac performance of diabetic rats. Science 1985; 227:1474–1477.

Hori C, Oka T. Vanadate enhances the stimulatory action of insulin on DNA synthesis in cultured mouse mammary gland. Biochim Biophys Acta 1980; 610:235–240.

Kadota S, Fantus IG, et al. Peroxide(s) of vanadium: a novel and potent insulin-mimetic agent which activates the insulin receptor kinase. Biochem Biophys Res Commun 1987a; 147:259–266.

Kadota S, Fantus IG, et al. Stimulation of insulin-like growth factor II receptor binding and insulin receptor kinase activity in rat adipocytes. Effects of vanadate and H_2O_2. J Biol Chem 1987b; 262:8252–8256.

Kahn CR. The molecular mechanism of insulin action. Annu Rev Med 1985; 36:429–451.

Kasuga M, Karlsson FA, et al. Insulin stimulates the phosphorylation of the 95,000-dalton subunit of its own receptor. Science 1982; 215:185–187.

Kasuga M, Fujita-Yamaguchi Y, et al. Tyrosine-specific protein kinase activity is associated with the purified insulin receptor. Proc Natl Acad Sci USA 1983; 80:2137–2141.

Kelly PA, Posner BI, et al. Studies of insulin, growth hormone and prolactin binding: ontogenesis, effects of sex and pregnancy. Endocrinology 1974; 95:532–539.

Kenner KA, Anyanwu E, et al. Protein-tyrosine phosphatase 1B is a negative regulator of insulin- and insulin-like growth factor-I-stimulated signaling. J Biol Chem 1996; 271:19810–19816.

Khan MN, Posner BI, et al. Intracellular hormone receptors: evidence for insulin and lactogen receptors in a unique vesicle sedimenting in lysosome fractions of rat liver. Proc Natl Acad Sci USA 1981; 78:4980–4984.

Khan MN, Savoie S, et al. Insulin and insulin receptor uptake into rat liver. Chloroquine action on receptor recycling. Diabetes 1985; 34:1025–1130.

Khan MN, Savoie S, et al. Characterization of rat liver endosomal fractions: in vivo activation of insulin-stimulable receptor kinase in these structures. J Biol Chem 1986; 261:8462–8472.

Khan MN, Baquiran G, et al. Internalization and activation of the rat liver insulin receptor kinase in vivo. J Biol Chem 1989; 264:12931–12940.

Kulas DT, Zhang WR, et al. Insulin receptor signaling is augmented by antisense inhibition of the protein tyrosine phosphatase LAR. J Biol Chem 1995; 270:2435–2438.

Lammers R, Bossenmaier B, et al. Differential activities of protein tyrosine phosphatases in intact cells. J Biol Chem 1993; 268:22456–22462.

McIntyre BS, Briski KP, et al. Effects of protein tyrosine phosphatase inhibitors on EGF- and insulin-dependent mammary epithelial cell growth. Proc Soc Exp Biol Med 1998; 217:180–187.

Moller NP, Moller KB, et al. Selective down-regulation of the insulin receptor signal by protein-tyrosine phosphatases alpha and epsilon. J Biol Chem 1995; 270:23126–23131.

Morgan DO, Ho L, et al. Insulin action is blocked by a monoclonal antibody that inhibits the insulin receptor kinase. Proc Natl Acad Sci USA 1986; 83:328–332.

Posner BI, Mierzwinski L, et al. Studies on amino acid levels and transport in the mechanically stressed rat heart. J Mol Cell Cardiol 1973; 5:221–233.

Posner BI, Kelly PA, et al. Studies of insulin, growth hormone and prolactin binding: tissue distribution, species variation and characterization. Endocrinology 1974; 95:521–531.

Posner BI, Kelly PA, et al. Prolactin receptors in rat liver: possible induction by prolactin. Science 1975; 188:57–59.

Posner BI, Patel B, et al. Uptake of insulin by plasmalemma and Golgi subcellular fractions of rat liver. J Biol Chem 1980; 255:735–741.

Posner BI, Patel BA, et al. Effect of chloroquine on the internalization of ^{125}I-insulin into subcellular fractions of rat liver. Evidence for an effect of chloroquine on Golgi elements. J Biol Chem 1982; 257:5789–5799.

Posner BI, Faure R, et al. Peroxovanadium compounds. A new class of potent phosphotyrosine phosphatase inhibitors which are insulin mimetics. J Biol Chem 1994; 269:4596–4604.

Posner BI, Khan MN, et al. Internalization of hormone receptor complexes: route and significance. Adv Exp Med Biol 1986; 205:185–201.

Powrie JK, Smith GD, et al. Mode of action of chloroquine in patients with non-insulin-dependent diabetes mellitus. Am J Physiol 1991; 260(6 Pt 1):E897–904.

Roth J. Peptide hormone binding to receptors: a review of direct studies in vitro. Metab Clin & Exp 1973; 22:1059–1073.

Shechter Y, Karlish SJ. Insulin-like stimulation of glucose oxidation in rat adipocytes by vanadyl (IV) ions. Nature 1980; 284:556–558.

Smith GD, Amos TA, et al. Effect of chloroquine on insulin and glucose homeostasis in normal subjects and patients with non-insulin-dependent diabetes mellitus. Bri Med J Clin Res Ed 1987; 294:465–467.

Stadtmauer LA, Rosen OM. Phosphorylation of exogenous substrates by the insulin receptor-associated protein kinase. J Biol Chem 1983; 258:6682–6685.

Stadtmauer L, Rosen OM. Increasing the cAMP content of IM-9 cells alters the phosphorylation state and protein kinase activity of the insulin receptor. J Biol Chem 1986; 261:3402–3407.

Takayama S, White MF, et al. Phorbol esters modulate insulin receptor phosphorylation and insulin action in cultured hepatoma cells. Proc Natl Acad Sci USA 1984; 81:7797–7801.

Tonks NK, Neel BG. From form to function: signaling by protein tyrosine phosphatases [see comments]. Cell 1996; 87:365–368.

Ullrich A, Bell JR, et al. Human insulin receptor and its relationship to the tyrosine kinase family of oncogenes. Nature 1985; 313:756–761.

White MF. The IRS-1 signaling system. Curr Opin Genet Dev 1994; 4:47–54.

Yale JF, Lachance D, et al. Hypoglycemic effects of peroxovanadium compounds in Sprague–Dawley and diabetic BB rats. Diabetes 1995; 44:1274–1279.

Yip CC, Moule ML, et al. Subunit structure of insulin receptor of rat adipocytes as demosntrated by photoaffinity labeling. Biochemistry 1982; 21:2940–2945.

Yu KT, Pessin JE, et al. Regulation of insulin receptor kinase by multisite phosphorylation. Biochimie 1985a; 67:1081–1093.

Yu KT, Werth DK, et al. src kinase catalyzes the phosphorylation and activation of the insulin receptor kinase. J Biol Chem 1985b; 260:5838–5846.

Zhang WR, Li PM, et al. Modulation of insulin signal transduction by eutopic overexpression of the receptor-type protein-tyrosine phosphatase LAR. Mol Endocrinol 1996; 10:575–584.

4
Growth Hormone Action
Signaling via a JAK/STAT-Coupled Receptor

David J. Waxman and Stuart J. Frank

CONTENTS

INTRODUCTION
PHYSIOLOGICAL, ENDOCRINE, AND CLINICAL ASPECTS OF GH ACTION
STRUCTURAL ASPECTS: GROWTH HORMONE, GH RECEPTOR AND GH BINDING PROTEIN
GH SIGNAL TRANSDUCTION—INITIAL EVENTS
GH SIGNAL TRANSDUCTION—ACTIVATION OF STAT TRANSCRIPTION FACTORS
OTHER GH-INDUCED INTRACELLULAR SIGNAL TRANSDUCTION PATHWAYS
MODULATORS AND ATTENUATORS OF GH SIGNALING
FATE OF GH-BOUND GH RECEPTOR AND MECHANISMS OF RECEPTOR DOWN-REGULATION
CONCLUDING REMARKS
SELECTED READINGS

1. INTRODUCTION

Among the many hormones and growth factors that collaborate to allow normal growth and development, growth hormone (GH) is one of the most important and extensively studied. GH has captured the interest of basic and clinical endocrinologists for many reasons. GH was recognized early in the history of modern endocrinology as a pituitary-derived peptide hormone that is a major positive regulator of growth. As a more sophisticated appreciation of hormonal physiology emerged, an important theory of hormone action, termed the "somatomedin hypothesis," was developed to explain the indirect growth promoting effects of GH; this concept spawned a burgeoning interest in insulin-like growth factors (IGFs) and related hormones.

Similarly, the field of neuroendocrinology and studies of the basic mechanisms of hormone secretion have blossomed through studies of the hypothalamic and pituitary regulation of GH production and release. These studies have led to a detailed understanding of GH releasing factors and their receptors, the role and regulation of the pulsatility of GH release, and the translation of this work to the clinical arena in terms of the diagnosis and treatment of GH deficiency.

Perhaps the most exciting area of molecular endocrinology and GH action that has developed, however, is the work carried out over the past decade related to the basic mechanisms of GH action. The cloning and characterization of GH and GH receptor (GHR) and their production using recombinant methods greatly facilitated crystallographic studies that provided critical molecular details of the interactions of GH with its plasma membrane bound receptor. This structural understanding in turn stimulated a keen interest in dissection at the biochemical level of the multiple intracellular signaling pathways induced by GH and the role of GHR in these processes. This exciting area has led to the unanticipated discovery that GHR is structurally and functionally related to the receptors for many other important hormones,

From: *Principles of Molecular Regulation* (P. M. Conn and A. R. Means, eds.), © Humana Press Inc., Totowa, NJ.

cytokines, and colony stimulating factors, a revelation that further broadens the impact of research into GH action.

In this chapter, we provide an overview of the clinical and basic aspects of GH physiology and describe the structural features of GH and GHR. Major emphasis is then placed on the molecular mechansims of GH signaling, including a detailed discussion of the major intracellular signalling pathways activated by GH and the interactions of GHR with key signaling proteins involved in those pathways. Finally, we discuss recent insights into how cellular GH signaling is modulated within cells and relate these findings to GH physiology.

2. PHYSIOLOGICAL, ENDOCRINE, AND CLINICAL ASPECTS OF GH ACTION

2.1. GH Stimulates Somatogenic Actions That Are Both Indirect and Direct

GH exerts somatogenic (growth-promoting) effects in both mammalian and nonmammalian species. GHRs are highly expressed in liver, but are also found in many other tissues; however, the physiological role of GH at many of these sites is only partially understood. According to the somatomedin hypothesis, much of the somatogenic activity of GH, including the promotion of postnatal longitudinal growth, is indirect, being mediated by IGF-1. IGF-1 is produced in and secreted by the liver in response to GH stimulation, and can interact with IGF-1 receptors in growth-responsive target cells and tissues. However, it is also clear that GH can act directly at the growth plate, for example, to collaborate with locally produced IGF-1 in promoting longitudinal bone growth. Studies of adipocyte differentiation indicate that GH (by promoting differentiation) and IGF-1 (by promoting clonal expansion) may collaborate in tissue formation at that site as well. Proliferation and/or hypertrophy at other nonhepatic cell types and tissues can similarly be promoted by GH acting either directly, or by local GH-induced IGF-1 production.

2.2. GH Exhibits a Broad Range of Metabolic Actions

GH exerts a wide range of metabolic effects on target cells and tissues; these include stimulation of fatty acid oxidation, amino acid uptake, and protein synthesis. The metabolic effects of GH are complex; they can be dependent on the dose and duration of GH exposure, the nature of the target cell or tissue, and the hormonal milieu of the tissue. Acute exposure of GH-starved cells or tissues to GH stimulates insulin-like effects, such as glucose uptake and lipogenesis. These insulinomimetic effects of GH are observed both in cell culture models and following GH infusion into normal human volunteers. Chronic GH exposure, however, is diabetogenic in that insulin-mediated glucose uptake, for example, is impaired. In humans, such diabetogenic actions of excessive GH can be seen most readily in patients with chronic GH elevation due to GH-secreting pituitary adenomas (acromegaly), some of whom manifest overt type 2 diabetes mellitus with a frequency and severity that correlates with the level of circulating GH. Moreover, acromegalics nearly always have detectable insulin resistance at both hepatic and extrahepatic sites, without associated changes in insulin receptor number or insulin binding characteristics. Dissection of the molecular mechanism(s) by which the polypeptide GH, essentially a single molecular species, can exert such a broad range of contextually specific and disparate effects presents a major challenge in this area of molecular endocrinology.

2.3. The Hypothalamo–Pituitary Axis Regulates GH Secretion

GH is primarily synthesized in the anterior pituitary gland, where its production is regulated by a complicated interplay of two hypothalamic peptides, GH-releasing hormone (GHRH) and somatostatin. GHRH and somatostatin are both synthesized in hypothalamic nuclei and are transported to the anterior pituitary somatotrophic cells by the hypothalamic–hypophysial portal blood vessels. GHRH acts on the somatotrophs to transcriptionally enhance GH production and to stimulate GH release. Somatostatin inhibits GH release, primarily by affecting the timing and amplitude of GH secretion rather than by regulating GH synthesis. Circulating IGF-1 also contributes to the regulation of plasma GH levels through its feedback inhibitory effects on the synthesis and secretion of GH by the pituitary (Fig. 1).

Further complexity regarding the regulation of pituitary GH secretion is indicated by the ability of synthetic GH secretagogues, such as GHRP-6, to promote pulsatile GH release from the anterior pituitary in a manner that is independent of pituitary GHRH and somatostatin receptors. A GH secretagogue receptor expressed in the pituitary and hypothalamus has been cloned, and this should facilitate studies of the physio-

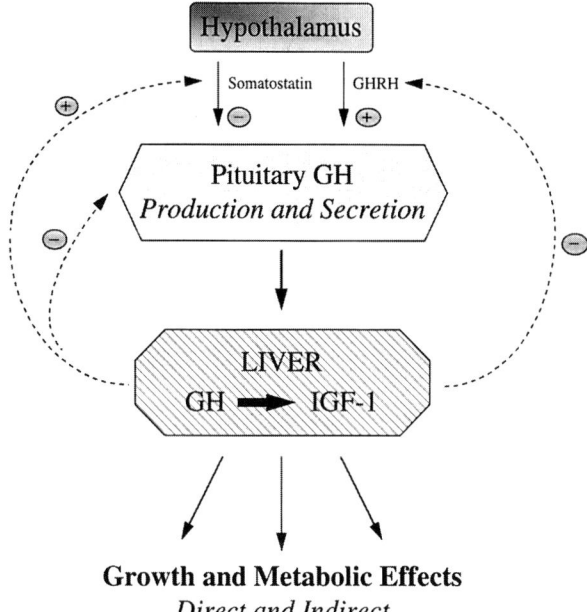

Fig. 1. Hypothalamic–pituitary GH axis. Pituitary GH production and secretion is regulated by the hypothalamic hormones somatostatin and GH releasing hormone (GHRH). GH and IGF-1 produced in the liver both feedback inhibit the pathway at the level of pituitary production, and at the level of somatostatin and GHRH secretion by the hypothalamus, as indicated.

logic and pharmacologic significance of these secretagogues, their role in regulation of plasma GH levels in normal individuals, and their development as potential therapeutic agents for some forms of GH deficiency.

While GH is produced primarily in the anterior pituitary gland, GH synthesis at extrapituitary sites (e.g., leukocytes, lymphocytes, thymocytes) has been described. It is unclear how GH secretion at these sites is regulated or whether the actions of such extrapituitary-derived GH are largely local (autocrine or paracrine) and/or to what degree they are mediated by IGF-1.

2.4. Humans and Rodents Both Exhibit Gender Differences in Plasma GH Profiles

Release of GH from the pituitary gland is not constant over time, but rather is intermittent, or pulsatile. GH secretion in humans begins at the end of the first trimester of gestation. Integrated daily GH secretion in humans is greatest during adolescence and declines with age thereafter. Notable sex differences in the neuroendocrine regulation of pituitary GH release have been observed in humans. Premenopausal women have two- to threefold higher daily GH secretion rates than men due to a two- to threefold higher GH pulse amplitude in women. However, both males and females exhibit the same GH pulse frequency. Normal young men experience 6–10 secretory bursts per day, the majority of which occur during sleep at night. Interpulse plasma GH levels are undetectable. Women exhibit higher serum GH peaks in the late follicular stage of the menstrual cycle, when serum estrogen levels are elevated. Estrogen is believed to be an important mediator of the sex differences in human plasma GH profiles, which in part originate in the reduced somatostatin inhibitory action that occurs in females. The reduced effectiveness of glucose ingestion at fully suppressing serum GH levels in females compared to males may also manifest this sex difference in somatostatin inhibitory activity. These sex differences in GH release are likely to be physiologically important, in view of the profound effects that sex-dependent GH patterns are demonstrated to have on linear skeletal growth and GH action in target tissues in rodent models (see below).

In comparison to humans, rodents display a dramatic sexual dimorphism in the pulsatility of GH secretion, which is directly regulated by the temporal pattern of hypothalamic hormone secretion. This temporal pattern is, in turn, sex dependent, and is established by gonadal steroid imprinting during the neonatal period. Male rats exhibit a highly regular pattern of plasma GH pulsatility that is characterized by high peaks of GH (~200 ng/mL) each ~3.5 h, followed by an interpulse period lasting up to ~2 h, during which circulating GH is essentially undetectable (1–2 ng/mL). By contrast, adult female rats are characterized by much more frequent, and irregular, pulses of pituitary GH secretion, resulting in sustained periods during which there is a near-continuous presence of GH in circulation at levels ≥ 20–40 ng/mL. In addition, young female rats have nocturnal surges of GH of high amplitude and short duration that are analogous to the high-amplitude nocturnal GH peaks seen in pubertal girls. These sex-dependent plasma GH profiles mediate the sex-dependent effects of GH on longitudinal bone growth and body weight beginning at puberty, as well as the sex-dependent expression in rodents of a large number of liver gene products, including hormone receptors, cytochromes P450, and other liver enzymes. Recent studies on GH signaling, described later in this chapter, have shed light on the cellular and molecular mechanisms through which the temporal pattern of GH stimulation dictates these sex-dependent patterns of liver gene expression.

GH STATE	CLINICAL SYNDROME	CAUSES
↑ GH ACTION	• ACROMEGALY (after puberty) • GIGANTISM (before puberty)	• GH-Secreting Pituitary Adenoma • Ectopic GHRH • Ectopic GH
↓ GH ACTION	• GH-Deficiency	• Isolated Gene Defect • Pituitary Failure • Hypothalamic Dysfuntion
	• Abnormal GH	• GH Mutations
	• GH Resistance (Laron Syndrome)	• GHR Mutations
	• "Normal Short Kids"	?

Fig. 2. Clinical syndromes related to GH pathology.

2.5. Several Clinical Disorders Involve GH Production or Its Action

The most striking clinical disorders related to the GH axis are those that manifest significantly increased or decreased net GH action (Fig. 2). Acromegaly, a syndrome generally caused by excessive GH secretion due to a somatotrophic anterior pituitary adenoma, is associated with excessive GH somatogenic activity, leading to bony and cartilaginous overgrowth in the face, jaw, hands, feet, and joints and visceral enlargement. Gigantism (excessive height) results when GH hypersecretion arises prior to the pubertal closure of the epiphyseal growth plate in the long bones. Conversely, decreased net GH action can usually be explained by congenital or early acquired GH deficiency. This may derive from GH gene deletion, somatotroph destruction, or dysregulation of GH secretion, and results in shortness of stature and abnormal skeletal and cartilaginous maturation that is most notable postnatally.

2.6. GH Resistance Can Result in Laron Syndrome

Laron syndrome is a rare disease that is similar in phenotype to GH deficiency. Circulating IGF-1 is low in Laron syndrome, as in the case of GH deficiency; however, in contrast to GH deficiency, circulating GH is present and often elevated in Laron patients. In most cases, Laron syndrome is associated with a variety of defects in GHR structure, function, or cell surface expression that serve as a basis for the observed hormone resistance. However, defects in GH signaling downstream of GHR are observed in some Laron patients. A mouse model of this syndrome (the "Laron mouse") has been obtained targeted disruption of the *GHR* gene. The GHR-deficient mouse mimics human Laron syndrome in that it displays marked postnatal growth retardation, proportionate dwarfism, greatly decreased serum IGF-1, and elevated serum GH concentrations.

Decreased net GH action attributable to a primary defect in IGF-1 production has been described in a single Lawn patient. This individual has a homozygous partial deletion in the *IGF-1* gene, and this results in a very low IGF-1 level and elevated circulating GH. In addition, there is marked prenatal and postnatal growth failure, much like that seen in mice with targeted homozygous disruption of the *IGF-1* gene.

These examples of decreased GH action caused by defects in GH secretion and/or GH sensitivity, or by defects in the *IGF-1* gene, illustrate two important features of the GH–IGF-1 axis: (1) although GH promotes IGF-1 secretion, the somatogenic effects of GH, in contrast to those of IGF-1, are not significantly manifested prenatally; and (2) as a consequence, some effects of IGF-1 on growth and development are apparently independent of GH. It is also becoming evident that the effects of GH on cellular proliferation, at least in some model systems, can be independent of IGF-1.

3. STRUCTURAL ASPECTS: GH, GH RECEPTOR AND GH BINDING PROTEIN

3.1. GH Is a Four-Helix Bundle Protein

Human GH (hGH) is a 191 amino acid (~22 kDa) product of the *hGH-N* gene and is primarily synthesized in the anterior pituitary gland. GH molecules from all species are ~50% α-helical in structure and contain two intrachain disulfide linkages (Fig. 3). X-ray crystallographic analyses of porcine GH and hGH indicate a structure composed of four antiparallel α-helical bundles. An alternatively spliced transcript of the *hGH-N* gene yields a 20-kDa form of hGH that has an internal deletion of amino acid residues 32–46 and represents ~16% of circulating GH. Other variants of GH (acetylated, phosphorylated, glycosylated, dimerized, and oligomerized GH) can be generated postranslationally but their functional signficance is unknown. As is discussed later, interaction of the 20 kDa form of hGH with GHR may differ mechanistically from that of 22 kDa hGH.

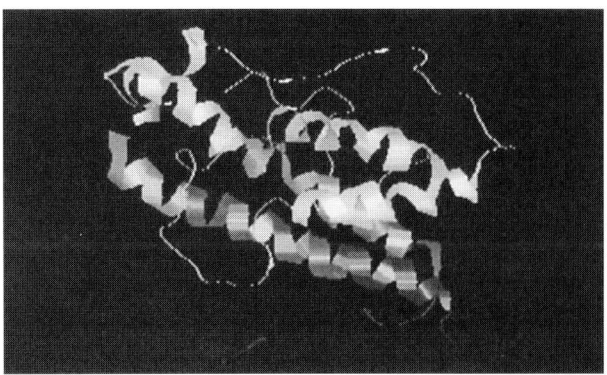

Fig. 3. Three-dimensional structure of hGH, shown as a ribbon model.

GH shares clear evolutionary and structural homology with prolactin, which is produced by lactotrophs in the anterior pituitary. As discussed later, GH and prolactin also share key features of their intracellular signaling pathways. While similar in three-dimensional structure, GH and prolactin share only ~24% amino acid sequence identity. GH and prolactin receptors are also similar in their overall topology and structure (see below), and human GH can bind to and activate prolactin receptors of lower species. However, the physiological effects of these two hormones, while overlapping, are quite distinct. Prolactin, for example, does not promote IGF-1 release.

3.2. GH Receptor Is a Single-Pass Transmembrane Protein

Binding sites for GH are found in the plasma membrane of many tissues. The cDNAs for the human and rabbit GHRs were first isolated in 1987 from liver, an abundant source of this receptor. GHRs from rodents (mouse and rat), ruminants (cow and sheep), pig, and chicken were cloned soon thereafter and exhibit significant homology and similar overall topologies. In the human and rabbit, GHR is a single membrane-spanning polypeptide of 620 residues with a ligand-binding extracellular domain of 246 residues, a short, single-pass transmembrane domain (~24 residues), and a large cytoplasmic domain of 350 residues (Fig. 4). Human and rabbit GHRs are the most similar to each other (84% identical), while the human and rodent GHRs are ~70% identical and the chicken and human receptors ~59% identical. Although predicted to be about 70 kDa in size based on amino acid sequence, GHRs of various species migrate heterogeneously at 110–140 kDa when resolved on denaturing (sodium dodecyl sulfate) gels under reducing conditions. The aberrant migration of GHR in part reflects

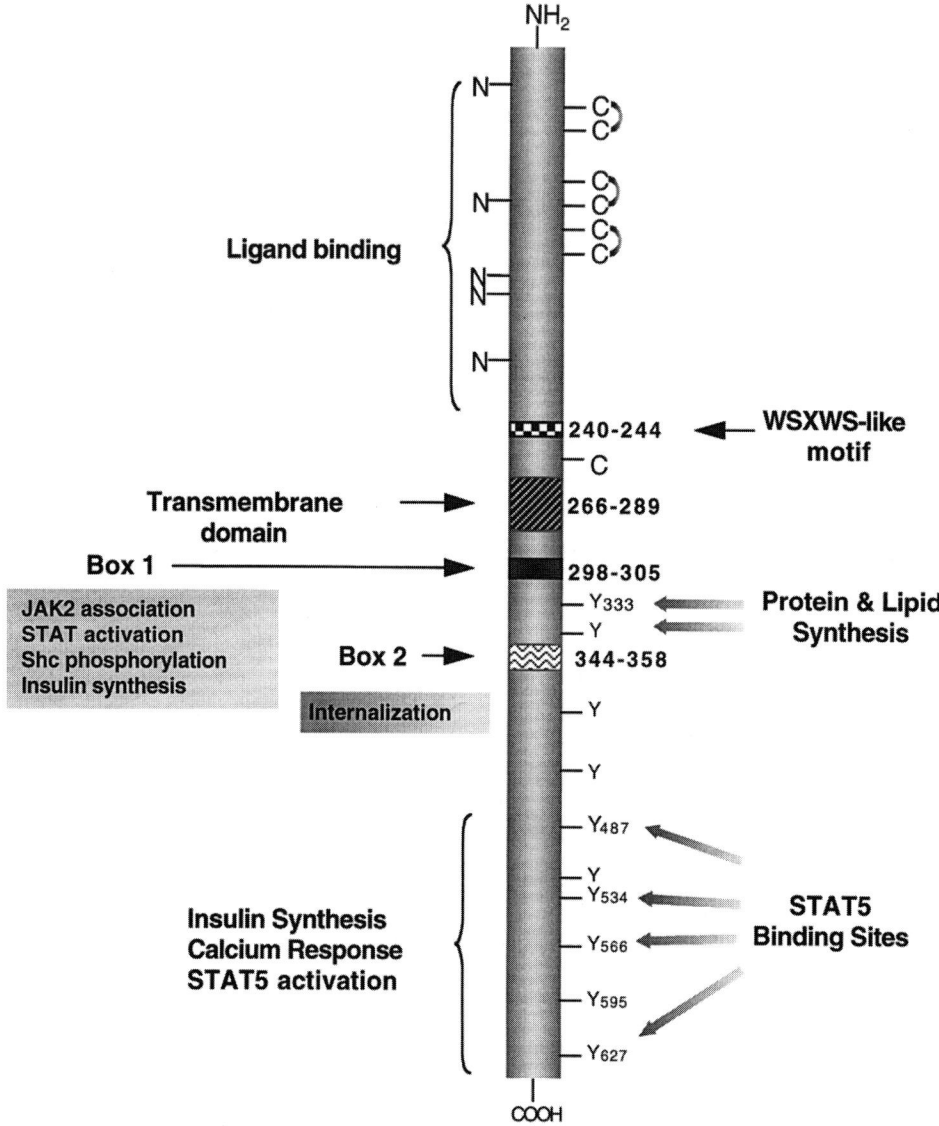

Fig. 4. Growth hormone receptor: key structural and functional features. Shown are the locations of five N-linked glycosylation sites (N) and seven cysteine residues (C) within the extracellular ligand binding domain of GHR. Ten tyrosines residues (Y) are present in the cytoplasmic region of rat GHR; the six tyrosines that are conserved across species are numbered. Also shown are the extracellular WSXWS-like motif and the intracellular Box 1 (proline-rich domain) and Box 2 regions. Regions of GHR required for various functions, including STAT5b binding are indicated using the rat GHR numbering system. (Adapted from Carter-Su et al. Annu Rev Biochem 1996; 58:187–207.)

the presence of five potential asparagine-linked glycosylation sites within the extracellular domain, several of which are glycosylated (Fig. 4). In addition, GHR can be modified by ubiquitination, a step that may target the receptor for internalization and degradation. However, cytoplasmic domain fragments of bacterially expressed GHR, which are neither glycosylated nor ubiquitinated, also migrate larger than expected, indicating that GHR binds sodium dodecyl sulfate differently than most proteins, even in the absence of posttranslational modifications.

3.3. GH Receptor Belongs to the Cytokine Receptor Superfamily

GHR contains several structural features that are conserved in other cell surface receptors and classify it within the cytokine/hematopoietin receptor superfamily. These include, in the extracellular domain, three pairs of conserved cysteine residues, which form intrachain disulfide linkages, a fibronectin-like domain, and a WSXWS-like motif, which in mammalian GHRs has the sequence YXXFS, where X represents a nonconserved amino acid (Fig. 4). Within the superfamily, GHR belongs to the type I cytokine receptor family, which includes the closely related prolactin receptor as well as receptors for various interleukins and colony-stimulating factors, erythropoietin, thrombopoietin, and leptin. The conserved cysteine residues and the WSXWS-like sequence of GHR help form and/or support the receptor's extracellular ligand binding pockets. Although GHR's WSXWS-like motif does not contact GH directly, it is essential for ligand binding and signal transduction, as demonstrated by alanine-scanning mutagenesis, and as indicated by the GH resistance phenotype seen in sex-linked dwarf chickens, where the motif's terminal serine is mutated to isoleucine. In addition, GHR and the other cytokine receptor family members all contain regions of homology in their cytoplasmic domains that confer similarities in their signaling mechanisms. One of these is termed Box 1, a proline-rich element eight amino acids long that is present in the receptors' proximal cytoplasmic domain (Fig. 4). The Box 1 element has the sequence ILPPVPVP in mammalian GHRs and is crucial for the association of receptor with Janus family tyrosine kinases (JAKs), as discussed later. A more loosely conserved cytoplasmic domain region of GHR containing seven acidic residues and ending with lysine is termed Box 2. Box 2 is located in the receptor's proximal cytoplasmic domain about 40 residues C-terminal to Box 1 and, like Box 1, plays an important role in potentiating GH signaling. Finally, each of the mammalian GHRs contains in its cytoplasmic domain six tyrosine residues conserved across species (Y^{333}, Y^{487}, Y^{534}, Y^{566}, Y^{595}, and Y^{627}, in the numbering system of the rodent receptors; Y^{314}, Y^{469}, Y^{516}, Y^{548}, Y^{577}, and Y^{609}, in the numbering system for the human and rabbit receptors), several of which can be phosphorylated following GH binding (*see* later).

3.4. Variants of GH Receptor Can Have a Variety of Physiological Effects

GHR variations of potential physiological importance are observed both between and within species. In rodents, for example, two alternatively spliced GHR transcripts of different size classes (4.2–4.7 and 1.0–1.4 kb) can be detected. The longer mRNA encodes the full-length GHR and the shorter message encodes a protein consisting of the GHR extracellular domain attached to a short hydrophilic peptide, which replaces the receptor's transmembrane and cytoplasmic domain. This shortened receptor is secreted and functions as a circulating high-affinity GH binding protein (GHBP). Other studies have described alternatively spliced GHR mRNAs that, because of frame shifting, predict the translation of a membrane-anchored GHR with a severely truncated cytoplasmic domain. These truncated GHRs have a significantly increased capacity to generate a soluble GHBP compared to full-length GHR, perhaps as a consequence of their markedly reduced internalization, and hence longer residence on the plasma membrane. In addition, the truncated GHR can heterodimerize with full-length GHR and thereby inhibit GH signaling via a dominant-negative inhibitory mechanism.

Mutations that confer pathophysiologically relevant alterations in GHR's extracellular domain can lead to changes in cell surface expression of GHR or its GH binding activity. This can result in severe GH resistance, as seen in Laron syndrome, or in a less severe deficiency of GH action associated with a condition referred to as idiopathic short stature (Fig. 2). A frameshift mutation that yields a severely truncated GHR with only seven cytoplasmic domain residues has been described as a cause of familial short stature. This autosomal dominant mutation is analogous to the cytoplasmic domain truncated GHR splice variants discussed earlier and is proposed to generate a dominant-negative GHR that results in GH insensitivity by forming inactive heterodimers with the wild-type receptor (*see* next section).

3.5. GH Induces a Functionally Important Dimerization of Its Receptor at the Cell Surface

The three-dimensional structure of GH is arranged as four antiparallel α-helical bundles without an obvious axis of symmetry (Fig. 3). By systematic mutagenesis of hGH and the extracellular domain of hGH, residues important for the high-affinity interaction been GH and its receptor have been mapped. Kinetic studies of the binding interactions between GH and GHBP and X-ray diffraction studies of the cocrystallized proteins indicate the formation of a trimolecular complex with a stoichiometry of one GH per two GHBP (i.e., two GHR) molecules. Despite its lack of an axis of symmetry, GH engages each receptor monomer at nearly identical contact points. By contrast, each of the two sites on GH that engages a GHR molecule (termed GH site 1 and site 2) is quite

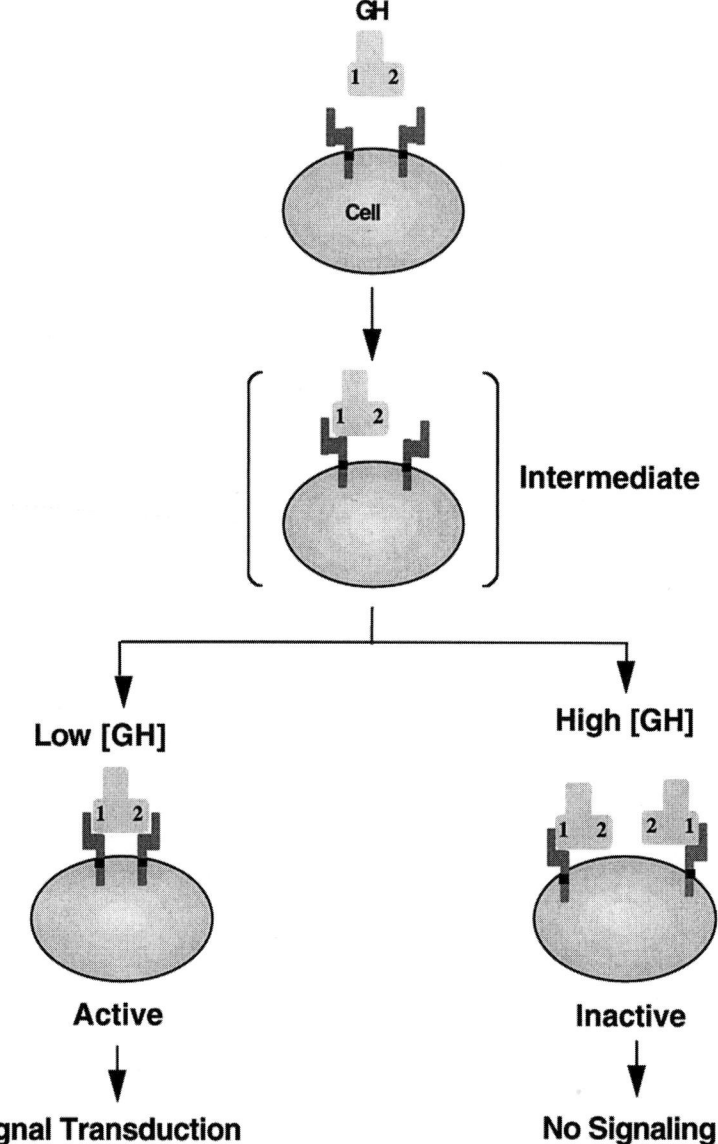

Fig. 5. Activation of GHR by sequential dimerization mechanism. GH is an asymmetric four-helix bundle protein that initially interacts via GH site 1 with a monomer of GHR to form the intermediate shown. At physiological GH concentrations, this initial step is followed by formation of the functional receptor dimer by interactions with GH site 2. At high GH concentrations, formation of the inactive monomeric complex dominates, and downstream signaling is prevented. (Adapted from Fuh et al. Science 1992; 256:1677–1680.)

different. Moreover, the binding reactions are sequential, such that the binding of one GHR molecule to site 1 of GH potentiates the binding of a second GHR molecule at GH site 2 (Fig. 5). Additional contact points between receptor monomers stabilize the tripartite complex.

GHR dimerization is necessary for receptor activation and effective signal transduction. This was originally demonstrated using growth factor-dependent hematopoietic cells that express a chimera between the extracellular domain of GHR and the cytoplasmic domain of a colony-stimulating factor receptor. These cells proliferate when the cell surface-expressed chimeric receptors are dimerized by treatment with GH or with antibodies directed to the extracellular domain of GHR, but not by their monovalent Fab fragments. Moreover, several cellular responses to GH, including GH-dependent cell proliferation, GH-dependent protein tyrosine phosphorylation, and GH-dependent lipogenesis are self-antagonized at high GH concentrations, consistent with GH binding to all available GHR molecules via site 1 in unproductive monomeric interactions. Similarly, mutants of GH with site-specific alterations of site 2 (e.g., *hGH–G120R*) are capable of interacting with GHR monomers via site 1, but are unable to promote receptor dimerization via site 2. These site 2 GH mutants are by themselves unable to promote signaling and can function as dominant negative antagonists of wild-type GH signaling both in vitro and in vivo. Such GH antagonists are currently being evaluated as therapeutic agents for acromegaly based on the hypothesis that interruption of the ability of GH to dimerize its receptors on target tissues may ameliorate the effects of excessive circulating GH. In a similar manner, some Laron syndrome patients have a GHR mutation (D152H), in which a charge change in the receptor's extracellular domain dimerization interface prevents GH-induced GHR dimerization. This finding further supports the conclusion that GHR dimerization is a key requirement for receptor activation and signaling.

GH treatment of intact cells promotes the appearance of a disulfide-linked, tyrosine phosphorylated high molecular weight form of GHR in human IM-9 B lymphocytes and in several other cell types. The disulfide-linked GHR form is not detected when cells are treated with a site 2 hGH mutant, such as G120K (a GH antagonist), yet G120K, by interacting with GHR via its intact site 1, antagonizes the ability of GH to promote GHR disulfide linkage. Thus, the disulfide-linked GHR form is a biochemically detectable correlate of dimerized GHR. The functional significance of the disulfide linkage *per se* is still unclear. Further evaluation of the mechanism by which it is formed (critical cysteine residues contributing to the disulfide linkage, for example) may facilitate studies of the signaling and trafficking of those GHR molecules that are engaged, dimerized, and activated by GH.

3.6. Plasma GH Is Bound to a GH-Binding Protein

A significant fraction (up to 50%) of circulating GH is complexed to a high-affinity GHBP in humans and other species. Although the precise physiological role of GHBP is still uncertain, GHBP can potentiate, or alternatively, may inhibit GH action, depending on the particular experimental paradigm and physiological situation. GHBP is identical in structure to the GHR ectodomain. Yet, in contrast to the 1:2 stoichiometry that characterizes GH: GHR complexes formed at the cell surface (Fig. 2), the majority of GHBP-associated GH that is present in circulation exists in the form of a 1:1 GH–GHBP complex. This difference in binding stoichiometry is likely due to the much lower concentration of GHBP in plasma (~1 n*M*) compared to the calculated GHR concentration at the cell surface (up to 60 n*M*). This large difference in GH binding site concentration would be expected to favor receptor dimer formation at the cell surface but not in plasma with the binding protein. Accordingly, the level of circulating GHBP is likely to modulate the concentration of free GH in plasma. GHBP can thus serve as a reservoir that facilitates delivery of GH from the circulation to its cell-surface receptor.

3.7. GH Binding Protein Can Be Formed By Multiple Mechanisms

In rodents, GHBP is largely derived from an alternatively spliced GHR mRNA that encodes the receptor ectodomain followed by a short hydrophilic tail in place of the transmembrane and cytoplasmic domains. In contrast, human and rabbit GHBPs are primarily formed by proteolytic shedding of the transmembrane ectodomain of GHR. The existence of two such distinct molecular mechanisms of GHBP generation highlights the likely biological importance of GHBP in GH physiology.

Mechanistic details of proteolytic shedding of human and rabbit GHRs have remained elusive. In IM-9 cells, a metalloprotease activity appears to be responsible for cleavage of GHR and the shedding of GHBP. Furthermore, this proteolysis is enhanced by activation of protein kinase C (PKC), a property shared with a number of other cell surface proteins that release their ectodomains in soluble form. These include tumor necrosis factor α, transforming growth factor α, and β-amyloid precursor protein. Identification of the enzyme(s) involved in cleavage of GHR from the cell surface will provide further insight into the regulation of GHBP formation and its physiological significance.

4. GH SIGNAL TRANSDUCTION—INITIAL EVENTS

Multiple biochemical events and intracellular signals can be detected shortly after GH-responsive target cells are stimulated with GH. Dimerization of GHR following GH binding initiates a series of protein phosphorylation cascades that lead to the activation of several enzymes and transcription factors, followed by the induction of gene expression (Fig. 6). In several instances, a detailed understanding of how these initial signaling events translate into the biological effects of GH is emerging. The remainder of this

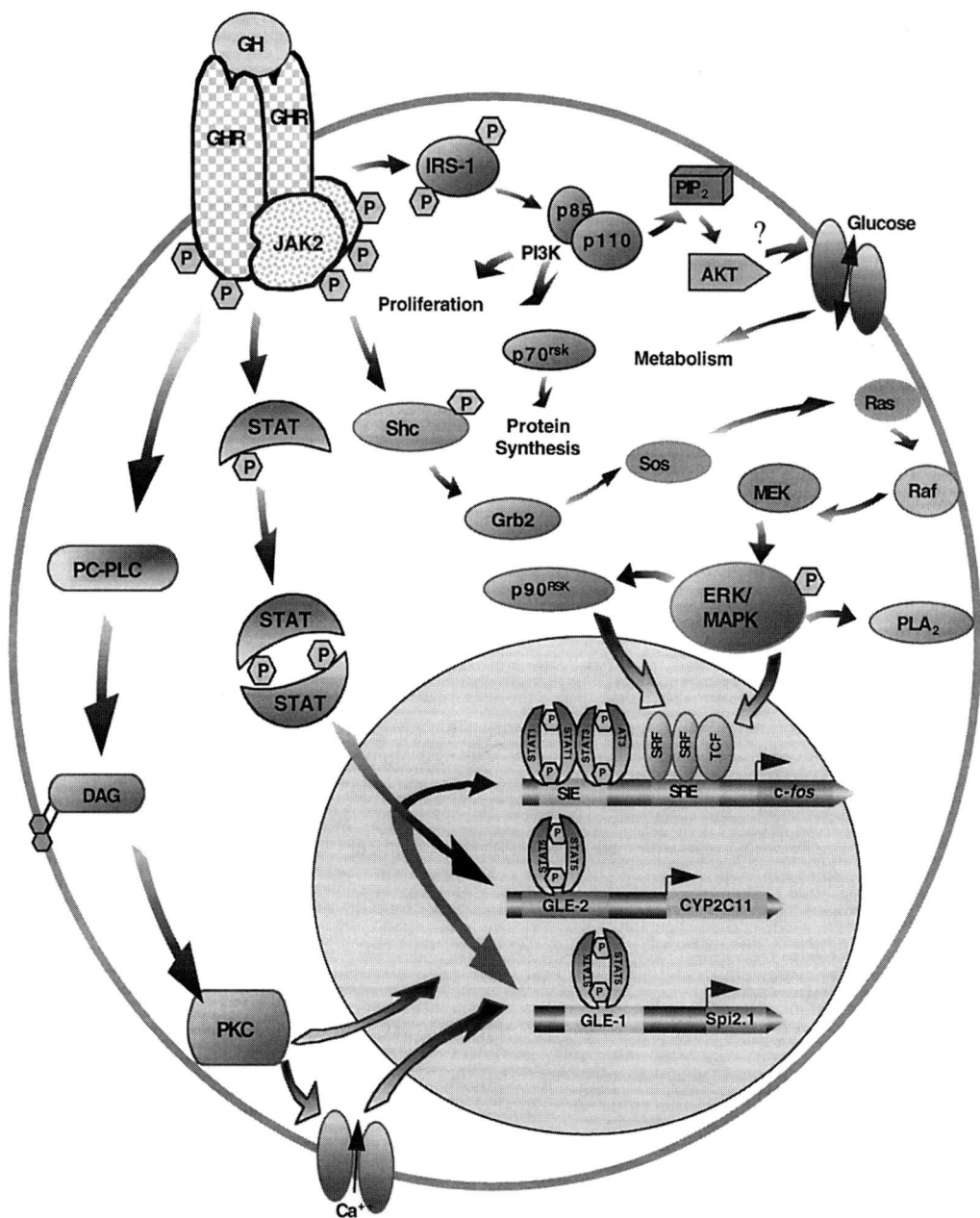

Fig. 6. Intracellular signaling pathways proposed to be activated by GH. GLE, interferon-γ-activated sequence (GAS)-like response element; PLA$_2$, phospholipase A$_2$; PLC, phospholipase C; PKC, protein kinase C; DAG, 1,2-diacylglycerol; SIE, Sis-induced element; SRE, serum response element; SRF, serum response factor; spi 2.1, serine protease inhibitor 2.1; CYP2C11, cytochrome P450 2C11; PI3K, phosphatidylinositol 3′-kinase; TCF, ternary complex factor; PIP2, phosphatidyl inositol 4,5-bisphosphate; AKT, protein kinase B. (Adapted from Carter-Su et al. Annu Rev Biochem 1996; 58:187–207.)

chapter discusses the intracellular signaling pathways activated by GH, the mechanisms by which GH signaling is modulated and/or attenuated, and the current status of studies aimed at understanding more clearly how GH signaling leads to the biological effects of GH.

As described later, a key early step in many GH signaling pathways involves GH-induced activation of a nonreceptor tyrosine kinase termed Janus kinase 2 (JAK2 kinase). Once activated by GH in this manner, JAK2 kinase catalyzes the tyrosine phosphorylation of other, nearby JAK2 molecules and of GHR. Four major signaling cascades are initiated following these primary tyrosine phosphorylation events (Fig.

6): (1) activation of signal transducer and activator of transcription (STAT) pathways leads to a direct stimulation of target gene transcription; (2) activation of the Ras–Raf–MEK–ERK pathway results in the activation of mitogen-activated protein kinase (MAP kinase), a dual specificity kinase whose activity is linked to intranuclear phosphorylation, c-*fos* transcription, and cell proliferation; (3) activation of phosphatidylinositol-3-kinase (PI-3-kinase), at least in part by tyrosine phosphorylation of insulin receptor substrates (IRS-1 and IRS-2) occurs and may be important for some of the metabolic and proliferative activities of GH; and (4) activation of PKC occurs and may have a substantial impact on the metabolic, differentiative, and proliferative actions of GH. In many instances, there is significant crosstalk between these four GH-activated pathways. GH signaling is thus characterized by intracellular pathways and events with multiple points of intersection and convergence, rather than linear paths leading to independent and exclusive biochemical endpoints and cellular events.

4.1. GH Activates the GH Receptor-Associated Tyrosine Kinase JAK2

GH-induced tyrosine phosphorylation of GHR was first observed by Carter-Su and colleagues in studies of murine 3T3–F442A fibroblasts and adipocytes. The absence of a consensus tyrosine kinase motif in the sequence of GHR's cytoplasmic domain indicated to these investigators that, rather than undergoing the classic autophosphorylation reaction that is well established for many other cell surface growth factor receptors, GHR is instead phosphorylated by a closely associated nonreceptor tyrosine kinase. In an important advance in our understanding of GH signaling, JAK2 kinase was later identified biochemically as a prominent GHR-associated tyrosine kinase. The rapidity and intensity of GH-induced JAK2 kinase activation and the tyrosine autophosphorylation of JAK2 seen in multiple cell types is consistent with JAK2 being a critical molecule in the initiation of GH signaling. Further, cells devoid of JAK2, but expressing GHR, do not respond to GH by tyrosine phosphorylation of STAT signaling molecules or Shc adaptor proteins important for MAP kinase activation (see below); by contrast, cells expressing JAK2 kinase respond to GH in these respects.

JAK2 is a member of the Janus family of mammalian tyrosine kinases, which is comprised of four structurally related (~36–47% identical) cytoplasmic tyrosine kinases—JAK1, JAK2, JAK3 and TYK3. All four kinases contain a catalytically inactive, kinase-like ("pseudokinase") domain located just N-terminal to the carboxy-terminal kinase domain. The JAK kinases are also typified by their propensity to physically and functionally couple to GHR, prolactin receptor and other cytokine receptor family members. Whereas JAK2 is the preferred kinase that associates with several receptors of the superfamily, including the receptors for GH, prolactin, and erythropoietin, other JAKs, or combinations of JAKs, associate with other cytokine receptor family members (e.g., JAK1 and TYK2 both associate with interferon α receptor, and JAK1 and JAK2 with interferon-γ receptor). This promiscuity of the JAK kinases for multiple receptors, coupled with the ubiquitous expression of three of the four JAKs (JAK3 being largely restricted to myeloid and lymphoid cells), dictates that the hormone- and cytokine-specificity for activation of downstream tyrosine kinase substrates (such as the STATs, *see* below) and target genes cannot be determined by the JAK kinase alone.

4.2. JAK2 Kinase Associates with GH Receptor via the Receptor's Proline-Rich Box 1 Sequence

GHR residues 271–390, corresponding to the membrane-proximal region of the receptor's cytoplasmic domain, can interact in vitro with JAK2 in cell extracts, as demonstrated by binding experiments using GHR-glutathione *S*-transferase (GST) fusion proteins that incorporate the Box 1 element of GHR, ^{298}ILPPVPVP^{305}. GHR Box 1 sequences are both necessary and sufficient for GHR to interact with JAK2, insofar as a GST fusion protein containing the Box 1 region alone is sufficient for weak, but specific, interaction with JAK2. Similarly, internal deletion of the Box 1 sequence prevents JAK2 activation and downstream JAK2-dependent signaling in cell-based reconstitution and transfection assays. JAK2 also requires Box 1 sequences for its physical and functional association with other cytokine receptor family members, such as the prolactin and erythropoietin receptors. A role for more distal cytoplasmic domain regions of GHR for stabilizing the receptor's interaction with JAK2 is, however, also apparent. The precise nature of the interaction of JAK2 with GHR Box 1 is unclear. Although Box 1 is proline-rich, it does not correspond to a typical proline-rich binding domain. Moreover, SH3 domains, which bind proline-rich sequences, have not been found in JAK2 kinase.

Coimmunoprecipation experiments demonstrate that interactions between GHR and JAK2 can be detected in the basal state, that is, in the absence of GH treatment and in the absence of tyrosine phosphorylation of either protein; however, the binding interaction is strengthened upon stimulation of cells with GH. The region of JAK2 most responsible for physical and functional interaction with the receptor is in the N-terminal ~20% of the molecule. The corresponding region is also important for JAK2's interaction with other cytokine receptors that it binds, such as the GM-CSF receptor β-chain and interferon-γ receptor, as well as for the interaction of the related JAK3 kinase with the interleukin-2 (IL-2) receptor γc chain.

Although JAK2 is the tyrosine kinase most tightly coupled to GHR activation, the participation of other tyrosine kinases in GH signaling is, by analogy to other cytokine receptors, quite possible. For example, GH induces a relatively low level of JAK1 tyrosine phosphoryation in certain cell types. However, since JAK1-deficient cells that express JAK2 are not impaired regarding GH-induced STAT and Shc tyrosine phosphorylation, while JAK2-deficient cells are unable to support these GH responses, JAK1 might not play a physiologically significant role in GH action. Recent evidence indicates that GH, like prolactin, can activate *src* and *fyn* kinases, two members of the src-family of nonreceptor tyrosine kinases. Details of this activation process and the significance of these kinases in GH signaling are important areas for future research.

4.3. JAK2 Is Activated by a Trans-Phosphorylation Reaction

Studies of the mechanism by which GH and other cytokines/growth factors activate JAK2 indicate that two C-terminal region tyrosine residues of JAK2, Y^{1007} and Y^{1008}, are sites of *trans-* or autophosphorylation, both in intact cells and in in vitro kinase assays. Y^{1007} and Y^{1008} are predicted to reside within the activation loop of the JAK2 kinase domain on the basis of sequence homology to the regulatory region of the insulin receptor tyrosine kinase. Mutagenesis of these two residues demonstrates that Y^{1007}, but not Y^{1008}, is essential for JAK2 kinase activation. Studies of the corresponding two tyrosine residues of JAK3 (Y^{980} and Y^{981}) indicate that Y^{980}, when phosphorylated, positively regulates JAK3 activity, while tyrosine phosphorylation of Y^{981} inhibits kinase activation.

These and other findings lead to a working model of early GH signaling in which JAK2 associates with GHR via interactions between the N-terminal portion of JAK2 and Box 1 of GHR. JAK2 is subsequently brought into close proximity with other JAK2 molecule(s) by GH-induced GHR dimerization. This GH-induced increase in JAK2–JAK2 proximity, together with a GH-induced stabilization of the GHR–JAK2 complex, enables JAK2 to utilize its low level basal kinase activity to *trans*-phosphorylate a nearby JAK2 molecule, for example, at Y^{1007}. This initial tyrosine phosphorylation reaction markedly enhances JAK2's intrinsic tyrosine kinase activity, enabling JAK2 to tyrosine phosphorylate the GHR cytoplasmic domain and other kinase substrates that associate with the GHR–JAK2 complex. In this model, which shares similarities with those proposed for other cytokine receptor–JAK systems, tyrosine phosphorylation of GHR and JAK2 at particular sites provides the opportunity for these molecules to bind, via interactions involving either SH2 domains or phosphotyrosine binding domains (PTBs), important signaling molecules (such as STATs), enzymes (kinases, phosphatases, etc.), or adaptor and docking molecules (e.g., Shc) that either further amplify or otherwise regulate intracellular signaling stimulated by GH. For instance, several specific tyrosine residues within the cytoplasmic tail of GHR, when phosphorylated, serve as docking sites for the SH2 domain of STAT5. This allows STAT5 to bind to the GHR–JAK2 complex, where it itself becomes tyrosine phosphorylated in a JAK2-dependent reaction that transforms the STAT from an inactive cytoplasmic protein to an active, nuclear transcription factor that can mediate some of GH's transcriptional responses.

5. GH SIGNAL TRANSDUCTION—ACTIVATION OF STAT TRANSCRIPTION FACTORS

5.1. Four STATs Can Be Activated by GH

Signal transducers and activators of transcription (STATs) are latent transcription factors that reside in the cytoplasm in an inactive form until converted to active, nuclear transcription factors by a tyrosine phosphorylation reaction. This tyrosine phosphorylation is JAK kinase-dependent, and in the case of GH signaling is generally presumed to be catalyzed by JAK2, rather than by an intermediate tyrosine kinase, such as fyn kinase. Of the seven known mammalian STATs (STATs 1–4, 5a, 5b, and 6), GH can activate four forms—STATs 1, 3, and 5a and 5b. STAT1 and STAT3 were originally discovered as the targets of activation by interferons and IL-6, respectively. STAT1 has a major role in mediating specific activation of target genes involved in antiviral immunity,

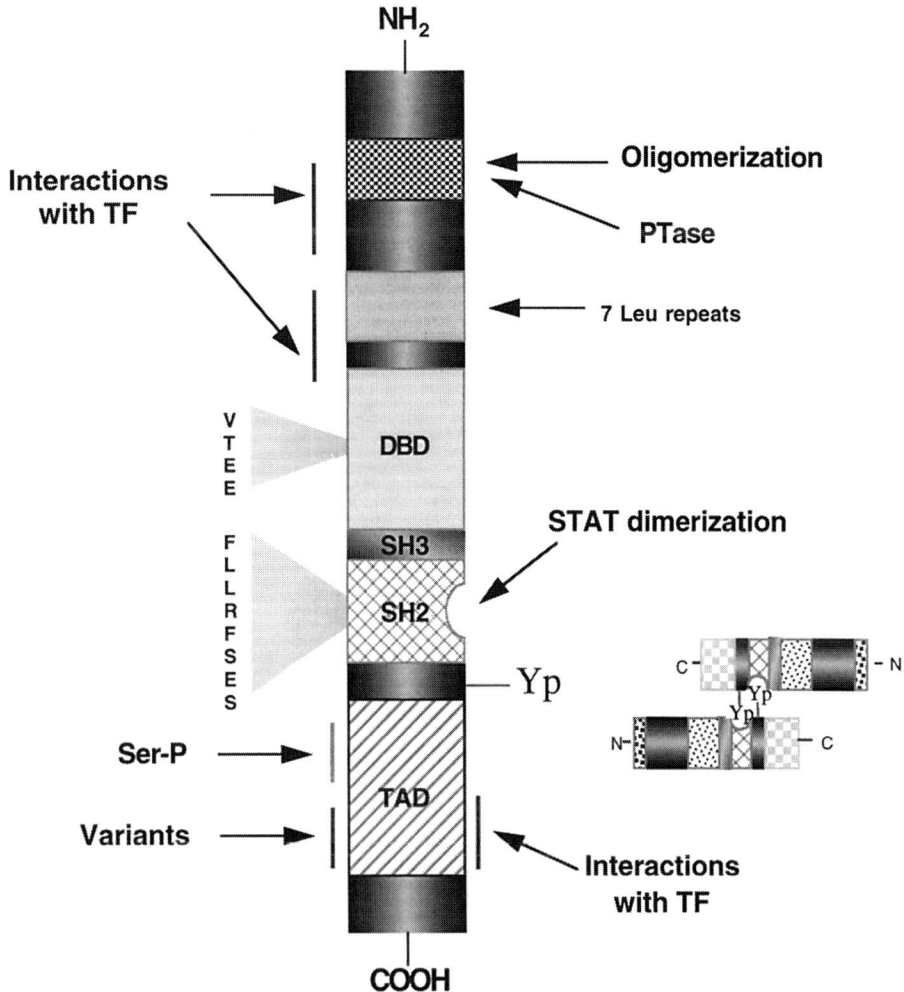

Fig. 7. Structural organization of GH-activated STATs. Functional regions that are homologous between STAT proteins include the DNA binding domain (DBD), the SH3-like and SH2 domains, and the C-terminal transactivation domain (TAD). Also shown are regions that interact with other transcription factors (TF), are important for STAT oligomerization or tyrosine phosphatase activity (PTase), as well as the C-terminal sites for phosphorylation on tyrosine and serine. (Adapted from Pellegrini and Dunsanter-Fourt. Eur J Biochem 1997; 248:615–633.) Shown at the right is a tyrosine-phosphorylated STAT-STATdimer.

while STAT3 recognition sites are often found in the regulatory regions of genes involved in the acute phase response, a response that is promoted by IL-6, leukemia inhibitory factor and other cytokines. As discussed in greater detail below, STAT5a plays a major role in mediating the effects of prolactin on mammary gland differentiation, while STAT5b plays a key role in GH action in the liver and other tissues. In each instance, activation of JAK2 kinase by dimerized GHR is a prerequisite for STAT activation.

Structural features common to the mammalian STAT proteins include a central DNA binding domain, an SH3-like domain, a highly conserved C-terminal SH2 domain, which is involved in binding the STAT to tyrosine phosphorylated signaling molecules, a single conserved tyrosine residue that is phosphorylated following activation of a receptor–JAK kinase complex (Tyr^{701} in the case of STAT1), and a *trans*-activation domain close to the STAT's extreme C-terminus (Fig. 7). Naturally ocurring, C-terminal truncated STATs can be formed by alternative splicing or by proteolysis, and these can exert dominant-negative effects with respect to STAT-dependent transcriptional responses.

STATs respond rapidly (within minutes) to stimulation of cells with GH by a sequence of events that involves both tyrosine and serine phosphorylation of the STAT protein, dimerization of the tyrosine phosphorylated STAT via its SH2 domain, translocation of the STAT to the nucleus, and finally, stimulation of target gene transcription (Fig. 8). This latter step is mediated by the binding of the activated, tyrosine

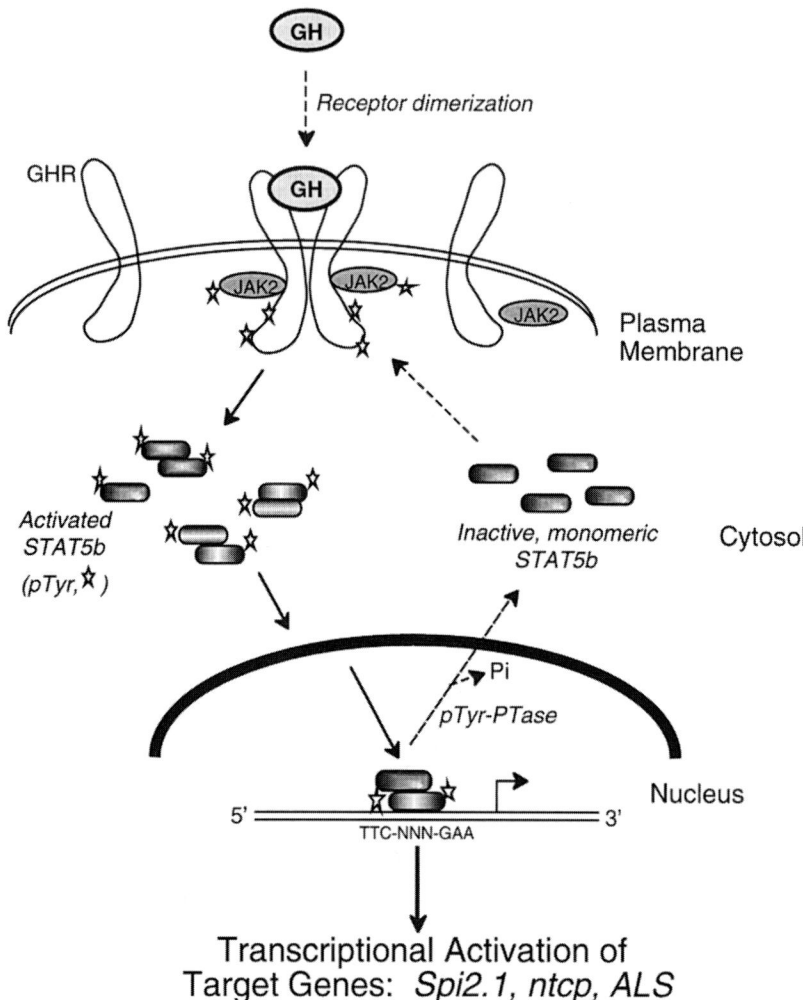

Fig. 8. STAT cycle: activation and deactivation of GH-responsive STATs. Plasma GH pulses activate STAT5b by a mechanism that involves dimerization of GHR at the cell surface, recruitment of JAK2 tyrosine kinase, followed by tyrosine phosphorylation of JAK2, GHR and then STAT5b. These initial steps are followed by STAT5b dimerization, nuclear translocation, and stimulation of the transcription of GH-induced target genes. Deactivation of the STAT occurs through the action of a tyrosine phosphatase (pTyr-PTase), which may be nuclear and allows the STAT to return to its monomeric cytoplasmic pool. (Adapted from Waxman DJ. Mechanisms and biological significance of pulsatile hormone secretion. Novartis Foundation Symposium 2000; 227:61–68.)

phosphorylated STAT as either a homodimeric or heterodimeric STAT complex to specific DNA enhancer elements. These "STAT response elements" are generally based on the 9-basepair palindromic sequence 5'-TT-N_5-AA-3' and can be found adjacent to several GH-inducible target genes. Moreover, the DNA-binding specificities of the four individual GH-activated STATs are somewhat distinct, and are based on the nucleotide sequences within and immediately adjacent to the 9 basepair palindrome. STAT transcriptional activity can further be modulated in a gene-specific manner through functional interactions with other transcription factors, such as c-Jun, the ubiquitous factor Sp1, and the glucocorticoid receptor, as well as with transcriptional coactivators and associated proteins, such as CBP/p300 and N-myc interacting protein (Nmi).

GH-induced activation of STATs 1 and 3 in cultured cells results in the assembly of complexes consisting of homo- and heterodimers of these two STATs bound to the c-*sis*-inducible element (SIE) of the c-*fos* promoter. This finding suggests that STAT1 and STAT3 may play a role in GH's induction of c-*fos*, an early GH-induced transcriptional response discussed below. However, targeted disruption of the *STAT1* gene in mice, while leading to defects in antiviral immunity, does not result in any overt disruption of growth; mice lacking STAT1 protein grow at the same rate as their wild-type littermates. Although more subtle effects of GH could well be mediated by STAT1, this has yet to be demonstrated. Targeted disruption of the *STAT3* gene leads to an embryonic lethal phenotype, precluding studies on the possible role of STAT3 in tissue-specific or whole-animal physiological responses to GH.

5.2. STAT5a and STAT5b Are Closely Related Molecules with Distinct Functions

STAT5 was originally identified by Groner and colleagues as a mammary growth factor that confers transcriptional responsiveness to prolactin in sheep. Subsequent studies led to the cloning from several species of two closely related STAT5 cDNAs, designated STAT5a and STAT5b, which are derived from separate genes and encode proteins that are >90% identical. Although generally quite similar in their properties, STAT5a and STAT5b exhibit differences with respect to their DNA binding specificities and in the sequences of their C-terminal transcription activation domain. They also exhibit differences in their tissue distribution and in their inhibitory crosstalk with the nuclear receptor PPARα, with STAT5b being more inhibitory than STAT5a to peroxisome proliferator signaling mediated by this nuclear receptor. In some tissues, such as mammary gland, STAT5a appears to dominate STAT5-dependent signaling events (stimulated by prolactin, in the case of mammary tissue), while in liver, STAT5b is apparently expressed at a ~10-fold higher level than STAT5a and is the key factor required for STAT5-dependent responses to male plasma GH pulses. However, a role for both STAT5a and STAT5b, perhaps acting as heterodimers, is also apparent from studies of the effects of STAT5a and STAT5b gene knockouts on the expression of certain GH-regulated genes in female mouse liver. Indeed, when activated by GH or other cytokines and growth factors, STAT5a and STAT5b form homodimeric and heterodimeric complexes that have distinct DNA binding activities and target gene specificities. Moreover, substitution of a single amino acid within STAT5's DNA binding domain (replacement of glycine 433 of STAT5b by glutamic acid, which is found at the corresponding position of STAT5a; and vice versa), is sufficient to interconvert the DNA binding specificities of the two STAT5 forms.

Studies carried out in tissue culture demonstrate that in addition to GH and prolactin, cytokines and growth factors as diverse as IL-2, -3, -5, and -15; erythropoietin; and epidermal growth factor (EGF) can each promote the activation of both STAT5 forms. Some cell type-specificity in the responses of STAT5a and STAT5b to interferons is apparent; however, the roles played by STAT5a and STAT5b in vivo in the physiological responses to many of these hormones and cytokines is still enigmatic.

In hepatocytes stimulated by GH, STAT5 promotes transcription of several liver-specific genes, including serine protease inhibitor 2.1 (Spi 2.1), the sodium-dependent bile acid cotransporter gene *ntcp*, and the gene encoding acid labile subunit (ALS), a serum protein that recruits IGF-1 into a complex with IGF binding protein-3. In each of these cases, GH-activated STAT5 interacts with a DNA enhancer sequence, termed a γ-activated sequence-like element (GLE), which has the consensus sequence 5′-TTC-NNN-GAA-3′ (Fig. 8). In several cases, STAT5-activated genes contain two adjacent GLEs, one of which may be too weak to bind STAT5 on its own, but can synergize with the nearby, more active STAT5 site. This arrangement facilitates the binding of two STAT5 dimers and an enhanced transcriptional response following GH stimulation. A similar arrangement of paired STAT5 binding sites is seen in several other STAT5-activatable genes. These include the IL-2 receptor gene and the gene that encodes cytokine-inducible SH2 protein 1 (CIS1), a STAT5-inducible early response factor that can contribute to the down-regulation of GH-induced STAT5 signaling discussed later in this chapter.

5.3. Specificity of STAT Activation Is Determined by Both GH Receptor and JAK2 Kinase

Detailed mapping studies have identified the contributions made by specific regions of GHR and JAK2 kinase to GH-induced STAT activation. GH activation of STATs 1 and 3 minimally requires a membrane-proximal receptor cytoplasmic domain region, which is limited to human GHR residues 271–317 in the case of STAT3, and thus maps to the same minimal region of GHR required for JAK2 activation. While GH activation of STATs 1 and 3 may involve interactions with JAK2, the nature of these interactions has not been established. It is clear, however, that GH activation of STATs 1 and 3 does not require interactions with any GHR cytoplasmic domain tyrosine residues. Thus, the SH2 domains of STATs 1 and 3 do not bind to the GHR–JAK2 complex by docking at one of the receptor's phosphotyrosine sites. In con-

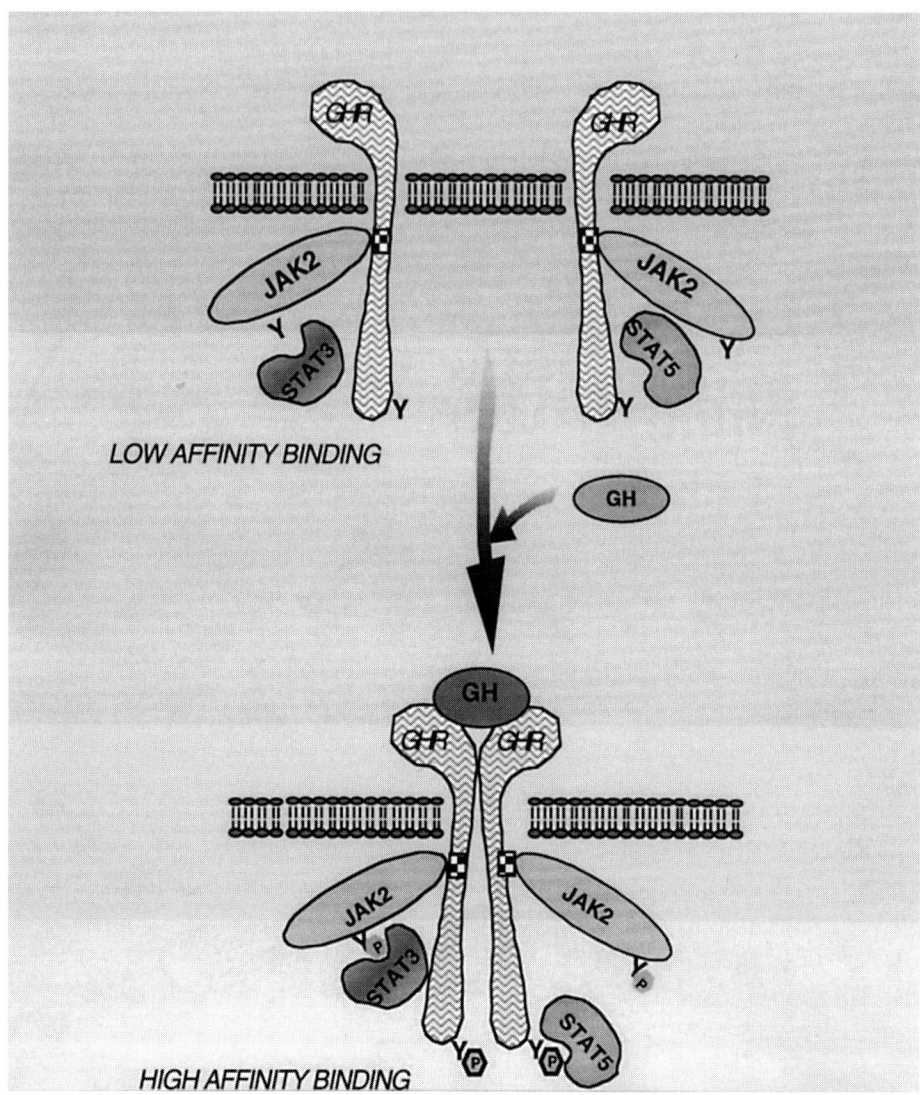

Fig. 9. STAT3 and STAT5 are activated by GH via distinct mechanisms. While GH activates both STAT3 and STAT5, STAT5 binds directly to GHR at any one of several tyrosine phosphorylation sites. By contrast, STAT3 preferentially binds to GH-activated JAK2 kinase, probably via its SH2 domain, as shown. (Adapted from Yi et al. Mol Endocrinol 1996; 10:1425–1443.)

trast, GH-induced activation of STAT5a and STAT5b primarily requires residues contained in the distal half of the GHR cytoplasmic domain, in addition to the Box 1 region of the receptor required for the initial JAK2 binding and activation steps (Fig. 9). Indeed, the presence of a single tyrosine at positions corresponding to rat GHR residues 487, 534, 566, or 627 (Fig. 4) is sufficient to allow for maximal or near-maximal STAT5a and STAT5b activation, as shown in studies where all but one of the six conserved cytoplasmic domain tyrosine residues are mutated to phenylalanine. This situation is in striking contrast to that described for the activation of STAT3 and STAT5 by IL-6: STAT3 is activated upon binding to specific phosphotyrosine residues on the gp130 subunit of the IL-6 receptor, while STAT5 can be activated by direct binding to the receptor-associated JAK1 kinase and in the absence of any gp130 tyrosine residues. Thus, STATs can be activated by two distinct mechanisms, one involving direct interactions with the receptor (as in the case of STAT5 with GHR and STAT3 with IL-6 receptor), and the other involving binding interactions with a JAK kinase (as in the case of STATs 1 and 3 with GH-activated JAK2 and STAT5 with IL-6-activated JAK1) (Fig. 9).

In addition to the four C-terminal GHR residues mentioned previously, GHR tyrosine residues 333 and/or 338 are also phosphorylated following GH treatment and can contribute to STAT5 tyrosine phosphorylation (Fig. 4). However, these membrane proximal GHR residues, as well as Tyrosine[487], are alone unable to mediate GH-stimulated, STAT5-dependent transcription of the Spi 2.1 promoter. By contrast, when phosphorylated, GHR tyrosines 534, 566, and 627 are individually sufficient to bind and activate STAT5 and thereby induce its transcriptional activity. Of interest is the apparent lack of amino acid sequence homology on the C-terminal side of each of the three GHR tyrosine residues associated with STAT5 transcriptional activation (Y^{534}FCEA, Y^{566}ITTE, Y^{627}VSTD), other than the presence of a hydrophobic residue immediately following the tyrosine phosphorylation site. The reason why GHR Tyr[487] can mediate STAT5 tyrosine phosphorylation, as reported in some studies, but is unable to activate STAT5-dependent transcription, as reported by others, is unclear.

Collectively, these findings suggest a model for GH-induced STAT activation in which STAT1 and STAT3 primarily interact with tyrosine phosphorylated residue(s) on JAK2 kinase, while STAT5b primarily interacts with the tyrosine phosphorylated GHR cytoplasmic domain (Fig. 9). Interactions between STAT5b and JAK2 are, however, also possible, as suggested by the ability of JAK2 to activate STAT5 in the absence of a cytokine receptor when both proteins are expressed in yeast, or when they are overexpressed by transient transfection in COS1 cells. A scheme similar to that of GH-induced STAT activation has been proposed for IL-6-induced STAT3 activation, in which a YXXQ sequence motif found naturally in the IL-6 receptor's gp130 subunit cytoplasmic domain is sufficient, when tyrosine phosphorylated, to mediate ligand-induced STAT3 binding and tyrosine phosphorylation. No such STAT3-binding motif is found in GHR, an observation consistent with the lack of STAT3 binding to tyrosine phosphorylated GHR cytoplasmic domain fragments, but two are present in JAK2 (Y^{266}TEQ and Y^{956}TSQ) and could potentially mediate both GH-induced STAT3 binding to JAK2 and STAT3 activation.

These differences in the mechanism by which GH activates STAT5b as compared to STAT1 and STAT3 are consistent with differences in the responsiveness of these STATs to GH pulses seen in hypophysectomized rat liver. Whereas liver STAT5b responds to physiological GH replacement doses given as intermittent pulses by repeated cycles of STAT5b tyrosine phosphorylation and nuclear translocation, STAT1 and STAT3 both require much higher GH doses for their initial response, and subsequently become desensitized to a second hormone pulse. The validity and physiological significance of these observations and their relationship to the proposed differential mechanism of GH-induced STAT1/STAT3 vs STAT5 activation require further study.

5.4. GH-Activated STATs Undergo Homo- and Heterodimerization

Independent of whether the binding of a STAT to the GHR-JAK2 complex occurs via interactions between its SH2 domain and a phosphotyrosine residue on GHR (as is proposed for STAT5) or on JAK2 (as proposed for STATs 1 and 3), once phosphorylated on tyrosine, the STAT must subsequently be released from the receptor–kinase complex and then form the dimers which migrate to the nucleus and stimulate target gene transcription. As noted previously, GH-activated STAT proteins can form both homodimeric and heterodimeric complexes via mutual interactions between the SH2 domain of one STAT molecule and the phosphotyrosyl residue of its dimeric partner. Not all potential STAT–STAT combinations have been observed, however. GH-activated STAT1 and STAT3 can form STAT1–STAT1 and STAT3–STAT3 homodimers, as well as STAT1–STAT3 heterodimers, each characterized by a distinct electrophoretic migration in a DNA binding (electrophoretic gel mobility shift) assay. Similarly, both homodimers and heterodimers invoving STAT5a and STAT5b can be observed following GH stimulation in liver. STAT5a–STAT5b heterodimers also form in mammary tissue and in Nb2 lymphocytes following prolactin stimulation. The activation by GH of multiple STATs that, in turn, form distinct complexes with unique DNA binding activities supports the proposed role of these hormone-activated transcription factors in mediating the pleiotropic effects that GH can have on gene transcription in target tissues.

Several models for STAT dimerization can be envisaged (Fig. 10). In the case of GH-activated STAT5, dimerization might be facilitated by the presence of two phosphorylated STAT monomers in close proximity within the same GHR–JAK2 receptor–kinase complex (pathway I). However, this model is not supported by the observation that IL-5 can activate STAT1 in cells expressing chimeric receptors engineered to contain only a single STAT-binding site within an active, dimerized receptor complex. Alternatively, a receptor-bound, phosphorylated STAT5 monomer might serve as a docking site to recruit a second STAT5 monomer, which then becomes tyrosine phosphorylated. Dimerization could then follow directly (pathway II). A third possibility is that, once phosphorylated on tyrosine, the STATs might have

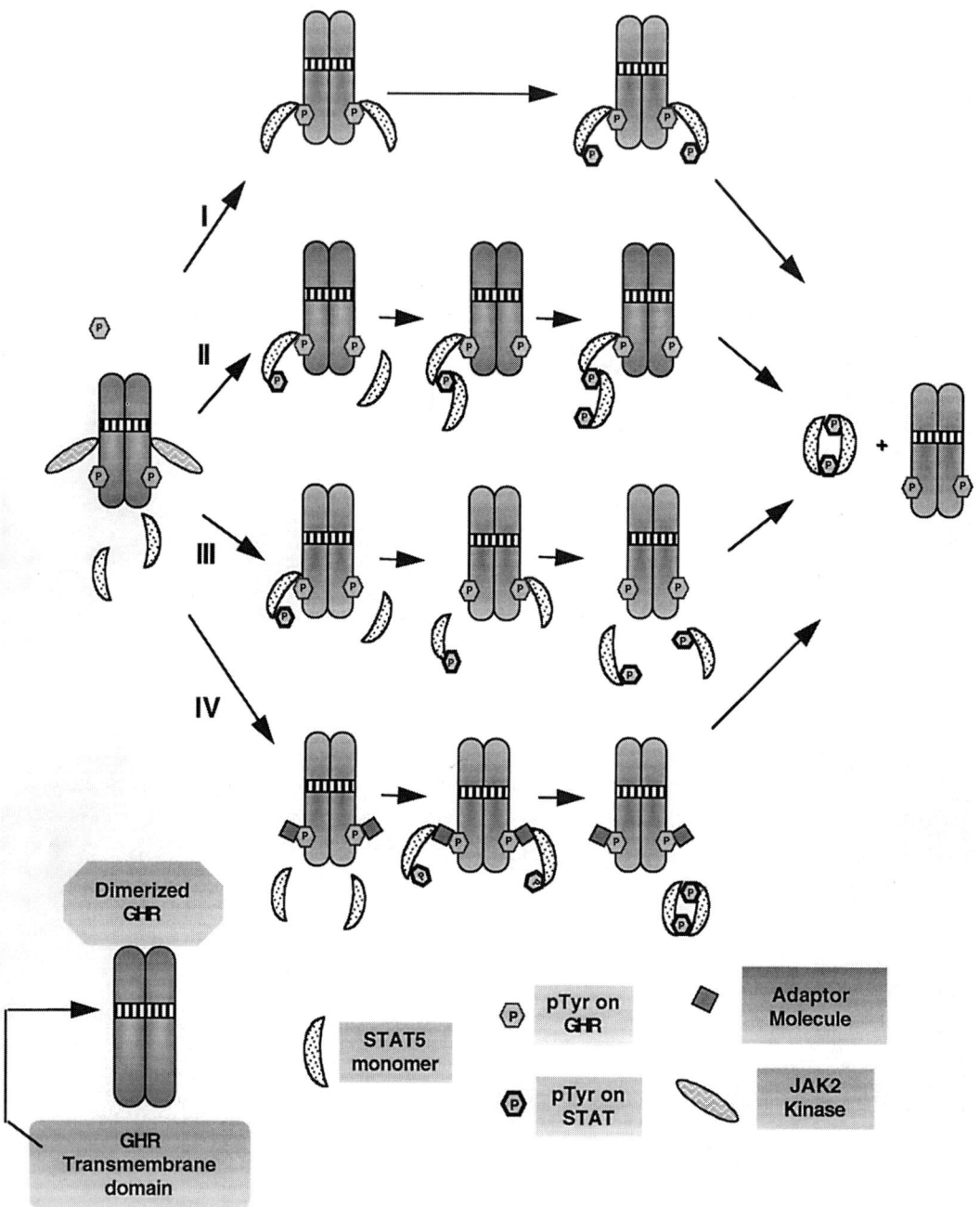

Fig. 10. Dimerization models for tyrosine phosphorylated STAT5. See text for details. (Modified from Behrmann et al. J Biol Chem 1997; 272:5249–5274.)

a reduced binding affinity for the receptor–kinase complex, allowing them to dissociate as monomers, followed by dimerization in the cytoplasm (pathway III). Finally, some of the observed interactions between GH-activated STATs and the receptor–kinase complex may be indirect and involve adaptor proteins, which could facilitate both recruitment of the STAT and its dimerization by one of the mechanisms discussed above (pathway IV).

5.5. STAT5b Plays a Key Role in the Sexually Dimorphic Responses to Plasma GH Secretory Patterns Seen in Rodent Liver

As discussed earlier in this chapter, pituitary GH secretion is regulated by neuroendocrine factors, which are in turn established by gonadal steroid imprinting, and this leads to sex differences in the temporal pattern of circulating GH. Gender differ-

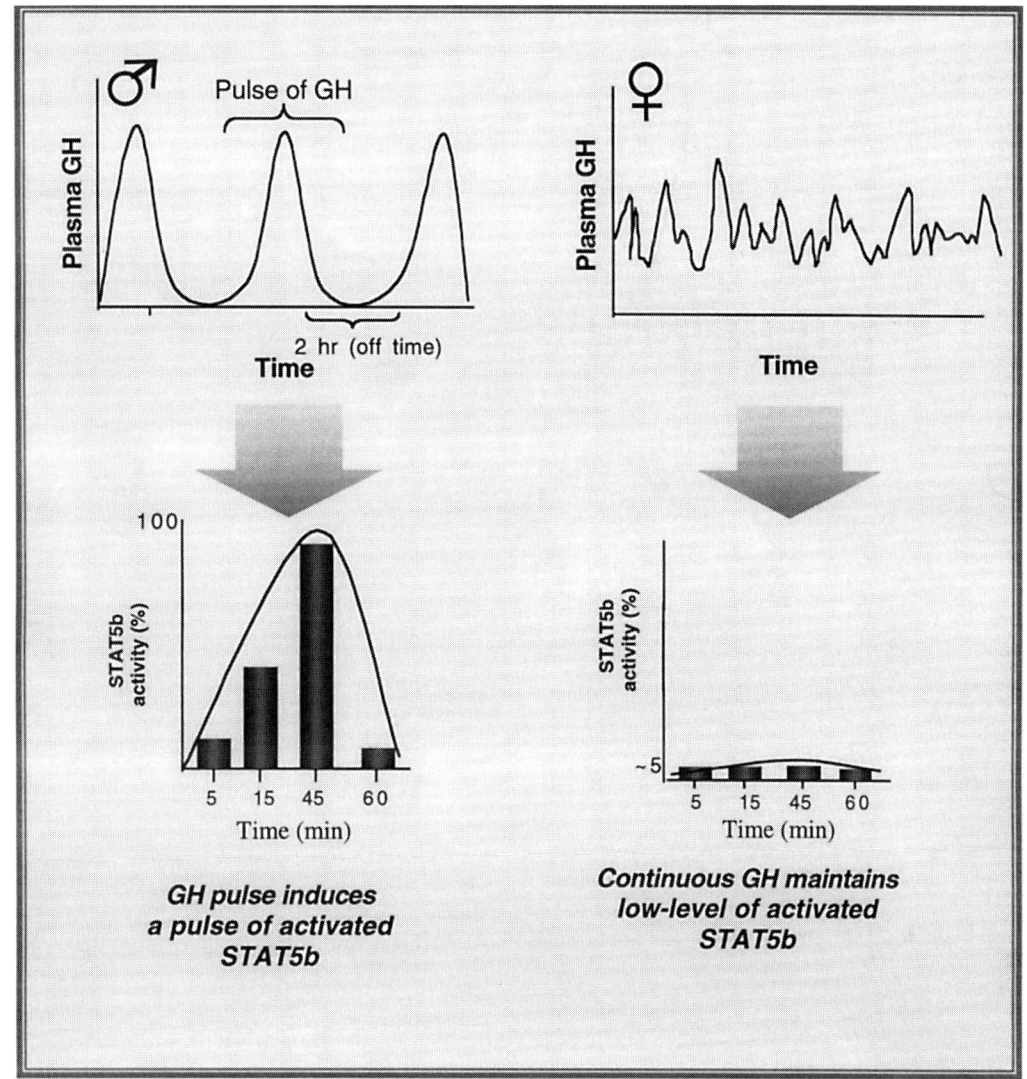

Fig. 11. DNA binding activity of STAT5b in response to gender-specific plasma GH pattern. Pulsatile GH secretion in male rats leads to intermittent activation of liver STAT5b and is hypothesized to stimulate transcription of male-expressed genes in the liver. By contrast, in females, the continuous presence of GH stimulates only low levels of STAT5b DNA binding activity. (Based in part on Choi HK, Waxman DJ. Endocrinology 1999; 140:5126–5135 and Waxman et al., 1995.)

ences in plasma GH profiles have been detected in humans, but are most dramatic in rodent species. Adult male rats are characterized by regular plasma pulses of GH spaced ~3.5 h apart, followed by a distinct period of time during which circulating GH is essentialy undetectable. In contrast, GH is secreted by the pituitary gland much more frequently in adult female rats, and this results in prolonged periods with a near-continuous presence of GH in circulation (Fig. 11). These sex differences in plasma GH patterns in turn lead to sexually dimorphic expression of a number of hepatic gene products. The best studied of these are the cytochrome P450 enzymes involved in hepatic steroid hydroxylation. In male rat liver, STAT5b is tyrosine phosphorylated in response to each successive plasma GH pulse, while in females the continuous pattern of GH exposure down-regulates the STAT5b activation pathway by >90% (Fig. 11). Moreover, the pulsatile pattern of plasma GH stimulation that characterizes males results in a repeated translocation of liver STAT5b from the cytoplasm to the nucleus, where the activated STAT5b homodimers that are formed are proposed to bind to STAT5-binding GLE sequences (STAT5 response elements) that have been found upstream of several GH pulse-induced, adult male-expressed cytochrome P450 genes.

Investigations of the cellular mechanisms of STAT5b activation and deactivation in the rat liver cell line CWSV-1 have revealed that STAT5b is recycled from the nucleus back to the cytoplasm at the conclusion of a GH pulse and then is reutilized (Fig. 8). Near-full responsiveness of STAT5b to succeeding GH pulses is achieved in these cells following a GH off-time of >2 h, similar to rat liver in vivo. The proposal that STAT5b is a key intracellular mediator of the stimulatory effects of GH pulses on male-specific liver P450 gene transcription is strongly supported by studies using a mouse *STAT5b* gene knockout model. Male *STAT5b$^{-/-}$* mice exhibit two striking phenotypes not seen in *STAT5b$^{-/-}$* females:

(1) A whole body growth rate deficiency becomes evident in STAT5b-deficient male mice at puberty. This growth deficiency and several other physiological characteristics of these mice during puberty and adulthood are reminiscent of Laron syndrome, discussed earlier in this chapter, although the growth deficiency of Laron patients becomes manifest well prior to puberty. These findings suggest that STAT5b plays a critical role in determining the male pattern of pubertal growth, which in rodents is known to be activated most efficiently by GH pulses; (2) There is a selective loss of male-specific liver cytochrome P450 gene expression in STAT5b-deficient mice. This loss of the male pattern of liver gene expression is coupled to a derepression of those P450 genes whose expression is normally limited to adult female mice.

Surprisingly, the closely related STAT5a protein is unable to substitute for STAT5b, despite its intrinsic responsiveness to GH activation by the same tyrosine phosphorylation/nuclear translocation mechanism. Furthermore, *STAT5a* gene disruption does not result in either the growth deficiency or the loss of male-specific gene expression phenotypes seen in *STAT5b* disrupted mice, although interesting effects on several GH-regulated, female-expressed liver P450 gene products are observed. Recent studies verify the hypothesis that *STAT5b$^{-/-}$* mice are, indeed, GH pulse-resistant, rather than displaying a phenotype that results from a perturbation of their endogenous pituitary GH secretory patterns.

In contrast to the repeated activation of STAT5b by male GH pulses, the continuous GH exposure pattern that is characteristic of females desensitizes the STAT5b activation pathway by a mechanism that likely involves enhanced dephosphorylation of both STAT5b and the GHR–JAK kinase signaling complex. Consequently, very little activated STAT5b transcription factor is present in the nucleus of adult female rats. Rather, in female rat liver nuclei, or following continuous GH treatment of male rats, there is enhanced expression of a novel non-STAT nuclear factor, termed GH-activated nuclear factor (GHNF). GHNF can bind to multiple sites along the 5′-flank of the female-expressed, continuous GH-activated rat liver P450 gene *CYP2C12*, and is greatly enriched in female as compared to male liver nuclei. The molecular nature of GHNF, and the precise role that it plays in the sexually dimorphic expression of *CYP2C12* and other continuous GH-activated liver genes is presently unknown.

5.6. Plasma GH Pulses Induce Repeat Cycles of STAT5b Activation and Deactivation

Data obtained by several laboratories studying GH signaling pathways supports the following sequence of events for the STAT5b activation cycle stimulated by a plasma GH pulse (see individual steps identified in Fig. 12): (1) GHR dimerizes at the cell surface upon binding GH. This leads to the activation of JAK2 tyrosine kinase, as discussed in detail above. (2) JAK2 phosphorylates GHR at tyrosine residues 534, 566, and 627, each of which can serve as a docking site for STAT5b. (3) STAT5b is recruited to one or more of these GHR phosphotyrosine docking sites via the STAT protein's SH2 domain. Once bound to GHR in this manner, STAT5b itself becomes phosphorylated on tyrosine residue 699 in a reaction that is likely to be catalyzed by JAK2 kinase. STAT5b also becomes phosphorylated on a C-terminal serine residue (Ser^{730}) in a step that can alter its DNA binding properties and perhaps also its transcriptional activity. (4) The tyrosine phosphorylated STAT5b homodimerizes via SH2 domain interactions; this step is closely followed by (5) nuclear translocation of the activated STAT transcription factor. It is uncertain whether the serine phosphorylation reaction precedes or follows STAT5b dimerization. (6) The homodimeric STAT complex binds to a DNA enhancer element comprised of TTC-NNN-GAA-containing sequences and stimulates target gene transcription. (7) STAT5b is most likely deactivated in the nucleus by a process that involves phosphotyrosine dephosphorylation as a key first step. This dephosphorylation reaction may be catalyzed by the SH2 domain-containing tyrosine phosphatase SHP-1, as discussed later in this chapter. (8) Signaling from GHR–JAK2 to STAT5b is subsequently terminated by a process that involves inhibition, inactivation, and/or internalization of the GHR–JAK2 signaling complex. Finally, (9) key signaling components, likely including

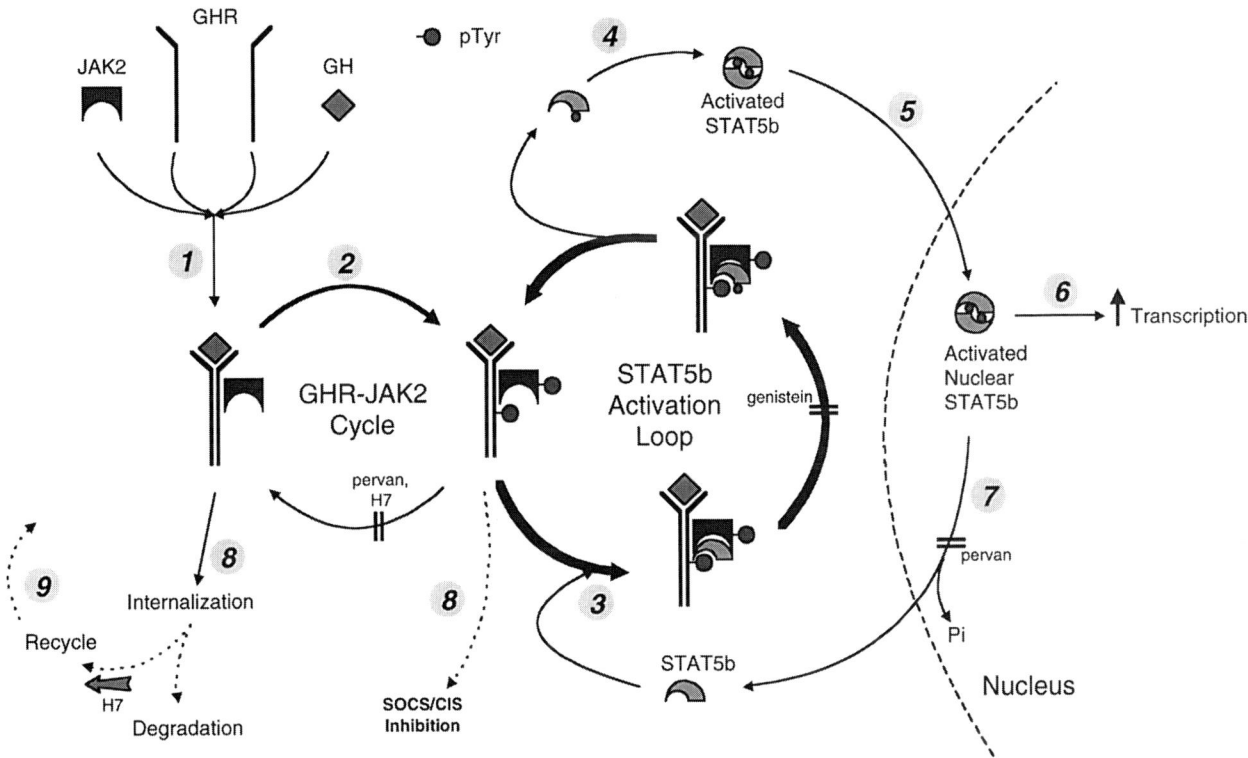

Fig. 12. GHR–JAK2 cycle and STAT5b activation loop. Shown are the proposed pathways for stimulation and termination of GH pulse-induced STAT5b signaling. See text for details. (Adapted from Gebert et al. Mol Endocrinol 1999; 13:38–56.)

STAT5b and perhaps also GHR and JAK2, reactivate and recycle back to the plasma membrane in time for the hepatocyte to respond to a subsequent GH pulse. This cycle of GH pulse-induced STAT5b activation, deactivation and recycling appears to be timed appropriately for liver cells to respond each ~3.5 h to an incoming plasma GH pulse with a strong STAT5b response. An interesting prediction of this model, which has yet to be tested experimentally, is that STAT5b transcriptional responses in liver are intermittent and closely follow each incoming plasma GH pulse.

Steps 1–6 of this proposed STAT5b cycle are fairly well understood at present. GH-induced serine phosphorylation of STAT5b (step 4) has been demonstrated for GH-activated STAT5b in liver tissue and in cultured liver cells, and has been confirmed for STAT5b activated by prolactin in Nb2 lymphocytes and for IL-2-activated STAT5 in T cells. This secondary phosphorylation of STAT5b is analogous to the serine phosphorylation of STAT3 at its consensus MAP kinase site, $PMS^{727}P$, which can greatly enhance the STAT's transcription activation potential (an analogous sequence, PSP, exists in both STAT5a and STAT5b). As discussed in later sections of this chapter, the cellular and biochemical events that underlie STAT5b dephosphorylation (step 7), termination of GHR–JAK2 signaling (step 8), and the resetting of GH signaling components during the plasma GH interpulse interval (step 9) are less well understood at present. These latter three steps are of prime importance for GH signaling, at least in hepatocytes, where they play a critical role in meeting the physiological requirement that STAT5b signaling undergo a rapid down-regulation at the conclusion of each plasma GH pulse.

6. OTHER GH-INDUCED INTRACELLULAR SIGNAL TRANSDUCTION PATHWAYS

6.1. GH Induces the Activation of MAP Kinases

MAP kinases play an important role in cell growth and differentiation, and can be activated by many cytokines and growth factors, including GH. GH can activate two cytosolic serine/threonine kinases belonging to the MAP kinases family, ERK1 and ERK2 (extracellular signal regulated kinases 1 and 2). GH-stimulated ERK tyrosine phosphorylation is accompanied by an increase in ERK enzymatic activity. ERKs catalyze serine/threonine phosphorylation associated with optimal activation of enzymes, such as the S6 serine/threonine kinases p70rsk and p90rsk, as well as transcription factors, such as c-Jun and some STATs. Consequently, the significance of GH-induced ERK activation lies in its potential to link multiple GH-activated signaling pathways (Fig. 6).

Mapping studies demonstrate that the membrane proximal Box 1 region of GHR's cytoplasmic domain, which is required for JAK2 activation, is both necessary and sufficient for GH activation of ERK activity. JAK2 mutants that either couple ineffectively to GHR or are catalytically inactive fail to support GH-induced ERK tyrosine phosphoryation. These findings firmly support the hypothesis that GH-induced JAK2 activation leads to ERK activation. Other studies demonstrate that, as is found for other cytokines and hormones that activate ERK, GH-activated JAK2 promotes tyrosine phosphorylation of the adaptor protein Shc, which in turn binds the small SH2–SH3 domain adaptor protein Grb2. Moreover, the tyrosine phosphorylated GHR cytoplasmic domain can interact with Shc, as well as several other adaptor proteins, including Grb10. These findings are supported by experiments using dominant-negative forms of Ras and Raf, which demonstrate a dependence on both Ras and Raf for GHR–JAK2 activation of ERKs.

The established Shc-Grb2-Sos-Ras-Raf signaling pathway of ERK activation can thus be activated by GH in a GHR-JAK2-dependent manner (Fig. 6). Time-course studies in cultured preadipocytes support the proposal that GH promotes assembly of a Shc–Grb2–Sos complex that activates Ras, thereby engaging the Raf–MEK–ERK pathway. These studies also demonstrate that GH-induced, MEK-dependent serine kinase activation negatively feeds back to regulate this process by causing Sos phosphorylation and dissociation of Shc–Grb2 complexes from Sos, thereby terminating further activation of ERK. Important unresolved questions include: (1) How does JAK2 kinase interact with the proximal members of this pathway to activate the Ras–Raf–MEK–ERK kinase pathway; and (2) Is this the major pathway that GH uses in target tissues to activate ERKs, or do other GH-stimulated pathways, such as those involving the activation of PI-3-kinase, protein kinase C (Fig. 6) or EGF receptor play a more central role in GH-induced ERK activation? GH-stimulated tyrosine phosphorylation of the EGF receptor is a novel pathway recently proposed to mediate ERK (and c-*fos*) activation in GH-treated liver cells. GH stimulates JAK2-dependent tyrosine phosporylation of the EGF receptor at tyrosine residue 1068 in a reaction that generates a docking site for Grb2, thereby coupling JAK2 to the Sos–Ras–Raf–MEK–ERK activation pathway. Interestingly, this crosstalk between the receptors for GH and EGF is independent of EGF receptor's intrinsic tyrosine kinase activity. In this case, therefore, the cytoplasmic domain of the EGF receptor, when phosphorylated following GH stimulation, primarily acts as a scaffold upon which Grb2-Sos complexes required for Ras–MAP kinase activation may assemble.

6.2. GH Induces Calcium Fluxes and the Activation of Protein Kinase C

GH stimulates increases in intracellular calcium in several GH-responsive cell types. In each instance, the increase can be accounted for, at least in part, by an influx of extracellular calcium. Inhibitor studies implicate voltage-dependent L-type calcium channels found in the plasma membrane in this process, and suggest a possible role for GH in their maintenance. The GH-induced increase in intracellular calcium is believed to underly the refractory nature of GH-stimulated cells to the insulin-like effects on lipogenesis and glucose metabolism of further acute GH exposure. Moreover, the increase in intracellular calcium may be required, in addition to other signals, for GH-induced *trans*-activation of the Spi 2.1 gene.

Reports vary regarding the signaling mechanisms that lead to this GH-induced calcium ion influx. CHO cells that stably express C-terminal truncated GHR constructs exhibit a requirement for GHR cytoplasmic domain sequences between residues 454 and 506 for GH-induced calcium influx. Most interesting, perhaps, is the observation that a GHR mutant internally deleted of the Box 1 sequence and incapable of JAK2 activation is fully capable of transducing a GH-dependent increase in intracellular calcium. Thus, unlike the pathways leading to activation of STAT proteins and the ERKs discussed previously, the GH-induced calcium response can apparently occur in the absence of JAK2-catalyzed tyrosine phosphorylation cascades.

GH induces increases in diacylglycerol, a potent activator of PKC, in various cell types (Fig. 6). In some cases the increase in diacylglycerol is accompanied by increased generation of inositol 1,4,5-trisphosphate (IP_3) (presumably via phospholipase C hydrolysis of phosphatidyl inositol 4,5-bisphosphate [PIP_2]), and in others it proceeds without IP_3 production (likely by phosphatidylcholine breakdown). A requirement of JAK2 kinase for GH-induced activation of PKC has not been established. Depletion of PKC activity by chronic exposure to the phorbol ester PMA or the use of PKC inhibitors has indicated a role for PKC in several GH responses. These include GH-induced MAP kinase activation, S6 kinase activation, stimulation of lipogenesis, activation of the transcription factors C/EBPβ and C/EBPδ and downregulation of cell surface GHR levels. Of interest, the GH-induced increase in calcium uptake seen in fat cells can be blocked by inhibitors of PKC or phospholipase C, but is stimulated by a diacylglycerol analog. These findings lead to the proposal that GH may promote direct PKC-catalyzed phosphorylation of the calcium channels or their regulator protein(s) (Fig. 6).

A uniform picture of the role of PKC in GH signaling has not yet emerged. Some investigators have concluded that GH-induced ERK tyrosine phosphorylation and activation is unlikely to require PKC activation, while others find that GH-induced ERK activation specifically involves the PKC$_\delta$ isoform. It is not clear what factors account for these discrepant findings. Mapping of the regions of GHR or JAK2 that might be required for PKC activation has not yet been reported.

6.3. GH Activates c-fos Gene Transcription

Activation of the c-*fos* protooncogene was one of the first GH-induced early gene activation events to be described. The general finding that c-*fos* is important in cell proliferation induced by GH and many other stimuli, and the knowledge that GH stimulates cell and tissue growth and other somatogenic effects emphasize the importance of the c-*fos* induction response. As described below, it is now clear that multiple GH-activated signaling pathways converge on promoter/enhancer elements upstream of the c-*fos* gene to stimulate its transcription.

c-*fos* gene transcription is activated by GH rapidly (detectable within 15 min after hormone treatment of target cells) and transiently (induction of c-*fos* signal is reversed within 1 h). The c-*fos* promoter contains enhancer elements that confer responsiveness to transcription stimulated by multiple growth factors. These elements include the c-*sis*-inducible element (SIE), a serum response element (SRE), an AP-1 site, and a cAMP response element. The effects of GH on c-*fos* transcription are primarily mediated via the SIE and SRE elements (Fig. 6). Transcription of a reporter gene driven by a high-affinity mutated version of the c-*fos* SIE (which binds STATs 1 and 3) can be activated by GH in cells that overexpress STAT3. However, a stronger GH-induced c-*fos* transcriptional response is mediated by the SRE, through which GH can synergize with other serum factors to exert its effects. In particular, GH stimulation of SRE-mediated c-*fos* transcription requires the collaboration of the serum response factor SRF with the ternary complex factor (TCF) Elk-1 (Fig. 6).

GH-induced c-*fos* transcription requires the membrane proximal half, but not the distal portion, of GHR's cytoplasmic domain. By inference, therefore, c-*fos* gene induction does not involve STAT5 activation or calcium ion influx, both of which require distal GHR cytoplasmic domain elements, as discussed earlier. Further, ERK-catalyzed serine phosphorylation of Elk-1, and ERK-mediated enhancement of SRF binding to the SRE (perhaps by phosphorylation of SRF) augment the GH-induced transcriptional response. Recent data indicate that Grb10, an adaptor protein capable of binding to the tyrosine phosphorylated GHR distal tail, can by as yet unknown mechanisms down-regulate GH-induced, SRE-mediated c-*fos* transcription. Thus, the intriguing possiblity is raised that the proximal GHR cytoplasmic domain may be required for SRE-mediated activation of c-*fos* while the distal GHR tail might regulate termination of that activation in a Grb10-dependent manner, and thereby insure the transient nature of c-*fos* induction noted above.

6.4. IRS Molecules Participate in GH Signaling

Insulin receptor substrates-1 and -2 (IRS-1 and -2) are nontransmembrane proteins of 160–180 kDa that become tyrosine phosphorylated at multiple sites in response to insulin, IGF-1, IL-4, and several other cytokines. Upon tyrosine phosphorylation, these proteins function as adaptors, providing docking sites for associations with signaling molecules such as the p85 subunit of PI-3-kinase, the adaptor Grb2, and the SH2 domain tyrosine phosphatase SHP-2. The IRS-mediated, insulin-induced activation of PI-3-kinase is of potential importance for insulin-induced glucose transport (Fig. 6). Targeted disruption of IRS-1 in mice leads to growth retardation and mild insulin resistance; however, knockout of the gene encoding IRS-2 results in a phenotype quite similar to human type 2 diabetes mellitus, with peripheral insulin resistance and inadequate pancreatic β-cell growth and insulin production to overcome the resistance.

GH promotes tyrosine phosphorylation of IRS-1 and IRS-2 and their association with PI-3-kinase in a broad range of GH-responsive cell types, suggesting a possible mechanism to account for some of the insulinomimetic and insulin-antagonistic metabolic effects of GH. Further, GH-induced lipid synthesis and GH's inhibition of noradrenaline-induced lipolysis in rat adipocytes are blocked by the PI-3-kinase inhibitor wortmannin. It is uncertain whether the requirement for GHR tyrosine residues 333 and/or 338 (which are phosphorylated in GH-treated cells) for GH-induced lipogenesis and protein synthesis is related to the activation of IRS-1 and/or IRS-2. While IRS proteins may have a role in potentiating GH-induced metabolic effects (given their likely importance in insulin-induced metabolic signaling), studies employing PI-3-kinase inhibitors and a dominant-negative form of the PI-3-kinase p85 subunit suggest that PI-3-kinase may not be involved in GH-mediated glucose transport in 3T3–L1 adipocytes. Thus, the linkage of IRS-1 and -2 to this important metabolic effect of GH is brought into question.

Recently, however, IRS-1 has been implicated as important in augmenting GH-induced cell proliferation. In vitro affinity precipitation experiments indicate that amino-terminal regions of IRS-1, which include the pleckstrin homology (PH), phosphotyrosine-binding (PTB), and *Shc* and *IRS-1 N*PXY binding ('SAIN') domains, specifically interact with JAK2, but that IRS-1 tyrosine phosphorylation and JAK2 tyrosine phosphorylation are not necessary for these interactions. This is particularly interesting in that the PTB and SAIN domains of IRS-1 are known to mediate the interaction of IRS-1 with the insulin receptor in a phosphotyrosine-dependent manner. Using IRS-1- and IRS-2-deficient 32D cells as a vehicle for reconstitution of IRS-1, it was further observed that cell proliferation and GH-induced activation of the MAP kinases ERK1 and ERK2 were both increased in cells transfected with IRS-1 in comparison to those lacking IRS-1. While the mechanism of IRS-1's enhancement of GH-induced ERK activation and the relationship of this activation to enhanced GH-induced proliferation are still unclear, these findings raise the possibility that IRS-1 may primarily modulate GH's proliferative effects, rather than its metabolic effects.

6.5. GH Induces Transcription of the IGF-1 Gene

Although IGF-1 has long been appreciated as a biologically important regulatory target of GH action, the mechanisms by which IGF-1 production is induced by GH have remained elusive. GH promotes hepatic *IGF-1* gene transcription in hypophysectomized rats within 15 min of a single systemic injection. The rapidity with which this transcriptional activation occurs suggests the involvement of some of the GHR-mediated early signaling events (e.g., tyrosine phosphorylation) discussed previously. Experiments to test this hypothesis, for example, using JAK2 kinase inhibitors such as genistein to block GH-induced IGF-1 transcription, have not been reported. Recent studies show, however, that GHR's cytoplasmic domain is necessary for GH-induced *trans*-activation of the IGF-1 promoter. In other studies, a set of constitutive nuclear protein binding sites has been localized to each of the two major IGF-1 promoters, but GH treatment does not elicit a detectable change in these DNA–protein interactions. A GH-inducible change in chromatin structure, manifested as a DNase I hypersensitivity site, has been mapped to the second intron of the *IGF-1* gene, but no discernible changes in DNA–protein interactions in that region accompany the hypersensitivity.

7. MODULATORS AND ATTENUATORS OF GH SIGNALING

GH-activated cell signaling, once engaged, can be both modulated and attenuated, and eventually is terminated. Examples of GH signal modulation have been noted previously with regards to GH-induced ERK activation and c-*fos* transcription. The ability of the JAK2-phosphorylated EGF receptor to augment GH-stimulated ERK activation by providing a docking site for Grb2 and the finding that Grb10, by interacting with the tyrosine phosphoryated GHR cytoplasmic domain, can dampen GH-induced c-*fos* activation provide additional examples. The termination of GH pulse-induced STAT5b signaling to the hepatocyte nucleus at the conclusion of a plasma GH pulse provides an example of signal termination that is physiologically important, in this case to permit the repeated activation of STAT5b by successive GH pulses that is essential to achieve a male pattern of pubertal growth and liver gene expression. In this section, the proposed roles of the protein tyrosine phosphatases SHP-1 and SHP-2 and the SOCS family of cytokine signaling attenuators in the modulation and termination of GH signaling are discussed. Additional aspects of the termination of GH signaling involving GHR down-regulation and desensitization are presented in the section that follows.

7.1. Protein Tyrosine Phosphatases Influence GH Signaling

As major components of GHR signaling are mediated by tyrosine phosphorylation cascades initiated by JAK2 kinase, significant interest has been focused on the role of protein tyrosine phosphatases as potential modulators of GH signaling. In the case of the GH pulse-activated GHR–JAK2 complex, tyrosine phosphatase activity makes an important contribution to the down-regulation of receptor–kinase activity (Fig. 12) beginning ~40 min after GH stimulation, as judged from experiments using tyrosine phosphatase inhibitors, such as pervanadate. The phosphatase that catalyzes dephosphorylation of the GHR–JAK2 complex has not been identified, but could be induced by new protein synthesis, as suggested by the finding that the protein synthesis inhibitor cycloheximide prolongs JAK2 signaling to STAT5b in liver cells.

7.2. Tyrosine Phosphatase SHP-1 Can Negatively Regulate JAK/STAT Signaling

The tyrosine phosphatase SHP-1 is an SH2 domain-containing tyrosine phosphatase that is important in the negative regulation of JAK2-induced signaling in the erythropoietin receptor system. Truncation of GHR's cytoplasmic tail to remove sequences between residues 520 and 540 (human GHR numbering system) results in prolonged JAK2 activation and enhanced activation of JAK2 substrates, such as STAT3 and IRS-1. One possible explanation of these findings is that cytoplasmic sequences of GHR serve to bind and recruit a tyrosine phosphatase, such as SHP-1, to the GHR–JAK2 complex. Indeed, mice deficient in SHP-1 exhibit delayed JAK2 dephosphorylation in liver extracts incubated in vitro. It has yet to be demonstrated, however, that SHP-1 directly catalyzes dephosphorylation of activated JAK2 kinase. Moreover, it is unclear whether GHR tyrosine phosphorylation is required for SHP-1 to be recruited to the GHR–JAK2 receptor–kinase complex.

Other studies indicate that SHP-1 catalytic activity is increased in liver cells by ~3–4-fold immediately following a pulse of GH and without an increase in SHP-1 protein. This activity increase occurs under conditions in which GH induces the association of SHP-1 with STAT5b and the translocation of both SHP-1 and STAT5b to the nucleus. As SHP-1 is known to be activated by the binding of tyrosine phosphorylated signaling molecules to one of its two SH2 domains, these findings suggest that the binding of tyrosine phosphorylated STAT5b to SHP-1 could serve to both activate the tyrosine phosphatase and to provide it with STAT5b substrate which it subsequently dephosphorylates.

SHP-1 could thus participate in two fundamentally distinct signal desensitization processes, both of which are required for efficient termination of STAT5b signaling. SHP-1-catalyzed dephosphorylation of cell surface GHR and/or JAK2 kinase could inhibit a catalytic cascade, such that one GHR or JAK2 dephosphorylation event prevents multiple downstream STAT5b activation events. By contrast, the dephosphorylation of GH-activated nuclear STAT5b by SHP-1 would represent a downstream inactivation event that is stoichiometric with respect to the STAT transcription factor. Further studies are required to verify or refute the proposed role of SHP-1 in tyrosine dephosphorylation of GHR–JAK2 and/or of GH-activated nuclear STAT5b.

7.3. Tyrosine Phosphatase SHP-2 Positively Regulates GH Signaling

In contrast to these data pointing to a negative regulatory role for SHP-1 in GH signaling, a related SH2 domain-containing tyrosine phosphatase, termed SHP-2, appears to exert positive effects on GH signaling. GH treatment of cultured preadipocytes leads to tyrosine phosphorylation of SHP-2 and the association of SHP-2 with the adaptor molecule Grb2 and with a transmembrane glycoprotein known as SIRPα. SIRPα and the closely related protein SHPS-1 interact via their intracellular phosphotyrosine residues with the SH2 domains of SHP-2, and in some cases SHP-1, and serve as substrates for the SHPs' tyrosine phosphatase activity. In the case of EGF and insulin signaling, SIRPs exert a negative role that is dependent on SIRP tyrosine phosphorylation. While SIRPs can apparently be tyrosine phosphorylated by JAK2 in response to GH, their role (positive or negative, direct or indirect) in GH signaling is uncertain.

One possibility is that SHP-2 positively regulates GH signaling leading to c-fos activation by augmenting activation of the Ras–Raf–MAP kinase (ERK) pathway. This is suggested by the correlation of GH-induced SHP-2 tyrosine phosphorylation with SHP-2's ability to associate with Grb2, and by the finding that overexpression of a catalytically-inactive SHP-2 mutant protein specifically inhibits GH-induced *trans*-activation of a c-*fos*-reporter gene construct. The mechanism for this positive effect of SHP-2, and whether SIRP is required for this effect, are important

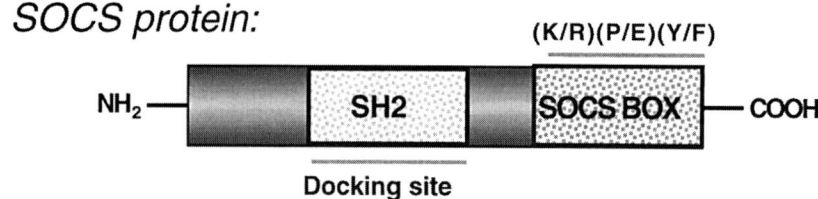

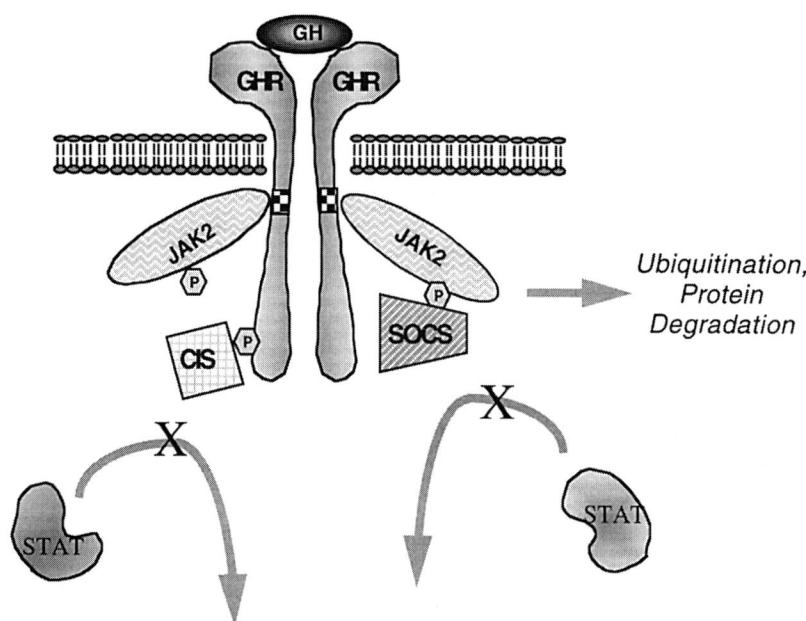

Fig. 13. SOCS/CIS protein inhibition of GH-induced JAK-STAT signaling. SOCS/CIS proteins (general structure shown at top) are synthesized as early response gene products following GH stimulation and are proposed to inhibit further signaling by binding to phosphotyrosine residues on GHR, as shown here for CIS, and/or JAK2, as shown for SOCS. SOCS/CIS protein binding may in turn facilitate ubiquitination and degradation of the receptor–JAK–SOCS/CIS complex, as described in the text. (Based in part on Ram PA, Waxman DJ. J Biol Chem; 1999, 274:35553–35561.)

open questions. Given that the tyrosine phosphatase activity of SHP-2 is required for it to augment this GH signal, it seems likely that dephosphorylation of particular cellular substrates, possibly including SIRP, may underly this action of SHP-2. By contrast, other studies show that SHP-2 can bind to JAK2 kinase in a reaction that does not require SHP-2's SH2 domains or its tyrosine phosphatase activity, but does not lead to JAK2 dephosphorylation.

7.4. SOCS Proteins Contribute to Down-Regulation of GH Signaling

The termination of GHR–JAK2 signaling to STAT5b, which occurs spontaneously in liver cells beginning ~40 min after GH pulse stimulation, is dependent on new protein synthesis. GH-activated STAT5b might therefore induce an "immediate early" response gene product(s) that down-regulates the STAT5b pathway by binding to the GHR–JAK2 signaling complex, thereby blocking the activation of additional STAT5b molecules. The desensitization of JAK2–STAT5b signaling in liver cells treated with GH continuously (i.e., mimicking the female rat plasma GH pattern) is consistent with a role for such an inhibitory factor in the longer term down-regulation of GH signaling. While a GH-inducible tyrosine phosphatase might be one such inhibitory factor, one or more of the recently described SOCS/CIS proteins (suppressor of cytokine signaling) could serve as an alternative inhibitory molecule (Fig. 13). SOCS/CIS proteins comprise a recently discovered family of receptor–JAK kinase signal inhibitory molecules that can be rapidly induced by a variety of cytokines via STAT-induced transcriptional mechanisms. The cytokine signaling inhibitory protein termed SOCS1 contains an SH2 domain through which it binds to, and thereby inhibits, the enzymatic activity of tyrosine phosphorylated (activated) JAK2 kinase. By contrast, CIS1, another member of the same cytokine inhibitor protein family, is a STAT5-inducible negative feedback regulator that binds to the tyrosine phosphorylated receptors for erythropoietin and IL-3. The receptor-bound CIS1 could thus block STAT5 docking and phosphorylation, in addition to all downstream STAT5-dependent transcriptional responses.

At least seven members of the SOCS/CIS protein family have been identified and shown to exhibit unique tissue-specific and time-dependent cytokine activation responses. Moreover, several SOCS/CIS mRNAs are rapidly induced in both preadipocytes and liver tissue in vivo in response to a pulse of GH. SOCS1 and SOCS3 are particularly effective at inhibiting GH-induced, STAT5-dependent transcriptional responses when constitutively expressed in COS cells, supporting the hypothesis that these proteins play an important role in the negative feedback regulation of GHR–JAK2 signaling to STAT5b. Potential mechanisms whereby SOCS/CIS proteins could contribute to termination of GH signaling might therefore include: (1) binding of the SOCS/CIS protein to tyrosine phosphorylated GHR or JAK2 to effect the initial down-regulation of receptor–kinase activity which first begins ~40 min into a GH pulse. The ~40 min requirement for GHR–JAK signal termination to begin may represent the time required for GH-activated STAT5b to induce sufficient SOCS/CIS protein to effect the observed GHR–JAK2 inhibition; (2) binding of the SOCS/CIS protein to GHR or JAK2 during the time interval between GH pulses, which could contribute to the decreased responsiveness of liver cells to a second hormone pulse that is observed during a ~2.5 h refractory time period between the pulses; and (3) binding of the SOCS/CIS protein to GHR or JAK2 in cells exposed to GH continuously, thereby contributing to the down-regulation of the GHR–JAK2–STAT5b signal transduction pathway that is characteristic of liver cells in female rats.

Individual SOCS/CIS proteins inhibit cytokine signaling via distinct mechanisms (Fig. 13). A key feature is the central SH2 domain of the SOCS protein, which can bind phosphotyrosine residues present in JAK kinases (in the case of SOCS1, and to a lesser extent SOCS3), thereby inhibiting JAK catalytic activity. In the case of other SOCS/CIS proteins, such as CIS1, the SOCS protein may bind to the receptor–kinase complex by means of interactions between the SOCS protein's SH2 domain and a phosphotyrosine residue in the cytokine receptor's cytoplasmic domain. In the case of GH signaling, this binding interaction between tyrosine phosphorylated GHR and a SOCS protein SH2 domain could inhibit STAT5 activation by competing directly with STAT5 for binding to the receptor. An alternative possibility, suggested by studies of the interaction of CIS1 with the erythropoietin receptor, is that once bound to the receptor–JAK2 complex, the SOCS/CIS protein undergoes ubiquitination. This, in turn, leads to efficient proteasomal degradation of the cell surface-bound SOCS/CIS–receptor–JAK2 complex, thereby preventing further JAK2-dependent signaling (Fig. 13). Further study is required to ascertain whether this proteasome-targeting inhibitory mechanism also applies to GHR and SOCS/CIS inhibition of GH-induced signaling.

8. FATE OF GH-BOUND GH RECEPTOR AND MECHANISMS OF RECEPTOR DOWN-REGULATION

Many biochemical changes, most notably activation of the tyrosine and serine phosphorylation cascades discussed above, occur intracellularly upon engagement and dimerization of GHR. In addition to its undergoing phosphorylation, however, GH-activated GHR also undergoes structural changes and changes in its intracellular routing in ways that ultimately alter its availability for further GH stimulation and activation at the cell surface. Down-regulation of GHR, defined as the loss of surface receptors after GH treatment, involves a combination of GH-induced receptor internalization, degradation and/or sequestration in intracellular compartments. In this section, we discuss aspects of GHR posttranslational modification and/or trafficking that relate to GH-induced receptor down-regulation and/or desensitization. These include GHR ubiquitination and endocytosis, nondissociable binding of GH, GHR detergent insolubility, and GHR accumulation in the nucleus.

8.1. GH Receptor Undergoes Endocytosis and Ubiquitination

As has been observed for many cell surface receptors, GHR internalization from the cell surface occurs in a broad range of cells. Both constitutive receptor internalization and ligand-induced GHR internalization occur. Structural determinants within GHR's cytoplasmic domain that mediate GH-induced receptor internalization have been identified through mutagenesis and transient transfection and reconstitution studies. In particular, GH-induced GHR internalization and down-regulation is abolished upon mutation of a critical cytoplasmic domain residue, Phe[346]. This observation is conceptually similar to findings with other cell suface proteins, such as low-density lipoprotein (LDL) receptor, in which an aromatic amino acid residue within the cytoplasmic domain is important for protein internalization.

Ubiquitination commonly targets proteins for degradation by the proteasome pathway. GHR is ubiquitinated on multiple lysine residues following GH treatment. This polyubiquitination step is an important determinant of GHR internalization and is linked to GHR endocytosis, as shown by the inhibition of GH-induced receptor internalization a 42°C but not 30°C in cells with a temperature-sensitive defect in ubiquitin conjugation. In contrast to other ubiquitinated proteins, however, degradation of the internalized, ubiquitinated GHR occurs in the endosomal/lysosomal compartment. Further, in the setting of a truncated GHR, clathrin-coated pit-mediated GHR internalization and lysosomal degradation can occur in a ubiquitin-independent manner. This ubiquitin-independent GHR internalization requires a dileucine motif within the receptor's proximal cytoplasmic domain, at residues 347 and 348 (rabbit GHR numbering system). The significance of the use of these ubiquitin and dileucine determinants for GH-induced GHR internalization is uncertain, but should be clarified by studies of receptor internalization in cell lines and tissues that homologously express GHR at physiological levels.

Studies carried out in cultured liver cells demonstrate that JAK2 signaling to STAT5b is prolonged by the proteasome inhibitor MG132, demonstrating that proteasome degradation is an important step in the termination of GHR–JAK2 signaling. Other studies show, however, that new protein synthesis is not required for these liver cells to respond to sequential GH pulses, suggesting that any proteasome degradation of GHR or JAK2 that may occur following a single GH pulse is perhaps limited to that portion of the cellular GHR and JAK2 pool which is then resident at the cell surface. Accordingly, the prolonged JAK2 signaling to STAT5b seen in proteasome inhibitor-treated cells could be a result of prolonged maintenance of the GHR–JAK2 complex at the cell surface, perhaps as a consequence of blocking the CIS protein-dependent receptor degradation mechanism discussed previously.

An alternative hypothesis is suggested by studies showing that GH-induced polyubiquitiration of GHR proceeds even when JAK2 kinase activity is inhibited. Moreover, even in the presence of active JAK2 kinase, those GHR molecules that become ubiquitinated are not tyrosine phosphorylated. This suggests that tyrosine phosphorylated GHR molecules may first be deactivated by phosphotyrosine dephosphorylation, and only then become targeted for internalization, as depicted in Fig. 12 (step 8). Phosphotyrosine dephosphorylation of GHR might thus deactivate the receptor and also facilitate its subsequent polyubiquitination and internalization. This receptor internalization process would, in turn, prevent further signaling to STAT5b until the time that GHR reappears on the cell surface (step 9). This latter event could involve both recycling of existing GHR and replacement of degraded receptors via new protein synthesis. Accordingly, proteasome inhibitors may stabilize the active, phosphorylated GHR–JAK2 signaling complex by an indirect mechanism, for example, by blocking degradation of a hypothetical tyrosine phosphatase inhibitor. Prolonged signaling to STAT5 from IL-2-activated JAK1 and JAK3 has also been observed in cells treated with proteasome inhibitors, where it is also speculated that a phosphotyrosine phosphatase inhibitor is the target of proteasome degradation.

8.2. GH Receptor May Be Subject to Nonproteolytic Processing and Internal Sequestration

In addition to receptor degradation, several other possible fates have been described for GHR. In IM-9 cells, which are notable for the long steady-state half-life of GHR (~8 h), the time course of GH-induced GHR down-regulation is similar to the time course observed for two other phenomena, the appearance of a nondissociable component of ^{125}I-hGH binding and the exocytic release of previously internalized hGH from the cell; this suggests that these three phenomena may be related. GH treatment of IM-9 cells also promotes the appearance of a disulfide-linked, tyrosine phosphorylated high molecular weight form of GHR that, as discussed earlier, may be represent a biochemical correlate of the dimerized receptor. Further, GH-induced appearance of a detergent-insoluble pool of GHRs, which is enriched in the disulfide-linked receptor form, can be detected in IM-9 cells with a time course similar to that observed for receptor down-regulation.

These observations indicate that receptor sequestration and/or nonproteolytic receptor processing are part of the itinerary of GH-bound GHRs. The intracellular location of the detergent-insoluble form of the GHR remains to be determined. It is also not yet known whether this receptor form specifically associates with particular membrane microdomains, such as caveolae, and/or subcellular structures, such as the cytoskeleton. Answers to these questions may reveal much about this apparently nondegradative routing of a subset of ligand-activated GHRs.

Finally, Lobie and colleagues have described the accumulation of GH in the nucleus, as well as transient GH-induced association of GHR with nuclear membranes. While the associations and functions of the nuclear-associated GH–GHR complex are still unclear, recent descriptions of the formation of GH-induced nuclear complexes comprised of key GH signaling molecules, such as JAK2 kinase, make it likely that nuclear signaling events may in part be mediated by this GH-induced subcellular routing of a portion of cellular GHR molecules. The GH-induced phenomena of GHR detergent insolubility and nuclear GHR accumulation have yet to be described in multiple cell lines and tissues, leaving open the possiblity that these GHR trafficking routes may be differentially accessed in different GH-responsive cells.

9. CONCLUDING REMARKS

In the present era of an expanding understanding of molecular and cellular biology, studies of basic aspects of hormone and cytokine action are flourishing. Studies of the cellular and molecular actions of GH and GHR function serve as an important paradigm in this arena. Important new directions for future research on GH action will continue to emerge as further details become known about the processes through which GH activates its receptor and initiates downstream signaling pathways leading to changes in gene expression with corresponding changes in cell physiology. Future areas of GH research will likely include studies of the control and regulatory mechanisms that modulate GH-activated signals, investigations of GH signaling as it affects and is affected by the actions of other growth factors and hormones ("crosstalk"), and identification of the determinants that allow GH to exert its differential effects in individual target cells and in a tissue-specific context. Unravelling of these issues will undoubtedly provide a more complete understanding of GH's effects on whole body physiology and human health.

ACKNOWLEDGMENTS

Preparation of this chapter was, in part, supported by NIH grants DK33765 (to D. J. W.) and DK46395 and a VA Merit Review award (to S. J. F.). The able assistance of Dr. Soo-Hee Park and Ms. Blossom Wu, Boston University, in figure preparation is gratefully acknowledged.

SELECTED READINGS

Alele J, Jiang J, Goldsmith JF, Yang X, Maheshwari HG, Black RA, Baumann G, Frank SJ. Blockade of growth hormone receptor shedding by a metalloprotease inhibitor. Endocrinology 1998; 139:1927–1935.

Argetsinger LS, Campbell GS, Yang X, Witthuhn BA, Silvennoinen O, Ihle JN, Carter-Su C. Identification of JAK2 as a growth hormone receptor-associated tyrosine kinase. Cell 1993; 74:237–244.

Darnell JE, Jr. STATs and gene regulation. Science 1997; 277:1630–1635.

de Vos AM, Ultsch M, Kossiakoff AA. Human growth hormone and extracellular domain of its receptor: crystal structure of the complex. Science 1992; 255:306–312.

Frank SJ, Yi W, Zhao Y, Goldsmith JF, Gilliland G, Jiang J, Sakai I, Kraft AS. Regions of the JAK2 tyrosine kinase required for coupling to the growth hormone receptor. J Biol Chem 1995; 270:14776–14785.

Gebert CA, Park SH, Waxman DJ. Termination of growth hormone pulse-induced STAT5b signaling. Mol Endocrinol 1999; 13:38–56.

Laron Z. Disorders of growth hormone resistance in childhood. Curr Opin Pediatr 1993; 4:474–480.

Leung DW, Spencer SA, Cachianes G, Hammonds RG, Collins C, Henzel WJ, Barnard R, Waters MJ, Wood WI. Growth hormone receptor and serum binding protein: purification, cloning and expression. Nature 1987; 330:537–543.

Udy GB, Towers RP, Snell RG, Wilkins RJ, Park SH, Ram PA, Waxman DJ, Davey HW. Requirement of STAT5b for sexual dimorphism of body growth rates and liver gene expression. Proc Natl Acad Sci USA 1997; 94:7239–7244.

Veldhuis JD. Gender differences in secretory activity of the human somatotropic (growth hormone) axis. Eur J Endocrinol 1996; 134:287–295.

Wakao H, Gouilleux F, Groner B. Mammary gland factor (MGF) is a novel member of the cytokine regulated transcription factor gene family and confers the prolactin response. EMBO J 1994; 13:2182–2191.

Waxman DJ, Ram PA, Park SH, Choi HK. Intermittent plasma growth hormone triggers tyrosine phosphorylation and nuclear translocation of a liver-expressed, Stat 5-related DNA binding protein. J Biol Chem 1995; 270:13262–13270.

Zhou Y, Xu BC, Maheshwari HG, He L, Reed M, Lozykowski M, Okada S, Cataldo L, Coschigamo K, Wagner TE, Baumann G, Kopchick JJ. A mammalian model for Laron syndrome produced by targeted disruption of the mouse growth hormone receptor/binding protein gene (the Laron mouse). Proc Natl Acad Sci USA 1997; 94:13215–13220.

5 Regulation of *Drosophila* Visual Transduction Through a Supramolecular Signaling Complex

Craig Montell

CONTENTS

INTRODUCTION
COMPARISON OF VERTEBRATE AND *DROSOPHILA* PHOTOTRANSDUCTION
ANATOMY OF THE *DROSOPHILA* COMPOUND EYE
GENETIC APPROACHES TO *DROSOPHILA* PHOTOTRANSDUCTION
CELL TYPE SPECIFIC EXPRESSION OF MULTIPLE OPSIN GENES
ACTIVATION AND REGULATION OF RHODOPSIN
DROSOPHILA VISION UTILIZES THE INOSITOL PHOSPHOLIPID SIGNALING SYSTEM
TRP AND TRPL REPRESENT THE ARCHETYPAL PHOSPHOLIPASE C-DEPENDENT ION CHANNELS
Ca^{2+} IS REQUIRED FOR DEACTIVATION OF PHOTOTRANSDUCTION
FORMATION OF A SIGNALPLEX THROUGH THE PDZ-CONTAINING PROTEIN INAD
POTENTIAL FUNCTIONS OF THE INAD SIGNALPLEX
PARALLELS BETWEEN *DROSOPHILA* PHOTOTRANSDUCTION AND MAMMALIAN SIGNALING
SELECTED READINGS

1. INTRODUCTION

The *Drosophila* phototransduction cascade has served as a paradigm for G protein-coupled signaling due in part to the availability of genetic approaches which offer the possibilty of identifying the key signaling components and characterizing the roles of such factors in vivo. Genetic screens have led to the identification of many of the proteins essential for *Drosophila* phototransduction, some of which were not previously known in other organisms. However, it now appears that each of the proteins critical for *Drosophila* vision is highly related to signaling proteins in vertebrates. As such, insights into *Drosophila* phototransduction appear to have general relevance to other G protein-coupled cascades.

In addition, the *Drosophila* visual cascade is among the fastest known G protein-coupled signaling cascades. Only ~10 ms elapse between the activation of the light receptor, rhodopsin, and opening of the light-sensitive channels. Moreover, the turn-off of the visual cascade is also very rapid (requires <100 ms). The speed of this cascade provides the opportunity to address the mechanisms underlying rapid G protein-coupled signaling in vivo.

During the last two decades, it has been widely assumed that the proteins that comprise signaling cascades interact through random stochastic collisions between activated proteins and downstream effectors. However, such a mechanism is difficult to reconcile

From: *Principles of Molecular Regulation* (P. M. Conn and A. R. Means, eds.), © Humana Press Inc., Totowa, NJ.

with the rapid kinetics of *Drosophila* phototransduction. Recent studies in a variety of systems have provided evidence that many of the components in signaling cascades are tightly linked. In the case of *Drosophila* phototransduction, it appears that most of the critical proteins are linked together in a massive signaling complex (signalplex). The identification of the signalplex has caused a reevaluation of the mechanisms by which *Drosophila* visual transduction, and perhaps other G protein signaling cascades, are activated and regulated.

2. COMPARISON OF VERTEBRATE AND *DROSOPHILA* PHOTOTRANSDUCTION

Both vertebrate and *Drosophila* visual transduction (phototransduction) are initiated by the photoactivation of the light receptor rhodopsin and subsequent activation of a heterotrimeric G protein (Fig. 1). In vertebrates, the effector for the G protein, which is referred to as transducin (*see* Chapter 1 for details on transducin), is a cGMP phosphodiesterase (PDE) which hydrolyzes 3′:5′-cyclic guanosine monophosphate (cGMP) to 5′-GMP (Fig. 1A; *see* Chapter 15 for details on phosphodiesterases). The reduction in cGMP levels results in the closing of the cGMP-gated ion channels. Thus, there is a significant current in the dark, which is carried by Na^+ and Ca^{2+}. This current is greatly reduced in the light, resulting in hyperpolarization of the photoreceptor cells.

In *Drosophila* photoreceptor cells, the effector for the G protein (G_q) is a phospholipase Cβ (PLC) which catalyzes the conversion of phosphatidylinositol-4,5-bisphosphate (PIP_2) to inositol-1,4,5-trisphosphate (IP_3) and diacylglycerol (DAG) (Fig. 1B). The second messenger(s) that activates the ion channels, TRP and TRPL, in *Drosophila* phototransduction has not been determined. However, in contrast to vertebrate vision, the light-responsive ion channels in *Drosophila*, which are permeable predominantly to Na^+ and Ca^{2+}, are closed in the dark and open in the light. Thus, stimulation of light results in depolarization in *Drosophila* photoreceptor cells.

Despite some major differences in the types of effector molecules used in vertebrate and *Drosophila* phototransduction, both types of visual systems display some important similarities. These include enormous signal amplification and responsiveness over a vast range of light intensities. In both systems, a single photon of light is sufficient to regulate large numbers of cation influx channels.

3. ANATOMY OF THE *DROSOPHILA* COMPOUND EYE

The adult compound eye is composed of approx 800 repeat units referred to as ommatidia (Fig. 2A). Each ommatidium is composed of 20 cells including 8 photoreceptor cells (Fig. 2B). Six of the photoreceptor cells, R1–6, extend the full depth of the retina and the remaining two are inner photoreceptor cells, R7 and R8, which occupy either the top or bottom halves of the retina, respectively. Consequently, only seven photoreceptor cells are present in any given plane of section. Each photoreceptor cell contains a specialized light sensing organelle consisting of ~60,000 microvilli. This organelle, referred to as a rhabdomere, is the site for light reception and phototransduction and provides an enormous surface area to concentrate large quantities of rhodopsin.

4. GENETIC APPROACHES TO *DROSOPHILA* PHOTOTRANSDUCTION

To identify the molecules and mechanisms underlying *Drosophila* phototransduction, electrophysiological, cell biological, and pharmacological approaches have been combined with genetic approaches. Many mutations affecting phototransduction have been identified using a relatively simple electrophysiological assay, electroretinogram recordings (ERGs). ERGs are extracellular recordings that measure the summed responses of all the cells in the retina to light. There are two main components of the ERG, the on- and off-transients, which emanate from activity postsynaptic to the photoreceptor cells in the optic lobes and the maintained component (or light coincident receptor potential, LCRP) which is principally a summed response of all the cells in the retina (Fig. 3A). Mutations have been identified which affect each of these two components; however, only those mutations that affect the LCRP are considered here (Table 1). Some of the mutations, such as *trp*, disrupted proteins that were not previously predicted to function in visual transduction and led to unexpected insights into *Drosophila* visual transduction. However, the screens did not saturate the genome for loci critical in phototransduction. Thus, mutations in other genes predicted to function in *Drosophila* visual

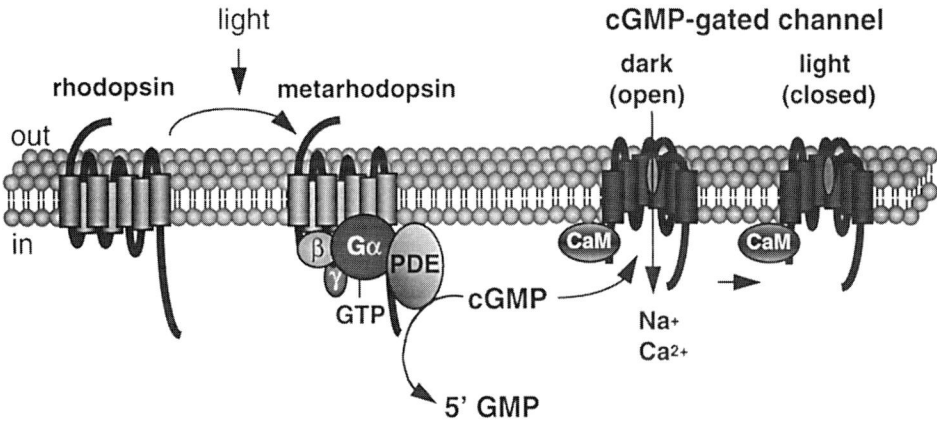

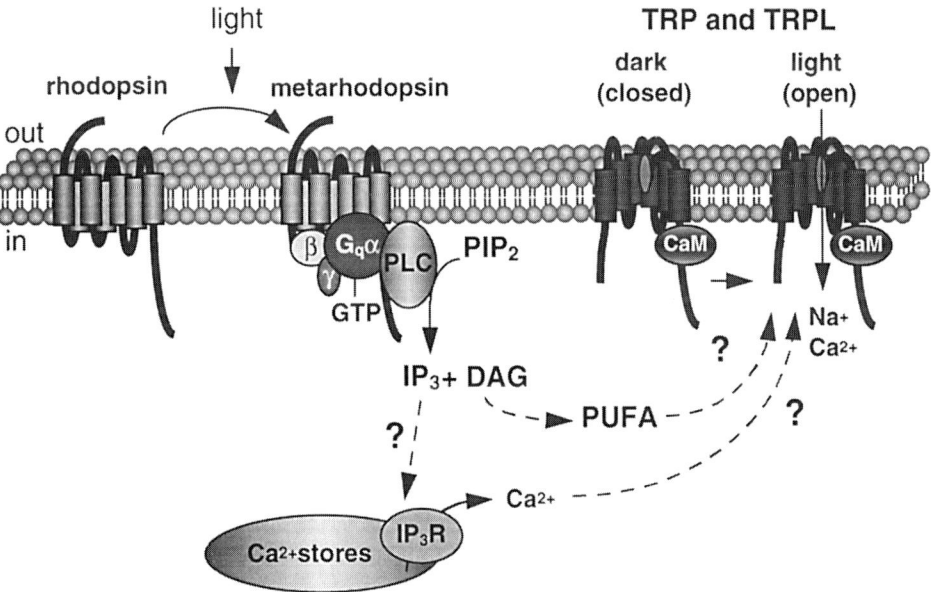

Fig. 1. Comparison between vertebrate and *Drosophila* phototransduction. (**A**) Vertebrate phototransduction. (**B**) Model of *Drosophila* phototransduction. *See* text for details. The question marks in (**B**) indicate the uncertainties as to whether TRP is activated by release of Ca^{2+} from the internal stores or by IP_3 and/or DAG through a mechanism independent of the stores. The extracellular (out) and cytoplasmic sides (in) of the membrane are indicated. 5′-GMP, 3′:5′-cyclic guanosine monophosphate; CaM, calmodulin; cGMP, 5′-guanosine monophosphate; DAG, diacylglycerol; Gα-, γ-, β-subunits of transducin; $G_q\alpha$, β and γ, subunits of G_q; IP_3, inositol-1,4,5-trisphosphate; PDE, phosphodiesterase; PIP_2, phosphatidylinositol-4,5-bisphosphate; PLC, phospholipase Cβ.

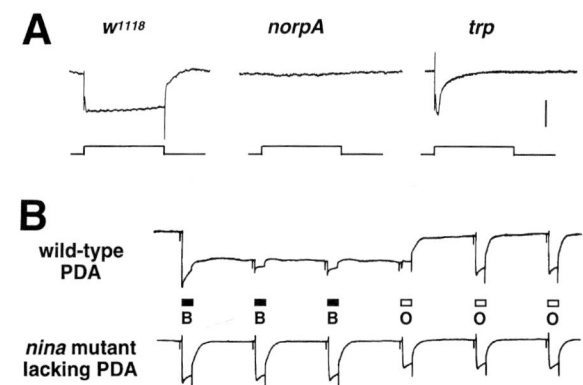

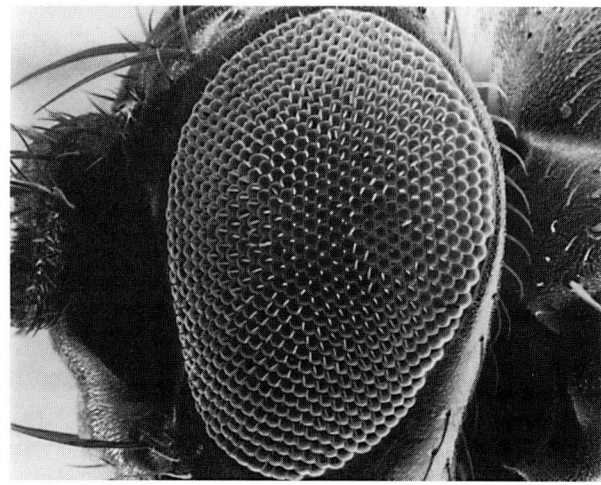

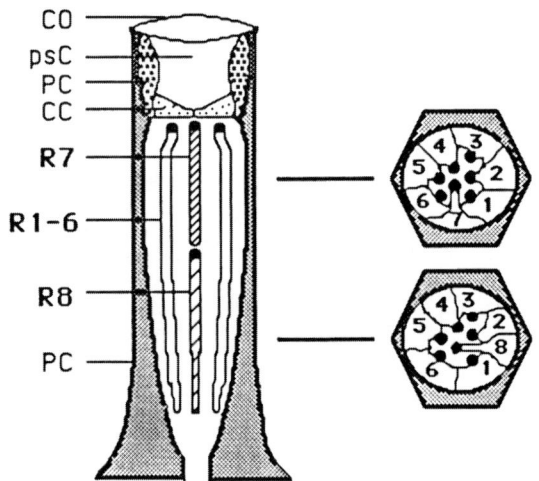

Fig. 2. Structure of *Drosophila* compound eye. (**A**) Scanning electron micrograph of the surface of the compound eye. Each eye contains ~800 ommatidia. (**B**) Schematic representation of a single ommatidium. Shown is an ommatidium at a 90° angle relative to the surface of the eye in the scanning electron micrograph. The ommatidia are ~0.1 mm in length. The corneal lens cell (CO) lies at the top of each ommatidium. Below the CO are the pseudocone (psC), pigment cells (PC), and cone cells (CC). The retina, which is comprised of the photoreceptor cells (R1–6, R7, and R8) and pigment cells (PC), is immediately beneath the cone layer. The microvillar portion of the photoreceptor cells, the rhabdomeres, of the R1–6, R7, and R8 photoreceptor cells are indicated by the columns without shading, with densely spaced diagonal lines and sparsely spaced diagonal lines, respectively. Indicated to the right are sections through the upper and lower portions of the ommatidia. In the cross-sections through the distal (*upper*) and proximal (*lower*) regions of the retina, shown to the *right*, the rhabdomeres are depicted as *black circles*. Note that the R7 and R8 cells occupy only the top and bottom portions of the retina respectively. Thus, only seven photoreceptor cells are present in any given plane.

Fig. 3. Examples of mutants with altered electroretinogram recordings. (**A**) *trp* and *norpA* show defects in the light coincident receptor potential (LCRP). *norpA* flies are unresponsive to light. The *trp* mutant is characterized by a transient (LCRP). Flies were dark adapted for 2 min and then exposed to a white-light stimulus. The duration of the light pulses was 5 s. The initiation and cessation of the light stimuli are indicated below and a 5 mV scale is included to the right. (**B**) *nina* (*neither-inactivation-nor-afterpotential*) mutants display reduced PDAs. Wild-type and nina mutant flies ($ninaD^{P245}$) were exposed to pulses of intense blue (B; 480 nm) and orange (O; 580 nm) light. The *black and white boxes* indicate the initiation and cessation of the blue and orange stimuli, respectively. The stimulus duration was 4 s and the calibration pulse before each response was 5 mV. (Adapted with permission by the Cambridge University Press from Stephenson et al., 1983.)

transduction were generated using genetic strategies that targeted specific loci for mutagenesis.

5. CELL TYPE SPECIFIC EXPRESSION OF MULTIPLE OPSIN GENES

Drosophila vision is polychromatic owing to the expression of a collection of six visual pigments, rhodopsins (Rh1–6), which respond to light with distinct absorption maxima (Table 2). Moreover, each photoreceptor cell in the compound eye expresses only a single rhodopsin isoform. The major rhodopsin, Rh1, is a blue pigment and is the only rhodopsin expressed in six of the photoreceptor cells, R1–6. A second rhodopsin, Rh2, is a violet-sensitive pigment that is not found in the compound eye but is expressed in three small light-sensing organs, ocelli, situated on the top of the fly head. Rh3 and Rh4 are restricted to nonoverlapping subsets of the ultraviolet (UV)-sensitive R7 cells. Both of these latter rhodopsins are UV photopigments with slightly different absorption maxima.

Approximately 30% of the R7 cells express Rh3

Table 1
Drosophila Phototransduction Loci That Have Been Characterized at the Molecular Level

Locus	Protein Product	Function in Photoresponse	Retinal Degeneration
arrestin 2	Binds to rhodopsin	Rhodopsin inactivation	Yes (LD)
cds	CDP–DAG synthase	Production of PIP_2 intermediate	Yes (LD)
$G_q\alpha$	α-Subunit of G protein	Activation	Not examined
$G\beta_q$	β-Subunit of G protein	Activation/deactivation	Not examined
inaC	Protein kinase C	Deactivation/adaptation	Not examined
inaD	PDZ-containing protein	Activation/deactivation	Yes (LD)
ninaA	Cyclophilin homolog	Rhodopsin transport	Yes (LI)
ninaC	Protein kinase/myosin	Deactivation/adaptation	Yes (LD)
ninaE	Rhodopsin	Light receptor	Yes (LI)
norpA	Phospholipase C	Activation/termination	Yes (LD)
rdgA	DAG kinase	Production of PIP_2 intermediate	Yes (LI)
rdgB	PI transfer protein	Recovery from light stimulation	Yes (LD)
rdgC	Protein phosphatase	Rhodopsin phosphatase	Yes (LD)
trp	Cation channel	Light-responsive channel	Yes (LD)
trpl	Cation channel	Light-responsive channel	Not examined

LD, light-dependent retinal degeneration; LI, light-independent retinal degeneration.

Table 2
Drosophila Rhodopsin Family

Rhodopsin	Expression Pattern	Pigment Type
Rh1	R1–6	Blue
Rh2	Ocelli	Violet
Rh3	R7 (~30%)	Ultraviolet
Rh4	R7 (~70%)	Ultraviolet
Rh5	R8 (~30%)	Blue
Rh6	R8 (~70%)	Green

and the remaining 70% express Rh4. The relative spatial distributions of the pair of R7 rhodopsin is devoid of any clear pattern and appears to be stochastic. The same appears to be the case for the remaining two rhodopsin isoforms, which are blue (Rh5) and green (Rh6) photopigments found exclusively in R8 cells. No clear pattern of Rh5 and Rh6 expressing cells can be discerned. However, the R7 and R8 opsins are coordinately expressed such that every R7 cell that contains Rh3 is situated above an R8 cell that expresses Rh5 and each Rh4 expressing R7 cells is located above an Rh6 expressing R8 cell. However, the molecular mechanisms that coordinate Rh3/Rh5 and Rh4/Rh6 expression remain to be identified.

In certain respects, the outer photoreceptor cells, R1–6, are functionally similar to the rods that populate the human retina. Human rods are very sensitive to light, are much more abundant than cones, and express only a single visual pigment that is important for night vision. Similarly, the Drosophila outer photoreceptor cells are also very sensitive to light, are the major class of photoreceptor cells, and express only one rhodopsin isoform. These features contrast with the inner photoreceptor cells, R7 and R8, which display some similarities to cone photoreceptor cells. Human cones are significantly fewer in number than rods and express three visual pigments—red, green and blue—that function to detect color during day vision. Moreover, human cones contribute to high visual acuity. The Drosophila R7 and R8 contain rhabdomeres that are not only fewer in number than the outer photoreceptor cells, but are also smaller in size. Consequently, the cross-sectional area of the rhabdomeres is ~10% that of the outer photoreceptor cells. Furthermore, the inner photoreceptor cells express four different visual pigments and may comprise a high acuity system.

6. ACTIVATION AND REGULATION OF RHODOPSIN

Rhodopsin consists of two components: an apoprotein moiety (opsin) consisting of seven transmembrane domains and a vitamin A derivative (3-hydroxy-11-*cis*-retinal in Drosophila and 11-*cis*-retinal in vertebrates) that binds to the seventh transmembrane segment (Fig. 4A). On exposure to light, there is a *cis* to *trans* isomerization around the 11–12 double bond in the retinal. This isomerization, which represents the only light-dependent event in phototransduction, results in an ensuing alteration in the conforma-

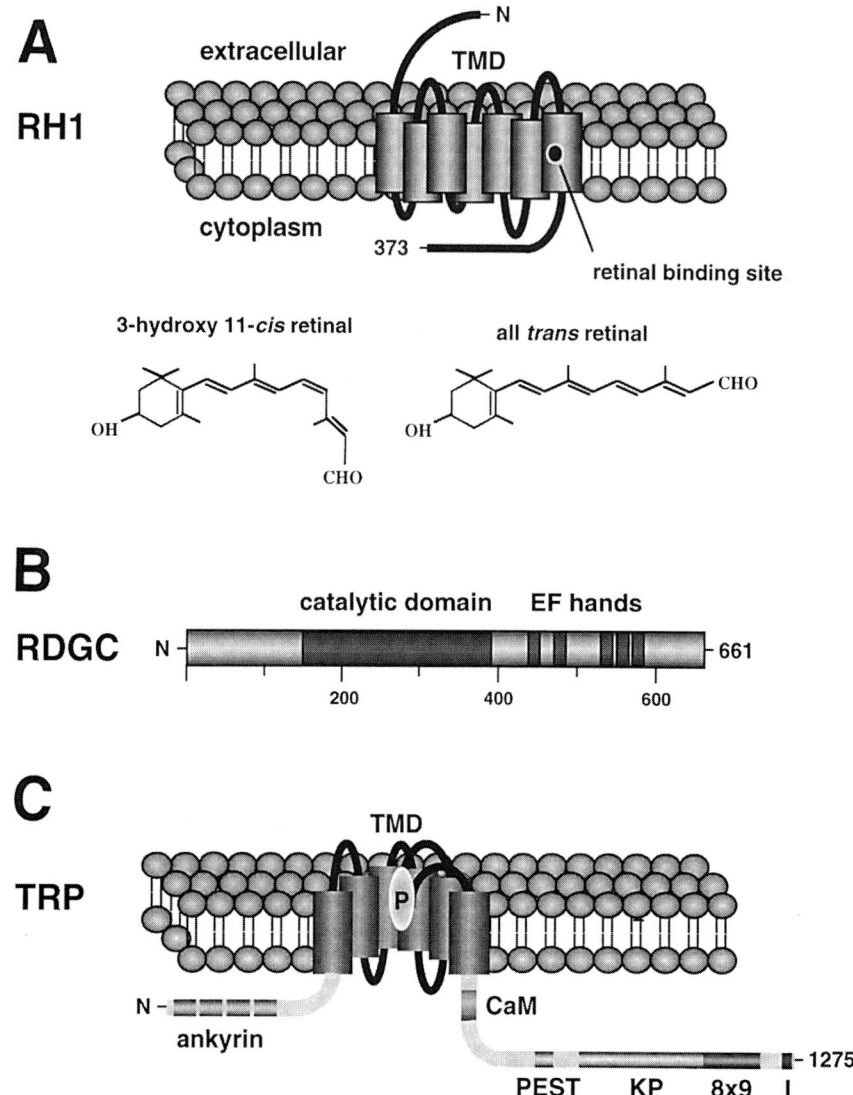

Fig. 4. Predicted structures of *Drosophila* rhodopsin, RDGC, and TRP. **(A)** The major *Drosophila* rhodopsin (Rh1). The position of the retinal binding site is indicated. The structures of the retinal chromophore in the *cis* and all-*trans* configurations are shown *below*. 373, the length of Rh1 in residues; N, N-terminus; TMD, seven putative transmembrane domains. **(B)** RDGC. 661, length of RDGC in amino acids; EF hands, the five putative Ca^{2+} binding motifs; N, N-terminus. **(C)** TRP. 1275, length of TRP in residues; 8 × 9, domain containing nine tandem copies of an eight amino acid repeat, D-K-D-K-K-P-G/A-D; ankryin, ankyrin repeats; CaM, calmodulin binding site; I, INAD binding site; KP, region containing multiple repeats of the dipeptide lysine and proline; P, putative pore-loop between TMDs 5 and 6; PEST, putative PEST degradation signal; TMD, six putative transmembrane domains.

tion of the opsin. The light activated rhodopsin, metarhodopsin, couples to and activates the G_q in a manner similar to that of other G protein-coupled receptors (*see* chapter 1).

In vertebrates, the light-activated rhodopsin, metarhodopsin II, is only semistable and dissociates into separate opsin and retinal components within minutes. However, the *Drosophila* Rh1 metarhodopsin is stable for at least 5 h and the retinal and opsin remain associated even after deactivation of the receptor. Stable light-activated metarhodopsin can be produced by exposing the photoreceptor cells to blue light. As a consequence, upon cessation of the blue light, a prolonged depolarization (PDA) is produced (Fig. 3B). The metarhodopsin can be converted back to the inactive state by exposure to orange light, thereby terminating the PDA. Although it is possible to generate a PDA in *Drosophila*, it is not normally observed. Ambient white light contains both blue and orange spectra. Therefore, the same steady white light stimulus that generates metarhodopsin also photoconverts the metarhodopsin back to rhodopsin. Moreover, a PDA can be generated only in flies that have white rather wild-type red eyes, as the pigment removes the majority of the blue portion of the spectrum.

The deactivation of metarhodopsin is very rapid and the fast kinetics of this process is essential to prevent excessive amplification which would occur in the event of prolonged coupling of the receptor with the G protein. Extensive analyses of vertebrate rhodopsin have demonstrated that deactivation is a multistep process that involves phosphorylation of the receptor and interaction with a rhodopsin binding protein referred to as arrestin (*see* chapter 1 for details on arrestins). Arrestin binding contributes to termination of the photoresponse by interfering with the interaction between metarhodopsin II and the G protein. The opsin–retinal complex subsequently dissociates and the opsin moiety is dephosphorylated. The receptor cycle is completed in the vertebrate retina by reassociation of the opsin with 11-*cis*-retinal.

Genetic studies in *Drosophila* have demonstrated that arrestin is also critical for deactivation of Rh1. Moreover, binding of *Drosophila* arrestin to rhodopsin is the rate-limiting step that determines the kinetics of receptor inactivation and the magnitude of signal amplification at the first step of the visual cascade. In contrast to vertebrates, phosphorylation of the receptor does not appear to be necessary for arrestin binding. Nevertheless, *Drosophila* rhodopsin undergoes light-dependent phosphorylation. This phosphorylation does appear to regulate rhodopsin activity, as mutations that eliminate the rhodopsin phosphatase, referred to as RDGC (Retinal Degeneration C; Fig. 4B), result in hyperphosphorylation of the receptor and defects in termination of the photoresponse. The mechanism by which dephosphorylation of the receptor contributes to deactivation remains to be elucidated.

Pak and co-workers have isolated many mutants with diminished PDAs (Fig. 3B). These mutants, referred to as *nina* (*neither-inactivation-nor-afterpotential*), each has reduced rhodopsin content. The lack of PDA in the *nina* mutants appears to be due to an excess of arrestin over metarhodopsin as a PDA is restored in *nina*; *arrestin* double mutants. Thus, the PDA generated in wild-type flies seems to result from a surplus of metarhodopsin relative to free arrestin.

7. *DROSOPHILA* VISION UTILIZES THE INOSITOL PHOSPHOLIPID SIGNALING SYSTEM

All known vertebrate and invertebrate rhodopsins couple to heterotrimeric G-proteins that consist of α-, β-, and γ-subunits (*see* chapter 1 for details on heterotrimeric G proteins). On light stimulation, the G protein binds to the receptor resulting in a GDP to GTP exchange and release of the βγ dimer. The α- and β-subunits that function in fly visual transduction ($G_{q\alpha}$ and $G_{\beta}e$, respectively) are highly enriched in photoreceptor cells. The effector for the activated G protein in *Drosophila* vision is a phospholipase Cβ encoded by the *norpA* locus. Genetic evidence that this is the case is that mutations that eliminate NORPA obliterate the light response. Furthermore, mutations in one of the enzymes involved in the synthesis of PIP_2, CDP-diacylglycerol synthase, eliminates the light response in photoreceptor cells that have been depleted of the PIP_2 pools by exposure to intense light. An unresolved issue in *Drosophila* phototransduction concerns the mechanism by which activation of PLC leads to opening of the light-sensitive ion channels TRP and TRPL. It is also unclear whether cGMP contributes to excitation in *Drosophila* vision. A cGMP-gated ion channel is expressed in the *Drosophila* retina, although it remains to be determined whether the channel is localized to the rhabdomeres. Furthermore, addition of a nonhydrolyzable analog of cGMP to *Drosophila* photoreceptors enhances the light response. However, genetic evidence supporting a role for cGMP in *Drosophila* phototransduction is currently lacking.

8. TRP AND TRPL REPRESENT THE ARCHETYPAL PHOSPHOLIPASE C-DEPENDENT ION CHANNELS

The major ion channel responsible for light-dependent cation influx is TRP, a protein with six putative transmembrane domains (Fig. 4C). *trp* mutant flies display a dramatic decrease in Ca^{2+} influx and moderate reduction in Na^+ influx. Moreover, the response in *trp* flies is only transient. The remaining influx in *trp* appears to be due to a second ion channel, TRPL, which is highly related to TRP. Flies that are missing both TRP and TRPL are completely unresponsive to light.

The mechanism whereby activation of PLC leads to opening of the light responsive ion channels has not been resolved. According to one model, TRP is activated indirectly by IP_3. The IP_3 binds to the IP_3 receptor, situated in the membrane of intracellular Ca^{2+} stores, resulting in Ca^{2+} release from the internal stores. Release of Ca^{2+} from the internal stores, which comprises part of the smooth endoplasmic reticulum, may then lead to activation of TRP. Consistent with the proposal that TRP is a store-operated channel (SOC) are in vitro studies demonstrating that cation influx can be activated by release of Ca^{2+} from internal stores after expressing TRP in tissue culture cells (Fig. 5). The link between Ca^{2+} release and opening of the TRP channels is not known; however, it does not appear to be Ca^{2+}, as store-operated cation influx can be induced in the presence of high intracellular levels of Ca^{2+} chelators.

In contrast to TRP, which can be activated in vitro by release of Ca^{2+} from internal stores, expression of TRPL in heterologous expression systems leads to a constitutive cation influx (Fig. 5). Thus, it appears that TRPL requires interaction with an additional subunit to confer store-operated activity. The additional subunit appears to be TRP, as TRP and TRPL bind directly and coexpression of two channels in vitro leads to a regulated SOC (Fig. 5) with characteristics distinct from either TRP or TRPL. TRP is significantly more abundant than TRPL in vivo and TRPL binds to TRP with higher affinity than to itself. Thus, it appears that the light-activated cation influx in photoreceptors is mediated by TRP homomultimers and TRP–TRPL heteromultimers. TRPL homomultimers may not exist in *Drosophila* photoreceptor cells. To account for the observation that the residual cation influx in *trp* mutant flies is light dependent rather than constitutive, it has been proposed that there exists a third TRP-related protein that is capable of associating with TRPL and forming a third class of regulated influx channels in photoreceptors. Recently, a third TRP-related protein, TRPγ, has been identified and shown to preferentially interact with TRPL. While TRPL and TRPγ homomultimers are constitutively active, TRPL-TRPγ heteromultimers produce a regulated phospholipase C-stimulated conductance with features distinct from that elicited by TRP homomultimers and TRP/TRPL heteromultimers.

Consistent with the in vitro studies indicating that TRP and TRP–TRPL are SOCs, are the observations that drugs that inhibit the IP_3 receptor block light-

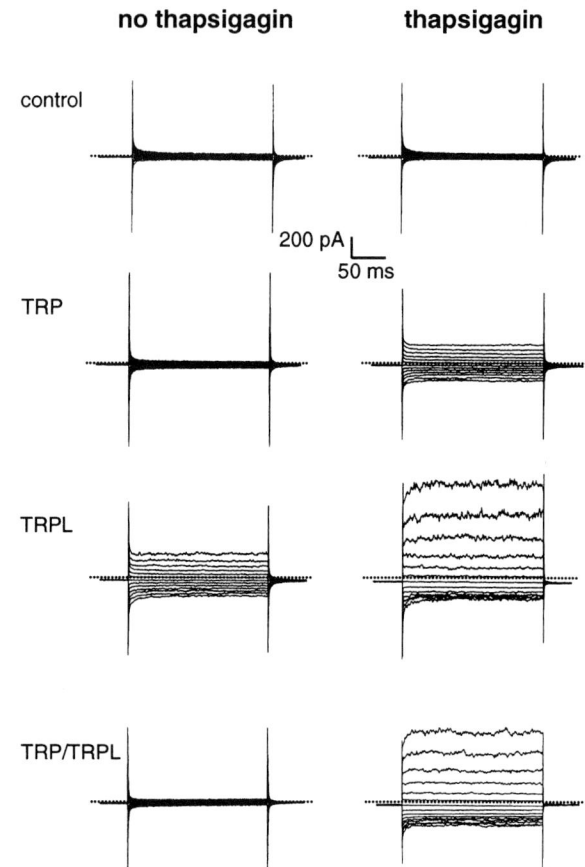

Fig. 5. TRP, TRPL, and TRP–TRPL currents in tissue culture cells before and after treatment with a drug, thapsigargin, that causes release of Ca^{2+} from the internal Ca^{2+} stores. Thapsigargin inhibits the Ca^{2+}-ATPase in the IP_3-sensitive Ca^{2+} stores and results in a net Ca^{2+} release from the internal stores via the leak current. The cDNA constructs, pTRP and pTRPL, were transfected either singly or in combination as indicated in a mammalian cell line, 293T cells. Na^+ currents were recorded either before or after depleting the intracellular Ca^{2+} stores by perfusing the cells with thapsigargin. Traces show currents produced using a set of step voltages from –100 to +80 mV in 15-mV increments. The holding potential was –10 mV. Dotted lines indicate 0 pA. (Adapted with permission by Cell Press from Xu et al., 1997.)

activated cation influx in vivo. Nevertheless, TRP and TRP–TRPL may not be activated through a store-operated mechanism in vivo. The very rapid, millisecond activation of the light responsive cation influx is difficult to reconcile with the large spatial separation between the stores and the phototransduction machinery (Fig. 6). The Ca^{2+} stores are excluded from the rhabdomeres, the organelle containing the proteins known to function in phototransduction. Thus, the kinetics of activation would presumably be retarded by the spatial separation between the stores and the

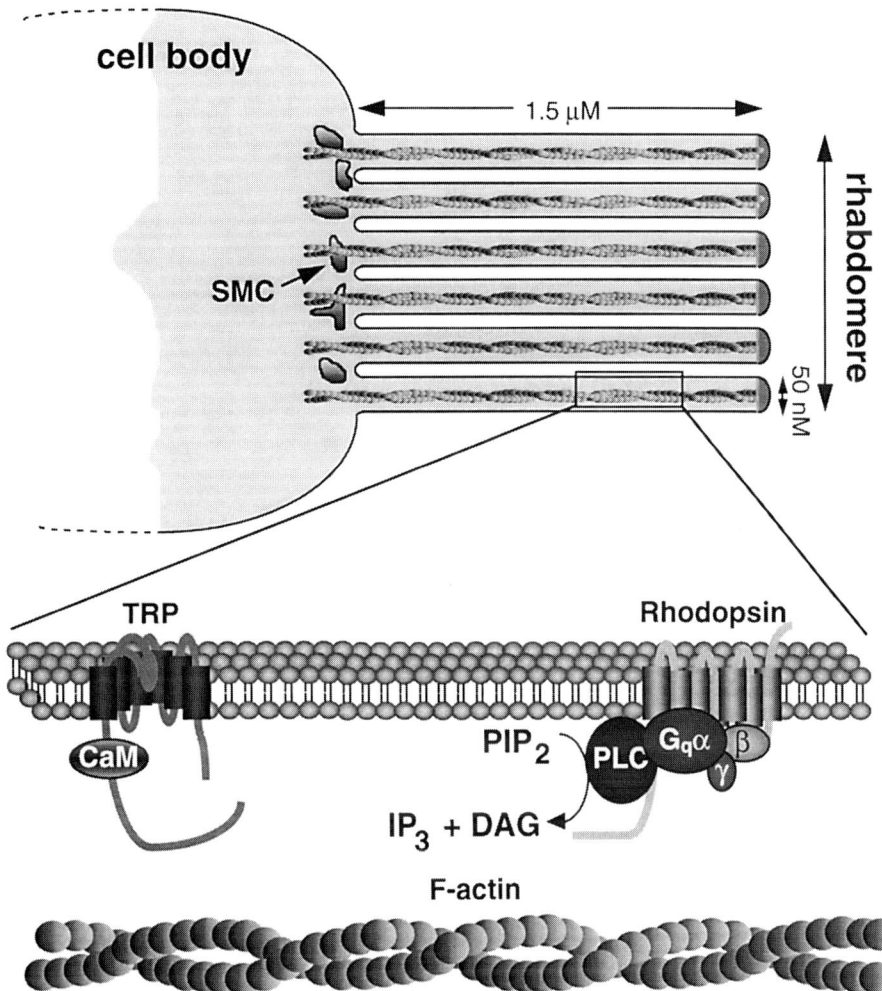

Fig. 6. The putative Ca^{2+} stores are spatially separated from most of the components critical in phototransduction. Shown is a cartoon of a cross-section through a single Drosophila photoreceptor cell. Photoreceptor cells contain a microvillar structure, referred to as a rhabdomere each of which (~50 nm in diameter and ~1.5 µm long) includes an actin core and the signaling proteins that function in phototransduction (some of which are illustrated below at higher magnification). Only six microvilli are included in the cartoon for clarity; however, a photoreceptor cell typically contains ~30 microvilli in a cross-section. Thus the width of a rhabdomeral cross-section at the distal end is similar to the length (~1.5 µm). The depth of the major photoreceptor cells (R1–6) is ~80 µm; therefore, there are ~50,000 microvilli/rhabdomere. Just below the base of the microvilli is the smooth endoplasmic reticulum, a presumed Ca^{2+} cisternae called the submicrovillar cisternae (SMC). (Reproduced with permission from Current Biology Ltd.; Montell, 1998.)

signaling proteins that mediate phototransduction. Experimental evidence suggesting that influx channels may not activated through a store-operated mechanism is that mutation of the *Drosophila* IP_3 receptor has no impact on the photoresponse. Furthermore, the light response cannot be mimicked by introduction of IP_3 into photoreceptor cells. Because activation of the light-dependent cation influx is PLC dependent, it is possible that DAG or a DAG metabolite is required, perhaps in combination with IP_3, for channel activation. Consistent with latter proposal is the observation that TRP and TRPL can be activated by polyunsaturated fatty acid metabolites of DAG. However, it is not known whether the effects of the polyunsuraged fatty acids are direct or indirect.

9. Ca^{2+} IS REQUIRED FOR TERMINATION OF PHOTOTRANSDUCTION

Deactivation of the phototransduction cascade, which is defined as the process of terminating the visual cascade following cessation of the light stimulus, is very rapid and occurs within ~100 ms. A critical messenger of deactivation is Ca^{2+} and it appears to function at multiple steps in the phototransduction cascade.

One of the mechanisms through which Ca^{2+} mediates rapid deactivation is through phosphorylation by the Ca^{2+}-dependent protein kinase, PKC. There are at least two distinct PKC isoforms expressed in the adult eye. Mutations that eliminate a retinal specific PKC, ePKC, result in a defect in deactivation as well as decreased adaptation. Light adaptation, which is also a Ca^{2+}-regulated process, allows photoreceptor cells to adjust their sensitivity to light over a wide range of light intensities.

Another important mediator of Ca^{2+}-dependent termination of the photoresponse is calmodulin. Calmodulin is a dumbbell-shaped protein (M_r ~17 kDa) that contains four Ca^{2+} binding sites. Binding of Ca^{2+} to calmodulin causes major conformational changes which in turn can affect its association and/or activity of associated calmodulin binding proteins (*see* chapter 10). In *Drosophila* photoreceptor cells, calmodulin is highly concentrated in the rhabdomeres where it is present at levels of ~0.5 mM. A severe reduction in the concentration of calmodulin results in a pronounced defect in deactivation.

The major calmodulin binding protein in the rhabdomeres is a protein, NINAC p174, which consists of an N-terminal serine/threonine kinase domain fused to a region homologous to the myosin heavy chain head (Fig. 7A). p174 contains two calmodulin binding sites, both of which are situated near the myosin head–tail junction. Mutations that disrupt the calmodulin binding sites or eliminate p174 cause mislocalization of calmodulin so that a majority is detected in the cell bodies (Fig. 7B). Furthermore, the NINAC–calmodulin interaction is required for normal termination of the photoresponse. The substrate(s) for the NINAC protein kinase and the molecular mechanism by which it functions in response termination remain to be identified.

Deactivation of rhodopsin may be a Ca^{2+} dependent phenomenon, as Ca^{2+} enhances the dephosphorylation of rhodopsin by RDGC. As described previously, mutations that eliminate RDGC result in hyperphosphorylation of rhodopsin and a defect in termination of phototransduction. The activity of arrestin may also be controlled by Ca^{2+}, as the major arrestin expressed in photoreceptor cells, ARR2, undergoes very rapid light-dependent phosphorylation by a Ca^{2+}–calmodulin-dependent protein kinase. Furthermore, ARR2 is not detectably phosphorylated in flies expressing a reduced concentration of calmodulin. However, the physiological significance of this phos-

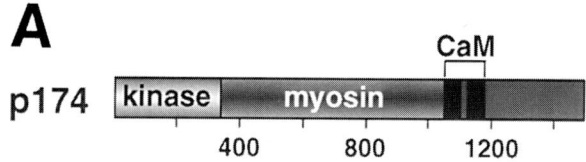

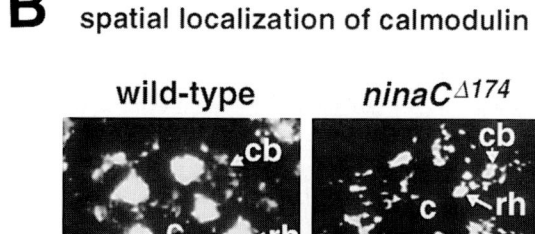

Fig. 7. Spatial distribution of calmodulin in the rhabdomeres is dependent on NINAC p174. (**A**) Structure of NINAC p174. A scale bar in amino acids is shown. CaM, two calmodulin binding sites; kinase, serine/threonine protein kinase domain; myosin, region similar to the head region of myosin heavy chains. (**B**) Spatial distribution of calmodulin in wild-type and transgenic flies that do not express NINAC p174 (*ninaC$^{\Delta 174}$*). Calmodulin was detected in cross-sections of adult ommatidia by indirect immunofluorescence using anti-calmodulin antibodies. c, central matrix; cb, cell bodies; rh, rhabdomeres.

phorylation has not been determined. The PLC encoded by the *norpA* locus appears to undergo negative feedback regulation by Ca^{2+}, as hydrolysis of PIP_2 is maximal at low Ca^{2+} concentrations (0.1 µM) and decreases in the presence of higher levels of Ca^{2+}.

The TRP and TRPL cation influx channels are also inactivated by Ca^{2+}. The TRPL ion channel is inactivated through a mechanism that involves calmodulin because the channel binds calmodulin and mutation of either of the two calmodulin binding sites causes a defect in inactivation. Inactivation of TRP may also result via Ca^{2+}–calmodulin, as TRP is a calmodulin binding protein. As of yet, the in vivo consequences of disrupting the TRP–calmodulin interaction have not been addressed. Thus, influx of Ca^{2+} through the light-dependent ion channels appears to modulate many proteins including the ion channels themselves.

10. FORMATION OF A SIGNALPLEX THROUGH THE PDZ-CONTAINING PROTEIN INAD

The rise in free intracellular Ca^{2+}, which occurs upon opening of TRP and TRPL, is highly localized

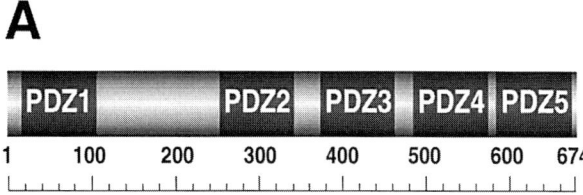

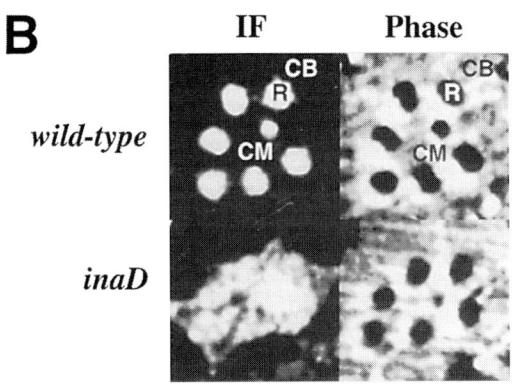

Fig. 8. TRP is mislocalized in *inaD* mutant photoreceptor cells. (A) PDZ domains in INAD. The positions of the five PDZ domains in INAD are indicated. A scale bar, in amino acids, is shown *below*. (B) Spatial localization of TRP in cross-sections of wild-type and *inaD* photoreceptor cells. Indirect immunofluorescent localization was performed using anti-TRP antibodies (IF). The phase contrast images are shown to the *right* (Phase). CB, cell bodies; CM, central matrix; R, rhabdomeres. (Adapted with permission by Cell Press from Chevesich et al., 1997.)

and may be restricted to a zone of only a few nanometers surrounding the channel pore. The limitation of the Ca^{2+} rise to microdomains presents a vexing question concerning the mechanism by which rhodopsin, PLC, PKC, and other signaling proteins are capable of sensing the increase in free Ca^{2+} resulting from the light-dependent Ca^{2+} influx. Such a problem would potentially occur if these signaling proteins were spatially separated from the ion channels by more than a few nanometers. An apparent resolution to this conundrum is that many of the proteins critical in *Drosophila* phototransduction are associated in a supramolecular signaling complex (signalplex).

The central player in the signalplex is INAD, a protein with five protein interaction domains referred to as PDZ domains (Fig. 8A). PDZ domains, which are named for the first three proteins recognized to contain these modules (*P*SD-95, *D*iscs large, *Z*O-1), are approx ~90–100 amino acids in length and are therefore similar in size to several other protein interaction motifs such as the Src homology 2 (SH2) and (SH3) domains. The structures of several PDZ mod-

ules have been determined in proteins distantly related to INAD and shown to consist of five-stranded antiparallel β-barrels flanked by three α-helices. The C-terminal three residues of PDZ binding proteins, which often conform to the consensus, Ser/Thr-X-Val (X refers to any amino acid) or consist of hydrophobic or aromatic side chains, insert into and interact with residues in a hydrophobic pocket of the PDZ module. Current evidence indicates that only a single target protein can interact at any given time with the hydrophobic pocket.

A minimum of seven proteins bind to INAD: Rh1, PLC, PKC, TRP, TRPL, NINAC, and calmodulin. Several of the INAD targets appear to depend on INAD for localization and/or retention in the rhabdomeres. While each of these INAD targets is highly enriched in the rhabdomeres of wild-type flies, mutations in INAD result in dramatic mislocalization of PKC, PLC, and TRP (Fig. 8B). Certain PDZ proteins expressed in vertebrates, *C. elegans,* and *Drosophila* also appear to function in spatial localization of interacting proteins. However, INAD is not required for rhabdomere localization of all of its targets, as the spatial distributions of NINAC, rhodopsin, and TRPL are indistinguishable between wild-type and *inaD* mutant flies.

With the exception of calmodulin, which binds to the linker region between PDZ1 and PDZ2, all of the INAD binding proteins associate with PDZ domains. Several of the targets interact with more than one domain and some PDZ domains associate with more than one protein. In contrast to most characterized PDZ–target protein interactions, some of the INAD binding proteins associate with INAD via internal sequences. In view of the large array of INAD target proteins and the multivalent nature of certain INAD–target proteins interactions, it would seem that INAD lacks the capacity to simultaneously nucleate seven or more interacting proteins. However, all of the interacting proteins do appear to be linked together in a large complex rather than to separate INAD monomers (Fig. 9).

The binding capacity of INAD may be increased by formation of INAD multimers because INAD is capable of forming homomultimers in vitro. The multimerization occurs through two PDZ domains, PDZ3 and PDZ4, resulting in the formation of large homomultimers. Although some of the target proteins also bind to PDZ3 and PDZ4, the homomultimerization and target binding are not mutually exclusive and can occur simultaneously. Thus, homomultimerization may increase the potential of INAD to coordinate

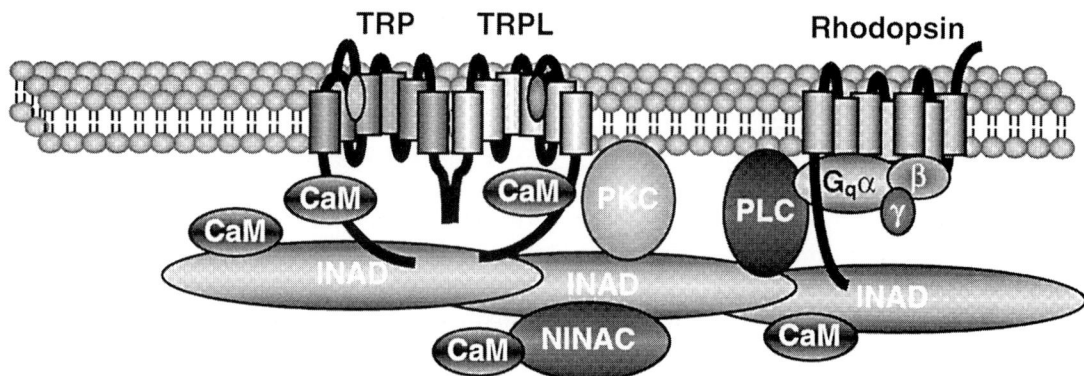

Fig. 9. Organization of the phototransduction cascade into a signalplex. Most components of the *Drosophila* visual signaling pathway are localized to the rhabdomeres. Within the rhabdomere, many components of the visual cascade bind directly to INAD. INAD is also capable of homomultimerization in vitro. CaM, calmodulin; PLC, phospholipase C; PKC, protein kinase C.

an array of signaling proteins. The observation that PDZ–PDZ and PDZ–target interactions are not mutually exclusive suggests that the two types of interactions occur through different interfaces on the PDZ module.

The organization of many of the signaling proteins required in *Drosophila* phototransduction into a supramolecular signaling complex differs from a formerly held concept that G protein-coupled signaling cascades consist primarily of free-floating components which associate with receptors as a consequence of random collisions. There is a growing body of evidence that many G protein-coupled cascades in vertebrate cells are also spatially organized. The critical question concerns the function(s) of the INAD signalplex during the photoresponse.

11. POTENTIAL FUNCTIONS OF THE INAD SIGNALPLEX

One potential function of the signalplex may be to facilitate rapid deactivation of the photoresponse mediated by the light-dependent rise in Ca^{2+}. As the Ca^{2+} influx is confined to microdomains, rhodopsin, PLC, and other proteins that are negatively regulated by Ca^{2+} may require close physical proximity to TRP and TRPL in order for these molecules to be subjected to the rise in Ca^{2+} levels. The signalplex would provide a potential mechanism by which multiple signaling proteins would be in close apposition to the mouth of the light-sensitive cation influx channels. Consistent with a proposed role of the signalplex in termination of the photoresponse, mutations in the INAD binding sites of several targets result in defects in termination. In some cases, the decreased rate of termination may be a secondary consequence of pronounced mislocalization of the target proteins rather than due to a requirement for binding INAD. However, mutation of the INAD binding site in the C-terminus of NINAC also results in a strong defect in inactivation despite any discernible effect of the mutation on the spatial distribution of NINAC. Thus, it appears that the signalplex does function in promoting rapid termination of the photoresponse.

A second potential function of the signalplex, which is not mutually exclusive with the first, is a role in activation of the phototransduction cascade. Opening of the TRP and TRPL channels occurs within milliseconds of light stimulation. Since light-dependent cation channels are spatially separated from the intracellular Ca^{2+} stores, gating of TRP and TRPL strictly through a store-operated mechanism may be too slow to achieve the rapid activation kinetics. Instead, the millisecond activation of TRP and TRP–TRPL may be mediated by production of a second messenger in the complex or through a direct, allosteric mechanism. The amplitude of the photoresponse is dramatically reduced in the null *inaD* allele, indicating that INAD does function in activation.

A third potential function of the signalplex may be in amplification of the phototransduction cascade. Both vertebrate and invertebrate vision are characterized by enormous signal amplification. A single photon of light is capable of opening hundreds of ion channels that give rise to the elementary unit of the light response referred to as a quantum bump. In contrast to vertebrate phototransduction, the amplification in *Drosophila* phototransduction does not appear to involve the G protein that binds to rhodopsin or the effector for the G protein (PDE or PLC in vertebrates and *Drosophila,* respectively). Instead, amplification appears to occur subsequent to activation of the PLC. The signalplex may function in amplification of the cascade, as the amplitude of the quantum bumps is reduced in *inaD* mutant flies.

12. PARALLELS BETWEEN *DROSOPHILA* PHOTOTRANSDUCTION AND MAMMALIAN SIGNALING

Mutation of most genes required for *Drosophila* phototransduction results in progressive atrophy of the photoreceptor cells. This phenomenon referred to as retinal degeneration is often, but not always, light dependent (*see* Table 1). In humans, more than 100 diseases have been identified that include degeneration of the photoreceptors. Several of these retinal diseases, which are collectively referred to as retinitis pigmentosa (RP), have a common etiology with *Drosophila*. Numerous alleles of the human rhodopsin gene have been identified as the cause of RP. However, the initial discovery that perturbation of rhodopsin leads to retinal degeneration was made in *Drosophila*. Other parallels between *Drosophila* retinal degeneration and RP include the findings that mutations in either the *Drosophila* or human light responsive ion channels, TRP and the α-subunit of the cGMP-gated channel respectively, result in degeneration of the photoreceptor cells.

The mechanisms underlying *Drosophila* phototransduction also have similarities with signaling cascades in mammalian systems. Inositol phospholipid signaling is widespread in mammalian cells and there are numerous examples in which stimulation of PLC is coupled to activation of cation influx. Both TRP and TRP–TRPL are activated in vitro through a store-operated mechanism, and store-operated Ca^{2+} entry has been detected in virtually every mammalian cell type specifically examined for this mode of Ca^{2+} entry. Moreover, store-operated Ca^{2+} entry has been implicated in a variety of processes ranging from mitogenesis in fibroblasts, osteoclast function, to T-cell activation. This latter possibility is strengthened by the correlation between a form of a severe immunodeficiency and a defect in store-operated Ca^{2+} entry in T cells. The vertebrate channels responsible for store-operated Ca^{2+} entry may be related to TRP, as numerous TRP homologs (TRPCs) have been identified in vertebrates and at least some of these channels appear to be activated through a store-operated mechanism.

There is a growing body of evidence that many G protein-coupled cascades in vertebrate cells are spatially organized. As appears to be the case for the INAD signalplex, compartmentalization of vertebrate cascades may also function in feedback regulation and in increasing the speed of activation. However, an additional function may be to achieve specificity by preventing unintended crosstalk between signaling cascades. This may be necessary, as many receptors that bind distinct agonists activate many of the same effector proteins.

SELECTED READINGS

Bloomquist BT, Shortridge RD, Schneuwly S, Perdew M, Montell C, Steller H, Rubin G, Pak WL. Isolation of a putative phospholipase C gene of *Drosophila*, *norpA*, and its role in phototransduction. Cell 1988; 54:723.

Hardie RC. Photolysis of caged Ca^{2+} facilitates and inactivates but does not directly excite light-sensitive channels in *Drosophila* photoreceptors. J Neurosci 1995; 15:889.

Hardie RC, Minke B. The *trp* gene is essential for a light-activated Ca^{2+} channel in *Drosophila* photoreceptors. Neuron 1992; 8:643.

Minke B, Selinger Z. The roles of *trp* and calcium in regulating photoreceptor function in *Drosophila*. Curr Opin Neurobiol 1996; 6:459.

Montell C. TRP trapped in fly signaling web. Curr Opin Neurobiol 1998; 8:389.

Pak WL. *Drosophila* in vision research. Invest Opthalmol Vis Sci 1995; 36:2340.

Scott K, Zuker C. Lights out: deactivation of the phototransduction cascade. Trends Biochem Sci 1997; 22:350.

Scott K, Sun Y, Beckingham K, Zuker CS. Calmodulin regulation of *Drosophila* light-activated channels and receptor function mediates termination of the light response in vivo. Cell 1997; 91:375.

Xu X-ZS, Li H-S, Guggino WB, Montell C. Coassembly of TRP and TRPL produces a distinct store-operated conductance. Cell 1997; 89:1155.

6

Dictyostelium
A Model Experimental System for Elucidating the Pathways and Mechanisms Controlling Chemotaxis

Chang Y. Chung and Richard A. Firtel

CONTENTS

INTRODUCTION
SIGNALING COMPONENTS IN THE REGULATION OF CHEMOTAXIS
REGULATION OF CYTOSKELETAL COMPONENTS IN CHEMOTAXIS
CONCLUDING REMARKS
SELECTED READINGS

1. INTRODUCTION

Chemotaxis, directed movement toward a chemoattractant agent, is involved in diverse biological responses, including wound healing in vertebrates, migration of tumor cells, axonal outgrowth to target cells, and aggregation leading to the formation of the multicellular organism in *Dictyostelium*. Chemotaxis is essential for the migration of polymorphonuclear leukocytes and macrophages to an inflammatory site. Amoeboid chemotaxis is also a key step in tumor metastasis, with morphological studies demonstrating that tumor cells exhibit leukocyte-like motility and psuedopod extension. The movement of cancer cells into and out of vascular channels has been shown to be stimulated and regulated by a number of cytokines, suggesting that chemotaxis is a key step in tumor metastasis. Axonal guidance in the developing nervous system may very well function via pathways that are analogous to chemotaxis of amoeboid cells. Higher concentrations of chemoattractant near target destinations of migratory neuroblasts appears to be involved in the regulation of the migration of neuroblasts while diffusible repulsive agents from a tissue may exclude axonal growth and help regulate the direction of the leading edge.

Chemotaxis is a fundamental process of cells and tissues and is based on the dynamics of the cytoskeleton. This process is activated by a variety of extracellular ligands. The binding of ligands to cell surface receptors leads to the directed reorganization of the actin and myosin cytoskeletons, pseudopod extension in the direction of the chemoattractant source, and cell movement via pathways that are thought to be highly conserved between mammals and *Dictyostelium*. *Dictyostelium* cells provide a powerful system to examine the role of cellular components to control coordinated cell movement because of the ability to apply genetic as well as cell biological approaches to study this evolutionarily conserved pathway. *Dictyostelium* cells, like neutrophils and macrophages, are motile and, in contrast to fibroblasts, lack stress fibers. In polymorphonuclear leukocytes, macrophages, and *Dictyostelium* cells, chemotaxis can be mediated through G protein-coupled cell surface receptors, and, in mammalian cells, the response is thought to function through G_i via release of $G\beta\gamma$-subunits. Conserved signaling components involved in the regulation of chemotaxis in *Dictyostelium* and neutrophils are shown in Fig. 1.

From: *Principles of Molecular Regulation* (P. M. Conn and A. R. Means, eds.), © Humana Press Inc., Totowa, NJ.

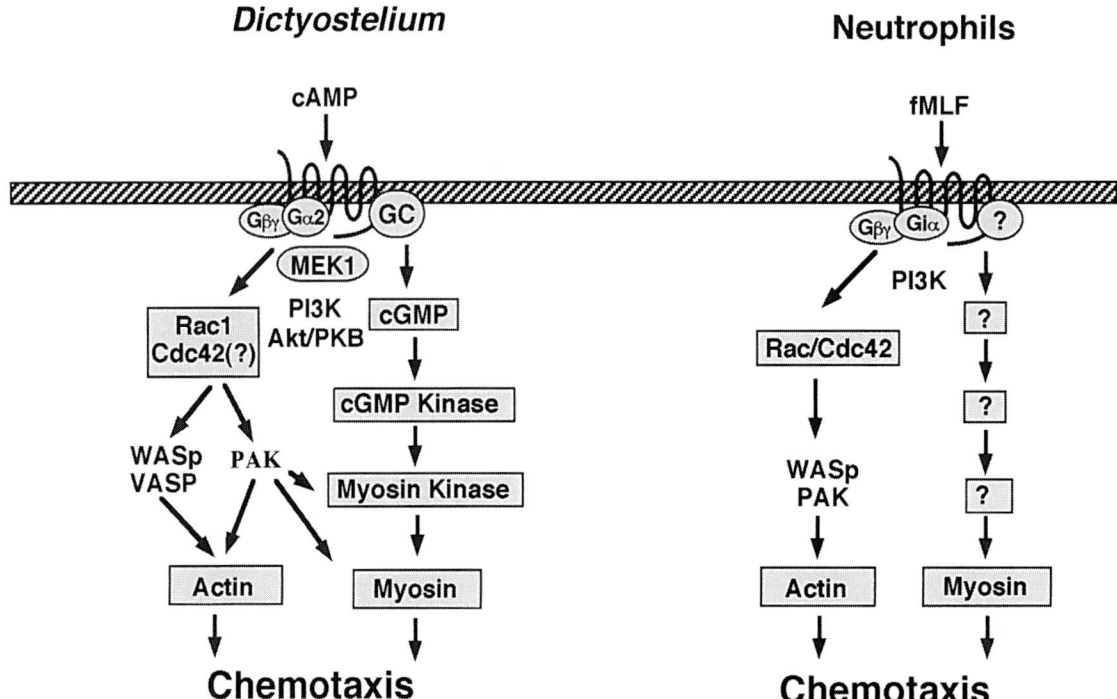

Fig. 1. Schematic diagram of signaling components involved in the regulation of chemotaxis in *Dictyostelium* and neutrophils. In both systems, chemotaxis can be mediated through G protein-coupled seven-transmembrane cell surface receptors. The binding of chemoattractants to the receptor activates a series of signaling events leading to the reorganization of the cytoskeleton. Directed reorganization of the actin and myosin cytoskeletons, pseudopod extension in the direction of the chemoattractant source, and cell movement via pathways are thought to be highly conserved between mammals and *Dictyostelium*.

Dictyostelium cells undergo a distinct developmental program in which up to 10^5 cells aggregate to form a multicellular organism. This is regulated by the chemoattractant cAMP, which is relayed from the center of an aggregation domain outward. Cells respond to the stimulus by chemotaxing up the cAMP gradient toward the cAMP source. This movement involves the activation of adenylyl cyclase and producing and secreting cAMP into the extracellular space, thus relaying the signal. Rises in intracellular cAMP, at the same time, activate PKA (cAMP-dependent protein kinase), whose function is also required for proper cell movement. Adaptation pathways make these responses transient so that a cell is responsive to the signal for only a short time. Cells closer to the aggregation center that have just responded to an outward moving wave of cAMP adapt and are not able to respond when cells produce and secrete cAMP. As a result, cells move inward while the cAMP wave moves outward, leading to the directionality of the cell movement. An extracellular phosphodiesterase hydrolyzes the cAMP, allowing cells to deadapt and become responsive to a new outward moving cAMP wave. At the height of aggregation, new cAMP waves are produced every 6 min. Approximately 20 cycles are required to produce a mound.

After mound formation, the multicellular organism undergoes morphogenesis and cell-type differentiation, which is also controlled by extracellular cAMP functioning through the same receptor, and can mediate both aggregation-stage pathways and those required for multicellular differentiation. Thus, for *Dictyostelium*, chemotaxis is an essential component of the developmental pathway. It provides an excellent biological system for identifying genes required for chemotaxis, as such genes would be defective in the ability to form aggregates and thus are easily identified in mutant screens. Most of the analysis of gene function described in the following sections results from the physiological and biochemical analysis of null mutants created by homologous recombination. Through gene discovery and the analysis of gene function in *Dictyostelium*, significant progress has been achieved that is directly applicable to understanding these processes in humans. These findings provide an appropriate basis for understanding disease states.

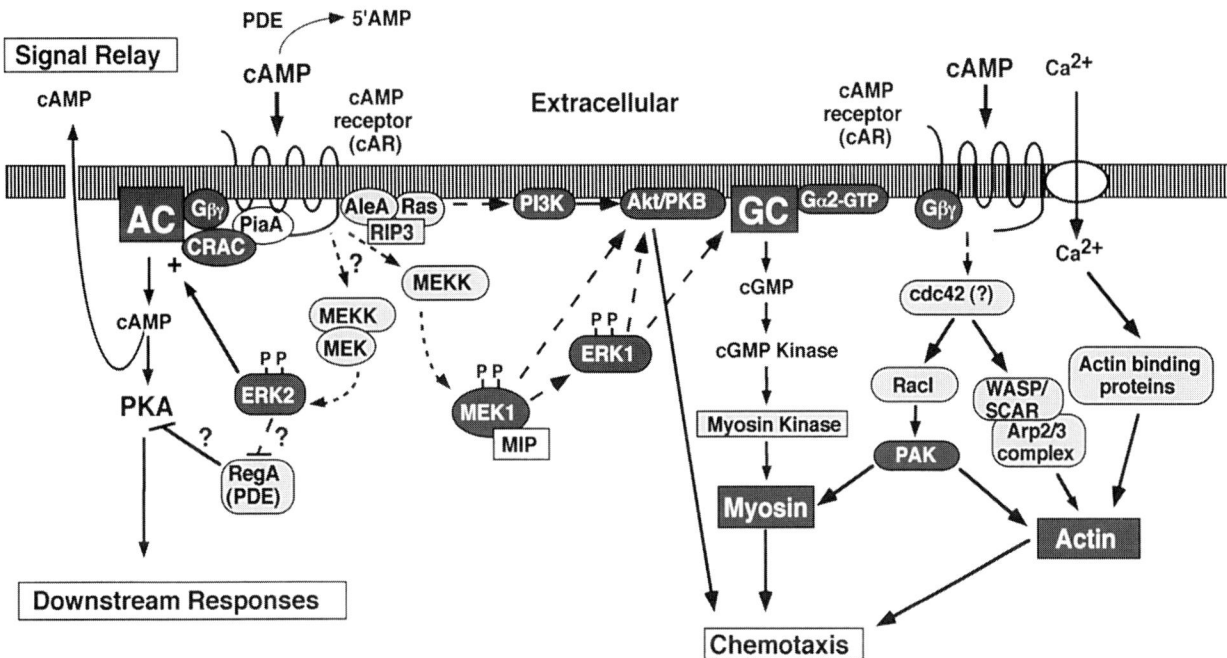

Fig. 2. Scheme of signal transduction involved in *Dictyostelium* chemotaxis. Binding of cAMP to the cell surface receptor induces (1) activation of adenylyl cyclase (ACA) resulting in the formation of cAMP for relay of the signal. cAMP activates cAMP-dependent kinase (PKA) leading to cellular responses such as chemotaxis and developmental gene expression. The activation of ACA requires Gβγ, CRAC, Pianissimo (PiaA), and ERK2 activity. (2) Activation of guanylyl cyclase (GC) leading to the formation of cGMP and the activation of cGMP-dependent kinase. Myosin II light and heavy chain kinases are regulated by cGMP, which regulates cytoskeletal organization during chemotaxis and morphogenesis. (3) Activation of Ras family proteins. Ras appears to activate PI3 kinases resulting in the activation of Akt/PKB which is required for the regulation of cytoskeleton and polarity during chemotaxis. Activation of Cdc42/Rac leads to the change of cytoskeletal organization via WASP regulating F-actin organization and PAK controlling myosin II filaments assembly. (4) Uptake of Ca^{2+} across the plasma membrane. Activities of many actin crosslinking and binding proteins are regulated by the Ca^{2+} concentration.

2. SIGNALING COMPONENTS IN THE REGULATION OF CHEMOTAXIS

cAMP functions through the interactions of serpentine, G protein-coupled receptors designated cARs. cAMP binding results in the activation of a series of signaling pathways, including: activation of adenylyl cyclase, which produces the cAMP signal; guanylyl cyclase, which mediates chemotaxis; Akt/PKB; phospholipase C; Ca^{2+} influx; and gene expression. Regulation of signaling components and interactions between them are illustrated in Fig. 2. Transduction of the chemotactic signal to second messengers induces rearrangement of cytoskeletal components, including the polymerization of actin and myosin II and the relocalization of myosins. The first step in the directed movement of cells toward a chemotactic source involves the extension of pseudopodia initiated by the focal nucleation and polymerization of actin at the leading edge of the cell, which includes many actin-binding proteins. Myosin II is also assembled into filaments and relocalizes to the posterior cortex. In this chapter, components involved in the transduction of chemotactic signals and their regulation are discussed. Changes and regulation of cytoskeletal components during chemotaxis are also reviewed.

2.1. Chemoattractant Receptors

Dictyostelium and neutrophils control chemotaxis through the activation of seven-transmembrane-domain G protein-coupled cAMP receptors. During the aggregation stage, chemotaxis that mediates the formation of the multicellular organism is controlled through cAMP receptors (cARs). Four cARs have been identified whose expression levels are tightly regulated throughout development. These receptors are approx 40% identical and 60% similar in sequence and share a great deal of homology within transmembrane and loop domains with metazoan G protein-coupled receptors. In addition to their expression pattern, the individual receptors also differ in their binding constant for cAMP. cAR1, which is preferentially expressed during aggregation and is required *in vivo* for aggregation-stage pathways, has the highest affinity and functions as the major receptor during aggregation and for the initial responses in the mound. *car1* null strains obtained by disrupting the cAR1 gene by homologous recombination do not aggregate normally but do show some response to cAMP due to the low level of expression of cAR3 at this stage. cAR3 is then maximally expressed during tight aggregate formation. cAR1 and cAR3 thus have partially overlapping functions. cAR2 and cAR4 are expressed in the prestalk cells in the multicellular stage of the development program and appear to regulate pattern formation and cellular differentiation by controlling the expression of prestalk genes.

2.2. Heterotrimeric G Proteins

Early studies showing that the affinity of the cAMP receptor was altered by guanine nucleotides and that cAMP increased GTPase activity strongly suggested the presence of heterotrimeric G proteins. Gα2 is a key player, as a null mutant of Gα2 does not aggregate. The *gα2* null mutant is unable to activate adenylyl or guanylyl cyclases and Akt/PKB, shows no chemotaxis, and does not respond to extracellular cAMP. Moreover, biochemical analyses strongly suggest that Gα2 may be the only Gα-subunit that interacts with any of the cAMP receptors. Seven other Gα-subunits have been identified and cloned, and deletion of most of these Gα-subunits results in subtle phenotypes. However, *gα4* null cells do not chemotax toward folic acid, the chemoattractant used by *Dictyostelium* cells during growth and early development to identify and locate food (bacteria). *gα4* null cells also exhibit developmental phenotypes and the gene is expressed from two promoters: one expressed during growth and one expressed during the mound stage. A single Gβ-subunit has been identified and the deletion of gene encoding Gβ results in the inability of cells to chemotax to any chemoattractant and to activate adenylyl and guanylyl cyclases and Akt/PKB. It is expected that the deletion of the Gβ-subunit would abrogate all signal transduction via heterotrimeric G proteins.

2.3. Activation and Regulation of Adenylyl Cyclase

The aggregation-stage *Dictyostelium* adenylyl cyclase (ACA) has a structure similar to that of mammalian adenylyl cyclases. cAMP levels are controlled by the activation of ACA in response to receptor stimulation and its subsequent degradation extracellularly by a membrane-associated and -secreted cyclic nucleotide phosphodiesterase (PDE) and intracellularly by a cAMP-dependent PDE, RegA, which is a two-component system (histidine kinase) response regulator. The activation of ACA by extracellular cAMP is mediated through cAR1 acting through the G protein containing the Gα2-subunit. Cells lacking cAR1, Gβ, or Gα2 cannot activate ACA in vivo but ACA can be activated in vitro by GTPγS in *gα2* null but not *gβ* null cells, suggesting that the pathway is mediated through G$\beta\gamma$-subunits. In *gα2* null cells, G$\beta\gamma$-subunits are presumably released from other heterotrimeric G proteins. ACA activation and cAMP accumulation also require the function of the mitogen-activated protein (MAP) kinase ERK2, as cAMP accumulation is absent in *erk2* null cells.

Two cytosolic proteins have been identified that are required for receptor activation of adenylyl cyclase, CRAC and pianissimo (PiaA). CRAC is a PH domain-containing protein that is proposed to interact with the Gβ-subunit and has been shown to translocate from cytosol to the plasma membrane in response to the chemoattractant. CRAC is required for the activation of adenylyl cyclase by Gβ, but its function in this process is not known. It is probable that the CRAC PH domain may interact with PtdIns(3,4,5)P$_3$ and/or PtdIns(3,4)P$_2$ produced in the expected activation of phosphatidylinositol 3-kinase (PI3 kinase) in response to chemoattractants, and the PI3 kinases PI3K1 and PI3K2 are required for receptor activation of *Dictyostelium* Akt/PKB. The transient, receptor-mediated translocation of the PH domain-containing protein CRAC to binding sites on the inner face of the plasma membrane has been demonstrated by using the GFP–CRAC fusion protein, which is thought to reflect a localized activation of the G protein-linked signaling system. This localized and transient activation of ACA, controlled by the transient recruitment of CRAC, may be a paradigm for general mechanisms of gradient sensing by G protein-linked chemotactic systems. Pianissimo (PiaA) is a 130-kDa cytosolic protein that is also required for the chemoattractant receptor and G protein-mediated activation of the adenylyl cyclase. Both PiaA and CRAC are essential for activation of the adenylyl cyclase, as lysates of *crac⁻ piaA⁻* double mutants require both proteins for reconstitution.

2.4. cGMP and Guanylyl Cyclase

Stimulation of cells with extracellular cAMP leads to a transient 10-fold increase of the cGMP level within 10 s after stimulus and returns to a basal level within 30 s. The role of cGMP in *Dictyostelium* chemotaxis was proposed from the phenotype of Streamer F (*stmF*) mutant, which is defective in aggregation. Streamer F mutant strains are defective in cGMP-phosphodiesterase activity and their chemotaxis is not as efficient as that of wild-type cells. They exhibit a prolonged increase in intracellular cGMP in response to stimulation with chemoattractants and the rate of chemotactic movement of *stmF* mutant cells is significantly reduced. This decrease in speed correlates with the prolonged duration of high intracellular cGMP concentration, suggesting that cGMP concentration must return to the basal level for efficient cell movement. During aggregation, guanylyl cyclase is regulated through cAR1, based on the observation that the cGMP response is greatly reduced in *car1* null mutant cells and absent in *car1/car3* double null mutant cells. cGMP is completely absent in *gβ* null cells, indicating the response functions through heterotrimeric G proteins. Gα2 and Gα4 mediate the cAMP- and folic acid-mediated activation of guanylyl cyclase, respectively, and disruption of either gene results in a loss of chemotaxis to the respective ligand. Guanylyl cyclase activity is strongly inhibited by Ca^{2+}, suggesting that the cGMP response can be controlled by Ca^{2+}-mediated inhibition of guanylyl cyclase. Analyses of the regulation of guanylyl cyclase in chemotactic mutants showed that a cGMP-binding protein mediates both stimulation and ATP-dependent inhibition of guanylyl cyclase. The positive and negative regulation of guanylyl cyclase by its product cGMP may explain how cells process the temporal and spatial information of chemotactic signals, which is necessary for sensing the direction of the chemoattractant.

2.5. Regulation by Ras

There are at least five *Dictyostelium Ras* genes (*B, C, D, G,* and *S*) in *Dictyostelium*, with *RasG* and *RasD* being closely related to mammalian Ras proteins. *RasC* and *RasS* presumably function in aggregation, as these genes are developmentally regulated and maximum levels of their transcripts are detected during aggregation, suggesting that the encoded proteins have distinct functions during aggregation. Cells overexpressing an activated *RasG* are unable to activate adenylyl cyclase, causing an inability to aggregate. *RasD* and *RasG* appear to have several regulatory functions during aggregation. Recent studies indicate that *Ras* is a negative regulator of cAMP receptor-mediated activation of the *Dictyostelium* MAP kinase ERK2. ERK2 activation is delayed and the level of activation is significantly reduced in cells overexpressing activated *RasD^{Q61L}*, whereas the activation is enhanced in cells overexpressing dominant negative *RasD^{S17N}* or *Ras^{D57Y}*. Activation of *Ras* is accomplished by guanine nucleotide exchange factors (*Ras*GEFs) which facilitate replacement of GDP bound to *Ras* with GTP. A *Dictyostelium* putative *Ras*GEF, Aimless (*AleA*), was identified and *aimless* null mutants were severely impaired in chemotaxis and activation of adenylyl cyclase. The motility of mutant cells appears normal, suggesting a defect in gradient sensing. Thus *aimless* null cells are defective in both the signal relay (activation of ACA) and chemotaxis pathways. ERK2 activation is enhanced in *aimless* null cells, which is consistent with enhanced ERK2 activity in cells expressing dominant negative *RasD*. Thus, *Ras* is involved in transduction signals from the cAMP receptor(s) to signaling pathways including activation of adenylyl cyclase and ERK2.

Another gene that is thought to function to control *Ras* regulation of aggregation is *RIP3* (*Ras*-interacting protein 3), which was identified in a two-hybrid screen. *RIP3* specifically interacts with *RasG* and *rip3* null cells show phenotypes that are indistinguishable from those of *aimless* null cells. These data strongly suggest a role for *Ras* pathways in the integration of the signal relay and chemotaxis pathways.

2.6. Role of Phosphatidylinositide 3-Kinases (PI3-kinases) and Protein Kinase B (Akt/PKB)

The *Dictyostelium* protein Akt/PKB (encoded by *pkbA*) is a structural homolog of mammalian Akt/PKB, having a conserved structure that includes an N-terminal PH domain, a kinase domain, and a C-terminal tail. During aggregation, *Dictyostelium* Akt/PKB is rapidly and transiently activated in response to chemoattractant signaling by cAMP with kinetics similar to receptor activation of guanylyl cyclase. Sites of phosphorylation that lead to Akt/PKB activation by upstream kinases in mammalian cells are conserved, and mutant analysis suggests that phosphorylation at these sites is required for activation. Akt/PKB activation is receptor and heterotrimeric G protein dependent, as determined by the analysis of gene knockout strains. Pertinent to the subject matter of this chapter, *pkbA* null cells exhibit defects in chemotaxis. The cells are unable to properly polarize when placed in a chemotactic gradient and move slowly. More detailed analysis reveals that the cells produce multiple lateral pseudopodia as well as pseudopodia at the leading edge. Analysis of mutants suggests that the cells "tumble" rather than move smoothly in the direction of the cAMP source. This leads to a defective chemotaxis and the inability to form a multicellular aggregate when cells do not have direct cell–cell contacts.

The PH domain-containing protein CRAC, which is required for receptor activation of the aggregation-stage adenylyl cyclase ACA, translocates to the plasma membrane in response to cAMP signaling. In polarized chemotaxing cells, the localization is to the leading edge, suggesting that the activation of adenylyl cyclase is localized under these physiological conditions. Interestingly, a gene fusion of the Akt/PH domain with GFP also rapidly translocates to the plasma membrane in response to cAMP signaling and localizes to the leading edge in chemotaxing cells. This suggests a general principle: in chemotaxing cells in which there is an ~10% difference in the chemotactic gradient between the front and back of the cells, receptor-mediated pathways are activated at the front and not laterally around the entire perimeter of cells. This phenomenon indicates there is a spatially localized activation of pathways, possibly the underlying mechanism by which cells produce a pseudopod only in the direction of the chemoattractant source. Many of the pathways required for preferential actin polymerization at the leading edge may result from a specific localized activation of the necessary signaling pathways at this point in the cell rather than randomly around the cell.

PI3-kinases have been implicated in controlling cell proliferation, actin cytoskeleton organization, and the regulation of vesicle trafficking between intracellular organelles. There are three genes in *Dictyostelium discoideum*, *DdPIK1*, *DdPIK2*, and *DdPIK3*, encoding proteins most closely related to the mammalian 110-kDa PI–3 kinase. PI3K1 and PI3K2 are most closely related to p110α, while PI3K3 is a member of the PI3Kγ subfamily. A strain in which the genes encoding both DdPIK1 and DdPIK are disrupted (*pi3k1/pi3k2* null cells) is unable to activate Akt/PKB in response to cAMP stimulation. Moreover, these cells are impaired in pinocytosis and also almost completely devoid of large postlysosomal vacuoles. The mutant cells contain numerous filopodia projecting laterally and vertically from the cell surface, and fluorescent microscopy indicates that these filopodia are enriched in F-actin, which accumulates in a cortical pattern in wild-type cells. Together, these results suggest that *Dictyostelium* DdPIK1 and DdPIK2 gene products regulate multiple steps in the endosomal pathway, and function in the regulation of cell shape and movement perhaps through changes in actin organization. The inability of *pi3k1/pi3k2* null cells to polarize properly might reflect the improper localization of Akt/PKB which is expected to localize to leading edge by the interaction of InsP3 produced by PIKs with the PH domain of PKB.

2.7. Ca^{2+} Influx by Ca^{2+} Uptake and Activation of Phospholipase C

Signal transduction in *Dictyostelium* for oriented movement and differentiation involves a fine tuning of the cytosolic Ca^{2+} concentration. cAMP binding to cAR1 results in a rise in intracellular Ca^{2+} levels within approx 5 s. This occurs directly by stimulating Ca^{2+} influx and indirectly by activating a phospholipase C. Chemoattractant binding to cARs appears to generate intracellular signals that induce activation and adaptation of the Ca^{2+} uptake response. Additional analyses on signaling mutants demonstrates that Ca^{2+} entry is not regulated by heterotrimeric G proteins or by G protein-mediated changes in intracellular cAMP or cGMP.

Chemoattractants stimulate the activation of a phospholipase C activity leading to the transient generation of inositol (3,4,5)-triphosphate (PIP_3) and diacylglycerol. A phosphatidylinositol-specific phospholipase C (Dd-PLC) has been identified and cloned in *Dictyostelium*. The C-terminal part of the protein shows strong homology to the mammalian PLCδ isoform. Expression of PLC in *Dictyostelium* cells results in increased basal levels of PIP_3 and enhanced PLC activity. Surprisingly, developmental gene regulation, cAMP-mediated chemotaxis, and activation of guanylyl and adenylyl cyclases are normal in Dd-PLC null cells. Although the cells lack PLC activity, PIP_3 is present at only slightly lower concentrations than in control cells, suggesting the presence of functional redundancy. PIP_3 induces Ca^{2+} mobilization from an NBD-Cl- and BHQ-sensitive compartment, which comprises the PIP_3 releasable pool. The role of diacylglycerol in chemotactic response and development of *Dictyostelium* is not known.

2.8. Function and Regulation of MAP Kinase Cascades

The MAP kinase cascade is a universal signaling unit used by a wide number of eukaryotic signaling pathways. The archetypal MAP kinase cascade is composed of members of MEK kinase (MAP kinase kinase kinase, MEKK), MEK (MAP kinase kinase), and MAP kinase (or ERK) families, which carry out a sequential phosphorylation reaction thereby serving to relay, amplify, and integrate diverse signals. MAP kinase is a serine/threonine-specific, proline-directed protein kinase. MAP kinase itself is activated by the upstream dual-specificity threonine and tyrosine protein kinase MEK via phosphorylation at tyrosine and threonine residues in the conserved TXY motif near the catalytic cleft of the kinase core. MEK in turn is activated by the further upstream serine/threonine protein kinase MEKK via phosphorylation at two serine or threonine residues in the activation loop.

A MAP kinase kinase (DdMEK1) required for proper aggregation in *Dictyostelium* has been identified. *mek1* null cells are unable to chemotax during aggregation in response to the chemoattractant cAMP. *mek1* null cells show highly impaired cAMP-mediated activation of guanylyl cyclase, suggesting that this may be the cause of the inability of *mek1* null cells to chemotax. The activation of the MAP kinase ERK2, which is essential for chemoattractant activation of adenylyl cyclase, is not affected in *ddmek1* null strains, indicating that DdMEK1 does not regulate ERK2, and at least two independent MAP kinase cascades control aggregation in *Dictyostelium*. Guanylyl cyclase is not constitutively active in *mek1* null cells expressing constitutively active MEK1 and can be activated in response to cAMP signaling. This suggests that MEK1 does not lie on a linear pathway between the receptor, G protein, and guanylyl cyclase, but its function is required for the ability of chemoattractants such as cAMP to activate guanylyl cyclase. MEK1 thus may function as a checkpoint in regulating chemotaxis.

Two MAP kinases, ERK1 and ERK2, have been identified. The function of ERK1 is not clear, but it may belong to a MAP kinase pathway including MEK1. ERK2 is required for aggregation. *erk2* null cells are defective in receptor-mediated cAMP production and thus do not produce sufficient cAMP for relaying the cAMP signal from cell to cell or the activation of PKA activity. ERK2 is rapidly and transiently activated in response to cAMP. This pathway requires cAR1, but neither Gα2 nor Gβ, suggesting that ERK2 is activated by a G protein-independent pathway. G protein-independent pathways also include Ca^{2+} influx and activation of STATa tyrosine phosphorylation and the transcription factor GBF. Thus, the cAR1 chemoattractant receptor is able to mediate distinct G protein-dependent and G protein-independent pathways. Activation of ERK2 by the chemoattractant appears to be negatively regulated by the Ras signaling pathway. Activated Ras results in reduced ERK2 activation, whereas disruption of putative RasGEF or expression of dominant negative Ras proteins have a more rapid, high, and extended activation. CRAC, a PH domain-containing protein required for adenylyl cyclase activation, is required for proper ERK2 adaptation. PKA overexpression results in a more rapid, high level of activation, whereas *pka* null cells show a lower level but more extended ERK2 activation. This suggests that PKA is required for both activation and adaptation of ERK2.

2.9. Protein Kinase C

Two protein kinase C-like enzymes have been suggested to exist in *Dictyostelium* due to the different cellular responses to two inhibitors specific for protein kinase C. One enzyme is preferentially sensitive to D-*erythro*-sphingosine, a diacylglycerol analog, and is required for growth. A second is preferentially inhibited by bisindolylmaleimide GF109203X and is required for chemotaxis. This activity is stimulated by diacylglycerol and phosphorylates a peptide substrate that is an efficient substrate for mammalian protein kinase Cs. This activity is a candidate for the effector of diacylglycerol generated during the aggregative phase of *Dictyostelium* development and defines a role for diacylglycerol in the chemotactic response.

2.10. Control of Chemotaxis by Rho Family Members and Downstream Effectors

The Rho family of small G proteins are key regulators of changes in the actin cytoskeleton. There are five Racs identified in *Dictyostelium*. Rac1B appears to regulate the cytoskeleton and cell movement during chemotaxis. Cells expressing constitutively active *DdRac1B^{61L}* (unable to hydrolyze GTP) exhibit an upregulated assembly of F-actin with multiple actin-enriched crowns, suggesting aberrant regulation of the cytoskeleton. In response to a chemoattractant,

$RacIB^{61L}$ cells polarize and move; however, the cells make many false turns and chemotaxis is inefficient. Cells expressing dominant negative $DdRacIB^{17N}$ do not have prominent F-actin-rich lamellipodia and $DdRacIB^{17N}$ cells do not polarize or move. Thus, DdRac1B regulates the general reorganization of the actin cytoskeleton. The activity of DdRac1B is negatively regulated by RacGAP (Rac GTPase activating protein). Cells lacking RacGAP show aberrant actin cytoskeletons and defective chemotaxis, very similar to cells expressing DdRac1B (Chung et al., submitted). The signaling pathway(s) activating Rac proteins are not known yet, although the activation of Rac proteins would be regulated by signaling pathways downstream from chemoattractant receptors. *Dictyostelium* DdRac1B is likely to regulate a signaling pathway downstream from chemoattractant receptors that control cell movement, migration, and directionality.

PAKs (p21-activated protein kinase) have been implicated in the morphological changes resulting from changes in the actin cytoskeleton associated with Rac and Cdc42. A *Dictyostelium* Ste20/PAK family kinase that phosphorylates severin, a Ca^{2+}-dependent F-actin fragmenting protein, was purified and cloned recently, indicating a direct signal transduction from the plasma membrane to the cytoskeleton by phosphorylating actin-binding proteins. A *Dictyostelium* myosin I heavy chain kinase (MIHCK) homologous to PAK and Ste20 has also been cloned and proposed to provide a direct link between Cdc42/Rac signaling pathways and motile processes requiring myosin I molecules. Another PAK homolog, PAKa, has been identified. PAKa contains a long N-terminal targeting and regulatory domain that includes an acidic domain, a potential polyproline SH3 binding domain, and a CRIB domain, which we show binds $DdRac1B^{GTP}$ and $HsCdc42^{GTP}$. *paka* null cells exhibit a defect in completing cytokinesis when they grow in suspension, suggesting a defect in the regulation of the actinomyosin cytoskeleton. In chemotaxis assays, *paka* null cells or wild-type cells expressing dominant negative PAKa produce many random, lateral pseudopodia, have a much higher frequency of making wrong turns than wild-type cells, and chemotax very inefficiently. PAKa appears to be required for the regulation of myosin II filament assembly in the posterior cortex, which is important to maintain polarity and restrict the formation of random, lateral pseudopodia. Genetic analyses suggest that PAKa may negatively regulate myosin II heavy chain kinase.

3. REGULATION OF CYTOSKELETAL COMPONENTS IN CHEMOTAXIS

3.1. Components of Cytoskeletal Organization and Changes During Cell Migration

Work in *Dictyostelium* has been at the forefront in identifying and understanding the components of the cytoskeleton and their function in controlling cell movement. These analyses have taken advantage of the ability to do gene knockouts to obtain null mutants and analyze the mutants via biochemical and physiological assays. The change of the cytoskeleton in a *Dictyostelium* cell on stimulation by a chemoattractant relies mainly on the equilibrium between monomeric and filamentous actin and numerous actin-binding proteins controlling the rigidity of the cytoskeleton.

3.1.1. ACTIN

Actin represents approximately five percent of the cellular proteins in *Dictyostelium*. In the resting cell, G- and F-actin are present in approximately equal amounts. The actin cytoskeleton undergoes dramatic rearrangements after chemotactic stimulation. Within a few seconds of stimulation of cells with a chemoattractant, levels of F-actin increase by 50–60% and it incorporates into the cytoskeleton. The assembly of F-actin returns to the resting level by 20 s after the chemoattractant stimulation, which correlates with the shape change of cells to round. After 30 s, a longer lasting increase of F-actin polymerization starts and lasts for several minutes, which represents the extension of new pseudopodia.

3.1.2. ACTIN-BINDING PROTEINS

The first step in the chemotaxis toward a chemoattractant gradient involves a regulated increase in actin nucleation activity that is correlated with an increase in actin polymerization occurring seconds after chemotactic stimulation at the leading edge of the cell. Actin-binding proteins regulate *in vitro* the assembly of actin into supramolecular structures.

3.1.2.1. Cofilin. Cofilin was identified and purified as an actin monomer binding protein of apparent molecular mass of 15,000 daltons from *Dictyostelium discoideum*. Cofilin has a depolymerizing activity of F-actin filaments in a pH-dependent manner and this activity is inhibited by phosphatidyl inositides. Two cofilin genes (*DCOF1* and *DCOF2*) were cloned. Gene disruption experiments suggest that *DCOF1* is essential for cell proliferation, whereas the disruption of *DCOF2* did not show any phenotypes. Cofilin is distributed diffusely throughout cytoplasm in vegetative cells, but localizes in ruffling membranes of the leading edge where the actin cytoskeleton is dramatically reorganizing. The overexpression of cofilin induces the formation of actin bundles beneath ruffling membranes and stimulates cell movement and membrane ruffling. This result is rather unexpected, as cofilin has F-actin severing activity. These changes suggest that cofilin may sever actin filaments *in vivo* and create many nucleation sites that induce growth and bundling of the filaments in the presence of cross-linking proteins to generate contractile systems involved in membrane ruffling and cell movement.

3.1.2.2. Aginactin and Cap32/34. A chemoattractant-regulated capping activity, called aginactin, from *Dictyostelium* may regulate changes in actin nucleation activity. Aginactin is a barbed-end capping activity and inhibits the rate and final extent of actin polymerization. Immunofluorescence staining indicates the localization of aginactin in F-actin-rich regions of the cell cortex and cell protrusions. Later studies demonstrated that the capping activity of aginactin is attributed to cap32/34 contained in aginactin. Cap32/34 is a highly conserved heterodimeric barbed-end capping protein. It binds to the fast growing ends of actin filaments and effectively inhibits filament elongation. Exposure of free barbed ends results in actin assembly, followed by entry of free capping protein into the actin cytoskeleton, which terminates, not initiates, the actin polymerization transiently.

3.1.2.3. ABP-50/EF1α. ABP-50 is the translation elongation factor 1 alpha (EF1α) and an actin filament binding and bundling protein. ABP-50 localizes with F-actin in surface extensions and exhibits a diffuse distribution throughout the cytosol in unstimulated cells. On addition of cAMP, ABP-50 becomes localized in the filopodia that are extended in response to stimulation within 90 s, suggesting that ABP-50 plays an important role in filopod extension. Crosslinked actin filaments bundled by ABP-50 are rotated by 90 degrees relative to each other, whereas other known crosslinking proteins require filaments to be parallel and unrotated. Bundles of actin ABP-50 would tend to exclude other actin bundling proteins. ABP-50 can thus regulate the state of the actin cytoskeleton as well as protein synthesis. At molar ratios present in the cytosol, the bundling activity of ABP-50 significantly blocks both polymerization and depolymerization of actin filaments and increases the final extent of actin polymer. The binding of ABP-50 to aminoacyl t-RNA is not pH dependent, but the interaction between ABP-50 and actin is dependent on cytoplasmic pH. Two pH-sensitive actin-binding sequences in ABP-50 are identified and are predicted to overlap with the aminoacyl-tRNA-binding sites. Thus, pH-regulated recruitment and release of ABP-50 from actin filaments in vivo may supply a highly localized concentration of ABP-50 to facilitate polypeptide elongation by the F-actin-associated translational apparatus.

3.1.2.4. ABP-120. The actin binding protein ABP-120 has been proposed to be required for crosslinking actin filaments in nascent pseudopods; this activity is important for the extension of pseudopodia in motile *Dictyostelium* amoebae. F-actin is incorporated into the Triton X-100-insoluble cytoskeleton at 30–50 s after cAMP stimulation, the time when ABP-120 is incorporated into the cytoskeleton and when pseudopods are extended after cAMP stimulation in wild-type cells. The *ABP-120* null mutant obtained by the homologous recombination exhibits profound defects in actin crosslinking, cytoskeletal structure, pseudopod number and size, cell motility, chemotaxis, and phagocytosis.

3.1.2.5. Talin. A *Dictyostelium* homolog of mammalian talin has been cloned. *Dictyostelium* talin is accumulated at the leading edge, where F-actin is enriched in polarized cells. Talin is cytoplasmic in unstimulated cells and translocates to the membrane cortex where chemoattractant receptors are strongly activated by local stimulation of cells with cAMP. The *talin* null mutant cells are capable of moving and responding to a chemoattractant, although they attach only loosely to a substrate via small areas of their surface, suggesting that talin might function as a membrane anchor.

3.1.2.6. Coronin. Coronin is an actin-binding protein identified in *Dictyostelium* and has an N-terminal domain with similarities to the β-subunits of G proteins (WD repeats) and a C-terminal domain with a high tendency for an α-helical structure. Coronin is a cytoplasmic actin-associated protein that is enriched at the leading edge of the cells and in projections of the cell surface called crowns. Coronin is a member of the WD-repeat family of proteins and interacts with actin–myosin complexes, suggesting coronin contributes to the dynamics of the actin system. The mutant cells lacking coronin grow slowly in suspension and become multinucleated. They also migrate more slowly than wild-type cells during chemotaxis.

3.1.2.7. Profilin. Profilin is a PIP_2-sensitive actin monomer sequestering protein. Two profilin isoforms (profilins I and II) have been purified and the genes encoding them cloned from *Dictyostelium*. Although both profilins contain a conserved lysine residue in the putative actin-binding region and can be crosslinked covalently to G-actin, the crosslinking efficiency of profilin II to actin is substantially higher than that of profilin I, resulting in a more efficient delay of the onset of elongation during the course of actin polymerization. Both profilin genes were disrupted and single mutants did not have any noticeable phenotype. However, cells lacking both profilins are 10 times larger than wild-type cells and cell motility is greatly impaired. Cells also show a broad rim of filamentous actin below the plasma membrane and the filamentous actin concentration in the cell is increased by about 60–70%, suggesting that profilin functions primarily as an actin-sequestering protein.

3.1.2.8. Ponticulin. Ponticulin, a 17-kDa integral glycoprotein, was purified as an F-actin binding protein. Ponticulin is abundant in the plasma membrane, constituting 0.4–1.0% of the total membrane protein. Ponticulin was proposed to be the major high-affinity link between the plasma membrane and the cortical actin network. Cells lacking ponticulin are deficient in high-affinity actin membrane binding. Even though cells are motile, *ponticulin* null cells are less efficient in chemotaxing to the chemoattractant due to the loss of positional stability of pseudopodia.

3.1.2.9. α-Actinin. α-Actinin is an actin-crosslinking protein. α-Actinin has been proposed to play a role in regulating actinomyosin-mediated contraction by promoting growth and/or crosslinking of actin filaments, based on the observation that it localizes with actin and myosin in newly formed projections filled with F-actin. Strains that are defective in α-actinin function chemotax and aggregate properly to a cAMP gradient, suggesting that there might be another crosslinking protein that is functionally redundant.

3.1.2.10. Severin. Reorganization of the actin cytoskeleton requires the fragmentation of existing F-actin filaments. Severin is a principal severing protein in *Dictyostelium*. Severin fragments F-actin filaments and caps the newly formed barbed ends. A *severin* null strain does not exhibit major defects in cell motility, which may reflect a functional redundancy. The severing activity of severin appears to be controlled by the intracellular Ca^{2+} concentration. In the presence of micromolar Ca^{2+}, severin cuts actin filaments, binds tightly to the fast-growing ends, and nucleates actin assembly. The activity of severin is inhibited by PIP_2 and other negatively charged phospholipids. Thus, severin regulates the actin cytoskeleton in response to the chemoattractant stimulus. Recently, a kinase homologous to mammalian PAKs was found to phosphorylate and regulate the function of severin.

3.1.2.11. WASp and SCAR. WAS (Wiskott–Aldrich syndrome) is an immunodeficiency syndrome caused by abnormal cytoskeleton organization in macrophages and neutrophils. The Wiskott–Aldrich syndrome protein, WASp, is an adapter protein implicated in the transmission of signals from tyrosine kinase receptors and small GTPases to the actin cytoskeleton. A gene homologous to mammalian WASp was identified. *Dictyostelium* WASp has five distinct domains. At the N-terminus, WASp has a PH domain and a CRIB domain interacting with Rac/Cdc42. WASp also has polyproline repeats that bind to the SH3 domain of proteins such as Nck and Fyn. The ARPH domain sequence is homologous to verprolin, which binds to F-actin and has actin filament severing activity. Thus, WASp might also have actin filament severing activity. At the C-terminus, WASp has series of D/E repeats which might form a coiled-coil structure and serve as an oligomerization motif. SCAR, a member of a new family of proteins related to WASp, was identified as a suppressor of the *cAR2* null phenotype. Cells lacking SCAR have reduced levels of F-actin staining during vegetative growth, and abnormal cell morphology and actin distribution during chemotaxis. A recent study indicates that WASp and SCAR interact with the Arp2/3 complex in mammalian cells. The actin-related proteins Arp2 and Arp3 are part of a seven-protein complex that is localized in the lamellipodia of a variety of cell types, and in actin-rich spots of unknown function. The Arp2/3 complex enhances actin nucleation and causes branching and crosslinking of actin filaments *in vitro*; *in vivo*, it is thought to drive the formation of lamellipodia and to be a control center for actin-based motility. Overexpression of SCAR or WASp in cells causes a disruption in the localization of the Arp2/3 complex and, concomitantly, induces a complete loss of lamellipodia and actin spots. These results suggest that WASp-related proteins may regulate the actin cytoskeleton through the Arp2/3 complex. Based on the sequence similarity, *Dictyostelium* WASp is expected to interact with the Arp2/3 complex and regulate the organization of the actin cytoskeleton.

3.1.3. MYOSIN II

Dictyostelium myosin II is the most abundant and best-studied myosin and is encoded by the *mhcA* gene. The molecular structure of *Dictyostelium* myosin II is very similar to that of mammalian myosin II. *Dictyostelium* myosin II exists as a hexameric complex composed of two 240-kDa heavy chains and two pairs (regulatory, 18 kDa and essential, 16 kDa) of light chains. The N-terminus of the heavy chain associates with a pair of light chains and forms a globular head domain that can be divided into the motor domain and neck domain. The motor domain contains Mg^{2+}-ATPase activity and actin-binding sites. The C-terminus tail region of myosin II is rich in sequences typical of α-helical coiled-coils. The coiled-coil domains of two heavy chains form a bipolar filament. Three threonine residues in the tail region of *Dictyostelium* myosin II heavy chain have been implicated in control of myosin filament formation. Alanine substitution of these sites causes substantial overassembly of myosin II in vivo. Similarly, aspartate substitution mimicking constitutive phosphorylation eliminates filament assembly in vitro and renders the myosin unable to drive any tested contractile event in vivo. These results demonstrate that heavy chain phosphorylation plays a key modulatory role in controlling myosin function in vivo.

Cells lacking functional myosin II were created by a gene knockout by homologous recombination and through the expression of antisense RNA. Neither strain can undergo cytokinesis and form large, multinucleated cells in suspension, which supports the idea that myosin II is a force-generating motor in the contractile ring or cleavage furrow. Surprisingly, these cells can chemotax and form aggregates. *myoII* null cells, however, are unable to develop past the mound stage. The weak aggregation defect suggests myosin II might not be required for pseudopodia protrusion. The ability of *myoII* null cells to complete cytokinesis on the substrate by traction-mediated cytofission reflects the functional motor system at the leading edge. However, based on the slower motility of *myoII* null cells on a substrate, it has been proposed that contractile forces generated by myosin II help the cell's rear edge to detach from the substratum and retract, allowing the cell to continue forward. Recent studies suggest that myosin II is required for retraction of the posterior end of the cell during cell movement. The regulation of myosin II polymerization occurs through receptor-mediated phosphorylation of myosin II heavy chain and myosin II light chain. Myosin II heavy chain is phosphorylated by two kinases, MHCKA and MHC–PKC, at threonine residues in the myosin II tail which lead to the disassembly of myosin filaments. Phosphorylation of residues in the myosin II tail is thought to lead to the disassembly of myosin II fibers, probably at the leading edge. In the posterior cortex, the phosphorylation of the myosin II tails might be inhibited by down-regulation of heavy chain kinase activity. Cells expressing a myosin II mutant in which the three mapped MHC-PKC phosphorylation sites are converted to Ala initially localize to the posterior cortex and form a C-shaped band, enabling the posterior of the cells to attain their wild-type shape. However, these cells do not maintain their shape and are unable to suppress the formation of lateral pseudopodia, suggesting that the assembly of myosin II at the posterior cortex might be important for maintaining cellular polarity and for preventing the formation of lateral pseudopodia. P21-activated protein kinase (PAKa) appears to be regulating the assembly of myosin II in the posterior cortex presumably via the regulation of myosin II heavy chain kinases.

The myosin essential light chain (ELC) was found to be required for myosin function in *Dictyostelium* from studies using cells expressing antisense RNA or a gene knockout. The *mlcE* null cells, when grown in suspension, exhibit the typical multinucleated phenotype observed in *myoII* heavy chain null cells. The aggregation of the *mlcE* null cells is delayed several hours and they never develop past the mound stage, similar to *myoII* null cells. The actin-activated ATPase activity of the myosin purified from the *mlcE* null cells is greatly reduced. This enzymatic defect of myosin probably results in the observed chemotactic defect of *mlcE* null cells. Despite the enzymatic defect, the localization of myosin in *mlcE* null cells is normal, suggesting that its phenotypic defects primarily arise from defective contractile function of myosin rather than its mislocalization. This supposition is strengthened by the observation that while *mlcE* null cells respond to chemoattractant with proper polarity, their movement is slower. During chemotaxis, the polarity toward the chemoattractant may depend primarily on proper localization of myosin II, while efficient motility requires proper contractile function of myosin II which can be regulated by the myosin light chain.

Phosphorylation of the 18-kDa regulatory light chain (RLC) is required for the full activity of myosin

II. A major RLC kinase (MLCKA) has been purified and cloned. *MLCKA* null cells undergo cytokinesis less efficiently than wild-type cells, but they undergo development and cap crosslinked surface receptors, processes that require the myosin heavy chain. A probable presence of additional light chain kinase(s) was indicated by the presence of phosphorylated regulatory light chain in *MLCKA* null cells.

3.1.4. UNCONVENTIONAL MYOSINS (MYOSIN IS)

Movement of a eukaryotic cell occurs by protrusion of lamellipodia and pseudopodia at the anterior and retraction at the posterior retraction of the cell body. Unconventional myosin Is appear to play roles in the protrusion of lamellipodia and pseudopodia. There are at least six distinct myosin Is, myoA–F. Each member of the myosin I family has a head domain at the N-terminus that is followed by a single IQ motif, a sequence implicated in the binding of light chains. C-terminal tails of myoB, C, and D have a polybasic domain, a GPA domain (rich in glycine, proline, and alanine or glutamine), and a Src homology (SH3) domain. myoA, E, and F lack the GPA and SH3 domains. Myosin IB was the first unconventional myosin identified and it has been shown by immunofluorescence microscopy that nonfilamentous myosin IB localizes at the leading edges of the lamellipodial projections of migrating *Dictyostelium* amoebae, which are devoid of myosin II, whereas filamentous myosin II is concentrated in the posterior of the cells. Myosin IC and ID also appear to colocalize in the actin-rich pseudopodia at the leading edge of migrating cells. Based on their locations and biochemical properties, actinomyosin I may contribute to the forces that cause extension at the leading edge of a motile cell, while the contraction of actinomyosin II at the rear squeezes the cell mass forward.

3.2. Regulation of Cytoskeletal Organization in Chemotaxis

A dramatic rise of the proportion of polymerized actin (F-actin) is one of the most striking changes after addition of extracellular cAMP to aggregation-competent *Dictyostelium* cells. A schematic diagram for the establishment of the leading edge by redistribution of actin filaments and components involved in the regulation of actin filaments is presented in Fig. 3. Localized actin assembly is required for the psuedopod formation during chemotaxis. Sites of actin filament assembly in the cell are regulated by the actin-binding proteins that control the localization, length, and stability of actin filaments. These actin-binding proteins are regulated by phosphorylation, Ca^{2+}, and phospholipids and thus are targets for the intracellular changes that occur upon stimulation of a cell with a chemoattractant. Chemotactic stimulation leads to an enrichment of ABP-120 with actin crosslinking activity in extending pseudopodia, suggesting that ABP-120 is involved in psuedopod extension. Talin stays in the cytoplasm and, within 30 s after chemotactic stimulus, translocates to the membrane cortex, where chemoattractant receptors are strongly activated by local stimulation of cells with cAMP. ABP-50 becomes localized in the filopodia that are extended in response to stimulation within 90 s, suggesting that ABP-50 plays an important role in filopod extension. The recruitment of ABP-50 at the leading edge seems to be regulated by the change of cytoplasmic pH. The activity of cap32/34 may be regulated by anionic phospholipids, as the preincubation of cap32/34 with phosphatidylinositol 4,5-bisphosphate (PIP_2) inhibits its actin binding and capping activities. The severing activity of a major severing protein, severin, appears to be controlled by the intracellular Ca^{2+} concentration. In the presence of micromolar Ca^{2+}, severin cuts actin filaments, binds tightly to the fast-growing ends, and nucleates actin assembly. The activity of severin is inhibited by PIP_2 and other negatively charged phospholipids. Scanning electron microscopy indicates that the double mutant disrupted in PI3K1 and PI3K2 contains no F-actin enriched pseudopodia, but has numerous filopodia projecting laterally and vertically from the cell surface. Fluorescent microscopy indicates that these filopodia are enriched in F-actin which accumulates in a cortical pattern in control cells. These results suggest that PI3K1 and PI3K2, possibly through modulation of the levels of $PI(3,4,5)P_3$ and $PI(3,4)P_2$, regulate the organization of actin filaments necessary for the extension of pseudopodia and the aggregation of cells into streams.

Immunofluorescence studies demonstrate that cytoskeletons composed of actin and myosin II rapidly reorganize in *Dictyostelium* cells that have been stimulated with the chemoattractant cAMP. The amounts of F-actin increase in the cortical region of cells after 5–10 s of stimulation by cAMP (the first peak). After a transient decrease in the amount of F-actin in the cortical region, the amounts of both actins increase in the cortical region after 25–35 s of stimulation (the second peak). A schematic diagram of signaling components involved in the regulation of myosin II filaments is presented in Fig. 4. Filaments of myosin II become associated within the cell membrane with increases in the mesh of actin filaments on the cell

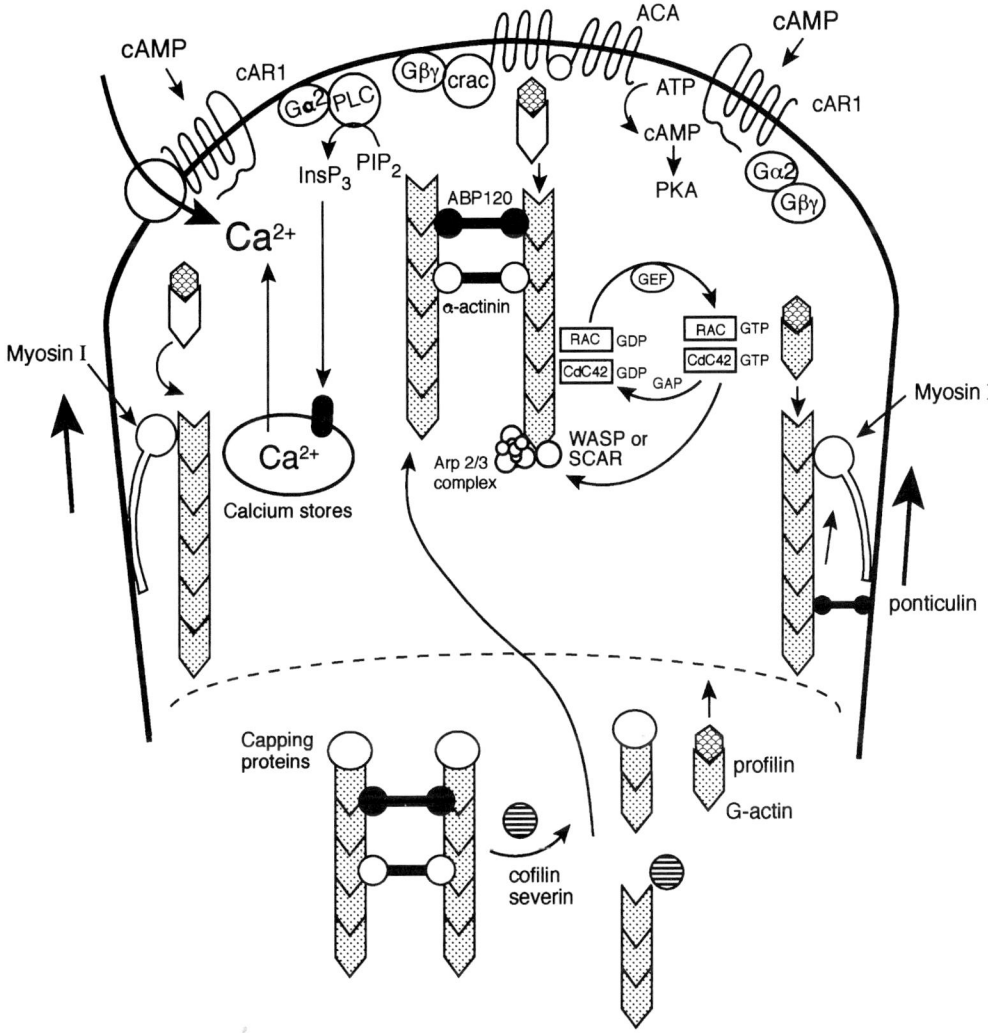

Fig. 3. Scheme for the establishment of the new leading edge. At the leading edge, the polymerization of actin filaments is stimulated by profilin. The interaction between between WASP or SCAR and the Arp2/3 complex has been suggested to be important for nucleation of actin filaments. New nucleation site can also be created by severing existing actin filaments by severing proteins such as cofilin and severin. Rac/Cdc42 proteins appear to regulate the activity of WASP. Actin crosslinking proteins stabilize the actin filaments into networks and they are regulated by intracellular pH and/or Ca^{2+} concentration. Ponticulian and talin are thought to link actin filaments to the leading edge of the plasma membrane. Myosin I function as a link between actin filaments and membrane and as a motor protein that is required for the extension of pseudopodice.

membrane at the time of the second peak. The amount of myosin II decreases thereafter, but F-actin stays localized in pseudopodia. The translocation of myosin II is suggested to be achieved by decreasing filament assembly in the cytoplasm and a concomitant increase of elongation and/or nucleation at the cortex. Many studies demonstrated that phosphorylations of three threonines near the C-terminus of myosin by heavy chain kinase inhibits myosin assembly. From the observation of cells expressing three myosin II mutants that cannot be regulated by the phosphorylation on the mapped light chain site, phosphorylation of the light chain is not required for the localization of myosin II to the furrow region of dividing cells and in the tail region of migrating cells. However, myosins that are deficient in heavy chain phosphorylation are distributed only in the cortical region of interphase cells and cells expressing this mutant myosin have a very slow rate of chemotactic migration. Cells expressing a myosin II mutant in which the three mapped MHC-PKC phosphorylation sites are converted to Ala initially localize to the posterior cortex and form a C-shaped band, enabling the posterior of the cells to attain their wild-type shape. However,

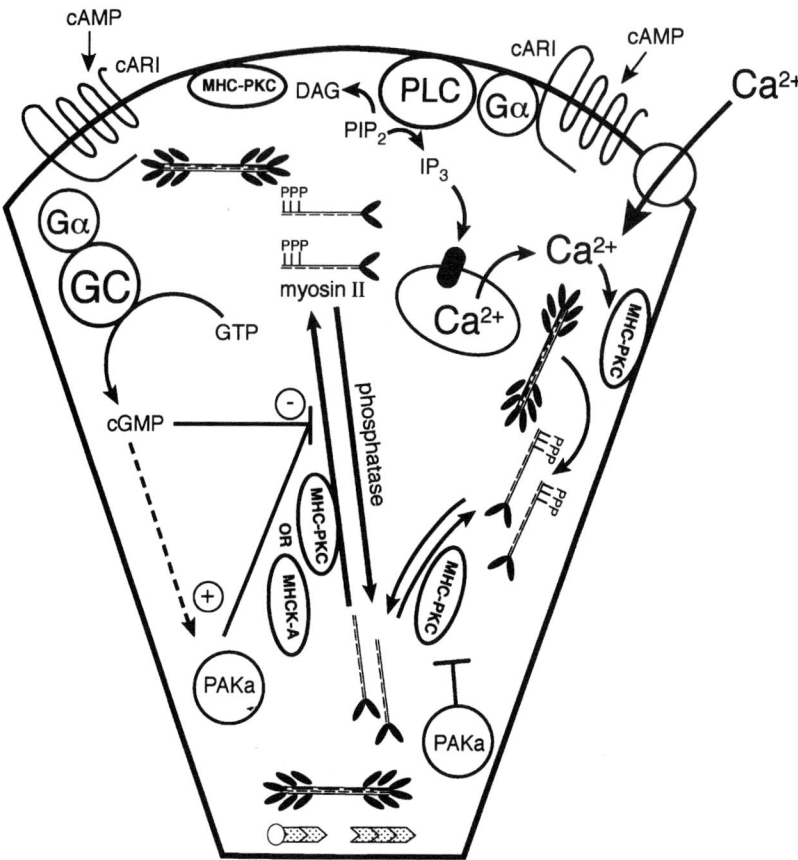

Fig. 4. Schematic diagram of the regulation of myosin II filaments assembly. The myosin II phosphorylated by MHC–PKC or MHCKA exists as a monomer while the myosin II molecules dephosphorylated by a phosphatase are in the form of parallel fibers that form thick filaments. Increase of intracellular Ca^{2+} concentration at the leading edge presumably recruits and activates the MHC–PKC, inducing the phosphorylation of heavy chains and disassembly into myosin II monomers. cGMP produced by the activation of GC inhibits heavy chain phosphorylation, thereby inducing an assembly of myosin II filaments and association of myosin II filaments into the cytoskeleton. Localized activation of PAKa also appears to be important for the regulation of the assembly of myosin II in the posterior part of the cell. The inhibition of phosphorylation is thought to be localized in the posterior cell body, resulting in the polarization of the migrating cell.

these cells do not maintain their shape and are unable to suppress the formation of lateral pseudopodia.

Binding of extracellular cAMP to cAR1 causes a transient phosphorylation of myosin heavy chain and light chain. Phosphorylation of myosin heavy and light chains is a key process regulating dynamic rearrangement of myosin thick filament in response to the chemoattractant. Two distinct myosin II heavy chain kinases have been identified. MHC–PKC was purified from the membrane fraction of developing cells and cloned. Sequence analysis of the cDNA revealed that the *Dictyostelium* MHC-PKC possesses all of the domains characteristic of members of the protein kinase C family. Cells lacking MHC–PKC exhibit substantial myosin II overassembly, as well as aberrant cell polarization and chemotaxis. Cells overexpressing the MHC–PKC contain highly phosphorylated MHC and exhibit impaired myosin II assembly and localization, no apparent cell polarization, and chemotaxis. From these observations, MHC–PKC appears to play an important role in the regulation of myosin II localization during chemotaxis. MHC–PKC has regulatory domains involved in interaction with Ca^{2+}, phospholipid, and diacylglycerol. Therefore, MHC–PKC is likely to be activated by Ca^{2+}, whose concentration rises in response to the chemoattractant. Another myosin II heavy chain kinase, MHCK-A, was purified and the gene was cloned. Analysis of the primary sequence reveals that the N-terminal amino acids form an α-helical coiled-coil domain and 300 residues of the C-terminus (residue 1146) form a WD repeat similar to the WD repeat of the β-subunit of heterotrimeric G proteins. Even though no part of the MHCK-A sequence displays similarity to the catalytic domain of conventional eukaryotic protein kinases, it autophosphorylates and phosphorylates *Dictyostelium* myosin II. Both growth phase and developed *mhckA* null cells show substantially reduced MHC kinase activity in crude lysates, as well as significant overassembly of myosin into the Triton-resistant cytoskeletal fractions. PAKa might phosphorylate and thereby inhibit the activity of MHC–PKC in the posterior cortex, as the assembly of myosin II filaments in the posterior cortex is absent in the *paka* null cell. This inhibition would occur in the posterior of cells where PAKa is localized, possibly allowing MHC–PKC to promote disassembly of myosin II fibers at the leading edge. Thus, the ability to sequentially polymerize and depolymerize myosin, possibly in response to the activation and adaptation of PAKa kinase activity, appears to be critical for cell movement.

cGMP regulates the association of myosin II with the cytoskeleton via inhibition of the phosphorylation of the myosin II heavy chain via a PKC activity. A mutant (KI-10) that is defective in chemotaxis and lacks the normal cAMP-induced cyclic GMP response exhibits a normal cAMP-induced cytoskeletal actin response but the cytoskeletal myosin II heavy chain response is abolished probably due to the lack of phosphorylation of myosin II heavy chain in response to cAMP. Cyclic GMP may play a role in the proper regulation of myosin function. Also, cGMP may be involved in regulating light chain phosphorylation in response to the chemotactic stimulus.

4. CONCLUDING REMARKS

How cells respond to environmental stimuli such as circulating hormones, growth factors, lymphokines, neurotransmitters, and stress (osmotic shock, UV, heat shock) has been a major question in understanding cell development, differentiation, and oncogenesis. Chemotaxis, a directed cell movement toward the chemoattractant gradient, plays an important role in responses to changes of environment. Chemotaxis involves the recognition of the concentration difference of the chemoattractant across the length of a cell, a relay of a signal through signaling pathways, and directional movement. Directed reorganization of the actin and myosin cytoskeletons, pseudopod extension in the direction of the chemoattractant source, and cell movement via pathways are thought to be highly conserved between mammals and *Dictyostelium*. G protein-coupled cAMP receptors are important for sensing the chemoattractant gradient. Relay of the signal through signaling pathways downstream of cAMP receptors induces a variety of changes in cellular components including the actinomyosin cytoskeleton, and these changes, in turn, induce chemotactic movement of cells. It is remarkable that these biochemical reactions including G protein-coupled receptors are quite similar in evolutionarily distant organisms. *Dictyostelium* cells thus provide a powerful system in which to examine the role of cellular components in controlling coordinated cell movement because of the ability to apply genetic and cell biological approaches to study this evolutionarily conserved path. A number of outstanding questions are still to be answered.

5. SELECTED READINGS

Arkowitz RA. Responding to attraction: chemotaxis and chemotropism in *Dictyostelium* and yeast. Trends Cell Biol 1999; 9:20–27.

Chen MY, Insall RH, Devreotes PN. Signaling through chemoattractant receptors in *Dictyostelium*. Trends Genet 1996; 12:52–57.

Condeelis J, Jones J, Segall JE. Chemotaxis of metastatic tumor cells: clues to mechanisms from the *Dictyostelium* paradigm. Cancer Metastas Rev 1992; 11:55–68.

Firtel RA. Interacting signaling pathways controlling multicellular development in *Dictyostelium*. Curr Opin Genet Dev 1996; 6:545–554.

Fukui Y. Toward a new concept of cell motility: cytoskeletal dynamics in amoeboid movement and cell division. Int Rev Cytol 1993; 144:85–127.

Gerisch G, Albrecht R, Heizer C, Hodgkinson S, Maniak M. Chemoattractant-controlled accumulation of coronin at the leading edge of *Dictyostelium* cells monitored using green fluorescent protein-coronin fusion protein. Curr Biol 1995; 5:1280–1285.

Kumagai A, Pupillo M, Gundersen R, Miake-Lye R, Devreotes PN, Firtel RA. Regulation and function of G α protein subunits in *Dictyostelium*. Cell 1989; 57:265–275.

Ma H, Gamper M, Parent C, Firtel RA. The *Dictyostelium* MAP kinase kinase DdMEK1 regulates chemotaxis and is essential for chemoattractant-mediated activation of guanylyl cyclase. Embo J 1997; 16:4317–4332.

Parent CA, Blacklock BJ, Froehlich WM, Murphy DB, Devreotes PN. G protein signaling events are activated at the leading edge of chemotactic cells. Cell 1998; 95:81–91.

Peracino B, Borleis J, Jin T, Westphal M, Schwartz JM, Wu LJ, Bracco E, Gerisch G, Devreotes P, Bozzaro S. G protein beta subunit-null mutants are impaired in phagocytosis and chemotaxis due to inappropriate regulation of the actin cytoskeleton. J Cell Biol 1998; 141:1529–1537.

Schleicher M, Andre B, Andreoli C, Eichinger L, Haugwitz M, Hofmann A, Karakesisoglou J, Stockelhuber M, Noegel AA.

Structure/function studies on cytoskeletal proteins in *Dictyostelium* amoebae as a paradigm. FEBS Lett 1995; 369:38–42.

Spudich JA, Finer J, Simmons B, Ruppel K, Patterson B, Uyeda T. Myosin structure and function. CSH Symp Quant Biol 1995; 60:783–791.

Tuxworth RI, Cheetham JL, Machesky LM, Spiegelmann GB, Weeks G, Insall RH. *Dictyostelium* RasG is required for normal motility and cytokinesis, but not growth. J Cell Biol 1997; 138:605–614.

van Haastert PJM, Kuwayama H. cGMP as second messenger during *Dictyostelium* chemotaxis. FEBS Lett 1997; 410: 25–28.

Westphal M, Jungbluth A, Heidecker M, Muhlbauer B, Heizer C, Schwartz JM, Marriott G, Gerisch G. Microfilament dynamics during cell movement and chemotaxis monitored using a GFP-actin fusion protein. Curr Biol 1997; 7:176–183.

7 gp130-Related Cytokines

Christoph J. Auernhammer and Shlomo Melmed

CONTENTS

INTRODUCTION
CYTOKINE AND CYTOKINE RECEPTOR FAMILIES AND THEIR GENES
GP130-SHARING CYTOKINES AND THEIR INTRACELLULAR SIGNALING CASCADE
SH2, SOCS, PIAS—INTRACELLULAR INHIBITORS OF THE JAK–STAT SIGNALING CASCADE
GP130-SHARING CYTOKINES—"CLASSIC" AND ENDOCRINE CYTOKINE ACTIONS
THE IL-6 CYTOKINE FAMILY—NEURO–IMMUNE–ENDOCRINE INTERFACE AND PHYSIOPATHOLOGICAL IMPLICATIONS
REFERENCES

1. INTRODUCTION

Cytokines are secreted multifunctional proteins, exerting local autocrine and paracrine, as well as systemic endocrine actions. Although originally thought to be derived from hematopoietic and immune cells, these molecules are secreted by a variety of different cell types and affect multiple target cells. Cytokine actions are characterized by biological pleiotropy and redundancy. Most cytokines affect different biological target cells and act in a cell type specific fashion. Cytokine families often exhibit similar or overlapping functions in a specific target tissue, as they use common receptor subunits and signaling cascades. Different factors, for example, interleukins (ILs), tumor necrosis factors, interferons (IFNs), colony-stimulating factors, chemokines, and others, have been categorized as cytokines, and the definition criteria are somewhat arbitrary. Cytokines act through specific receptors and use common signaling pathways.

This chapter provides a general overview on the large family of four α-helix bundle cytokines and class I cytokine receptors. Cytokine signaling and classical hematopoietic as well as neuro–immuno–endocrine actions of cytokines are extensively discussed for the IL-6 cytokine family, which share the common gp130 receptor subunit and act through the Jak–STAT signaling cascade.

2. CYTOKINE AND CYTOKINE RECEPTOR FAMILIES AND THEIR GENES

Cytokines and cytokine receptors are classified according to: (1) primary or tertiary protein structure, (2) use of common cytokine receptor subunits, and (3) common biological actions.

2.1. Four α-Helix Bundle Cytokines

Based on their tertiary protein structure, the largest family of cytokines are four α-helix bundle cytokines (Bazan, 1991; Sprang and Bazan, 1993; Walter, 1997). Although exhibiting only a low degree of homology in their primary structures, four α-helix bundle cytokines exhibit high homology in their tertiary protein structures and functional receptor epitopes. Four α-helix bundle cytokines are subdivided into short-chain and long-chain cytokines, as their helices are constituted of approx 15 or 25 residues, respectively. The tertiary structure, from the N- to the C-terminus, consists of helices A, B, C, and D,

From: *Principles of Molecular Regulation* (P. M. Conn and A. R. Means, eds.), © Humana Press Inc., Totowa, NJ.

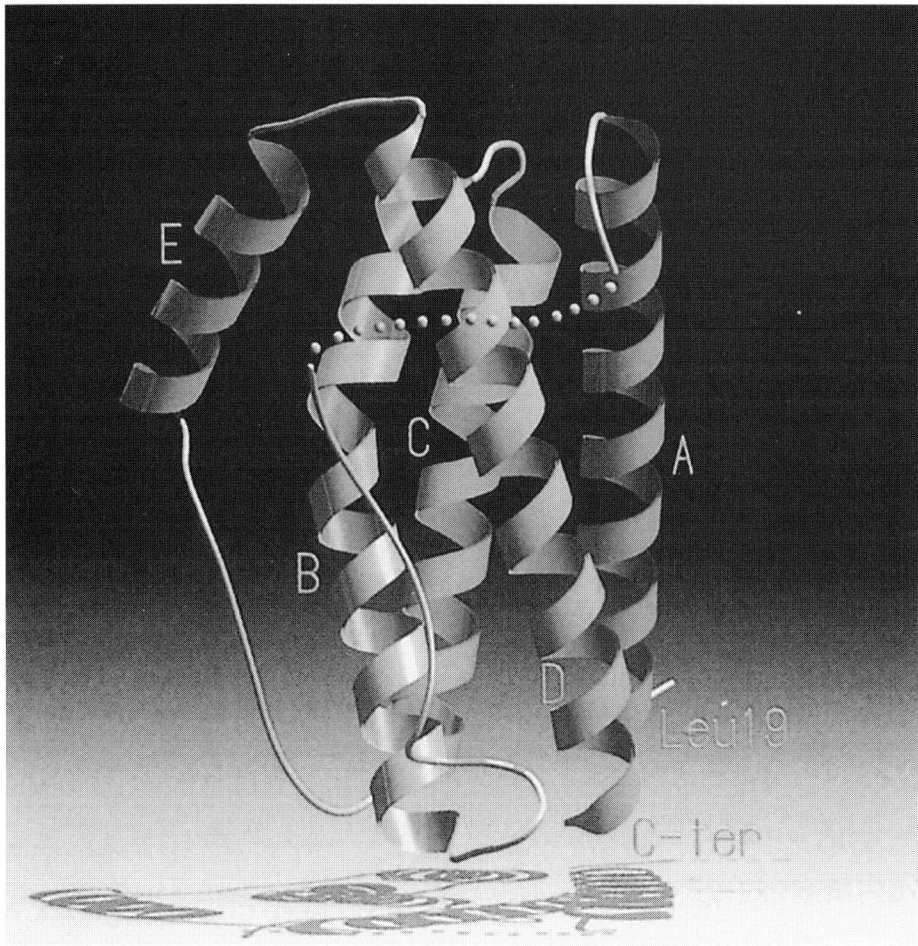

Fig. 1. Comparison of the tertiary structure of long-chain four α-helix bundle cytokines GH (de Vos, 1992), IL-6 (Somers, 1997), LIF (Smith DK, 1998), leptin (Zhang, 1997).

in up–up–down–down topology (Bazan, 1991; Sprang and Bazan, 1993; Walter, 1997). Thus far, crystal structures have been determined for the short-chain four α-helix bundle cytokines IL-2, IL-4, IL-3, IL-5, GM-CSF, and M-CSF, as well as the long-chain four α-helix bundle cytokines IL-6, LIF, CNTF, GH, G-CSF, and leptin (Somers et al., 1997; Zhang et al., 1997, for review see Simpson et al., 1997). Several other cytokines are predicted to fold in a similar way (Fig. 1).

2.2. Class I Cytokine Receptors

The hematopoietic or class I cytokine receptor superfamily (Bazan, 1990; Johnston et al., 1996; Walter, 1997) is characterized by structural and sequence similarities in their extracellular region containing cytokine binding domains (CBD), a single transmembrane domain, and an intracellular domain of variable length, lacking endogenous kinase activity. Each CBD spans ~200 residues, consisting of two fibronectin type III (FBT-III) molecules with four conserved cystein residues in the N-terminal and a Trp-Ser-X-Trp-Ser (WSXWS) motif in the C-terminal part. In addition, more FBT-III and immunoglobulin (Ig)-like domains can comprise the extracellular region (Bazan, 1990; Johnston et al., 1996; Walter, 1997). Subfamilies of the class I cytokine receptor superfamily consist of a cytokine-specific α-chain and a common β-chain heterodimer (IL-3R, IL-5R, GM-CSFR), a common γ-chain heterodimer (IL-2R, IL-4R, IL-7R, IL-9R, IL-15R), or a common gp130 receptor subunit homo/heterodimer (IL-6R, IL-11R, LIFR, OSMR, CNTFR, CT-1R). Alternatively, homodimerization of the cytokine specific receptor subunit (GHR, PRLR, EpoR, G-CSFR) has been observed (Johnston et al., 1996; Somers et al., 1997; Walter, 1997).

IFNα/βR and IFNγR constitute the class II cytokine receptor family, which lacks a WSXWS motif but exhibits conserved cystein residues (Pestka et al., 1997; Ransohoff, 1998).

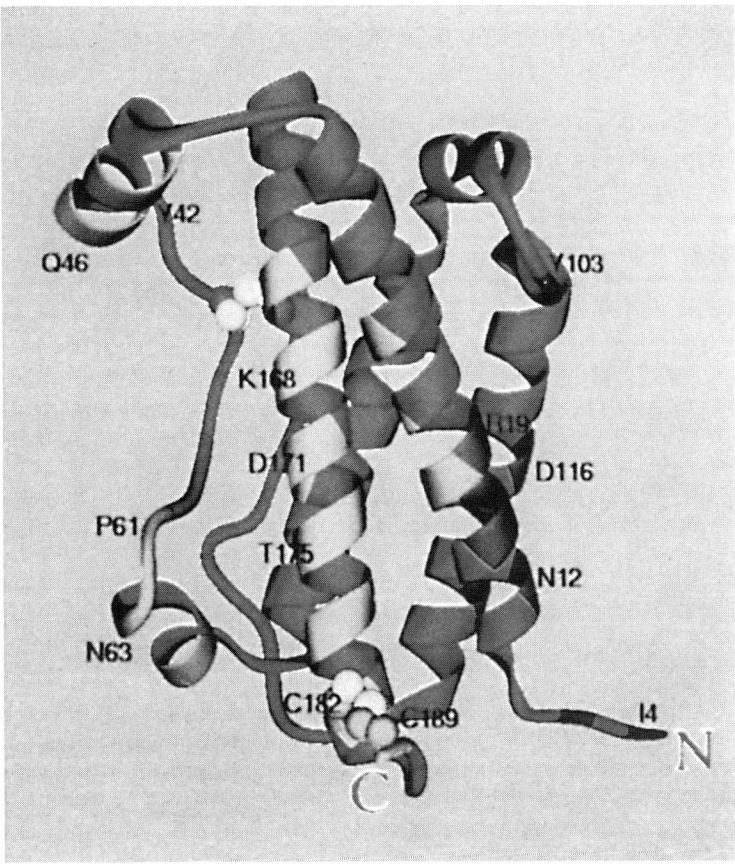

Fig. 2. Schematic diagram of the IL-6 cytokine family and their receptors. (Adapted from Hirano, 1998.)

3. gp130-SHARING CYTOKINES AND THEIR INTRACELLULAR SIGNALING CASCADE

3.1. gp130-Sharing Cytokines

The IL-6 cytokine family is part of the class I cytokine receptor superfamily and characterized by their receptors sharing the common gp130 receptor subunit (Fig. 2). The IL-6 cytokine family consists of IL-6, IL-11, LIF, OSM, CNTF, and CT-1. Binding of IL-6 or IL-11 to their specific receptor subunits is believed to result in homodimerization of gp130, while LIF, OSM, CNTF, and CT-1 cause gp130 heterodimerization (for extended review see Hirano, 1998; Simpson et al., 1997; Taga, 1997; Nakashima and Taga, 1998; Taupin et al., 1998). Human gp130 is an ubiquitously expressed ~100-kDa protein, comprising a 22 amino acid signal sequence, a 597 amino acid extracellular domain, a 22 amino acid transmembrane domain, and a 277 amino acid cytoplasmic domain (Hibi, 1990). The extracellular gp130 domain consists of an N-terminal Ig-like domain, a CBD and three membrane-proximal FBT-III modules (Hibi, 1990). The crystal structure of the cytokine binding domain of human gp130 has recently been resolved (Bravo et al., 1998).

IL-6 forms a hexameric IL-6–IL-6R–gp130 complex with a 2:2:2 stoichiometry (Murakami et al., 1993; Paonessa et al., 1995; Menziani et al., 1997; Simpson et al., 1997) (Fig. 3). A similar model has also been suggested for IL-11 (Dahmen et al., 1998; Barton et al., 1999). Thus, both IL-6 and IL-11 induce homodimerization of gp130.

IL-6 has three receptor binding regions on distinct epitopes, named binding site I, II, and III, respectively (Simpson et al., 1997). Formation of the hexameric IL-6–IL-6R–gp130 complex starts with binding of site I to the IL-6R and binding of site II to the cytokine binding domain of gp130, forming a IL-6–IL-6R–gp130 heterotrimer with 1:1:1 stoichiometry (Simpson et al., 1997). Subsequently, site III binds the Ig-like domain of another gp130 molecule, causing homodimerization of gp130 and formation of the hexameric complex IL-6–IL-6R–gp130 complex with a 2:2:2 stoichiometry (Simpson et al., 1997; Ham-

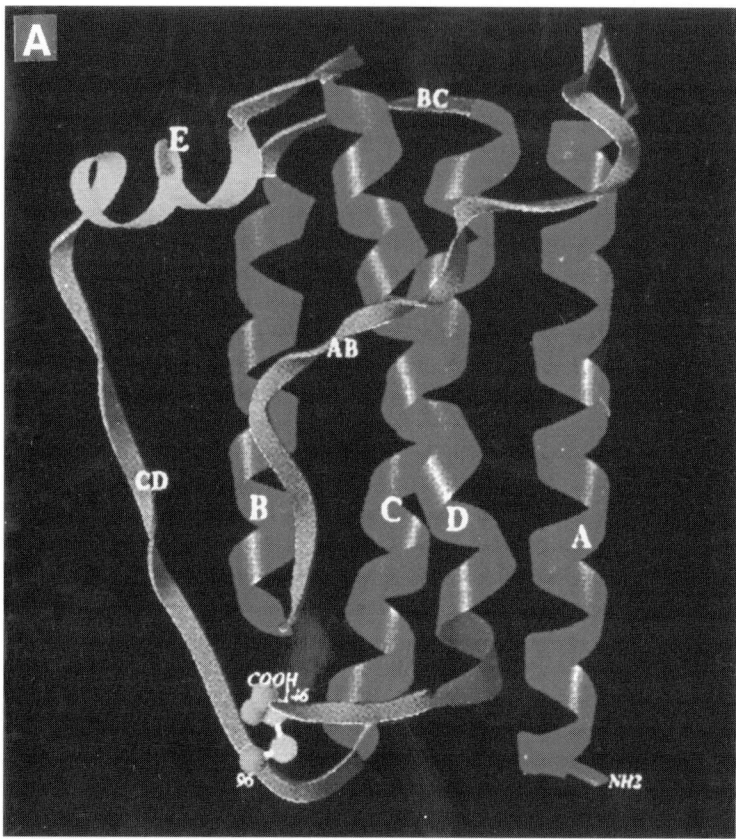

Fig. 3. Model of the hexameric IL-6:IL-6R:gp130 complex with 2:2:2 stoichiometry. (Reprinted with permission from Simpson, Protein Sci 1997; 6:929–955.)

macher et al., 1998). Based on data derived from crystal structure analysis of IL-6 (Somers et al., 1997), a fourth binding site enabling a low-affinity IL-6–IL-6 interaction has been suggested. Dimeric IL-6 (IL-6D) presenting an increased affinity for the IL-6R but a decreased ability to couple with gp130 (Ward et al., 1996), as well as several IL-6 mutants unable to homodimerize gp130 (de Hon, 1994; Savino, 1994; Ehlers, 1995), have been shown to act as specific IL-6 antagonists. Fusion proteins of IL-6 antagonists with a defective binding site II, or binding site III covalently linked to the soluble form of the human IL-6R, bind to gp130 with their respective intact binding site, but are unable to induce gp130 homodimerization. Due to their binding to gp130, these antagonists inhibit gp130 homo- as well as heterodimerization and act as general inhibitors of gp130-mediated cytokine signaling (Renne et al., 1998).

IL-11 also exhibits three distinct receptor binding epitopes (Barton et al., 1999) and IL-11 and IL-6 show overlapping binding motifs on gp130 (Dahmen et al., 1998), suggesting a very similar mechanism of receptor complex formation for IL-11 and IL-6, respectively. Site I of mIL-11 binds to the IL-11R, with Arg^{169} being essential for this interaction of IL-11 with the IL-11R. Similarly, site I of hIL-6 harbors an important Arg at position 179. Site II and site III of mIL-11 are binding sites for gp130. Mutation of Tyr^{147} in the site III binding site of mIL-11 does not alter binding affinity of the molecule for the IL-11R, but inhibits its biological activity. These data strengthen the concept of IL-11, similar to IL-6, forming a hexameric complex with a homodimeric gp130 (Barton et al., 1999). However, others have suggested formation of a pentameric receptor complex involving IL-11–IL-11R–gp130 in a 2:2:1 stoichiometry, without induction of a gp130 homodimer (Neddermann et al., 1996).

The LIFR is a multifunctional receptor subunit, which heterodimerizes with gp130 after ligand binding of LIF (Hudson et al., 1996) or OSM (Gearing et al., 1992). In addition, the LIFR forms a tripartite complex with gp130 and a respective third cytokine specific receptor subunit after ligand binding of CNTF (Davis et al., 1993a; Stahl et al., 1994; McDonald et al., 1995; Panayotatos et al., 1995; Di Marco et al.,

1996) or CT-1 (Pennica et al., 1995; Robledo et al., 1997; Arzt, 1999). Thus, LIF, OSM, CNTF, and CT-1 induce heterodimerization of gp130.

While the LIFR alone exhibits a relatively low binding affinity ($K_d \sim 1 \times 10^{-9}$ M) for LIF, the subsequent association with gp130 forms a high affinity ($K_d \sim 0.1 \times 10^{-10}$ M) complex. Similar to IL-6, LIF exhibits three potential receptor binding sites. However, in contrast to IL-6, only site II binds to the CBD of gp130, but the Ig-like domain of gp130 is not required for LIF signaling (Hammacher et al., 1998). Instead, LIF site III, and to a lower extent site I, both seem to be important for LIFR binding. Several amino acid residues at binding site III have been demonstrated to be essential for LIF binding to its receptor. These residues are highly conserved in LIF, OSM, CNTF, and CT-1, providing a possible explanation for their ability to share the LIFR. The requirement of a LIFR–gp130 heterodimer in LIF, OSM, CNTF, and CT-1 signaling is also demonstrated by mutated human LIF, exhibiting reduced binding affinity to gp130 but unchanged affinity for LIFRα, is acting as a general LIFR antagonist (Vernallis et al., 1997).

The human (Mosley et al., 1996) and murine OSMR (Lindberg et al., 1998; Tanaka et al., 1999) have recently been cloned. While hOSM binds to the LIFR (OSMR type I) as well as to the specific OSMR (OSMR type II) (Gearing et al., 1992; Thoma et al., 1994; Auguste et al., 1997; Kuropatwinski et al., 1997), mOSM (Yoshimura et al., 1996) seems unable to signal through the mLIFR and is an exclusive inducer of OSMR type II signaling (Tanaka et al., 1999). Murine M1 leukemic cells respond to mLIF and hOSM equally well, but 30- to 100-fold higher concentrations of mOSM are required to obtain a similar response. NIH3T3 cells respond to mOSM, but not to mLIF or hOSM, respectively (Yoshimura et al., 1996).

Soluble specific receptors—consisting of the extracellular receptor domain created by alternative splicing—have thus far been described for IL-6 (sIL-6R) (Muller-Newen et al., 1998; Peters et al., 1998; Nishimura et al., 1998), IL-11 (sIL-11R) (Baumann et al., 1996; Karow et al., 1996; Curtis et al., 1997), CNTF (sCNTFR) (Davis et al., 1993b; Panayotataos et al., 1994), and LIF (sLIFR) (Layton et al., 1992; Tomida, 1997; Hui et al., 1998; Zhang et al., 1998), respectively. Soluble cytokine receptors can act as antagonists (Layton et al., 1992; Curtis et al., 1997; Hui et al., 1998; Muller-Newen et al., 1998) as well as agonists (Davis et al., 1993b; Panayotataos et al., 1994; Baumann et al., 1996; Karow et al., 1996; Nishimura et al., 1998; Peters et al., 1998). As sIL-6R, sIL-11R, and sCNTFR have been demonstrated to act as agonists, a new concept of cytokine action has been proposed. While the common receptor subunit gp130 is ubiquitously expressed on many cell types, complexes of a respective cytokine and its specific soluble receptor component might mediate cytokine signaling to cell types, lacking the specific receptor component. This model could explain the diversity of cytokine action (Hirano, 1998).

Soluble forms of gp130 with a molecular mass of 90–110 kDa exist in human serum and have been suggested to arise by proteolytic cleavage, rather than alternative splicing. Soluble gp130 has also been demonstrated to act as an antagonist of IL-6 and LIF signaling, respectively (Narazaki et al., 1993; Modrell et al., 1994; Montero-Julian et al., 1997; Muller-Newen et al., 1998; Zhang et al., 1998).

Members of the IL-6 cytokine family sharing the common gp130 receptor subunit explains some of their biological effects. First, sharing of the gp130 receptor subunit might cause overlapping and partially redundant biological effects of different members of the IL-6 cytokine family (Table 1). Cytokine specific effects in different tissues and cell types are determined by expression of the rsepective specific receptor subunit. Even if the cytokine specific transmembrane receptor subunit is not expressed on a specific cell type, soluble receptors might enable signal transduction mediated by the almost ubiquitously expressed gp130. Second, sharing gp130 might be a limiting step, preventing additive or synergistic action of several members of the IL-6 cytokine family.

3.2. Jak–STAT Pathway

The Jak–STAT signaling cascade is a common signaling pathway for type I and type II cytokine receptors. While the respective cytokine receptor subunits lack intrinsic kinase activity, receptor-associated cytoplasmic Janus kinases (Jaks) exhibit tyrosine kinase activity after ligand binding to the receptor. Jaks associate with the cytoplasmic receptor subunits in the absence of ligand, but are autophosphorylated and activated only after ligand binding and hetero/homodimerization of the receptor complex. Activation of receptor-associated Jaks by autophosphorylation is followed by tyrosine phosphorylation of the cytoplasmatic receptor domain and signal transducers and activators of transcription (STAT) proteins. Using LIFR signaling as an example (for extended review see Auernhammer, 2000), the 74 membrane-proximal amino acids of the LIFR are sufficient for binding of

Table 1
IL-6 Cytokine Family: Overlapping Biological Effects

Cytokine-Mediated effect	IL-6	IL-11	LIF	OSM	CNTF	CT-1
Maintenance of ES cell pluripotentiality	±*	−	+	+	+	+
Macrophage-differentiation in M1 cells	+	−	+	+	±*	+
Growth promotion of myeloma cells	+	+	+	+	+	nd
Promotion of thrombopoiesis	+	+	+	+	nd	nd
Induction of hepatic acute phase proteins	+	+	+	+	+	+
Induction of ACTH secretion in vivo	+	+	+	+	+	+
Induction of ACTH secretion in vivo	±*	+	+	+	nd	nd
Neural differentiation	+	+	+	+	+	+
Induction of bone loss/osteoclast formation	+	+	+	+	nd	nd
Induction of cardiac hypertrophy in vitro	±*	+	+	+	±	+

Adapted from Taga, 1997; Nakashima, 1998.

Jak1 and Jak2 (Stahl et al., 1994). Ligand binding and specific receptor subunit complex formation activates Jak kinase activity, followed by phosphorylation of gp130 and the LIFR. Phosphorylated tyrosine residues on LIFR and gp130 provide specific docking sites for the SH2 domains of STAT proteins (Stahl et al., 1994), causing receptor association and subsequent phosphorylation of STAT1, STAT3, or STAT5a. C-terminal truncation of the cytoplasmic gp130 or LIFRβ domain, respectively, revealed that the membrane-proximal Box 1 and Box 2 regions are not sufficient for STAT3 phosphorylation (Stahl et al., 1995). In contrast, a consensus sequence YXXQ (Stahl et al., 1995), located on several membrane-distal locations in the cytoplasmic domains of gp130 and LIFRβ, respectively, is required for STAT3 acquisition to the receptor and subsequent phosphorylation.

Four different members of the Jak family, named Jak1, Jak2, Jak3, and Tyk2 (Johnston, 1996; Carter-Su and Smit, 1998; Moutoussamy, 1998) have been described. The general structure of all Janus kinases consists of seven Janus homology (JH) domains, JH1 exhibiting the kinase domain (Johnston et al., 1996, Carter-Su and Smit, 1998; Moutoussamy et al., 1998). Binding of STATs to tyrosine phosphorylated residues of the cytoplasmic receptor subunit causes a closer sterical association with Jak kinases, which is suggested to cause tyrosine phosphorylation of the STATs. Phosphorylation of C-terminal tyrosine sites in STAT3 (Tyr^{705}) and STAT1 (Tyr^{701}) causes the SH2 domains to enable homo- or heterodimerization of STAT3–STAT3, STAT1–STAT3, or STAT1–STAT1, respectively (Darnell, 1997; Horvath and Darnell, 1997). Crystal structure analysis of STAT3 and STAT1 homodimers demonstrates that the SH2 domain of one STAT monomer binds to the C-terminal phosphotyrosine of the other, thus enabling homodimerization (Becker et al., 1998; Chen et al., 1998) (Fig. 4). The dimerized STAT complexes are translocated to the nucleus and their DNA binding domain (amino acids 400–500) allows binding to specific DNA STAT binding elements (SBEs), causing transcriptional activation (Lamb et al., 1995; Darnell, 1997; Horvath and Darnell, 1997). Thus far, seven STAT proteins, namely STAT1α and STAT1β, STAT2, STAT3, STAT4, STAT5, and STAT6 are known (Darnell, 1997; Horvath and Darnell, 1997). STAT1, STAT3, and STAT5 are commonly involved in the signaling cascade of multiple cytokines, while signaling through STAT2, STAT4, and STAT6 is relatively exclusive (see Table 2).

3.3 Ras–MAPK

Besides the Jak–STAT cascade, several cytokines and growth factors, including GH, PRL, and IL-6 cytokine family members, also activate the Ras–mitogen-activated protein kinase (MAPK) signaling cascade.

In contrast to its inhibitory functions on gp130-mediated STAT activation, SHP2 has also been shown to be essential for LIFR/gp130-mediated activation of MAPK (Schiemann et al., 1995; Schaper et al., 1998). Using chimeric receptor models, mutation of Y118 in the cytoplasmic domains of gp130 inhibits ligand-induced MAPK activity (Schiemann et al., 1995). Y118 of gp130 comprises part of a YXXV module, which has been shown to be essential for SHP2 activation (Boulton et al., 1995; Fukada et al., 1996; Byon et al., 1997; Stofega et al., 1998). In addition, coexpression of a dominant-negative SHP2 form blocked gp130- and LIFR-induced MAPK activity by ~70% (Schiemann et al., 1995). These data

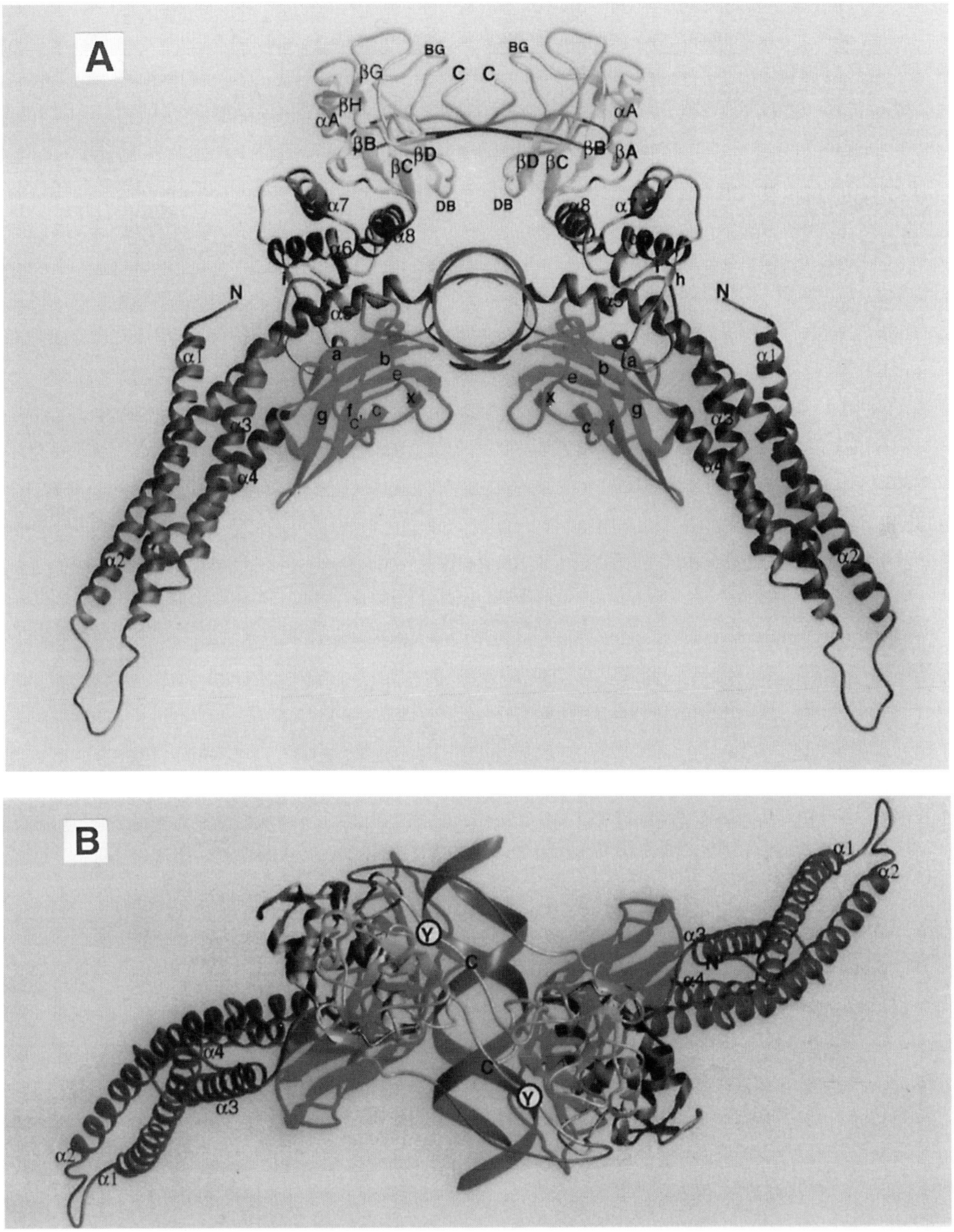

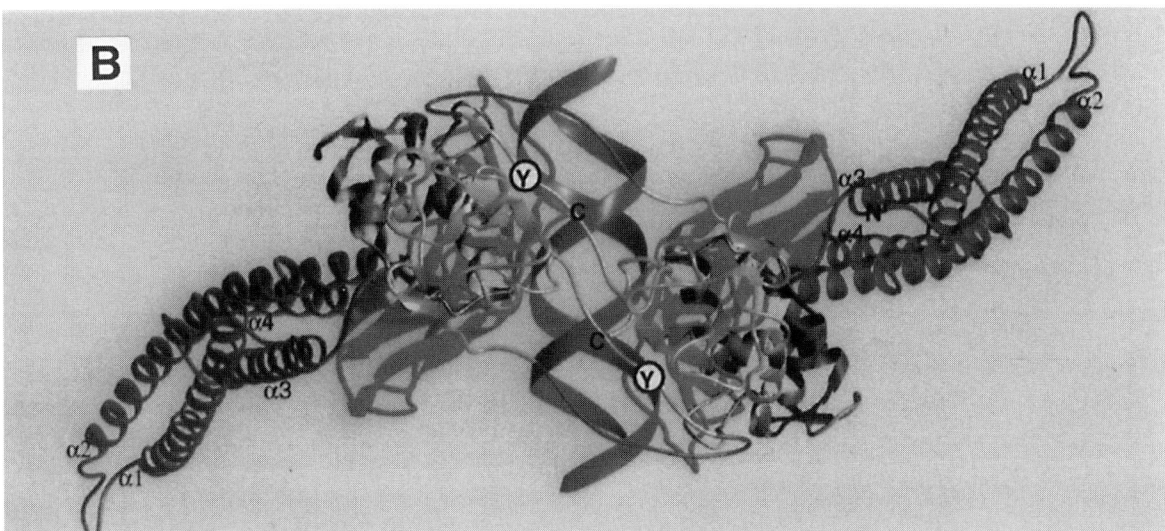

Fig. 4. Ribbon diagram of the STAT3β homodimer–DNA complex. (From Becker 1998.) Views are shown **(A)**, along the DNA axis; **(B)**, from the top.

Table 2
Cytokines–Cytokine Receptors–Jak/STAT Signaling System

4α-Helix Bundle Cytokines			Cytokine Receptor Subunit		Signaling System		
Name	Chain	Class	Specific	Common	Receptor	Jaks	STATs
IL-2	Short		IL-2Rα and IL-2Rβ		γc receptors		
IL-2	Short		IL-2Rα and IL-2Rβ		IL-2	Jak1, Jak3	STAT1, STAT3, STAT5
IL-4	Short	I	IL-4R	γc	IL-4	Jak1, Jak3	STAT3, STAT6
IL-7	Short		IL-7R		IL-7	Jak1, Jak3	STAT1, STAT3, STAT5
IL-9	Short		IL-9R		IL-9	Jak1, Jak3	STAT3, STAT5
					βc receptors		
IL-3	Short		IL-3R		IL-3	Jak2	STAT1, STAT5
IL-5	Short	I	IL-5R	βc	IL-5	Jak2	STAT1, STAT5
GMC-CSF	Short		GM-CSFR		GM-CSF	Jak2, Jak1	STAT1, STAT5, STAT3
					gp130 receptors		
IL-6	Long		IL-6R		IL-6	Jak1, Jak2, Tyk2	STAT1, STAT3
IL-11	Long		IL-11R		IL-11	Jak1, Jak2, Tyk2	STAT1, STAT3
LIF	Long	I	LIFR	gp130	LIF	Jak1, Jak2, Tyk2	STAT1, STAT3, STAT5
OSM	Long		LIFR or OSM		OSM	Jak1, Jak2, Tyk2	STAT1, STAT3, STAT5
CNTF	Long		LIFR and CNTFR		CNTF	Jak1, Jak2, Tyk2	STAT1, STAT3
CT-1	Long		LIFR and CT-1R		CT-1	Jak1, Jak2, Tyk2	STAT3
					homodimeric hormone receptors		
GH	Long		GHR		GH	Jak2	STAT1, STAT3, STAT5
PRL	Long	I	PRLR		PRL	Jak2	STAT1, STAT5
Epo	Long		EpoR		EPO	Jak2	STAT5
IFN-α/β	Long	II	IFN-α/βR		IFN-α/βR		STAT1, STAT2
IFN-γ	Short		IFN-γR		IFN-γR		STAT1

Adapted from Johnston, 1996; Liu, 1998; Nakashima, 1998.

suggest MAPK activation by LIFR and gp130 is in part mediated through activation of SHP2. Using the same chimeric receptor model, coexpression of a dominant negative Jak2 or a dominant negative Ras partially inhibited or completely abrogated G-CSF-stimulated MAPK activity, respectively, indicating a convergence of the Jak–STAT and Ras–MAPK pathways at the level of Ras (Schiemann et al., 1995). IL-6 or LIF-induced MAPK activity is inhibited by the PI-3 kinase inhibitor wortmannin (Oh et al., 1998; Takahashi-Tezuka et al., 1998). Thus, PI-3 kinase seems to be an essential mediator of LIF- and IL-6-induced MAPK activation.

Bidirectional interactions of the Jak–STAT and the Ras–MAPK pathway, respectively are in fact suggested by several lines of evidence. IL-6- or LIF-induced secondary serine phosphorylation of STAT3 (Boulton et al., 1995; Zhang et al., 1995) is dependent on MAPK activity (Stephens et al., 1998), while Erk2 and STAT3 have been shown to be associated *in vivo* (Jain et al., 1998). MAPK-induced secondary serine phosphorylation of STAT3 has been suggested to either enhance (Lamb et al., 1995), not affect (Zhang et al., 1995), or inhibit (Chung et al., 1997a) STAT3 dimerization, nuclear translocation, and DNA binding. Recent in vitro results have shown that overexpression of Erk2 or Mek1, subsequently activating Erk1/2, inhibits STAT3 tyrosine phosphorylation and activation (Jain et al., 1998; Sengupta et al., 1998). Mutation of the major serine phosphorylation site Ser[727] on STAT3 did not prevent the inhibitory effect of MAPK on IL-6-induced STAT3 activation, while Jak1 and Jak2 kinase activity were both inhibited by MAPK. Thus, reduced Jak kinase activity rather than serine phosphorylation of STAT3 might be responsible for MAPK-mediated inhibition of IL-6-induced STAT3 activation, while serine phosphorylation of STAT3 appears to be only a coincidental phenomenon of MAPK activity (Sengupta et al., 1998). Although the biological significance of MAPK-induced serine phosphorylation of STAT3 remains to be further elucidated and might be differentially regulated in different cell types, close interaction with the Jak–STAT and Ras–MAPK pathways is now apparent.

3.4. Others

Besides the Jak–STAT signaling cascade, several other cytoplasmic tyrosine kinases are activated by IL-6, including Btk (Matsuda et al., 1995b), Tec (Matsuda et al., 1995b), Fes (Hallek et al., 1997), p59Fyn (Ernst et al., 1996), p56/59Hck (Ernst et al., 1996), and p56Lyn (Ernst et al., 1996).

Insulin receptor substrate (IRS) proteins are adaptor proteins with multiple tyrosine phosphorylation sites, serving as docking sites for SH2 domains of various proteins. IRS proteins are involved in signaling of insulin and various cytokines including LIF (Argetsinger et al., 1995, 1996; Matsuda et al., 1995; Carter-Su and Smit, 1998). Phosphorylated IRS1 or IRS2 associate with various proteins, including the p85 regulatory subunit of phosphatidylinositol 3′-kinase, Grb2, or protein tyrosine phosphatase SHP2, respectively (Carter-Su and Smit, 1998).

As all of the above mentioned molecules might be involved in gp130-induced signaling cascades, their respective functional roles merit further investigation.

4. SH2, SOCS, PIAS—INTRACELLULAR INHIBITORS OF THE JAK–STAT SIGNALING CASCADE

Several intracellular inhibitors of the Jak–STAT signaling pathway act as negative feedback mediators, limiting the activation of the Jak–STAT cascade. Besides SH2-containing protein tyrosine phosphatase (SHP2), two new protein families, namely, SOCS (Suppressor Of Cytokine Signaling) and PIAS (Protein Inhibitor of Activated Stat) inhibit the Jak–STAT cascade at different levels.

4.1. SHP-2

The cytosolic protein tyrosine phosphatase SHP2 associates with various tyrosine phosphorylated proteins by its highly specific tandem N-terminal SH2 domain (Ottinger et al., 1998), resulting in conformational changes and activation of the central phosphatase domain (Byon et al., 1997; Barford and Neel, 1998; Stein-Gerlach et al., 1998). gp130-sharing cytokines induce SHP2 association with the gp130 receptor subunit (Stofega et al., 1998) and stimulate SHP2 tyrosine phosphorylation in various cell lines (Megeney et al., 1996; Stofega et al., 1998). Chimeric receptor models demonstrate the presence of a membrane-proximal tyrosine phosphorylation site in the cytoplasmic domain of the gp130 receptor (Y118) which is essential for tyrosine phosphorylation of SHP2 (Stahl et al., 1995; Fukada et al., 1996; Symes et al., 1997; Kim et al., 1998). Jak1-deficient cell lines exhibit significantly reduced SHP2 phosphorylation SH2 after stimulation with IL-6-soluble–IL-6R complexes (Schaper et al., 1998). In contrast, IL-6-induced SHP2 phosphorylation is not impaired in Jak2 or Tyk2-deficient cell lines (Schaper et al., 1998), indicating that Jak1, but not Jak2 or Tyk2, is required for tyrosine phosphorylation of SHP2.

Negative feedback regulation of SHP2 on gp130-mediated STAT activation has been suggested by several recent studies (Symes et al., 1997; Qu and Feng, 1998; Servide et al., 1998). Abrogation of SHP2 activation by overexpression of a dominant negative SHP2 variant (Symes et al., 1997; Servide et al., 1998) or a mutated gp130 subunit lacking the cytoplasmic binding site for SHP2 (Symes et al., 1997), respectively, significantly increased CNTF-stimulated nuclear transport and DNA binding of STAT dimers in different cell models. LIF-induced pluripotent stem cell proliferation, which has previously been shown to be STAT3 dependent (Boeuf et al., 1997; Niwa et al., 1998), is enhanced in embryonic stem cells with a homozygous inactivating mutation of SHP2 in comparison to wild-type cells (Qu and Feng, 1998). In contrast to its inhibitory functions on gp130-mediated STAT activation, SHP2 has also been shown to be essential for gp130-mediated activation of MAPK.

4.2. SOCS

SOCS proteins are a new family of molecules, named Suppressors Of Cytokine Signaling (Starr et al., 1997). The nomenclature is not uniform and members of this protein family have also been termed CIS (Cytokine-Inducible SH2 containing protein) (Yoshimura et al., 1995; Masuhara et al., 1997), SSI (STAT-induced STAT Inhibitors) (Minamoto et al., 1997; Naka et al., 1997), and JAB (Jak Binding Protein) (Endo et al., 1997), respectively. Thus far, eight members of this protein family, called CIS/CIS1 (Yoshimura et al., 1995; Masuhara et al., 1997), SOCS1/JAB/SSI1 (Endo et al., 1997; Naka et al., 1997; Starr et al., 1997), SOCS2/CIS2/SSI2 (Masuhara et al., 1997; Minamoto et al., 1997; Starr et al., 1997), SOCS3/CIS3/SSI3 (Masuhara et al., 1997; Minamoto et al., 1997; Starr et al., 1997), and SOCS4–SOCS7 (Hilton et al., 1998) have been described, exhibiting a common protein structure with a variable N-terminal

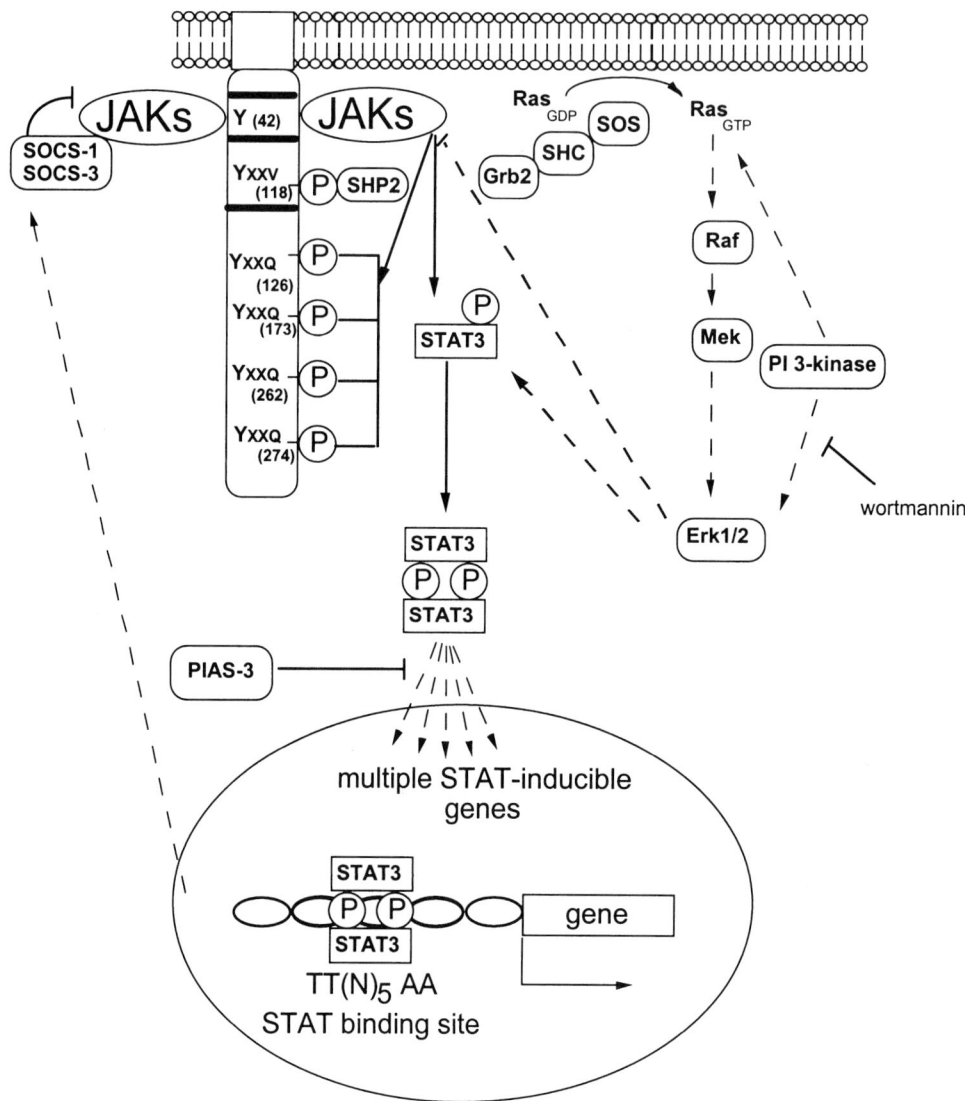

Fig. 5. Model of gp130 mediated signalling pathways (adapted from Hirano 1997, Vojtek 1998).

region, a central SH2 domain, and a C-terminal SOCS-box motif.

SOCS1 and SOCS3 bind to the JH1 domain of Jak2 and inhibit Jak2 activity (Fig. 5) (Endo et al., 1997; Masuhara et al., 1997; Ohya et al., 1997; Suzuki et al., 1998; Nicholson et al., 1999; Yasukawa et al., 1999). Data on the inhibitory action of SOCS3 on Jak kinase activity *in vitro* are somewhat arbitrary (Suzuki et al., 1998; Nicholson et al., 1999; Yasukawa et al., 1999), probably due to the much weaker Jak binding affinity and inhibitory action of SOCS3, in comparison to SOCS1 (Masuhara et al., 1997; Song and Shuai, 1998; Suzuki et al., 1998; Yasukawa et al., 1999). Both the N-terminal pre-SH2 domain and the SH2 domain are essential for the inhibitory action of SOCS1 on cytokine-induced Jak–STAT signaling (Narazaki et al., 1998; Nicholson et al., 1999; Yasukawa et al., 1999). The C-terminal SOCS-box is not directly required for the inhibitory function of SOCS proteins (Narazaki et al., 1998); however, conflicting data indicate a role for the SOCS box in SOCS protein stability and proteasomal degradation (Kamura et al., 1998; Narazaki et al., 1998; Zhang et al., 1999). SOCS1 has been demonstrated to specifically bind to the tyrosine residue (Y1007) in the activation loop of Jak2, whose phosphorylation is required for activation of kinase activity (Yasukawa et al., 1999). SOCS1 also associates with other kinases, including Jak1, Jak3, Tyk2, and Tec and to inhibit their activation (Ohya et al., 1997; Sakamoto et al., 1998).

SOCS1 and/or SOCS3 inhibit the signaling cascade of several Jak–STAT-dependent cytokines, including IL-6 (Naka et al., 1997; Starr et al., 1997), LIF (Masuhara et al., 1997; Minamoto et al., 1997; Naka et al., 1997; Starr et al., 1997; Auernhammer et al., 1998; Yasukawa et al., 1999), OSM (Starr et al., 1997), IL-11 (Auernhammer et al., 1999), IL-2 (Endo et al., 1997), IL-3 (Endo et al., 1997), Epo (Endo et al., 1997; Yasukawa et al., 1999), GH (Adams et al., 1998), PRL (Helman et al., 1999), leptin (Bjorbaek et al., 1998), and IFNs (Starr et al., 1997; Sakamoto et al., 1998; Song and Shuai, 1998). CIS inhibits the signaling of Epo and IL-3, respectively (Yoshimura et al., 1995). Conflicting data exist on the role of SOCS2 in the inhibition of LIF signaling (Masuhara et al., 1997; Minamoto et al., 1997). The physiologic role and function of the other SOCS proteins, including SOCS4–SOCS7, has not yet been elucidated.

SOCS protein expression is stimulated by multiple cytokines in a tissue and cell type specific manner (Endo et al., 1997; Naka et al., 1997; Starr et al., 1997; Adams et al., 1998; Auernhammer et al., 1998, 1999; Bjoerbaek et al., 1998; Sakamoto et al., 1998; Helman et al., 1998). Several findings indicate a negative autoregulatory feedback of CIS, SOCS1, and SOCS3 on their own cytokine-induced gene expression, thus limiting their cellular accumulation. Cytokine-induced SOCS1 and SOCS3 gene expression is STAT3 dependent (Naka et al., 1997; Auernhammer et al., 1999), while both proteins inhibit cytokine-induced STAT3 phosphorylation. Characterization of the murine SOCS3 promoter revealed an essential STAT1/STAT3 binding element, located at nt −72 to −64, to be responsible for these findings (Auernhammer et al., 1999). Similarly, the cytokine-induced CIS expression is STAT5 dependent as evidenced by the presence of essential STAT5 binding elements in its promoter region (Matsumoto et al., 1997; Verdier et al., 1998), while CIS inhibits cytokine-induced STAT5 phosphorylation (Yoshimura et al., 1995).

4.3. PIAS

Another family of negative regulators of STAT signaling, termed protein inhibitors of activated STAT (PIAS), has recently been described (Chung et al., 1997a; Liu et al., 1998). PIAS1 and PIAS3 interact directly with both activated STAT1 and STAT3. Association of PIAS with STATs occurs only after ligand-induced activation of the respective STAT proteins, and inhibits DNA binding of the respective STAT homo- or heterodimers. In contrast to SOCS proteins that inhibit Jak activity and subsequently prevent STATs phosphorylation and activation, PIAS interacts directly with activated STAT proteins and inhibits their binding to specific DNA sequences. Thus, the Jak–STAT signaling cascade may be negatively regulated at several different levels.

4. gp130-SHARING CYTOKINES— "CLASSIC" AND ENDOCRINE CYTOKINE ACTIONS

4.1. Hematopoietic and Immune Effects

Targeted disruption or postnatal inactivation of gp130 in mice results in hematological disorders, indicating an essential role of gp130-sharing cytokines in hematopoiesis. Homozygous gp30 knockout mice (gp130$^{-/-}$) die from embryonic stage 12.5E through term with cardiac hypoplasia, and also possess greatly reduced hematopoietic progenitor cells. Postnatal inactivation of gp130 in mice results in a reduced number of peripheral blood platelets.

Megakaryocyte proliferation and platelet production in vivo and/or in vitro is costimulated in concert with other growth factors by IL-6, IL-11, LIF, OSM, CNTF, and CT-1, respectively (for review see Nakashima and Taga, 1998). gp130-sharing cytokines stimulate myeloma cell survival and proliferation (for review see Klein, 1998; Treon and Anderson, 1998).

4.2. Neuro–Immune–Endocrine Regulation of the HPA Axis

Besides IL-1 and IL-2, several gp130-sharing cytokines are also important neuro–immuno–endocrine modulators of HPA axis function (for review see Besedovsky and del Ray, 1996; Ray and Melmed, 1997; Nussdorfer and Mazzocchi, 1998; Ehrhart-Bornstein et al., 1998; Arzt et al., 1999; Auernhammer and Melmed, 2000). In vivo administration of IL-6 (Besedovsky and del Rey, 1996; or LIF (Akita et al., 1996; Chesnokova et al., 1998) stimulates ACTH secretion in rodents as well as nonhuman or human primates, respectively. Intracerebroventricular injection, but not peripheral injection, of CNTF causes elevation of murine corticosterone serum levels (Meazza et al., 1997). Peripheral coadministration of CNTF plus sCNTFR in mice has been suggested to activate the HPA axis (Demitri et al., 1998). Peripheral coadministration of each respective gp130-sharing cytokine, namely IL-6, IL-11, LIF, OSM, CNTF, and CT-1, together with IL-1 in mice results in increased corticosterone serum levels in comparison to IL-1 alone (Benigni et al., 1996). These data suggest a role for all gp130-sharing cytokines in the neuro–immuno–endocrine modulation of the HPA axis. Thus, IL-6 and LIF are the best characterized members of the g130-sharing cytokine family modulating HPA axis function and development.

In human pituitary tissue, expression of the gp130 receptor subunit (Shimon et al., 1997) and specific IL-6R (Shimon et al., 1997), LIFR (Shimon et al., 1997), and IL-11R (Auernhammer et al., 1999) has been demonstrated. In primary cultures derived from human fetal pituitaries of 16–31 wk of gestation, recombinant human IL-6, LIF, and OSM treatments induced ACTH secretion either alone or in synergistic action with CRH (Shimon et al., 1997). In murine corticotroph AtT-20 cells, LIF, OSM, and IL-11 induce *POMC* gene expression and ACTH secretion (Akita et al., 1995; Ray et al., 1996; Auernhammer et al., 1999). LIF-induced *POMC* gene expression and ACTH secretion involve tyrosine phosphorylation and activation of Jak2 (Bousquet, unpublished data), gp130 (Ray et al., 1996), STAT3 mutants (Ray et al., 1996, 1998), and STAT1 (Ray et al., 1996, 1998), respectively. LIF-induced ACTH secretion can be inhibited by gp130-directed antibodies (Ray et al., 1996; Shimon et al., 1997) as well as overexpression of dominant negative STAT3 mutants (Bousquet and Melmed, 1999), verifying the essential role of the Jak–STAT signaling cascade in LIF-induced corticotroph cell activation. Corticotroph SOCS3 is a cytokine-inducible negative feedback regulator of LIF- and IL-11-induced *POMC* gene expression and ACTH secretion, by inhibiting the Jak–STAT signaling cascade (Auernhammer et al., 1998).

Both IL-6 and LIF are expressed in rodent pituitary folliculostellate cells, whereas IL-6 is not present in human folliculotellate cells (for review *see* Renner et al., 1998; Tsagarakis et al., 1998). IL-6 and LIF are produced by various pituitary hormone-secreting adenomas. Human LIF and IL-6 are also produced by different hormone-secreting cells of the normal pituitary (for review *see* Ray et al., 1998; Tsagarakis et al., 1998; Schwartz et al., 1999), suggesting autocrine and paracrine effects. Pituitary expression of LIF (Wang et al., 1996; Auernhammer et al., 1998) and IL-6 (reviewed by Tsagarakis et al., 1998) is stimulated by systemic administration of LPS or IL-1. Although IL-6-knockout and LIF-knockout mice exhibit normally elevated corticosterone levels in response to short-term stress and inflammatory stimuli (Chesnokova et al., 1998; Kozak et al., 1998; Manfredi et al., 1998), LIF-knockout mice show a blunted response of the HPA axis, particularly ACTH secretion, to various long-term stress and inflammatory stimuli (Akita et al., 1996a; Chesnokova et al., 1998).

In addition to their effects on HPA axis function, gp130-sharing cytokines are also involved in pituitary development as well as pituitary tumorigenesis. LIF induces corticotroph differentiation and ACTH secretion (Stefana et al., 1997; Li et al., 1997), while it inhibits corticotroph cell proliferation (Stefana et al., 1996). Transgenic mice overexpressing LIF directed by a pituitary-targeted αGSU-promoter exhibit a Cushingoid phenotype with elevated basal corticosterone serum levels and pituitary corticotroph cell hyperplasia (Yano et al., 1998). Furthermore, these animals are dwarfs and their pituitaries exhibit a reduced number of thyrotroph, gonadotroph, lactotroph, and somatotroph cells. This is probably due to diminished expression of the homeobox genes *Lhx3* and *Pit-1* and subsequent diminished expression of *Lhx3* and *Pit1*-dependent pituitary cell lines (Yano et al., 1998).

5. THE IL-6 CYTOKINE FAMILY— NEURO–IMMUNE–ENDOCRINE INTERFACE AND PHYSIOPATHOLOGICAL IMPLICATIONS

The gp130-related cytokines thus participate in both hematopoietic regulation as well as signaling molecules for the neuro–immune–endocrine interface. Furthermore, they play critical roles in regulation of uterine implantation; and in bone, thyroid, and pancreatic function. These protean metabolic functions reflect the critical role of these ubiquitous regulatory molecules.

REFERENCES

Adams TE, Hansen JA, Starr R, Nicola NA, Hilton DJ, Billestrup N. Growth hormone preferentially induces the rapid, transient expression of SOCS-3, a novel inhibitor of cytokine receptor signaling. J Biol Chem 1998; 273:1285–1287.

Akita S, Webster J, Ren SG, Takino H, Said J, Zand O, Melmed S. Human and murine pituitary expression of leukemia inhibitory factor. Novel intrapituitary regulation of adrenocorticotropin hormone synthesis and secretion. J Clin Invest 1995; 95:1288–1298.

Akita S, Conn PM, Melmed S. Leukemia inhibitory factor (LIF) induces acute adrenocorticotrophic hormone (ACTH) secretion in fetal rhesus macaque primates: a novel dynamic test of pituitary function. J Clin Endocrinol Metab 1996a; 81:4170–4178.

Akita S, Malkin J, Melmed S. Disrupted murine leukemia inhibitory factor (LIF) gene attenuates adrenocorticotrophic hormone (ACTH) secretion. Endocrinology 1996b; 137:3140–3143.

Akita S, Readhead C, Stefaneanu L, Fine J, Tampanaru-Sarmesiu A, Kovacs K, Melmed S. Pituitary-directed leukemia inhibitory factor transgene forms Rathke's cleft cysts and impairs adult pituitary function. A model for human pituitary Rathke's cysts. J Clin Invest 1997; 99:2462–2469.

Arce V, Garces A, de Bovis B, Filippi P, Henderson C, Pettmann B, deLapeyriere O. Cardiotrophin-1 requires LIFRbeta to promote survival of mouse motoneurons purified by a novel technique. J Neurosci Res 1999; 55:119–126.

Argetsinger LS, Hsu GW, Myers MG Jr, Billestrup N, White MF, Carter-Su C. Growth hormone, interferon-gamma, and leukemia inhibitory factor promoted tyrosyl phosphorylation of insulin receptor substrate-1. J Biol Chem 1995; 270:14685–14692.

Argetsinger LS, Norstedt G, Billestrup N, White MF, Carter-Su C. Growth hormone, interferon-gamma, and leukemia inhibitory factor utilize insulin receptor substrate-2 in intracellular signaling. J Biol Chem 1996; 271:29415–29421.

Arzt E, Pereda MP, Castro CP, Pagotto U, Renner U, Stalla GK. Pathophysiological role of the cytokine network in the anterior pituitary gland. Front neuroendocrinol 1999; 20:71–95.

Auernhammer CJ, Bousquet C, Melmed S. Autoregulation of pituitary corticotroph SOCS-3 expression—characterization of the murine SOCS-3 promoter. Proc Natl Acad Sci USA, in press.

Auernhammer CJ, Chesnokova V, Bousquet C, Melmed S. Pituitary corticotroph SOCS-3: novel intracellular regulation of leukemia-inhibitory factor-mediated proopiomelanocortin gene expression and adrenocorticotropin secretion. Mol Endocrinol 1998; 12:954–961.

Auernhammer CJ, Melmed S. Interleukin-11 stimulates POMC gene expression and ACTH secretion in corticotroph cells: evidence for a redundant cytokine network in the HPA axis. Endocrinology.

Auernhammer CJ, Melmed S. Leukemia inhibitory factor—neuroimmune modulator of endocrine function. Review, manuscript in preparation.

Auguste P, Guillet C, Fourcin M, Olivier C, Veziers J, Pouplard-Barthelaix A, Gascan H. Signaling of type II oncostatin M receptor. J Biol Chem 1997; 272:15760–15764.

Balhoff JP, Stephens JM. Highly specific and quantitative activation of STATs in 3T3-L1 adipocytes. Biochem Biophys Res Commun 1998; 247:894–900.

Barford D, Neel BG. Revealing mechanisms for SH2 domain mediated regulation of the protein tyrosine phosphatase SHP-2. Structure 1998; 6:249–254.

Barton VA, Hudson KR, Heath JK. Identification of three distinct receptor binding sites of murine interleukin-11. J Biol Chem 1999; 274:5755–5761.

Baumann H, Wang Y, Morella KK, Lai CF, Dams H, Hilton DJ, Hawley RG, Mackiewicz A. Complex of the soluble IL-11 receptor and IL-11 acts as IL-6-type cytokine in hepatic and nonhepatic cells. J Immunol 1996; 157:284–290.

Bazan JF. Structural design and molecular evolution of a cytokine receptor superfamily. Proc Natl Acad Sci USA 1990; 87:6934–6938.

Bazan JF. Neuropoietic cytokines in the hematopoietic fold. Neuron 1991; 7:197–208.

Becker S, Groner B, Muller CW. Three-dimensional structure of the Stat3beta homodimer bound to DNA. Nature 1998; 394:145–151.

Benigni F, Fantuzzi G, Sacco S, Sironi M, Pozzi P, Dinarello CA, Sipe JD, Poli V, Cappelletti M, Paonessa G, Pennica D, Panayotatos N, Ghezzi P. Six different cytokines that share GP130 as a receptor subunit, induce serum amyloid A and potentiate the induction of interleukin-6 and the activation of the hypothalamus-pituitary-adrenal axis by interleukin-1. Blood 1996; 87:1851–1854.

Besedovsky HO, del Rey A. Immune-neuro-endocrine interactions: facts and hypotheses. Endocrinol Rev 1996; 17:64–102.

Betz UAK, Bloch W, van den Broek M, Yoshida K, Taga T, Kishimoto T, Addicks K, Rajewsky K, Muller W. Postnatally induced inactivation of gp130 in mice results in neurological, cardiac, hematopoietic, immunological, hepatic, and pulmonary defects. J Exp Med 1998; 188:1955–1965.

Bjorbaek C, Elmquist JK, Frantz JD, Shoelson SE, Flier JS. Identification of SOCS-3 as a potential mediator of central leptin resistance. Mol Cell 1998; 1:619–625.

Boeuf H, Hauss C, Graeve FD, Baran N, Kedinger C. Leukemia inhibitory factor-dependent transcriptional activation in embryonic stem cells. J Cell Biol 1997; 138:1207–1217.

Bole-Feysot C, Goffin V, Edery M, Binart N, Kelly PA. Prolactin (PRL) and its receptor: actions, signal transduction pathways and phenotypes observed in PRL receptor knockout mice. Endocrinol Rev 1998; 19:225–268.

Boulton TG, Zhong Z, Wen Z, Darnell JE Jr, Stahl N, Yancopoulos GD. STAT3 activation by cytokines utilizing gp130 and related transducers involves a secondary modification requiring an H7-sensitive kinase. Proc Natl Acad Sci USA 1995; 92:6915–6919.

Bousquet C, Melmed S. Critical role for STAT3 in murine pituitary ACTH leukemia inhibitory factor (LIF) signaling. J Biol Chem 1999; 274:10723.

Bousquet C, Ray DW, Melmed S. A common pro-opiomelanocortin-binding element mediates leukemia inhibitory factor and corticotropin-releasing hormone transcriptional synergy. J Biol Chem 1997; 272:10551–10557.

Bravo J, Staunton D, Heath JK, Jones EY. Crystal structure of a cytokine-binding region of gp130. EMBO J 1998; 17:1665–1674.

Burstein SA, Mei RL, Henthorn J, Friese P, Turner K. Leukemia inhibitory factor and interleukin-11 promote maturation of murine and human megakaryocytes in vitro. J Cell Physiol 1992; 153:305–312.

Byon JC, Kenner KA, Kusari AB, Kusari J. Regulation of growth factor-induced signaling by protein-tyrosine-phosphatases. Proc Soc Exp Biol Med 1997; 216:1–20.

Carter-Su C, Smit LS. Signaling via JAK tyrosine kinases: growth hormone receptor as a model (Review). Recent Prog Horm Res 1998; 53:61–82.

Chen X, Vinkemeier U, Zhao Y, Jeruzalmi D, Darnell JE Jr, Kuriyan J. Crystal structure of a tyrosine phosphorylated STAT-1 dimer bound to DNA. Cell 1998; 93:827–839.

Chesnokova V, Auernhammer CJ, Melmed S. Murine leukemia inhibitory factor gene disruption attenuates the hypothalamo-pituitary-adrenal axis stress response. Endocrinology 1998; 139:2209–2216.

Chung CD, Liao J, Liu B, Rao X, Jay P, Berta P, Shuai K. Specific inhibition of Stat3 signal transduction by PIAS3. Science 1997a; 278:1803–1805.

Chung J, Uchida E, Grammer TC, Blenis J. STAT3 serine phosphorylation by ERK dependent and -independent pathways negatively modulates its tyrosine phosphorylation. Mol Cell Biol 1997b; 17:6508–6516.

Curfs JHAJ, Meis JFGM, Hoogkamp-Korstanje JAA. A primer on cytokines sources, effects, and inducers. Clin Microbiol Rev 1997; 10:742–780.

Curtis DJ, Hilton DJ, Roberts B, Murray L, Nicola N, Begley CG. Recombinant soluble interleukin-11 (IL-11) receptor alpha-chain can act as an IL-11 antagonist. Blood 1997; 90:4403–4412.

Dahmen H, Horsten U, Kuster A, Jacques Y, Minvielle S, Kerr IM, Ciliberto G, Paonessa G, Heinrich PC, Muller-Newen G. Activation of the signal transducer gp130 by interleukin-11 and interleukin-6 is mediated by similar molecular interactions. Biochem J 1998; 331 (Pt 3):695–702.

Darnell JE Jr. STATs and gene regulation. Science 1997; 277:1630–1635.

Davis S, Aldrich TH, Ip NY, Stahl N, Scherer S, Farruggella T, DiStefano PS, Curtis R, Panayotatos N, Gascan H, et al. Released form of CNTF receptor alpha component as a soluble mediator of CNTF rsponses. Science 1993a; 259:1736–1739.

Davis S, Aldrich TH, Stahl N, Pan L, Taga T, Kishimoto T, Ip NY, Yancopoulos GD. LIFR beta and gp130 as heterodimerizing signal transducers of the tripartite CNTF receptor. Science 1993b; 260:1805–1808.

Demitri MT, Benigni F, Meazza C, Zinetti M, Fratelli M, Villa P, Acheson A, Panayotatos N, Ghezzi P. Protective effect of ciliary neurotrophic factor (CNTF) in a model of endotoxic shock: action mechanisms and role of CNTF receptor alpha. J Inflamm 1998; 48:47–55.

Di Marco A, Gloaguen I, Graziani R, Paonessa G, Saggio I, Hudson KR, Laufer R. Identification of ciliary neurotrophic factor (CNTF) residues essential for leukemia inhibitory factor receptor binding and generation of CNTF receptor antagonists. Proc Natl Acad Sci USA 1996; 93:9247–9252.

Endo TA, Masuhara M, Yokouchi M, Suzuki R, Sakamoto H, Mitsui K, Matsumoto A, Tanimura S, Ohtsubo M, Misawa H, Miyazaki T, Leonor N, Taniguchi T, Fujita T, Kanakura Y, Komiya S, Yoshimura A. A new protein containing an SH2 domain that inhibits JAK kinases. Nature 1997; 387:921–924.

Erhart-Bornstein M, Hinson JP, Bornstein SR, Scherbaum WA, Vinson GP. Intraadrenal interactions in the regulation of adrenocortical steroidogenesis. Endocrinol Rev 1989; 19:101–143.

Ernst M, Oates A, Dunn AR. Gp130-mediated signal transduction in embryonic stem cells involves activation of Jak and Ras/mitogen-activated protein kinase pathways. J Biol Chem 1996; 271:30136–30143.

Fukada T, Hibi M, Yamanaka Y, Takahashi-Tezuka M, Fujitani Y, Yamaguchi T, Nakajima K, Hirano T. Two signals are necessary for cell proliferation induced by a cytokine receptor gp130: involvement of STAT3 in anti-apoptosis. Immunity 1996; 5:449–460.

Gearing DP, Comeau MR, Friend DJ, Gimpel SD, Thut CJ, McGourty J, Brasher KK, King JA, Gillis S, Mosley B, et al. The IL-6 signal transducer, gp130: an oncostatin M receptor and affinity converter for the LIF receptor. Science 1992; 255:1434–1437.

Hallek M, Neumann C, Schaffer M, Danhauser-Riedl S, von Bubnoff N, de Vos G, Druker BJ, Yasukawa K, Griffin JD, Emmerich B. Signal transduction of interleukin-6 involves tyrosine phosphorylation of multiple cytosolic proteins and activation of Src-family kinases Fyn, Hck, and Lyn in multiple myeloma cell lines. Exp Hematol 1997; 25:1367–1377.

Hammacher A, Richardson RT, Layton JE, Smith DK, Angus LJ, Hilton DJ, Nicola NA, Wijdenes J, Simpson RJ. The immunoglobulin-like module of gp130 is required for signaling by interleukin-6, but not by leukemia inhibitory factor. J Biol Chem 1998; 273:22701–22707.

Helman D, Sandowski Y, Cohen Y, Matsumoto A, Yoshimura A, Merchav S, Gertler A. Cytokine-inducible SH2 protein (CIS3) and JAK2 binding protein (JAB) abolish prolactin receptor-mediated STAT5 signaling. FEBS Lett 1998; 441:287–291.

Hibi M, Murakami M, Saito M, Hirano T, Taga T, Kishimoto T. Molecular cloning and expression of an IL-6 signal transducer, gp130. Cell 1990; 63:1149–1157.

Hilton DJ, Hilton AA, Raicevic A, Rakar S, Harrison-Smith M, Gough NM, Begley CG, Metcalf D, Nicola NA, Willson TA. Cloning of a murine IL-11 receptor alpha-chain; requirement for gp130 for high affinity binding and signal transduction. EMBO J 1994; 13:4765–4775.

Hilton DJ, Richardson RT, Alexander WS, Viney EM, Willson TA, Sprigg NS, Starr R, Nicholson SE, Metcalf D, Nicola NA. Twenty proteins containing a C-terminal SOCS box form five structural classes. Proc Natl Acad Sci USA 1998; 95:114–119.

Hinds MG, Maurer T, Zhang JG, Nicola NA, Norton RS. Solution structure of leukemia inhibitory factor. J Biol Chem 1998; 273:13738–13745.

Hirano T. Interleukin 6 and its receptor: ten years later. Int Rev Immunol 1998; 16:249–284.

Horsten U, Muller-Newen G, Gerhartz C, Wollmer A, Wijdenes J, Heinrich PC, Grotzinger J. Molecular modeling-guided mutagenesis of the extracellular part of gp130 leads to the identification of contact sites in the interleukin-6 (IL-6). IL-6 receptor.gp130 complex. J Biol Chem 1997; 272:23748–23757.

Horvath CM, Darnell JE. The state of the STATs: recent developments in the study of signal transduction to the nucleus. Curr Opin Cell Biol 1997; 9:233–239.

Horvath CM, Wen Z, Darnell JE Jr. A STAT protein domain that determines DNA sequence recognition suggests a novel DNA-binding domain. Genes Dev 1995; 9:984–994.

Hudson KR, Vernallis AB, Heath JK. Characterization of the receptor binding sites of human leukemia inhibitory factor and creation of antagonists. J Biol Chem 1996; 271:11971–11978.

Hui W, Bell MC, Carroll GJ, Layton MJ. Modulation of cartilage proteoglycan metabolism by LIF binding protein. Cytokine 1998; 10:220–226.

Ichihara M, Hara T, Kim H, Murate T, Miyajima A. Oncostatin M and leukemia inhibitory factor do not use the same functional receptor in mice. Blood 1997; 90:165–173. Published erratum appears in Blood 90(5):2120.

Jain N, Zhang T, Fong SL, Lim CP, Cao X. Repression of Stat3 activity by activation of mitogen-activated protein kinase. Oncogene 1998; 17:3157–3167.

Jenab S, Morris PL. Testicular leukemia inhibitory factor (LIF) and LIF receptor mediate phosphorylation of signal transducers and activators of transcription (STAT)-3 and STAT-1 and induce c-*fos* transcription and activator protein-1 activation in rat Sertoli but not germ cells. Endocrinology 1998; 139:1883–1890.

Johnston JA, Bacon CM, Riedy MC, O'Shea JJ. Signaling by

IL-2 and related cytokines: Jak, STATs, and relationship to immunodeficiency. J Leukocyte Biol 1996; 60:441–452.

Juge-Morineau N, Francois S, Puthier D, Godard A, Bataille R, Amiot M. The gp130 family cytokines IL-6, LIF and OSM but not IL-11 can reverse the anti-proliferative effect of dexamethasone on human myeloma cells. Br J Haematol 1995; 90:707–710.

Kamura T, Sat S, Haque D, Liu L, Kaelin Jr WG, Conaway RC, Conaway JW. The elongin BC complex interacts with the conserved SOCS-box motif present in members of the SOCS, ras, WD-40 repeat, and ankyrin repeat families. Gen Develop 1998; 12:3872–3881.

Karow J, Hudson KR, Hall MA, Vernallis AB, Taylor JA, Gossler A, Heath JK. Mediation of interleukin-11-dependent biological responses by a soluble form of the interleukin-11 receptor. Biochem J 1996; 318:489–495.

Kim H, Hawley TS, Hawley RG, Baumann H. Protein tyrosine phosphatase 2 (SHP-2) moderates signaling by gp130 but is not required for the induction of acute-phase plasma protein genes in hepatic cells. Mol Cell Biol 1998; 18:1525–1533.

Klein B. Update of gp130 cytokines in multiple myeloma. Curr Opin Hematol 1998; 5:186–191.

Kodama H, Fukuda K, Pan J, Makino S, Baba A, Hori S, Ogawa S. Leukemia inhibitory factor, a potent cardiac hypertrophic cytokine, activates the JAK/STAT pathway in rat cardiomyocytes. Circ Res 1997; 81:656–663.

Kozak W, Kluger MJ, Soszynski D, Conn CA, Rudolph K, Leon LR, Zheng H. IL-6 and IL-1 beta in fever. Studies using cytokine-deficient (knockout) mice. Ann NY Acad Sci 1998; 856:33–47.

Kunisada K, Hirota H, Fujio Y, Matsui H, Tani Y, Yamauchi-Takihara K, Kishimoto T. Activation of JAK-STAT and MAP kinases by leukemia inhibitory factor through gp130 in cardiac myocytes. Circulation 1996; 94:2626–2632. (Published erratum appears in Circulation 1997 Apr 1;95 [7]:1975).

Kuropatwinski KK, De Imus C, Gearing D, Baumann H, Mosley B. Influence of subunit combinations on signaling by receptors for oncostatin M, leukemia inhibitory factor, and interleukin-6. Biol Chem 1997; 272:15135–15144.

Lamb P, Seidel HM, Haslam J, Milocco L, Kessler LV, Stein RB, Rosen J. STAT protein complexes activated by interferon-gamma and gp130 signaling molecules differ in their sequence preferences and transcriptional induction properties. Nucleic Acids Res 1995; 23:3283–3289.

Layton MJ, Cross BA, Metcalf D, Ward LD, Simpson RJ, Nicola NA. A major binding protein for leukemia inhibitory factor in normal mouse serum: identification as a soluble form of the cellular receptor. Proc Natl Acad Sci USA 1992; 89:8616–8620.

Li QL, Yano H, Ren SG, Li X, Friedman TC, Melmed S. Leukemia inhibitory factor (LIF) modulates pro-opiomelanocortin (POMC) gene regulation in stably transfected AtT-20 cells overexpressing LIF. Endocrine 1997; 7:325–330.

Lindberg RA, Juan TS, Welcher AA, Sun Y, Cupples R, Guthrie B, Fletcher FA. Cloning and characterization of a specific receptor for mouse oncostatin M. Mol Cell Biol 1998; 18:3357–3367.

Liu B, Liao J, Rao X, Kushner SA, Chung CD, Chang DD, Shuai K. Inhibition of Stat1-mediated gene activation by PIAS1. Proc Natl Acad Sci USA 1998; 95:10626–10631.

Manfredi B, Sacerdote P, Gaspani L, Poli V, Panerai AE. IL-6 knock-out mice show modified basal immune functions, but normal immune responses to stress. Brain Behav Immun 1998; 12:201–211.

Masuhara M, Sakamoto H, Matsumoto A, Suzuki R, Yasukawa H, Mitsui K, Wakioka T, Tanimura S, Sasaki A, Misawa H, Yokouchi M, Ohtsubo M, Yoshimura A. Cloning and characterization of novel CIS family genes. Biochem Biophys Res Commun 1997; 239:439–446.

Matsuda T, Fukada T, Takahashi-Tezuka M, Okuyama Y, Fujitani Y, Hanazono Y, Hirai H, Hirano T. Activation of Fes tyrosine kinase by gp130, an interleukin-6 family cytokine signal transducer, and their association. J Biol Chem 1995a; 270:11037–11039.

Matsuda T, Takahashi-Tezuka M, Fukada T, Okuyama Y, Fujitani Y, Tsukada S, Mano H, Hirai H, Witte ON, Hirano T. Association and activation of Btk and Tec tyrosine kinases by gp130, a signal transducer of the interleukin-6 family of cytokines. Blood 1995b; 85:627–633.

Matsumoto A, Masuhara M, Mitsui K, Yokouchi M, Ohtsubo M, Misawa H, Miyajima A, Yoshimura A. Cis, a cytokine inducible SH2 protein, is a target of the JAK-STAT5 pathway and modulates STAT5 activation. Blood 1997; 89:3148–3154.

Mayer P, Giessler K, Ward M, Metcalf D. Recombinant human leukemia inhibitory factor induces acute phase proteins and raises blood platelet counts in nonhuman primates. Blood 1993; 81:3226–3233.

McDonald NQ, Panayotatos N, Hendrickson WA. Crystal structure of dimeric human ciliary neurotrophicfactor determined by MAD phasing. EMBO J 1995; 14:2689–2699.

Meazza C, Di Marco A, Fruscella P, Gloaguen I, Laufer R, Sironi M, Sipe JD, Villa P, Romano M, Ghezzi P. Centrally mediated inhibition of local inflammation by ciliary neurotrophic factor. Neuroimmunomodulation 1997; 4:271–276.

Megeney LA, Perry RL, LeCouter JE, Rudnicki MA. bFGF and LIF signaling activates STAT3 in proliferating myoblasts. Dev Genet 1996; 19:139–145.

Menziani MC, Fanelli F, De Benedetti PG. Theoretical investigation of IL-6 multiprotein receptor assembly. Proteins 1997; 29:528–544.

Metcalf D, Hilton D, Nicola NA. Leukemia inhibitory factor can potentiate murine megakaryocyte production in vitro. Blood 1991; 77:2150–2153.

Minamoto S, Ikegame K, Ueno K, Narazaki M, Naka T, Yamamoto H, Matsumoto T, Saito H, Hosoe S, Kishimoto T. Cloning and functional analysis of new members of STAT induced STAT inhibitor (SSI) family: SSI-2 and SSI-3. Biochem Biophys Res Commun 1997; 237:79–83.

Modrell B, Liu J, Miller H, Shoyab M. LIF and OM directly interact with a soluble form of gp130, the IL-6 receptor signal transducing subunit. Growth Factors 1994; 11:81–91.

Montero-Julian FA, Brailly H, Sautes C, Joyeux I, Dorval T, Mosseri V, Yasukawa K, Wijdenes J, Adler A, Gorin I, Fridman WH, Tartour E. Characterization of soluble gp130 released by melanoma cell lines: a polyvalent antagonist of cytokines from the interleukin 6 family. Clin Cancer Res 1997; 3:1443–1451.

Moreau JF, Donaldson DD, Bennett F, Witek-Giannotti J, Clark SC, Wong GG. Leukaemia inhibitory factor is identical to the myeloid growth factor human interleukin for DA cells. Nature 1988; 336:690–692.

Mosley B, De Imus C, Friend D, Boiani N, Thoma B, Park LS, Cosman D. Dual Oncostatin M (OSM) Receptors. Cloning and characterization of an alternative signaling subunit conferring osm-specific receptor activation. J Biol Chem 1996; 271: 32635–32643.

Moutoussamy S, Kelly PA, Finidori J. Growth-hormone-receptor and cytokine-receptor-family signaling. Eur J Biochem 1998; 255:1–11.

Muller-Newen G, Kuster A, Hemmann U, Keul R, Horsten U, Martens A, Graeve L, Wijdenes J, Heinrich PC. Soluble IL-6 receptor potentiates the antagonistic activity of soluble gp130 on IL-6 responses. J Immunol 1998; 161:6347–6355.

Murakami M, Hibi M, Nakagawa N, Nakagawa T, Yasukawa K, Yamanishi K, Taga T, Kishimoto T. IL-6-induced homodimerization of gp130 and associated activation of a tyrosine kinase. Science 1993; 260:1808–1810.

Naka T, Narazaki M, Hirata M, Matsumoto T, Minamoto S, Aono A, Nishimoto N, Kajita T, Taga T, Yoshizaki K, Akira S, Kishimoto T. Structure and function of a new STAT-induced STAT inhibitor. Nature 1997; 387:924–928.

Nakashima K, Taga T. gp130 and the IL-6 family of cytokines: signaling mechanisms and thrombopoietic activities. Seminars in Hematol 1998; 35:210–221.

Narazaki M, Yasukawa K, Saito T, Ohsugi Y, Fukui H, Koishihara Y, Yancopoulos GD, Taga T, Kishimoto T. Soluble forms of the interleukin-6 signal-transducing receptor component gp130 in human serum possessing a potential to inhibit signals through membrane-anchored gp130. Blood 1993; 82:1120–1126.

Narazaki M, Fujimoto M, Matumoto T, Morita Y, Saito H, Kajita T, Yoshizaki K, Naka T, Kishimoto T. Three distinct domains of SSI-1/SOCS-1/JAB protein are required for its suppression of interleukin-6 signaling. Proc Natl Acad Sci 1998; 95:13130–13134.

Neddermann P, Graziani R, Ciliberto G, Paonessa G. Functional expression of soluble human interleukin-11 (IL-11) receptor alpha and stoichiometry of in vitro IL-11 receptor complexes with gp130. J Biol Chem 1996; 271:30986–30991.

Nicholson SE, Willson TA, Farley A, Starr R, Zhang JG, Baca M, Alexander WS, Metcalf D, Hilton DJ, Nicola NA. Mutational analyses of the SOCS proteins suggest a dual domain requirement but distinct mechanisms for inhibition of LIF and IL-6 signal transduction. EMBO J 1999; 18:375–385.

Nishimoto N, Ogata A, Shima Y, Tani Y, Ogawa H, Nakagawa M, Sugiyama H, Yoshizaki K, Kishimoto T. Oncostatin M, leukemia inhibitory factor, and interleukin 6 induce the proliferation of human plasmacytoma cells via the common signal transducer, gp130. J Exp Med 1994; 179:1343–1347.

Nishimura R, Moriyama K, Yasukawa K, Mundy GR, Yoneda T. Combination of interleukin-6 and soluble interleukin-6 receptors induces differentiation and activation of JAK-STAT and MAP kinase pathways in MG-63 human osteoblastic cells. J Bone Miner Res 1998; 13:777–785.

Niwa H, Burdon T, Chambers I, Smith A. Self-renewal of pluripotent embryonic stem cells is mediated via activation of STAT3. Genes Dev 1998; 12:2048–2060.

Nussdorfer GG, Mazzocchi G. Immune-endocrine interactions in the mammalian adrenal gland: facts and hypotheses. Internat Rev Cytol 1998; 183:143–183.

Oh H, Fujio Y, Kunisada K, Hirota H, Matsui H, Kishimoto T, Yamauchi-Takihara K. Activation of phosphatidylinositol 3-kinase through glycoprotein 130 induces protein kinase B and p70 S6 kinase phosphorylation in cardiac myocytes. J Biol Chem 1998; 273:9703–9710.

Ohya K, Kajigaya S, Yamashita Y, Miyazato A, Hatake K, Miura Y, Ikeda U, Shimada K, Ozawa K, Mano H. SOCS-1/JAB/ SSI-1 can bind to and suppress Tec protein-tyrosine kinase. J Biol Chem 1997; 272:27178–27182.

Ottinger EA, Botfield MC, Shoelson SE. Tandem SH2 domains confer high specificity in tyrosine kinase signaling. J Biol Chem 1998; 273:729–735.

Panayotatos N, Everdeen D, Liten A, Somogyi R, Acheson A. Recombinant human CNTF receptor alpha: production, binding stoichiometry, and characterization of its activity as a diffusible factor. Biochemistry 1994; 33:5813–5818.

Panayotatos N, Radziejewska E, Acheson A, Somogyi R, Thadani A, Hendrickson WA, McDonald NQ. Localization of functional receptor epitopes on the structure of ciliary neurotrophic factor indicates a conserved, function-related epitope topography among helical cytokines. J Biol Chem 1995; 270:14007–14014.

Paonessa G, Graziani R, De Serio A, Savino R, Ciapponi L, Lahm A, Salvati AL, Toniatti C, Ciliberto G. Two distinct and independent sites on IL-6 trigger gp130 dimer formation and signalling. EMBO J 1995; 14:1942–1951.

Pennica D, Shaw KJ, Swanson TA, Moore MW, Shelton DL, Zioncheck KA, Rosenthal A, Taga T, Paoni NF, Wood WI. Cardiotrophin-1. Biological activities and binding to the leukemia inhibitory factor receptor/gp130 signaling complex. J Biol Chem 1995; 270:10915–10922.

Pennica D, Swanson TA, Shaw KJ, Kuang WJ, Gray CL, Beatty BG, Wood WI. Human cardiotrophin: protein and gene structure, biological and binding activities, and chromosomal localization. Cytokine 1996; 8:183–189.

Pestka S, Kotenko SV, Muthukumaran G, Izotova LS, Cook JR, Garotta G. The interferon gamma (IFN-gamma) receptor: a paradigm for the multichain cytokine receptor. Cytokine Growth Factor Rev 1997; 8:189–206.

Peters M, Muller AM, Rose-John S. Interleukin-6 and soluble interleukin-6 receptor: direct stimulation of gp130 and hematopoiesis. Blood 1998; 92:3495–3504.

Piekorz RP, Nemetz C, Hocke GM. Members of the family of IL-6-type cytokines activate Stat5a in various cell types. Biochem Biophys Res Commun 1997; 236:438–443.

Piekorz RP, Rinke R, Gouilleux F, Neumann B, Groner B, Hocke GM. Modulation of the activation status of Stat5a during LIF-induced differentiation of M1 myeloid leukemia cells. Biochim Biophys Acta 1998; 1402:313–323.

Portier M, Zhang XG, Ursule E, Lees D, Jourdan M, Bataille R, Klein B. Cytokine gene expression in human multiple myeloma. Br J Haematol 1993; 85:514–520.

Qu CK, Feng GS. Shp-2 has a positive regulatory role in ES cell differentiation and proliferation. Oncogene 1998; 17:433–439.

Ransohoff RM. Cellular responses to interferons and other cytokines: the JAK-STAT paradigm. N Engl J Med 1998; 338:616–618.

Ray D, Melmed S. Pituitary cytokine and growth factor expression and action. Endocrinol Rev 1997; 18:206–228.

Ray DW, Ren SG, Melmed S. Leukemia inhibitory factor (LIF) stimulates proopiomelanocortin (POMC) expression in a corticotroph cell line. Role of STAT pathway. J Clin Invest 1996; 97:1852–1859.

Ray DW, Ren SG, Melmed S. Leukemia inhibitory factor regulates proopiomelanocortin transcription. Ann NY Acad Sci 1998; 840:162–173.

Renne C, Kallen KJ, Mullberg J, Jostock T, Grotzinger J, Rose-John S. A new type of cytokine receptor antagonist directly targeting gp130. J Biol Chem 1998; 273:27213–27219.

Renner U, Gloddek J, Pereda MP, Arzt E, Stalla GK. Regulation and role of intrapituitary IL-6 production by folliculostellate cells. Domest Anim Endocrinol 1998; 15:353–362.

Robledo O, Fourcin M, Chevalier S, Guillet C, Auguste P, Pouplard-Barthelaix A, Pennica D, Gascan H. Signaling of the cardiotrophin-1 receptor. Evidence for a third receptor component. J Biol Chem 1997; 272:4855–4863.

Sakamoto H, Yasukawa H, Masuhara M, Tanimura S, Sasaki A, Yuge K, Ohtsubo M, Ohtsuka A, Fujita T, Ohta T, Furukawa Y, Iwase S, Yamada H, Yoshimura A. A Janus kinase inhibitor, JAB, is an interferon-g-inducible gene and confers resistance to interferons. Blood 1998; 92:1668–1676.

Schaper F, Gendo C, Eck M, Schmitz J, Grimm C, Anhuf D, Kerr IM, Heinrich PC. Activation of the protein tyrosine phosphatase SHP2 via the interleukin-6 signal transducing receptor protein gp130 requires tyrosine kinase jak1 and limits acute-phase protein expression. Biochem J 1998; 335:557–565.

Schiemann WP, Nathanson NM. Involvement of protein kinase C during activation of the mitogen-activated protein kinase cascade by leukemia inhibitory factor. Evidence for participation of multiple signaling pathways. J Biol Chem 1998a; 269:6376–6382.

Schiemann WP, Nathanson NM. Raf-1 independent stimulation of mitogen-activated protein kinase by leukemia inhibitory factor in 3T3-L1 cells. Oncogene 1998b; 16:2671–2679.

Schiemann WP, Graves LM, Baumann H, Morella KK, Gearing DP, Nielsen MD, Krebs EG, Nathanson NM. Phosphorylation of the human leukemia inhibitory factor (LIF) receptor by mitogen-activated protein kinase and the regulation of LIF receptor function by heterologous receptor activation. Proc Natl Acad Sci USA 1995; 92:5361–5365.

Schwartz J, Ray DW, Perez FM. Leukemia inhibitory factor as an intrapituitary mediator of ACTH secretion. Neuroendocrinology 1999; 69:34–43.

Sengupta TK, Talbot ES, Scherle PA, Ivashkiv LB. Rapid inhibition of interleukin-6 signaling and Stat3 activation mediated by mitogen-activated protein kinases. Proc Natl Acad Sci USA 1998; 95:11107–11112.

Servidei T, Aoki Y, Lewis SE, Symes A, Fink JS, Reeves SA. Coordinate regulation of STAT signaling and c-fos expression by the tyrosine phosphatase SHP-2. J Biol Chem 1998; 273:6233–6241.

Shimon I, Yan X, Ray DW, Melmed S. Cytokine-dependent gp130 receptor subunit regulates human fetal pituitary adrenocorticotropin hormone and growth hormone secretion. J Clin Invest 1997; 100:357–363.

Simpson RJ, Hammacher A, Smith DK, Matthews JM, Ward LD. Interleukin-6: structure–function relationships. Protein 1997; 6:929–955.

Somers W, Stahl M, Seehra JS. 1.9 A crystal structure of interleukin 6: implications for a novel mode of receptor dimerization and signaling. EMBO J 1997; 16:989–997.

Song MM, Shuai K. The suppressor of cytokine signaling (SOCS) 1 and SOCS3 but not SOCS2 proteins inhibit interferon-mediated antiviral and antiproliferative activities. J Biol Chem 1998; 273:35056–35062.

Sprang S, Bazan J. Cytokine structural taxonomy and mechanisms of receptor engagement. Curr Opin Stryct Biol 1993; 3:815–827.

Stahl N, Yancopoulos GD. The tripartite CNTF receptor complex: activation and signaling involves components shared with other cytokines. J Neurobiol 1994; 25:1454–1466.

Stahl N, Boulton TG, Farruggella T, Ip NY, Davis S, Witthuhn BA, Quelle FW, Silvennoinen O, Barbieri G, Pellegrini S, et al. Association and activation of Jak-Tyk kinases by CNTF-LIF-OSM-IL-6 beta receptor components. Science 1994; 263:92–95.

Stahl N, Farruggella TJ, Boulton TG, Zhong Z, Darnell JE Jr, Yancopoulos GD. Choice of STATs and other substrates specified by modular tyrosine-based motifs in cytokine receptors. Science 1995; 267:1349–1353.

Starr R, Willson TA, Viney EM, Murray LJ, Rayner JR, Jenkins BJ, Gonda TJ, Alexander WS, Metcalf D, Nicola NA, Hilton DJ. A family of cytokine-inducible inhibitors of signalling. Nature 1997; 387:917–921.

Stefana B, Ray DW, Melmed S. Leukemia inhibitory factor induces differentiation of pituitary corticotroph function: an immuno-neuroendocrine phenotypic switch. Proc Natl Acad Sci USA 1996; 93:12502–12506.

Stein-Gerlach M, Wallasch C, Ullrich A. SHP-2, SH2-containing protein tyrosine phosphatase-2. Int J Biochem Cell Biol 1998; 30:559–566.

Stephens JM, Lumpkin SJ, Fishman JB. Activation of signal transducers and activators of transcription 1 and 3 by leukemia inhibitory factor, oncostatin-M, and interferon-gamma in adipocytes. J Biol Chem 1998; 273:31408–31416.

Stofega MR, Wang H, Ullrich A, Carter-Su C. Growth hormone regulation of SIRP and SHP-2 tyrosyl phosphorylation and association. J Biol Chem 1998; 273:7112–7117.

Suzuki R, Sakamoto H, Yasukawa H, Masuhara M, Wakioka T, Sasaki A, Yuge K, Komiya S, Inoue A, Yoshimura A. CIS3 and JAB have different regulatory roles in interleukin-6 mediated differentiation and STAT3 activation in M1 leukemia cells. Oncogene 1998; 17:2271–2278.

Symes A, Stahl N, Reeves SA, Farruggella T, Servidei T, Gearan T, Yancopoulos G, Fink JS. The protein tyrosine phosphatase SHP-2 negatively regulates ciliary neurotrophic factor induction of gene expression. Curr Biol 1997; 7:697–700.

Taga T. GP130 and the interleukin-6 family of cytokines. Annu Rev Immunol 1997; 15:797–819.

Takahashi-Tezuka M, Yoshida Y, Fukada T, Ohtani T, Yamanaka Y, Nishida K, Nakajima K, Hibi M, Hirano T. Gab1 acts as an adapter molecule linking the cytokine receptor gp130 to ERK mitogen-activated protein kinase. Mol Cell Biol 1998; 18:4109–4117.

Tanaka M, Hara T, Copeland NG, Gilbert DJ, Jenkins NA, Miyajima A. Reconstitution of the functional oncostatin M (OSM) receptor: molecular cloning of the mouse OSM receptor β subunit. Blood 1999; 93:804–815.

Taupin JL, Pitard V, Dechanet J, Miossec V, Gualde N, Moreau JF. Leukemia inhibitory factor: part of a large ingathering family. Int Rev Immunol 1998; 16:397–426.

Thoma B, Bird TA, Friend DJ, Gearing DP, Dower SK. Oncostatin M and leukemia inhibitory factor trigger overlapping and different signals through partially shared receptor complexes. J Biol Chem 1994; 269:6215–6222.

Tomida M. Presence of mRNAs encoding the soluble D-factor/LIF receptor in human choriocarcinoma cells and production of the soluble receptor. Biochem Biophys Res Commun 1997; 232:427–431.

Treon SP, Anderson KC. Interleukin-6 in multiple myeloma and related plasma cell dyscrasias. Curr Opin Hematol 1998; 5:42–48.

Tsagarakis S, Kontogeorgos G, Kovacs K. The role of cytokines in the normal and neoplastic pituitary. Crit Rev Oncol Hematol 1998; 28:73–90.

Turnbull AV, Rivier CL. Regulation of the hypothalamic-pituitary-adrenal axis by cytokines: actions and mechanisms of action. Physiol Rev 1999; 79:1–71.

Verdier F, Rabionet R, Gouilleux F, Reisenherz-Huss C, Varlet P, Muller O, Mayeux P, Lacombe C, Gisselbrecht S, Chretien S. A sequence of the CIS gene promoter interacts preferentially with two associated STAT5A dimers: a distinct biochemical difference between STAT5A and STAT5B. Mol Cell Biol 1998; 18:5852–5860.

Vernallis AB, Hudson KR, Heath JK. An antagonist for the leukemia inhibitory factor receptor inhibits leukemia inhibitory factor, cardiotrophin-1, ciliary neurotrophic factor, and oncostatin M. Biol Chem 1997; 272:26947–26952.

Vojtek AB, Der CJ. Increasing complexity of the Ras signaling pathway. J Biol Chem 1998; 273:19925–19928.

Walter MR. Structural biology of cytokines, their receptors, and signaling complexes: implications for the immune and neuroendocrine circuit. Chem Immunol 1997; 69:76–98.

Wang Z, Ren SG, Melmed S. Hypothalamic and pituitary leukemia inhibitory factor gene expression in vivo: a novel endotoxin-inducible neuro-endocrine interface. Endocrinology 1996; 137:2947–2953.

Ward LD, Hammacher A, Howlett GJ, Matthews JM, Fabri L, Moritz RL, Nice EC, Weinstock J, Simpson RJ. Influence of interleukin-6 (IL-6) dimerization on formation of the high affinity hexameric IL-6 receptor complex. Biol Chem 1996; 271:20138–20144.

White MF. The IRS-signaling system: a network of docking proteins that mediate insulin and cytokine action. Recent Prog Horm Res 1998; 53:119–138.

Yano H, Readhead C, Nakashima M, Ren SG, Melmed S. Pituitary-directed leukemia inhibitory factor transgene causes Cushing's syndrome: neuro-immune-endocrine modulation of pituitary development. Mol Endocrinol 1998; 12:1708–1720.

Yasukawa H, Misawa H, Sakamoto H, Masuhara M, Sasaki A, Wakioka T, Ohtsuka S, Imaizumi T, Matsuda T, Ihle JN, Yoshimura A. The Jak-binding protein JAB inhibits Janus tyrosine kinase activity through binding in the activation loop. EMBO J 1999; 18:1309–1320.

Yoshida K, Taga T, Saito M, Suematsu S, Kumanogoh A, Tanaka T, Fujiwara H, Hirata M, Yamagami T, Nakahata T, Hirabayashi T, Yoneda Y, Tanaka K, Wang WZ, Mori C, Shiota K, Yoshida N, Kishimoto T. Targeted disruption of gp130, a common signal transducer for the interleukin 6 family of cytokines, leads to myocardial and hematological disorders. Proc Natl Acad Sci USA 1996; 93:407–411.

Yoshimura A, Ohkubo T, Kiguchi T, Jenkins NA, Gilbert DJ, Copeland NG, Hara T, Miyajima A. A novel cytokine-inducible gene CIS encodes an SH2-containing protein that binds to tyrosine-phosphorylated interleukin 3 and erythropoietin receptors. EMBO J 1995; 14:2816–2826.

Yoshimura A, Ichihara M, Kinjyo I, Moriyama M, Copeland NG, Gilbert DJ, Jenkins NA, Hara T, Miyajima A. Mouse oncostatin M: an immediate early gene induced by multiple cytokines through the JAK-STAT5 pathway. EMBO J 1996; 15:1055–1063.

Zhang F, Basinski MB, Beals JM, Briggs SL, Churgay LM, Clawson DK, DiMarchi RD, Furman TC, Hale JE, Hsiung HM, Schoner BE, Smith DP, Zhang XY, Wery JP, Schevitz RW. Crystal structure of the obese protein leptin-E100. Nature 1997; 387:206–209.

Zhang JG, Zhang Y, Owczarek CM, Ward LD, Moritz RL, Simpson RJ, Yasukawa K, Nicola NA. Identification and characterization of two distinct truncated forms of gp130 and a soluble form of leukemia inhibitory factor receptor alpha-chain in normal human urine and plasma. J Biol Chem 1998; 273:10798–10805.

Zhang JG, Farley A, Nicholson SE, Willson TA, Zugaro LM, Simpson RJ, Moritz RL, Cary D, Richardson R, Hausmann G, Kile BJ, Kent SBH, Alexander WS, Metcalf D, Hilton DJ, Nicola NA, Baca M. The conserved SOCS box motif in suppressors of cytokine signaling binds to eingin B and C and may couple boumd proteins to proteasomal degradation. Proc Natl Acad Sci USA 1999; 96:2071–2076.

Zhang X, Blenis J, Li HC, Schindler C, Chen-Kiang S. Requirement of serine phosphorylation for formation of STAT-promoter complexes. Science 1995; 267:1990–1994.

PART II
Signaling Mechanisms Mediated by Ion Channels, Calcium, and Lipids

8   Ion Channels

Deborah J. Nelson and Harry A. Fozzard

CONTENTS

INTRODUCTION
BIOPHYSICS AND STRUCTURAL MOTIFS
BIOLOGICAL ROLES
MODULATION
EVOLUTION OF ION CHANNELS
SUMMARY
SELECTED READINGS

1. INTRODUCTION

Ion channels are intrinsic membrane proteins or protein complexes that form permeation pathways for the movement of ions to move down an energy gradient from one side of the membrane to the other. They are found in both cell surface membranes and intracellular compartment membranes, and in all plant and animal cells. On the cell surface channels may be few in number, and some small cells may have only one or two channels per cell. Alternatively, in some parts of nerve cells, half of the membrane surface is composed of ion channel proteins. Some channel types have specialized protein structural motifs that identify channel families and assist in understanding function. These structural motifs have been extremely well preserved throughout evolution.

The function of ion channels is to permit regulated passage of ions from one compartment to another, with transfer of charge determined by the electrical and chemical gradients between the compartments. Some channels are highly selective for one ion, discriminating between similarly sized and charged ions many thousand-fold, and other channels are poorly selective or selective only by size. The flow of ions through channels is regulated by "gates," which open or shut the channel's pore in response to the binding of extracellular ligands, neurotransmitters and other hormones, or cytoplasmic second messengers such as G proteins or Ca^{2+}. Others are gated by physical processes, especially membrane electrical field, stretch, or temperature. As proteins they can be compared to enzymes, but typically their speed of ion translocation and the electrical consequences of bulk charge movement emphasize that very different mechanisms exist in the function of these remarkable molecular machines. Some of the functions of channels can be investigated at the single-molecule level, because of the sensitivity of the electrical methods used to detect current from single channels.

The differential sensitivity of ion channels to both extracellular and intracellular conditions enables channels to control a vast array of biological processes. The most dramatic is perhaps their roles in mediating excitability in the nervous system and muscle, with the consequent manifestation of such higher order mammalian phenomena as thinking and memory. But the fundamental process of establishing a membrane potential means that channels regulate membrane transport of all charged chemical species. They act as sensors for detection of light, sound, touch, smell, movement, and most other interactions with the world outside of a multicellular organism. They mediate Ca^{2+} release for muscle contraction or

From: *Principles of Molecular Regulation* (P. M. Conn and A. R. Means, eds.), © Humana Press Inc., Totowa, NJ.

other means of motility. They regulate secretion of endocrine and exocrine secretory cells. And they are moderators of cell communication in embryogenesis and in mature tissues. In turn, they are carefully modulated by covalent and noncovalent second messenger systems to integrate cell responses.

Because channels are found in plants and primitive bacteria, it is clear that they evolved early and, for the most part, in their present form. Their role in regulating development of multicellular organisms is critical, if rather poorly understood. In this chapter we describe the characteristic protein structural motifs of channels and what is known of the structural basis of their function. We survey the multiple roles of channels in biological regulation and identify ways they themselves are regulated. Finally, we speculate on future progress in this field.

2. BIOPHYSICS AND STRUCTURAL MOTIFS

2.1. Biophysics

2.1.1. PERMEATION

The primary function of an ion channel is to facilitate translocation of selected ions from one side of the impermeable membrane barrier to the other, utilizing energy stored in the electrochemical transmembrane gradients. Ions are single atoms or small molecules that have an excess or deficiency of electrons, resulting in a negative or a positive charge. The ions interact electrically with the polar water molecules to form ion–water complexes. The channel must provide a path for these ions to pass the low dielectric membrane lipid bilayer, which is about 50 Å thick. Most channels are more or less selective for a particular ion such as K^+ or for a particular class of ions such as monovalent cations or anions. This selectivity property requires some intimate interaction of the channel protein with the ion seeking passage. The energy responsible for net transfer of ions (current) is found in the electrical and chemical gradients for that ion between the electrolyte solutions bathing the two sides of the membrane.

The permeation path (Fig. 1) provides a pore either wide enough to accommodate hydrated ions or with a pore lining of amino acids that can substitute for waters of hydration. The reason this is necessary is that water is held with very high energy, so transport requiring dehydration without energetically equivalent substitution for the water molecules would be

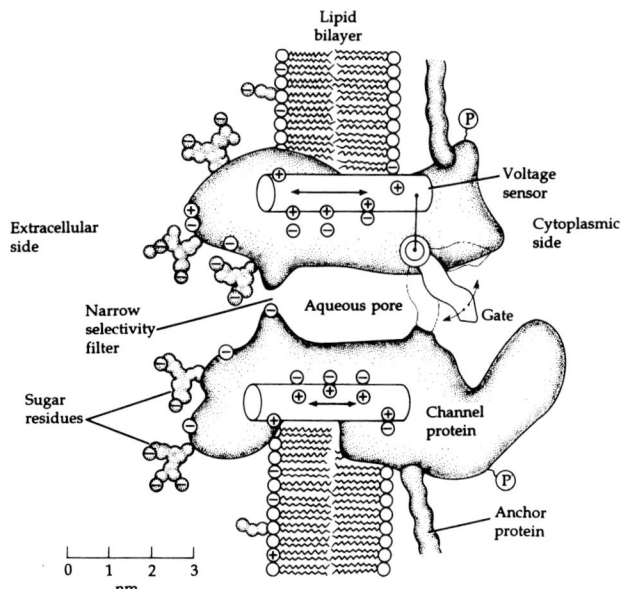

Fig. 1. Cartoon of the biophysicist's concept of a voltage-gated ion channel. The channel is an intrinsic membrane protein that is often glycosylated on the outside and may be anchored on the inside by the cytoskeleton. It creates an aqueous pore for ion permeation, with a narrow region that selects for the ions that may permeate. A gate opens or shuts the pore in response to a change in the membrane potential, detected by a voltage sensor. Not shown here are binding sites for drug or toxin interaction. The protein may also be modulated by phosphorylation. (From Hille B. Ion Channels of Excitable Membranes, 2nd edit., Sunderland, MA: Sinauer Associates, 1992).

impossible. However, channels pass ions at rates of 10–100 million ions per second, requiring very short dwell times in the pore. Most of the 50 Å-long pore is indeed wide enough to provide an almost normal water environment for the ions. However, highly selective channels have short 5–10 Å-long ion binding sites that are only 3–5 Å wide, and these are the sites of the selectivity process. Interaction at these sites must be brief, or the channel could not sustain high flow.

Four mechanisms are employed by channels to achieve selectivity. The first mechanism is for the channel to act as a sieve. The channel pore may simply be too small for an ion, so that it cannot pass through. All channels have a sievelike function, but this mechanism cannot select for a larger ion over a smaller one, for example, for K^+ (radius 1.33 Å) over Na^+ (radius 0.95 Å). The second mechanism is electric field repulsion. Cation-selective channels tend to have negatively charged amino acids in the pore lining, thereby creating a negative field that repels anions while attracting cations. The third mechanism is replacement of waters of hydration by oxygens in the pore lining. Na^+ and Ca^{2+} channels have a selectivity filter composed in part of carboxylic amino acids, which supply oxygens to replace those of water. K^+ channels use the carbonyl oxygens of their protein backbone for this purpose. Because these tethered oxygens can substitute in an energetically equivalent manner only if they are located at the correct distance, channels can select between ions with only small differences in their radii. The fourth mechanism is electrostatic displacement of a bound ion, pushing it through the channel. The high energy of hydration of divalent ions means that their binding to channel oxygens is much higher affinity. As long as they are bound, they block the channel to passage of any other ion. Only a second divalent ion with similar affinity can displace the blocking ion, allowing one to complete its passage through the pore in a "billiard ball" fashion. These four selectivity mechanisms resemble substrate specificity of enzymes, Selectivity is the property that allows channels to separate charge. For example, a Na^+ channel discriminates between Na^+ and Cl^-. So independent movement of Na^+ separates charge and creates potential differences. Remember that there are 6×10^{23} Na^+-positive charges in a mole of NaCl (Avogadro's number), and this represents about 10^6 coulombs of charge (Faraday's number).

The energy for net transfer of ions is found in the electrochemical gradients established across the membrane. Although channels are just as important for ion movement between intracellular compartments such as mitochondria and endoplasmic reticulum, this discussion focuses on the surface membrane. The Na,K pump establishes a high K^+ concentration (e.g., 140 mM) within the cytoplasm, but outside [K^+] is much lower (e.g., 4 mM). K^+ exits the cell through channels down its chemical concentration gradient. The anion partners of K^+ (necessary to approach electroneutrality) in the cell are mostly large negatively charged molecules that are too large to pass channels ("fixed anions"), so exit of K^+ leaves the cytoplasm negative. This negative field attracts K^+ back into the cell, and an equilibrium is approached, defined as the Nernst potential for K^+, and this is primarily responsible for the stable electrical difference across all cell membranes (the cell's resting potential). Intracellular Na^+ is about 5–10 mM compared to extracellular Na^+ of about 140 mM, so the total Na^+ electrochemical difference from the resting potential is approx 130 mV, providing a large energy source for net entry of Na^+ through open channels or to drive coupled transporters. In each case, ions will have net movement down their electrochemical gradients when a channel is open that allows their passage.

Ca^{2+} requires a special discussion, because of its critical roles in cytoplasmic signaling. The cell membrane has an ATP-dependent Ca^{2+} pump and/or a Na/Ca exchanger (which uses the energy in the Na^+ gradient to transport Ca^{2+} out of the cytoplasm, lowering it from outside levels of about 2 mM to near 100 nM in the cell. This results in a huge electrochemical gradient for Ca^{2+} entry if Ca^{2+}-permeable channels are open that permit its passage. Under those conditions, the membrane potential can have a huge effect on the amount of Ca^{2+} entering. Also, because of the low cytoplasmic concentration, the net flow of Ca^{2+} through channels can raise the cytoplasmic concentration to micromolar levels, initiating many of the cytoplasmic processes controlled by cytoplasmic Ca^{2+}. Cytoplasmic Ca^{2+} is regulated by surface and endoplasmic reticulum channels.

2.1.2. Gating

Although some channels remain permanently open, most open only intermittently. The process of opening and closing of channels is called gating (Fig. 1), and it is particularly susceptible to modulation by other cell systems. The two principal mechanisms of gating are ligand binding and membrane electric field. Binding of a signaling molecule to the channel complex acts allosterically to open an occluded permeation path. Other channels have protein conformations responsive to the membrane's electric field. Some channels have multiple gates, so that ligand binding or voltage change can initiate both opening and closing of channels, with the processes following different time courses. Individual channels are either open or shut before interaction with gating signals. Binding of ligand or an electric field then alters the probability that the channel will open or shut, so the channels behave stochastically. Whole cell currents are channel population responses, resulting from time-dependent recruitment of channels to the open state or to closure. Some channels have several open conformations with different permeation rates called subconductance states, providing another mechanism for modulation.

Gating mechanisms have been studied more in voltage-gated channels, probably because of the early pioneering work of Hodgkin and Huxley describing Na^+ and K^+ channels as the ionic mechanism for the nerve action potential. A major concept for channel gating is the "ball and chain" model for inactivation. This proposes that a protein plug has a binding site at the inner mouth of the pore, and when bound it occludes the pore. This plug is attached by an amino acid chain, which keeps the plug near the pore. Somehow, ligand binding or changes in electric field can release the "safety" on the chain, allowing the ball to block the pore. This has been shown unequivocally to be the mechanism for fast inactivation of Shaker type K^+ channels. A similar mechanism exists for Na^+ channels, except that the ball is tethered at both ends, producing a "hinged lid" blocking mechanism. A third mechanism is called C-type inactivation. It includes a conformational change in the outside mouth of the pore by a mechanism as yet unresolved.

In contrast to inactivation, activation (channel opening) is less well understood, and this is a current frontier in channel biophysics. One of the best models is that of the nicotinic acetylcholine (ACh) receptor channel in nerve and neuromuscular synapses. Binding of two ACh molecules opens the channel by rotating two pore-lining α-helices that have a kink near their intracellular ends. The rotation moves the kinked helices apart, opening the pore. Similar mechanisms are suspected for voltage-gated channels.

Although not exactly gating, another mechanism for regulating channel opening is found in channels that show rectification, a non-ohmic behavior of channels. Rectification is the ability of some channels to conduct only in one direction or much better in that direction than in the other. One mechanism seen in inward rectifier K^+ channel and in the N-methyl-D-aspartate (NMDA) type of glutamate receptor channels is Mg^{2+} block. These channels have a Mg^{2+} binding site within the pore. Depending on the electric field, Mg^{2+} can access and bind to the site, thereby blocking permeation. A second similar mechanism in inward rectifier channels is the entry of the small cytoplasmic polycationic molecules spermine and spermidine into the pore occluding it.

2.2. Structural Motifs

2.2.1. General Channel Structure

The general organization of ion channel proteins is similar to other intrinsic membrane proteins such as transporters and receptors. The long sequential chain of amino acids is woven through the membrane multiple times, creating separate extracellular and intracellular regions of the molecule. The N-terminus is usually in the cytoplasm, but the location of the C-terminus is variable. The extracellular region often has multiple sites for glycosylation. In contrast to soluble proteins, the ion channel three-dimensional structure is hydrophobic on the outside, where it contacts the membrane lipid.

It has gradually become apparent that there are several structural motifs that characterize ion channels. The neurotransmitter-gated channels are usually composed of five subunits, each with three or four transmembrane segments, arranged around a large, poorly selective central pore. Voltage-gated channels have four subunits or four covalently connected domains that surround a pore, with each subunit or domain containing six transmembrane segments (Fig. 2d,f). Another large family of inward rectifier channels has a similar four-subunit structure, but with smaller subunits containing only two transmembrane segments each (Fig. 2b). These channels are usually K^+ selective or cation selective, and they are gated or modulated by mostly cytoplasmic mediators, including G proteins, ATP or cyclic nucleotides, and/or Ca^{2+}. Channels in this structural group function as sensory receptors by responding to extracellular molecules or physical processes such as stretch or temperature.

The five subunits of the neurotransmitter gated channels may be identical, forming a homopentamer, but are most often composed of different but similar subunits. For example, the nicotinic ACh receptor is composed of two α-subunits, and one each of β-, γ-, and δ-subunits. Each subunit has a similar membrane topology of four transmembrane segments The opportunity to mix different subunits from a family of channel proteins provides a rich variety of channels to suit various functions. The voltage-gated Na^+ and Ca^{2+} channels have a four-domain α-subunit, equivalent to covalently linking four subunits for the K^+ channel. These four domains define the pore, so different types of channels are not different mixtures of subunits, but they must be different gene products. The four-subunit K^+ and nonselective channels all have substantial subfamilies that can assemble to make heterotetramers, providing a rich variety of functional channels.

Voltage-gated channels and the two-transmembrane segment family of channels typically have additional subunits that do not contribute directly to the pore. For example, the brain Na^+ channel has two β-subunits, each with only one transmembrane segment (Fig. 2a). The Ca^{2+} channel has three or four subunits, at least one of which is transmembrane. K^+ channels often have β-subunits, and their stoichiometry is now being studied. The roles of these extra subunits are

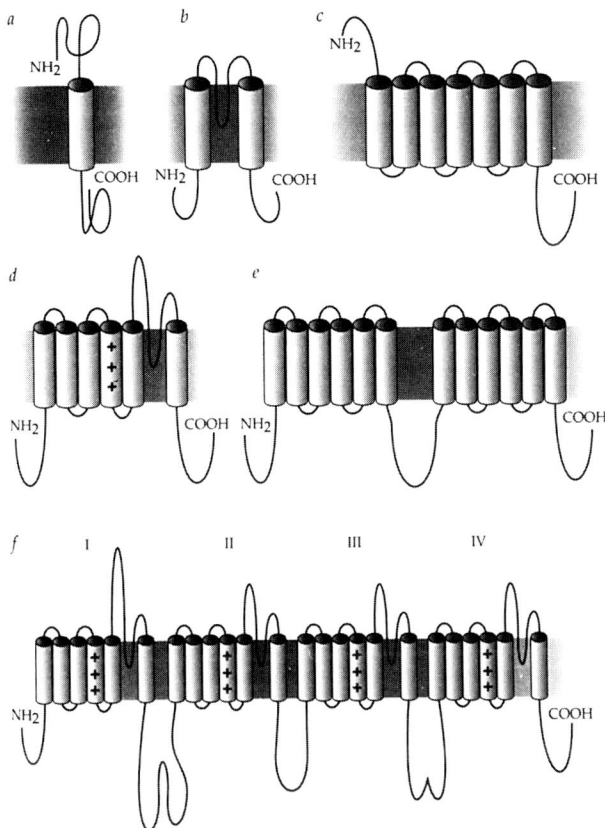

Fig. 2. Membrane topology of several ion channels and membrane transporters. (a) Typical supplemental subunit with single transmembrane segment. (b) Two-transmembrane segment structure of the inwardly rectifying K^+ channel family. Note that the extracellular segment is folded back into the membrane to form a P loop. (c) A seven-transmembrane segment protein typical of the muscarinic G protein-coupled receptor. (d) The six-transmembrane segment pattern of the voltage-gated K^+ channel subunit. The channel requires four such subunits to form the pore. The fourth segment (S4) is highly charged and represents part of the voltage sensor. (e) 12-transmembrane segment structure for several membrane transporters and the CFTR channel. (f) 24-transmembrane segment protein, organized into four covalently linked homologous sets of six segments, each similar to the voltage-gated K^+ channel subunit. This protein forms a pseudotetramer-subunit to create a pore Na^+ and Ca^{2+} channels. (From Fozzard HA. In Haber E (ed). Molecular Cardiovascular Medicine. New York: Scientific American, 1995, pp. 211–224.)

varied, including stabilizing the pore-containing complex during transport and insertion into the membrane, anchoring the channel in functionally important locations to make bipolar cells, or modulating the channel's gating. The roles of these nonpore subunits are only now being studied.

Although many channel amino acid sequences have been known for some time, and their membrane topology inferred, the detailed three-dimensional structure necessary to understand these beautiful machines has been out of reach. The proteins are too large for solution by nuclear magnetic resonance (NMR) analysis, although some soluble fragments can be studied. X-ray analysis awaits development of useful crystals. Extensive electron density studies of two-dimensional arrays of nicotinic ACh receptors have reached 5 Å resolution, providing a general organizational scheme. The transmembrane (pore-containing) piece of a bacterial two-transmembrane K^+ selective channel has been crystalized and analyzed to about 2 Å resolution (Fig. 3). Valuable structure–function information has been derived from tedious experiments using chimeras and point-mutated channels expressed in heterologous systems and studied with electrophysiological techniques. More direct physical studies with electron paramagnetic resonance spectroscopy (EPR) and fluorescence energy coupling have just begun. Extrapolation of these clues by molecular modeling using rules of protein structure derived from X-ray analysis of soluble proteins has been invaluable in creating an intriguing picture of the channels. The brief review that follows must be considered a first glimpse into this rapidly developing field.

2.2.2. THE PORE

Two structural concepts have been developed to describe the pore, one for the nicotinic ACh receptor (hopefully as a model for the five-subunit channel family) and the other for voltage-gated channels. In both, the bulk of the water-filled pore is lined with α-helices. For the nicotinic ACh receptor, the second of the four transmembrane segments (M2) lines the pore, and several rings of charged and hydrophobic residues contributed by each of the five subunits determine permeation properties. Voltage-gated channels have segments called P loops, located on the outside of the protein between the fifth and sixth transmembrane segments of each subunit or of each of the four domains (Fig. 3). These P loops are folded back into the helix-lined pore to form the outer mouth of the channel. For channels composed of four identical subunits the P loops are identical, but the P loops differ in heterotetramers and in the four-domain Na^+ and Ca^{2+} channels.

The P loop was proposed as a critical component for selectivity of channels because of its very high homology within channels of similar selectivity. For example, the P loops of K^+ channels from plants, bacteria, and humans all contain a common motif of

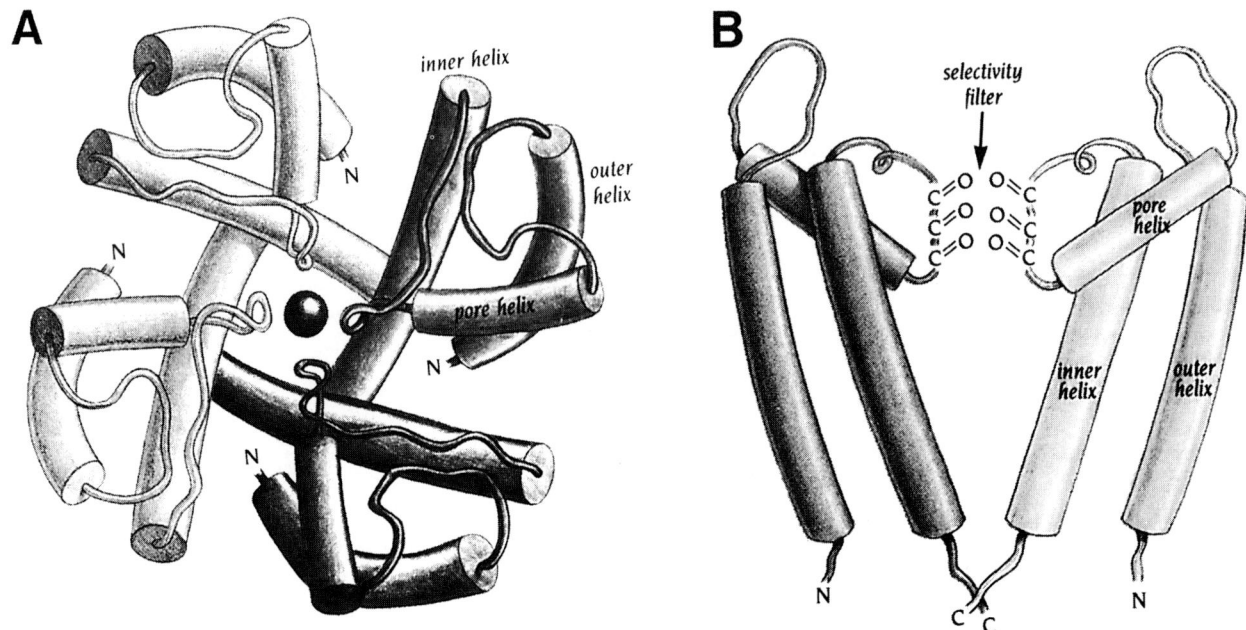

Fig. 3. Crystal structure of the KcsA channel, formed by four subunits of the topology shown in Fig. 2b. **(A)** View from above the membrane. The transmembrane helices are shown as cylinders, and the P loops as strands. Four P loops comprise a short pore helix and a random coil strand to fold into the center pore, surrounding an unhydrated K^+ ion. **(B)** Side view of two of the subunits of this K^+-selective channel. The P loop residues Gly-Tyr-Gly supply at least three carbonyl oxygens from each subunit to the critical selectivity region. (From Branden C, Tooze J. Introduction to Protein Structure, 2nd ed. New York: Garland, 1999, Chapter 12.)

a pair of glycines surrounding an aromatic residue (typically Gly-Tyr-Gly). The central role of P loops in K^+ channels was first demonstrated by switch of permeation properties by chimeras: the permeation property accompanies the P loop inserted into another channel type with different gating properties. Point mutations demonstrated that K^+ selectivity required the Gly-Tyr-Gly motif, and structural models predicted that the residues of this sequence faced away from the pore, presenting their backbone carbonyl oxygens to the lumen. X-ray analysis of the bacterial K^+ channel transmembrane piece confirmed this prediction that the selectivity of K^+ channels is determined by carbonyl oxygens in the P loop (Fig. 3). Several structurally related channels that are only cation selective have similar P loops, but with modification of the critical Gly-Tyr-Gly motif that is necessary for K^+ selectivity.

Voltage-gated Na^+ and Ca^{2+} channels have highly conserved rings of amino acid residues at the deepest penetration of the extracellular P loops into the pore. The conserved ring is estimated to be 30% into the membrane electric field. The Ca^{2+} channel P loop ring has four carboxyls, and mutation of these residues alters permeation and selectivity of the channel. The conserved ring of the Na^+ channel P loop is composed of Asp, Glu, Lys, and Ala (usually called DEKA) (Fig. 4). Mutation of these residues affects permeation and selectivity, and the Na^+ channel can be made to show the permeation and selectivity properties of the Ca^{2+} channel by mutational substitution of a carboxyl for the lysine. Although the physical–chemical properties of selectivity remain obscure, it is clear that carboxyl oxygens play an important role, in contrast to the carbonyl oxygens of the K^+-selective channels.

In spite of the almost universal presence of the DEKA motif for selectivity of the voltage-gated Na^+ channels, the amiloride-sensitive epithelial Na^+ channel does not appear to contain the motif. Several other channel types also represent challenges to our understanding of selectivity. The proton channel probably provides a channel path for protons without the customary water-filled pore. The aquaporin channel allows translocation of water without accompanying ions. Finally, anion channels have specific permeability sequences for anions, and included in the pore are several positively charged residues. The details of anion channel selectivity have not been resolved.

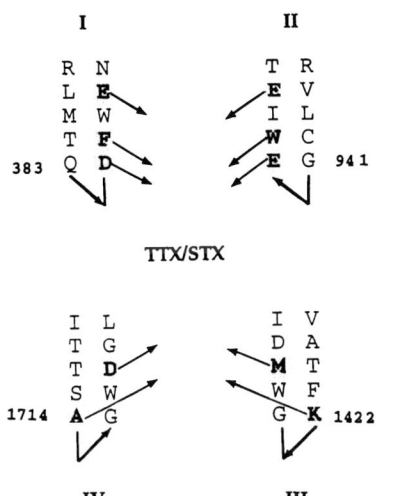

Fig. 4. Illustration of the outer vestibule and selectivity region of the Na⁺ channel. The four P loops of the most common brain Na⁺ channel are shown, one from each of the four domains (I–IV). The amino acids are identified by their single-letter code. The residues facing the central pore are identified with *arrows*. The selectivity ring is the DEKA set seen at the bottom of the loops within the turn. TTX/STX indicates the binding sites of tetrodotoxin and saxitoxin. (From Lipkind, Fozzard, Biophys J 1994; 66:1–13).

2.2.3. SENSORS

We have already referred to the ability of bound ligands and membrane electric field to change the equilibrium between various closed and open conformations of the channel. Little is known about the allosteric events responsible for ligand-dependent conformational changes, except by extrapolation from allosteric events in enzymes. However, the key components of electric field sensing have been located and several mechanisms have been proposed. Stretch and temperature are important modulators, but no details are yet available about their mechanisms.

Good progress is being made in locating the binding sites for ligands, especially the ACh binding site in the nicotinic channel and the glutamate and γ-aminobutyric acid (GABA) binding sites in neurotransmitter gated channels. The cystic fibrosis transmembrane conductance regulator (CFTR) Cl⁻ channel is gated in part by ATP and it has two cytoplasmic nucleotide binding folds. These sites have been shown to have different and critical functions in the channel gating. One very active area of study is the location of α and βγ G protein binding sites on the Ca^{2+} channel or on the G protein coupled K⁺ channels.

The most thoroughly characterized channel binding sites are those of specific toxins. The very high affinity Na⁺ channel specific tetrodotoxin and saxitoxin bind within the Na⁺ channel's outer mouth, called the vestibule, with specific interactions of several kilocalories per mole with residues of the DEKA selectivity filter residues and at least four other vestibule residues (Fig. 4). μ-Conotoxin, a toxin from sea snails, also can bind to the same region of the Na⁺ channel vestibule, and similar conotoxins bind to the vestibule of the N-type Ca^{2+} channel. Scorpion and sea anemone toxins bind to the domain IV S3-S4 extracellular loop and disrupt that voltage sensor in Na⁺ channels. Local anesthetic antiarrhythmic drugs bind partly to two aromatic residues on the domain IV S6, just below the Na⁺ channel selectivity filter. Several scorpion toxins block the voltage-gated K⁺ channels, and their interactions with the K⁺ channel's outer vestibule have been characterized in detail.

One of the earliest suggestions about structure–function was the location of the voltage sensor. The Na⁺ channel amino acid sequence showed four homologous domains of six putative transmembrane segments (Fig. 2). The fourth segment (S4) of each domain had the surprising motif of a positively charged amino acid (usually arginine) in every third position of the presumed transmembrane helix. This placed positive charges within the membrane where they could be influenced by the field, and this was proposed as the structural basis of the voltage sensor. Subsequently the same motif has been found in the sequence of every significantly voltage-sensitive channel. For example, each of the four domains of the Ca^{2+} channel α_1-subunit has the same S4 motif. For K⁺ channels composed of four subunits, each of the subunits has its S4 motif. Mutational studies have amply proved that these S4 segments are essential parts of the voltage sensor, but adjacent residues in other segments are also involved in this very complicated mechanism. These channels are very sensitive to electric field, often with an e-fold change in channel open probability with 5–6 mV change. Consequently, the arginines in the S4 structures must not be rigidly fixed by high-energy salt bridges, because that would require larger energies to displace the charges. Therefore, it seems likely that the S4 segments are surrounded by a water-filled pocket. Indeed, several mutational studies with cysteine substitutions have shown that the charge locations within the S4 can be accessed from the cytoplasm during one gating state of the channel and then from the outside solution in another state. Either the charged site translocates across the entire membrane or one access path is shut and another opened. In either case, the local electric field will change dramatically, and could plausibly set in motion the conformational changes required for opening or shutting of the pore. The mechanisms of temperature and stretch gating are only speculative at this time.

There is substantial evidence that channels are not floating freely in the membrane, but are distributed on the membrane in very specific fixed locations. The most obvious examples of this are the polarity found in epithelial cells and the composition of synapses in the nervous system. One obvious mechanism for anchoring of channels in specific locations is that they are coupled to the cytoskeleton. Several channel proteins have consensus binding sites for ankyrin in their cytoplasmic regions, and the Na^+ channel can be isolated in association with ankyrin. Interventions that are known to alter the integrity of the cytoskeleton can influence single-channel gating behavior. A particularly interesting stretch mechanism is proposed for the cochlear hair cells, where sound-generated movement of the hair opens channels that transduce the sound. Temperature could also act through the cytoskeleton by affecting assembly of the cytoskeletal complexes. The gating mechanisms of these channels involved in sensory functions are a promising area of cell biology.

3. BIOLOGICAL ROLES

The advent of high-resolution electrophysiological techniques has allowed exploration of the diversity of roles ion channels play in transducing external stimuli into functional changes at the single-cell level. Although a number of channels are designed to fit unique functional niches, for example, in sensing nucleotide levels or dynamic changes in membrane tension stretch, many cells use the same channel to control a number of functional roles ranging from salt and water transport in epithelia to mitogenesis in the immune system.

3.1. Resting Membrane Potential

All cells maintain a membrane resting membrane potential via a combination of both active transport processes, which establish ionic gradients between the intracellular and extracellular environments, and the gating of ion channels, which allows for the passive flow of ions down their respective gradients. The movement of K^+ down its electrical–chemical gradient is ubiquitously important in determining the resting membrane potential in almost all cell types.

The molecular species of K^+ channel that is utilized to mediate the passive flow of K^+ may be different from cell to cell. In excitable cells such as nerve and muscle, the outwardly rectifying K^+ channel family plays the predominant role. Activation of these channels is enhanced at depolarized membrane potentials following the regenerative, voltage-dependent Na^+ channel activation. Unlike Na channels, members of the outwardly rectifying K^+ channel family inactivate slowly, and therefore are the predominant conductance during membrane potential repolarization following the initiation of the action potential during the initial increase in Na^+ conductance. Tight control of K^+ permeability is clearly important in excitable cells, which signal to one another via action potential generation. Nonexcitable cells, those that do not fire action potentials, including myelinating oligodendrocytes and circulating macrophages, use the inwardly rectifying K^+ channels (Kir 1–6) to determine the resting potential.

3.2. Secretion

Intracellular Ca^{2+} is the key element in the mechanisms regulating secretion of hormones and neurotranmitters. Voltage-dependent Ca^{2+} channels localized at nerve terminals open in response to action potential depolarization, producing a localized elevation in intracellular Ca^{2+} (Ca_i). The transient and localized increase in Ca_i activates several Ca^{2+}-dependent proteins that are then able to trigger the movement and fusion of vesicles containing peptides or neuroactive components. Exocytosis of the vesicle then releases the neurotransmitter contents into the extracellular milieu. The entry of Ca^{2+} into cells can originate from pathways other than voltage-dependent Ca^{2+} channels. Stretch-activated ion channels are permeable to monovalent and divalent cations when the cell is mechanically perturbed. Such is the case with cardiac cells that secrete natriuretic peptides, which are important hormones in the control of blood pressure and solute and water excretion. Increases in intraatrial pressure in the beating heart trigger the opening of Ca^{2+}-permeable mechanosensitive channels, thereby up-regulating secretion not only by atrial distention but also by natriuretic peptide mRNA expression.

3.3. Metabolic Sensors

The ATP-sensitive K channel (K_{ATP}) found in cardiac and smooth muscle as well as insulin secreting pancreatic β-cells is tightly coupled to a 17-transmembrane domain, ATP-binding protein, the sulfonylurea receptor (SUR). The K^+ ionophore (Kir 6.0) is a member of the inwardly rectifying K⁺ channel superfamily. Kir 6.X isoforms are unique within the superfamily in that they do not express channel function without coexpression of the SUR receptor. Physical association of the two proteins suggests that the SUR subunit directly regulates K_{ATP} channel activity by direct interaction with the Kir subunit. Although the precise location of these protein–protein interactions are still a subject of investigation, the complex is uniquely designed to sense changes in the intracellular ATP/ADP ratio coupling glucose metabolism to electrical activity leading to insulin secretion in the pancreatic β-cell and hyperpolarization in the ischemic brain and heart. Increases in the ATP/ADP ratio closes K_{ATP} channels and conversely a decrease in the ratio opens them. In the pancreatic β-cell, the role of K_{ATP} is pivotal. Circulating serum glucose is transported into the β-cell through the activity of the glucose transporter at the plasma membrane. Glucose is metabolized within the cell resulting in an increase in cytoplasmic ATP. An increase in the ATP/ADP ratio signals the closing of the hyperpolarizing K⁺ conductance, resulting in cellular depolarization and the opening of voltage-dependent Ca^{2+} channels. The influx of extracellular Ca^{2+} leads to subsequent vesicular release of insulin and thereby glucose utilization in the periphery.

3.4. Volume Regulation

When cells are exposed to a hypotonic environment they passively swell and then spontaneously shrink over a period of minutes to return to their normal volume. The return to normal volume, the regulatory volume decrease, is mediated through the activation of both K⁺ and Cl⁻ channels. Movement of salt through the two conductance pathways obligates the osmotic flow of water out of the cell, causing it to shrink and eventually return to normal volume. Anion channels involved in this response may allow for the permeation of large cellular osmolytes including taurine, further contributing to the loss in cellular osmotic load. The passive movement of water that accompanies solute efflux from the cell may be through a member of the family of aquaporin water channel proteins. These proteins contain six membrane-spanning domains and possess a common structural motif forming a single water-selective aqueous channel that spans the plasma membrane. Water reabsorption in the kidney is under the control of the peptide hormone vasopressin and appears to involve alterations in the expression and trafficking of aquaporin channels through G protein activation and an elevation of cAMP.

3.5. Growth and Proliferation

Functional K⁺ channels have been shown to be a requirement for activation and proliferation of T and B cells in the immune system by mitogenic stimuli. Expression of a specific isoform of outwardly rectifying K⁺ channels (Kv1.3) similar to those found in nerve and muscle is correlated with the degree of cell proliferation. In addition, the density of K⁺ channels in actively cycling cell subsets found in the thymus is 10- to 20 fold higher than that found in quiescent cells. Conversely, mitogenesis is inhibited in a dose-dependent manner by compounds that block K⁺ channel activation. The link between an enhancement of nuclear transcriptional activity and membrane hyperpolarization is mediated through transduction of a second channel. Lymphocytes, like almost all nonexcitable cells, do not express voltage-dependent Ca^{2+} channels. Transient increases in Ca_i used in a number of cytoplasmic signaling pathways, including the response to mitogens, is brought about by the release of Ca^{2+} from internal stores through the opening of Ca^{2+}-selective, inositol triphosphate sensitive channels in internal membranes as well as the influx of Ca^{2+} through a voltage-insensitive Ca^{2+}-permeable pathway localized in the plasma membrane. The plasma membrane conductance pathway is sensitive to the state of stored Ca^{2+} in the intracellular compartment through a mechanism that is actively under study and may involve both direct and indirect cytoplasmic signaling pathways. Importantly, however, the passive divalent ion influx through this pathway is not gated by voltage but rather is augmented at hyperpolarized membrane potentials. Thus, the increase in K⁺ channel expression seen following the exposure of lymphocytes to mitogenic stimuli provides for a stable hyperpolarized membrane potential supporting Ca^{2+} influx through store-operated Ca^{2+} channels.

3.6. Sensory Transduction

Ion channels are critical elements in most sensory transduction systems serving to convert changes in stimulus intensity to changes in membrane potential. Cyclic nucleotide gated (CNG) channels act as signal transducers in both phototransduction and olfaction. These channels are structurally related to the voltage-gated K⁺ channels but are activated by the direct binding of either cGMP (photoreceptors) or cAMP (olfactory receptors) to a cytoplasmic site on the C-terminus of each of the subunits that make up the channels. Like their cousins the voltage-gated K+ channels, CNG channels are heterotetramers of four subunits arranged around a central ion conduction pore. The possibility for binding of multiple ligands serves to steepen the relationship between stimulus intensity and response, ensuring efficient neuronal signaling. Thus, in the case of the CNG channels, binding of a single nucleotide significantly increases open channel probability with maximal open state probability achieved only when four ligands are bound. CNG channels are nonselective cation channels. The influx of cations through CNG channels leads to cellular depolarization and subsequent increases in Ca_i, initiating a cascade of cellular transduction events that is cell type specific.

3.7. Cell–Cell Communication

Intercellular communication to synchronize the response of populations of cells to external stimuli is mediated by intercellular channels present in gap junctions. Cellular processes that make use of this conductance pathway include coordinated excitation of cardiac and smooth muscle as well as the transmission of excitatory signals at electrical synapses that do not employ neurotransmitters. These channels are encoded by a multigene family called the connexins, forming structural proteins that span two plasma membranes. Connexins form single-membrane channels (connexons) as oligomeric complexes, which align between the two cells to form a continuous pathway. Typically, each connexon is composed of six connexins that are arranged concentrically around the central pore. The interconnecting complex, which can be composed of different connexons, is then termed the gap junction. Gap junctions are relatively nonselective, allowing for the passage of signaling molecules such as the cyclic nucleotides cell to cell.

4. MODULATION

Ion channels are integrated into a number of intracellular signaling cascades. They serve as targets for phosphorylation by cytoplasmic kinase, as well as gates for the influx of calcium, linking a number of signaling cascades. The activity of ion channels is linked to G proteins both directly and indirectly through the activation of a variety of seven transmembrane receptors. Ion channels can activate or gate neighboring populations of channels by allowing the influx of Ca^{2+} and they may regulate the expression of other populations of channels within the same cell.

Thus, ion channels serve as targets for cellular regulation while themselves serving as regulatory elements.

4.1. Phosphorylation

Ion channels function as effective, cell-surface signal transducers by means of their interaction with cellular protein kinases. Coded within each protein's primary structure are consensus sequences that serve as phosphorylation sites for a number of cellular kinases including the serine threonine kinases as well as tyrosine kinases. Channel phosphorylation may gate the channel, may shift the voltage dependence of gating, or may facilitate interaction with other regulatory proteins including the cytoskeleton. Specific protein–protein interaction between signaling proteins and ion channels are mediated by modular binding domains including the Src homology 3 domain of the Src tryosine kinase family, PDZ domains, and guanylate kinase-like domains.

The CFTR chloride channel is localized to the apical membranes of epithelial cells lining secretory tubules and organs including the sweat duct, trachea, and pancreas. The channel is gated directly via cAMP-dependent phosphorylation of its cytoplasmic regulatory domain via the activity of protein kinase A. The phosphorylation-dependent gating of CFTR is arguably one of the most active areas in the study of ion channel modulation by cytoplasmic kinase not only because of its complexity but also because of the contribution of the channel to the movement of salt and water transport in the lung. The most common mutation in the molecule with results in the disease of cystic fibrosis (CF) is a single amino acid deletion in the cytoplasmic first nucleotide binding fold of the protein (Δ508) causing incomplete processing of the protein, resulting in the loss of channel targeting to the apical membrane. The absence of the phosphorylation-dependent anion conductance of the affected epithelia, especially in the lungs, results in the loss of chloride movement that drives water flow and the formation of a dilute aqueous saline secretion. Although many organs are affected by the disease, the most devastating effect of the loss of chloride conductance is seen in the lungs. The normal mucus present in the lungs is severely dehydrated in CF patients. The thick immobile mucus effectively traps bacteria, and this infection produces the inflammatory response that eventually destroys the lungs and causes death.

Ions channels serve as targets of cytoplasmic kinase phosphorylation as well as interact with kinases participating in complex cellular signaling cascades. An example of the interaction of an ion channel in such a signaling cascade is the α-amino-3-hydroxy-5-methyl-4-isoxazole propionic acid (AMPA) channel of the glutaminergic receptor neurotransmitter family. The AMPA receptor on the postsynaptic membrane in a number of central nervous system synapses relays signals to the nucleus by means of its interaction with the Src-family protein tyrosine kinase, Lyn. Recent studies indicate that the AMPA receptor acts not only as a ligand-gated ion channel in normal synaptic transmission but also as a sort of postsynaptic sensor that recognizes only the strong stimulation from the presynaptic to the postsynaptic membrane. Thus, Lyn-mediated signaling through the AMPA receptor may contribute to reshaping synaptic strength through a process that involves gene expression and the induction of intracellular signaling pathways that mediate release of neurotrophins.

4.2. G Protein Linked Receptor Activation

Activation of a single G protein-linked receptor may generate a multiplicity of effects at the level of cellular electrophysiological responses. Receptor activation of membrane phospholipase C results in the liberation of the second messenger inositol-1,4,5-triphosphate, which then mobilizes intracellular free Ca^{2+} through the activation of Ca^{2+}-permeable channels in the endoplasmic reticulum. Reduction in the level of Ca^{2+} within the intracellular stores signals the opening of a Ca^{2+}-selective influx pathway in the plasma membrane through a signal transduction element that is yet to be elucidated. An increase in the levels of cytoplasmic diacylglycerol induces the translocation of protein kinase C to the plasma membrane, which may phosphorylate both anion and cation channels giving rise to the activation of Cl^- channels in some cells. Transient increases in intracellular Ca^{2+} may lead to the activation of calcium-dependent K^+ channels via direct binding to cytoplasmic sites. It also leads to the activation of the multifunction intracellular Ca^{2+}–calmodulin-dependent kinase, which may phosphorylate plasma membrane Cl^- channels and induce activation. G protein subunits released during receptor activation may also activate ion channels directly as is the case for the inwardly rectifying K^+ channels in the heart and brain. These channels are gated through the binding of the G protein βγ complex to a cytoplasmic C-terminal site of the channel directly. Conversely, the same G protein βγ complex could bind to a cytoplasmic site on a voltage-dependent Ca^{2+} channel, augmenting the channel's activation in response to voltage during a second voltage stimulus. The formation of free fatty acids such as arachidonic acid as a result of phospholipase C activation may activate K^+ channels directly or indirectly through the opening of a Ca^{2+} influx pathway, as well as by the release of Ca^{2+} from intracellular stores.

4.3. Clustering of Ion Channels with Regulatory Elements at Specific Subcellular Sites

The assemblage of ion channels and receptors at specific sites such as the neural synapse is crucial for signaling. However, the mechanisms by which proteins are targeted to and clustered at these sites is still poorly understood. Compartmentalization of ion channels with receptors at the synapse is established via protein–protein interactions with the submembranous postsynaptic density-95 protein, a protein containing PDZ domains. PDZ domains are found in a large number of multifunctional proteins, where they mediate complex protein–protein interactions. There is a short consensus sequence at the C-terminus of membrane proteins that is critical for the interaction with PDZ domains consisting of the amino acids (D/E)(S/T)XV. With the recognition that channels may exist in a multiprotein complex via interactions with a PDZ-containing protein scaffold comes the possibility that ion channel modulation is not determined solely by the random diffusion of regulatory elements in the vicinity of the channel but rather by the concerted positioning of regulatory elements in close proximity with the channel itself. The targeting of ion channels and receptors to discrete plasma membrane domains is then critical to cellular function. It is clear that identification of proteins that associate with ion channels, either directly or via scaffolding proteins, will facilitate our understanding of how membrane kinases and phosphatases efficiently control ion channel activation and, therefore. fine tune cellular function.

4.4. Channel Interaction with Membrane Phospholipids

Membrane phospholipids have recently been identified as significant regulatory elements in ion channel activation. The activity of the ATP-sensitive, inwardly rectifying K^+ channels expressed in insulin-secreting β cells in vascular smooth muscle and heart is dependent upon both intracellular levels of ATP and coexpression of the sulfonylurea receptor. The activity of K_{ATP} is inhibited by ATP binding to the tightly linked sulfonylurea receptor and, therefore, highly sensitive to cellular metabolism. Although the mechanism of channel activation/inhibition by ATP is similar in channels expressed in all three tissues, their sensitivity to ATP varies widely. Recently, it has been shown that the inhibitory effect of ATP on channel activation is dependent on the membrane phospholipids phosphatidylinositol-4,5-bisphosphate and phosphatidylinositol-4-phosphate. Membrane phospholipid composition may vary from tissue to tissue. The profound effects of membrane phospholipids on K_{ATP} activity may explain the divergence in ATP sensitivity between tissues and furthermore provide a fine-tuning mechanism for activity of the channel in the determination of excitability in neurons and vascular smooth muscle and secretion of insulin in the pancreas.

5. EVOLUTION OF ION CHANNELS

5.1. Ion Channels Are Ancient Proteins

The remarkable similarity of structure and function of ion channels in biology emphasizes their general roles in cells. Consequently, it is not surprising that the genes and proteins are highly conserved. Prokaryotes (archaebacteria and eubacteria) have several types of ion channels that are surely ancestors of mammalian channels. Although the functions of bacterial ion channels are not well understood yet, bacteria do need to regulate their ionic environment to maintain their metabolism and to regulate their volume. Channels in bacteria show the characteristic four-subunit motif for formation of a transmembrane pore, and they use a typical P loop motif for K^+ selectivity. Indeed, the Streptomyces K^+ channel has a three-dimensional structure very much like human K^+ channels.

When eukaryotes evolved about 1400 million years ago, a dramatic increase in the type and variety of ion channels occurred. K^+ channels developed the S4 voltage sensor motif, probably by a mosaic process, and channels acquired a variety of ligand switches, including those for amino acids. Presumably by gene duplication, eukaryotes developed the covalently linked four-subunit motif of Ca^{2+} channels, maintaining the S4 voltage sensor and perfecting a new type of P loop selectivity filter. A third type of newly developed channel in plant and animal cells are the Cl^--selective channels, some of which are gated by intracellular Ca^{2+}. Essential to electrical excitability, the cells also developed a variety of special ion co- and countertransporters.

Eukaryotic cells depend on Ca^{2+} as a major intracellular messenger for modulation and for motility. In combination with the new set of ion channel pumps, the dual ion channel system of K^+ channels and voltage-dependent Ca^{2+} channels mediated important cellular behavior, including excitability.

The next large evolutionary step was to multicellular organisms—fungi, plants, and animals. Now cells differentiated to perform special functions using the new diversity of channels. For several types of K^+ and Ca^{2+} channels it is not easy to distinguish recordings obtained from mammalian cells, algae, and plants. But the new biological need was for fast communication systems, and several types of nervous systems evolved. Although action potentials are achievable in single-cell organisms, these large organisms needed fast conduction for integration of their movements. This was difficult to achieve with Ca^{2+} channels because the large currents required for fast conduction would flood the cells with Ca^{2+}, producing unwanted signals or injury to the cells. This new fast communication function was achieved by development of the Na^+ channel, with its high influx of the relatively nontoxic Na^+. Eukaryotic cells have invested heavily in the ability to achieve large transmembrane Na^+ gradients, including the rapidly conducted action potentials and a variety of transport systems dependent on the Na^+ gradient. The most primitive Na^+ channel so far has been found in the jellyfish, which is more than 60% identical in its amino acid structure and has all of the typical kinetic and most of the pharmacologic properties of mammalian Na^+ channels. These multicellular organisms also developed a variety of ligand-gated ion channels for synaptic transmission.

5.2. Prokaryotic Origins of Ion Channels

Bacteria depend strongly on intrinsic membrane transport proteins to maintain their environment and metabolism. Several types of ligand binding motifs that are used in eukaryotic ligand gated ion channels probably are derived from these transport systems, including those for amino acids, cyclic nucleotides, and ATP. Some bacteria apparently have no channels, but others have several types of channels. The two common channel motifs are a six-transmembrane segment (6 TM), 1 P loop type, and a 2 TM, 1 P loop type. They may be coupled to a sequence that contains a NAD^+ binding site (presumably so that gating could respond to the cell's redox state) or without that site. One without the NAD^+ site does have a critical histidine that allows pH gating. It is not clear whether the NAD$^+$ binding channel preceded or followed in development. Although it is plausible to think of the 6 TM/1 P channel as the ancestor of eukaryotic K$^+$ channels, the first four transmembrane segments of the 6 TM/1 P type do not show any relationship to the eukaryotic ones. Specifically, no prokaryotic channel has the S4 voltage sensor motif. Perhaps the 2 TM/1 P prokaryotic channel represents the eukaryotic ancestor, and subsequent development added the new four transmembrane segments. Bacterial cells also show genes for several of the 2 P type, including both a 6 TM/1 P + 2 TM/1 P type and a 2 TM/1 P + 2 TM/1 P type. In any case, the P loop with its characteristic Gly-Tyr-Gly signature sequence for K$^+$ selectivity is a critical conserved element between prokaryotes and eukaryotes.

5.3. Eukaryotic Four-Subunit Channels

According to analysis of sequences of the residues surrounding the K$^+$ selectivity signature sequence, several K$^+$ channel families developed quite early in eukaryotic evolution, including the 2 TM segment K$_{IR}$ family, the Kv family of voltage-dependent channels, the eag family, and the tandem 2 P loop families. The 6 TM segment cyclic nucleotide gated chanels are closely related, although they have become nonselective for Na$^+$ and K$^+$. We know less about other nonselective channels such as the volume- and stretch-activated channels, although there is some suggestion that they may have been derived from early 2 P loop ancestors.

The four-domain Ca^{2+} channel is presumed to have been formed by two gene duplication steps. Evidence for this includes the resemblance between domains I and III and between II and IV, suggesting that the channel existed in a two-domain form for some time. No present-day Ca^{2+} channels have been found with that structure, but several intrinsic membrane proteins have a pair of 6 TM segment domains. It is also assumed that the Na$^+$ channel is derived from some primitive Ca^{2+} channel, perhaps the T-type channel. However, it also could have been derived from a separate pair of gene duplications, and may have also developed in the two-domain form first. The selectivity filter structure of the Na$^+$ channel shows more variation than that of the Ca^{2+} channel.

As expected from such ancient genes, a variety of isoforms of Na$^+$ and Ca^{2+} channels has been found. One of their distinguishing characteristics is the extent to which they are sensitive to marine toxins—snail conotoxins for the Ca^{2+} channel and to tetrodotoxin (TTX) and saxitoxin (STX) for the Na$^+$ channel. It is plausible that there was evolutionary value in resistance to these toxins by fish and molluscs that depend on toxin-producing organisms for their food source, but our nervous system has inherited a mixture of types.

5.4. Eukaryotic Five-Subunit Channels

The highly diverse neurotransmitter-gated family of ion channels appeared sometime after the evolution of multicellular animals. They are unique in their dependence on five subunits to form the pore. Each of the five subunits has a 4 TM motif, with the second segment lining the pore. Because it has five subunits, the pore is larger and less selective than the 4 TM type. Some of these channels are composed of five identical subunits, but most have several types of subunits. The original evolutionary step was in selectivity for either cations (acetylcholine- or 5-hydroxytryptamine gated) or anions (glycine- or γ-aminobutyric acid gated). These channels are found in nematodes and insects, as well as in humans. Although not found as such in single-cell organisms, they appear to have inherited their ligand binding domains from bacterial amino acid binding ancestor genes.

5.5. Glutamate-Gated Channels

In contrast to the other amino acid gated channels, the important glutamate gated channels must be unrelated to the five-subunit type described earlier. They now appear to be composed of four subunits that each contain only three transmembrane segments. The extracellular N-terminal domain contains the glutamate binding site. After one membrane crossing, there is a structure like the K$^+$ channel P loop, although with loss of its selectivity between K$^+$ and Na$^+$. This P loop is infolded from the cytoplasmic side of the membrane, like an upside down K$^+$ channel. The P loop is followed by two transmembrane segments and finally a C-terminal tail that contains modulation sites. This structure has raised the suggestion that this family is derived from a mosaic of the gene for a bacterial glutamate receptor, the gene for an inward rectifier K$^+$ channel with loss of selectivity, and the regulatory part of some other gene.

5.6. Other Channels

Little information is yet available about the evolution of other important channels, including the six-subunit gap junctions and the intracellular four-subunit Ca^{2+}-release channels such as the ryanodine receptor or the IP_3 receptor. Much more information is needed about the amino acid structures of volume- and stretch-gated channels, before any insight can be obtained. No clues are yet available about the voltage-gated Cl^- channels or the amiloride-sensitive Na^+ channels.

In summary, bacteria developed two critical preserved motifs—the four-subunit motif for pore formation and the delicate pore structure needed for K^+ selectivity. They also developed the 2 P loop channel structure that is prominant in eukaryotic nonselective channels. Single-celled eukaryotic fungi, plants, and animals further expanded the types of K^+ channels and developed the S4 voltage sensor motif. They also generated the four-domain Ca^{2+} channel. Multicellular organisms required electrically coordinated nervous systems. For this they developed voltage-dependent Na^+ channels and amino acid gated five-subunit channels. When the various genome projects are completed, a rich opportunity will be available to explore in much greater detail the evolution of ion channels.

6. SUMMARY

Ion channels are defined by a set of critical functions: a transmembrane hydrophilic path for ions, selectivity for specific ions, and gating in response to a membrane electric field or to a host of extracellular or intracellular ligands. Through these functions, they mediate a huge range of biological functions, including electrical signaling by action potentials in nerve, muscle, and endocrine cells; regulation of cell volume; response to temperature, light, taste, sound, and pressure; and the processes involved in embryogenesis and cell activation. A combination of powerful and sensitive electrophysiological tools, such as the patch clamp, with molecular biology has facilitated progress toward understanding the structure–function of these remarkable molecular machines and sensors for a full range of physical and chemical stimuli.

SELECTED READINGS

Anderson O, Koeppke R. Molecular determinants of channel function. Physiol Rev (Suppl) 1992; 72:S89.

Babenko AP, Aguilar-Bryan L, Bryan J. A view of SUR/K_{IR}6.X., K_{ATP} channels. Annu Rev Physiol 1998; 60:667–687.

Branden C, Tooze J. Introduction to Protein Structure, 2nd edit., New York: Garland, 1999, Chapter 12.

Calaghan SC, White E. The role of calcium in the response of cardiac muscle to stretch. Prog Biophys Mol Biol 1999; 71:59–90.

Catterall WA. Cellular and molecular biology of voltage-gated channels. Physiol Rev (Suppl) 1992; 72:S15.

Dibas AI, Mia AJ, Yorio T. Aquaporins (water channels): role in vasopressin-activated water transport. Proc Soc Exp Biol Med 1998; 219:183–199.

Doyle DA, Cabral JM, Pfuetzner RA, Kuo A, Gulbis JM, Cohen SL, Chait BT, MacKinnon R. Structure of the potassium channel: molecular basis of K^+ conduction and selectivity. Science 1998; 280:69.

Fozzard HA. Ion channels and cardiac function. In Haber E (ed). Molecular Cardiovascular Medicine. New York: Scientific American 1995: pp. 211–224.

Fozzard HA, Hanck DA. Structure and function of voltage-dependent sodium channels. Physiol Rev 1996; 76:887.

Gadsby DC, Nairn AC. Control of CFTR channel gating by phosphorylation and nucleotide hydrolysis. Physiol Rev 1999; 79:S77–S107.

Gelfand EW, Mills GB, Cheung RK, Lee JW, Grinstein S. Transmembrane ion fluxes during activation of human T lymphocytes: role of Ca^{2+}, Na^+/H^+ exchange and phospholipid turnover. Immunol Rev 1987; 95:59–87.

Hille B. Ion Channels of Excitable Membranes, 2nd edit., Sunderland, MA: Sinauer Associates, 1992.

Kerschbaum HH, Cahalan MD. Single-channel recording of a store-operated Ca^{2+} channels in Jurkat T lymphocytes. Science 1999; 283:836–839.

Ranganathan R, Ross EM. PDZ domain proteins: scaffolds for signaling complexes. Curr Biol 1997; 7:R770–773.

Strong M, Chandy KG, Gutman GA. Molecular evolution of voltage-sensitive ion channel genes: on the origins of electrical excitability. Mol Biol Evol 1993; 10:221.

Tokuda M, Hatase O. Regulation of neuronal plasticity in the central nervous system by phosphorylation and dephosphorylation. Mol Neurobiol 1998;17:137–156.

Yamada M, Inanobe A, Kurachi Y. G protein regulation of potassium ion channels. Pharmacol Rev 1998; 50:723–760.

Yeager M. Structure of cardiac gap junction intercellular channels. J Struct Biol 1998; 121:231–245.

9 Calcium Ions as Intracellular Messengers
Pathways and Actions

Stanko S. Stojilkovic with Melanija Tomić, Taka-aki Koshimizu, and Fredrick Van Goor

Contents

INTRODUCTION
LIGAND-GATED CATION CHANNELS
VOLTAGE-GATED CALCIUM CHANNELS
INTRACELLULAR CALCIUM RELEASE CHANNELS
STORE-OPERATED CALCIUM CHANNELS
CALCIUM-MOBILIZING RECEPTORS AND V_m PATHWAY
TEMPORAL, SPATIAL, AND CELL-SPECIFIC CALCIUM SIGNALING PARADIGMS
INTERCELLULAR CALCIUM WAVES
CALCIUM-CONTROLLED CELLULAR MACROMOLECULES
CALCIUM-CONTROLLED CELLULAR PROCESSES
CONCLUSION
SELECTED READINGS

1. INTRODUCTION

Calcium acts as both an extracellular (first) and an intracellular (second) messenger to regulate a diverse array of cellular functions, from cell division and differentiation to cell death. Accordingly, Ca^{2+} concentration in the plasma, as well as in the intercellular and intracellular environment, is under tight neuronal and hormonal control. The availability of Ca^{2+} for controlling Ca^{2+} concentration in the plasma and extracellular environment is secured from both diet and the bone matrix, thereby preventing any risk of Ca^{2+} deprivation. Changes in the extracellular free Ca^{2+} concentration ($[Ca^{2+}]_e$) signal the activation of Ca^{2+}-sensing receptors, which act to restore basal $[Ca^{2+}]_e$. Although these receptors are found in several tissues, those expressed in parathyroid cells are intimately involved in the control of $[Ca^{2+}]_e$. This chapter focuses on the role of Ca^{2+} as an intracellular messenger and the pathways that regulate its intracellular concentration. All selected references are review articles that can direct the readers to more specific aspects of Ca^{2+} signaling. Also, several chapters in this textbook are critical for understanding some of the aspects of Ca^{2+} signaling. This includes, but is not limited to, chapters on Ca^{2+} and Ca^{2+}-binding protein-mediated pathways, G protein-coupled receptors, phospholipases, adenylyl cyclases, protein kinase, ion channels, and apoptosis.

From: *Principles of Molecular Regulation* (P. M. Conn and A. R. Means, eds.), © Humana Press Inc., Totowa, NJ.

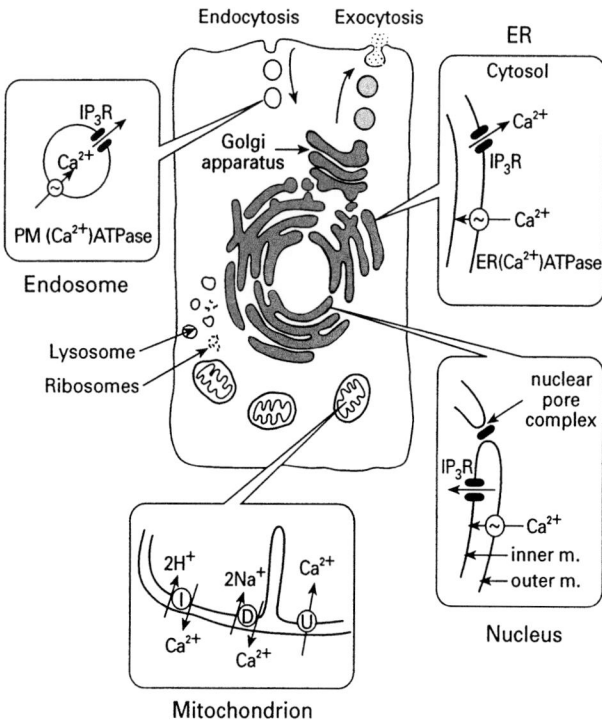

Fig. 1. Schematic representation of intracellular calcium releasable pools. PM, plasma membrane; ER, endoplasmic reticulum; IP$_3$R, inositol 1,4,5-trisphosphate receptor; m, membrane.

1.1. Calcium as an Intracellular Messenger

The generation and control of intracellular Ca^{2+} signals depends on the dynamic relationships among extracellular, cytosolic, and intraorganelle Ca^{2+} concentrations. For cells, the extracellular medium, with about 1.8 mM free Ca^{2+}, represents an unlimited and tightly controlled reservoir of Ca^{2+}. Many intracellular organelles also contain high concentrations of Ca^{2+}, which allows them to serve as an additional, but limited, reservoir. These intracellular organelles include the endoplasmic reticulum (ER)/sarcoplasmic reticulum (SR), Golgi network, mitochondria, endosomes, lysosomes, and secretory vesicles (Fig. 1). The total Ca^{2+} concentration, that is, concentration of free and bound Ca^{2+}, in intracellular organelles is in the millimolar concentration range. In contrast, the cytosol contains only 10–20% of the total intracellular Ca^{2+}, much of which is bound to soluble cytosolic proteins and membranes, making it inaccessible to many Ca^{2+}-controlled cellular processes. In unstimulated cells, the intracellular free calcium concentration ([Ca^{2+}]$_i$) is between 50 and 200 nM, whereas in stimulated cells, the [Ca^{2+}]$_i$ can increase to micromolar concentrations.

More than 100 neurotransmitters and hormones, acting on their respective receptors or receptor channels, relay their signal by altering [Ca^{2+}]$_i$. For the generation of intracellular Ca^{2+} signals, cells depend on the large electrochemical gradient of Ca^{2+} across the plasma membrane (termed *Ca^{2+} influx* or *Ca^{2+} entry*), or across ER membrane (termed *Ca^{2+} mobilization* or *Ca^{2+} release*). There are several pathways controlling Ca^{2+} influx. Ligand-gated receptor channels, including nicotinic acetylcholine receptor channels (nAChRs), glutamate receptor channels (GluRs), and ATP receptor channels (P2XRs), require the binding of an extracellular ligand, such as neurotransmitters, to open the channel pore. Many of these channels are permeable to Ca^{2+}, but also conduct other ions to control membrane excitability. In contrast, voltage-gated Ca^{2+} channels (VGCCs) are highly selective for Ca^{2+} and are opened by a decrease in the electrical potential across the plasma membrane (V_m), which is termed membrane depolarization. As many different ionic channels act in concert to control membrane potential, the voltage-gated Ca^{2+} entry pathway will be referred to as the V_m-pathway. In typical excitable cells, this pathway is typified by the firing of Ca^{2+}-dependent action potentials (APs).

Although the V_m-pathway provides a sufficient Ca^{2+} signal for the activation of many cellular functions, the Ca^{2+}-mobilizing pathway elicits more complex Ca^{2+} signals that are necessary for the selective control of cellular functions. Intracellular Ca^{2+} sources also provide a mechanism for the rapid propagation of Ca^{2+} signals within the cell, which is important in the control of intraorganelle function. Calcium mobilization occurs in many nonexcitable and excitable cell types, and is mediated through Ca^{2+} channels that are associated with intracellular organelles. These channels belong to two major classes: inositol 1,4,5-trisphosphate (IP$_3$)-sensitive receptor channels (IP$_3$Rs) and ryanodine-sensitive calcium release channels (RyRs). IP$_3$Rs are intracellular ligand-gated Ca^{2+} channels expressed in the ER membrane and are activated by the binding of IP$_3$. RyRs are also express in the ER and SR membrane and are sensitive to the plant alkaloid ryanodine, from which they are named. At the present time, it is unknown whether or not they operate as ligand-gated channels.

Calcium influx and mobilization pathways are frequently integrated to control the [Ca^{2+}]$_i$. For example, VGCCs are functionally connected with RyRs in skeletal and cardiac muscle cells, as well as in many neural and other excitable cells. Action potential-driven Ca^{2+} influx through VGCCs signals the release of intracellular Ca^{2+} from RyRs, a process known as *Ca^{2+}-induced Ca^{2+} release*. IP_3-controlled Ca^{2+} release is also commonly associated with Ca^{2+} influx through voltage-gated and other Ca^{2+}-conducting channels. In such cells, Ca^{2+} release predominates during the early phase of agonist stimulation, whereas Ca^{2+} entry is essential during sustained agonist stimulation. In nonexcitable cells, Ca^{2+} mobilization is usually accompanied by activation of Ca^{2+} entry mechanism(s) other than VGCCs. These include *store-operated Ca^{2+} channels* (SOCCs) and *nonselective Ca^{2+}-conducting channels*. In excitable cells, these channels act in conjunction with VGCCs to regulate Ca^{2+} entry.

The role of plasma membrane receptors in the control of Ca^{2+} influx and mobilization is critical. For example, modulation of the V_m pathway is predominantly controlled by G_s- and G_i/G_o-coupled receptors directly, by α- and/or β/γ-subunits, or indirectly, by intracellular messengers activated by these pathways. The control of the Ca^{2+}-mobilization-dependent signaling system is more diverse than that of the V_m-signaling pathway, as it often involves many biochemical steps and multiple second messenger molecules. The phospholipase C-controlled production of IP_3 is essential in the initiation of Ca^{2+} release through IP_3Rs, and this family of enzymes is controlled by several G protein-coupled receptors, as well as by some members of tyrosine kinase receptors. These two classes of receptors are known as *calcium-mobilizing* receptors, as they initiate the release of Ca^{2+} from intercellular Ca^{2+} stores (for review *see* Chapter 13 in this volume).

Single cells do not act in isolation, but rather in coordination to control tissue function. At the tissue level, endogenous and receptor-controlled Ca^{2+} signals are frequently synchronized, leading to intercellular Ca^{2+} waves. In many cellular networks, gap junction channels mediate the propagation of intercellular Ca^{2+} signals. These channels physically connect the cytoplasm of neighboring cells to allow the direct transfer of ions and small molecules, such as Ca^{2+}, IP_3, and cADP rybose, from one cell to another. In excitable cells, such connection also provides electrical coupling between cells. The propagation of intercellular Ca^{2+} waves can also occur by the extracellular diffusion of messenger molecules over short distances, which spreads the Ca^{2+} signal to neighboring cells by activation of ligand-gated receptors and/or receptor channels.

1.2. Patterns of Ca^{2+} Signaling and Cellular Functions

Spontaneous and agonist-induced Ca^{2+} signals can exhibit different temporal and spatial patterns. Whereas some cells respond to agonist stimulation with nonoscillatory rapid and transient rises, others respond with oscillatory increases in $[Ca^{2+}]_i$. Both the V_m pathway and the Ca^{2+}-mobilization pathway can create oscillatory Ca^{2+} responses, the first named the *plasma membrane oscillator*, and the second the *ER oscillator*. For some cellular functions, such as Ca^{2+}-dependent excitability, the oscillatory nature of Ca^{2+} signals is essential. For example, Ca^{2+} oscillations provide the frequency coding of the extracellular stimuli, which informs the cell of the intensity of hormone stimulation. Calcium oscillations also protect the cell from the toxic effects of sustained elevations in $[Ca^{2+}]_i$. For others, periodic Ca^{2+} release is not imperative, but contributes additional features in the control of cellular functions such as exocytosis.

In addition to different temporal patterns, distinct spatial patterns can occur in response to agonist activation. Calcium signals can be *localized* within the cytoplasm, such as those driven by electrical activity, but can also be propagated within the cell (intracellular Ca^{2+} wave) or between cells (intercellular Ca^{2+} wave) to produce more global Ca^{2+} signals. The spatial and temporal aspects of Ca^{2+} signals vary depending on a number of factors, which include receptor and cell type, distribution of plasma membrane receptors and channels, intracellular distribution of Ca^{2+} release channels and ER (Ca^{2+})ATPase, and the distribution of Ca^{2+}-buffering proteins within the cytosol. In addition to the ER, the mitochondria and the nucleus participate in intracellular redistribution of Ca^{2+}. Although other cytoplasmic organelles are also equipped with the mechanism for Ca^{2+} influx and release, it is unlikely that in agonist-stimulated cells they play an important role in the dynamic intracellular Ca^{2+} equilibrium.

The complexity of the temporal and spatial organization of Ca^{2+} signals is in accord with the numerous functions of intracellular Ca^{2+} functions. Calcium signaling is required for many intracellular Ca^{2+} storage compartment functions, including control of mitochondrial metabolism, "packaging" in the trans-Golgi network, exocytosis of secretory vesicles, and control of gene expression in the nucleus. The activity of several enzymes is also dependent on $[Ca^{2+}]_i$. These include calmodulin, protein kinase C, phospholipase

C, and phospholipase D. Cytosolic calcium is an active participant in the control of IP$_3$Rs and RyRs, as well as in the control of several calcium, potassium, sodium, chloride, and nonselective cationic channels. Calcium has also been implicated in the control of capacitative calcium entry and cytoskeletal functions. Finally, Ca^{2+} serves as a signal for cell fertilization, differentiation, and death.

1.3. Intracellular Ca^{2+} Homeostasis

In addition to plasma membrane and intraorganelle Ca^{2+} channels, Ca^{2+} pumps in the plasma membrane and ER membrane are essential for the control and maintenance of [Ca^{2+}]$_i$. Calcium removal from the cytosol is required to stop many Ca^{2+}-regulated events and to allow for additional or repetitive stimulation. Because [Ca^{2+}]$_i$ in activated cells is in the nanomolar to low-micromolar concentration range and intraluminal Ca^{2+} is in a high-micromolar to low-millimolar concentration range, Ca^{2+} uptake into intracellular organelles and Ca^{2+} efflux across the plasma membrane proceed against its concentration gradient and thus require energy. Such transport occurs through three well-characterized pathways, Ca^{2+}-ATPase, ion exchangers, and the Ca^{2+} electrogenic uniport. Ca^{2+}-ATPases and ion exchangers are expressed in both the SR and the ER, as well as in the plasma membrane. Calcium electrogenic uniport operates in mitochondria and utilizes the mitochondrial membrane potential (Fig. 1).

Plasma membrane Ca^{2+}-ATPases are expressed in both excitable and nonexcitable cells and provide an effective mechanism for removal of cytosolic Ca^{2+}. The ER Ca^{2+}-ATPases function as pumps in a manner comparable to the plasma membrane Ca^{2+}-ATPases, actively transporting Ca^{2+} from the cytosol into the internal stores. A variety of tissue-specialized isoforms of the SR Ca^{2+}-ATPase (SERCA) have been identified. All have a high affinity for Ca^{2+} (0.1–0.4 µM) and transport it with a second-order, sigmoidal [Ca^{2+}]$_i$ dependence. These characteristics of the ER Ca^{2+}-ATPase are appropriate for maintaining low [Ca^{2+}]$_i$. There are several specific SERCA inhibitors, such as thapsigargin and 2,5-di(*tert*-butyl)-1,4-benzohydroquinone (BHQ). These inhibitors can be used in experimental conditions to deplete Ca^{2+} stores and to increase [Ca^{2+}]$_i$.

Intracellular Ca^{2+} buffers are also important in the control of Ca^{2+} signals and cellular Ca^{2+} homeostasis. For example, the diffusion rate of Ca^{2+} is dependent on cytosolic and ER calcium buffers. The concentration of Ca^{2+} buffers in the cytosol is in the range of 100–300 µM, and is in the millimolar range in the ER. The cytosol contains slowly diffusible Ca^{2+}-binding proteins that rapidly buffer cytosolic Ca^{2+}. About 25% of the cytosolic Ca^{2+} buffers are mobile, which is important for propagation of intracellular Ca^{2+} signals. The ratio between free and bound Ca^{2+} in cytosol is about 1:100. It is possible that the role of the Ca^{2+} buffers is significant near the ER Ca^{2+} release channels, and that their saturation is required for the initiation of Ca^{2+} spikes and waves. In other words, the balance between Ca^{2+} release from the ER and its binding to soluble proteins determines the spreading of localized Ca^{2+} signals.

After Ca^{2+} is transported into the ER by the Ca^{2+}-ATPases, the majority of it is bound to intraluminal Ca^{2+} buffers. This allows large quantities of Ca^{2+} to be accumulated and quickly mobilized when the release mechanism is activated. Calsequestrin and calreticulin are such intraluminal Ca^{2+} binding proteins. These proteins have the basic properties required to play a dynamic Ca^{2+} storage function; they have a high capacity and a low affinity for Ca^{2+}. By buffering luminal free Ca^{2+} concentrations, calsequestrin and calreticulin protect the accumulation of the insoluble calcium phosphate precipitates. They also protect several endomembrane functions from the toxic effects of high luminal Ca^{2+} concentration.

2. LIGAND-GATED CATION CHANNELS

Extracellular ligand-gated receptor channels differ from other Ca^{2+}-conducting plasma membrane channels (*see* below) in three respects. First, their activation depends on the delivery and binding of a ligand to the extracellular domain of these receptor channels. Second, termination of their activities requires removal of the ligand, which is usually mediated by a specific pathway for ligand degradation and/or uptake. Third, they are not highly selective for Ca^{2+}, but conduct other ions as well. Because ligand-gated channels are generally activated by neurotransmitters, they are also known as *neurotransmitter-controlled channels*. However, some of these channels can be controlled through hormonal transmission, that is, a ligand for a particular receptor channel can be colocalized with hormones in secretory vesicles and cosecreted upon cell activation. Once released, the ligand can act in an autocrine and/or paracrine manner.

There are two classes of ligand-gated channels, the *excitatory cation-selective receptor channels,* operated by acetylcholine, glutamate, 5-hydroxytrypta-

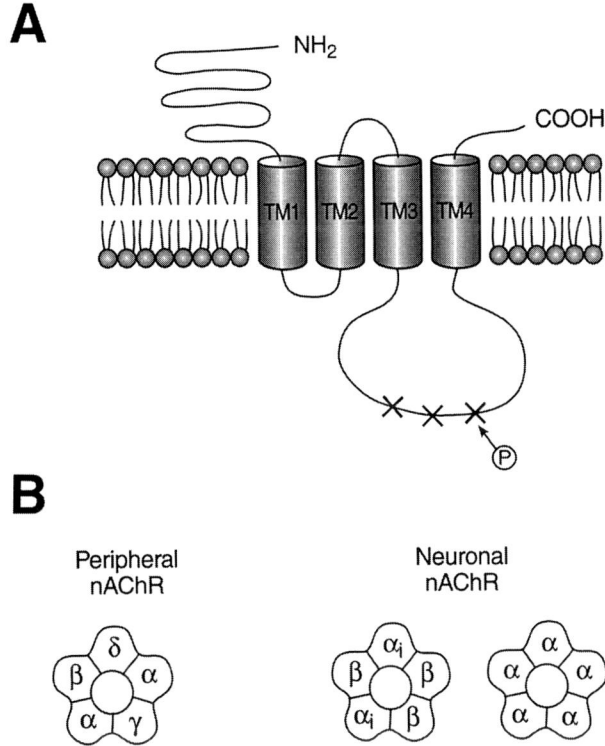

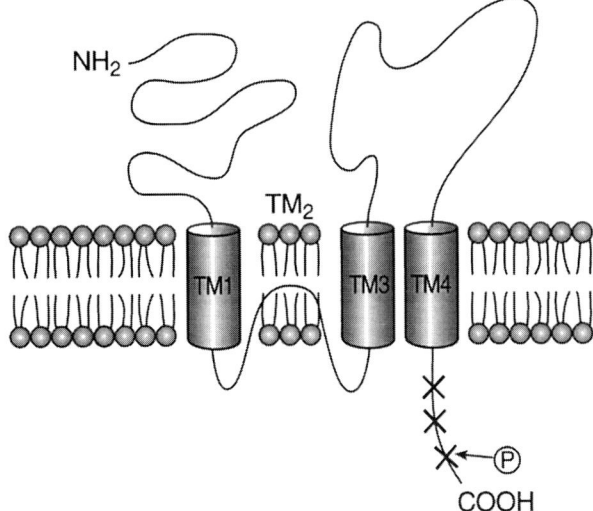

Fig. 2. Structural organization of muscle and neuronal nicotinic channels (nAChRs). **(A)** Putative transmembrane organization of nAChRs. TM, transmembrane domain. **(B)** Front view of the models of muscle and neuronal nAChRs. X, the potential phosphorylation sites.

Fig. 3. Transmembrane topology of glutamate ion channels. TM, transmembrane domain. TM_2 does not cross the membrane, even though it contributes to the lining of the pore. X, the potential phosphorylation sites.

mine (5-HT), and adenosine 5′-triphosphate (ATP), and the *inhibitory anion-selective receptor channels*, activated by γ-aminobutyric acid (GABA) and glycine. Structural information obtained by cDNA cloning of ligand-gated receptor channels has led to the identification of several families of evolutionary related proteins. The 5-HT_3, GABA, and glycine receptor channels possess structural features similar to the nicotinic acetylcholine receptor channel (nAChRs); hence these receptors can be grouped as one family. These channels are composed of five subunits (pentamers), each of which contributes to the ionic pore. Each subunit has a large extracellular NH_2 (N)-terminal region followed by four hydrophobic putative membrane-spanning segments and an extracellular COOH (C)-terminus (Fig. 2).

Information about the structure of nAChRs was used as a guideline for the examination of other ligand-gated channels, including glutamate receptor channels. Although it was initially believed that these channels are also composed of four transmembrane segments, the second hydrophobic segment seems to make up a loop, which inserts into the membrane from the intracellular site, but does not span the membrane. Furthermore, the entire region between M3 and M4, which was previously believed to be intracellular, is extracellular. Thus, the N-terminus is exctracellularly located, whereas the C-terminus is intracellularly located, and is regulated by signaling molecules, including the kinases (Fig. 3). At the present time, it is not clear if these receptor channels are composed of four or five subunits.

The ATP-gated channels, termed *P2X receptor channels,* have only two putative transmembrane domains with the N and C-terminus facing the cytoplasm (Fig. 4). Seven cDNAs have been expressed individually in different expression systems. As with nicotinic and glutamate channels, functional diversity of P2X channels is generated by subunit multimerization. It is believed that functional channels are tetramers and that both hetero- and homotetramers account for the native receptor channels. In addition to ATP channels, there are several other two-transmembrane-domain channels that are not ligand gated, including FMRFamide-gated sodium channels, epithelial sodium channels, inward rectifier potassium channels, and mechanosensitive channels.

Interestingly, many ligand-gated channels share common agonists with G protein-coupled receptors (Table 1). For example, there are two types of acetylcholine receptors that can be distinguished by their pharmacology. The nicotinic acetylcholine receptor binds nicotine and is a cation-selective receptor channel. On the other hand, the muscarinic acetylcho-

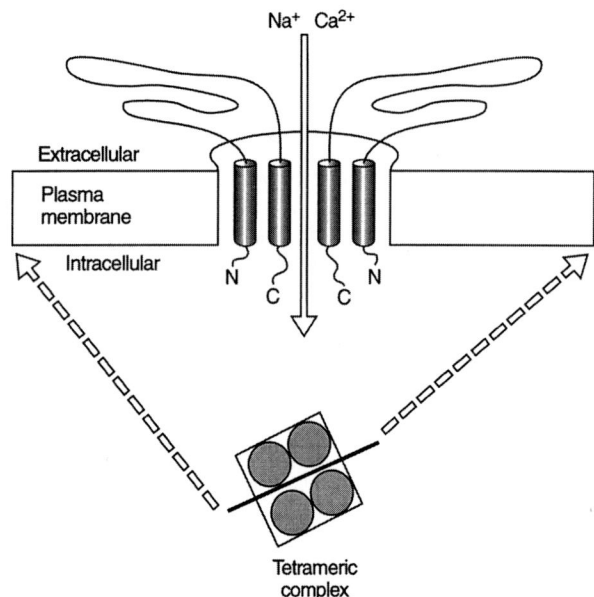

Fig. 4. Topological model of purinergic receptor channels. Quaternary structure of the channels may by tetrameric, by analogy with the structure of inwardly rectifying potassium channel, which also contains two transmembrane domains.

Table 1
Neurotransmitters as Common Agonists for Extracellular Ligand-Gated Receptor Channels (Ionotropic Receptors) and G Protein-Coupled Receptors (Metabotropic Receptors)

Ligand	Receptor Channels	Receptors
ACh	Nicotinic: muscle and neuronal	Muscarinic: M1–M5
ATP	$P2X_{1-7}$	$P2Y_{1-8}$
GABA	$GABA_A$	$GABA_B$
Glutamate	AMPA, kainate, NMDA	$MGlyR_1$-$mGluR_7$
5-HT	$5\text{-}HT_3$	$5\text{-}HT_{1A-F}$
		$5\text{-}HT_{2A-C}$
		$5\text{-}HT_4$
		$5\text{-}HT_{5A,B}$
		$5\text{-}HT_6$
		$5\text{-}HT_7$

line receptor binds muscarine and is coupled to a G protein-dependent signaling pathway. Receptors that operate as channels are also known as *ionotropic receptors* and those coupled to G protein-dependent pathways as *metabotropic receptors*. Thus, receptor subtype expressed in a particular tissue maintains tissue specificity of a common agonist function. For details on the mechanism of metabotropic receptor-mediated Ca^{2+} signaling *see* Section 4.

The pathways for delivery and degradation of ligands are also complex and specific for a particular receptor channel, as illustrated here by ATP secretion and degradation. ATP is stored along with acetylcholine in a variety of peripheral and central synaptic vesicles, and with noradrenaline in vesicles of sympathetic nerve terminals. It has also been detected in synaptosomes from the mammalian brain and neuromuscular junction, and the electric organ of electric ray. ATP is also cosecreted from chromaffin cells in a receptor-controlled and Ca^{2+}-dependent manner. It is likely that ATP is cosecreted with neurotransmitters and hormones from many other cell types; however, its secretion may be masked by the activity of endogenous ectonucleotidases. Ectonucleotidase are plasma membrane-bound enzymes that dephosphorylate extracellular ATP, ADP, and AMP to adenosine in a sequential manner (Fig. 5). Some of the cells express only ectoATPase, whereas in others ectoADPase and ectoAMPase (5'-nucleotidase) are also operative. It has also been reported that the enzyme ATP-diphosphohydrolase controls degradation of both ATP and ADP. The operation of ectonucleotidase in a particular cell type is consistent with the hypothesis that ATP is cosecreted by target cells, and that a mechanism for its degradation is needed to terminate the signal upon the removal of agonist.

2.1. Nicotinic Acetylcholine Receptor Channels

2.1.1. STRUCTURAL DIVERSITY

The native nAChR was initially identified as a pentamer protein of about 300,000 MW from the fish electrical organ. Recombinant DNA studies revealed close homologies between nAChR subunit sequences derived from electrical organ and skeletal muscle tissue, as well as between muscle and neuronal nAChRs. Five peripheral nAChR subunits, labeled as α_1, β_1, δ, γ, and ϵ, and 10 neuronal nAChR subunits, labeled as α_2–α_9 and β_2–β_4, were identified (Fig. 6A). In both muscle and neuronal nAChRs, the large N-terminal domain contains the ligand binding sites. The transmembrane segment M2 forms the wall of the ion channel and the variable C-terminal domain faces the cytoplasm and is subject to regulation by phosphorylation (Fig. 2A). Both peripheral and neuronal receptors form heteropentamers that form barrel-like structures. In muscle and electrical organ, nAChRs are composed of four subunits: α, β, γ or ϵ, and δ. The neuronal AchRs assemble according to the general 2α, 3β stoichiometry, with possibly more than one

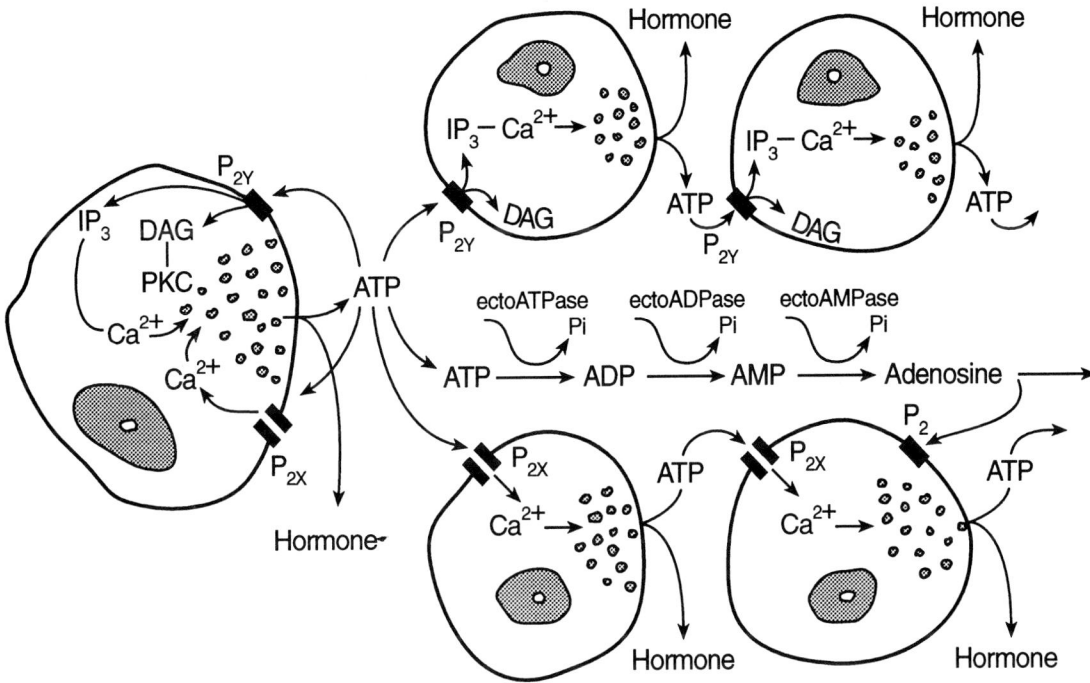

Fig. 5. Schematic representation of purinergic system for self-amplification of Ca^{2+} signaling and hormone secretion, as well as cell-to-cell spreading of Ca^{2+} signals and ATP degradation. This signaling system is operative only in tissue in which ATP is cosecreted with neurotransmitters or hormones.

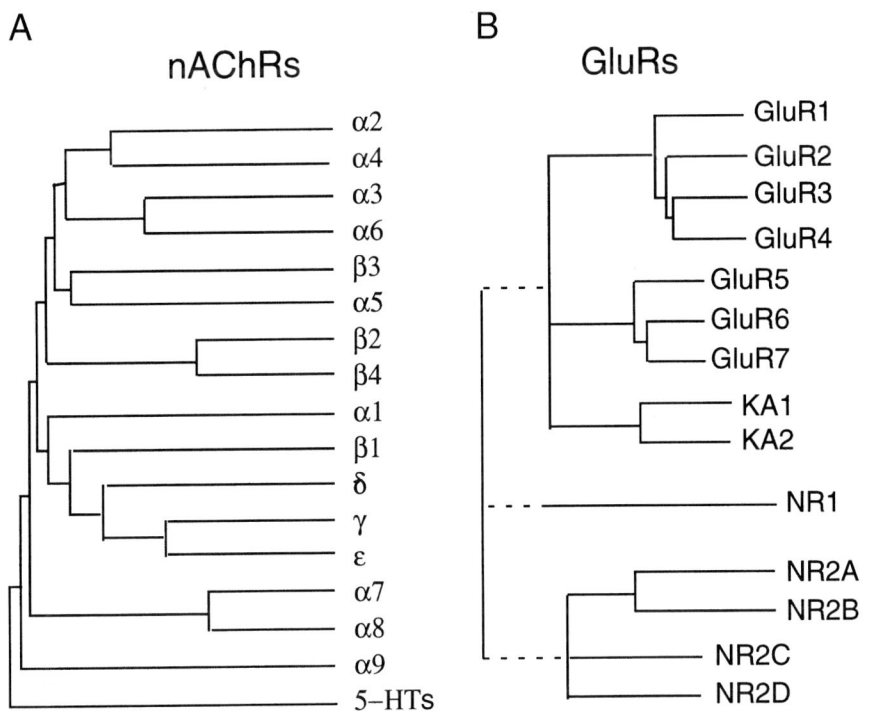

Fig. 6. Simplified dendrograms of the members of the nicotinic and glutamate receptor-channel families. **(A)** Nicotinic receptor family. **(B)** Glutamate receptor family. (Derived from Changeux JP [Brain Res Rev 1998; 26:198] and Ozawa S [Prog Neurobiol 1998; 54:581] with permission.)

α-subunit class within a pentamer. Figure 2B illustrates the models for peripheral and neuronal nAChRs. In reconstitution experiments, neuronal subunits also form functional homooligomeric channels. In addition, multiple combinations of both α- and β-subunits from two or more different subtypes form a wide range of functional heterooligomers. These multiple combinations of nAChR subunits possess distinct pharmacological and physiological properties. Also, the subunit composition determines the rate of desensitization. Furthermore, the distribution of neuronal nAChR subunits within the brains of adult animals varies. The expression of these subunits during embryonic and postnatal development is highly specific for a particular subunit.

2.1.2. Functional Diversity

Neuronal nAChRs exhibit a higher permeability to Ca^{2+} than muscle and electrical organ receptors, owing to the multiple combinatorial possibilities available for the assembly of the various subunits into heterooligomers. For example, the α_7/α_8-subunit-based homoologomers are highly permeable for Ca^{2+} compared to Na^+, whereas subunits α_2–α_6 and β_2–β_4 from heterooligomeric channels have similar permeabilities for these cations. The specific amino acids important for Ca^{2+} permeability and selectivity of neuronal nAChRs are located in the M2 domain. The muscle nAChRs are the least permeable to Ca^{2+}, with the estimated percentage of the inward current carried by Ca^{2+} being only 2%. For these channels, receptor activation and plasma membrane depolarization are the primary initiators of excitation–contraction coupling. Upon binding of acetylcholine, nAchR channels open to allow Na^+ to flow through the channel. The resulting depolarization opens VGCCs and initiates Ca^{2+} release from SR through RyRs. In skeletal muscle, activation of VGCCs in the T-tubule plasma membrane is the primary signal that activates intracellular Ca^{2+} release channels and ultimately stimulates muscular contraction (*see* Sections 4. and 7.).

2.1.3. Physiology and Pathology

The role of nAChRs in neuromuscular coupling is well established. In contrast, although the nAchRs are found in most parts of the brain, their functional significance is not well characterized. Both Ca^{2+} influxes through activated nAChRs and Ca^{2+} potentiation account for the physiological actions of these channels. Calcium potentiation is a process in which Ca^{2+} influx through one channel regulates the efficacy of other ligand-gated channels, leading to the modulation of membrane excitability in neurons, as well as their ability to integrate synaptic and paracrine signals. Furthermore, nAChR-dependent Ca^{2+} signals enhance protein kinase activity in myotubes, leading to phosphorylation of the nAChR γ-subunit. Because this process is dependent on Ca^{2+} influx, it can be considered as autoregulation of phosphorylation by nAChRs. Recent results indicate that point mutation in this receptor may abolish desensitization, increase the affinity for agonists, and convert the effects of competitive antagonists into the agonist responses. Such mutations also occur spontaneously in humans and may be involved in diseases such as congenital myasthenia or frontal lobe epilepsy. Nicotinic agonists might also serve as therapeutic agents for Alzheimer's disease, schizophrenia, and Tourette's syndrome.

2.2. Glutamate Receptor Channels

Glutamate receptor channels (GluRs) are traditionally divided into three major subtypes: (1) α-amino-3-hydroxy-5-methyl-4-isoxazole propionic acid (AMPA), (2) kainate, and (3) *N*-methyl-D-asparate (NMDA) receptor channels. Because there are no pharmacological agents that can clearly distinguish between AMPA and kainate receptors, they are also termed as non-NMDA receptors. Molecular cloning revealed numerous subunits for each receptor group. For NMDA receptor channels, NR1 and NR2A to NR2D subunits have been established. For the non-NMDA receptor channels, GluR1–GluR4 denote the AMPA-sensitive family, whereas GluR5–GluR7, KA1, and KA2 denote the kinate subclass. The phylogenic trees of these subunits are shown in Fig. 6B. In addition to these 14 subunits, two cDNAs for β-subunits have been cloned. A single δ-subunit also exists and belongs to the GluR type subunit, but the function of this particular subunit is unknown. Finally, the molecular diversity of GluRs is further increased by variants created by alternative splicing and RNA editing. Alternative splicing leads to subunit variants with different C-termini, additional or substituted sequences in N-termini, or additional or substituted sequences in the extracellular segment between M3 and M4. All GluR subunits contain four hydrophobic segments, labeled M1–M4, but their M2 segment forms a pore-loop structure, entering and exiting the cell membrane from the intracellular side (Fig. 3). The M2 amino acids line the inner channel pore, and specific residues in this segment determine the ion selectivity of the channel.

2.2.1. NMDA Receptor Channels

NMDA receptors are heteromeric channels formed by NR1 and NR2 types of subunits. For the NR1 subunit, the existence of eight splice variants was reported. In addition to the specific structure and pharmacology, NMDA channels have a very different excitation behavior than those not activated by NMDA. These channels are both ligand and voltage gated. Full activation of the NMDA receptor requires application of two ligands, L-glutamate and glycine. The NMDA receptors become fully activated by glutamate only after their Mg^{2+} block has been relieved by membrane depolarization. The NMDA receptor exhibits low binding affinity sites for Ca^{2+}, which results in a low selectivity among cations. However, because of the lower affinity binding, Ca^{2+} moves through the pore rapidly. Their kinetics is much slower, resulting in a large Ca^{2+} influx and long-term metabolic or structural changes.

The importance of Ca^{2+} signals generated by NMDA receptor channel activity is well established. Some of the most important functions of the nervous system, such as synaptic plasticity, are dependent on the behavior of NMDA receptor channels and Ca^{2+} influx through these channels. For example, tetanic stimulation-induced long-term potentiation of synaptic transmission in the hippocampus has been extensively studied as a model of activity-dependent changes in synaptic efficacy. This process may provide the physiological basis for information storage in the brain. It is believed that Ca^{2+} enters the cell via NMDA receptor channels and triggers the processes leading to a sequence of events that culminate in enhanced synaptic efficacy. The relationship between learning ability and NMDA receptor activities in hippocampus is also examined by employing the specific receptor channel antagonists, or by genetically engineering mouse models, which lack a particular subtype of NMDA receptor channels. The results of these investigations support the idea that the NMDA channel-dependent synaptic plasticity may represent the cellular basis of certain forms of learning.

2.2.2. AMPA and Kainate Receptors

Immunoprecipitation studies, using subunit type specific antibodies, and experiments with compounds that differentially affect the desensitization properties of these channels, suggest that kainate and AMPA receptor subunits do not form mixed channel complexes. However, both types of receptors can be expressed in the same neuron. Native AMPA receptors are either homomeric or heteromeric oligomers composed of these multiple subunits. AMPA receptors in mammalian CNS differ considerably with respect to gating kinetics and Ca^{2+} permeability. Although AMPA channels had generally been considered permeable only to Na^+ and K^+, some native AMPA receptors display a substantial permeability to Ca^{2+}, and a weaker selectivity among the divalent cations compared to NMDA channels. Accordingly, it has been found that homomeric receptors assembled from GluR2 subunits exhibit low permeability to Ca^{2+}. In contrast, homomeric GluR1, GluR3, or GluR4 subunits are highly permeable to Ca^{2+}. Such subtypes of receptor channels are expressed by cultured hippocampus and cerebellar Purkinje neurons. The rapid kinetics of AMPA receptors is suitable for fast neurotransmission. The Ca^{2+} permeable AMPA receptors are involved in the excitatory synaptic transmission in hippocampal and neocortical nonpyramidal neurons. It is believed that Ca^{2+} influx through these channels plays a significant role in modulating long-term synaptic functions.

Until recently our knowledge of the functional properties of the kainate receptors was limited. The major reason was related to difficulty in pharmacologically dissociating AMPA and kainate receptor-mediated responses. Also, kainate receptors are rapidly desensitizing, whereas AMPA receptors are slowly desensitizing, that is, in cells expressing both receptors, the native kainate receptor-mediated responses were difficult to separate from long-lasting AMPA responses. Therefore, the channel properties have been examined mostly in recombinant kainate receptors. Homomeric GluR6 receptor channels show substantial Ca^{2+} permeability. In contrast to AMPA receptors, the RNA editing in the M2 region does not significantly change the Ca^{2+} permeability. A number of recent studies have indicated the role of these receptors at neuronal synapses. Both presynaptic and postsynaptic localization of these receptors has been suggested. Depending on the subtype of receptors and the localization, these channels may exhibit stimulatory or inhibitory action.

2.2.3. Glutamate Receptors and Diseases

Excessive activation of glutamate receptors by glutamate or glutamate analogs causes neuronal damage, a phenomenon referred to as excitotoxicity. It is also likely that NMDA receptors play an important role in glutamate excitotoxicity, a process thought to be involved in stroke, Parkinson's, Huntington's, and Alzheimer's diseases, as well as in schizophrenia and epilepsy. One of the major pathways leading from glutamate receptor activation to neuronal death involves excessive Ca^{2+} influx. Overactivation of the glutamate receptors, particularly NMDA receptors, has been implicated in a number of neurological disorders, including ischemia, hypoglycemia, seizure, and mechanical trauma. In the case of cerebellar granule cells, glutamate toxicity resulted in early necrotic cell death in a population of cells, and delayed apotosis in the others. On the other hand, controlled Ca^{2+} influx through the NMDA receptor blocked apoptosis induced by serum withdrawal in NG108 neuroblastoma cells. This is consistent with a view that NMDA receptors may or may not induce apoptosis, depending on the amplitude and duration of Ca^{2+} signals.

2.3. Purinergic Receptor Channels

ATP-gated receptors, also known as *purinoreceptors*, belong to two major groups: G protein-coupled receptors (P2YRs), whose activation causes Ca^{2+} release from intracellular stores, and plasma membrane channels (P2XRs), whose activation promotes cation influx (Table 1). In contrast to nAChRs and GluRs, P2XRs have been identified in both excitable and nonexcitable cells. Seven channels from this family have been cloned and named $P2X_1$ to $P2X_7$. In addition, several spliced forms of $P2X_2Rs$ were reported and it is possible that other forms also express spliced and/or edited mRNAs. Each P2XR subunit appears to have intracellular N- and C-termini, and two putative transmembrane domains (M1 and M2) with a large hydrophilic loop of about 300 residues (Fig. 5). There is 35–50% identity and 50–65% similarity in amino acid sequences between pairs of P2XRs, whereas the recombinant receptors characterized differ among themselves with respect to the action of ATP analogs, the desensitization rate, and the effectiveness of antagonists.

2.3.1. STRUCTURAL AND FUNCTIONAL SPECIFICITY OF P2X CHANNELS

The $P2X_1R$ and $P2X_3R$ are cation-selective ion channels, with a relatively high Ca^{2+} permeability. The pharmacology of $P2X_1R$ resembles that of classical P2X channels in smooth muscle, whereas the properties of $P2X_3R$ expressed in oocyte- and HEK293 cells are similar to those observed in rat sensory neurons from the dorsal root ganglion. Both channels are sensitive to α,β-methylene-ATP and the P2 channel antagonist suramin. Another group of P2XRs is composed of the $P2X_2$, $P2X_4$, $P2X_5$, and $P2X_6$ channels. All these channels have a low sensitivity to α,β-methylene-ATP, and two of them, the $P2X_2R$ and $P2X_5R$, are sensitive to suramin. The ligand-selectivity profile of cloned $P2X_2R$ resembles those found in PC12 cells and several other neuronal types. Northern blot analysis suggests that the $P2X_2R$ is expressed in the pituitary, brain, spinal cord, intestine, and vas deferens. Functional $P2X_2Rs$ and several of its spliced forms were found in pituitary somatotroph populations. Other subpopulations of anterior pituitary cells express other P2XR subtypes. Unique to $P2X_5R$ is the small current, which is only 5–10% of that observed in other channels. The message for this channel is present in lymph node, spleen, and brain tissue. $P2X_4R$ RNA is expressed in the spinal cord, lung, thymus, bladder, adrenal, testis, and brain, with a staining pattern that closely mirrored the $P2X_6R$. The $P2X_7R$ shows low or no sensitivity to suramin and pyridoxalphosphate-6-azophenyl-2′,4′-disulfonic acid, and is insensitive to the $P2X_1$ and $P2X_3$ agonist α,β-methylene-ATP. The protein of this channel is the largest among P2XRs. In contrast to other channels from this family, it represents a bifunctional molecule; it operates as a channel, but also permeabilizes cells, forming large pores in the plasma membrane. Another distinguishing characteristic of $P2X_7Rs$ is their high sensitivity to BzATP.

2.3.2. INACTIVATION PROPERTIES OF P2XRs

In addition to ligand-selectivity profiles and antagonist sensitivity of P2XRs, they also differ with respect to their desensitization rates. Based on difference in the desensitization kinetics, these channels are divided into two groups: rapidly desensitizing channels ($P2X_1R$ and $P2X_3R$) and slowly desensitizing channels ($P2X_2R$, $P2X_6R$, and $P2X_7R$). It is unclear which group $P2X_4Rs$ belong to, as they exhibit fast and complete desensitization when expressed in oocytes, and slow desensitization when expressed in HEK293 cells. Two experimental approaches have been used to study the molecular mechanism underlying P2XR desensitization, the first being construction of chimeric channels between slowly and rapidly desensitizing subtypes and the second being coexpression of both types. Chimeric studies suggest that the responsible domain for desensitization is localized within the two transmembrane regions of $P2X_1R$ and $P2X_3R$. Cotransfection of the expression plasmids for these two channel types was also found to yield a P2XR with altered desensitization and agonist selectivity properties, indicating that such channels are presumably heteropolymers. Recently, a new view about P2XR desensitization has emerged. The $P2X_2R$ splice variant, termed $P2X_{2-2}R$, lacks a stretch of the

C-terminal amino acids of the $P2X_2R$ molecule and encodes a functional channel that desensitizes faster than the wild-type. This observation suggests the importance of the spliced segment for prolonged Ca^{2+} influx through wild-type channels. The presence of specific residues in the C-terminus of $P2X_2R$, with two conserved amino acids contributing significantly to the development of a sustained Ca^{2+} influx through these channels, may indicate a common mechanism of desensitization for the P2XR family. The Lys^{375} residue is common to all four slowly desensitizing channels, $P2X_2R$, $P2X_5R$, $P2X_6R$, and $P2X_7R$, whereas Pro^{374} is present only in three of them, $P2X_2R$, $P2X_6R$, and $P2X_7R$. In contrast, the rapidly desensitizing $P2X_1R$ and $P2X_3R$ do not contain these two amino acids. The removal of the Pro^{373}–Pro^{376} segment by splicing resulted in the rapidly desensitizing $P2X_{2-2}R$. Certainly, future experiments with insertion of this segment into rapidly desensitizing channels and its deletion from slow-desensitizing P2XRs will provide important insight into the functional regulation of P2XR by their C-termini.

2.3.3. Physiology

In contrast to the well-characterized structure and pharmacology of P2XRs, their physiological significance is not well understood. In general, Ca^{2+} is a charge carrier through these channels, although the permeability of Ca^{2+} vs Na^+ varies widely among different cell types. Thus, these channels can serve as Ca^{2+} influx channels. These channels also increase $[Ca^{2+}]_i$ by influencing Ca^{2+} entry through other Ca^{2+}-selective channels. For example, activation of ATP-gated channels leads to plasma membrane depolarization, which opens VGCCs to increase Ca^{2+} entry. The direct or indirect increase in $[Ca^{2+}]_i$ by ATP-gated Ca^{2+} channels may be important for the sustained IP_3-induced and ER-derived Ca^{2+} spiking in some cells. In addition to stimulating intracellular Ca^{2+} signals, the paracrine actions of ATP can generate the cell-to-cell spread of Ca^{2+} signals in mast and glial cells in the absence of gap-junctional communication (*see* Section 8). The best-characterized agonist role of ATP is in synaptic transmission from sympathetic nerves, where ATP acts as a cotransmitter with noradrenaline. ATP has also been implicated in parasympathetic, sensory, and somatic neuromuscular transmission. About 40% of hypothalamic neurons in culture respond to ATP by a rapid increase in $[Ca^{2+}]_i$ owing to Ca^{2+} entry through P2XRs. ATP-induced Ca^{2+} signals are sufficient to trigger hormone release in several secretory cell types. For example, in gonadotrophs, chromaffin cells, β-pancreatic cells, and PC12 cells, ATP induces an inward current, and increases Ca^{2+} influx and hormone release. In a subpopulation of pituitary cells, as well as in a subpopulation of chromaffin and insulin-secreting cells, ATP also releases $[Ca^{2+}]_i$ from internal stores bathed in Ca^{2+}-deficient medium. Finally, ATP is cosecreted during agonist- and depolarization-induced secretion of catecholamines and gonadotropins. In the pineal gland, ATP potentiates the effect of noradrenaline in N′-acetyl-5-hydroxytryptamine production.

3. VOLTAGE-GATED CALCIUM CHANNELS

3.1. Classification, Structure, and Distribution of VGCCs

Central to the operation of the V_m-dependent Ca^{2+} signaling pathway is the incorporation of VGCCs into the electrical membrane activity. These channels are large multisubunit proteins that span the plasma membrane to provide a Ca^{2+}-selective pathway into the cytosol. At rest, the channel remains closed, blocking Ca^{2+} entry into the cell. In response to membrane depolarization, the molecular "gates" of the channel open and Ca^{2+} is free to flow through the aqueous pore of the channel. The voltage-dependent opening of ionic channels is called *activation*. The direction of Ca^{2+} flow through the channel pore is determined by the electrochemical gradient for Ca^{2+}, which under normal conditions drives Ca^{2+} from the extracellular milieu into the cytosol. After only a few milliseconds, or in as much as several hundred milliseconds, the channel closes and the flow of Ca^{2+} is again blocked. The closing of an ionic channel is called *inactivation*. Following inactivation, the channel returns to its resting state until the next membrane depolarization triggers the whole process over again (*see also* the chapter by H. A. Fozzard and D. J. Nelson in this volume).

The Ca^{2+} selectivity and voltage sensitivity of these channels are common features among two major groups of VGCCs, which are separated by their sensitivity to changes in V_m. The first group of channels requires only weak membrane depolarization to open. Consequently, they are activated at relatively hyperpolarized membrane potentials and are known as *low-voltage activated* (LVA) Ca^{2+} channels. The most distinguishing feature of this group, however, is their rapid inactivation and requirement for strong membrane hyperpolarization to bring them out of steady-

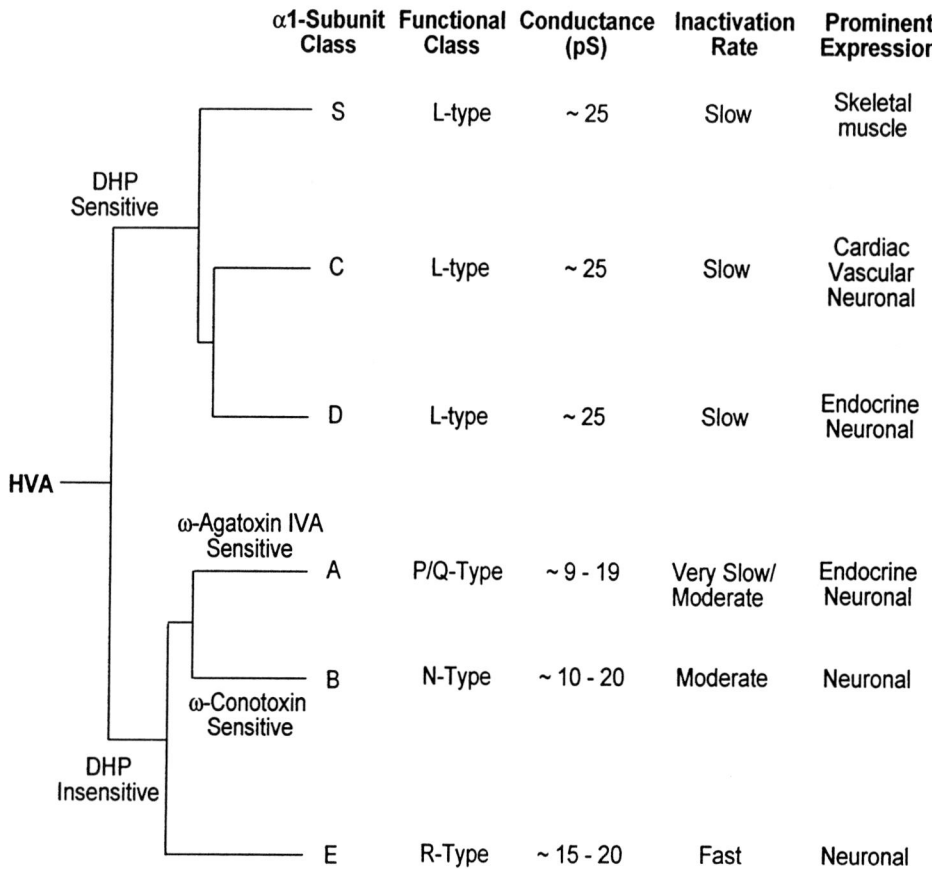

Fig. 7. Structural relationship and functional characteristics among α_1-subunits of high-voltage-activated calcium channels. (Derived from Wheeler DB [Prog Brain Res 1995; 105:65] with permission.)

state inactivation. Owing to the rapid inactivation of these channels, they are often referred to as transient or T-type Ca^{2+} channels. The second group of VGCCs requires moderate to strong membrane depolarization to open. They are therefore activated at relatively depolarized membrane potentials and are known as *high-voltage activated* (HVA) Ca^{2+} channels. Compared to LVA channels, most HVA channels inactivate slowly, or exhibit little to no inactivation during sustained membrane depolarization. Among this group, biophysical and pharmacological studies have identified multiple subtypes that can be distinguished by their ion selectivity, single channel conductance, pharmacology, metabolic regulation, and tissue localization. Based on these criteria, five HVA Ca^{2+} channel subtypes have been identified to date: L-, N-, P-, Q-, and R-type Ca^{2+} channels (Fig. 7).

Consistent with the functional studies, molecular cloning has identified several genes that encode different VGCC subtypes. The first Ca^{2+} channel was purified from skeletal muscle, as it is a highly enriched source of L-type Ca^{2+} channels. Purification of the channel identified five subunits, including a large α_1- (200–260 kDa) subunit and four smaller ancillary subunits: α_2, β, γ, and δ. The α_1-subunit consists of four homologous repeats, each one composed of six transmembrane segments (Fig. 8). Located within the α_1-subunit are the voltage sensor, gating machinery, channel pore, and multiple protein kinase A and cAMP-dependent-kinase phosphorylation sites. Since the first α_1-subunit was cloned from skeletal muscle, at least seven isoforms have been identified, including the α_{1S}, α_{1C}, α_{1D}, α_{1B}, α_{1A}, α_{1E}, and α_{1G}. The α_{1S}-, α_{1C}-, and α_{1D}-subunits make up the L-type Ca^{2+} channels in skeletal muscle, cardiac muscle, and neurons, respectively. The α_{1B} subunit is associated with N-type channels in neurons. The α_{1A}-subunit is associated with both P- and Q-type neuronal channels, and the α_{1E}-subunit with the R-type channels (Fig. 7). Finally, the α_{1G}-subunit has been linked to T-type Ca^{2+} channels. Mutations in the α_1-subunit of VGCCs are the underlying defects in a growing number of human disorders, including hypokalemic periodic paralysis, hemiplegic migraine, episodic ataxia type

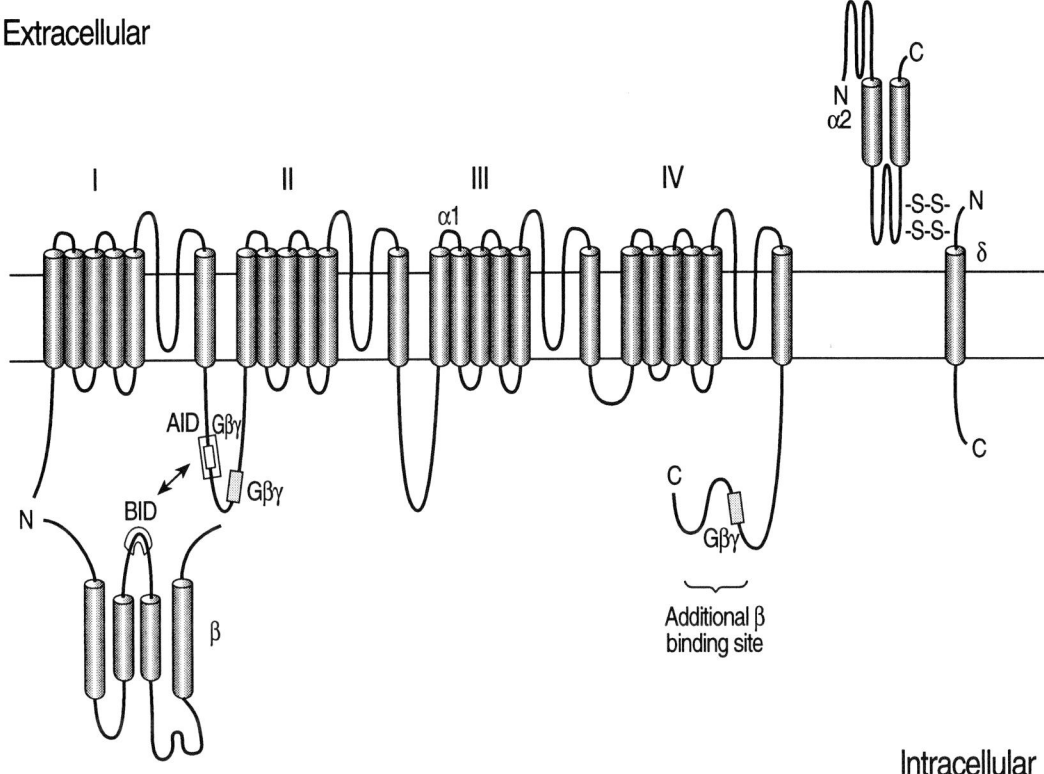

Fig. 8. Schematic representation of voltage-gated calcium channel complex. Diagram indicates the putative transmembrane topologies of the α_1-subunit, as well as α_2-, β-, and δ-subunits. The binding sites for the Gβ/γ dimer are also shown. (Derived from Dolphin AC [J Physiol 1998; 506:3] with permission.)

2, and cerebellar ataxia. Although the α_1-subunit can function as a VGCC when expressed alone, the ancillary subunits have a substantial impact on current amplitude, voltage dependence, and activation/inactivation kinetics. In addition, the β_1-subunit is a target for second-messenger-mediated modulation of VGCCs.

Coexpression of several Ca^{2+} channel subtypes in a single cell is common in neurons and endocrine cells. For example, both T- and L-type Ca^{2+} channels are expressed in pituitary lactotrophs, gonadotrophs, corticotrophs, somatotrophs, and melanotrophs. In sensory neurons, T-type and L-type Ca^{2+} channels are coexpressed with N-type Ca^{2+} channels. In other neurons, P/Q-type Ca^{2+} channels are also found in conjunction with other VGCC subtypes. Although multiple Ca^{2+} channel subtypes may be coexpressed in the same cell, they are often distributed nonuniformly in different regions. In inferior olivary neurons, HVA Ca^{2+} channels are found mostly, but not exclusively, in the dendrites, whereas LVA Ca^{2+} channels are found predominately in the cell body. The distribution of HVA Ca^{2+} channel subtypes within the same cell may also be nonuniform. In neurons, extensive expression of the α_{1B}- and α_{1A}-subunit of the N- and P/Q-type Ca^{2+} channels has been found in the dendritic shafts and presynaptic nerve terminals, but not the cell body. Conversely, the α_{1C}- and α_{1D}-subunit of L-type Ca^{2+} channels were found predominately in the soma and proximal dendrites. The α_{1E}-subunit of the R-type Ca^{2+} channel was found predominantly in the cell body of central nervous system (CNS) neurons. The nonuniform distribution of Ca^{2+} channel subtypes likely reflects their different functional roles. For example, the slow inactivation kinetics of N- and P/Q-type Ca^{2+} channels make them ideal for prolonging AP duration, allowing enough Ca^{2+} entry to stimulate exocytosis in the presynaptic nerve terminals.

3.2. Action Potential-Driven Ca^{2+} Influx

Voltage-gated Ca^{2+} channels serve two major functions in excitable cells; one is to generate and/or shape APs and the other is to allow Ca^{2+} influx during the transient depolarization. The voltage dependency of these Ca^{2+} channels endows them with the ability to

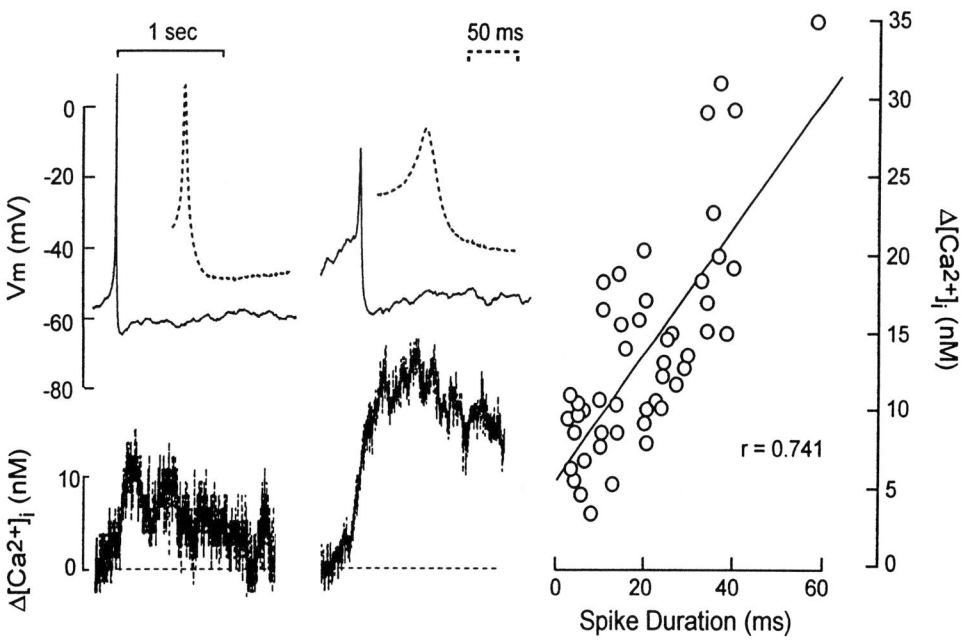

Fig. 9. Action potential driven Ca^{2+} influx in GT1 cells. **(Left panels)** Expanded time scale of action potentials and their associated $[Ca^{2+}]_i$ transients in cells exhibiting narrow *(left panel)* and broad *(right panel)* spiking. **(Right panel)** Relationship between AP duration and the maximum amplitude of the $[Ca^{2+}]_i$ transients.

substitute for Na^+ channels in the generation of APs. Membrane depolarization increases the probability that VGCCs will open, so that a small membrane depolarization (driven by LVA Ca^{2+} current or some other depolarizing current) increases the probability for VGCCs to open. The resulting activation of VGCCs and associated Ca^{2+} influx further depolarizes the membrane and activates more Ca^{2+} channels. Ultimately, such a positive feedback action of Ca^{2+} entry on V_m leads to the threshold level of depolarization and the "all-or-none" firing of APs. The resulting Ca^{2+} entry through VGCCs during the AP causes a transient increase in $[Ca^{2+}]_i$. This can be observed as discrete $[Ca^{2+}]_i$ transients that follow each AP (Fig. 9).

In some cells, VGCCs account entirely for the upstroke of an AP. Such cells include cardiac pacemaker cells of the sinoatrial node, vertebrate smooth muscle cells, invertebrate muscle cells, and some pituitary and neuroendocrine cell types. In other cells, voltage-gated Na^+ and Ca^{2+} channels contribute to the upstroke of the AP. The contribution of voltage-gated Na^+ and Ca^{2+} channels to the AP in a single cell type can also vary depending on the level of hyperpolarization reached during the interpulse interval. For example, in GnRH-secreting neurons, the availability of Na^+ channels and LVA Ca^{2+} channels is dependent on the inactivation status of the channels, which is determined by the level of hyperpolarization reached during the interpulse interval. In more hyperpolarized cells, there is little steady-state inactivation of these channels, allowing them to participate in AP generation, and resulting in sharp, high-amplitude spikes with a limited capacity to drive Ca^{2+} (Fig. 9). In cells with more depolarized interpulse V_m, there is an increasing proportion of Na^+ channels under steady-state inactivation. This prevents them from contributing to the APs, resulting in broad spikes of lower amplitude, and with an increased capacity to drive Ca^{2+} (Fig. 9). The increase in AP duration is likely due to the activation of fewer voltage-gated K^+ channels owing to the smaller amplitude of the APs. Therefore, in GnRH-secreting neurons, the availability of tetrodotoxin-sensitive Na^+ channels to contribute to AP firing can influence its amplitude and duration, which in turn regulates voltage-gated Ca^{2+} entry.

3.3. Spontaneous Pacemaking and $[Ca^{2+}]_i$ Transients

The biophysical properties of each VGCC subtype and its association with other ionic channels determine its role in generating and shaping APs. The low activation range of T-type channels allows them to trigger rebound APs following pronounced membrane hyperpolariztions. This membrane hyperpolarization is necessary to remove the steady-state inactivation of the T-type channels. Such behavior allows "inhibitory" presynaptic neurons to transiently remove the steady-state inactivation of T-type channels, which triggers the rebound firing of AP bursts in thalamic neurons and ventrobasal relay neurons. Conversely, the slow inactivation rates of HVA channels make them well suited for extending AP duration, allowing for prolonged Ca^{2+} influx. The weak inactivation of some HVA channels can also enable them to act as pacemaker channels in the control AP frequency. Finally, Ca^{2+} entry through VGCCs itself can influence their participation in the control of AP spiking. For example, the inactivation rate of L-type Ca^{2+} channels is augmented by the increase in $[Ca^{2+}]_i$ following Ca^{2+} entry. Therefore, unless Ca^{2+} is sufficiently removed from around the channel pore by intercellular Ca^{2+} buffers and/or plasma membrane Ca^{2+} pumps, L-type Ca^{2+} channel-dependent APs will be abolished. In addition, some VGCCs are coupled to the activation of Ca^{2+}-controled K^+ channels, which help to repolarize the membrane in response to Ca^{2+} entering the cells during the AP. This suggests that Ca^{2+}-activated K^+ channels can act as a negative feedback pathway to limit Ca^{2+} influx during AP spiking. Also, in some cells activation of Ca^{2+}-controlled K^+ channels mediates the burst-like firing of APs.

Extracellular Ca entry though VGCCs during AP firing contributes to the level of $[Ca^{2+}]_i$ in unstimulated cells. In GnRH-secreting neurons, each AP drives a transient increase in $[Ca^{2+}]_i$, the magnitude of which is determined by the AP amplitude and duration (Fig. 9). In pituitary somatotrophs and lactotrophs, spontaneous AP activity and the associated $[Ca^{2+}]_i$ fluctuations contribute to basal growth hormone (GH) and prolactin release, respectively. In these and other cells, an increase in the AP frequency is followed by increases in the amplitude and duration of AP-driven $[Ca^{2+}]_i$ signals. This leads to the accumulation of $[Ca^{2+}]_i$, which eventually reaches a new steady-state plateau level. The level of this plateau is proportional to the AP frequency. Both spontaneous AP firing and the associated $[Ca^{2+}]_i$ fluctuations are abolished in the absence of extracellular Ca^{2+} and by the addition of L-type calcium channel blockers. The spiking amplitude of spontaneous $[Ca^{2+}]_i$ fluctuations in cultured cells ranges from 50 to 700 nM. In view of the differences in the amplitudes of the spontaneous Ca^{2+} spikes and the irregularity in their frequency, they are often termed fluctuations rather than oscillations.

3.4. Modulation of AP-Driven Ca^{2+} Influx

Hormones and neurotransmitters activate plasma membrane receptors to modulate AP-driven Ca^{2+} entry through VGCCs. In quiescent cells, receptor activation can initiate AP firing to drive Ca^{2+} influx through VGCCs. In cells spontaneous active cells, modulation of AP-driven Ca^{2+} entry is characterized by changes in AP amplitude, duration, or firing frequency; all of which alter Ca^{2+} influx. This leads to modulation of $[Ca^{2+}]_i$ and Ca^{2+}-regulated events, including enzyme activity, secretion, and gene expression. Receptor-mediated stimulation and inhibition of AP firing can be regulated by direct actions on VGCCs and/or indirectly through other plasma-membrane ionic channels, including Na^+ channels, K^+ channels, and ligand-gated channels. In addition, several different intercellular signaling pathways can mediate the modulation of channel activity by receptor activation (Fig. 10). The impact of these channels and their second messenger pathways on membrane excitability is summarized in the following sections.

3.4.1. Voltage-Gated Ca^{2+} Channels

Facilitation of Ca^{2+} influx through VGCCs can occur by phosphorylation of the α_1-subunit or the ancillary subunits that regulate channel activity. Putative phosphorylation sites have been identified on both the α_1- and β-subunits of VGCCs. In cardiac myocytes, activation of protein kinase A (PKA) augments Ca^{2+} influx through VGCCs. The effects of PKA phosphorylation have been attributed to an increase in the open probability of the channel being open and to an increase in the mean open time. Such actions would increase AP duration and associated Ca^{2+} influx. Activation of G_s-coupled receptors by GH-releasing hormone in pituitary somatotrophs, as well as in several other cell types operated by adenylyl-cyclase-coupled receptors, also facilitates voltage-gated Ca^{2+} entry. In contrast, activation of G_i/G_o-coupled receptors frequently leads to inhibition of pacemaker activity and voltage-gated Ca^{2+} influx. For example, in pituitary melanotrophs and lactotrophs, activation of dopamine D_2 receptors inhibits VGCCs, which reduces $[Ca^{2+}]_i$ and hormone secretion. Similarly, activation of somatostatin receptors in somatotrophs inhibits VGCCs to reduce $[Ca^{2+}]_i$ and GH secretion. Inhibition of VGCCs can occur by two pathways. The first pathway is the fast membrane delimited pathway, in which the $\beta\gamma$ dimer of the G_i/G_o protein is a direct intermediate between the plasma membrane receptor and VGCCs. Several

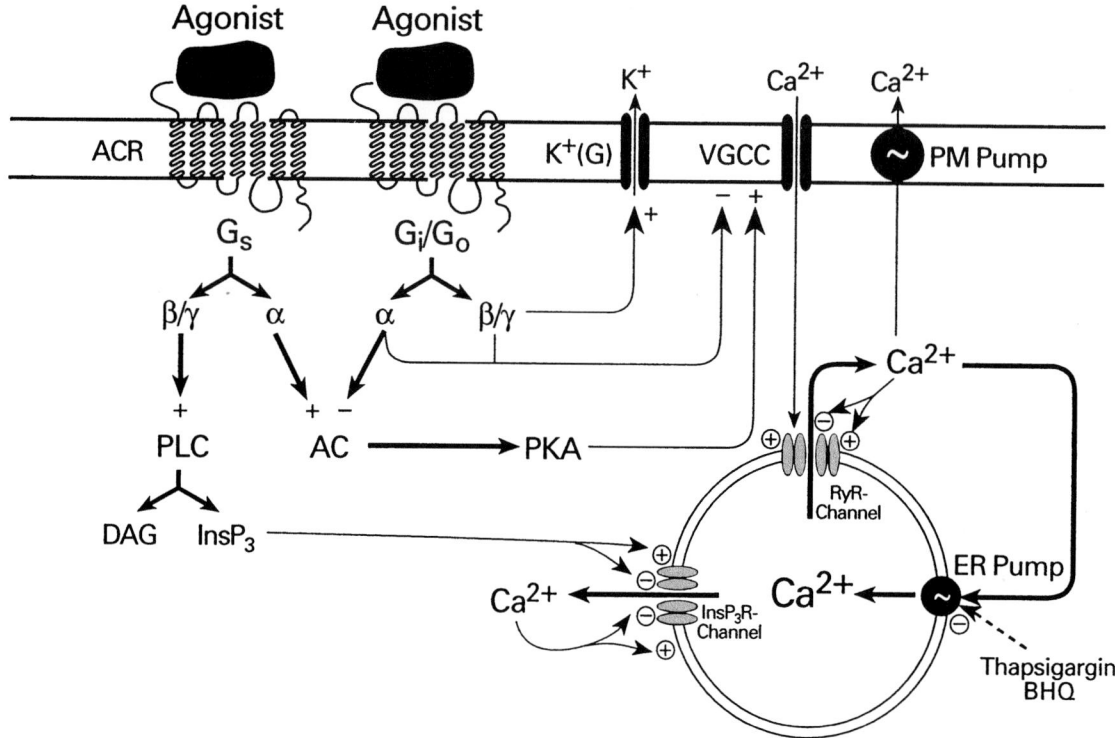

Fig. 10. Adenylyl cyclase coupled receptors and spontaneous electrical activity. Voltage-gated Ca^{2+} entry facilitates Ca^{2+} release through RyRs expressed in endopasmic reticulum (ER). In some cells, ER also express IP₃Rs. In addition to "classical" Ca^{2+}-mobilizing receptors, G_s-coupled receptors can also stimulated phospholipase C, leading to production of IP_3 and facilitation of Ca^{2+} release from ER. Adenylyl cyclase coupled receptors (ACR) can also modulate voltage-gated Ca^{2+} influx by the direct actions of G subunits on plasma membrane channels, as well as through protein kinase A (PKA)-dependent phosphorylation of these channels. PLC, phospholipase C; AC, adenylyl cyclase; $K^+(G)$, G protein-regulated inward rectifier potassium channels.

aspects of this pathway have been characterized. The essential structural elements required for G protein modulation are found in VGCCs, supporting the direct action of G proteins. Inhibition of the Ca^{2+} channel mediated by the β/γ dimer is a time- and voltage-dependent process. As a fraction of the total Ca^{2+} channels open much more slowly and require a larger depolarization to open, this produces a slowing of the activation kinetics and a reduction in the current amplitude. In the second pathway, diffusion of cytoplasmic messengers mediates the inhibitory actions of the G protein coupled receptor. In some cells, both membrane delimited and second messenger pathways can act to inhibit voltage-gated Ca^{2+} channels.

3.4.2. Na⁺ Channels

These channels can be classified into two general groups: tetrodotoxin (TTX)-sensitive and -insensitive. TTX-sensitive Na^+ channels are characterized by their fast activation and inactivation kinetics and are susceptible to steady-state inactivation. In many excitable cell types, these channels act in conjunction with VGCCs to generate the upstroke of the AP and to control the firing amplitude. Therefore, modulation of TTX-sensitive Na^+ channels can alter membrane excitability to influence AP-driven Ca^{2+} entry. Unlike other ionic channels, few receptors or intracellular messenger pathways have been demonstrated to modulate TTX-sensitive Na^+ channels. However, one such pathway includes the dopamine D_2 receptor in frog melanotrophs, which decreases the Na^+ current amplitude. In other cells, activation of PKA reduces the amplitude of TTX-sensitive Na^+ currents. Accordingly, PKA phosphorylation sites have been identified in these channels. In rat and human pituitary adenoma cells, pituitary adenylate cyclase activating polypeptide (PACAP) activates TTX-sensitive Na^+ channels via an adenylate cyclase–PKA pathway to increase hormone secretion.

TTX-insensitive channels exhibit slow activation and inactivation in response to membrane depolarization. In some cardiac cells, neurons, and endocrine cells, these channels provide the background depolarizing current that is necessary to counteract the hyperpolarizing drive of K⁺ channels, such as the inward rectifying K⁺ channels (K_{ir}). This keeps the V_m near the threshold for AP firing. Accordingly, removal of extracellular Na⁺ hyperpolarizes the membrane and abolishes AP-driven Ca^{2+} entry in these cells. The importance of TTX-insensitive Na⁺ channels in controlling the V_m makes them excellent targets for the regulation of membrane excitability. In pituitary somatotrophs, for example, GH releasing hormone activates TTX-insensitive Na⁺ channels via a cAMP/PKA signaling pathway to depolarize the membrane. This leads to an increase in AP frequency and AP-driven Ca^{2+} entry.

3.4.3. INWARD RECTIFYING K⁺ CHANNELS

When K_{ir} were first described 50 yr ago they were called anomalous rectifying K⁺ channels. This was because, unlike other ionic channels, their conductance increased with membrane hyperpolarization. Consequently, these channels are open during the interpulse interval and are closed during the AP. Owing to a substantial outward current through these channels during the interpulse interval, modulation of these currents greatly impacts membrane excitability. Inhibition of the K_{ir} depolarizes the membrane to increase excitability, whereas activation hyperpolarizes the membrane to decrease excitability. These channels are susceptible to modulation by a wide range of hormones and neurotransmitters that exert their actions via several different intracellular messenger pathways, including G proteins, intercellular ATP, pH, protein kinase C (PKC), and cAMP. In pituitary corticotrophs, corticotropin-releasing factor inhibits the K_{ir} via cAMP-dependent pathways to increase AP frequency and associated Ca^{2+} entry. Thyrotropin-releasing hormone can also inhibit K_{ir} channels in GH_3 cells and pituitary lactotrophs to increase membrane excitability. In contrast, in pituitary melanotrophs and lactotrophs, dopamine activates K_{ir} channels to hyperpolarize the membrane, causing a cessation of AP firing and a decrease in $[Ca^{2+}]_i$ and hormone secretion. Like dopamine, somatostatin stimulates K_{ir} channels to reduce membrane excitability and GH secretion in pituitary somatotrophs. The stimulatory actions of dopamine and somatostain on K_{ir} are consistent with the expression of the G protein-regulated K_{ir-3} family of channels in pituitary cells (*see* Fig. 11 and the chapter by H. A. Fozzard and D. J. Nelson in this volume).

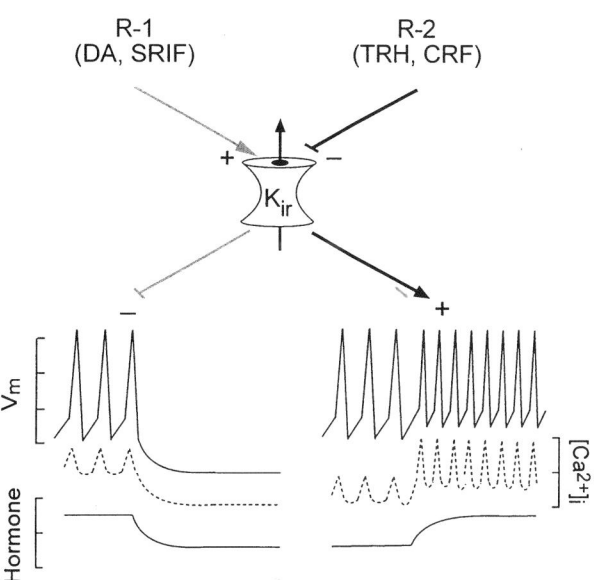

Fig. 11. Receptor-controlled inward rectifier potassium current (K_{ir}) and membrane excitability. A number of receptors negatively coupled to adenylyl cyclase (R_1) stimulate K_{ir}, channels (subtype 3), leading to membrane hyperpolarization and abolition of spontaneous electrical activity (**gray lines**). This leads to a decrease in $[Ca^{2+}]_i$ and a concomitant decrease in hormone secretion. Calcium-mobilizing receptors and those positively coupled to adenylyl cyclase (R_2) inhibit K_{ir} channels (subtypes 1 and 2), leading to depolarization of cells and an increase in action potential frequency, which leads to increase in $[Ca^{2+}]_i$ and the augmentation of hormone secretion.

3.4.4. LIGAND-GATED CHANNELS

The influx of Na⁺ and Ca^{2+} through ligand-gated channels depolarizes the membrane to increase AP firing frequency and associated Ca^{2+} entry. The Ca^{2+} permeability of these channels also allows them to drive Ca^{2+} into the cell. Thus, these channels act as a Ca^{2+} influx pathway and as a modulator of AP-driven Ca^{2+} influx. In pituitary gonadotrophs, for example, activation of P2XRs drives Ca^{2+} into the cell to initiate AP firing in quiescent cells or increase AP frequency and associated Ca^{2+} entry in spontaneously active cells. Because ATP is cosecreted with other neurotransmitters or hormones, it may bind to its receptors to amplify agonist-induced Ca^{2+} signals and secretory responses by further increasing AP-driven Ca^{2+} entry. Activation of $GABA_A$ receptor channels in GnRH-secreting neurons also depolarizes the membrane to increase membrane excitability. Unlike activation of P2XRs, however, activation of $GABA_A$ receptor channels increases the membrane permeability to chloride ions, which diffuse out of the cell to depolarize the membrane. In other cells, $GABA_A$ receptors decrease excitability due to hyperpolarization of plasma membrane. Thus, ligand-gated anionic and cationic receptor channels can modulate membrane excitability indirectly, by altering voltage-gated Ca^{2+} influx.

4. INTRACELLULAR CALCIUM RELEASE CHANNELS

4.1. Inositol 1,4,5-Trisphosphate Receptor Channels

The lack of expression and operation of a V_m-dependent Ca^{2+} signaling pathway in nonexcitable cells makes mobilization of Ca^{2+} from intracellular stores the major pathway for Ca^{2+} signaling in these cells. However, this pathway is not unique to nonexcitable cells, as it is also operative in excitable cells. In both cell types, Ca^{2+} mobilization is triggered by activation of two classes of receptors: seven membrane domain receptors coupled to G proteins and tyrosine kinase plasma membrane receptors. Calcium-mobilizing receptors that are coupled to G_q/G_{11}, as well as several receptors coupled to G_s and G_i, activate phospholipase Cβ, whereas tyrosine kinase receptors activate phospholipase Cγ. Both enzymes hydrolyze the membrane-associated phosphatidylinositol 4,5-bisphosphate to increase the production of IP_3 and diacylglycerol (*see* chapter by Exton in this volume). Diacylglycerol remains in the plasma membrane, where it acts on PKC. In contrast, IP_3 rapidly diffuses into the cytosol to release Ca^{2+} from a fraction of the nonmitochondrial stores containing the specific intracellular receptors for IP_3.

4.1.1. STRUCTURE

Purification and functional reconstitution of IP_3Rs demonstrate that the binding sites for IP_3 are on the same protein that makes up the Ca^{2+} release channels. This channel is composed of four similar subunits that are noncovalently associated to form a four-leaf cloverlike structure, the center of which forms the Ca^{2+}-selective channel (Fig. 12). The IP_3 binding sites are located within the first 788 residues of the N-terminus of each subunit. Complete cDNA sequences of three distinct IP_3R encoding genes have been determined. Most cells express multiple isoforms of IP_3Rs, which suggests that different isoforms have different functions. Investigation of the single channel function of type 1, type 2, and type 3 IP_3R revealed isoform-specific properties in terms of their sensitivity to IP_3 and Ca^{2+}. IP_3Rs are present in almost all cells and are

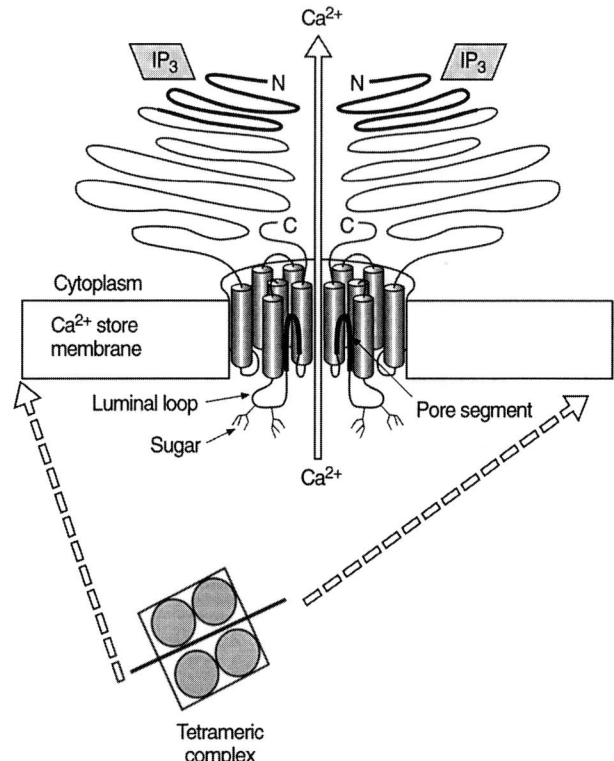

Fig. 12. Topological model of IP_3 receptor channels expressed in the endoplasmic reticulum membrane. (Derived from Mikoshiba K [Curr Opin Neurobiol 1997; 7:339].)

localized in the ER membrane, nuclear membrane, and possibly the plasma membrane in some cell types. Functionally reconstituted purified IP_3Rs respond to IP_3 with an increase in the open probability due to a large conformational change. The release of Ca^{2+} is electrically compensated by an inward potassium flux.

4.1.2. REGULATION

Cytosolic Ca^{2+} is the major messenger controlling IP_3R gating. In the presence of stimulatory concentrations of IP_3, type 1 and type 2 IP_3Rs respond to increases in $[Ca^{2+}]_i$ in a biphasic manner; Ca^{2+} increases IP_3R activity at low concentrations and inhibits it at higher concentrations. Conversely, the type 3 IP_3R open probability increases monotonically as the concentration of cytosolic Ca^{2+} increases. In both cases, the binding of IP_3 to residues within the N-terminal domain of IP_3Rs is required for cytosolic Ca^{2+} to exhibit its messenger function. It is probable that binding of IP_3 to a single subunit of these tetramer channels causes a conformational change that exposes the Ca^{2+} binding site(s) of the channel to $[Ca^{2+}]_i$. This, in turn, leads to a slight elevation in $[Ca^{2+}]_i$, that converts IP_3Rs from a state of low affinity for IP_3 to a state of high affinity, an action that affects the ability of channels to bind IP_3 and to further respond to Ca^{2+}. The rate of activation of IP_3Rs is controlled by the rate of Ca^{2+} release, which is determined by IP_3 concentration, that is, saturation of binding sites of IP_3Rs. Once the channels are fully activated, additional elevation of $[Ca^{2+}]_i$ does not change the affinity of the receptor to IP_3, but decreases the conductance of the channel, that is, inactivates the channel. The rate of $[Ca^{2+}]_i$-dependent inactivation is relatively slow and is not well characterized. Several other factors also modulate the activity of IP_3Rs, including luminal calcium, PKA, PKC, calcium/calmodulin-dependent protein kinase II, adenine nucleotides, and pH.

4.1.3. Pharmacology

In addition to IP_3, there are other highly potent agonists for IP_3Rs: some inositol tetrakisphosphates and few *myo-*, *chiro-*, and *scyllo-*inositol trisphosphate and trisphosphorothioate analogs. Adenophostins are structurally distinct from IP_3, yet exhibit 10- to 100-fold greater potency at the IP_3Rs. Two inhibitors, heparin and decavanadate, competitively inhibit IP_3 binding to IP_3Rs; neither inhibitor is highly specific. Also, caffeine, a RyR stimulator, inhibits IP_3Rs. A series of xestospongins have been identified as noncompetitive IP_3R antagonists. Another compound reported to be a functional and cell-permeable IP_3R antagonist is 2-aminoethoxydiphenyl borate. A monoclonal antibody against the C-terminus of the IP_3R molecule also affects Ca^{2+} mobilization.

4.1.4. ER Calcium Excitability

Several models for oscillatory Ca^{2+} release from the ER have been developed. Some of these models are based on fluctuations in IP_3 production (*see* Section 9). Others are analogs to the Hodgkin–Huxley model for the electrical excitability of the plasma membrane. In the later models, the first step is IP_3 production and its binding to IP_3Rs, resulting in a small release of Ca^{2+} from the ER. Once released into the cytosol, Ca^{2+} initially exerts a positive feedback effect on the IP_3Rs. This augments Ca^{2+} release, leading to a large Ca^{2+} pulse or wave. The *ER calcium excitability* refers to the ability of such small Ca^{2+} increment to trigger a large calcium pulse. When $[Ca^{2+}]_i$ reaches a critical level, Ca^{2+} begins to exert a negative feedback effect on the IP_3R channel. During this refractory period, $[Ca^{2+}]_i$ undergoes a delicate redistribution between the free and bound forms, which is influenced by ER Ca^{2+}-ATPase, the reuptake of Ca^{2+} by mitochondria, and an exclusion of Ca^{2+} from the cell by plasma membrane Ca^{2+}-ATPase and exchangers. When the coordinate actions of these cycles bring $[Ca^{2+}]_i$ to lower facilitatory levels, the process is repeated and another transient occurs (Fig. 13). Thus, like plasma membrane potential, the $[Ca^{2+}]_i$ represents the excitation variable in the Ca^{2+} excitability of the ER membrane. The analogy between the two types of membrane-associated excitabilities goes beyond these general features. For example, the positive and negative effects of Ca^{2+} on IP_3Rs are analogous to the voltage-gated activation and inactivation of the TTX-sensitive Na^+ channel.

4.2. Ryanodine Receptor Channels

Ryanodine receptor channels (RyRs) were originally identified as Ca^{2+} release channels expressed in the SR of skeletal muscle fibers and cardiac myocytes, where they play a central role in excitation–contraction coupling. These channels are also expressed in neurons, chromaffin cells, sea urchin eggs, and several nonexcitable cell types. Mammalian tissues express three isoforms: RyR_1 is expressed predominantly in skeletal muscle, RyR_2 is expressed in cardiac muscle, and RyR_3 has a wide tissue distribution, including the nonexcitable cells. RyR_1 and RyR_2 channels display a 66% identity, whereas the RyR_3 channel is much shorter. Like the IP_3Rs, RyRs are tetramers, with a large N-terminal region forming heads, and a C-terminal region that forms the Ca^{2+}-selective channel. Although these channels are frequently coexpressed with IP_3Rs, the physiological importance of their coexpression and their variable density within the cells are still largely unclear.

4.2.1. Regulation

RyRs are the largest known ion channels and are susceptible to many different modulators, including cytosolic calcium, V_m, and several intracellular messengers. As in the regulation of IP_3Rs, cytosolic Ca^{2+} is a major regulator of RyRs; at low concentrations, Ca^{2+} promotes release, whereas higher concentrations are inhibitory. However, inhibition of RyRs by high $[Ca^{2+}]_i$ is somewhat controversial, as it requires $[Ca^{2+}]_i$ to be in the millimolar concentration range, which is not reached under physiological conditions. The ability of Ca^{2+} to stimulate its release from the ER/SR via RyRs is called "Ca^{2+}-induced Ca^{2+} release." This process is of fundamental importance for coordinating the elementary Ca^{2+} release events into Ca^{2+} spikes and waves (*see* Section 6). Unlike IP_3Rs, RyRs can release Ca^{2+} in response to an increase in $[Ca^{2+}]_i$, without any other change in the concentration of sec-

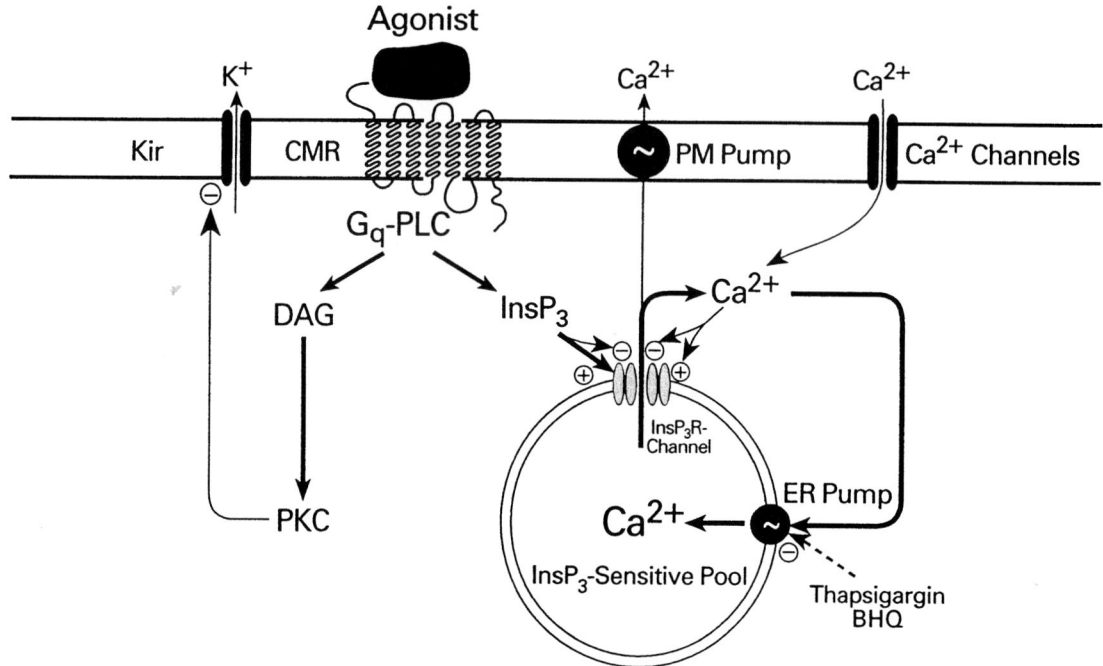

Fig. 13. Endoplasmic reticulum calcium excitability. Inositol 1,4,5-trisphosphate and Ca^{2+} act as coagonists in the control of IP_3R permeability. When the receptor is occupied by IP_3, low concentrations of calcium facilitate Ca^{2+} release from ER and subsequently the elevated $[Ca^{2+}]_i$ inhibit IP_3Rs. During the prolonged occupation, IP_3 may also inhibit IP_3Rs. PKC, protein kinase C; CMR, calcium mobilizing receptors; Kir, inward rectifier potassium channels.

ond messengers. This is crucial for excitation–contraction coupling. For example, in cardiac cells, Ca^{2+} entry through dihydropyridine-sensitive channels activates SR RyRs to induce a further increase in $[Ca^{2+}]_i$. In skeletal muscle cells, the dihydropyridine receptors act primarily as voltage sensors to directly activate RyRs in response to V_m depolarization.

In addition to Ca^{2+}, there are numerous other endogenous modulators of RyRs. These include ER/SR Ca^{2+}, cytosolic pH, Mg^{2+} and other cations, Cl^- and other anions, nucleotides, cyclic adenosine 5′-diphosphate ribose (cADP-R), several protein kinases, and calmodulin and other Ca^{2+}-binding proteins. Of these, cADP-R deserves special attention. This molecule is present in many cell types and is generated from NAD by two enzymes. In parallel to the $[Ca^{2+}]_i$ response, the intracellular concentration of cADP-R also increases in response to agonist stimulation. Although injection of cADP-R is accompanied by release of Ca^{2+} from intracellular stores, the mechanism and the site of cADP-R action are controversial. It was thought that cADP-R directly activates RyRs, but recent studies made its direct action on RyR_1 and RyR_2 unlikely. Furthermore, modulation of RyRs by cADP-R was reported by some, and the lack of its effect by other investigators. Finally, the finding that cADP-R binds to two proteins of 100 and 140 kDa raised the possibility that these proteins might act as intermediators between cADP-R and RyRs, or may even form a novel type of channel.

4.2.2. PHARMACOLOGY

Caffeine is a standard pharmacological tool for activation of RyRs. It acts on RyRs in both intact cells and isolated channels. Another pharmacological agent, dantrolene, has biphasic effects on RyR_1; in nanomolar concentrations it increases the open probability of these channels and in micromolar concentrations it inactivates the channel. Ryanodine also activates and inhibits RyRs, depending on its concentration. At 1–10 μM, ryanodine locks the channel in a subconductance stage, which slows the channel opening and closing. At higher concentrations, ryanodine inhibits RyRs, and this action is mimicked by ruthenium red. Both compounds do not alter IP_3Rs, indicating that they are relatively specific for RyRs. On the other hand, the IP_3Rs blocker heparin activates RyRs in a Ca^{2+}-dependent manner. Thus, although IP_3Rs and RyRs are structurally and functionally distinct, the use of caffeine and heparin as pharmacological tools may lead to misleading conclusions about the receptor types expressed in a particular cell.

4.2.3. THE RYANODINE RECEPTOR AND DISEASE

The role of RyRs in several diseases is relatively well characterized. The best example is malignant hyperthermia, in which mutations in RyR_1 gene are related to a specific clinical syndrome. Ryanodine receptor modulators, such as dantrolene, are useful in the management of these patients. These receptors have also been implicated in myocardial protection. Alterations in RyRs are involved in the pathogenesis of heart failure, cardiomyopathies, and degenerative skeletal muscle diseases. As RyRs are also expressed in nonmuscle tissues, it is likely that these receptors play a role in other diseases.

5. STORE-OPERATED CALCIUM CHANNELS

It is generally accepted that activation of IP_3-dependent Ca^{2+} mobilization is associated with depletion of the ER calcium pool, and that Ca^{2+} entry through Ca^{2+}-conducting channels is essential for sustaining agonist-induced Ca^{2+} spiking and refilling the intracellular Ca^{2+} pools upon termination of receptor activation. The two phases of Ca^{2+} signaling during sustained activation of phospholipase C-coupled receptors can be easily dissected by comparing the $[Ca^{2+}]_i$ responses of cell populations to a Ca^{2+}-mobilizing agonist in the presence or absence of extracellular Ca^{2+}. Receptor activation results in a transient increase in $[Ca^{2+}]_i$ in cells bathed in Ca^{2+}-deficient medium, while in Ca^{2+}-containing medium the initial spike phase is accompanied by an additional plateau component. In nonexcitable cells, voltage-insensitive Ca^{2+} entry completely accounts for sustained Ca^{2+} influx. In excitable cells, however, both voltage-sensitive and -insensitive Ca^{2+} influxes account for the plateau $[Ca^{2+}]_i$ response (Fig. 14). The mechanism of activation of voltage-insensitive Ca^{2+} influx pathways has been termed *capacitative calcium entry*.

5.1. Capacitative Calcium Entry

The term capacitative Ca^{2+} entry, by analogy with a capacitor in an electrical circuit, implies that intracellular Ca^{2+} stores prevent entry when they are charged (filled by Ca^{2+}), but promote entry as soon as the stored Ca^{2+} is discharged (released) (Fig. 15). The similarities in the properties of this entry within different cell types, including excitable cells, suggest a common mechanism. In addition to Ca^{2+}-mobilizing agonists, capacitative Ca^{2+} entry can be activated by injection of IP_3 or its nonmetabolized forms into the cell, inhibition of the Ca^{2+}-ATPase by thapsigargin, discharge of the intracellular content by calcium ionophores, or prolonged incubation of cells in Ca^{2+}-deficient medium. Injection of heparin, an IP_3R inhibitor, completely blocks agonist- and IP_3-induced Ca^{2+} mobilization and capacitative Ca^{2+} entry. Thus, a decrease in ER Ca^{2+} content in a receptor or nonreceptor manner is an effective signal for Ca^{2+} influx. Because depletion of the ER Ca^{2+} stores is followed by the influx of Ca^{2+} into the cell, the channels involved in such influx were termed (SOCCs). At the present time, the nature of these channels and the mechanism of their regulation in response to store depletion are unknown. Three types of channels have been suggested to mediate capacitative Ca^{2+} entry: calcium release activated (CRAC) channels, Ca^{2+}-activated nonselective channels (CAN), and transient receptor potential protein (TRP) channels.

5.1.1. CRAC CHANNELS

The ion currents associated with capacitative Ca^{2+} entry have been extensively studied in nonexcitable cells. Although the low unitary cord conductance of such Ca^{2+} entry (24 fS, in contrast to VGCCs, whose conductance is between 7 pS and 30 pS) is compatible with the involvement of a carrier in Ca^{2+} influx, noise analysis is more consistent with the involement of a channel. CRAC channels are blocked by several trivalent and bivalent ions in the following order: $La^{3+} > Zn^{2+} > Cd^{2+} > Be^{2+} = Co^{2+} = Mn^{2+} > Ni^{2+} > Sr^{2+} > Ba^{2+}$. Like VGCCs and IP_3Rs, CRAC channels are sensitive to $[Ca^{2+}]_i$; both calcium-dependent inactivation and activation of these channels have been observed. Initially, it has been suggested that CRAC channels are IP_3Rs expressed in the plasma membrane. This hypothesis is supported by the presence of an immunospecific IP_3R expressed in the plasma membrane of T lymphocytes and endothelial cells, as well as the ability of IP_3 and/or $Ins(1,3,4,5)P_4$ to activate Ca^{2+} channels on the cytoplasmic surface. However, there are two findings that contradict this hypothesis. First, the single cord conductivity of IP_3Rs is higher than that of CRAC channels. Second, as Ca^{2+} entry in lymphocytes can be activated by thapsigargin (i.e., without an increase in inositol phosphate production), these channels should be regulated by a mechanism other than IP_3, or by dual regulation, by inositol phosphates and some other messenger(s) generated by the depletion of intracellular Ca^{2+} pools. The mechanism that store depletion signals for activation of CRAC channels is not clear. The presence of a diffusible cytoplasmic messenger, termed CIF, has been proposed. Also, it has been suggested that these

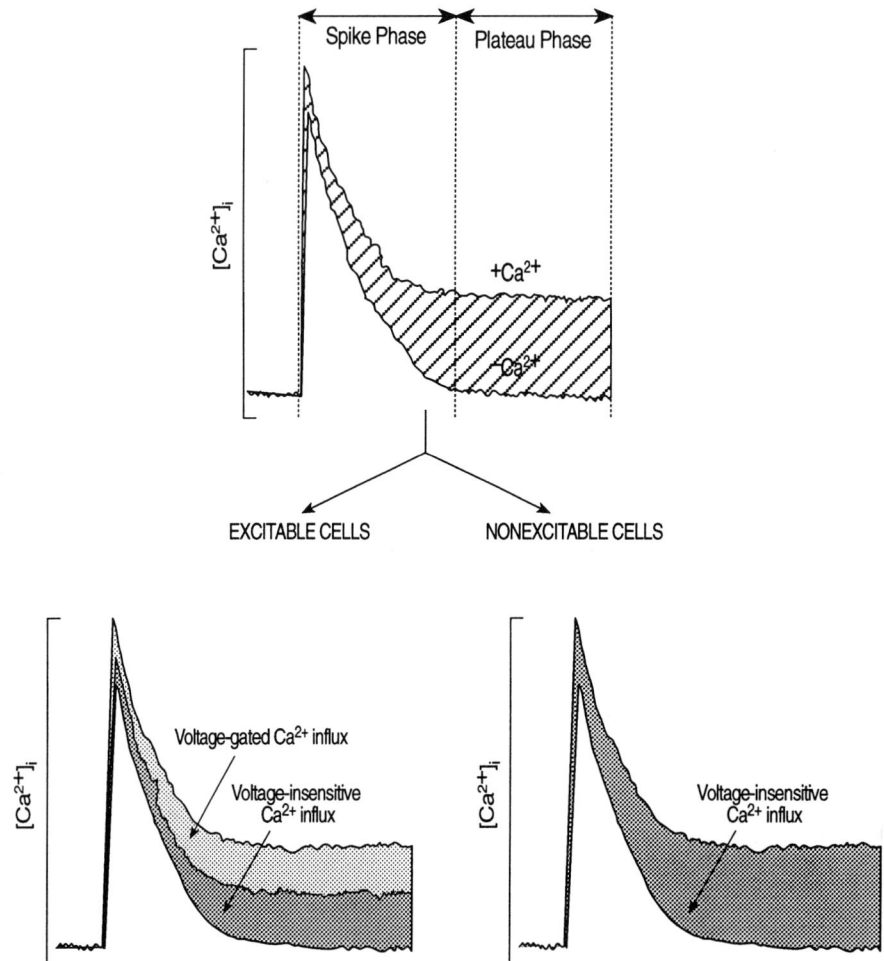

Fig. 14. Components of the agonist-induced calcium signals in excitable and nonexcitable cells. Calcium-mobilizing receptor-induced and IP_3-mediated Ca^{2+} signals can be resolved into a largely extracellular Ca^{2+}-independent spike component and an influx-dependent plateau component. Whereas in nonexcitable cells, voltage-insensitive Ca^{2+} influx accounts for the plateau response, in excitable cells both voltage-sensitive and insensitive Ca^{2+} influxes participate in the sustained plateau response. Voltage-sensitive Ca^{2+} influx can be dissected further by utilizing specific inhibitors of VGCCs (*see* Fig. 7).

channels form a direct link with IP_3Rs when ER Ca^{2+} store is depleted.

5.1.2. NONSELECTIVE CATIONIC CHANNELS

Several published observations support the view that the signal for capacitative Ca^{2+} entry can also regulate Ca^{2+} influx through calcium-conducting nonselective cation channels. Endothelial cells from human umbilical cord express a nonselective cation channel, with a permeation ratio of $K^+/Na^+/Ca^{2+}$ = 1:0.9:0.2, and with a single-channel conductance of about 27 pS. In these cells, application of histamine induces Ca^{2+} transients and an ionic current with a reversal potential near 0 mV. The amplitude of this current closely correlates with the amplitude of the concomitant Ca^{2+} transients, suggesting that calcium influx through histamine-activated nonselective cation channels is responsible for Ca^{2+} spiking. A 50 pS nonselective cation channel is expressed in mast cells and contributes to the sustained increase in Ca^{2+} following receptor activation. However, this form of Ca^{2+} entry is not blockable by dialyzing cells with the IP_3R channel inhibitor heparin, which is necessary to demonstrate the involvement of capacitative Ca^{2+} influx. The depletion of intracellular Ca^{2+} stores by thapsigargin also activates a Ca^{2+}-conducting nonselective cation current in mouse pancreatic acinar cells, demonstrating more directly that capacitative Ca^{2+} entry can be associated with different plasma membrane channels.

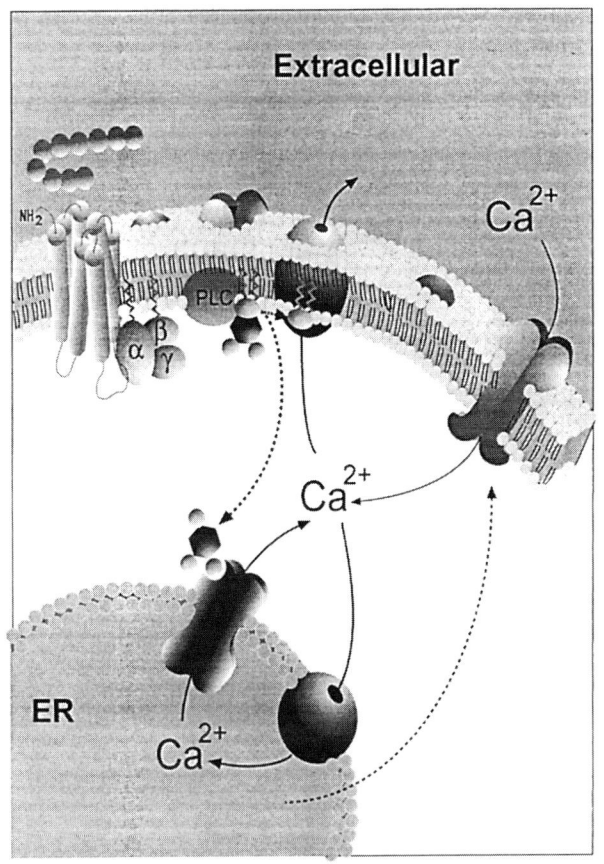

Fig. 15. Store depletion-activated Ca^{2+} influx through voltage-insensitive Ca^{2+} channels. IP_3-induced Ca^{2+} release from the ER leads to depletion of the ER calcium store. This in turn activates voltage-insensitive Ca^{2+} influx through store-operated Ca^{2+} channels via a yet uncharacterized pathway.

5.1.3. TRANSIENT RECEPTOR POTENTIAL PROTEIN CHANNELS

The other proposal is that SOCCs have a structure similar to transient receptor potential (TRP) protein from *Drosophila* mutants, whose photoreceptors are incapable of eliciting a sustained Ca^{2+} response. The TRP and related TRPL gene products have been found to display significant amino acid sequence similarity to the α-subunit of L-type VGCCs, but lack the specific basic residues in the S4 region, which are responsible for the voltage sensitivity of VGCCs. In insects, photoreceptor signaling utilizes a phospholipase C-coupled system, implicating that the TRP gene product may operate as a Ca^{2+} channel activated by emptying the intracellular Ca^{2+} pools. In accord with this, heterologous expression of TRP_3 cDNA in eukaryotic cells led to the formation of ion channels that were activated by the depletion of intracellular Ca^{2+} stores.

However, the conductances of TRP channels and CRAC channels are not the same. Unlike the TRP_3, TRPL gene product forms nonselective cation channels that are not coupled to store depletion. The cDNAs of mammalian homologs of TRP/TRPL proteins have been recently cloned. One group of these channels encodes a Ca^{2+}-permeable nonselective cation channel, which is insensitive to store depletion. However, a bovine homolog of TRP/TRPL known as bTRP4 represents the first member of another group of cDNAs, which encodes a store-operated Ca^{2+}-conducting channel. Some excitable cells express TRP5, which forms a functional store-operated and Ca^{2+}-conducting channel, suggesting that SOCCs, like VGCCs and ligand-gated channels, represent a family of related proteins. Moreover, it strengthens the hypothesis that capacitative Ca^{2+} entry is operative in excitable cells as well.

6. CALCIUM-MOBILIZING RECEPTORS AND V_m PATHWAY

Agonist-stimulated release of Ca^{2+} from the ER modulates plasma membrane excitability in neuronal and endocrine cells by altering the V_m. For example, activation of Ca^{2+}-mobilizing receptors in GnRH-secreting neurons, pituitary lactotrophs, and GH cell lines leads to a biphasic increase in $[Ca^{2+}]_i$, consisting of a transient spike phase, followed by a sustained plateau phase. During the spike phase, AP firing is abolished due to a transient hyperpolarization of the membrane. This is followed by a sustained depolarization, resulting in an increase in AP frequency and associated Ca^{2+} entry through VGCCs. The increase in AP-driven Ca^{2+} entry participates in the generation of the plateau rise in $[Ca^{2+}]_i$ that is necessary to refill the ER Ca^{2+} stores during sustained receptor activation (Fig. 16). Pituitary gonadotrophs exhibit more complex interactions between $[Ca^{2+}]_i$ and membrane potentials owing to the nature of the Ca^{2+} release process. Activation of GnRH Ca^{2+}-mobilizing receptors in these cells stimulates an oscillatory rise in $[Ca^{2+}]_i$. Each spike increase in $[Ca^{2+}]_i$ is associated with a transient membrane hyperpolarization that is followed by the firing of one or more rebound APs that drive Ca^{2+} into the cell. The participation of AP-driven Ca^{2+} influx is minor during the initial phase of GnRH stimulation and intracellular Ca^{2+} mobilization, but is critical for the maintenance of the Ca^{2+} signal during sustained receptor activation. As discussed in Subheading 6.1., Ca^{2+}-controlled channels, as well as

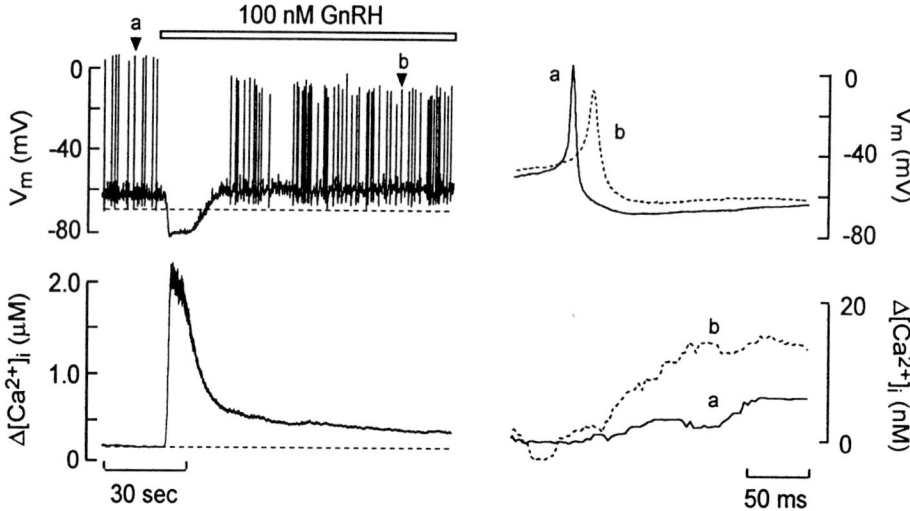

Fig. 16. Agonist (GnRH)-induced modulation of electrical activity in spontaneously active cells. (**Left panels**) Simultaneous measurements of electrical activity and [Ca^{2+}]$_i$ in single GT1 neurons. (**Right panels**) Expanded time scale of single action potentials and associated [Ca^{2+}]$_i$ transients from the records on left labeled as **a** and **b**. Note the increase in AP frequency and duration, which lead to enchanced Ca^{2+} entry during agonist activation.

other channels such as CAN, M-channels, and SOCCs, may synchronize the V_m and Ca^{2+} release process during activation of Ca^{2+}-mobilzing receptors.

6.1. Ca^{2+}-Activated K$^+$ Channels

In many excitable cells expressing Ca^{2+}-mobilizing receptors, agonist-induced release of Ca^{2+} from the ER opens Ca^{2+}-activated K$^+$ channels that hyperpolarize the membrane. In GnRH-secreting neurons (Figs. 16 and 17), pituitary lactotrophs, and GH cell lines, the spike increase in [Ca^{2+}]$_i$ activates apamin-sensitive K$^+$ channels to induce transient membrane hyperpolarization. Similarly, the repetitive membrane hyperpolarization in pituitary gonadotrophs is mediated by the transient activation of apamin-sensitive K$^+$ channels by the oscillatory Ca^{2+} release (Fig. 17). There are at least three reasons why Ca^{2+}-activated K$^+$ channels are incorporated into the Ca^{2+} signaling pathway. First, activation of these channels may relieve the steady-state inactivation of voltage-gated Na$^+$ and Ca^{2+} channels, which stimulates or enhances AP generation in some cells. This is exemplified in male rat gonadotrophs, in which oscillatory activation of Ca^{2+}-activated K$^+$ channels by GnRH is necessary to remove the steady-state inactivation of Na$^+$/Ca^{2+} channels allowing for the initiation of AP spiking and voltage-gated Ca^{2+} influx. Second, activation of Ca^{2+}-activated K$^+$ channels may prevent a lethal increase in [Ca^{2+}]$_i$ by limiting voltage-gated Ca^{2+} influx. For example, in hypothalamic GT1 neurons, Ca^{2+}-activated K$^+$ channels prevent AP-driven Ca^{2+} entry during the high spike increase in [Ca^{2+}]$_i$ and limit it during the lower plateau increase in [Ca^{2+}]$_i$. Lastly, activation of Ca^{2+}-activated K$^+$ channels and the resulting membrane hyperpolarization may serve to synchronize electrical activity in cell networks to generate pulsatile hormone secretion.

6.2. Nonselective Ca^{2+} Activated Cationic Channels

In some neuronal and nonneuronal cells, an increase in [Ca^{2+}]$_i$ leads to the activation of plasma membrane associated CAN channels. Several features of CAN channels make them excellent candidates for mediating the synchronization of the V_m pathway with the Ca^{2+}-mobilizing pathway. First, CAN channels are opened by an increase in [Ca^{2+}]$_i$, which occurs in response to activation of Ca^{2+}-mobilizing receptors. Second, most CAN channels are permeable to Na$^+$, K$^+$, and Ca^{2+}, giving them a reversal potential of approx 0 mV. Therefore, activation of these channels would depolarize the membrane to increase AP frequency and associated voltage-gated Ca^{2+} entry. Third, they do not undergo voltage- or Ca^{2+}-dependent inactivation and thus are capable of maintaining sustained membrane depolarizations in response to an increase in [Ca^{2+}]$_i$. In addition to increasing voltage-gated Ca^{2+} entry, the permeability of most CAN channels to Ca^{2+} provides an additional Ca entry pathway. Also, CAN channels can indirectly increase [Ca^{2+}]$_i$ by the reversal of the plasma membrane Na$^+$/Ca^{2+} exchanger.

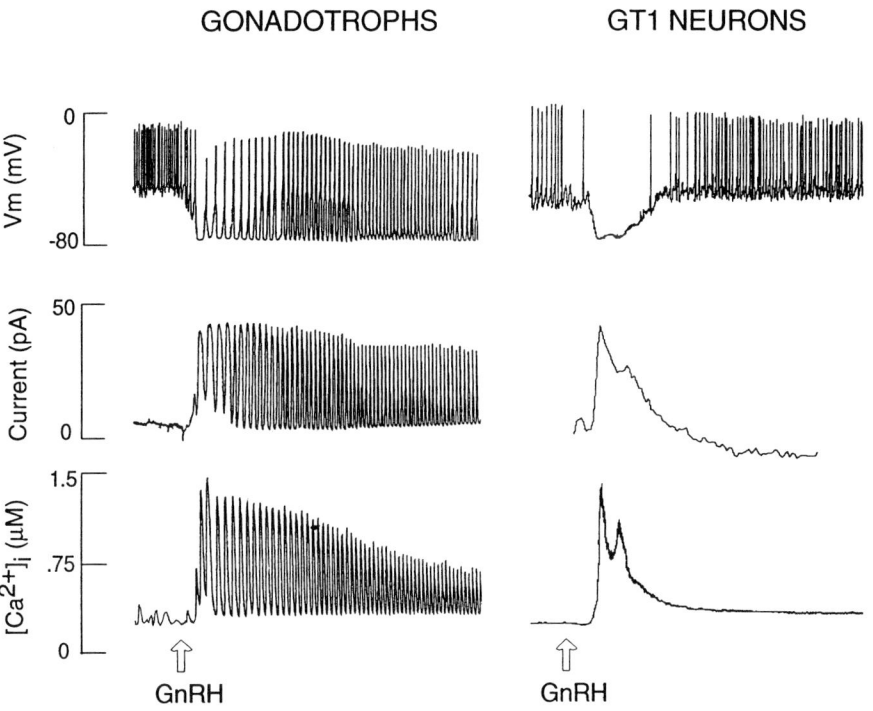

Fig. 17. Agonist-induced calcium signaling and electrical activity in gonadotrophs and GT1 neurons. In gonadotrophs, activation of GnRH calcium-mobilizing receptors is associated with baseline Ca^{2+} oscillations (**left panel,** *bottom tracing*) and periodic activation of apamin-sensitive potassium (SK) channels (**middle panel**). SK channels transiently hyperpolarize the membrane. When $[Ca^{2+}]_i$ is at the baseline level, SK channels are not activated, and gonadotrophs are able to fire one to five action potentials. In GT1 neurons, GnRH induces a spike and plateau type of $[Ca^{2+}]_i$ response (**right panel,** *bottom tracing*) that transiently activates SK channels (**middle panel**). This leads to a transient hyperpolarization of membrane, followed by sustained depolarization of cells and increase in the firing frequency.

6.3. M-Channels

The M-channel driven current is a time- and voltage-dependent K⁺ current that slowly activates when a cell is depolarized toward the AP firing threshold. Once activated, this current has the unique characteristic of sustained activation, making it ideal for opposing membrane depolarization to reduce membrane excitability and associated voltage-gated Ca^{2+} entry. Receptor-mediated inhibition of this current depolarizes the membrane and increases the input resistance, which augments AP frequency and associated voltage-gated Ca^{2+} entry. Because there can be a considerable current through these channels at depolarized membrane potentials, suppression of this current may also broaden APs to prolong the duration of Ca^{2+} influx in some cells. Inhibition of this current was first observed in response to activation of muscarinic receptors in bullfrog sympathetic ganglion neurons, hence the name M-type current. Since then, it has been demonstrated that activation of a variety of receptors suppresses this current, including GnRH, bradykinin, opioids, substance P, ATP, adrenergic, TRH, and angiotensin II receptors. As these receptors are typically coupled to the phospholipase C pathway, it is likely that IP_3, 1,2-diacylglycerol (DAG), or one of their metabolites, and/or an increase in $[Ca^{2+}]_i$ mediates the inhibition of M-currents. Moreover, the link between activation of these Ca^{2+}-mobilizing receptors and inhibition of M-currents may provide a mechanism to increase AP-driven Ca^{2+} entry to refill the ER Ca^{2+} stores.

6.4. Store-Operated Ca^{2+} Channels

Until recently, it was believed that excitable cells do not express SOCCs. However, recent evidence suggests that excitable cell also express SOCCs and that these channels act as a Ca^{2+} influx pathway and as pacemaker channels to modulate AP-driven Ca^{2+} entry. In smooth muscle cells, store depletion with the ER Ca^{2+}-ATPase inhibitor thapsigargin activates Ca^{2+} influx that was not blocked by VGCC inhibitors. Similarly, depletion of the ER Ca^{2+} stores by the activation of Ca^{2+}-mobilizing receptors or thapsigargin activates a voltage-independent Ca^{2+}-permeable channel in N1E-115 neuroblastoma cells and GnRH-secreting neurons. Moreover, in GnRH-secreting neurons, activation of SOCCs depolarizes the membrane to increase AP frequency and duration, resulting in enhanced Ca^{2+} influx through VGCCs (Figs. 16 and 17). Activation of Ca^{2+}-mobilizing receptors in GH_3 cells may also activate SOCCs to regulate membrane excitability. Consistent with the presence of SOCCs in electrically excitable cells, the *trp5* gene product has been identified in neurons. Expression of *trp5* cDNA into cells leads to the formation of an ion channel that is activated by store depletion and is primarily permeable to Ca^{2+}.

7. TEMPORAL, SPATIAL, AND CELL-SPECIFIC CALCIUM SIGNALING PARADIGMS

7.1. Local and Global Calcium Signals

Analysis of images obtained by confocal microscopy of individual cells loaded with Ca^{2+}-sensitive fluorescent dyes makes possible the visualization of elementary events in Ca^{2+} signaling. These events occur in the vicinity of a single channel or a small group of channels and consist of a brief opening of the channels and a concomitant rise in local $[Ca^{2+}]_i$ of about 2 μm in diameter. Further Ca^{2+} release is inhibited, presumably due to Ca^{2+} inhibition of ER channels, leading to a decline in local $[Ca^{2+}]_i$ due to a passive diffusion and reuptake of Ca^{2+}. These localized Ca^{2+} signals have been termed *sparks* (observed in cardiac myocytes, skeletal and smooth muscles, the cells expressing RyRs), and *puffs* (observed in cells expressing IP_3Rs). Both phenomena involve concerted action of several channels within a single release site. In IP_3-stimulated cells, smaller local signals, termed *blips*, were also observed and may underlie Ca^{2+} release from an individual channel. At subthreshold agonist concentrations, the two major factors for the generation of localized Ca^{2+} signals are the site of action of the primary stimulus and the density of Ca^{2+}-mobilizing receptors in the plasma membrane. The focal and discrete injection of an agonist close to one region of the cell leads to a localized rise in $[Ca^{2+}]_i$ near the region of stimulation. Similarly, the initial rise of $[Ca^{2+}]_i$ in oocytes occurs at the point of sperm entry. In pancreatic acinar cells, the IP_3Rs are predominantly localized in a zone with secretory granules, providing localized Ca^{2+} signals independent from the regions of stimulation by agonist.

AP-driven Ca^{2+} entry forms several types of localized increase in $[Ca^{2+}]_i$ that are close to the plasma membrane. At the mouth of an open VGCC, $[Ca^{2+}]_i$ can reach hundreds of micromolar and such a signal is analogous to the blips generated by the activation of a single IP_3Rs. These highly localized peaks are transient, with a time scale faster than that of Ca^{2+} binding to its buffers, and are relevant for colocalized processes, such as exocytosis. Conversely, the synchronized activity of several VGCCs can generate bumps or quantum emission domains, which are localized Ca^{2+} signals that are analogous to sparks and puffs.

These elementary events may be arrested and remain local Ca^{2+} signals, or they may trigger the spread of Ca^{2+} throughout the cytoplasm to form global Ca^{2+} signals. Both, Ca^{2+} release and Ca^{2+} influx can make the transition from local to global signals. Single IP_3R-derived Ca^{2+} blips are isolated events initiated by very low levels of IP_3, and participate in the control of the resting $[Ca^{2+}]_i$. Further increase in IP_3 concentrations generates puffs, which are the primary sites of a rise in $[Ca^{2+}]_i$ that spreads or does not spread throughout the cytosol, depending on the temporal coordination of a sufficient number of those elementary events, that is, on agonist concentration and distribution of IP_3Rs. At threshold agonist concentration, puffs serve as the initiation sites for the onset of Ca^{2+} waves. Calcium diffusing away from a puff site is sufficient to trigger propagation of the signal.

Depending on the cell type, Ca^{2+} waves propagate at 5–100 μm/s. The pattern of IP_3Rs distribution and the size of cells are critical for wave propagation. For example, the density of IP_3Rs in astrocytes (about 100 μm in diameter) varies within the cell, creating a nonlinear propagation of Ca^{2+} waves, with several focal loci where the wave is amplified. Smaller cells, such as pituitary gonadotrophs, usually have only one focal point, which serves as a "pacemaker" to generate Ca^{2+} waves. The priming of cells with IP_3 is required for wave initiation, but Ca^{2+} serves as a positive as well as a negative feedback element. Thus, wave propagation is the manifestation of the ER calcium excitability, where local increases in $[Ca^{2+}]_i$ initiates a self-propagated signal in which Ca^{2+} serves as a diffusible factor.

Depolarization-induced global signals result from the synchronized activity of VGCCs and RyRs, where the opening of RyRs is tightly coupled to AP firing. In skeletal muscle, the dihydropyridine receptor serves as the voltage sensor and is directly coupled to the RyR$_1$. The voltage sensor responds to membrane depolarization by a change in conformation, which in turn activates RyRs. Such coupling can activate all RyRs, leading to an explosive release of Ca^{2+} and subsequently to muscle contraction. In cardiac cells, voltage-gated Ca^{2+} entry leads to local [Ca^{2+}]$_i$ increase, which is then greatly amplified by activation of a small group of RyRs through a Ca^{2+}-induced Ca^{2+} release mechanism. Thus, the interconnected voltage sensor and RyRs represent a functional unit that generates Ca^{2+} sparks, which do not activate the silent neighboring units because of their low sensitivity to [Ca^{2+}]$_i$. However, when the cardiac cells are overloaded with Ca^{2+}, the neighboring units start to communicate with each other, leading to a regenerative Ca^{2+} wave.

7.2. Pathway- and Cell-Specificity of Calcium Signaling

Global Ca^{2+} signals can take different forms depending on the pathways for Ca^{2+} delivery and the cell types in which a particular pathway is expressed. Figure 18 illustrates the patterns of Ca^{2+} signals generated by V_m-dependent and IP$_3$-dependent pathways in pituitary gonadotrophs. Both pathways generate oscillatory Ca^{2+} signals. In these cells, agonist-induced Ca^{2+} oscillations are higher in amplitude (1–2 μm) and have more variable frequencies, which are determined by agonist concentrations. In contrast, the amplitudes of spontaneous V_m-derived Ca^{2+} oscillations reached only 50–200 nm and there is no obvious modulation in their frequencies. In other cell types, such as somatotrophs, spontaneous V_m-derived oscillations are higher in amplitude, comparable to that induced by Ca^{2+}-mobilizing agonists.

There is an enormous heterogeneity of [Ca^{2+}]$_i$ signals from one cell type to another. Among cells operated by Ca^{2+}-mobilizing receptors, some respond by the generation of sinusoidal oscillations, in which agonist concentration regulates the amplitude, but not the frequency of Ca^{2+} transients. This pattern of oscillatory Ca^{2+} signaling has been observed in excitable and nonexcitable cell types. In other cells, discrete [Ca^{2+}]$_i$ transients or spikes can be observed above a constant baseline level, and this pattern of [Ca^{2+}]$_i$ oscillations is referred to as *baseline spiking*. Here, the agonist concentration regulates the frequency but not the spike amplitude. Interestingly, the upstroke of the [Ca^{2+}]$_i$ transient is sometimes preceded by a more gradual increase in [Ca^{2+}]$_i$, which is reminiscent of the slow membrane depolarization observed during

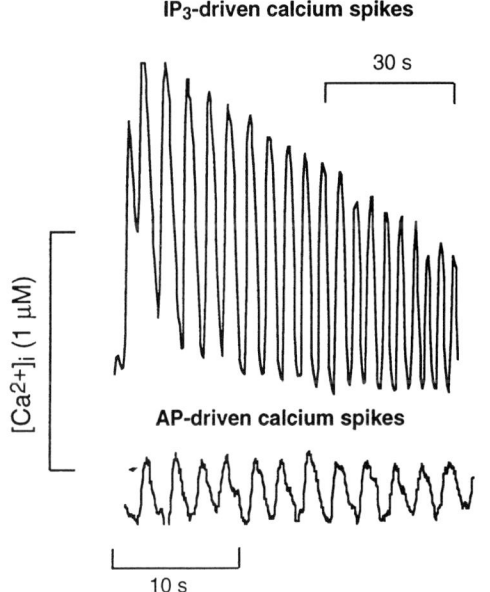

Fig. 18. The patterns of Ca^{2+} signaling in pituitary gonadotrophs. **Bottom tracing** illustrates spontaneous and extracellular Ca^{2+}-dependent fluctuations in [Ca^{2+}]$_i$, that are typical of the plasma membrane oscillator. **Upper tracing** illustrates IP$_3$-induced [Ca^{2+}]$_i$ oscillations that are dependent on Ca^{2+} mobilization from the intracellular stores (ER oscillator).

the pacemaker potential in electrically excitable cells. This is consistent with the view that Ca^{2+} participates in the autocatalytic process that leads to the "all or none" firing of an [Ca^{2+}]$_i$ transient. In some cells, both oscillatory patterns are observed depending on the agonist concentration. The frequency of intracellular Ca^{2+} oscillations can also vary from cell to cell. For example, in pituitary gonadotrophs the frequency of Ca^{2+} spiking is between 5 to 30 per minute, whereas in hepatocytes it is on the order of one spike per minute.

Not all cells expressing Ca^{2+}-mobilizing receptors respond to agonist simulation by the generation of [Ca^{2+}]$_i$ oscillations. In many cells, monophasic or biphasic nonoscillatory Ca^{2+} signals are observed in response to agonist stimulation. Monophasic [Ca^{2+}]$_i$ responses are characterized by the generation of a single spike increase in [Ca^{2+}]$_i$. Conversely, biphasic [Ca^{2+}]$_i$ signals can be characterized by the generation of a single transient increase in [Ca^{2+}]$_i$ followed by a sustained increase in [Ca^{2+}]$_i$ of lower amplitude. Alternatively, they may be characterized by bidirectional changes in [Ca^{2+}]$_i$, in which a single spike increase in [Ca^{2+}]$_i$ is followed by a sustained decrease in [Ca^{2+}]$_i$ below basal levels. Moreover, the pattern of the [Ca^{2+}]$_i$ response in different cell types all expressing the same Ca^{2+}-mobilizing receptor can also vary considerably. For example, parathyroid cells, Leydig cells, GT1 neurons, pituitary lactotrophs, and gonadotrophs all express the endothelin-A receptor, but respond to its activation with very different Ca^{2+} signals. In Leydig cells, endothelin generates a monophasic increase in $[Ca^{2+}]_i$. In GT1 neurons, endothelin stimulates a biphasic increase in $[Ca^{2+}]_i$, consisting of a transient increase in $[Ca^{2+}]_i$ that is followed by a lower amplitude sustained increase in $[Ca^{2+}]_i$. Conversely, in lactotrophs, activation of endothelin receptors stimulates a bidirectional change in $[Ca^{2+}]_i$, consisting of a spike increase in $[Ca^{2+}]_i$ that is followed by a sustained and prominent decrease in $[Ca^{2+}]_i$. Lastly, in pituitary gonadotrophs, endothelin stimulates $[Ca^{2+}]_i$ oscillations similar to those observed by activation of GnRH receptors (Fig. 19).

7.3. Intraorganelle Calcium Signaling

The propagation of Ca^{2+} signals within the cell is required for adequate regulation of cytosolic and membrane-associated cellular functions, such as control of the channels on the plasma and ER membranes, cytosolic enzyme activities, protein folding, etc. Moreover, Ca^{2+} plays an important role in the function of cellular organelles (*see* Section 10). Except for the ER and SR, where the mechanism of Ca^{2+} release into and uptake from the cytosol is quite well established (as discussed in Section 4), Ca^{2+} fluxes across membranes of other organelles are much less understood. New results are emerging especially on nuclear and mitochondrial Ca^{2+} transport and regulation. The assessment of $[Ca^{2+}]$ in the organelles is made possible with recent advancements in experimental techniques. These include confocal microscopy, development of specially designed fluorescent Ca^{2+} indicators that would preferably load into specific organelles, and, most importantly, development of aeqorins, Ca^{2+}-sensitive photoproteins that can be localized to desired organelles by the addition of protein targeting sequences.

7.3.1. NUCLEUS

The cell nucleus is surrounded by the nuclear envelope, which is made up of two membranes that enclose the perinuclear space and is continuous with the ER membrane (Fig. 1). The nuclear membranes share many features with that of the ER, including expression of IP_3Rs and Ca^{2+}-ATPase in the outer membrane, and IP_3 and cADP-ribose regulated channels in the inner nuclear membrane. As the inositol lipid cycle

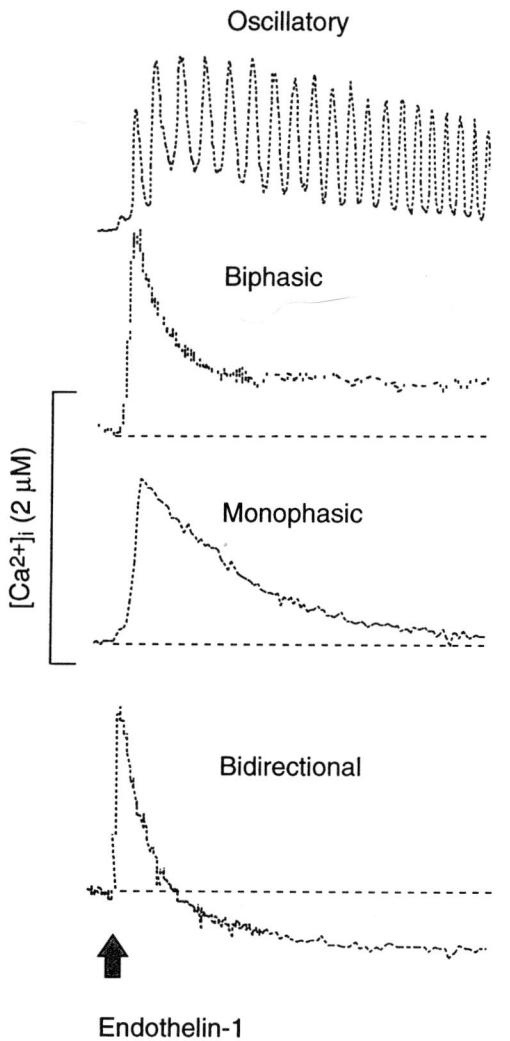

Fig. 19. Cell specificity of agonist-induced Ca^{2+} signals. From **top** to **bottom**: gonadotrophs, GT1 neurons, Leydig cells, and lactotrophs. All these cells express ET_A receptors.

also takes place in the nucleus, Ca^{2+} can enter the nucleus directly from perinuclear stores, but can also diffuse from the cytosol through the nuclear pore complex. The nuclear pore complex is a large multiprotein structure and is the only known pathway for a direct exchange between the cytosplasm and the nucleoplasm. Although research on the control of transport through the pore complex is far from complete, it seems clear that the opening of the pore is regulated by the state of depletion in the perinuclear Ca^{2+} stores. These findings, and the identification of Ca^{2+} binding proteins in the nucleus, suggest that nuclear $[Ca^{2+}]$ is regulated independently of $[Ca^{2+}]_i$. However, a direct comparison of $[Ca^{2+}]_i$ and nuclear $[Ca^{2+}]$ is still controversial owing to technical shortcomings of intracellular Ca^{2+} indicators. Furthermore, there are cell-to-cell differences in the apparent nuclear $[Ca^{2+}]$ that are primarily based on the size of the cytoplasm. Cells whose nuclear volume is large compared to the cytoplasm have shown different results from those in which the cytoplasm is larger than the nucleoplasm. Although nuclear $[Ca^{2+}]$ and $[Ca^{2+}]_i$ may be regulated independently, Ca^{2+} waves generated in the cytoplasm can spread into the nucleoplasm of several cell types. Furthermore, cytoplasmic Ca^{2+} puffs can cause transient increase in nucleoplasmic $[Ca^{2+}]$. In HeLa cells, such puffs originate primarily in the vicinity of the nucleus, they are propagated through the entire nucleus, and Ca^{2+} is delivered to regions of the cytoplasm that were not otherwise within the reach of diffusion of the initial puff. Thus, the diffusion of Ca^{2+} through the nucleus may play an important role in the formation of global Ca^{2+} signals.

7.3.2. Mitochondria

These organelles possess their own sophisticated Ca^{2+} efflux/influx system that is unique compared to the mechanisms taking place at the membranes of the ER and nucleus. Experiments on isolated mitochondrial preparations made possible the identification and characterization of several components of the Ca^{2+} transport systems across the mitochondrial membrane (Fig. 1). Calcium influx occurs via a Ca^{2+} uniporter that is very fast and does not require energy for its operation. The energy released by the passage of electrons along the respiratory chain of mitochondria generates pH and voltage gradients across the inner membrane. The matrix of mitochondria is alkaline and the membrane is negatively charged to about -160 mV. Thus, the membrane potential attracts positive ions into the matrix and pushes OH^- ions out, reinforcing the effects of the pH gradient. The very low permeability of the inner membrane to cations prevents massive Na^+ and K^+ fluxes. In contrast, Ca^{2+} entry is facilitated, as the inner membrane contains a transport protein that efficiently moves Ca^{2+} into the matrix using the voltage gradient as the driving force.

In contrast to mitochondrial Ca^{2+} influx pathways, Ca^{2+} efflux from the mitochondria via two electroneutral antiporters, $2Na^+/Ca^{2+}$ and $2H^+/Ca^{2+}$, requires energy in the form of ATP hydrolysis. Sodium-dependent transport dominates in muscle and neuronal tissues, and sodium-independent transport dominates in the liver and kidney. The collapse of mitochondrial membrane potential is associated with a rapid Ca^{2+} release, but it is unclear whether this process ever takes place under physiological conditions. The role of the permeability transition pore in Ca^{2+} influx appears more likely. This pore is a voltage-dependent channel activated by cytosolic and matrix Ca^{2+} and is inhibited by matrix H^+. In its low-conductance mode, it is an important element in the control of intracellular Ca^{2+} signaling, whereas in the high-conductance mode it promotes apoptosis (see Section 10). Like RyRs in the SR, opening of the permeability transition pore in mitochondria can be initiated by cytosolic Ca^{2+}, leading to Ca^{2+}-induced Ca^{2+} release. The signals can further propagate from an activated mitochondrion to another in vitro and in vivo, forming Ca^{2+} waves. Thus, mitochondria may serve as amplification sites for Ca^{2+} signals originating from the ER. Accordingly, amplification sites for Ca^{2+} waves in some cell types have been colocalized with mitochondria.

Mitochondria also act as Ca^{2+} reservoirs following ER Ca^{2+} release, by taking up Ca^{2+} released into the cytosplasm. This process can buffer $[Ca^{2+}]_i$ and modulate mitochondrial metabolism. The coupling of cytosolic and mitochondrial Ca^{2+} pathways was observed during both oscillating and nonoscillating Ca^{2+} signaling patterns. IP_3-mediated release of Ca^{2+}, for example, results in a rapid and large (several-fold higher than $[Ca^{2+}]_i$) increase in mitochondrial Ca^{2+} concentration. However, other mechanisms of elevating $[Ca^{2+}]_i$, such as Ca^{2+} influx, are much less effective in elevating mitochondrial $[Ca^{2+}]$, despite a similar change in global $[Ca^{2+}]_i$. These findings cannot be explained by the normal function of the Ca^{2+} uniporter, unless morphological considerations are taken into account: close contacts of mitochondria and IP_3Rs on the ER membrane enable the exposure of mitochondrial surfaces to higher $[Ca^{2+}]$ than the concentration in the bulk of the cytosol. In addition, the phenomenon of fast and large mitochondrial uptake might be a consequence of a rapid uptake mode, a new mechanism of Ca^{2+} uptake into the mitochondria. In some cells, however, Ca^{2+} influx through the plasma membrane was found to significantly influence mitochondrial $[Ca^{2+}]$ and function.

Mitochondrial Ca^{2+} handling may also influence intracellular Ca^{2+} signaling by modulating Ca^{2+} fluxes through plasma and ER membranes. For example, by buffering $[Ca^{2+}]_i$ during voltage-gated Ca^{2+} influx and/or Ca^{2+} release from ER, mitochondria may modulate the open probability of IP_3Rs. Also, the mitochondrial Ca^{2+} pathway modulates ATP production, which in turn alters the dynamics of plasma membrane and ER

Ca^{2+} release and uptake. Therefore, the mitochondrial Ca^{2+} handling system constitutes an active element in the formation of Ca^{2+} waves and oscillations, and it can also regulate transient $[Ca^{2+}]_i$ changes, induced by either V_m-controlled or Ca^{2+}-mobilizing processes.

7.3.3. Golgi Apparatus

The Golgi apparatus has recently been identified as an IP_3-sensitive Ca^{2+} store. Its importance in cellular Ca^{2+} homeostasis is still unknown.

7.3.4. Endosomes

There are indications that endocytosis causes substantial import of Ca^{2+} into the cytosol due to loss from early endosomes. Furthermore, caveolae, the small surface invaginations that eventually become vesicles, contain plasma membrane Ca^{2+}-ATPase and IP_3R channels, and thus may participate in the control of intracellular Ca^{2+} signaling (Fig. 1). Additional experiments are required to characterize the endocytotic mechanism for Ca^{2+} signaling and its significance to oscillatory Ca^{2+} response.

7.3.5. Secretory Granules and Lysosomes

Secretory granules and lysosomes may also work as dynamic Ca^{2+} storage compartments, as they contain high Ca^{2+} concentrations. These compartments may play an important role in exocytosis. However, the information on their participation in Ca^{2+} signaling is controversial.

8. INTERCELLULAR CALCIUM WAVES

In addition to intracellular Ca^{2+} signals, waves of $[Ca^{2+}]_i$ can flow through connected cells to coordinate the cellular activity of a tissue. Intercellular Ca^{2+} waves are characterized by an increase in $[Ca^{2+}]_i$ that crosses cell borders into neighboring cells through direct or indirect contact with the stimulated cell. Intercellular Ca^{2+} waves can be initiated in vitro by electrical or mechanical stimulation of a single cell, or by focal application of hormones or neurotransmitters. Once initiated, they travel complex routes for several hundred micrometers at rate of 15 μm/s to 25 μm/s. Intercellular Ca^{2+} waves have been observed in dissociated cell and organotypic cultures of many tissues, including the heart, brain, pituitary, kidney, liver, mammary gland, and lens epithelium. They have also been observed *in situ* using glial cells in acutely isolated rat retinas. Two pathways mediate intercellular Ca^{2+} waves—the movement of cytosolic messengers between connected cells via gap junctions and extracellular diffusion of messenger molecules.

8.1. Gap Junctions and Ca^{2+} Wave Propagation

The movement of cytoplasmic messengers between connected cells is mediated by gap junctions, which are clusters of intercellular channels that form a cytoplasmic bridge between adjacent cells to allow for the cell-to-cell transfer of ions, metabolites, and small messenger molecules. This includes Ca^{2+}, ATP, cAMP, cADP-ribose, and IP_3. Vertebrate intercellular channels are made up of a multigene family of conserved proteins called *connexins*. To date, at least 20 connexin genes have been identified. Connexins are made up of four hydrophobic transmembrane domains with the N- and C-termini located in the cytoplasm. In the plasma membrane, six connexin subunits assemble in a circle to form hemichannels called *connexons*. Connexons can contain a single type of connexin (homomeric) or multiple connexins (heteromeric) to form the hemichannel pore. When two connexons from adjacent cells come together, they form an intercellular channel that spans the gap between the two cells. Two identical connexons or different connexons can join to form either homotypic or heterotypic intercellular channels, respectively. The presence of heteromeric connexins and heterotypic intercellular channels can produce a diverse group of structurally distinct intercellular channels, with different permeabilities and/or functions. A variety of other factors, including membrane potential, Ca^{2+}, pH, and phosphorylation of channels can also alter gap junction channels. Several neurotransmitters and hormones, such as dopamine, acetylcholine, GABA, and estrogens have also been found to alter intercellular channel activity.

Gap junctions are found in a variety of cell types and serve many functional roles, including normal and abnormal brain functions, wound healing, ciliary beating, bile flow, insulin secretion, follicular growth, oogenesis, ovulation, and bioluminescence. Mutations or defective productions in gap junction channels have been linked to a number of diseases, including Charcot–Marie–Tooth X-linked neuropathy, malignancy, infertility, cataractogenesis, and deafness.

At the cellular level, several lines of evidence implicate gap junction channels in mediating the propagation of intercellular Ca^{2+} waves. In airway epithelial cells, osteoblastic cells, endothelial cells, and astrocytes, the specific gap junction channel antagonists, octanol and halothane, inhibit the propagation of intercellular Ca^{2+} waves. These antagonists also prevented intercellular Ca^{2+} waves between *Obelia* photocytes and their support cells and between follicular cells and oocytes. In several cell lines, transfection of cDNAs encoding gap junction proteins enabled them to propagate intercellular Ca^{2+} waves when they were not previously able to do so. For example, intercellular Ca^{2+} waves were observed in C6 glioma and HEK293 cells lines only after transfection with the cDNA for connexin 43. Agents that modify gap junction channels, such as H^+, $[Ca^{2+}]_i$, and PKC, also alter the propagation of intercellular Ca^{2+} waves. Finally, the injection of specific antibodies against gap junction channels inhibited wave propagation in airway epithelial cells.

Calcium used for the generation of intercellular Ca^{2+} waves via gap junctions can come from two sources, intracellular Ca^{2+} stores and/or the intercellular space. In epithelial cells, aortic endothelial cells and mammary epithelial cells, removal of extracellular Ca^{2+} did not prevent the spread of a mechanically induced intercellular Ca^{2+} wave. Conversely, depletion of intracellular Ca^{2+} stores with thapsigargin prevents intercellular Ca^{2+} wave propagation. This indicates that intracellular Ca^{2+} stores are primarily responsible for the generation of Ca^{2+} waves in these cells. In other cells, removal of extracellular Ca^{2+} prevented Ca^{2+} wave propagation. Moreover, in some cells, the source of Ca^{2+} depends on the trigger for activation of intercellular Ca^{2+} waves. In glial cells, for example, glutamate-triggered Ca^{2+} waves are dependent on extracellular Ca^{2+} entry, whereas focal mechanical stimulation is dependent only on intracellular Ca^{2+} stores. In addition, unlike glutamate-stimulated waves, mechanically stimulated waves exhibit short delays at cell borders. These results demonstrate that different pathways for the generation of intercellular Ca^{2+} waves can exist in a single cell type.

Due to the nonselectivity of gap junctions to ions and small molecules, several diffusible second messenger molecules are potential candidates for mediating the propagation of intercellular Ca^{2+} waves via gap junctions. Because many Ca^{2+}-mobilizing receptors stimulate the generation of intercellular Ca^{2+} waves, two obvious candidates are Ca^{2+} and IP_3. One model for the propagation of Ca^{2+} waves suggests that there is a regenerative interaction between $[Ca^{2+}]_i$ and the release of additional Ca^{2+} from intracellular Ca^{2+} stores. Thus, Ca^{2+} diffuses short distances to stimulate the autocatalytic release of Ca^{2+} from intracellular stores to propagate the intercellular Ca^{2+} wave. Consistent with the involvement of Ca^{2+} diffusion in the generation of the intercellular Ca^{2+} wave, addition of exogenous Ca^{2+} buffers to the cytoplasm of glial cells slows or abolishes intercellular Ca^{2+} waves. One uncertainty with this model, however, is that it remains unknown if Ca^{2+} diffuses through gap junctions to stimulate the release of Ca^{2+} in the adjoining cell. A further complication is that in some cells bathed in Ca^{2+}-deficient medium, there is no increase in $[Ca^{2+}]_i$ in the stimulated cell, yet a Ca^{2+} wave occurs in adjacent cells. This indicates that an increase in $[Ca^{2+}]_i$ in the stimulated cell is not required for the initiation of intercellular Ca^{2+} waves. Thus, Ca^{2+} is not the cytosolic messenger that propagates Ca^{2+} waves through gap junctions in these cells.

Another model for the propagation of intercellular Ca^{2+} waves involves the regenerative production of IP_3. In this model, it is suggested that the rise in $[Ca^{2+}]_i$ stimulated by IP_3 activates phospholipase C to generate additional IP_3. Consistent with the involvement of IP_3 in the generation of intercellular Ca^{2+} waves, inhibiting IP_3 synthesis or blocking its receptors eliminates the propagation of Ca^{2+} waves. In addition, inhibition of phospholipase C abolishes mechanically stimulated waves in some cells, pointing to its involvement in the regenerative generation of IP_3. Mathematical modeling simulations indicate that intercellular wave propagation cannot be explained by the simple diffusion of IP_3, as the required concentration for diffusion is much higher than that reached by agonist stimulation. This further supports the hypothesis that regenerative IP_3 production mediates the propagation of intercellular Ca^{2+} waves. As IP_3 molecules are small enough to pass through gap junctions, the simplest hypothesis to explain IP_3-mediated intercellular Ca^{2+} waves is that IP_3 is the intercellular diffusible messenger that synchronizes cellular activity by the production of intercellular Ca^{2+} waves.

In addition to IP_3 and Ca^{2+}, cADP-R is also a good candidate for mediating intercellular Ca^{2+} waves. Like IP_3 and Ca^{2+}, molecules of cADP-R are small enough to diffuse through gap junctions, indicating that they could trigger the intercellular Ca^{2+} wave in adjoining cells. In addition, like IP_3, cADP-R stimulates Ca^{2+} release in a variety of cell types and can act as a global messenger within single cells. Accordingly, in sheep lens cells, injection of cADP-R stimulates the production of intercellular Ca^{2+} waves.

8.2. Extracellular Messenger-Mediated Ca^{2+} Wave Propagation

The second pathway mediating Ca^{2+} wave propagation along interconnected cells involves a diffusible extracellular messenger released from the stimulated cell. In rat basophilic leukemia cells, hepatocyte neuroepithelioma cells, and astrocytes, ATP has been identified as the extracellular messenger. In some cells, ATP is cosecreted with hormones or neurotransmitters in response to receptor stimulation. Following its release, ATP diffuses short distances to stimulate purinergic receptors in neighboring cells. This leads to an increase in $[Ca^{2+}]_i$ and further ATP secretion, which then acts on the next cell and so on to generate intercellular Ca^{2+} waves (Fig. 5). Although the action of ATP is restricted to short distances, as it is degraded rapidly by several ecto-ATPases, this autocatalytic action of ATP allows for the propagation of Ca^{2+} waves over several hundred micrometers. Moreover, the actions of ATP on its neighboring cells are not restricted to a single cell type, as it may act on purinoreceptors in other cell types to control their activity. For example, pituitary gonadotrophs, lactotrophs, and somatotrophs all express purinergic receptors, which may be activated in response to receptor-induced ATP secretion from a single cell type. Therefore, extracellular-mediated Ca^{2+} waves can propagate through networks of multiple cell types in a single tissue to regulate its function. The involvement of extracellular messengers does not exclude the involvement of gap junctions. In rat liver epithelial cells and astrocytes, for example, diffusible extracellular messenger molecules and gap junctions act in concert to mediate wave propagation. It has also been suggested that gap junctions potentiate ATP release leading to facilitation of Ca^{2+} wave propagation.

9. CALCIUM CONTROLLED CELLULAR MACROMOLECULES

9.1. Calcium-Controlled Enzymes

Calcium can indirectly, by binding to specific intracellular proteins, control the activity of a number of enzymes. In all nonmuscle and smooth muscle cells, calmodulin is the predominant Ca^{2+} binding protein. A related Ca^{2+} binding protein, known as troponin C, dominates in skeletal muscle. The calcium–calmodulin complex controls more than 20 intracellular enzymes, which participate in regulation of several intracellular functions, such as cyclic nucleotide and glycogen metabolism, secretion, motility, Ca^{2+} transport, and cell cycle. Many of these enzymes are inhibited in an intramolecular manner, and the calcium–calmodulin complex releases this inhibition. Activated calmodulin can bind directly to the effector protein and modulate its activity, or indirectly through a family of calcium–calmodulin-dependent protein kinases and phosphatases. The kinase family includes myosin light chain kinase, phosphorylase kinase, calmodulin kinase I, calmodulin kinase II, EF-2 kinase (which is also known as calmodulin kinase III), and recently discovered calmodulin kinase IV. Calmodulin kinase II is known as multifunctional Ca^{2+}–calmodulin-dependent protein kinase, while other members of this family are more specialized or dedicated to the regulation of a specific function. For details on calmodulin and the enzymes controlled by this Ca^{2+} binding protein see the chapter by A. R. Means in this volume.

A number of enzymes that participate in intracellular signaling are directly controlled by cytosolic Ca^{2+}. For example, mammalian cells express a family of several closely related PKC isozymes. The classical group (A) consists of four isozymes—α, βI, βII, and γ-PKC, all single polypeptides containing four conserved regions (C1–4) and five variable regions (V1–5). The C2 region represents the calcium-binding domain. Another group (B) consists of four enzymes: δ, ε, η, θ. This group lacks the C2 region and thus does not require calcium for activation. The third group (C), composed of two atypical PKC enzymes, ζ and λ, is also insensitive to cytosolic calcium. For details on Ca^{2+}-dependent regulation of PKC isozymes see the chapter by A. C. Newton in this volume.

Cytosolic Ca^{2+} also participates in the control of phospholipases C, D, and A_2. In several neuronal cell types, depolarization-induced Ca^{2+} entry leads to a modest increase in the activity of phospholipase C, as estimated by increases in IP_3 production. Such observations concur with in vitro studies of the effects of Ca^{2+} on phospholipase C activity. However, the calcium dependence of IP_3 production is more complex than was initially proposed. In a majority of cells, G protein-mediated activation of phospholipase C is a prerequisite for the development of sensitivity to Ca^{2+}. Furthermore, the two calcium pathways activated by the G protein–phospholipase C complex show different efficiencies in the sensitization of IP_3 production. While the early Ca^{2+} mobilization-dependent phase is relatively inefficient, the sustained Ca^{2+} entry-dependent pathway is essential for the activity of the G protein–phospholipase C signaling system.

In contrast to Ca^{2+}, PKC may negatively modulate phospholipase C. The sensitivity of phospholipase C to $[Ca^{2+}]_i$ and PKC was employed in two models for $[Ca^{2+}]_i$ oscillations. In one model, Ca^{2+} activates PKC, which in turn generates an oscillatory Ca^{2+} signal through transient inhibition of phospholipase C. In other models, elevated $[Ca^{2+}]_i$ stimulates phospholipase C. Thus, in both models the IP_3 production is oscillatory. Although a recent study measuring diacylglycerol in single agonist-stimulated cells is consistent with this view, two lines of evidence argue against the requirement for IP_3 to oscillate in order to generate $[Ca^{2+}]_i$ oscillations. First, in many cell types, injection of nonmetabolized IP_3 is sufficient to create oscillatory Ca^{2+} responses, demonstrating that oscillations in IP_3 are not essential for oscillations in $[Ca^{2+}]_i$. Second, no obvious modulation of phospholipase C activity has been observed in several cell types that were treated with PKC activators, or in PKC-depleted cells treated with agonists.

The roles of PKC and Ca^{2+} in the control of an enzyme are not limited to phospholipase C. PKC stimulates phospholipase D activity and this action is facilitated by voltage-gated Ca^{2+} influx. Phospholipase A_2 shows a similar pattern of regulation by PKC and calcium, a feature potentially important in understanding the crosstalk among multiple signal activated phospholipases. For detailed description of phospholipases and their regulation, including the role of calcium, see the chapter by J. H. Exton in this volume.

Several members of the adenylyl cyclase family of enzymes are also sensitive to changes in $[Ca^{2+}]_i$. The AC5 and AC6 forms of these enzymes are inhibited by elevation in $[Ca^{2+}]_i$. This inhibition does not appear to involve calmodulin, and is also observed in vitro with isolated AC5. Calcium-dependent inhibition is removed by a deletion of 112 amino acids in the C region of the central cytoplasmic loop. The physiological significance of such inhibitory effects of Ca^{2+} is not known, but it may serve as the mechanism for the generation of the oscillatory cAMP production in response to the oscillatory Ca^{2+} signals. The activity of three other members of this family of enzymes—AC1, AC3, and AC8—is facilitated by the rise in $[Ca^{2+}]_i$, and this action of Ca^{2+} is mediated by Ca^{2+}–calmodulin complex. The amino acids 495–522 of AC1 are involved in Ca^{2+}–calmodulin stimulation. It is also possible that calmodulin kinase mediates the stimulatory action of Ca^{2+} on AC3 activation. For details on these enzymes and their regulation by Ca^{2+} see the chapter by L. Birnbaumer in this volume.

9.2. Calcium-Controlled Channels

A rise in $[Ca^{2+}]_i$ is an effective signal for the activation and/or inhibition of several plasma and ER membrane calcium channels. The role of Ca^{2+} in control of IP_3Rs and RyRs is discussed in Section 4. Like RyRs and IP_3Rs, SOCCs are also sensitive to $[Ca^{2+}]_i$ and both calcium-dependent activation and inactivation of these channels have been observed. The conductivity of L-type calcium channels is also altered by $[Ca^{2+}]_i$. An increase in $[Ca^{2+}]_i$ inhibits L-type channel activity via a recently identified Ca^{2+} binding motif. In addition to Ca^{2+}-selective channels, other ionic channels are controlled or modulated by $[Ca^{2+}]_i$. These include several potassium, chloride, and nonspecific channels, all of which are known as Ca^{2+}-activated channels. The family of Ca^{2+}-activated potassium channels (I_{K-Ca}) is composed of at least three members: apamin-sensitive Ca^{2+}-activated small conductance K^+ channels (SK), apamin-resistant small conductance I_{K-Ca}, and carybdotoxin-sensitive large conductance (BK) I_{K-Ca} channels. In agonist-stimulated pituitary cells and hypothalamic neurons, the predominant mediators of the Ca^{2+}-dependent plasma membrane conductances are apamin-sensitive SK channels. These channels show a higher sensitivity to Ca^{2+} than other Ca^{2+}-activated K^+ currents, rendering them appropriate for controlling the interspike interval in firing cells by producing long hyperpolarizing periods following bursts of APs, as well as spike frequency adaptation during depolarizing pulses. Consistent with these observations, an increase in the AP spiking frequency follows the application of apamin in several cell types. Also, apamin increases secretion of several hormones, suggesting that facilitation of firing frequency results in elevation in $[Ca^{2+}]_i$. Some pituitary and hypothalamic cells also express BK channels. The intrinsic properties of these channels are appropriate for the regulation of AP duration and the associated Ca^{2+} entry through VGCCs.

Calcium-activated chloride channels are commonly present in many neurons and secretory cells. In addition to Cl^-, these channels are permeable to several other small anions. The opening of these channels is controlled by $[Ca^{2+}]_i$ and depolarization. In neurons, these channels may play a role similar to that of SK channels in pituitary cells. Moreover, because of its sensitivity to $[Ca^{2+}]_i$, the current through these channels is frequently employed as an assay for monitoring changes in $[Ca^{2+}]_i$.

CAN channels are expressed in many excitable and nonexcitable cells. These channels are activated by an increase in $[Ca^{2+}]_i$ due to Ca^{2+} influx and/or release of Ca^{2+} from intracellular stores. They also provide a route for Ca^{2+} entry by two mechanisms. First, in a manner comparable to glutamate receptor channels and purinergic receptor channels, CAN channels depolarize cells, leading to activation of

VGCCs and the subsequent influx of Ca^{2+}. Second, many CAN channels are permeable to Ca^{2+}, and thus can serve as calcium-permeant channels. They are voltage insensitive and have a single channel conductance of about 30 pS. The structure of these channels is unknown, but may relate to the *trp* gene products.

10. CALCIUM-CONTROLLED CELLULAR PROCESSES

10.1. Fertilization, Development, and Apoptosis

Calcium plays important roles in both fertilization and embryonic development. At fertilization, the sperm interacts with the egg to trigger Ca^{2+} oscillations. These oscillations last for several hours, and may be critical in triggering the enzymes involved in the cell division cycle. The mechanism of these oscillations is not well defined, but it is likely that IP_3 is involved in their generation. Also, a prolonged period of Ca^{2+} signaling, similar to that during fertilization, is an important growth signal for many cells, including normal immune cells and cancer cells exhibiting unlimited growth. During development, cells also use Ca^{2+} signaling pathways to regulate development. For example, the rise in $[Ca^{2+}]_i$ contributes to body axis formation, organ development, cell migration, and formation of neuronal circuits. Both pathways for calcium signaling, Ca^{2+} influx and Ca^{2+} mobilization, are utilized during the development.

The finding that alterations in intracellular Ca^{2+} homeostasis are commonly observed in necrosis and apoptosis is consistent with a view that elevated $[Ca^{2+}]_i$, especially if maintained for long periods, can be cytotoxic. The necrosis is usually accompanied by a prolonged nonoscillatory elevations in $[Ca^{2+}]_i$. In contrast, the apoptosis represents a more orderly program of cell death. A wide variety of candidate molecules could be involved in Ca^{2+}-sensitive apoptotic process. These include calmodulin-dependent kinase/phosphatase, Ca^{2+}-sensitive protease such as calpain and nuclear scafford protease, Ca^{2+}-activated endonuclease, and transglutaminase. For details on Ca^{2+} signals and apoptosis see the chapter by R. B. Evens-Storms and J. A. Cidlowski in this volume.

10.2. Exocytosis

In both endocrine cells and neurons, regulated exocytosis is triggered by an elevation in $[Ca^{2+}]_i$. A rise in $[Ca^{2+}]_i$ is a critical step in the process of exocytosis, which is accomplished by the rapid, Ca^{2+}-regulated fusion of neurotransmitter- or hormone-filled vesicles with the plasma membrane. Once the fusion is completed, the components of secretory vesicles are selectively recovered by endocytosis. The secretory vesicles are regenerated by a recycling pathway, and the final stages of this pathway include the recruitment, docking, and ATP-dependent priming of the secretory vesicles, in the preparation of the next cycle of exocytosis. The distinct factors are required for sequential docking, ATP-dependent priming, and Ca^{2+}-dependent triggering (fusion) reaction. For example, polyphospholipids may contribute to the priming, but not the fusion, whereas Ca^{2+} is required for both steps in exocytosis.

The specific pathways involved in neurotransmission and hormone secretion differ, presumably owing to the different vesicle types expressed in neurons and endocrine cells. One of the major difference is related to the $[Ca^{2+}]_i$ dependence of fusion. In neurons, there is a nonlinear dependence of the secretory rate on the level of $[Ca^{2+}]_i$, within the 50–200 μM concentration range. At synapses between nerve cells, such high concentrations of $[Ca^{2+}]_i$ occur during APs, and are localized in the active zones, which contain both vesicles and clusters of VGCCs. Thus, Ca^{2+} microdomains are critical for driving neurotransmission. In endocrine cells, the coupling of Ca^{2+} channels to the secretory machinery is less integrated, and the submembrane shell of Ca^{2+} is only 2–5 μM.

Although exocytosis is a highly specialized process, the molecules involved in this process are related to those that mediate the targeting and fusion of transport vesicles in other intracellular membrane trafficking pathways. Among the identified cytosolic Ca^{2+} sensors, synaptotagmin I has been the most thoroughly characterized. Synaptotagmin I is a presynaptic vesicle protein that has two repeats of cytoplasmic calcium domain (termed C2A and C2B), each sharing homology with the C2 regulatory domain of PKC. These domains mediate Ca^{2+}-regulated interaction to liposomes and plasma membrane protein syntaxin, suggesting the underlying mechanism for synaptic vesicular docking and fusion to plasma membrane. Genetic analysis of *Drosophila* mutant and null mice at the synaptotagmin I allele confirmed this hypothesis. More than 50 C2 domain sequences have been identified, several of which have been implicated in exocytosis. In addition to those found in synaptogamin I, these include synaptogamin isoforms, rabphilin, doc2, and munc13-1.

The relationship between the pattern of Ca^{2+} signaling and the extent of hormonal secretion is very complex. Two questions regarding Ca^{2+} signaling and secretion merit close attention. These are: the importance of the oscillatory vs nonoscillatory Ca^{2+} response to secretion, and the requirement for the frequency-coded signal to control exocytosis. For example, in gonadotrophs GnRH induces oscillatory Ca^{2+} responses at low to intermediate concentrations, and a biphasic response at high concentrations. Measurements of Ca^{2+} response and LH secretion in individual gonadotrophs, using a reverse hemolytic plaque assay, suggest that gonadotrophs require a spike/plateau Ca^{2+} signal, rather than an oscillatory signal to secrete gonadotropins. However, simultaneous measurements of LH release by capacitance measurements and Ca^{2+} signaling in oscillatory gonadotrophs demonstrate that each Ca^{2+} spike induces a burst of exocytosis. Also, oscillatory $[Ca^{2+}]_i$ elevations induced by photolysis of caged IP_3 trigger exocytosis in these cells. The concentration-dependent regulation of secretion through frequency and/or amplitude coding of Ca^{2+} signals is still not well defined. A reasonable correlation is found between the frequency of spiking and GH release in spontaneously secreting somatotrophs. However, the amplitude of Ca^{2+} spikes may also play an important role in secretion in these cells. Finally, the sensitivity of the secretory mechanism to calcium during initial and sustained secretion differs significantly.

10.3. Nuclear Functions

The transcriptional activity of several genes is controlled by $[Ca^{2+}]_i$. Calcium itself can modulate transcriptional activity, but it can also act through individual Ca^{2+}-sensitive proteins that serve as molecular decoders of the intracellular Ca^{2+} signals. For example, both Ca^{2+} and cAMP can be involved in the regulation of c-*fos* transcription via calcium- and cAMP-dependent phosphorylation of the dimmeric transcription factor, CREB (Fig. 20). It is interesting that Ca^{2+} exhibits stimulatory and inhibitory actions on c-*fos* expression that depend on $[Ca^{2+}]_i$. The $[Ca^{2+}]_i$ dose dependence of c-*fos* transcription and mRNA accumulation is bell shaped, with facilitation at low and inhibition at high $[Ca^{2+}]_i$. Recent evidence has shown that c-*fos* gene transcription is modulated by a Ca^{2+}-dependent block to transcriptional elongation. The finding that these cells respond to agonist stimulation with nonoscillatory amplitude-modulated Ca^{2+} signals suggests that the amplitude of the Ca^{2+} response is the signal for the dual action on transcriptional activity. It is also likely that different genes require different $[Ca^{2+}]_i$ thresholds for their activation.

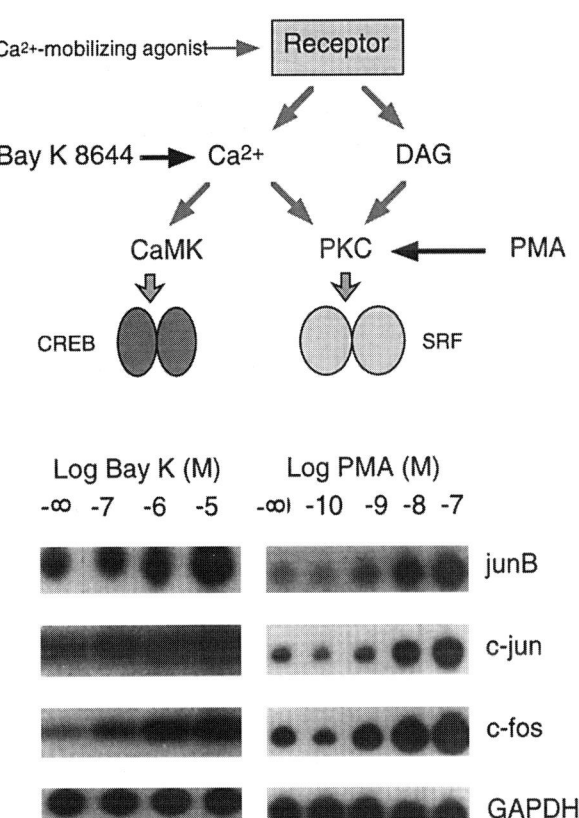

Fig. 20. Mechanism of early response gene induction in excitable cells operated by Ca^{2+}-mobilzing receptors. In accord with the role of calmodulin kinase and protein kinase C in early response gene induction, the calcium channel activator, Bay K 8644, and the protein kinase C activator, PMA, can substitute for calcium-mobilizing agonist in stimulation of early gene expression.

In lymphocytes, for example, transcription of pro-inflammatory genes is selective, depending on the amplitude of Ca^{2+} signals. It has also been suggested that $[Ca^{2+}]_i$ oscillations can reduce the effective Ca^{2+} threshold for activating a particular transcription factor. The frequency of spiking can also serve as a determinant for a selective activation of responsible transcription factors. For example, IP_3-induced oscillations with a 1-min interval showed more gene expression in T lymphocytes than those responding with higher or lower frequency, or sustained nonoscillatory plateau.

10.4. Mitochondrial Functions

Mitochondria not only serve as a transient reservoir for Ca^{2+} during agonist stimulation, but Ca^{2+} influx into these organelles is also required for their functions. Calcium regulates several mitochondrial dehydrogenase enzymes involved in oxidative ATP synthesis, including pyruvate dehydrogenase, isocitrate dehydrogenase, and α-ketoglutaride dehydrogenase. Calcium regulates these enzymes in the 0.2–2 μM concentration range. It has been suggested that increases and decreases in [Ca^{2+}]$_i$ are translated into parallel changes in the concentration of mitochondrial Ca^{2+}. Accordingly, V_m-dependent and calcium mobilizing-dependent mitochondrial [Ca^{2+}] spiking occurs in intact hepatocytes. Furthermore, each Ca^{2+} spike in cytosol is sufficient to cause a transient increase in mitochondrial [Ca^{2+}], which is then associated with a transient activation of Ca^{2+}-sensitive mitochondrial dehydrogenases. This action also triggers the sustained activation of mitochondrial metabolism as long as the spiking frequency is higher than 0.5/min. In contrast, a sustained low-amplitude [Ca^{2+}]$_i$ response and the slow, partial elevations in [Ca^{2+}]$_i$ are ineffective in increasing mitochondrial [Ca^{2+}] and dehydrogenase activities. These observations are further supported by the finding in chromaffin cells and isolated mitochondria, in which an increase in [Ca^{2+}]$_i$ to about 500 nM is required for activating the calcium electrogenic uniport.

10.5. Brain

Neurons provide an example of how local and global Ca^{2+} signals have been used to control different functions. The existence of local and global signals in neurons is in accord with the discrete morphological and functional regions of these cells, which are composed of the dendritic tree and its spines, the cell body containing the nucleus, the axon, and the synaptic endings. The connection of neuronal excitability with Ca^{2+} mobilization is another feature of these cells that is of great functional significance. Such coupling is needed for the propagation of Ca^{2+} signals within the cell. Thus, both electrical activity and Ca^{2+} signals can be localized, but also propagated along and/or within the cells. Such intimate connection of electrical activity and Ca^{2+} signals at the local and global level may account for the complex neuronal functions. For example, propagation of APs along the axon represents a global electrical signal combined with local Ca^{2+} signals. As discussed in Section 3., voltage-gated Ca^{2+} influx is intimately involved in generation and propagation of APs. Modification in the amplitude and duration of APs is critical for Ca^{2+} influx and subplasma membrane Ca^{2+} concentrations. Such localized Ca^{2+} signals control neurotransmission and plasma membrane channel activities. On the other hand, the local electrical activity within the dendritic tree functions as a biomicroprocessor, and the associated local Ca^{2+} signals probably control short-term memory. To make the memory more permanent, propagation of these localized Ca^{2+} signals to the nucleus is required, and is achieved by generation of global Ca^{2+} signals through Ca^{2+}-induced Ca^{2+} release. Certainly, many steps in these processes are incompletely defined.

11. CONCLUSION

As stressed frequently by Berridge, Ca^{2+} is one of the most versatile and universal messengers. This simple ion participates in cellular function during the birth, life, and death of a cell. Accordingly calcium regulates a diverse array of cellular processes, from muscle contraction to memory storage. In addition to its intracellular messenger functions, Ca^{2+} also acts as a first messenger, participating in cellular Ca^{2+} homeostasis and several other processes, by controlling a number of extracellular Ca^{2+}-sensing receptors. In contrast to other intracellular messengers being generated by a single pathway, Ca^{2+} signals can be generated by activation of many different ionic channels and other pathways expressed in the plasma membrane, ER, mitochondria, nucleus, etc. This results in the incredible diversity and versatility of Ca^{2+} signals, which differ among themselves in term of their spatial/temporal organization and amplitude coding. Spatial organization includes not only local and global Ca^{2+} signals within the cytosol, but also the intraorganelle Ca^{2+} signals. Such a variety of patterns of Ca^{2+} signals and pathways involved in their generation provides the framework for understanding the universality of Ca^{2+} as the agent that signals for a new life at the stage of fertilization and is used during embryonic development. The differentiation of cells can also be characterized by the selection of the components of Ca^{2+} signaling tools that a particular cell type selects to best fit its cellular functions. All differentiated cells can reuse the same pattern of Ca^{2+} signaling over and over again, but the ability to switch the pattern of signaling to one that will activate the program that leads to cell death is preserved.

SELECTED READINGS

Berridge MJ. Capacitative calcium entry. Biochem J 1995; 312:1.

Berridge MJ, Bootman MD, Lipp P. Calcium—a life and death signal. Nature 1998; 395:645.

Brown EM, Vassilev PM, Hebert Sc. Calcium ions and extracellular messengers. Cell 1995; 83:679.

De Camilli P, Takei K. Molecular mechanisms in synaptic vesicle exocytosis and recycling. Neuron 1996; 16:481.

Dermietzel R. Gap junction wiring: a "new" principle in cell-to-cell communication in the nervous system. Brain Res Rev 1998; 26:176.

Dolphin AC. Mechanisms of modulation of voltage-dependent calcium channels by G proteins. J Physiol 1998; 506:3.

Galzi JL, Changeux JP. Neuronal nicotinic receptors: molecular organization and regulations. Neuropharmacology 1995; 6:563.

Li YX, Keizer J, Stojilkovic SS, Rinzel J. Ca^{2+} excitability of the ER membrane: an explanation for IP_3-induced Ca^{2+} oscillations. Am J Physiol 1995; 26:C1079.

North RA, Barnard EA. Nucleotide receptors. Curr Opin Neurobiol 1997; 7:346.

Ozawa S, Kamiya H, Tsuzuki K. Glutamate receptors in the mammalian central nervous system. Prog Neurobiol 1998; 54:581.

Pozan T, Rizzuto R, Volpe P, Meldolesi J. Molecular and cellular physiology of intarcellular calcium stores. Physiol Rev 1994; 74:595.

Putney JW Jr, Bird G St J. The inositol phosphate-calcium signaling systems in nonexcitable cells. Endocrinol Rev 1993; 14:610.

Stojilkovic SS, Catt KJ. Calcium oscillations in anterior pituitary cells. Endocrinol Rev 1992; 13:256.

Taylor CW. Inositol trisphosphate receptors: Ca^{2+}-modulated intracellular Ca^{2+} channels. Biochim Biophys Acta 1998; 1436:19.

Thomas AP, Bird GS, Hajnoczky G, Robb-Gaspers LD, Putney JW Jr. Spatial and temporal aspects of cellular calcium signaling. FASEB J 1996; 10:1505.

Zucchi R, Ronca-Testoni S. The sarcoplasmic reticulum Ca^{2+} channel/ryanodine receptor: modulation by endogenous effectors, drugs and disease states. Pharmacol Rev 1997; 49:1.

10 Calcium and Calmodulin-Mediated Regulatory Mechanisms

Anthony R. Means

Contents
Regulation of Calcium Homeostasis
Calmodulin as an Intracellular Calcium Receptor
Physiological Processes Regulated by Calmodulin
Perspectives
Selected Readings

1. REGULATION OF CALCIUM HOMEOSTASIS

1.1. In Serum

Calcium is one of the five most abundant elements on the planet, yet it serves a remarkable array of regulatory intracellular functions. In fact, calcium can be a potent intracellular poison. Perhaps it is for this reason that cells have evolved so many mechanisms to keep the intracellular calcium concentration low. Even in the bloodstream, the levels of calcium are tightly controlled by the actions of three hormones. Calcitonin from the C cells in the thyroid, parathyroid hormone (PTH) from the chief cells of the parathyroid, and the active form of vitamin D (calcitrol) combine to keep the level of circulating calcium within a narrow concentration range. An increase in calcium causes release of calcitonin whereas a decrease in calcium causes release of PTH. Elevated serum calcium acts as a hormone on cells of the parathyroid by binding to a calcium-sensing receptor. This receptor is a member of the seven transmembrane family of receptors and is coupled by G_q to the activation of phospholipase C (PLC). The action of PLC results in inhibition of the release of PTH, which in turn leads to an increase in plasma calcium (Fig. 1). The increase in plasma calcium is due to mobilization of calcium from bone, increased calcium absorption, and stimulation of the synthesis of vitamin D_3 (calcitrol) by the kidney. Calcitrol, in turn, increases calcium absorption from the gut and synergizes with PTH in mobilizing calcium from the bone. Together these hormone-mediated events work to normalize serum calcium levels.

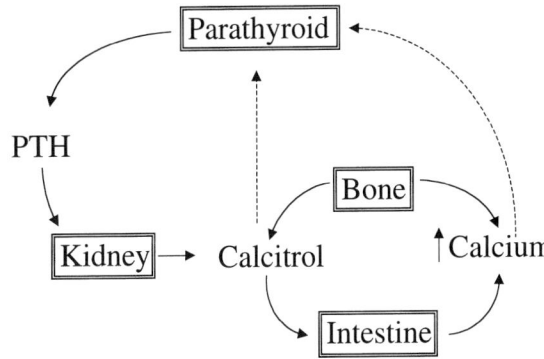

Fig. 1. Regulation of calcium homeostasis in serum. PTH is parathyroid hormone. *Solid lines* to *arrowheads* signify positive regulation, whereas *dashed lines* to *arrowheads* signify negative regulation. Calcitrol is an active form of vitamin D.

From: *Principles of Molecular Regulation* (P. M. Conn and A. R. Means, eds.), © Humana Press Inc., Totowa, NJ.

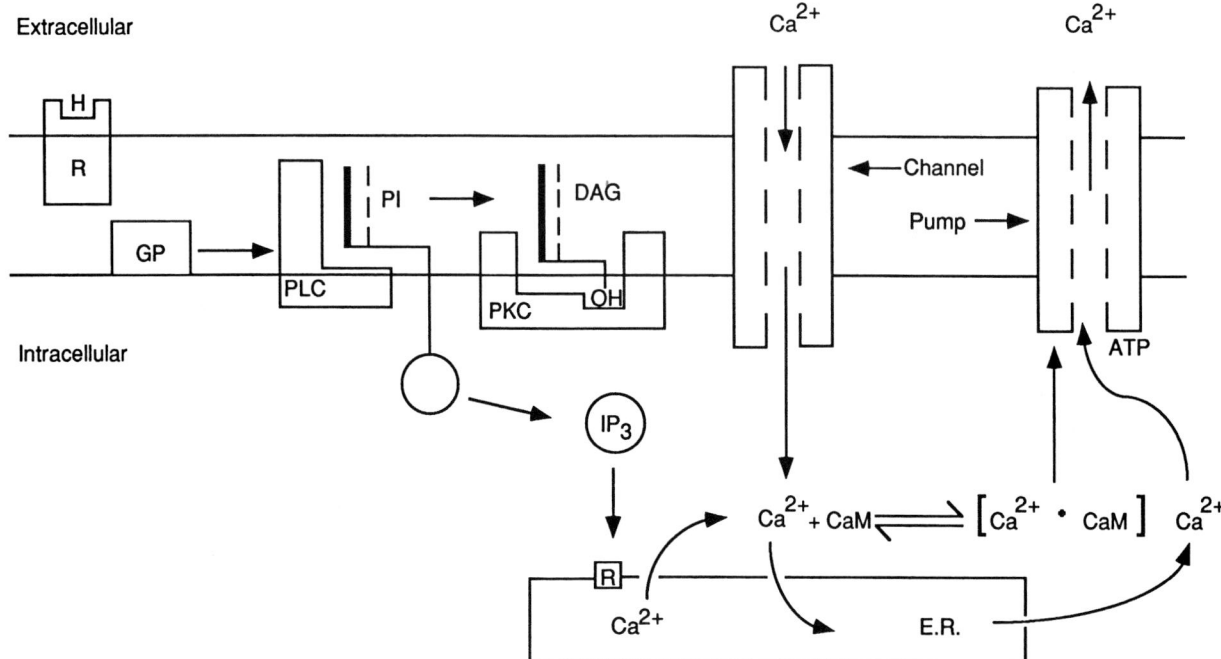

Fig. 2. Regulation of intracellular calcium homeostasis. Hormones (H) can interact with members of the seven transmembrane family of receptors (R). The R is coupled to heterotrimeric G proteins, which can couple to and activate phospholipase C (PLC) in the membrane. PLC hydrolyzes phosphatidylinositol (PI) into diacylglycerol (DAG) and inositol trisphosphate (IP_3). DAG activates protein kinase C (PKC) at the plasma membrane. IP_3 is a diffusible messenger that interacts with a receptor (R) on the endoplasmic reticulum (ER). The R is a Ca^{2+} release channel that releases Ca^{2+} into the cytoplasm. Ca^{2+} can also enter the cytoplasm through a number of plasma membrane Ca^{2+} channels (Channel). One type of channel is voltage sensitive. Another type is activated by depletion of the ER Ca^{2+} pool. Intracellular Ca^{2+} can be resequestered into the ER, removed from the cell through Ca^{2+} release channels in the plasma membrane (Pump), or bound to intracellular Ca^{2+}-binding proteins such as calmodulin (CaM). The Ca^{2+} Pump is also a Ca^{2+}–calmodulin-binding ATPase, and Ca^{2+}–calmodulin is necessary to activate the pump.

1.2. In Nonexcitable Mammalian Cells

Intricate mechanisms within a cell also combine to keep resting calcium concentrations in the nanomolar range (Fig. 2). Calcium concentrations in cells usually increase only transiently. This is because calcium does not diffuse very far as many proteins and organelles exist to prevent free diffusion. A number of growth factors and G protein-coupled receptors activate PLC, which is metabolized to release the water-soluble messenger inositol 1,4,5-trisphosphate (IP_3). IP_3 then interacts with a receptor on the endoplasmic reticulum (ER), resulting in the release of calcium from ER into cytoplasm. This release can activate a plasma membrane calcium channel known as Icrac or TRPL resulting in what has been termed capacitative influx (see chapter 5 in this volume). Calcium can also be bound by a number of calcium binding proteins, resequestered into the ER, removed from the cell via the action of a calcium-pumping ATPase or taken up into mitochondria. The calcium-pumping ATPase is activated by calcium when bound to its primary intracellular receptor, calmodulin. Thus, as we will see, calmodulin is not only involved in mediating signal transduction pathways involving calcium, but also participates in the control of intracellular calcium homeostasis.

1.3. Nuclear Calcium

Depletion of the intracellular stores of calcium both from the ER and the nuclear envelope can have profound influences on cell metabolism. Pharmacological agents such as thapsigargin inactivate the ER calcium pump; thus, although calcium is rapidly released from the ER, the ER calcium compartment cannot refill. This sends signals to the cell that result in inhibition of nuclear transport, inhibition of protein synthesis, and arrest of the cell cycle. In many cells, the eventual consequence of this inability to refill the stores with calcium is apoptosis. Regarding the nuclear entry of calcium and other molecules, it has been suggested that this entry process is controlled by the status of the nuclear calcium store. When nuclear

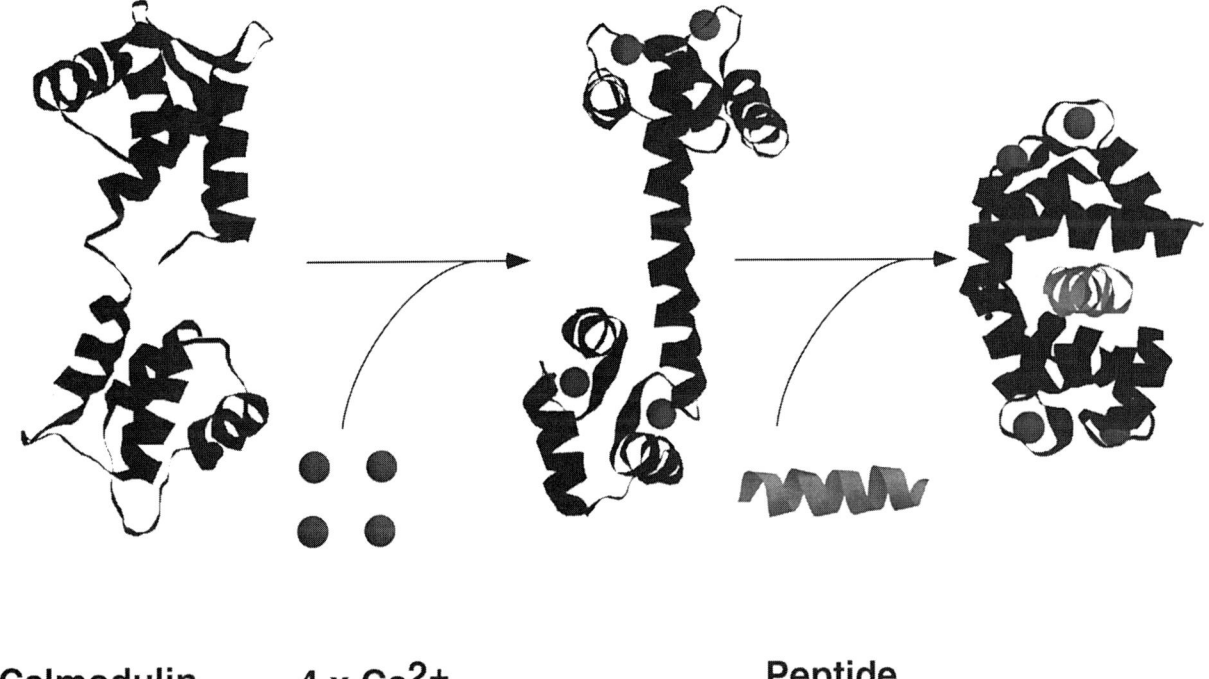

Calmodulin   **4 × Ca²⁺**   **Peptide**

Fig. 3. Structural conformations of calmodulin. All three structures are ribbon diagrams of three-dimensional structures. **(Left)** Calmodulin in the absence of bound calcium. **(Middle)** Calmodulin in the presence of four bound calcium ions. **(Right)** Ca^{2+}-bound calmodulin interacting with a target peptide.

stores of calcium are filled, the nuclear pore complexes are open and intermediate-sized molecules (>10 kDa) can diffuse freely across them. However, depletion of this calcium pool in response to messengers such as IP_3 results in blockade of the nuclear pore by a central plug. This prevents the import of intermediate-sized molecules but allows ions and small molecules to cross. It was hypothesized that a sensor of the filling state of the nuclear calcium pool initiates the conformational change of the nuclear pore complex.

There are many roles for calcium in the nucleus including transcriptional regulatory pathways. Considerable controversy exists as to whether cytoplasmic and nuclear calcium levels are regulated differentially. In mammalian cells, hormonal stimuli appear to evoke release of elementary units of calcium called puffs. When the stimulus is applied at or near the plasma membrane, calcium transients are observed in the cytoplasm but not in the nucleus. However, if the stimulus occurs in a perinuclear region of the cytoplasm, calcium levels rise in and are propagated across the nucleus as a wave. Thus, while there are differential requirements for calcium rises in cytoplasm vs nucleus, it is likely that the waves are initiated in the cytoplasm. Regardless, most of the nuclear events that have been clearly shown to require calcium are mediated by calcium binding to calmodulin.

2. CALMODULIN AS AN INTRACELLULAR CALCIUM RECEPTOR

2.1. Calcium Binding to Calmodulin

Calmodulin is a 148 amino acid protein that is essential in all eukaryotic cells. It binds four calcium ions with micromolar affinity using a high-affinity Ca^{2+} binding motif called an EF-hand. The protein is comprised of N-terminal and C-terminal globular domains that are separated by an eight-turn, solvent-exposed α-helix. The two Ca^{2+} binding sites in the N-terminal region are of slightly higher affinity than are the two sites in the C-terminal domain. However, the binding of calcium to all sites is highly cooperative. There is considerable similarity between how calcium causes changes in the structure and function of calmodulin and how steroid hormones cause changes in the structure and function of their cognate receptors. Thus, calcium binds calmodulin with micromolar affinity and causes a conformational change that increases the hydrophobicity of the protein and stabilizes the central α-helix (Fig. 3). When in the calcium-bound conformation, calmodulin interacts with calmodulin-binding segments in target proteins. The affinity of the calcium–calmodulin complexes with target peptide sequences is nanomolar. The three order of magnitude difference between calcium binding to calmodulin and the complex binding to a target protein measured *in vitro* is largely overcome when all three components of the ternary complex are present together. Practically, this means that the apparent calcium concentration required to stimulate the activity of a calcium-dependent enzyme is in the nanomolar rather than micromolar range. As the increases in intracellular calcium in response to hormonal stimuli are frequently also within this same nanomolar concentration range, the activation of enzymes within a cell will typically follow first-order kinetics.

2.2. Calmodulin Binding to Target Proteins

Calmodulin binds to and regulates a remarkable array of cellular proteins. Some of these targets require Ca^{2+} binding to calmodulin whereas others bind calmodulin independently of Ca^{2+}. More than 30 molecular targets for Ca^{2+}–calmodulin have been identified, many of which are proteins involved in signal transduction pathways. These molecules include protein kinases and phosphatases, Ca^{2+}-pumping ATPases, adenylyl cyclases, cyclic nucleotide phosphodiesterases, proteases, nitric oxide synthases, small G proteins related to Ras, membrane receptors, and ion channels. Whereas calmodulin will bind to a wide variety of amino acid motifs within these target proteins, there are some common themes (*see* Fig. 4a). First, the calmodulin binding segment has the potential to form an amphipathic α-helix in solution. Thus, calmodulin binding peptides are typically 16–18 amino acids in length and generally contain both basic and hydrophobic amino acids. Second, the N-terminus of the target peptide usually begins with two or three basic amino acids, followed by a bulky hydrophobic amino acid at position 5. At 8–12 amino acids C-terminal to amino acid 5 is a second bulky hydrophobic amino acid. These two hydrophobic residues (usually Trp, Phe, Leu, or Ile) serve to anchor the hydrophobic pockets that each contain two bound calcium ions and constitute either end of calmodulin. Finally, a basic residue usually ends the calmodulin binding sequence.

Additionally, calmodulin can bind to target proteins in a manner that is independent of Ca^{2+}. One type of such interaction is represented by a 23 amino acid segment called an IQ motif. The consensus sequence for this motif is IQXXXRGXXXR (where X is any amino acid) and it is most frequently found in tandem repeats within proteins. The IQ motif was identified in the X-ray structure of the regulatory domain of scallop myosin as the region responsible for binding the essential and regulatory myosin light chains which are also Ca^{2+}-binding proteins. The first calmodulin binding protein found to contain such a binding region was also a myosin, but of the nonconventional type found in vertebrate brush border epithelial cells. In this type of myosin, calmodulin plays the role of the regulatory light chains. A number of other proteins are now known to contain IQ domains and bind calmodulin independently of Ca^{2+}. One example is a protein present in neuronal growth cone membranes called GAP-43 or neuromodulin. In this case, addition of Ca^{2+} causes calmodulin to be released from neuromodulin. It has been suggested that the released Ca^{2+}–calmodulin complex can now interact with and activate other membrane-associated proteins such as the calmodulin-dependent forms of adenylyl cyclase. Even though a number of proteins that interact with calmodulin in the absence of Ca^{2+} have been identified, the most mechanistic insights have been generated from the study of calmodulin binding to targets in a manner that requires Ca^{2+} binding to calmodulin. Thus, it is these types of interactions that are discussed in detail in this chapter.

The three-dimensional structures of Ca^{2+}–calmodulin bound to peptides derived from several target enzymes have been solved. The interaction of the peptide with calmodulin results in remarkable conformational changes of both the peptide and calmodulin. As predicted from the hydrophobicity analysis, the peptide assumes a α-helical conformation throughout its length. Completely unexpectedly, however, calmodulin bends and twists around its central helix to completely engulf the helical peptide (Fig. 3). Remarkably, there are about 180 contacts formed between calmodulin and the peptide with more than 80% of these contacts involving van der Waals forces. A comparison of three such structures revealed how the calmodulin central helix would unwind in order to position the two hydrophobic domains optimally to affect recognition of different target enzymes. Thus, the central helix serves as a variable expansion joint that allows different relative positioning of the lobes as unique target enzymes are recognized. A major determinant of the final structure is the extent of separation of the two bulky hydrophobic residues at either end of the target peptide.

The functional significance of the different confor-

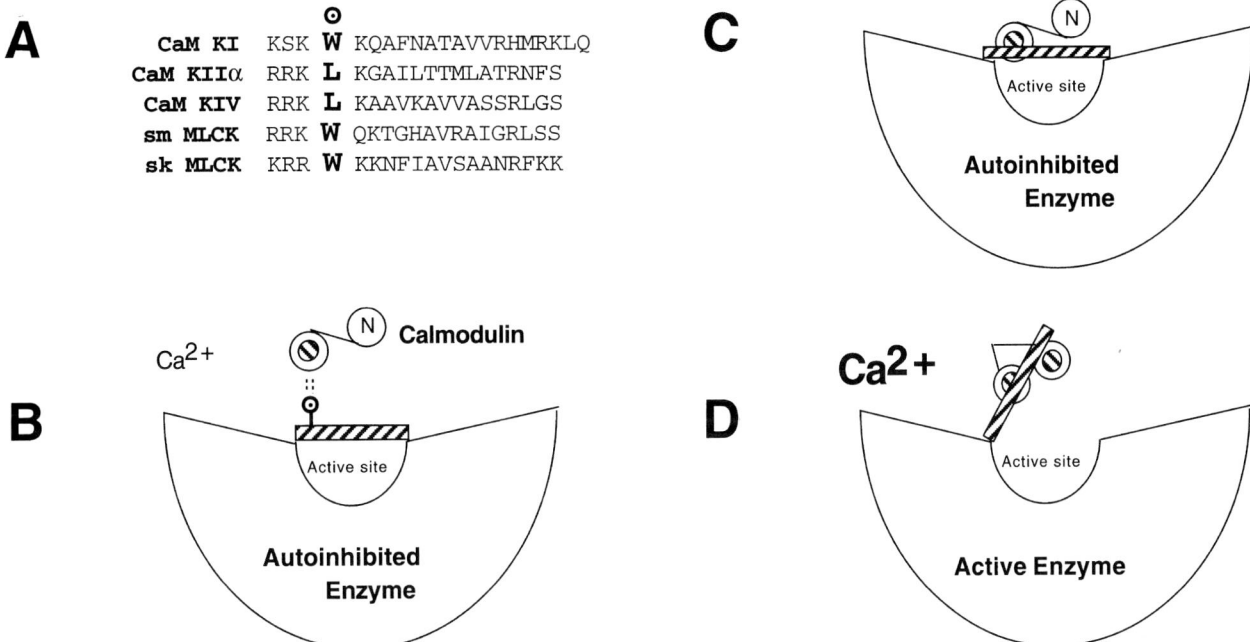

Fig. 4. Regulation of protein kinase activity by Ca^{2+}–calmodulin binding. **(a)** The amino acid sequence of the calmodulin binding regions of five calmodulin-dependent protein kinases in single letter code for amino acids. The hydrophobic amino acid critical in making the initial interaction with the C-terminal half of calmodulin is in **bold**. **(b)** Ca^{2+} binds to the two sites in the C-terminus of calmodulin, causing a conformational change in the protein such that the C-terminal half of the protein can now make contact with the hydrophobic residue in the N-terminus of the calmodulin binding sequence. **(c)** The initial interaction between calmodulin and the W of the enzyme promotes additional contacts between the C-terminal lobe of calmodulin and the N-terminal portion of the calmodulin binding domain of the enzyme. These interactions do not activate the enzyme. **(d)** Additional Ca^{2+} binds to the two sites in the N-terminus of calmodulin. Together the interactions promote a bending and twisting of calmodulin around its central helix. These conformational changes allow multiple interactions to form between the N-terminus of calmodulin and the C-terminus of the calmodulin binding region. The final conformation results in removal of the autoinhibitory domain from the active site of the enzyme. The enzyme can now phosphorylate its substrates.

mations of calmodulin when bound to target peptides was demonstrated by careful analysis of the contacts made between calmodulin and its target sequence from smooth muscle myosin light chain kinase (smMLCK) and calmodulin kinase II (CaMKII). It was known from mutagenesis studies that the C-terminal Leu residue in the smMLCK sequence was crucial for calmodulin-dependent activation of the enzyme. This Leu made contacts with the side chains of four calmodulin residues. Interestingly none of the four calmodulin residues made contacts with the peptide derived from CaMKII, leading to the hypothesis that mutation of the four residues would result in a calmodulin that could activate CaMKII but not smMLCK. The four residues were changed to Ala and this hypothesis was proven experimentally. Thus, the conformation of calmodulin when bound to a target sequence results in exposing a surface that participates in activation of the enzyme in an enzyme-specific manner. This remarkable ability to assume different conformations underlies the ability of calmodulin to specifically regulate such a large number of enzymes that contain slightly to highly variable calmodulin binding sequences.

2.3. Autoinhibition and Activation of Calmodulin-Dependent Enzymes

2.3.1. Myosin Light Chain Kinases (MLCKs)

In general, the mode of activation of calmodulin-dependent protein kinases is similar to that described for the MLCKs. This generalized activation mechanism is depicted in Fig. 4. The interaction between calmodulin and its target sequences within MLCK is antiparallel. That is, the C-terminal region of calmodulin binds the N-terminal portion of the calmodulin binding sequence. A bulky hydrophobic residue is positioned at the same point in the N-terminal region of the calmodulin binding domain and is represented for five different enzymes in Fig. 4a. The C-terminus of calmodulin recognizes this hydrophobic residue, which provides the first site of interaction between the two proteins (Fig. 4b). This interaction is sufficient to initiate the bending and twisting around the central helix, which allows formation of many additional contacts between the C-terminus of calmodulin and the N-terminus of the calmodulin binding site. In addition, this conformational change promotes contacts to be made between the N-terminus of calmodulin and the C-terminus of the calmodulin binding site (Fig. 4c). This final conformation provides a surface that results in removing the autoinhibitory region of the enzyme from its active site (Fig. 4d). The surface formed by the N-terminus of calmodulin when bound to a target enzyme is determined by the sequence of the calmodulin binding domain and varies from enzyme to enzyme.

As depicted in Fig. 4 for MLCK, virtually all calmodulin-dependent enzymes are completely inactive in the absence of calmodulin. However, they can be activated by limited proteolysis due to the removal of the autoregulatory segment that contains the calmodulin binding region. The fact that some proteolytic fragments could be generated that neither bound calmodulin nor demonstrated activity, yet the enzymes could be activated by an additional proteolytic digestion, led to the suggestion that calmodulin-dependent enzymes were autoinhibited. In the case of the MLCK enzymes, the autoinhibition is of the "pseudosubstrate" type. This means that a portion of the enzyme, which resembles its substrate but cannot be phosphorylated, occludes the active site. Binding of Ca^{2+}–calmodulin physically removes this inhibitory segment and allows access of the protein and/or ATP substrates. Three types of experimental evidence support this mechanism. First, it was predicted, based on the sequence similarity between a portion of smMLCK and the normal myosin light chain substrate of this enzyme, that a His residue in the autoinhibitory region of the enzyme would lie in the site occupied by the Ser that is phosphorylated when the myosin light chain substrate binds to smMLCK. Indeed conversion of the His to a Ser, in the context of a consensus recognition site for smMLCK, resulted in intramolecular phosphorylation of the substituted Ser that activated the enzyme independent of Ca^{2+}–calmodulin association. Moreover, binding of Ca^{2+}–calmodulin to the mutant smMLCK prevented the intramolecular phosphorylation, presumably by removing the Ser from the active site. Second, synthetic peptide analogs of the autoinhibitory regions of calmodulin-dependent enzymes are enzyme-specific competitive inhibitors relative to protein substrates. These autoinhibitory peptides do not bind calmodulin although the autoinhibitory region and calmodulin binding region frequently overlap and thus share some amino acids in common (e.g., the C-terminal few amino acids of the smMLCK autoinhibitory peptide and the N-terminal few amino acids of the calmodulin binding peptide are identical). Third, the X-ray structures of three calmodulin-dependent protein kinases have been solved. Two of the enzymes, *C. elegans* twitchin and vertebrate titin, are homologs of the MLCKs whereas the other enzyme is vertebrate CaMKI. All three enzymes were crystallized in the autoinhibited form and clearly show the autoinhibitory segment occluding the catalytic center of the respective enzyme in a manner consistent with the pseudosubstrate hypothesis. In all three enzymes the protein substrate binding site is blocked, whereas in CaMKI the ATP binding pocket is also deformed by the autoinhibitory segment. The details and generality of these "intrasteric" autoinhibitory mechanisms are thoroughly discussed in the chapter by Kemp and colleagues in this volume.

2.3.2. INSIGHTS FROM THE CaM KINASE I STRUCTURE

Elucidation of the X-ray structure of CaMKI clarified the importance of the hydrophobic residue at the fifth position of a calmodulin binding domain. This amino acid is essential for calmodulin binding to and activation of target enzymes. In the case of CaMKI this residue is a Trp, and it lies on an entirely solvent-exposed surface of the protein. So, as depicted generically in Fig. 4, the initial residue on CaMKI that would interact with calmodulin would be the Trp (W). This interaction with the calmodulin C-terminal globular domain would initiate the large conformational change in calmodulin that results in engulfment of the calmodulin binding sequence and a sequence specific activation of the enzyme by exposure of the active site. The corollary to this mechanism is that the C-terminal domain of calmodulin binds very similarly in all known three-dimensional structures—it is the N-terminus that differs in the bound state. So, as stated previously, the N-terminal conformation of calmodulin must participate in both target enzyme binding and activation.

2.3.3. CaM KINASE II

Many calmodulin-dependent enzymes are regulated exclusively by the interaction with Ca^{2+}–calmodulin. For example, smMLCK becomes active when Ca^{2+}–calmodulin binds to it in a manner that depends on an acute increase in the intracellular Ca^{2+}

concentration, but is rapidly inactivated on removal of calmodulin, which occurs when the Ca^{2+} levels fall. However, some Ca^{2+}–calmodulin-dependent protein kinases are also subject to regulation by phosphorylation. The prototypical case is that of CaMKII. This enzyme, the subunits of which are encoded by one of several genes (α, β, γ, δ), is a multimer of identical (or dissimilar) subunits. The association domain is at the C-terminus and the holoenzyme is arranged like a wheel with the association domains in the center and the catalytic domains around the periphery. One subunit phosphorylates an adjacent subunit on a single Thr residue that is located between the C-terminal end of the catalytic domain and the N-terminal residue of the calmodulin binding domain (Thr^{286} in CaMKIIα) only when the adjacent subunits are both complexed with Ca^{2+}–calmodulin. Thus, calmodulin not only relieves autoinhibition of one subunit but also exposes the Thr of the adjacent subunit to its catalytic activity. The autophosphorylated subunit then "traps" calmodulin by slowing its dissociation rate by 1000-fold relative to the nonphosphorylated form. Once phosphorylated, a subunit becomes autonomous of Ca^{2+}–calmodulin. In other words, even when calmodulin is removed by chelation of its Ca^{2+} ions, the protein kinase remains active. Thus, the enzyme is subject to gain of function regulation in response to graded increases in the frequency and/or amplitude of intracellular Ca^{2+} transients, but loss of function requires both a decrease in Ca^{2+} concentration and dephosphorylation of Thr^{286} by the action of a protein phosphatase such as PP2A. Perhaps the best studied physiological roles of CaMKII are in the nervous system where it is involved in various aspects of learning and memory as well as in the control of the activity of several ion channels.

2.3.4. CaM Kinases I and IV

More recently, two other members of the calmodulin-dependent protein kinase family have been shown to be regulated by both Ca^{2+}–calmodulin binding and phosphorylation. Like CaMKII, these enzymes, CaMKI and CaMKIV, are also phosphorylated on a single Thr residue. However, these two enzymes are both monomeric and their phosphorylation is catalyzed by a distinct group of enzymes called CaMK kinases that are themselves Ca^{2+}–calmodulin binding proteins. In addition, the location of the phosphorylated residue in CaMKI and IV is different than in CaMKII. The phosphorylated residue in CaMKI and IV is positioned in the catalytic domain in the region between the protein kinase homology subdomains VII

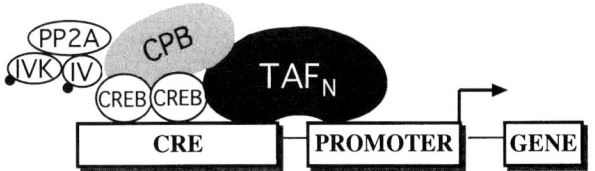

Fig. 5. Schematic representation of the association of CaMKIV and PP2A with a gene promoter in the nucleus of a cell. The CRE binds CREB, which serves as a transcription factor. CREB is phosphorylated on Ser^{133} by CAMKIV (IV) after IV is phosphorylated by its CaM kinase kinase (IVK). Both IV and IVK require calmodulin binding for activation. Calmodulin is represented by the solid dots. This sequence of events is initiated by either a rise in the intranuclear concentration of Ca^{2+}, or by a translocation of Ca^{2+}–calmodulin into the nucleus. Phosphorylation of CREB on Ser^{133} allows binding of the coactivator CBP. Binding of CBP stabilizes the transcription unit by increasing the affinity of the basal transcription machinery (TAFn) with the promoter. PP2A can dephosphorylate and inactivate IV. The inactivation of IV is one of the steps required to reverse the entire process.

and VIII. This region is also known as the "activation loop" and is phosphorylated in many other protein Ser/Thr kinases such as protein kinase A (PKA), some forms of protein kinase C (PKC), the mitogen-activated protein (MAP) kinases, and the cyclin-dependent kinases. Ca^{2+}–calmodulin binding is required for activation loop phosphorylation. In the case of CaMKI, the phosphorylated enzyme remains entirely dependent on Ca^{2+}–calmodulin for activity. Indeed, phosphorylation sensitizes the enzyme to reactivation upon a secondary rise in Ca^{2+} by causing a three-fold decrease in the amount of calmodulin required for half-maximal enzyme activation (K_{cam}). Calmodulin binding to CaMKIV not only allows phosphorylation of the active loop Thr by a CaMK kinase but also promotes intramolecular autophosphorylation of two Ser residues in the extreme N-terminus of the protein. The autophosphorylation relieves a novel form of autoinhibition. Once phosphorylated, CaMKIV activity becomes independent of Ca^{2+}–calmodulin, a property shared with CaMKII.

Whereas CaMKI is present in the cytoplasm of all or most tissues, CaMKIV distribution is limited to the brain, testis, bone marrow, skin, and T lymphocytes. CaMKIV is predominantly found in the nucleus where it exists in a high affinity, stoichiometric complex with the heterotrimeric protein Ser/Thr phosphatase, PP2A (Fig. 5). The phosphatase can dephosphorylate and therefore inactivate CaMKIV. This mechanism seems to operate in T lymphocytes to rapidly inhibit CaMKIV activation in response to occupancy of the

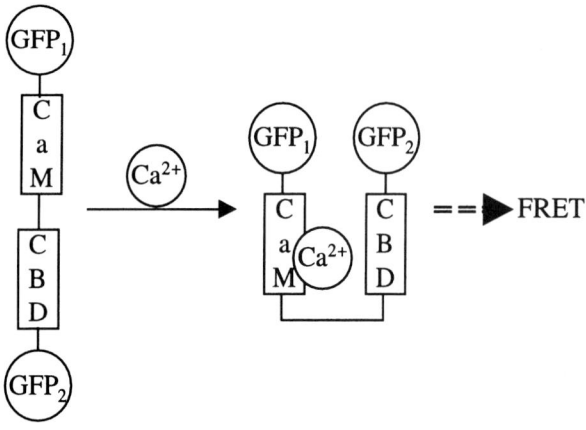

Fig. 6. Intracellular Ca²⁺ indicators based on the structure–function relationships of calmodulin interaction with a binding peptide. Calmodulin (CaM) is linked to a calmodulin binding domain peptide sequence (CBD) through a spacer (the *solid line* between CaM and CBD). One molecule of the green fluorescent protein (GFP) is linked to CaM and another to CBD. The light emission properties of the two GFP molecules are different. Calcium (Ca²⁺) binding to CaM initiates a structural change that brings the two molecules of GFP into close proximity. This conformation change results in fluorescent energy transfer (FRET) between the two molecules of GFP. Changes in the intracellular Ca²⁺ concentration can be calculated from the FRET.

T cell receptor in the face of elevated nuclear levels of Ca²⁺ that are needed to ensure expression of cytokine genes that are essential for T-cell proliferation. Targets of CaMKIV include transcription factors such as CREB and CREMt, as well as the coactivator proteins p300/CBP. Some evidence exists to suggest that CaMKIV may be required for the T-cell production of several cytokines such as interleukin (IL)-2, IL-4, and IL-5. In addition, it facilitates transcription by several orphan members of the steroid receptor superfamily such as RORα. It has also been reported that CaMKIV is required for the latent phase of long-term potentiation in pyramidal neurons of the hippocampus. Thus, both CaMKII and CaMKIV seem to be involved in learning and memory in the same cells but by different mechanisms, as detailed later in this chapter.

2.4. Calmodulin-Based Intracellular Calcium Indicators

It has also been possible to take advantage of the remarkable conformational changes that occur when calmodulin binds a target peptide to create a new type of fluorescent indicator for Ca²⁺ called cameleons (Fig. 6). The indicator is composed of two molecules of the green fluorescent protein (GFP) each with a

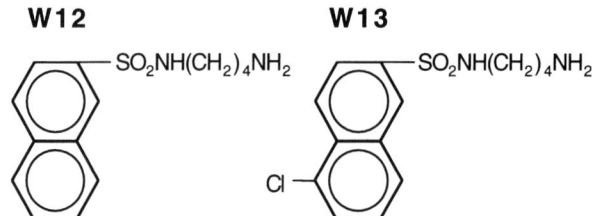

Fig. 7. Structure of the naphthalenesulfonamide calmodulin antagonist W13 and its less active analog W12. W13 binds to calmodulin in a Ca²⁺-dependent fashion and acts as a competitive antagonist of the interaction of Ca²⁺–calmodulin with its Ca²⁺-dependent protein targets. Although W13 and W12 differ by only a single Cl⁻, W13 is about six times more potent that W12 as a calmodulin antagonist.

different fluorescence emission. Calmodulin is ligated between the two GFPs and, after addition of a linker peptide, to a calmodulin binding target peptide sequence. Upon Ca²⁺ binding, calmodulin will interact with the target peptide, thereby bringing the GFPs at either end of the complex in much closer proximity. This increases the fluorescence resonance energy transfer (FRET) between the two GFP molecules which is sufficient to monitor localized changes in the calcium concentration in individual living cells.

2.5. Methods for Inhibiting Calmodulin Action in Cells

2.5.1. Pharmacological Agents

As calmodulin is involved in interacting with more than 30 separate proteins and enzymes, and Ca²⁺ is one of the two most frequently used second messengers in cells, a number of strategies have been developed to identify calmodulin-dependent signal transduction pathways. The initially used and most frequently employed inhibitory agents are the small-molecule calmodulin antagonists. The first such agent championed as a calmodulin antagonist was chlorpromazine. This substance effectively serves as a competitive calmodulin antagonist *in vitro*, but has a variety of targets in cells. Perhaps the most popular chemical inhibitors are members of the naphthalenesulfonimide family, called "W" compounds (Fig. 7). The double saturated ring structure of these molecules mimics the bulky hydrophobic residues that anchor peptide binding to the N- and C-terminal domains of calmodulin and the aliphatic side chain can make interactions with the calmodulin central helix. Indeed, structural studies have shown that these drugs cause conformational changes in calmodulin similar to those caused by the peptide analogs of the calmodulin domains of known calmodulin-dependent enzymes. The most efficacious and selective of these W compounds is W13. It has an IC_{50} of about 50 μM as a competitive inhibitor of calmodulin-dependent enzymes. Substitution of the single Cl^- on carbon 5 with an H^+ decreases the IC_{50} of this analog, called W12, to about 300 μM. As both compounds are quite hydrophobic, they enter cells with reasonable facility. When designing studies in cells it is imperative to use both compounds as, at similar concentrations, W12 is a good control for W13. It is also important to carry out dose–response studies because, as calmodulin is an essential gene product in all cells, it is possible to kill cells with calmodulin antagonists. For example, it is known that calmodulin is required for cell proliferation. Thirty micrograms per milliliter of W13 causes complete arrest of chinese hamster ovary (CHO) cell proliferation but 100% of the cells recover after the drug is removed from the culture medium. The same concentration of W12 has no growth inhibitory effects. However, increasing the W13 concentration to only 35 μg/mL results in 60% cell death and 60 μg/mL produces 99% lethality. Obviously if the cell dies, it is impossible to determine if calmodulin was important for any individual physiologically relevant event. Thus, as is the case for all pharmacological agents, caution and careful experimental design are critical if effects of these drugs are to be appropriately interpreted.

2.5.2. Calmodulin Binding Peptides

2.5.2.1. Secretion. One of the most specific ways to inhibit calmodulin is to employ a peptide antagonist modeled on the calmodulin binding site of a Ca^{2+}-calmodulin-dependent enzyme. A frequently used peptide is from smooth muscle myosin light chain kinase (smMLCK). The sequence of this peptide antagonist is: ARRKWQKTGHAVRAIGRL. A peptide in which the Trp (W) at position 5 and the Leu (L) at position 18 are converted to Ala (A) serves as a reasonable control. The latter peptide retains the potential to form an α-helical and carries the same charge distribution as the peptide antagonist. *In vitro*, the K_i of the peptide antagonist is 1 nM whereas that of the control peptide is 30–100 nM. The peptide antagonist has been used effectively in permeabilized cells, has been microinjected into frog eggs and has been used in both *Drosophila* and mammalian expression vectors. An example of how these peptides were used in cells to determine if a secretory process was calmodulin dependent was to examine the mechanism responsible for Ca^{2+}-dependent release of von Willdebrand Factor from endothelial cells in response to thrombin. These cells can be permeabilized with saponin yet retain thrombin-dependent secretion. The peptide antagonist (100 μM) completely inhibited secretion whereas equivalent concentrations of the control peptide or a peptide derived from PKC were without effect. It is not clear why such a high concentration of the peptide was required but calmodulin concentrations in cells can be several micromolar and the concentration of calmodulin binding equivalents several times higher. Fortunately, the peptide could be washed out of the permeabilized cells, to show that the cells were not permanently damaged by the calmodulin antagonist.

2.5.2.2. Activation of Frog Oocytes. In the frog, *Xenopus laevis*, progesterone is the physiological stimulant for mature oocytes to undergo meiosis. However, the cell division cycle arrests the resulting eggs at metaphase of the second reduction division. Fertilization provides the signal necessary to release the arrest, which allows the cells to undergo anaphase and cytokinesis, thus completing mitosis. Multiple rounds of mitosis then ensue to continue embryonic development. A rise in intracellular calcium is both the usual fertilization-induced signal and is sufficient for resumption of meiosis in oocytes arrested in meiosis II. It was shown that the smMLCK peptide antagonist would block the response to elevated Ca^{2+} whereas the control peptide would not. These results implicate calmodulin as an obligate intermediate in the Ca^{2+}-initiated signal transduction cascade leading to the destruction of cyclin B by ubiquitin-mediated proteolysis and completion of meiosis. Subsequently, the single calmodulin target was identified as CaMKII and its substrate suggested to be a component of the Anaphase Promoting Complex, a E3 ubiquitin ligase, that marks proteins such as cyclin B for ubiquitin-dependent proteolysis via the 26S proteasome.

2.5.2.3. Axon Guidance and Lung Development. In *Drosophila*, the smMLCK antagonist peptide was used to investigate the role of calmodulin in the growth of pioneer neurons that initiate specific neuronal pathways in the fly nervous system. A promoter expressed only in these neurons was ligated to a cDNA encoding the motor portion of kinesin to ensure movement of the transgene product into the growth cone. The antagonist peptide was attached to the C-terminus of the kinesin domain. Disruption of the calmodulin-signaling pathway by the antagonist peptide resulted in either stalling of axon extensions or errors in axon guidance. These phenotypic consequences occurred only when the antagonist peptide was directed to the growth cone. A similar strategy was used in the mouse to target expression of a concatemerized MLCK antagonist peptide to the lung using the lung cell specific surfactant promoter. This peptide was targeted to the nucleus due to the creation of a bipartite nuclear localization sequence. Mice expressing this peptide died within 15 min after birth and demonstrated extensive cyanosis coupled with little or no movement. Histological examination of the lung revealed markedly dilated cysts. Thus, even when expressed as a part of a fusion protein *in vivo*, the antagonist peptide is sufficiently potent to specifically disrupt calmodulin signaling either in the cytoplasm or in the nucleus.

2.5.3. Calmodulin Antisense RNA

A third way to antagonize calmodulin in a cell is to utilize antisense RNA expression vectors. Calmodulin can be present in quite high concentrations in cells (micromolar). The calmodulin protein concentration of G_1 cells (1×) doubles at the G_1/S boundary in proliferating cells (2×), remains at that level until the cell divides, and then each daughter cell receives half of the calmodulin. Thus calmodulin is a very stable protein and it is extremely difficult to decrease the concentration to <50% in a mammalian cell. However, as the 1× concentration is required for survival and the 2× concentration is required for entry into and progression through DNA synthesis, a 50% reduction can have considerable biological consequences. Therefore, the antisense approach is a useful alternative to consider, particularly in exponentially growing cells in culture. This strategy was first used in a mammalian cell line to examine the effects of decreasing the concentration of calmodulin on cell proliferation. Although mammals contain three calmodulin genes that differ in nucleic acid sequence but encode the identical protein, the region of the three mRNAs surrounding the ATG that initiates translation are remarkably similar. This sequence similarity allows design of a single antisense RNA that will compete with all three mRNAs to an extent that results in about a 40% reduction in cell calmodulin levels. Remarkably this decrease was sufficient to completely halt cell cycle progression. The transgene, regulated by an inducible metallothionein promoter, was stably integrated into the cells and clonal lines established. Induction of the transgene by Zn^{2+} resulted in a decrease in calmodulin but activation of the promoter was transient. Thus, calmodulin levels were restored within 8 h, which coincided with resumption of cell proliferation. In subsequent studies antisense oligonucleotides have been used in neuronal cell lines to selectively decrease individual calmodulin mRNAs and question the functional consequences.

2.5.4. Calmodulin Mutant Proteins

2.5.4.1. Insulin Secretion. Generation of calmodulin mutant proteins has also been used successfully to assess specific functions of calmodulin *in vitro* and *in vivo*. As mentioned previously, the two globular ends of calmodulin that contain the pairs of Ca^{2+} binding sites are separated by a long central helix. The flexibility of this helix is essential for calmodulin to properly associate with its various target proteins to ensure a biological response. When the central helix is shortened by eight amino acids, a protein is produced that binds Ca^{2+} identically to the authentic calmodulin but cannot bind to or activate a variety of calmodulin-dependent enzymes. Such a mutant can be used to differentiate between the effects of calmodulin as a regulator of target proteins and as a Ca^{2+} binding protein (Fig. 8). Creation of transgenic mice that overexpress calmodulin in the pancreatic β cells results in an early onset, nonimmune diabetes. Although the pancreas contains plenty of insulin, it cannot be released in response to glucose because of a metabolic defect that impairs the ability to produce ATP as a product of glucose metabolism. Because the ATP concentration does not rise sufficiently, the ATP-dependent K^+ channel in the plasma membrane cannot be closed. Therefore, the membrane does not depolarize, Ca^{2+} channels do not open, and insulin is not secreted. Bypassing the metabolic effects of glucose by the use of sulfonylureas to initiate membrane depolarization or agents that increase the intracellular release of Ca^{2+} such as carbachol allows normal biphasic insulin release in these mice. On the other hand, expression of the calmodulin mutant that binds Ca^{2+} but cannot interact with target proteins (CaM-8) results in the inability to open the voltage-dependent Ca^{2+} in response to glucose or sulfonylureas although glucose metabolism, ATP production, and membrane depolarization are normal. As Ca^{2+} release from intracellular stores causes normal insulin secretion, it was suggested that the calmodulin mutant might directly interfere with the activity of the calcium channels. Mice expressing CaM-8 were also diabetic but due to a defect quite different than the one responsible for diabetes resulting from overexpression of calmodulin. This suggests that calmodulin action vs excess Ca^{2+} binding can result in similar phenotypic consequences but by completely different mechanisms.

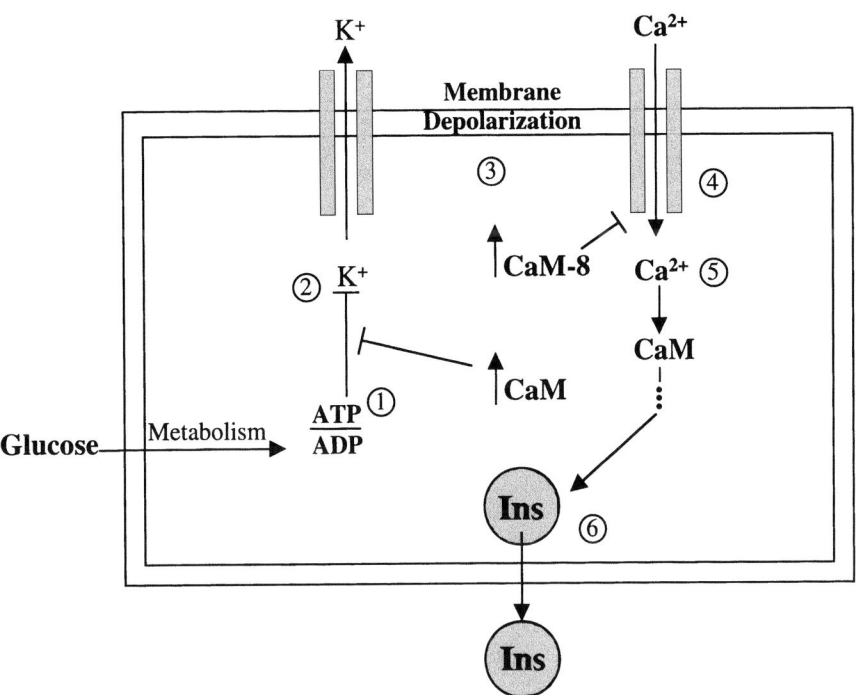

Fig. 8. Glucose-mediated insulin secretion from pancreatic β cells. Glucose enters the cell and is metabolized, which changes the intracellular ATP/ADP ratio (1). ATP then interacts with an ATP-dependent K⁺ channel, which inactivates the channel (2). Inactivation of the K⁺ channel leads to depolarization of the plasma membrane (3) which activates voltage-dependent Ca²⁺ channels (4). Activation of the Ca²⁺ channels allows Ca²⁺ to enter the cell from the extracellular space (5). The increase in intracellular Ca²⁺ activates calmodulin (CaM) which initiates a series of reactions that result in insulin secretion (6). Overexpression of the calmodulin mutant CaM-8 CaM blocks insulin secretion by inhibiting appropriate regulation of the Ca²⁺ channels. Overexpression of the CaM also blocks insulin secretion, but in this case the inhibition is owing to the inability to inactivate the K⁺ channels.

2.5.4.2. Enzyme-Specific Antagonists. It is also possible to produce a mutant calmodulin that will serve as an enzyme-selective calmodulin antagonist. This was first demonstrated by forming chimeras between calmodulin and its structural homolog, troponin C, by exchanging one or more of the four Ca²⁺ binding EF hands. Such mutant proteins would activate some calmodulin-dependent enzymes normally, partially activate others, and fail to activate yet others. Analysis of the residues that differed between the chimera that failed to activate smMLCK and calmodulin allowed a "designer" calmodulin antagonist to be produced. This protein, containing only three mutations in the first domain, was a potent antagonist of smMLCK (K_i ~38 nM) but activated several other enzymes similarly to wild-type calmodulin. The mutant protein bound smMLCK normally but could not produce a surface appropriate for enzyme activation. This finding reinforced the concept that the structure of calmodulin, when bound to an enzyme, presented a specific surface that aided in activation of that particular enzyme.

3. PHYSIOLOGICAL PROCESSES REGULATED BY CALMODULIN

3.1. Regulation of the Metaphase/Anaphase Transition in Meiosis

An understanding of the individual roles of the functional domains of calmodulin-dependent enzymes has allowed the generation of a number of specific biological reagents to assess the roles of these enzymes in cellular processes. One of the best examples of the use of such reagents is in the identification of CaMKII as the primary target of the Ca²⁺ signal that results from the fertilization of *Xenopus laevis* eggs. Fully developed *Xenopus* oocytes are arrested in prophase of meiosis I. Progesterone treatment results in release of this block and continuation of meiosis until metaphase of meiosis II. A block again occurs at this stage of meiosis and is held until fertilization occurs. Entry of sperm causes a release of intracellular Ca²⁺ initiated at the point of egg penetration by the sperm, which proceeds as a wave across

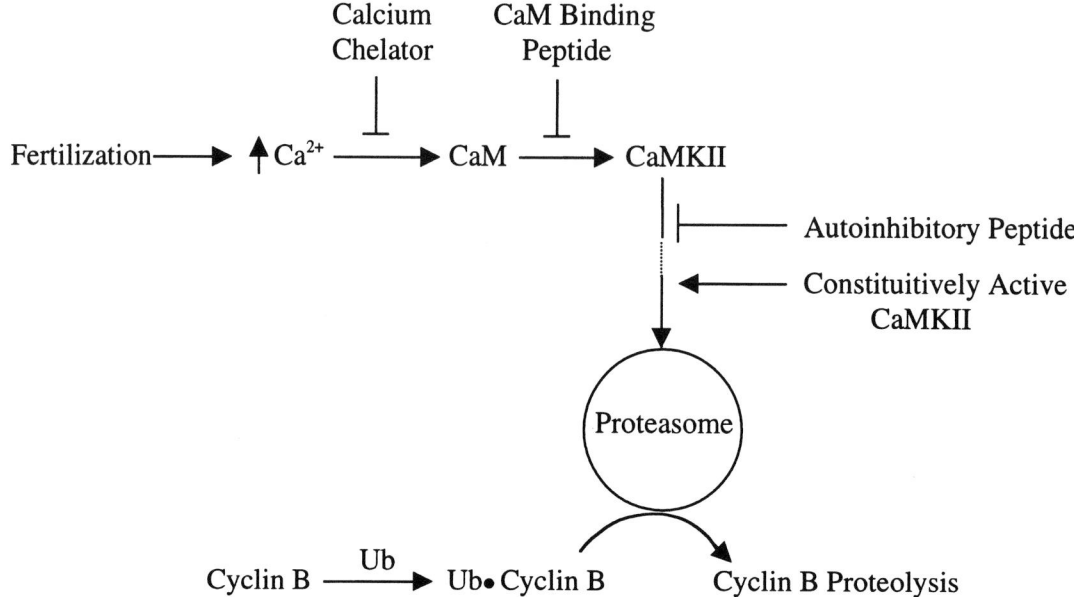

Fig. 9. The sequence of Ca^{2+}-dependent steps involved in activation of *Xenopus* oocytes upon fertilization. Progesterone action in *Xenopus* eggs results in the initiation of meiosis. Meiosis continues until metaphase of the second meiotic division at where an arrest occurs. At the arrest point, the cells contain high levels of the active form of cyclin B/cdc2 (also known as MPF). This arrest is relieved by fertilization, meiosis is completed, and mitotic divisions ensue. Fertilization produces an increase in intracellular Ca^{2+}, which binds calmodulin (CaM) and activates CaMKII. Activation of CaMKII results, in turn, in activation of the 26S proteasome. Activation of the proteasome leads to proteolysis of ubiquinated proteins (Ub is ubiquitin), one of which is cyclin B. Destruction of cyclin B inactivates cdc2. Inactivation of cdc2 allows exit from meiosis to occur. The figure also shows the points at which specific reagents have been used to inhibit (⊥) or stimulate (↑) the process.

the entire cell leading to the completion of meiosis. Mitotic divisions and embryogenesis then ensue.

Much is known about the signaling pathways that are important for *Xenopus* oocyte activation and fertilization. Rather than acting through the traditional progesterone receptor, this steroid hormone, which is clearly the physiological stimulus, acts at the level of the plasma membrane to decrease the activity of adenylyl cyclase by a completely unknown mechanism. This results in a decrease in cAMP levels and PKA activity which are required for the activation of the protooncogene, c-*mos*. C-*mos* aids in the activation of Maturation Promoting Factor (MPF), which is comprised of cyclin B and the cyclin-dependent protein kinase, cdc2. MPF leads to the breakdown of the germinal vesicle surrounding the nucleus (GVBD) and progression of meiosis. As mentioned previously, meiosis is halted at metaphase II owing to formation of a complex known as Cytostatic Factor (CSF), of which c-*mos* is also a component. Cells arrest with high levels of MPF activity.

The *Xenopus* system is amenable to manipulation because it is relatively simple to microinject potential activators or inhibitors and score GVBD. The pathway described previously was largely determined in this manner. For instance, cyclic nucleotide phosphodiesterase (PDE), the regulatory subunits of PKA, the PKA inhibitory protein PKI, c-*mos*, and an active form of MPF are all activators of the oocyte. In contrast, activators of adenylyl cyclase, cAMP, inhibitors of PDE, or the catalytic subunit of PKA are inhibitors of activation. A similar strategy has been used to define the pathway by which fertilization leads to resumption of the second meiotic division (Fig. 9). The initial experiments had demonstrated that fertilization results in a rapid increase in Ca^{2+} which, in turn, leads to inhibition of CSF and MPF, characterized by measuring the proteolysis of cyclin B, and resumption of meiosis. Cyclin B degradation and inactivation of MPF could be blocked by microinjection into mature eggs of calcium chelators such as BAPTA or EGTA and calmodulin was implicated because pharmacological inhibitors of this protein such as W7 were also inhibitory. To prove that calmodulin was the target of Ca^{2+}, it was demonstrated that microinjection of the smMLCK antagonist peptide would inhibit resumption of meiosis whereas the control peptide would not. To investigate the calmodulin-dependent enzyme involved, eggs were injected with truncated, constitutively active fragments of such enzymes and it was shown that the relevant portion of CaMKII readily activated the cyclin B destruction machinery even in the absence of Ca^{2+} or in the presence of the calmodulin antagonist peptide. The effects of the truncated form of CaMKII were completely inhibited by coinjection of a synthetic autoinhibitory peptide derived from CaMKII. The strong implication of these studies was that CaMKII was the calmodulin-dependent enzyme activated by the Ca^{2+} wave in response to fertilization. To prove this hypothesis, endogenous CaMKII was immunoprecipitated from eggs as a function of time after fertilization. Activation of the enzyme was shown to occur temporally between the increase in Ca^{2+} and the proteolysis of cyclin B. This enzyme activation could also be blocked by either the calmodulin antagonist or pseudosubstrate peptide. This is a demonstration of how structural and sequence information generated by biochemical and molecular technology can suggest an experimental strategy to dissect the components of a signaling pathway in a cell. Such an approach has been applied in numerous experimental paradigms to elucidate the role of calmodulin and specific calmodulin-dependent enzymes in many other cell processes.

3.2. Activation of T Lymphocytes

Perhaps one of the most highly publicized roles of calmodulin and a specific calmodulin-dependent enzyme is in the activation of T lymphocytes (Fig. 10). The enzyme in this case is the Ca^{2+}–calmodulin-dependent protein Ser/Thr phosphatase, calcineurin, and the reason for the publicity is the fact that this enzyme is the target of the immunosuppressive drugs cyclosporin A and FK-506. However, this cell system is equally illustrative of how calmodulin can be involved in regulation of transcription by multiple pathways. Engagement of the T cell receptor (TCR) causes a number of events to occur very rapidly at or close to the plasma membrane. One such event is the activation of phospholipase C to generate IP_3 which, in turn, binds to its receptor on the endoplasmic reticulum membrane to stimulate release of Ca^{2+} from this intracellular store. The free Ca^{2+} concentration is rapidly increased from the resting level of about 100 nM, achieves a near micromolar maximum by 2 min and then decreases to a new baseline level of approx 400 nM by 5 min. The rise in Ca^{2+} results in increased binding to calmodulin which activates the Ca^{2+}–calmodulin-dependent protein phosphatase, calcineurin. Calcineurin exists in the cytoplasm in a complex that includes a transcription factor, the Nuclear Factor of Activated T cells (NFATc). Calcineurin dephosphorylates NFATc, which allows it to enter the nucleus and interact with the promoter regions of cytokine genes such as the one encoding IL-2. Cyclosporin A binds cyclophilin and FK506 binds to FKBP12. Either drug/protein interacts with and inhibits the activation of calcineurin. Thus, NFATc remains phosphorylated and in the cytoplasm. Reagents used to decipher this pathway included Ca^{2+} chelating agents, Ca^{2+}–calmodulin-independent fragments of calcineurin, pharmacological and peptide antagonists of calmodulin, and forms of calcineurin that are mutated so that the phosphatase activity is silenced.

The IL-2 promoter is a complicated one and requires, among other factors, binding of both AP1 as well as NFATc for transcriptional activation. AP1 is composed of heterodimers of members of the c-*fos* and c-*jun* families, which are immediate early genes. Calcium is required for the activation of immediate early genes in T cells and this effect of Ca^{2+} is still apparent in cells treated with cyclosporin A or FK-506, so is independent of calcineurin. However, this immediate early gene activation is inhibited by calmodulin antagonists (in cells exposed to ionomycin, which increases the release of Ca^{2+} from intracellular stores), implicating the existence of a second calmodulin target. One of the transcription factors required for activation of the *fos* and *jun* genes is CREB. As CaMKIV is present in T cells, is a nuclear protein and will phosphorylate CREB on Ser^{133}, which is required for its transactivating function, CaMKIV seemed a reasonable candidate for the additional Ca^{2+}–calmodulin target. This idea was tested using a series of reagents made possible by understanding the structure and regulation of CaMKIV. Perhaps the most convincing series of experiments involved creation of a transgenic mouse that expressed a catalytically inactive form of CaMKIV specifically in T cells when resident in the thymus. Such cells could not phosphorylate CREB, activate the immediate early genes, or produce IL-2. However, splenic T cells that had lost the transgene product due to normal protein turnover had regained the ability to respond normally to agonists of the TCR. The importance of CaMKIV in T-cell activation was subsequently confirmed by analysis of the phenotypic consequences of silencing the gene in mice. Thus calmodulin-dependent enzymes can affect transcriptional responses either by regulating the translocation of a transcription factor from cytoplasm to nucleus or by direct phosphorylation of transcription factors within the nucleus.

Interestingly, calmodulin kinases can also negatively regulate transcription of genes that require

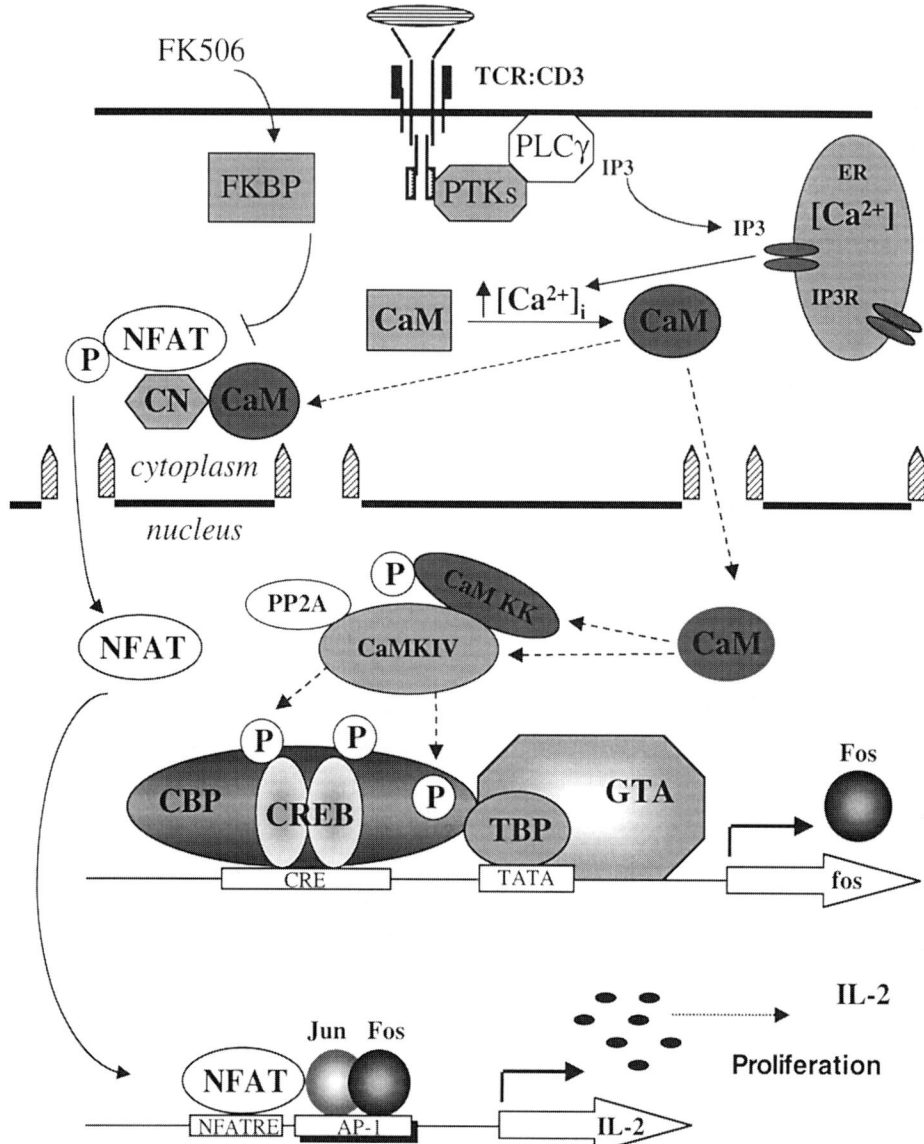

Fig. 10. Roles for Ca^{2+}–calmodulin in the activation of T lymphocytes. Engagement of the T cell receptor (TCR:CD3) stimulates the activity of several protein tyrosine kinases (PTKs). Two essential PTKs are called lck and ZAP-70. These PTKs activate phospholipase Cγ (PLCγ, which, in turn, results in the production of inositol trisphosphate (IP$_3$). IP$_3$ interacts with its receptor on the endoplasmic reticulum (ER) to release Ca^{2+} from this Ca^{2+} store. The Ca^{2+} interacts with calmodulin (CaM). Ca^{2+}–calmodulin can initiate two sequences of reactions that are required for T-cell activation. One sequence of events is controlled by the Ca^{2+}–calmodulin-dependent phosphatase, calcineurin (CN). Binding of Ca^{2+}–calmodulin to CN activates the phosphatase, which dephosphorylates a cytoplasmic component of the transcription factor, Nuclear Factor of Activated T cells (NFAT). This allows NFAT to translocate from cytoplasm to nucleus where it binds to a response element (NFATRE) required for transcriptional activation of the gene encoding the cytokine IL-2. The immunosuppressive drug, FK506, exerts its action by inhibiting the activation of CN. FK506 enters the cell by diffusion and binds to its receptor FKBP. The FK506–FKBP complex then interacts with the complex containing CN and NFAT. The other series of reactions involving Ca^{2+}–calmodulin leads to activation of genes encoding immediate early genes such as those of the c-*fos* and c-*jun* families. The sequence of reactions involve phosphorylation of CREB by CaMKIV and are detailed in Fig. 5. Fos and Jun together form a transcription factor called AP-1. NFAT and AP-1 act in tandem to regulate IL-2 gene expression. IL-2 is synthesized in the T cell, exits the cell, and interacts with its own cell surface receptor to initiate proliferation of the T cell. (TATA is a sequence of nucleotides in the basal promoter of the c-*fos* gene that binds to the TATA binding protein transcription factor, TBP; GTA refers to the complex of proteins that constitute the general transcription apparatus.)

CREB for activation. The activation of CREB requires its phosphorylation on Ser^{133}. This modification is absolutely required for the association of CREB with the coactivators p300/CBP and can be catalyzed by a number of protein kinases including PKA, p90rsk, PKB (AKT), or CaMKIV, I or II. All of these protein kinases phosphorylate only Ser^{133} except CaMKII also phosphorylates Ser^{142}. Initially, it was surprising to find that phosphorylation of Ser^{142} was both dominant and inhibitory. However, it is currently believed that Ser^{142} phosphorylation prevents an effective association between CREB and p300/CBP. This action of CaMKII in T cells results in inhibition of activation due to abrogation of IL-2 production which is presumably a consequence of phosphorylation of CREB on Ser^{142}. However, more recently it has been shown that CaMKII can also antagonize transcription of virtually any gene that requires p300/CBP as a coactivator. As many of these genes do not require CREB, it follows that another target for CaMKII must exist which, when phosphorylated, results in transcriptional inhibition in a CBP–p300-dependent manner.

3.3. Regulation of Cell Proliferation

Based on the information presented in this chapter, it should come as no surprise that calmodulin is an essential protein in all eukaryotic cells. Deletion of the unique calmodulin gene in the genetically tractable organisms *Saccharomyces cerevisiae, Schizosaccharomyces pombe, Aspergillus nidulans,* and *Drosophila melanogaster* is lethal. A common phenotypic consequence of the absence of calmodulin is inhibition of cell proliferation. Calmodulin is required for quiescent cells to enter the cell cycle in response to a mitogenic stimulus and is also essential for exponentially growing cells at the G_1/S, G_2/M, and metaphase/anaphase transitions. As an extensive review on this subject has recently appeared, it will be dealt with here only in an abbreviated fashion.

3.3.1. VERTEBRATE CELLS

In mammalian cells in culture, calmodulin is required at two points during entry of quiescent cells into the cell cycle (Fig. 11). The first is within the first 1–2 h after addition of a mitogen. Cells can be reversibly arrested at this point by the use of anticalmodulin drugs or the calcineurin antagonists cyclosporin A or FK-506, but not by the calmodulin kinase inhibitor KN-93. This has led to the suggestion that the calmodulin target at this point is calcineurin. The second block occurs much closer to the G_1/S

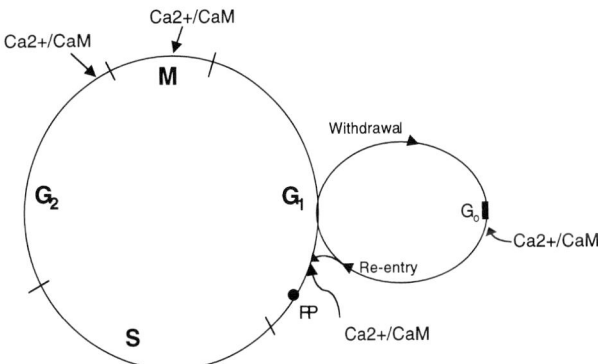

Fig. 11. Regulation of the cell cycle by Ca^{2+}–calmodulin (Ca^{2+}/CaM). The phases of the cell cycle that require Ca^{2+}/CaM are indicated by *arrows*. G_0 refers to quiescent cells. RP indicates the restriction point in G_1. After a cell passes the RP, it is committed to undergo another round of cell division. The decision to exit the cell cycle and enter G_0 is made before cells pass the RP.

boundary and can be inhibited by anticalmodulin drugs or KN-93 but not cyclosporin A or FK-506. Indeed, cells arrested at the first point by FK-506 can be released from this block into KN-93 and arrest before entry into DNA synthesis. These results are compatible with a CaM kinase being required for the G_1/S transition. Unfortunately, at this time, nothing is known of the targets for the action of either calcineurin or CaM kinase during cell cycle reentry. On the other hand, overexpression of calmodulin results in a shortened cell cycle due to a decrease in the length of G_1. The only other proteins suggested to shorten G_1 when overexpressed are cyclins D1 and E, c-*myc,* and the transcription factor E2F. Interestingly, the suggestion has been made that calmodulin is important both for the activity of the cyclin D1–cdk4 complex and maintaining the nuclear localization of cyclin D1 during late G_1. Thus, calmodulin and the cyclin-dependent pathways may be linked at G_1/S.

Certainly there is a link between calmodulin and cdk-dependent pathways at G_2/M and in the transition from metaphase to anaphase during mitosis. In mammalian cells depletion of calmodulin or extracellular Ca^{2+} causes an arrest in G_2. Overexpression of a constitutively active fragment of CaMKII also causes an arrest in G_2 whereas the CaM kinase inhibitor KN-93 results in a G_2 block. These studies suggest that the timing of an event or events catalyzed by CaMKII is critical for executing the G_2 to M transition. As mentioned earlier, the critical calmodulin-dependent enzyme for the metaphase to anaphase transition is also a CaMKII, at least in *Xenopus* eggs. Therefore, it seems likely that a CaM kinase is required both for the entry into and progression through mitosis.

3.3.2. ASPERGILLUS NIDULANS

The studies in vertebrate systems regarding the roles of calmodulin in the cell cycle have been complemented by analyses in the genetically tractable filamentous fungus *Aspergillus nidulans*. As mentioned previously, the calmodulin gene is essential for the nuclear division cycle in *Aspergillus*. Similar to mammalian cells, calmodulin is required for the entry of spores into the nuclear division cycle, the passage of cells from G_1/S and the transition from G_2 into mitosis. Each of these cell cycle requirements also needs Ca^{2+} and the Ca^{2+} binding properties of calmodulin. Because of these features, *Aspergillus* may be a better model for understanding roles for calmodulin in the cell cycle than is the yeast *S. cerevisiae*. Although the calmodulin gene is essential in yeast, the essential functions do not require Ca^{2+} and can be maintained by a mutant form of calmodulin that does not bind Ca^{2+} at all. Three essential calmodulin targets have been identified in yeast and, as expected from the lack of Ca^{2+} requirements of calmodulin, these targets bind calmodulin independent of calcium. On the other hand, although yeast has several genes encoding CaM kinases and calcineurin, these genes are not essential for the cell cycle but only for the recovery from the G_1 arrest induced by treatment with pheromone.

The results in yeast are in contrast to those in *Aspergillus* (Fig. 12). First, as mentioned previously, the calmodulin gene is essential. A strain of *Aspergillus* conditional for expression of calmodulin can be rescued by expression of vertebrate but not yeast calmodulin. Calmodulin that has been mutated so it cannot bind Ca^{2+} is also unable to rescue the strain. The single calcineurin gene is essential and required for passage from G_1 to S. Three CaM kinase genes have been isolated from *Aspergillus*. One, which is the most similar to mammalian CaMKII although it is a monomeric enzyme, is required for the G_2/M transition. The other two are homologs of CaMKI/IV and the CaMK kinase that phosphorylates and activates CaMKI/IV and both are required for the entry of quiescent spores into DNA synthesis in response to a germination stimulus. The requirements for CaM kinases at both the G_0/S and G_2/M transitions occur prior to the activation of the cyclin-dependent kinases. At the G_2 arrest point, NIMA, a protein kinase required for entry into mitosis in *Aspergillus*, and the cdc25 phosphatase that dephosphorylates and acti-

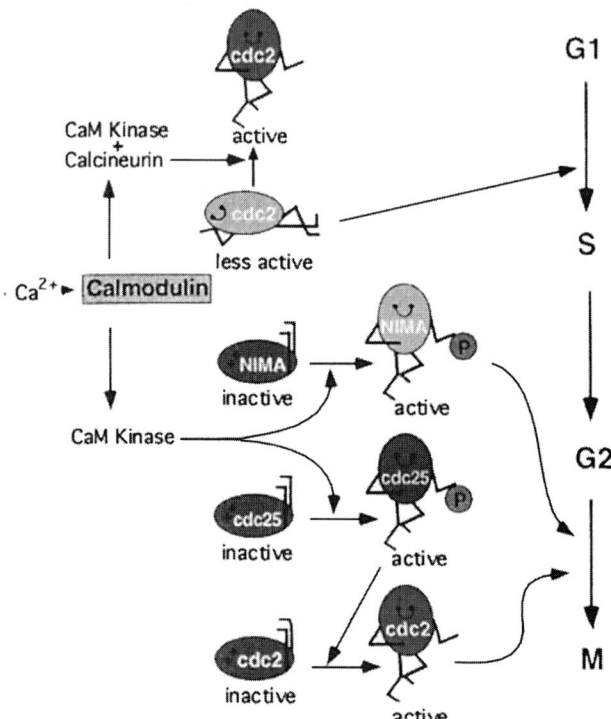

Fig. 12. Regulation of the cell cycle by Ca^{2+}–calmodulin in the filamentous fungus *Aspergillus nidulans*. One cyclin and one cyclin-dependent kinase (cdc2) are sufficient to direct cells from G_1 to S and from G_2 to M in this organism. The Ca^{2+}–camodulin-dependent enzymes involved in the G_1 to S transition are calcineurin and a CaM kinase. Both of these enzymes are required to appropriately activate cdc2. A CaM kinase is also required in regulation of the G_2 to M transition. This CaM kinase helps to activate the protein kinase NIMA, which is essential for the G_2 to M transition in *Aspergillus*. In addition, the CaM kinase is required for the proper activation of the protein phosphatase cdc25. Cdc25 then dephosphorylates and activates cdc2, another protein kinase that is essential for the G_2 to M transition.

vates the cyclin B/cdc2 enzyme are inactive when Ca^{2+}, calmodulin, or CaM kinase concentrations are held low. At the G_0/S block, the cyclin B–cdc2 complex does not become phosphorylated and is therefore inactive when Ca^{2+}–calmodulin or the CaMKI/IV homolog concentrations are held low. Thus, like in the mammalian cells, the calmodulin-dependent pathways required for cell cycle progression in *Aspergillus* appear to be linked to the cyclin–cdc2 pathways that control the cell cycle transitions. However, in no case has the target or targets of the calmodulin-dependent enzymes been clearly defined.

3.4. Avoidance Behaviors

One organism in which the physiological target for a Ca^{2+}–calmodulin-dependent behavior has been identified is *Paramecium*. This free-swimming organism exhibits an "avoidance reaction" by which it can reverse its ciliary motion, stop, turn, and begin swimming in a new direction. This behavior is initiated by a Ca^{2+}-based action potential that, in turn, causes a transient increase in intracellular Ca^{2+}. The action potential is owing to the sequential opening and closing of two ion channels. Both channels require Ca^{2+} but one demonstrates an inward current carried by Na^+ whereas the other reveals an outward current carried by K^+. A screen for mutants in this avoidance behavior resulted in generation of *Paramecium* that could only swim backward, and thus were overactive mutants, as well as those that could not execute the avoidance reaction at all and continued to swim forward. Both types of altered behavior were found to result from point mutations in the calmodulin gene. Analysis of the several mutants that only swam backward revealed mutations in the N-terminal half of calmodulin that affected the Na^+ current whereas the mutants that showed no avoidance reaction were the result of mutations in the C-terminal half of calmodulin and affected the K^+ current. Thus, the calmodulin targets were Ca^{2+}-dependent Na^+ channels and Ca^{2+}-dependent K^+ channels, respectively. All mutant behaviors could be rescued by microinjection of wild-type calmodulin. More recently, the calmodulin gene has been deleted in *Drosophila*. Whereas maternal calmodulin was sufficient to support embryogenesis, the larva die within 2 d after hatching. Null larva show an avoidance reaction to food that is strikingly similar to that seen in *Paramecium*. In the calmodulin-deficient fly, the locomotion activity elicited by the presence of food is also spontaneous backward movement. This suggests that roles for calmodulin in regulation of membrane excitability may well be evolutionarily conserved. Indeed, such roles have also been shown to be the case in the mammalian nervous system.

3.5. Learning and Memory

One of the most studied behaviors that requires Ca^{2+} and calmodulin in mammals is long-term potentiation (LTP) in hippocampal neurons. LTP is an electrical activity at the synapse that is important for spatial memory and is the most often used assay for learning and memory. Calcium is essential for LTP. This response can be divided into a short-term Ca^{2+} requiring step (seconds), which is independent of gene expression or protein synthesis, and a longer term Ca^{2+} requiring step (minutes), which involves activation of the immediate early genes of the fos and jun families and protein synthesis. The short-term response requires CaMKII as has been found in *Aplysia* and mice from which the gene encoding CaMKIIα has been silenced. Although CaMKII is one of the most abundant proteins in the postsynaptic membrane and is clearly activated by synaptic Ca^{2+} entry, its target required for the first phase of LTP remains unknown. The second phase of LTP, which prolongs synaptic output, appears to require nuclear events that involve two calmodulin-dependent enzymes, calcineurin and CaMKIV. It has been hypothesized that an increase of Ca^{2+} through the synapse leads to an increase in nuclear Ca^{2+} (and possibly calmodulin) which activates a CaMK kinase to increase the activity of CaMKIV. In turn, CaMKIV phosphorylates CREB, and possibly p300/CBP which are reactions required for the transcription of the immediate early genes. One way to reverse this response is to dephosphorylate CaMKIV (via the associated PP2A) and CREB (via protein phosphatase 1, PP1). Calcineurin has been suggested to accelerate the dephosphorylation of CREB, perhaps by dephosphorylating the inhibitory subunit of PP1 (inhibitor 1) which inhibits PP1 only when phosphorylated. Such a scenario would ensure that a brief stimulus would result in a large but transient phosphorylation of CREB, whereas a prolonged stimulus would result in CREB remaining phosphorylated for a longer time, possibly due to an accumulation of superoxide which inhibits calcineurin. Support for this possibility comes from the fact that, in the presence of pharmacological inhibitors of calcineurin (such as FK-506), even a brief stimulus results in an increase in the length of time that CREB remains phosphorylated. The supposition is that the duration of CREB phosphorylation will determine whether or not CREB-requiring genes are transcribed.

4. PERSPECTIVES

Calmodulin is one of the most versatile proteins known. It is present in all cells and is essential for cell survival. Many have suggested that calmodulin is simply a "housekeeping" protein because of its pleiotropic properties. However, the information provided in this chapter demonstrates that the roles for calmodulin can be very specific and differ from cell to cell. In a smooth muscle cell, MLCK is the rate-limiting enzyme in the Ca^{2+}-mediated pathway that controls contraction. In T lymphocytes, both calcineurin and CaMKIV are essential for the activation of the cells by antigens. In hippocampal neurons, three calmodulin-dependent enzymes—calcineurin, CaMKII, and CaMKIV—combine to fine-tune the synaptic output and thus contribute to learning and memory. In red blood cells, the primary calmodulin-dependent enzyme is the plasma membrane ATPase that maintains intracellular Ca^{2+} homeostasis and thereby the flexibility of this important cell. In unicellular organisms, calmodulin and some of its specific targets are essential for cell proliferation. Thus, the ways in which a given cell uses calmodulin as an intracellular Ca^{2+} receptor depends on its complement of calmodulin binding proteins. This is analogous to how a cell responds to cAMP. Whereas all cells contain an adenylyl cyclase to generate cAMP and PKA to mediate its actions, the PKA substrates vary considerably from cell to cell. So, calmodulin is much more important than are most ubiquitously expressed proteins in that it is essential for many pathways that use Ca^{2+} to initiate the cascade of events that culminate in a physiologically important response. Indeed, many of the actions of calmodulin within a cell depend on Ca^{2+} signaling which, in turn, is regulated by a wide variety of hormones and growth factors. Amazingly, although we know a great deal about the structure and functions of calmodulin, there is much left to learn. New functions are discovered at an alarming pace and unraveling the sequence of steps that underscore each function will keep a large number of researchers occupied for a long time to come.

SELECTED READINGS

Deisseroth K, Heist EK, Tsien, RW. Translocation of calmodulin to the nucleus supports CREB phosphorylation in hippocampal neurons. Nature 1998; 392:198.

DeKonick P, Schulman H. Sensitivity of CaM kinase II to the frequency of Ca^{2+} oscillations. Science 1998; 279:227.

Hu S-H, Parker MW, Lei JY, Wilce MCJ, Benian GM, Kemp BE. Insights into autoregulation from the crystal structure of twitchin kinase. Nature 1994; 369:581.

Ikura M. Calcium binding and conformational response in EF-hand proteins. TIBS 1996; 21:14.

Lorca,T, Cruzalegui FH, Fesquet D, Cavadore J-C, Mery J, Means A, Doree M. Calmodulin-dependent protein kinase II mediates inactivation of MPF and CSF upon fertilization of *Xenopus* eggs. Nature 1993; 366:270.

Mayford M, Bach EB, Huang Y-Y, Wang L, Hawkins RD, Kandel ER. Control of memory formation through regulated expression of a CaMKII transgene. Science 1996; 274:1678.

Meador WE, Means AR, Quiocho FA. Modulation of calmodulin plasticity in molecular recognition on the basis of X-ray structures. Science 1993; 262:1718.

Miyawaki A, Liopis J, Helm R, McCaffery JM, Adams JA, Ikura M, Tsien RY. Fluorescent indicators for Ca^{2+} based on green fluorescent proteins and calmodulin. Nature 1997; 388:882.

Saimi Y, Kung C. Ion channel regulation by calmodulin binding. FEBS Lett 1994; 350:155.

VanBerkum MF, Goodman CS. Targeted disruption of Ca^{2+}–calmodulin signaling in *Drosophila* growth cones leads to stalls in axon extension and errors in axon guidance. Neuron 1995; 14:43.

Westphal RS, Anderson KA, Means AR, Wadzinski BE. A signaling complex of Ca^{2+}–calmodulin-dependent protein kinase IV and protein phosphatase 2A. Science 1998; 280:1258.

11 Protein Kinase C

Alexandra C. Newton

Contents
> Introduction
> Background
> Structure of Protein Kinase C
> Regulation of Protein Kinase C
> Function of Protein Kinase C
> Summary
> Selected Readings

1. INTRODUCTION

Phosphorylation is a universal language used by cells to relay information between and within cells. One family of kinases that plays a key role in transducing information is the protein kinase C family. Members of this family interpret information from signals that result in phospholipid hydrolysis; they communicate to substrates throughout the cell, causing both short-term and long-term changes in cell function. This chapter focuses on the molecular mechanisms of protein kinase C, from the protein to the cell biological level.

2. BACKGROUND

2.1. Discovery of Lipid Second Messengers and Protein Kinase C

Almost half a century ago, Hokin and Hokin unearthed the first clue that cells use lipids to transduce information: they discovered that cholinergic stimulation of pancreatic slices promoted the incorporation of ^{32}P from radiolabeled ATP into phospholipids. This initiated a flurry of research into stimulus-dependent lipid turnover and it became quickly apparent that the phosphoinositides were hydrolyzed in response to a large number of diverse extracellular signals (with the noted incorporation of radiolabeled phosphate reflecting recycling of the hydrolyzed phospholipids). However, the effector enzyme coupled to this lipid signaling pathway remained elusive for another 25 yr.

In the meantime, research into the molecular basis for cancer had identified a potent tumor promoter, whose mechanism remained unknown, but that induced dramatic biological responses in virtually every biological system examined. It had been known over the millennia that the milky sap exuded by plants of the Euphorbiaccae family is a potent irritant and croton oil, derived from *Croton tiglium*, was used medicinally as a counterirritant and cathartic. In the late 1960s, the active ingredient was identified as a family of diesters of the tetracyclic diterpene phorbol, with varying acyl chains at the C-12 and C-13 positions. The most potent compound was phorbol 12-myristate 13-acetate (PMA, also referred to as TPA for 12-*O*-tetradecanoyl phorbol 13-acetate). Figure 1 shows the structure of PMA and the plant, *Croton tiglium*, from which it is isolated. The lipophilicity of PMA causes it to partition readily into membranes and this had made it difficult for researchers to identify whether the molecule bound specifically to cells. A breakthrough occurred when Peter Blumberg and co-workers synthesized a less hydrophobic phorbol ester,

From: *Principles of Molecular Regulation* (P. M. Conn and A. R. Means, eds.), © Humana Press Inc., Totowa, NJ.

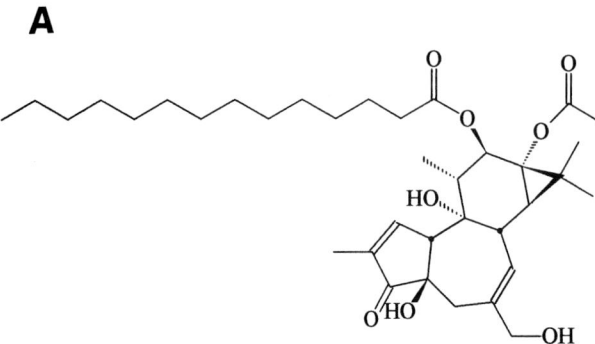

Fig. 1. Structure of phorbol myristate acetate **(top)** and the plant from which it is isolated, Croton tiglium **(bottom).** (Courtesy of P. Blumberg, NIH.)

phorbol 12,13-dibutyrate (PDBu), which they predicted would be sufficiently water-soluble to overcome the high nonspecific membrane intercalation of PMA but would still retain its "pharmacophore." In 1980, they successfully used this molecule to show that phorbol esters have a saturable, high-affinity binding site in the membrane fraction of cells; affinities on the order of 10 nM and approx 10^5 binding sites per cell were measured. Brain showed uniquely high levels of binding and the receptors were associated with a specific phospholipid, phosphatidylserine.

This question remained: What is the identity of the phorbol ester receptor? Given its high expression in brain, had it already been detected and characterized through one of its other functions?

In the early 1970s, Yasutomi Nishizuka and co-workers were working on the newly discovered protein kinase G. Their research revealed that a constitutively active fragment of protein kinase G, insensitive to cGMP, was generated on proteolytic cleavage. This led them to hypothesize that a regulatory module that constrained catalysis had been released, a hypothesis that proved seminal in the discovery of protein kinase C. Shortly thereafter, his group began purifying a kinase that was constitutively active, which they named protein kinase M (M for its only known requirement, Mg^{2+}). They noted that much higher levels of this novel kinase were obtained from stocks of frozen rat brain compared with brains from freshly slaughtered rats. Taking advantage of their experience with protein kinase G, they hypothesized that they were examining a proteolytically generated kinase domain and thus set out to identify the "proenzyme." In 1977 Nishizuka's group reported the discovery of this proenzyme. They called it protein kinase C because it was activated by the Ca^{2+}-dependent protease, calpain. It soon became evident that activity of the proenzyme was dependent on anionic phospholipids, in particular phosphatidylserine. Curiously, a crude extract of phospholipids from brain membranes was the most effective at activating the kinase, leading Nishizuka's group to search for an impurity that could be an essential activator. The identification of diacylglycerol as the impurity responsible for this stimulation led Nishizuka to make the conceptual breakthrough that this enzyme might be the target for diacylglycerol as a second messenger. These hallmark findings heralded the beginning of research on one of the most described enzymes in the literature in the past two decades. Attesting to the widespread interest in this kinase, a review by Nishizuka in the mid-1980s was the most quoted article in all the sciences in that decade.

Interest in protein kinase C catapulted to the forefront of signal transduction when the enzyme was identified as the phorbol ester receptor. The similar subcellular localization and association with phosphatidylserine of the phorbol ester receptor and protein kinase C, together with the rapid effects of phorbol esters on protein kinase C in cells (see Section 2.2.), beckoned a direct comparison. Monique Castagna, who had spent the previous summer in Blumberg's laboratory where the phorbol ester receptor was being characterized, traveled from France to Japan to do the experiment. In 1982, Castagna, Nishizuka, and co-workers published the first article showing that phorbol esters directly bind and activate protein kinase C; the following year, several groups showed that the phorbol ester receptor and protein kinase C copurified.

2.2. Translocation of Protein Kinase C

With the discovery that protein kinase C is directly activated by phorbol esters, an avalanche of reports on the effects of phorbol esters on cells marked the 1980s. The one unifying theme, first reported by Anderson and co-workers, is that phorbol ester stimulation results in rapid redistribution of protein kinase C from the cytosol to the membrane. This membrane translocation has continued to serve as the hallmark for protein kinase C activation.

The past two decades have provided enormous insight into the molecular mechanisms of protein kinase C. This family of kinases serves as a paradigm for how protein function is exquisitely regulated by macromolecular and intradomain interactions.

3. STRUCTURE OF PROTEIN KINASE C

Cloning of protein kinase C in the mid-1980s revealed a family of related enzymes, with 11 mammalian isozymes identified to date (Fig. 2). They all share a common primary structure: a single polypeptide with the N-terminal half containing regulatory modules and the C-terminal moiety containing the kinase domain. Four major subclasses have been characterized: the conventional protein kinase Cs (α, γ, and the alternatively spliced βI and βII), the novel protein kinase Cs (δ, ϵ, η/L, θ), the atypical protein kinase Cs (ζ and ι/λ), and protein kinase μ (protein kinase D). They differ primarily in the N-terminal modules that regulate the kinase domain. They do, however, share a common regulatory mechanism: almost all isozymes use two membrane-targetting modules to bring the kinase to the membrane, an interaction that results in release of an autoinhibitory (pseudosubstrate) domain from the substrate binding cavity. The existence of multiple isozymes of protein kinase C likely provides breadth, while maintaining specificity, in the regulation of cellular function by lipid second messengers.

The regulatory moiety is separated from the catalytic domain by a "hinge" that becomes proteolytically labile when protein kinase C binds membranes. Cleavage at the hinge liberates a constitutively active kinase domain, freed of constraints imposed by the regulatory domain. It is this proteolytically generated kinase domain (protein kinase M) that led to the discovery of protein kinase C.

3.1. Regulatory Moiety

The regulatory moiety ranges in size from 20 to 70 kDa and contains two types of critical regulatory determinants: a pseudosubstrate domain (see Chapter 17 in this volume) that binds the substrate binding cavity and acts as an autoinhibitory module, and two membrane-targetting modules (typically C1 and C2) that recruit protein kinase C to membranes upon generation of diacylglycerol. The binding to membranes by the membrane-targetting modules provides the energy to release the pseudosubstrate from the active site.

3.1.1. PSEUDOSUBSTRATE

A pseudosubstrate motif has been identified in all the protein kinase Cs except protein kinase Cμ (human homolog, protein kinase D). It comprises a stretch of basic amino acids with an Ala at the position of the phosphoacceptor site. Peptides based on this sequence are effective competitive inhibitors of protein kinase C; ones with a Ser replacing the Ala at the phosphoacceptor position are good substrates.

The pseudosubstrate is a key molecular switch in the regulation of the protein kinase C isozymes. It has been proposed that, in the absence of cofactors, it binds and sterically blocks the substrate binding cavity. Activation results from release of the pseudosubstrate from the substrate binding cavity. Consistent with this hypothesis, the pseudosubstrate is extremely proteolytically labile when protein kinase C is active and resistant to proteolysis when the enzyme is inactive. Indeed, this proteolytic sensitivity has served as a useful conformational probe of protein kinase C. Furthermore, an antibody to the pseudosubstrate activates protein kinase C in the absence of cofactors, presumably by binding the pseudosubstrate with high affinity and displacing it from the active site.

Mutants in which the pseudosubstrate has been removed are unstable, suggesting that this portion of the molecule is important to allow the proper folding of protein kinase C. Mutants in which some of the basic residues are replaced with neutral or acidic ones have higher basal activity, again consistent with the notion that decreasing the affinity of the pseudosub-

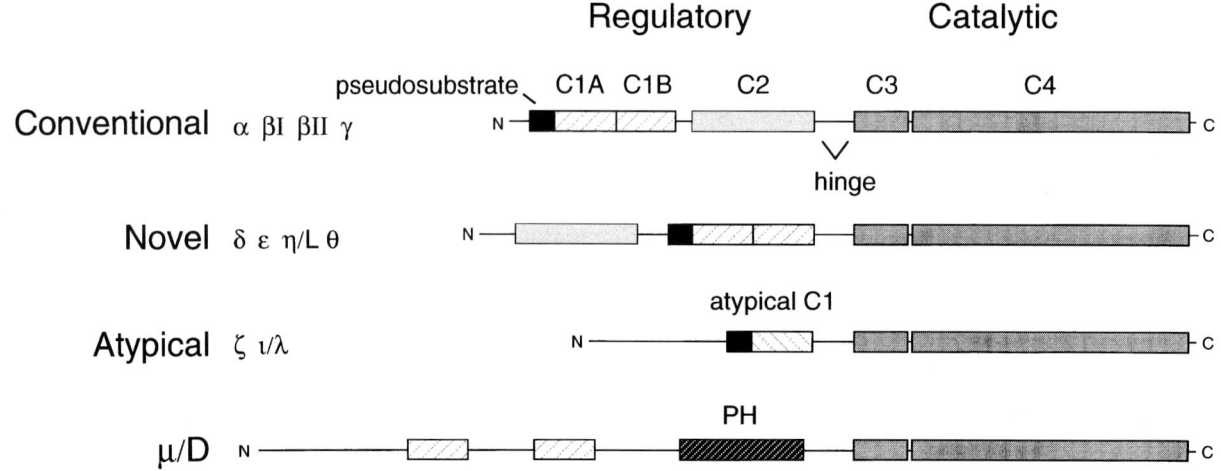

Fig. 2. Primary structure of members of the protein kinase C family showing subclasses and members within each subclass. The N-terminal moiety contains membrane-targetting modules: the C1A and C1B domains which bind diacylglycerol or phorbol esters in all but the atypical protein kinase Cs (*hatched box*); the C2 domain which binds anionic lipids and, for conventional protein kinase Cs, Ca^{2+} (*light gray box*); the PH domain which binds phosphoinositides (*dark hatched box*). Also present in the regulatory domain is the autoinhibitory pseudosubstrate (*black box*). The C-terminal half contains the kinase domain (*dark gray boxes*) which comprises the ATP binding lobe (C3) and substrate binding lobe (C4). (Adapted from BBA 1998; 1376:155–172.)

strate for the active site promotes substrate binding and catalysis.

Molecular modeling of the protein kinase Cs based on the crystal structure of protein kinase A (*see* Chapter 16 in this volume) suggests that the basic pseudosubstrate is maintained in the substrate binding cavity by an electrostatic gate: the substrate binding cavity is highly negatively charged. This strong electronegative potential results, in part, from a conserved cluster of acidic residues (EGEDEDE in the conventional protein kinase Cs) surrounding the substrate binding cavity.

3.1.2. MEMBRANE-TARGETING MODULES

3.1.2.1. C1 Domain. Elucidation of both nuclear magnetic resonance (NMR) and crystal structures of the C1 domain have revealed a globular domain containing a hydrophilic ligand-binding pocket formed by two pulled apart β sheets (Fig. 3A). Two zinc atoms help maintain the fold of the domain by coordinating Cys and His residues at opposite ends of the primary sequence of the domain.

An important finding that emerged from structural studies is that ligand binding does not alter the conformation of the C1 domain. Rather, ligand binding dramatically alters the surface properties of the module: ligand binding caps the hydrophilic ligand binding pocket such that the top third of the module displays a contiguous hydrophobic surface. Thus, the targeting of the C1 domain to membranes results from ligand binding altering the properties of the domain such that it has a high affinity for membranes.

Curiously, atypical protein kinase Cs do not respond to phorbol esters or diacylglycerol. Structural studies have provided a molecular basis for this lack of ligand binding: one face of the ligand binding pocket is missing. Whether the atypical C1 domains serve a separate function, such as promoting protein–protein interactions, remains to be established.

3.1.2.2. C2 Domain. The C2 domain serves as a membrane-targetting module in a number of signaling proteins, such as cytosolic phospholipase A_2 and phospholipase C. Usually, this module binds anionic lipids in a Ca^{2+}-dependent manner, although some C2 domains bind neutral lipids and others are not regulated by Ca^{2+}. Elucidation of both crystal and NMR structures has revealed a β-sheet-rich domain with a novel Ca^{2+}-binding pocket: loops formed by sequences at opposite ends of the primary structure that come together to form an aspartate-lined mouth (Fig. 3B). The actual Ca^{2+} binding site contains three loops, each contributing aspartates important in binding Ca^{2+}. This site coordinates 2–3 Ca^{2+} ions; curiously, the coordination sphere is not complete for any of the Ca^{2+} ions, suggesting that interactions with lipid headgroups could complete the coordination.

Mutagenesis throughout the C2 domain has local-

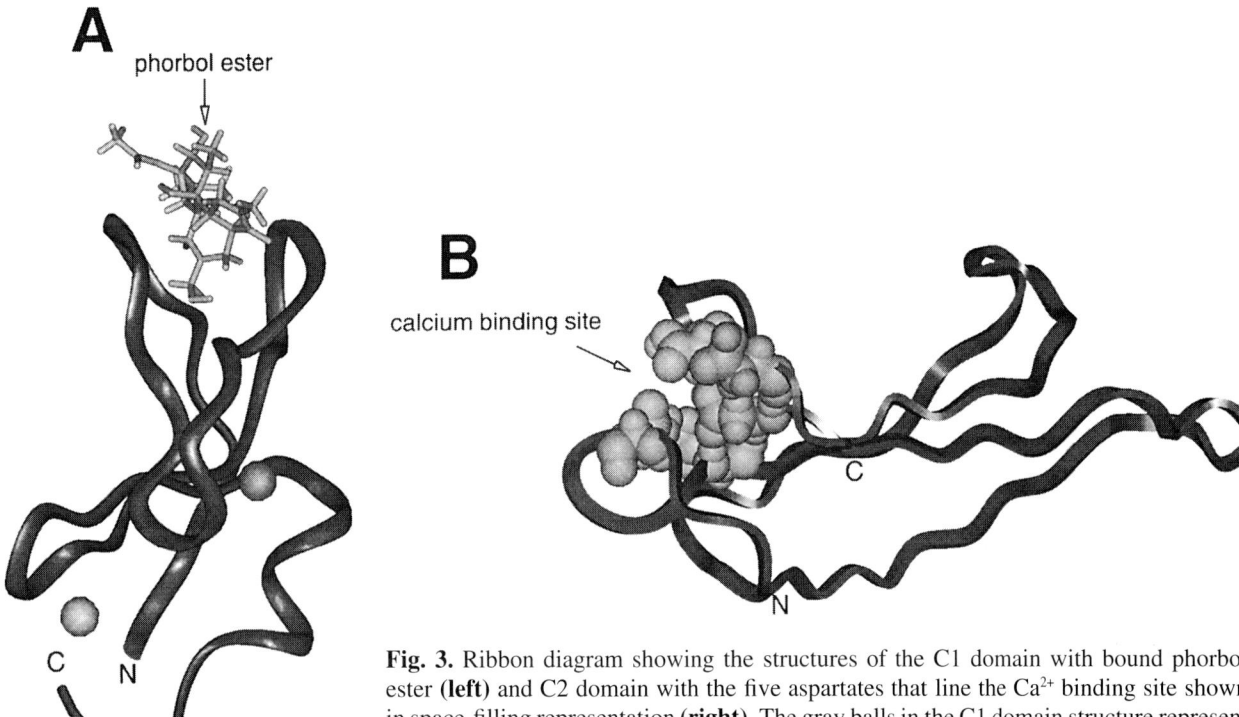

Fig. 3. Ribbon diagram showing the structures of the C1 domain with bound phorbol ester (**left**) and C2 domain with the five aspartates that line the Ca^{2+} binding site shown in space-filling representation (**right**). The gray balls in the C1 domain structure represent two zinc atoms. Adapted from BBA 1998; 1376:155–172.

ized the membrane-interacting portion to the Ca^{2+} binding region, so that the domain approaches the membrane "jaws first." Ca^{2+} binding is essential for this interaction: mutagenesis to neutralize the Ca^{2+} binding site, thereby mimicking the effect of Ca^{2+} on the electrostatic potential of the domain, does not promote membrane binding. Rather, it dramatically decreases the affinity of Ca^{2+} and hence the affinity for membranes; however Ca^{2+} binding is still required for the module to bind membranes. Thus, Ca^{2+} does not regulate the domain by acting as an "electrostatic switch" (i.e., charge neutralization does not trigger membrane binding). Rather, the cation may serve as a bridge to lipid headgroups.

3.1.3. KINASE DOMAIN

The catalytic domain is approx 45 kDa and sequence alignment reveals 40% identity with the sequence of protein kinase A (*see* chapter by Corbin in this volume). This sequence is also highly related to the recently discovered Akt kinase (protein kinase B) and the protein kinase C-related kinase, PRK (protein kinase N). Although protein kinase C has been refractory to crystallography so far, its kinase core has been modeled based on that of protein kinase A. Not surprisingly, residues that maintain the fold or are important for catalysis are uniformly conserved, with differences in residues found primarily on the surface of the protein. The protein kinase A sequence ends with a hydrophobic residue, phenylalanine, that tucks into the protein. All protein kinase Cs have this phenylalanine, which is then immediately followed by a conserved phosphorylation site (hydrophobic phosphorylation motif, *see* Section 3.2.) followed by a short sequence unique to each isozyme.

The activity of the kinase core is regulated by two independent mechanisms: phosphorylation, which is required to render the kinase catalytically competent, and the pseudosubstrate (*see* above), whose removal from the active site is required for substrate binding. The regulation by phosphorylation is discussed in Section 3.2.

The maximal rate of catalysis of protein kinase C is on the order of 10 reactions per second, similar to that of protein kinase A. Unlike protein kinase A, it does not have a strict "consensus sequence" for substrate phosphorylation, although regions surrounding phosphorylation sites in physiological substrates are generally basic. Studies using an oriented peptide library to determine optimal peptide sequences phosphorylated by protein kinase C have confirmed the importance of basic residues at either N- or C-terminal ends of the phosphoacceptor site, showing particular importance of basic residues at the P-3 position. In addition, a hydrophobic residue at the P+1 position markedly enhances recognition by protein kinase C. Nonetheless, the selectivity for specific residues is modest and differences between preferences for isozymes are subtle for most isozymes. The modest substrate selectivity surrounding the phosphorylation site of protein kinase C substrates in vitro underscores the importance of location and accessibility in determining substrates in vivo.

Similar to other kinase family members, including tyrosine kinases, a loop near the entrance to the active site regulates the active site of protein kinase C. In addition, the C-terminus serves a key regulatory role. This region of the kinase domain has traditionally been referred to as a variable one among protein kinase C isozymes, but it is, in fact, quite conserved: importantly, it contains two motifs that are regulated by phosphorylation. The activation loop and C-terminus are discussed in Section 3.2.

4. REGULATION OF PROTEIN KINASE C

4.1. Regulation of Protein Kinase C by Lipid Cofactors

The activation of protein kinase C depends on the release of the pseudosubstrate from the substrate binding cavity. The role of cofactors is to allow this conformational change to take place. Specifically, the binding of ligands to both the C1 and C2 domains (for conventional and novel protein kinase Cs) provides the energy to break the electrostatic interactions between the basic pseudosubstrate and acidic substrate binding cavity. Why have two membrane-targetting modules? It turns out that protein kinase C can be recruited to membranes by either module alone. However, the binding mediated by either the C1 or the C2 domain alone is too weak to provide the energy to release the pseudosubstrate. Both domains must be membrane-bound for the high-affinity interaction that results in release of the pseudosubstrate. Thus, nature has ensured specificity in the activation of protein kinase C by membrane binding by taking advantage of the "synergism" (i.e., binding constants multiply) of having two points of membrane attachment. It also ensures reversibility: metabolism of diacylglycerol results in the loss of one point of attachment, with the C2 interaction being too weak to activate the protein. A large number of signaling molecules use two membrane anchors to achieve high-affinity, reversible membrane attachment—combinations of pleckstrin homology, C1, C2, myristoylation, and basic sequences are used to target such proteins to membranes.

4.1.1. PHOSPHATIDYLSERINE AND OTHER ANIONIC LIPIDS

Protein kinase C is optimally active in the presence of phosphatidylserine, diacylglycerol, and Ca^{2+}. Of these three cofactors, phosphatidylserine is essential for maximal activity. It is the interaction with phosphatidylserine that seems to be the major one responsible for release of the pseudosubstrate from the substrate binding cavity.

In the absence of diacylglycerol, protein kinase C binds all anionic membranes with equal, but relatively low, affinity. This interaction depends only on the net charge of the membrane and is driven primarily by electrostatic forces. The affinity of full-length protein kinase C for anionic membranes and that of constructs of isolated C2 domains is approximately equal, consistent with the binding to anionic lipids being mediated primarily by the C2 domain in the full-length enzyme (see Fig. 4). Thus, protein kinase C can be recruited to membranes by the C2 domain alone.

Diacylglycerol induces a remarkable increase in the membrane affinity of protein kinase C which is selective for phosphatidylserine. That is, the presence of a C1 domain ligand in membranes increases the affinity of protein kinase C for anionic lipids, as would be expected given the protein is now tethered to membranes by two domains, but there is an additional order of magnitude increase in the membrane association constant that occurs if phosphatidylserine is present in membranes. Detailed studies with enantiomeric lipids have revealed that the interaction with phosphatidylserine is specific for the sn-1,2-phosphatidyl-L-serine configuration. This unusual specificity in a protein–lipid interaction argues strongly for the presence of specific determinants on protein kinase C that specifically recognize the molecular shape of the L-serine headgroup.

4.1.2. DIACYLGLYCEROL AND PHORBOL ESTERS

Diacylglycerol and phorbol esters regulate protein kinase C by increasing the enzyme's membrane affinity—they are essentially "molecular glue" that retains the C1 domain on the membrane. Both ligands regulate protein kinase C by this same mechanism; however, the two can have very different biological effects for two reasons. First, diacylglycerol is rapidly metabolized, providing transient activation of protein kinase C, whereas phorbol esters are not readily metabolized and hence result in constitutive activation of the kinase. Second, phorbol esters are several orders of magnitude more potent than diacylglycerol in recruiting protein kinase C to membranes. The remarkable

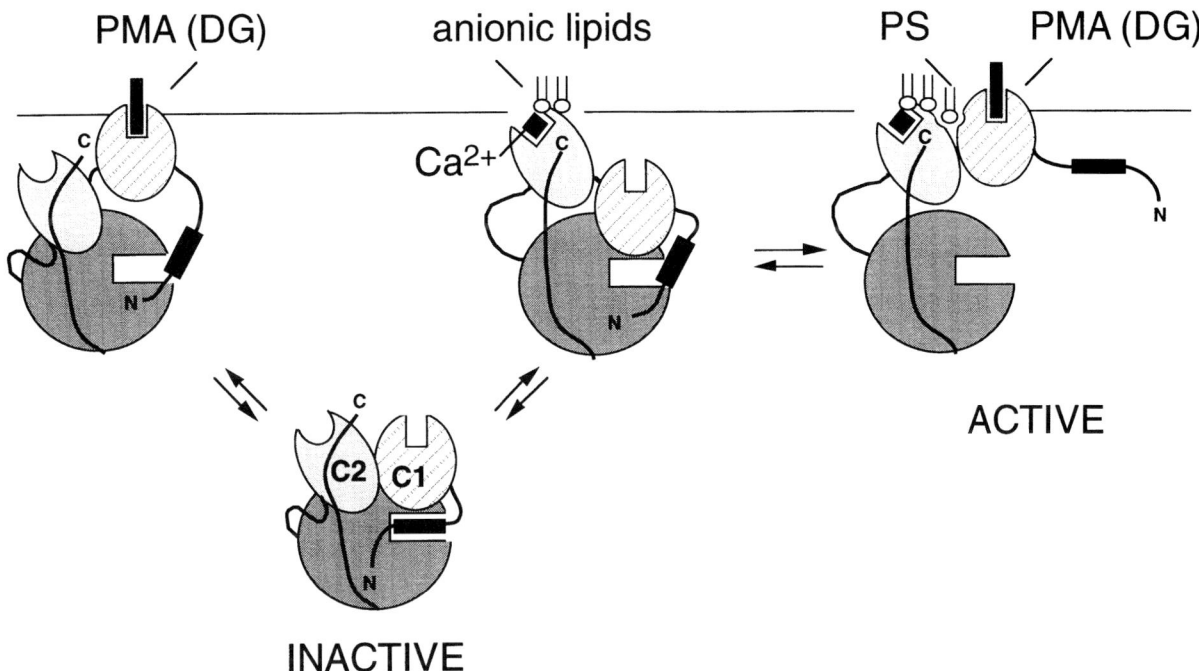

Fig. 4. Model for the regulation of protein kinase C by its two membrane-targetting modules. In its inactive conformation, the pseudosubstrate (*black rectangle*) binds the active site/substrate binding cavity (*open rectangle*) of the kinase domain (*large gray circle*), thus sterically blocking substrate binding. The enzyme can be recruited to membranes by the C1 domain alone (*upper left*), an interaction that is driven by binding of phorbol esters or diacylglycerol to this domain (note only one C1 domain is shown for clarity). It can also be recruited to the membrane by the C2 domain alone; this interaction requires anionic lipids and, for conventional protein kinase Cs, Ca^{2+} (*black square*). Membrane binding by either domain alone does not provide sufficient energy to release the pseudosubstrate from the active site. This requires the interaction of both the C1 and C2 domains with membranes **(right)**. Once the C1 domain is bound to membranes, the enzyme displays a 10-fold increase in affinity for phosphatidylserine relative to other anionic lipids, suggesting the presence of a specific binding site for this lipid. The energy supplied by the binding to diacylglycerol and phosphatidylserine helps break electrostatic interactions between the basic pseudosubstrate and acidic substrate binding cavity, resulting in release of the pseudosubstrate and activation of protein kinase C. Membrane binding involving the C2, but not C1, domain results in a conformational change that exposes the hinge region separating the regulatory and catalytic moieties (S-shaped line in soluble conformation). Several lines of evidence suggest that the C-terminus of the enzyme, which contains key phosphorylation sites (*see* Fig. 5), interfaces with the C2 domain. (Adapted from BBA 1998; 1376:155–172.)

potency of phorbol esters is illustrated by the finding that 1 mol% phorbol esters (i.e., one molecule phorbol esters per 99 lipid molecules in a membrane) increases the membrane affinity of protein kinase C by three orders of magnitude. As a result of this high-affinity binding, phorbol esters are able to recruit protein kinase C to neutral (phosphatidylcholine) membranes by the C1 domain alone, in the absence of any C2 domain interactions (*see* Fig. 4).

Most protein kinase C isozymes contain two C1 domains (*see* Fig. 2), and there has been debate as to whether both contribute to the in vivo function of protein kinase C. In vitro measurements are consistent with one molecule of phorbol ester binding per molecule of protein kinase C, although recent studies with a fluorescent analog of phorbol esters suggest that the stoichiometry may be 2:1. Importantly, studies in which one or the other C1 domain was deleted have established that the two domains are not functionally equivalent in vivo: the second C1 domain (C1B) mediates the phorbol ester-dependent translocation of protein kinase Cδ in NIH 3T3 cells. Curiously, constructs of isolated C1A or C1B domains of most isozymes bind phorbol esters with similar affinity in vitro, suggesting that orientation or accessibility underlies the functional nonequivalence of the domains in the full-length protein.

One of the early discoveries from the Nishizuka lab regarding protein kinase C was the observation that diacylglycerol and Ca^{2+} "synergize." That is, the more of one ligand that was present, the less of the other was required for maximal activation. This apparent synergism does not result from allosteric interactions between the binding sites for each ligand in the C1 and C2 domains. Protein kinase C can be recruited to membranes via the C1 domain alone (in the absence of anionic lipids) and this interaction is insensitive to Ca^{2+}. Rather, the apparent synergism arises because each ligand, independently, increases protein kinase C's affinity for anionic lipids.

4.1.3. Calcium

The conventional protein kinase Cs are marked by their unique requirement for Ca^{2+}. The role of this cation is to facilitate the interaction of the C2 domain with anionic membranes. The requirement for Ca^{2+} is not, however, absolute. Conventional protein kinase Cs can be recruited to membranes and maximally activated provided sufficiently high levels of diacylglycerol or phosphatidylserine are present. One way to think about the activation of protein kinase C is that a certain amount of binding energy is required to activate the enzyme (protein kinase C bound to membranes with apparent binding constants of 10^5 M^{-1} or greater is maximally active). Relatively high concentrations of diacylglycerol will cause the C1 domain to contribute significantly to this binding energy, and relatively high concentrations of Ca^{2+} or anionic lipids will cause the C2 domain to contribute significantly. Thus, high enough concentrations of either diacylglycerol or Ca^{2+} can abrogate the need for the other second messenger.

Is the ability to maximally activate conventional protein kinase Cs in the absence of Ca^{2+} physiologically relevant? Phorbol esters, without addition of Ca^{2+} and ionophore, can effectively activate conventional protein kinase C isozymes in many cell types. Whether there are conditions in which diacylglycerol, which is about 1000-fold less potent on a molar basis than phorbol esters in activating protein kinase C, can activate conventional protein kinase Cs in the absence of Ca^{2+} has not been resolved. It is certainly possible that the presence of domains enriched in diacylglycerol could promote sufficiently tight membrane binding of conventional protein kinase Cs that Ca^{2+} would not be required for activation.

4.1.4. Other Lipids and Lipid Products

Protein kinase C activity is directly modulated by a number of other lipids and lipid products, in addition to diacylglycerol and phosphatidylserine. Almost all isozymes are activated by fatty acids, in particular arachidonic acid. Activity is also sensitive to highly negatively charged lipids such as the polyphosphoinositides. Specific activation of the atypical protein kinase Cs by phosphatidylinositol-3,4,5-trisphosphate has been reported, although this lipid has also been shown to activate other isozymes as well. Lysophosphatidic acid has been shown to activate the protein kinase C isozymes. These acidic lipids may modulate protein kinase C activity by helping to recruit the enzyme to membranes, presumably by binding the C2 domain.

Protein kinase C activity can also be regulated by components of the ceramide signaling pathway. Sphingosine, a basic hydrolytic product of sphingomyelin, directly inhibits protein kinase C isozymes, presumably by neutralizing the charge of anionic lipid activators.

The activity of protein kinase C depends on its phosphorylation state, and at least two lipids regulate enzymes that control this phosphorylation. Ceramide, which plays a crucial role in apoptosis, has been reported to indirectly inhibit protein kinase C by modulating the activity of phosphatases that regulate the phosphorylation state of protein kinase C. In addition, phosphatidylinositol-3,4,5-trisphosphate activates the heterologous kinase that regulates the protein kinase Cs (see Section 3.2.1.).

4.1.5. Membrane Structural Properties

The maximal rate of catalysis of protein kinase C depends on lipid structural properties rather than membrane structural properties in vitro. For example, the enzyme is maximally activated by phosphatidylserine and diacylglycerol dispersed in Triton X-100 mixed micelles or present in membrane bilayers where the diluting lipid is phosphatidylcholine.

Nonetheless, the activity of the enzyme can be exquisitely sensitive to membrane structural properties such as headgroup packing, surface curvature, the presence of membrane intercalators such as anesthetics and alcohols, and the presence of nonbilayer forming lipids. For example, lipids such as phosphatidylethanolamine that have a propensity to form nonlamellar membrane structures increase the activity of the enzyme. Activity also requires acyl chain unsaturation somewhere in the bilayer, not necessarily on phosphatidylserine. The common thread in many of the effects of membrane modulators on protein kinase C activity appears to be that activity is sensitive to defects in headgroup packing. Such defects may facilitate the intercalation of protein kinase C into the hydrophobic core of the membrane (e.g., C1 domain) or facilitate specific headgroup–protein interactions.

4.2. Regulation of Protein Kinase C by Phosphorylation

For protein kinase C to be competent to respond to second messengers, it must first be processed by three ordered phosphorylations. These phosphorylations regulate both the subcellular localization and catalytic competence of the enzyme. Specifically, the enzyme is regulated by two phosphorylation switches that are conserved separately or together among various members of the kinase superfamily (see Fig. 5 and 6).

Pulse-chase experiments have revealed that newly synthesized protein kinase C associates with a membranous fraction that has both detergent-soluble and detergent-insoluble components. This species is first phosphorylated on its activation loop, a loop near the entrance to the active site, by another kinase. Phosphorylation at this position has been proposed to correctly align residues in the active site for catalysis, with the immediate consequence being two ordered autophosphorylations at the C-terminus. These C-terminal phosphorylations have been proposed to lock protein kinase C in a catalytically competent conformation, so that phosphate at the activation loop is no longer required, and to release mature protein kinase C into the cytosol. It is this phosphorylated (mature) species that is competent to respond to second messengers.

4.2.1. ACTIVATION LOOP PHOSPHORYLATION SWITCH

The activation loop is phosphorylated by a heterologous kinase. A flurry of interest in this area has identified the recently discovered phosphoinositide-dependent kinase, PDK-1, as regulating this phosphorylation site in vivo. This kinase specifically phosphorylates the activation loop of conventional, novel, and atypical protein kinase Cs.

For the conventional protein kinase Cs, phosphorylation at the activation loop triggers the maturation of protein kinase C. Overexpression of a dominant-negative (kinase-inactive) PDK-1 in COS cells was shown to result in the accumulation of unphosphorylated protein kinase C, whereas overexpression of a kinase active PDK-1 was shown to increase the fraction of mature (i.e., phosphorylated at C-terminal sites) protein kinase C. In vitro studies established that the phosphorylation of the activation loop triggers the autophosphorylation of the C-terminal sites. However, once the enzyme has been phosphorylated at these latter positions, phosphate at the activation loop is no longer required for function. Furthermore, phosphate at the activation loop does not seem to regulate the function of the mature enzyme: the mature enzyme's basal and maximal cofactor-stimulated activity is the same whether or not phosphate occupies the activation loop. Thus, the role of the activation loop phosphate, at least for the conventional protein kinase Cs, is to trigger the maturation of protein kinase C by phosphorylation and not to regulate the activity of the mature form.

PDK-1, like protein kinase C, translocates to membranes upon activation. Specifically, generation of

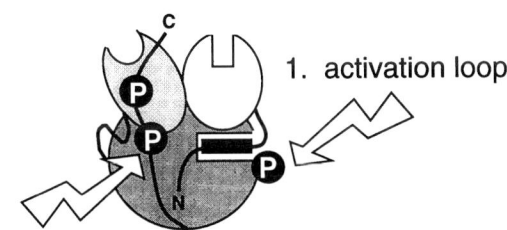

Fig. 5. Protein kinase C is regulated by two phosphorylation switches, one at the activation loop and one at the C-terminus. The former is phosphorylated by a heterologous kinase, the phosphoinositide-dependent kinase 1 (PDK-1) or related kinase. This phosphorylation is required to process catalytically competent protein kinase C; its role appears to be to correctly align residues for catalysis to promote the phosphorylation of the two C-terminal sites. Once the C-terminal sites are phosphorylated, phosphate on the activation loop is no longer required for the catalytic function of the enzyme. At least for the conventional protein kinase Cs, the C-terminal sites appear to be modified by autophosphorylation. Phosphate at these latter two positions locks the enzyme in a catalytically competent conformation and releases the enzyme into the cytosol.

```
               activation loop           turn motif              hydrophobic motif
PKC βII   496  VTTKTFCGTPDYIAPE     628  NFDRFFTRHPPVLTPPDQEVIRNIDQS----EFEGFSFVNSEFLKPEVKS
     S6K  225  TVTHTFCGTIEYMAPE     358  QFDSKFTRQTPVDSPDDSTLSESANQV----FLGFTYVAP...
Akt/PKBα  304  ATMKTFCGTPEYLAPE     437  YFDEEFTAQMITITPPDQDDSMECVDSERRPHFPQFSYSASGTA
     PKA  193  GRTWTLCGTPEYLAPE     326  NFDDYEEEEIRV-SINEKCGK---------EFTEF
```

Fig. 6. The phosphorylation switches of protein kinase C are conserved among other members of the kinase superfamily. Alignment of sequences surrounding the activation loop phosphorylation switch and the two C-terminal phosphorylation site of protein kinase CβII, S6 kinase, Akt (protein kinase B), and protein kinase A.

phosphatidylinositol-3,4,5-trisphosphate recruits PDK-1 to membranes by binding to the PH domain on this kinase. Lipid binding does not appear to be crucial for the activity of the enzyme because it can phosphorylate substrates such as S6 kinase in the complete absence of lipids. Importantly, the phosphorylation by PDK-1 of the conventional protein kinase Cs does not appear to be sensitive to regulators of either kinase in vitro. Thus, membrane localization is not required for the activation loop regulation of the conventional protein kinase Cs. The ability to be phosphorylated by PDK-1 in the absence of signals from the phosphoinositide-3-kinase pathway may explain the high level of activation loop phosphorylation of the conventional protein kinase Cs found typically in cells.

The atypical protein kinase Cs are also phosphorylated by PDK-1, but this phosphorylation differs from that of the conventional protein kinase Cs in two important ways. First, in contrast to the conventional protein kinase Cs, the phosphorylation by PDK-1 of the atypical protein kinase Cs is sufficient to activate the enzyme, in the apparent absence of cofactors for these isozymes. Second, the phosphorylation of atypical protein kinase Cs by PDK-1 is stimulated by phosphatidylinositol-3,4,5-trisphosphate. Thus, membrane localization is important for the phosphorylation and activation of the atypical protein kinase Cs. Most of the protein kinase Cζ in cells is not phosphorylated at the activation loop, with ligands that activate PI-3-kinase causing a dramatic increase in activation loop phosphorylation.

In summary, activation loop phosphorylation serves two different functions for two subclasses of protein kinase Cs: for the conventional protein kinase Cs it triggers the processing of the enzyme into a form that can later respond to second messengers, and for the atypical protein kinase Cs, it triggers the direct, stimulus-dependent activation of the enzyme.

4.2.2. C-Terminal Phosphorylation Switch

Following phosphorylation at the activation loop, protein kinase C becomes rapidly phosphorylated at two C-terminal positions. These positions are found separately or together in a number of other kinases (*see* Fig. 6). The phosphate on the "turn motif" in protein kinase A is anchored at the very top of the upper lobe of the kinase domain. By analogy, this phosphorylation on protein kinase C has been proposed to tether the C-terminus away from the active site. The structure surrounding the hydrophobic phosphorylation motif has not been solved, although the spacing of hydrophobic and polar residues suggests it may form an amphipathic helix.

For the conventional protein kinase Cs, both the turn motif and hydrophobic motif are modified by autophosphorylation. Expression of kinase-inactive protein kinase Cs, either as a result of mutation of the activation loop switch or mutation of active site residues, prevents phosphorylation of the C-terminal sites. These sites are also effectively autophosphorylated in vitro. There has been some speculation that a heterologous kinase could modify the hydrophobic phosphorylation motif in the novel protein kinase Cs. No such kinase exists for the atypical protein kinase Cs because they contain a Glu instead of a phosphorylatable residue in their hydrophobic motif.

Biochemical and mutagenesis studies have established that the phosphate on the turn motif is required for stability and catalytic competence of protein kinase C. Mutation of the Thr at this position to an Ala results in inactivatable protein that accumulates in the detergent-insoluble fraction of cells (note that mutation of just this Thr to Ala results in compensating phosphorylations at nearby residues; these must also be mutated to Ala to address the function of phosphate on the turn motif). Furthermore, dephosphorylation of this position abolishes protein kinase C activity; dephosphorylation of the activation loop and hydrophobic phosphorylation motif has no significant effect on protein kinase C activity.

4.3. Regulation of Protein Kinase C by Anchoring Proteins

Soon after the discovery that protein kinase C actually comprises a family of proteins, it became apparent that numerous isozymes were present in specific cell types, that specific isozymes localized to specific intracellular locations, and that these locations were different in different cell types. Attempts to discern the function of specific isozymes from in vitro studies were not particularly fruitful because the cofactor regulation and substrate specificity of the various isozymes are not strikingly different in vitro. Thus, much interest has focused on how the function of the various isozymes is regulated in the actual cell.

An attractive possibility to account for isozyme selectivity is the targeting of protein kinase C to specific intracellular locations by protein–protein interactions. Enormous experimental support for this hypothesis has accumulated in recent years. Most striking is an example from the *Drosophila* visual system: here, mutants lacking the scaffold protein InaD are defective in visual transduction because signaling proteins, including protein kinase C and phospholipase C, become mislocalized. Isozyme specificity arising from protein–protein interactions makes structural sense: the surfaces of each isozyme is unique, thus providing unique determinants to partner binding proteins that can provide the basis for isozyme-selective function.

A wide variety of protein kinase C interacting proteins have been identified by yeast two hybrid analysis, interaction cloning (in which a cDNA expression library is probed with protein kinase C), and overlay assays (a modification of Western blot analysis where protein kinase C, rather than antibody, is used to probe bands). Binding proteins have been proposed to tether the various forms of protein kinase C (unphosphorylated vs mature), the various activation state of protein kinase C (inactive vs active), and the various isozymes of protein kinase C. Such binding proteins could potentially localize protein kinase C either near its substrates or near its cofactors (e.g., pool of diacylglycerol).

One of the first characterized binding proteins for protein kinase C are the RACKS (Receptors for Activated C Kinase). These proteins have been proposed to bind the active conformation of protein kinase C, maintaining it near its substrates. For some of these proteins, the motif on protein kinase C that mediates the interaction has been identified and microinjection into oocytes of peptides based on this motif disrupts the stimulus-triggered translocation of protein kinase C.

Another class of binding proteins are the STICKs (Substrates That Interact with C Kinase). These proteins have been proposed to tether inactive protein kinase C, with activation of the kinase causing phosphorylation of the STICK and release from this anchor.

Multiscaffold proteins have also been identified that bind protein kinase C. The anchoring protein AKAP 79 binds at least three signaling molecules: protein kinase A, calcineurin, and protein kinase C. This protein localizes these signaling molecules to postsynaptic densities, a localization that, at least for protein kinase A, has been shown to be required for the phosphorylation of specific ion channels.

4.4. Downregulation of Protein Kinase C

Treatment of cells with phorbol esters results in a rapid redistribution of most isozymes, typically to the plasma membrane. As mentioned previously, this translocation serves as hallmark for the activation of protein kinase C. In addition, phorbol esters have another profound effect on all protein kinase Cs except the atypical ones: prolonged treatment with phorbol esters results in disappearance of protein kinase C from the detergent-soluble fraction of cells. This disappearance has been termed down-regulation and for many years was thought to result exclusively from proteolytic degradation of the enzyme. The hinge of membrane-bound protein kinase C is extremely labile proteolytically, so that prolonged membrane binding would be expected to result in proteolysis of the enzyme.

Recently, phorbol ester treatment has also been shown to result in dephosphorylation of protein kinase C, an event that is coupled to the redistribution of the enzyme to the detergent-insoluble fraction of cells. Whether the dephosphorylated protein is specifically sequestered by an anchoring protein, whether this species can be recycled by phosphorylation, and whether it is this species that is preferentially proteolyzed (or perhaps protected from proteolysis) are questions that are currently under study.

5. FUNCTION OF PROTEIN KINASE C

The presence of multiple isozymes of protein kinase C within single cell types, the broad substrate specificity in vitro, and the multitude of effects of phorbol esters in vivo have confounded elucidating the precise function of protein kinase C isozymes. What is clear, however, is that the protein kinase C isozymes are involved in a plethora of cellular functions, from short-term changes such as altering membrane permeability to long-term changes such as memory and learning.

5.1. Substrates

Treatment of cells with phorbol esters, as well as in vitro phosphorylation assays, have identified an abundance of protein kinase C substrates. The enzyme phosphorylates a large number of proteins, in particular membrane proteins (e.g., receptors, ion channels), cytoskeletal proteins (e.g., vimentin, myosin), and amphipathic proteins that cycle on and off the membrane (e.g., src, annexins). The promiscuity of the kinase in in vitro phosphorylation assays has, unfortunately, impaired the identification of physiologically relevant substrates. Clearly, correct subcellular location is one of the key determinants in marking a protein for phosphorylation by protein kinase C in vivo, so that the ability to be phosphorylated in vitro is not a reliable indicator of whether a specific protein is a protein kinase C substrate.

One of the best characterized in vivo substrates of protein kinase C is the MARCKS (Myristoylated Alanine Rich C Kinase Substrate) protein. In most cells, the appearance of an 80-kDa phosphoprotein on sodium dodecyl sulfate-polyacrylamide gel electrophoresis (SDS-PAGE) analysis of cell lysates is a hallmark "readout" for the stimulation of protein kinase C. This ubiquitous protein has a basic amphipathic helix that helps tether it to membranes, in conjunction with the binding energy provided by an N-terminal myristate. Activation of protein kinase C results in phosphorylation within this basic membrane tether, resulting in release of MARCKS into the cytosol. Although the regulation of MARCKS' membrane interaction by protein kinase C has been elegantly characterized, the precise cellular function of MARCKS is not clear. Generation of knockout mice has revealed that it plays an essential role in neural tube development. However, the lethality of this defect has restricted the information that can be obtained from knockout studies. Studies with cells grown in culture have revealed that MARCKS regulates cellular architecture: specifically, its phosphorylation by protein kinase C is required to allow actin-driven cell movement in processes such as cell adhesion.

Another major protein kinase C substrate, in this case present in cells of hematopoietic origin, is pleckstrin. Similar to MARCKS, this is also an amphipathic protein whose phosphorylation is involved in regulating the actin cytoskeleton.

5.2. Examples of Specific Isozyme Functions

An abundance of functions have been ascribed to the protein kinase C family in general, and to specific isozymes. Biochemical analyses have shown, for example, that protein kinase Cε regulates the actin-based cytoskeleton, with a defined actin-binding motif identified between the C1A and C1B domains of this isozyme. As another example, protein kinase CβII specifically phosphorylates the nuclear protein, lamin B, in human erythroleukemia cells, a phosphorylation that is required for proliferation of these cells.

The generation of knockout mice defective in specific protein kinase C isozymes is beginning to shed light on some of the physiological functions of protein kinase C. Disruption of the gene for protein kinase Cγ results in mice that appear relatively normal, with only modest effects on learning and memory, indicating this isozyme either is not essential for viability or its function can be compensated by other isozymes.

However, one striking difference between knockout and wild-type mice has been characterized: mice lacking protein kinase Cγ display reduced responses to nonnoxious pain stimuli following painful stimulation such as resulting from nerve injury, reduction in a phenomenon referred to as neuropathic pain. Studies with knockout mice in protein kinase Cε have also implicated this isozyme as a potential target for pain and, also, anxiety, for example, mice lacking this isozyme display less anxiety in response to threatening situations. Targeted disruption of the gene encoding protein kinase Cβ results in mice with an impaired immune response, with analysis of B cells from these mice revealing that the β isozymes are involved in B-cell activation. However, the molecular basis for many of the physiological differences observed in knockout mice is largely unresolved.

The protein kinase C isozymes do have some unifying regulatory roles. For example, a large number of both G protein-coupled receptors and tyrosine kinase receptors are desensitized to incoming information as a result of direct phosphorylation by protein kinase C. For example, members of the muscarinic receptor family, the β-adrenergic receptor, and the visual receptor, rhodopsin, are phosphorylated to reduce the coupling of these proteins to the relevant G protein. This type of desensitization complements that by G protein receptor kinases (GRKs), kinases that specifically phosphorylate the active conformation of the receptor. Typically, protein kinase Cs modify both liganded and nonliganded forms of the receptor, a modification that tends to be slower and has been proposed to be important for desensitizing cells to low levels of stimulation.

The uncontrolled activity of protein kinase C has long been implicated in tumorogenesis given the profound effects of phorbol esters in tumor promotion. Recently, high levels of expression of protein kinase CβII have been implicated in the first steps in colon cancer.

5.3. Chemical Intervention of Protein Kinase C Signaling

The use of selective inhibitors for protein kinase C isozymes to determine isozyme-specific functions has been met with limited success. A variety of approaches have been taken to develop inhibitors. Classical approaches have been based on the rationale design of molecules that compete for substrates or ligands. For example, a large number of inhibitors have been developed that recognize the active site,

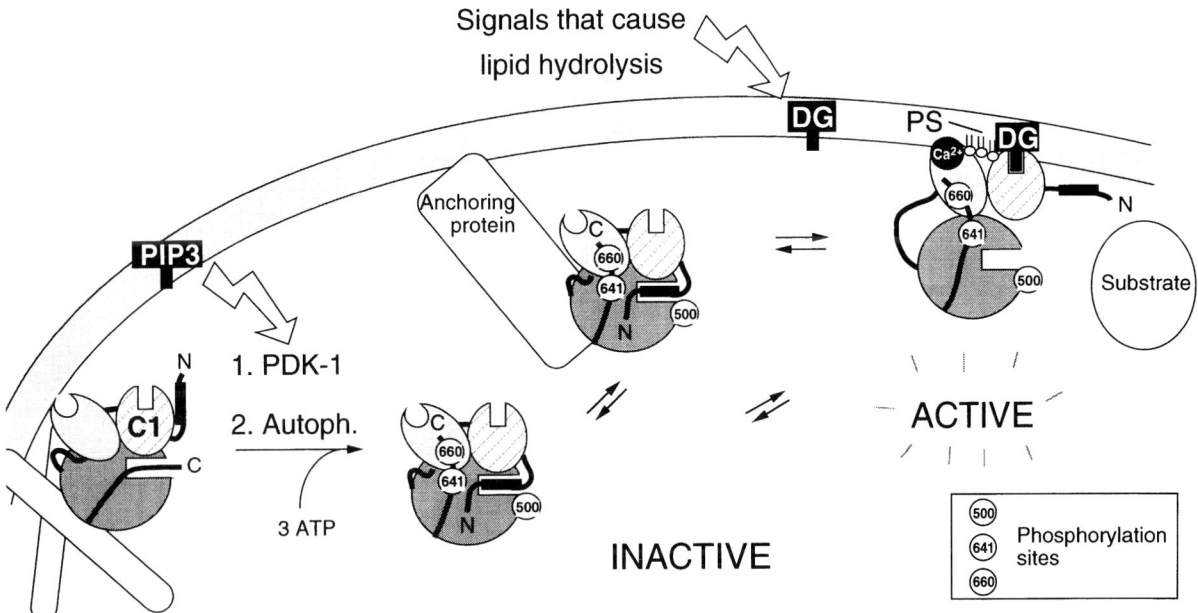

Fig. 7. Model summarizing the regulation of protein kinase C by phosphorylation, targetting proteins, and cofactors. Newly synthesized protein kinase C associates with a particulate fraction of cells that likely comprises both cytoskeletal and membrane components **(far left)**. In this conformation, the pseudosubstrate is out of the active site (as assessed by its proteolytic sensitivity), and it is hypothesized that the C-terminus may, instead, occupy the active site. The first step in the post-translational modification of protein kinase C is phosphorylation at the activation loop by PDK-1 or a related kinase. This phosphorylation correctly aligns residues for catalysis, triggering the phosphorylation of the two C-terminal sites (*see* Figs. 5 and 6). Phosphorylation at these sites is accompanied by release of protein kinase C into the cytosol, where anchoring proteins can localize specific isozymes to defined intracellular sites **(upper middle)**. Generation of diacylglycerol provides the allosteric switch to activate protein kinase C by recruiting it to membranes. The coordinate interaction of the C1 and C2 domains with membranes provides the energy to release the pseudosubstrate from the active site, allowing substrate binding and catalysis. (Adapted from Curr Opin Cell Biol 1997; 9:161–167.)

such as staurosporine, balanol, and other ATP analogs. Others have been made that compete with substrates such as ones based on the pseudosubstrate sequence. An "antagonist" to the C1 domain has been difficult to design because C1 ligands do not operate allosterically; rather, they regulate protein kinase C by altering the surface hydrophobicity of the domain to promote membrane binding.

Yet another approach is to take advantage of combinatorial approaches to generate molecules that specifically bind and inhibit a particular isozyme. For example, screening a library of RNA aptamers revealed one aptamer sequence that specifically inhibited protein kinase CβII and not other isozymes. Presumably an aptamer was selected that bound a unique surface on this particular isozyme.

Antisense technology to probe for specific isozyme function has also been met with moderate success. For example, antisense oligonucleotides to protein kinase Cα have been used to dramatically reduce the expression of this isozyme without affecting the expression of other isozymes, allowing the effects of protein kinase Cα on cellular adhesion to be studied.

6. SUMMARY

The protein kinase C family of enzymes are exquisitely regulated by phosphorylation, by cofactors, and by protein–protein interactions (Fig. 7). Signals from the phosphoinositide-3-kinase pathway and the diacylglycerol pathway each provide separate switches to regulate the function of protein kinase C. The former is required to phosphorylate the protein and thus render it catalytically competent, and the latter is required to recruit protein kinase C to membranes to allow pseudosubstrate release. In addition, the location of the enzyme is fine-tuned by specific binding proteins. With the advent of new technologies for studying specific isozymes, the next decade should enrich our understanding of protein kinase C by continuing to unveil the physiological role of this key signal transducer.

SELECTED READINGS

Avruch J. PDK1—a kinase at the hub of things. Curr Biol 1999; 9, R93–R96.

Blumberg PM. In vitro studies on the mode of action of the phorbol esters, potent tumor promoters: part 1. Crit Rev Toxicol 1980; 8:153.

Keranen LM, Dutil EM, Newton AC. Protein kinase C is regulated *in vivo* by three functionally distinct phosphorylations. Curr Biol 1995; 5:1394–1403.

Leitges M, Schmedt C, Guinamard R, Davoust J, Schaal S, Stabel S, Tarakhovsky A. Immunodeficiency in protein kinase Cβ-deficient mice. Science 1996; 273:788–791.

Mochly-Rosen D. Localization of protein kinases by anchoring proteins: a theme in signal transduction. Science 1995; 268:247.

Newton AC. Regulation of protein kinase C. Curr Opin Cell Biol 1997; 9:161–167.

Nishikawa K, Toker A, Johannes F-J, Songyang Z, Cantley LC. Determination of the specific substrate sequence motifs of protein kinase C isozymes. J Biol Chem 1997; 272:952–960.

Nishizuka Y. The role of protein kinase C in cell surface signal transduction and tumour promotion. Nature 1984; 308:693.

Orr JW, Keranen LM, Newton AC. Reversible exposure of the pseudosubstrate domain of protein kinase C by phosphatidylserine and diacylglycerol. J Biol Chem 1992; 267:15263–15266.

Parker PJ, Coussens L, Totty N, Rhee L, Young S, Chen E, Stabel S, Waterfield MD, Ullrich A. The complete primary structure of protein kinase C—the major phorbol ester receptor. Science 1986; 233:859–866.

Szallasi Z, Bogi K, Gohari S, Biro T, Acs P, Blumberg PM. Nonequivalent roles for the first and second zinc fingers of protein kinase Cδ. Effect of their mutation on phorbol ester-induced translocation in NIH 3T3 cells. J Biol Chem 1996; 271:18299–18301.

Tsunoda S, Sierralta J, Sun Y, Bodner R, Suzuki E, Becker A, Socolich M, Zuker CS. A multivalent PDZ-domain containing protein assembles signalling complexes in a G-protein-coupled cascade. Nature 1997; 388:243–249.

Zhang G, Kazanietz MG, Blumberg PM, Hurley JH. Crystal structure of the Cys2 activator-binding domain of protein kinase Cδ in complex with phorbol ester. Cell 1995; 81:917–924.

12 Nitric Oxide

Randy Krainock and Sean Murphy

CONTENTS

INTRODUCTION
CHEMISTRY AND REACTIVITY
REGULATION OF THE NOS GENES AND THEIR PRODUCTS
MOLECULAR TARGETS FOR NO
ROLES FOR NO
REAL AND POTENTIAL THERAPEUTICS
SELECTED READINGS

1. INTRODUCTION

The award of the 1998 Nobel prize for physiology and medicine, while controversial in terms of those excluded, was expected in terms of subject matter. Nitric oxide (NO), a somewhat reactive and highly diffusible gas, participates in numerous biological processes. Perhaps most importantly, NO is a signal molecule involved in the regulation of blood flow, in neurotransmission, and in the immune response.

Twenty years ago, while it was known that vasodilating drugs such as nitroglycerine produced NO, that NO could activate soluble guanylyl cyclase and so increase cGMP, and that stimulation of the endothelium lining blood vessels releases a relaxing factor onto underlying smooth muscle, it took 10 yr to put these together. This was coincident with the realization that activated macrophages produce NO oxidation products when engaged in the destruction of microorganisms. The last ten yr have seen not only the cloning and sequencing of cDNAs for NO synthases but also a resurrection of nitrogen chemistry. In addition, the roles evoked for NO have multiplied. There continues to be debate about precisely how NO is produced, and whether it is actually NO that mediates the many functions attributed to it.

From: *Principles of Molecular Regulation* (P. M. Conn and A. R. Means, eds.), © Humana Press Inc., Totowa, NJ.

Most famously, and from an increased appreciation of the mechanics of NO-induced vasodilation has emerged Viagra, one of the most publicized drugs to ever come on the market. This selective inhibitor of type 5 phosphodiesterase (PDE) is now widely prescribed as a remedy for sexual dysfunction. However, changes in therapeutic intervention have not been limited to Viagra. Introducing NO is useful in pulmonary dysfunction, in restenosis, and in wound healing. Blocking NO production via NOS inhibitors has potential for the treatment of ischemia and also for septic shock.

2. CHEMISTRY AND REACTIVITY

It seems that nearly all eukaryotic cells, as well as some simpler organisms, are capable of synthesizing NO (with L-citrulline as the coproduct) from the five-electron oxidation of L-arginine. This reaction is catalyzed by nitric oxide synthases (NOS) which employ components familiar from other enzymes but in novel and somewhat unconventional ways.

As a radical, NO reacts rapidly with other species containing unpaired electrons (Fig. 1). For example, NO binds directly to the heme group of soluble guanylyl cyclase (sGC), a major target for the physiological effects of NO, stimulating cGMP formation in a cell by many hundreds of fold. Further, NO reacts

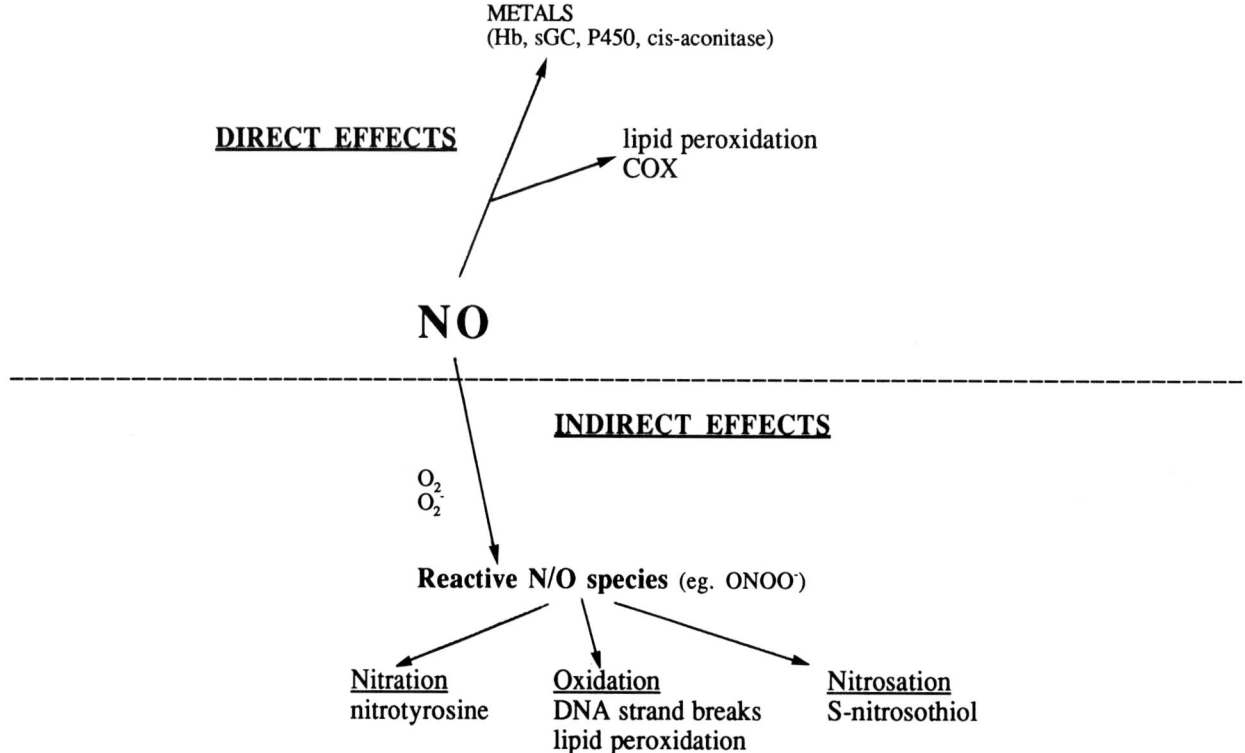

Fig. 1. The chemical biology of nitric oxide.

with other metalloproteins such as *cis*-aconitase, thus regulating cellular iron metabolism. Local NO concentrations of >1 μM permit indirect reactions. NO reacts with O_2, forming an intermediate species (N_2O_3) that can efficiently nitrosate thiols and amines. Outcompeting superoxide dismutase (SOD), NO reacts with O_2^- to produce peroxynitrite ($ONOO^-$) which, if protonated, decomposes rapidly (seconds) to a hydroxyl radical. This $ONOO^-$ reacts rapidly with CO_2 (acting as a catalyst) in a complex manner that produces several short-lived reaction intermediates, such as NO_2 and CO_3^-. These intermediates are probably responsible for many of the reported toxic effects of NO. For example, $ONOO^-$ nitrates and thereby inactivates manganese SOD, an enzyme that dismutes superoxide.

3. REGULATION OF THE NOS GENES AND THEIR PRODUCTS

To date, three major isoforms of NOS have been described and these are products of distinct genes. The cDNAs share 50–60% homology at the nucleotide and amino acid levels. Two of these gene products, NOS-1 and NOS-3, are constitutively expressed enzymes that are calcium–calmodulin dependent and produce small amounts of NO in response to transient elevations in intracellular calcium level. Both NOS-1 and NOS-3 are functional only as dimers (head-to-head), exhibit alternative splice products, display interactive domains in the N-terminal region, and depend upon calcium transients for activity. To date, the understanding of transcriptional regulation of these isoforms is rudimentary. For NOS-1, first characterized in the nervous system, activators of the gene are mechanical injury, hypoxia, and sex steroids. For NOS-3, responsible for the production of endothelium-derived relaxing factor, transcriptional activation follows shear stress, hypoxia, sex steroids, and growth factors.

While the NOS-2 isoform is also active only as a dimer, it is soluble and functions independently of a rise in intracellular calcium. This isoform is not normally expressed but can be transcriptionally induced in response to particular stimuli (among many others; proinflammatory cytokines, bacterial endotoxins, hypoxia, viral coat proteins). Unlike the constitutive isoforms, NOS-2 has calmodulin bound at all times, maintaining the enzyme in a tonically active state capable of producing a large and continuous flux

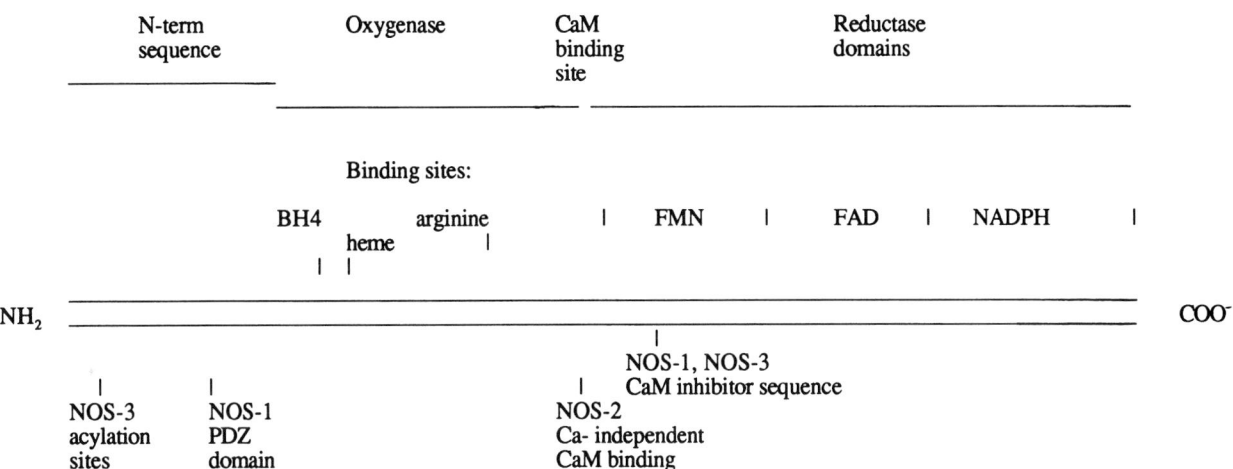

Fig. 2. Functional elements in the NOS sequence. BH4, tetrahydrobiopterin; CaM, calmodulin; FAD, flavin adenine dinucleotide; FMN, flavin mononucleotide; NADPH, nicotinamide adenine dinucleotide phosphate; PDZ, protein interaction domain.

of NO. The relatively large amount of NO generated by NOS-2 over a sustained period is implicated not only in the pathogenesis of various disorders but also in the response to infectious organisms. This isoform has been variously named ("inducible," "immunologic," "macNOS"), terms that are somewhat misleading because NOS-2 has also been reported in a wide variety of normal and neoplastic cell types, in some cases constitutively. The first 500 amino acids comprise the catalytic center of the enzyme and deletion of residues 66–114 denies dimerization and therefore a NOS function. The promoter region of the NOS-2 gene displays a wide variety of potential response elements and it appears from in vitro studies that the NFκB site in the proximal region is important for transcriptional activation.

Each NOS isoform has similar catalytic domains: a C-terminal reductase (homologous with cytochrome P450 reductase) exhibiting binding sites for flavins and NADPH, and an N-terminal oxygenase domain that contains bound heme and the site for H4 biopterin (see Fig. 2). The reductase domain transfers electrons from NADPH to heme. Calmodulin binds just at the N-terminal side of the reductase. Each isoenzyme has a different N-terminal extension that is not essential for catalysis but functions in intracellular localization of that particular protein. For example, in the N-terminal region of NOS-1 there is a PDZ domain that mediates subcellular targetting of the enzyme. In the CNS, this PDZ domain targets the enzyme to postsynaptic sites by binding to postsynaptic density proteins. In skeletal muscle, NOS-1 is targetted to the endplate owing to association with the dystrophin glycoprotein complex. In both cases, this strong association couples the enzyme to the source of calcium, resulting in enzyme activation. By analogy, NOS-3 is targeted to plasmalemmal caveolae and this occurs through dual acylation by myristate (irreversible) and palmitate (reversible). Caveolae are sites for the sequestration of signaling molecules, again leading to NOS activation.

Phosphorylation of the NOS isoforms allows both crosstalk between NO and other signaling pathways and also posttranslational modification. Purified protein kinases phosphorylate all three isoforms, generally inhibiting NOS activity. Both NOS-1 and NOS-3 have associated regulatory proteins, abbreviated as PIN and ENAP-1, respectively. Transcript stability is another means by which protein expression can be regulated. On activation of the NOS-2 gene, the transcript is very unstable with a $t_{1/2}$ of 2–3 h. This can be shortened by agents such as transforming growth factor. Clearly there are proteins coinduced that are responsible for this RNA instability as $t_{1/2}$ is prolonged by inhibitors of transcription and also protein synthesis.

The regulation of dimer assembly is of interest, as this leads to a functional NOS. The degree to which H4 biopterin is required depends on the isoform and the cell type in which the enzyme is expressed. Whether arginine and/or heme limits intracellular NOS assembly is still unclear. Certainly, in activated cells, NOS-2 exists as a mixture of dimer and monomer, implying either that dimers are unstable or that assembly is limited. Cell culture studies point to the former and that accumulating NO limits dimerization, perhaps by preventing heme insertion into the protein.

4. MOLECULAR TARGETS FOR NO

As depicted in Fig. 1, NO can interact with a wide variety of biomolecules. The following survey aims at being comprehensive but not exhaustive.

4.1. DNA

The chemical alteration of DNA underlies a variety of pathological states. NO can potentially damage DNA through reactive nitrogen species (e.g., $ONOO^-$, N_2O_3), inhibition of DNA repair processes, or increased production of genotoxic (e.g., alkylating) agents. A species such as N_2O_3 causes mutations in cells, chemically altering DNA. Deamination of cytosine, adenine, and guanine results in conversion to uracil, hypoxanthine, and xanthine, respectively. As well as modification, NO can form or modulate the activity of carcinogens. Nitrosamines are metabolized to alkylating species that lesion DNA.

4.2. Proteins

The ability to nitrosate and to nitrate key amino acids in proteins (transcriptional regulators, enzymes, receptors) explains many of the actions of NO in terms of gene regulation and alterations in cell signaling pathways.

4.2.1. TRANSCRIPTION FACTORS

Transcription factors are regulatory proteins that bind to specific DNA sequences and recruit RNA polymerase. Transcriptional activators display sequences specific for DNA binding and also a transactivation domain. A number of transcriptional activators have been shown to be regulated by NO, either posttranslationally through direct modification of the protein, or indirect regulation via alteration in the rate of their own transcription. In so doing, NO may either turn off constitutively expressed genes, activate transcriptionally regulated genes, or prevent their activation. The best examples to date are the transcription factors NFκB and AP-1, whose activation can be blocked by NO and these proteins can also be directly nitrosated, with a reduction in promoter association. These two transcription factors are involved in the regulation of expression of a very large number of genes. Serine substitution of a cysteine residue at the N-terminal region of the p50 NFκB subunit reduces DNA binding. Specific binding of NFκB is also inhibited reversibly by NO, and these effects appear to be mediated by nitrosation of the same cysteine residue. There is also good evidence that NO stabilizes the NFκB inhibitory protein IκBα and activates its transcription, thus inhibiting NFκB translocation from the cytoplasm to the nucleus. AP-1 is a dimeric complex of Jun–Jun or Jun–Fos proteins. The key to dimer formation resides in two cysteine residues in the leucine zipper and basic regions of Fos and Jun, and nitrosation of Cys^{252} in the DNA binding domain decreases AP-1 binding to DNA.

4.2.2. ENZYMES

Critical signaling proteins can be influenced by NO, functioning at the transcriptional and/or posttranscriptional level either as an activator (poly[ADP-ribose] synthetase, p21ras, sGC) or inhibitor (adenylyl cyclase type I, protein kinase C, cytochrome P450, nitric oxide synthases, lipoxygenase).

The heme protein cyclooxygenase (COX) is inhibited by NO at high concentrations but activated by low concentrations. The mitochondrial respiratory chain is susceptible. NO inhibits cytochrome c oxidase (complex IV) in a reaction that is reversible and competitive with oxygen, while $ONOO^-$ irreversibly inhibits respiratory complex I–III as glutathione levels decrease.

Nitric oxide can have diverse effects on cell death, initiating or protecting against apoptosis depending on the cell type, NO concentration, and redox environment. The family of protein-cleaving enzymes known as caspases are targets for NO, which inhibits their activity in a reversible manner.

4.2.3. CYTOKINES

Exposing cells to NO, coupled with the use of NOS inhibitors, has revealed complex effects on the production of cytokines. Some members of the caspase family participate in the maturation of cytokines. Precursors of interleukin-1β (IL-1β) and IL-18 are cleaved by caspase-1, and NO reversibly blocks this process, thereby acting as an antiinflammatory agent. However, NO can be proinflammatory, increasing release of other cytokines such as tumor necrosis factor-α (TNF-α) and IL-6.

Chemokines, small chemoattractant cytokines, can be quite specific in determining the immigration patterns of lymphocytes and macrophages to certain inflammatory regions. There appears to be an interesting crosstalk between NO and chemokines in that the expression of IL-8 and MCP-1 is inhibited by NO, whereas MCP-1 can prevent the expression of NOS-2.

4.2.4. MEMBRANE RECEPTORS

Excessive activation by glutamate of the *N*-methyl-D-aspartate (NMDA) receptor in mature neurons mediates cell death through calcium entry and downstream events that may or may not involve the activation of NOS-1. In addition, NO in its various redox states can down-regulate NMDA receptor activity at modulatory sites consisting of critical cysteine sulfhydryl or thiol groups.

The induction of MHC class II by interferon-γ (IFN-γ) is blocked by NO donors and NO inhibits the induction of adhesion molecules such as intercellular adhesion molecule (ICAM) and VCAM-1 via effects on NFκB.

4.2.5. STRUCTURAL PROTEINS

The nitration of tyrosine residues in proteins is a convenient marker of reactive nitrogen-centered oxidants being produced. Whether it is $ONOO^-$ or another oxidation product that is responsible for this nitration in vivo is the subject of debate. If these tyrosine residues are sites for regulation via PTK-mediated phosphorylation, then functional activity of the target protein can be affected.

Important structural proteins such as neurofilament L show evidence of nitration in diseases of the nervous system, such as amyotropic lateral sclerosis. Nitration converts a negatively charged hydrophilic residue and so dirupts assembly into polymers, so important for axonal integrity.

5. ROLES FOR NO

5.1. Nervous System

Survival, differentiation, and the removal of excess cells via apoptosis are essential for the proper development of the nervous system. A role for NO in these processes is suggested by the transient high expression of NOS found within specific brain and spinal cord sites during the maturation process. Interestingly, NO serves to activate pathways of programmed cell death as well as pathways that promote cell survival and differentiation. The balance between NO-induced death and protection rest on the quantity and the redox state of NO and the cellular targets of NO present at the site of production. Apoptotic pathways leading to cell death may be activated by damage to cellular structures, particularly the mitochondria, by excessive levels of NO or $ONOO^-$. Direct upregulation of the tumor suppressor p53 by NO may lead to activation of proapoptotic caspases and a cell death cascade. On the other hand, interaction of NO with NMDA receptors and NO-mediated regulation of intracellular calcium levels may serve as a protective mechanism against cell death. Cell survival and differentiation is enhanced through the NO activation of p21WAF1, a cyclin-dependent kinase inhibitor. In addition, cell maturation is promoted through a cGMP-independent NO activation of the protooncogene *p21ras*.

Further functions of NO in the developing nervous system include roles in the maturation of the retinotectal system and the formation of ocular dominance patterns within the visual cortex. Olfactory neurons may require NO to establish appropriate synapses and for the creation of odor memory. Neurite targeting and growth appear to be regulated by NO interactions with cytoskeletal proteins.

NO acts as a neurotransmitter for nonadrenergic noncholinergic (NANC) synapses found in peripheral tissues. Often released as a cotransmitter with neuropeptides, NO activates sGC with the subsequent production of cGMP. Elevations in cellular cGMP levels induce relaxation of smooth muscle. Activity of the NANC innervation of the respiratory tract is a component of ventilation/perfusion *(V/Q)* matching. NO generated by nitrergic innervation and the respiratory epithelium mediate the relaxation of the bronchiolar and vascular smooth muscle. In males, stimulation of NOS-containing neurons of the pelvic plexus results in the vasodilation of cavernosum vessels and penile erection. Similarly in the female, relaxation of uterine smooth muscle during pregnancy is modulated by NANC neurons. Gastrointestinal peristalsis is regulated in part by NO release from neurons located in the submucosa and myenteric plexus, which stimulates smooth muscle relaxation. Intestinal absorption and secretion of fluid and electrolytes may be mediated in a similar NANC–NO-dependent fashion.

Within the central and peripheral nervous systems,

the function of NO as a neurotransmitter is linked to glutamate receptor activity. Activation of glutamate receptors, specifically the NMDA receptors, results in NO production due to an increase of intracellular calcium levels and subsequent stimulation of NOS. As a permeable molecule, NO may act on enzymes, transcription factors, or structural proteins of the NO-producing cell or diffuse to act within neighboring cells. NO may also alter the neurotransmitter release in targeted neurons via the phosphorylation of synaptic vesicle proteins by cGMP-dependent kinase (PKG), a downstream consequence of sGC activity. The function of the NMDA receptor and calcium channel activity may be regulated in a similar fashion by NO-stimulated phosphorylation or nitrosylation of receptor regulatory sites. The neurotransmitter function of NO may play a role in tolerance to ethanol-induced motor deficits as well as the alcohol withdrawal syndrome. Nociception and hyperalgesia may be modulated by NO, as might tolerance to the antinociceptive actions of opioids. Autonomic nervous system activity is modulated by NO including the sinoatrial and atrioventricular nodes of the heart. Skeletal muscle and some associated neurons contain NOS. The strength of muscle contraction is modulated by NO effects on energy production in fast twitch fibers.

The neurotransmitter activity of NO may be involved in long-term potentiation (LTP), a model of learning and memory in the hippocampus. The process of LTP is initiated by the activation of postsynaptic NMDA receptors with the subsequent production of NO. NO produced by the postsynaptic neuron acts as a retrograde messenger activating sGC in the presynaptic neuron and effecting changes in the amount or ease of release of neurotransmitter during a future depolarization. This NO signaling from post- to presynpatic neuron results in an increase in the synaptic strength or efficiency. Glutamate, α-amino-3-hydroxy-5-methyl-4-isoxazole propionic acid (AMPA) receptor activation, and NO are components of long-term depression (LTD), a similar model of learning and memory in the cerebellum. In the model of LTD, however, NO acts as an anteriograde messenger and synaptic strength and efficiency are reduced.

NO plays a dual role in ischemic events in the brain. Neuron death that is a result of reduced blood flow and hypoxia initiates a chain reaction of excitotoxicity in neighboring cells. Cells damaged by hypoxia, trauma, or infection release excessive levels of the excitatory NMDA receptor ligand glutamate. The ensuing hyperactivation of the NMDA–NO signaling pathway initiates destructive cellular events within surrounding neurons. The increased release of NO may cause direct cellular damage or, more likely interact with O_2^- to form the very reactive $ONOO^-$. DNA damage caused by NO or $ONOO^-$ activates poly (ADP-ribose) synthetase (PARS). PARS is a DNA repair enzyme that utilizes NAD and through the repair process may exhaust cellular energy stores resulting cell death. Inhibition of oxidative phosphorylation through interactions of NO or $ONOO^-$ with the mitochondrial respiratory complexes may result not only in the loss of cellular energy production but also in enhanced free radical formation and further oxidative stress. The function of the glycolytic enzyme glyceraldehyde-3-phosphate dehydrogenase (G3PDH) may be reduced following nitrosylation by nitrogen reactive molecules. NO produced during excitotoxic events may induce loss of cellular energy production, increased formation of free radicals, or activation of the p53 apoptotic pathway resulting in neuronal death.

NOS-1 is considered the primary source of toxic NO in acute ischemic events. Delayed damage observed in ischemia may be due to cytokine induction of NOS-2 in glial cells that are observed at the infarct boundary. Counter to these damaging effects, NO produced by vascular endothelial NOS-3 is neuroprotective. Relaxation of vascular smooth muscle, stimulated by endothelial-derived NO, acts to limit the effects of ischemia by maintaining blood flow to the tissues.

Several degenerative brain diseases are considered to result from the death of critical neuronal populations due to the overabundance of free radicals or oxidative stress. The precise etiologies for the presence of increased free radicals and other highly reactive molecules believed to be at work in these disorders are not known. Increased generation of damaging oxidative species by the activation of enzymes such as xanthine oxidase and NOS may play crucial roles in the observed pathology. Equally damaging is the loss of protective antioxidant pathways such as glutathione peroxidase or superoxide dismutase.

Amyotrophic lateral sclerosis (ALS) is a progressive and fatal loss of motor neurons in the central nervous system. A mutation of superoxide dismutase-1 (SOD-1) is thought to play a major part in the etiology of this disease. The SOD-1 mutation may be a gain of function that utilizes $ONOO^-$ to produce a nitrogen reactive species that may damage vital proteins of the motor neurons resulting in cell death. The neurodegeneration of Alzheimer's disease (AD) results in progressive memory loss and dementia. Glutamate activation of NMDA receptor, subsequent NO production, and a resulting excitotoxicity may be mechanisms of neuronal death in AD. Huntington's disease may result from a similar excitotoxic mechanism. NO may play a role in the oxidative stress believed to result in the loss of nigrostriatal dopaminergic neurons which is characteristic of Parkinson's disease (PD). Evidence from experiments utilizing the dopaminergic specific toxin 1-methyl-4-phenyl-1,2,3,6-tetrahydropyridine (MPTP) as a model for PD supports a role for NO in the pathogenesis of PD. Multiple sclerosis (MS) is the progressive demyelination of central nervous system axons. The etiology of MS is not known. Oligodendrocytes may be damaged by an immune-mediated increase in NO and ONOO$^-$ production and therefore be unable to synthesize myelin. Human immunodeficiency virus (HIV) coat protein gp120 may precipitate the dementia that is associated with acquired immunodeficiency syndrome (AIDS). Excessive levels of NO generated by gp120-stimulated excitotoxicity and through a direct gp120 activation of NOS-2 may play a key role in the neuron loss noted in AIDS dementia.

5.2. Immune System

Bacterial endotoxins, exotoxins, and host cell derived cytokines induce the expression of NOS-2, resulting in the production of high levels of NO. When so stimulated, cells of the immune system (macrophages/monocytes, neutrophils, T lymphocytes, microglia) as well as nonimmune cells (astrocytes, neurons, cardiac myocytes, pancreatic cells) express NOS-2. Viruses, bacteria, fungi, and parasites are activators of NOS-2 as well as the targets of cytotoxic NO effects. Actions of NO and reactive nitrogen molecules damage nucleic acid and vital structures in the target organism, resulting in stasis or death. NO released during the immune response is nonspecific and may potentially interact with or damage host cells. The signs of inflammatory diseases such as asthma, arthritis, and colitis may be associated with excessive immune-derived NO production. Septic shock results from massive activation of NOS in response to multiple inflammatory mediators released from immune cells in response to bacterial invasion. Vasodilation, hypotension, and myocardial depression are a direct result of the high level of NO that is generated in septic shock. Elevated production of NO resulting from an activated immune system may play a role in carcinogenesis and metastasis as well as tumor cell killing. Diabetes mellitus may result from apoptosis of pancreatic cells stimulated by NO released from activated T cells and macrophages. Death of cells may result from NO activation of the p53 pathway or from NO induced cell expression of the proapoptotic Fas receptor.

5.3. Cardiovascular System

Normal blood pressure and maintenance of blood flow to body tissues requires a balance between vasoconstrictor and vasodilator substances. Vascular endothelial cells, by virtue of their production and release of multiple vasoactive molecules, are central to cardiovascular function. NOS-3 is abundant within the vascular endothelium and generates a basal level of NO that by counteracting vasoconstrictor influence serves to regulate vascular tone. Stimulation of the endothelial cell receptors by shear stress, or vasoactive substances such as ATP, 5-hydroxytryptamine, and bradykinin, activates NOS-3. A primary target for endothelial-derived NO is vascular smooth muscle sGC. The activation of the NO–sGC–cGMP signaling cascade results in vascular smooth muscle relaxation. Vessel wall architecture is regulated in part by the NO inhibitory effect on vascular smooth muscle proliferation and migration. Monocyte and leukocyte adhesion to vessel walls is modulated by an NO-induced reduction of endothelial cell expression of the adhesion molecules ICAM-1 and VCAM-1. Platelet adhesion and aggregation are blocked by an NO-mediated reduction in the expression of the glycoproteins IIb/IIIa and P-selectin. Vascular endothelial permeability may be altered by NO. Together with modulation of cardiac myocyte contractility, the effects of NO on the vasculature and cellular blood components are vital for the maintenance of the circulatory system.

Dysfunction of the vascular endothelial cell regulation of blood vessel tone is a feature of hypertension. Whether this functional disruption is a cause or a consequence of elevated blood pressure is not known. In either case, the vasoactive effects of endothelial derived NO are attenuated. One explanation for this phenomenon is a decrease of the endothelial production of NO. The drop in NO release may be due to a reduced availability of or a decreased NOS affinity for the substrate L-arginine. A similar effect may occur during a reduced availability of the NOS cofactor tetrahydrobiopterin. Stimulated endothelial cell production of NO may be limited owing to a reduction in the levels of bradykinin. Bradykinin is a vasodilator that activates endothelial cell receptors to stimulate NOS activity. In hypertension, levels of the bradykinin inactivator angiotensin converting enzyme (ACE) are increased. Therefore, the ACE metabolism of bradykinin blocks a stimulus for NO production. A second mechanism for the reduced NO effects observed in hypertension is the sequestering of available NO through its reaction with superoxide. The reaction between NO and O_2^- is highly efficient, which when O_2^- is in excess, may limit the amount of NO available to interact with vascular smooth muscle sGC. In addition, NOS may produce O_2^- if L-arginine is limited, further reducing the levels of free NO. Endothelial NOS-mediated vascular relaxation may not be able to overcome the effects of a possible excessive release of vasoconstrictors such as angiotensin II (AngII) and endothelin-1 (ET-1). AngII also sequesters available NO by stimulating the cellular production of O_2^-. Regardless of the exact cause, loss of NO actions allows for the vasoconstrictor substances to act unopposed, leading to smooth muscle contraction, leukocyte and platelet adhesion, and vessel wall remodeling. Loss of NO function in the vasculature, if not the initial cause, certainly aids in sustaining hypertension.

Hypertension is a risk factor for the development of atherosclerosis. Therefore, the mechanisms responsible for the loss of endothelial function discussed above are also likely to play a role in the vessel wall remodeling (plaques) that are the hallmark of atherosclerosis. In addition to the functional changes observed in hypertension, the vascular effects of NO are further altered by elevated levels of serum cholesterol and low-density lipoproteins (LDL) of atherosclerosis. NOS-3 protein expression in vascular endothelial cell levels is lowered owing to a cytokine-mediated reduction in the transcription of the *NOS-3* gene. Oxidized LDL is believed to be the stimulus for the release of the downregulatory cytokines from immune cells. Production of endothelial-derived NO may also be reduced by the actions of elevated blood cholesterol. NO production may be reduced via the cholesterol-stimulated increase of the endogenous NOS inhibitor asymmetric dimethylarginine (ADMA). The diversion of NO by reactions with superoxide, similar to that described for hypertension, reduce the level of NO available to act within normal cellular pathways. The G protein receptor signaling pathways of endothelial cells used by vasodilators substances to activate NOS may be disrupted by oxidized LDL. Through these mechanisms, NO-mediated endothelial cell function is inhibited. This decrease in function allows for the unregulated adhesion and infiltration of leukocytes into the vessel wall.

Adherent migrating macrophages engulf oxidized LDL to become foam cells, a common finding in atherosclerotic plaques. Vascular smooth muscle proliferation and migration is unchecked. NO reaction with O_2^-, generating reactive molecules such as $ONOO^-$, may damage or alter the normal function of vessel wall components. Cytokine expression by infiltrating immune cells may stimulate the expression of NOS-2 in smooth muscle cells with high levels of NO produced. Depending on the redox environment, the induced NO may serve to cause further vascular damage or protect and promote healing of the cells of the atherosclerotic plaque.

Restoration of endothelial function by the transfection of NOS cDNA into the vascular endothelial cells is a possible strategy for treatment of both hypertension and atherosclerosis. Interestingly, estrogen replacement may serve to preserve vascular function in postmenopausal women. Estrogen appears to enhance endothelial NO production. Local dysfunction of the endothelial NO component of vascular tone regulation may also play a role in the occurrence of angina or migraine.

6. REAL AND POTENTIAL THERAPEUTICS

Broadly, there are three strategies: introducing NO (or amplifying its downstream effects, in the case of Viagra), blocking NO production via NOS inhibitors, and scavenging NO oxidation products.

Inhaled NO is utilized as a therapeutic agent in pathological states in which pulmonary arterial hypertension is present. Indeed, the use of NO in patients with severe adult respiratory distress syndrome (ARDS) is now commonplace. Inhaled NO (5–80 ppm) results in improved hemodynamics and oxygenation status in a variety of pathological states. In addition, inhaled NO extends far beyond the pulmonary site and affects peripheral microvascular beds, making it a potential means of treating reperfusion injury associated with trauma. Major potential toxicities of inhaled NO, such as pulmonary edema, are related to the formation of NO_2 (a strong oxidizer). There is also the risk of methemoglobinemia, but this is rarely a problem as NO delivery is confined to 0.5–4%.

With its vasoprotective properties, local NO delivery is a promising approach for the prevention of intimal hyperplasia associated with vascular injury and atheroscleosis. In addition, adequate rates of NO production are essential for normal wound healing as it promotes angiogenesis and endothelial and epithelial cell proliferation/migration. From animal studies, viral transfer of the *NOS-2* gene appears to be an ideal strategy for restenosis and in the promotion of excisional wound healing.

Another potentially therapeutic application of NO lies in its ability to inhibit caspases and thus to block apoptosis. Programmed cell death contributes to dysfunction or failure in many organs, including the heart and liver, resulting in the need for radical therapies such as transplantation. The generation of molecules capable of delivering NO to a specific organ without causing widespread systemic effects is currently underway. On the other hand, chronic inflammation and infection are risk factors for cancer, situations in which the *NOS-2* gene is persistently expressed. The inactivation of tumor suppressor genes and the activation of oncogenes involve damage to DNA, and NO can deaminate purines and pyrimidines and also cause oxidative damage to DNA.

Most of the current NOS inhibitors have no appreciable isoform selectivity. Recently, however, several compounds with selectivity have been reported, such as L-N^6-(1-iminoethyl)lysine and *N*-(3-(aminomethyl)benzyl)acetamidine for NOS-2, and also pteridine antagonists. The main indications for NOS-1 inhibitors would be disease states involving brain ischemia, but there is a much wider spectrum for NOS-2 inhibitors (septic shock, inflammatory and infectious diseases, transplantation). While their development is at an early stage, scavengers of ONOO$^-$ (more appropriately, with the intermediates derived from the reaction of ONOO$^-$/CO$_2$) may prove useful in protecting tissues from the effects of reperfusion injury.

SELECTED READINGS

Brune B, von Knethen A, Sandau KB. Nitric oxide and its role in apoptosis. Eur J Pharmacol 1998; 351:261.

Cannon RO. Role of nitric oxide in cardiovascular disease. Clin Chem 1998; 44:1809.

Christopherson KS, Bredt DS. Nitric oxide in excitable tissues. J Clin Invest 1997; 100:2424.

Forstermann U, Boissel J-P, Kleinert H. Expressional control of the constitutive isoforms of nitric oxide synthase. FASEB J 1988; 12:773.

Harrison DG. Cellular and molecular mechanisms of endothelial cell dysfunction. J Clin Invest 1997; 100:2153.

Knight JA. Reactive oxygen species and the neurodegenerative disorders. Ann Clin Lab Sci 1997; 27:11.

Loihl AK, Murphy S. Expression of NOS-2 in glia associated with CNS pathology. In Mize R, Friedlander M, Dawson T, Dawson V (eds). Nitric Oxide and Other Diffusible Signals in Brain Development, Plasticity, and Disease. Amsterdam: Elsevier, 1998: 253.

MacMicking J, Xie Q-W, Nathan C. Nitric oxide and macrophage function. Annu Rev Immunol 1997; 15:323.

Mayer B, Hemmens B. Biosynthesis and action of NO in mammalian cells. Trends Biochem Sci 1997; 22:477.

Mayer B, Andrew P. Nitric oxide synthases: catalytic function and progress toward selective inhibition. Naunyn Schmiedebergs Arch Pharmacol 1998; 358:127.

Merrill JE, Murphy S. Regulation of gene expression in the nervous system by reactive oxygen and nitrogen species. Metab Brain Dis 1997; 12:97.

Michel T, Feron O. Nitric oxide synthases: which, where, how, and why? J Clin Invest 1997; 100:2146.

Nathan C. Inducible nitric oxide synthase. J Clin Invest 1997; 100:2417.

Squadrito GL, Pryor WA. Oxidative chemistry of nitric oxide. Free Rad Biol Med 1998; 25:392.

Stuehr DJ. Structure–function aspects of the nitric oxide synthases. Annu Rev Pharmacol Toxicol 1997; 37:339.

Wink DA, Mitchell JB. Chemical biology of nitric oxide. Free Rad Biol Med 1998; 25:434.

13 Phospholipases

J. H. Exton

CONTENTS

INTRODUCTION
PHOSPHOLIPASE A_2
PHOSPHOLIPASE A_1
PHOSPHOLIPASE C
PHOSPHOLIPASE D
SELECTED READINGS

1. INTRODUCTION

Phospholipids are essential components of all cell membranes, but are also hydrolyzed by phospholipases to remodel the membrane lipids or to yield molecules that are important regulators of cellular processes. As shown in Fig. 1, there are four types of phospholipases that hydrolyze phospholipids at specific bonds. Phospholipases A_1 and A_2 release fatty acids from the *sn*-1 and *sn*-2 positions of the glycerol backbone of phospholipids, yielding the corresponding lysophospholipids, whereas phospholipases C and D cleave the phosphodiester bond at different sites. Phospholipase C yields 1,2-diacylglycerol and the phosphorylated head group, whereas phospholipase D yields phosphatidic acid and the nonphosphorylated head group. Phospholipases occur in different isoforms, many of which are highly regulated by extracellular and intracellular signals. The regulated isoforms play major roles in the control of cell functions.

2. PHOSPHOLIPASE A_2

2.1. Low Molecular Mass Phospholipase A_2 Isoforms

Phospholipase A_2 (PLA$_2$) exists in many isoforms. In general, these may be grouped into those of low molecular mass (~14 kDa) and those of higher mass (~85 kDa). The small PLA$_2$ isozymes are secreted, and are found in insect and snake venoms and pancreatic juice, or are released into areas of inflammation, for example, synovial fluid. They all require millimolar Ca^{2+} concentrations for catalysis and have a high disulfide bond content. Some forms have been crystallized and they have a rigid structure with about 50% α-helical content. Catalysis is by a proton relay system involving a conserved His/Asp pair. Ca^{2+}

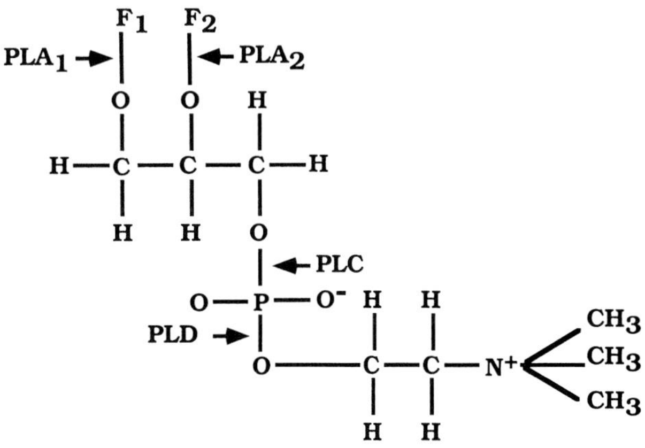

Fig. 1. Sites of hydrolysis of phosphatidylcholine. Indicated are the bonds hydrolyzed by phospholipase A_1 (PLA$_1$), phospholipase A_2 (PLA$_2$), phospholipase C (PLC), and phospholipase D (PLD).

From: *Principles of Molecular Regulation* (P. M. Conn and A. R. Means, eds.), © Humana Press Inc., Totowa, NJ.

bound to a conserved Ca^{2+}-binding loop stabilizes the transition state. The small, secreted PLA_2 isozymes have been classified into three groups. Group I and II isozymes contain highly conserved amino acids and have a closely similar exon–intron structure, implying they are members of an evolutionarily conserved family. Group III isozymes have rearranged sequence homologies, but their catalytic residues are identical to those in groups I and II. Another type (group V) of small PLA_2 has recently been described. The enzyme is present in mammalian tissues and is involved in eicosanoid formation in mast cells and macrophages. It is also secreted from macrophages and participates in signaling in these cells.

2.2. Cytosolic PLA_2 Isoforms

The large PLA_2 isozymes are cytosolic and are classified as group IV enzymes. They comprise a Ca^{2+}-dependent 85-kDa form and a Ca^{2+}-independent 80-kDa form. They show no sequence similarity to the low molecular mass PLA_2 isozymes, and utilize a mechanism of catalysis that is different from that employed by those isozymes. It involves Ser as a nucleophile and is similar to that utilized by many serine esterases and neutral lipases, which have a similar Gly-X-Ser-X Gly motif.

2.2.1. PROPERTIES OF Ca^{2+}-DEPENDENT CYTOSOLIC PLA_2

Ca^{2+}-dependent cytosolic PLA_2 ($cPLA_2$) is widely distributed in human tissues, but is not present in mature T and B lymphocytes. It shows little discrimination between phospholipids with different head groups or with different linkages (1-acyl or 1-O-alkyl) at the sn-1 position. Thus, its cellular substrates appear to be those in greatest abundance in cell membranes, that is, phosphatidylcholine (PC) > phosphatidylethanolamine (PE) > phosphatidylserine (PS) > phosphatidylinositol (PI). Unlike the other PLA_2 isoforms, $cPLA_2$ preferentially hydrolyzes arachidonic acid at the sn-2 position of its phospholipid substrates. This is important because this fatty acid is further metabolized to a variety of physiologically active eicosanoids. The enzyme contains a Gly-Leu-Ser^{228}-Gly-Ser motif, with Ser^{228} being essential for catalysis. $cPLA_2$ is stimulated by micromolar Ca^{2+}, which binds to a Ca^{2+}-lipid-binding (CaLB) domain in the N-terminus of the enzyme (Fig. 2). CaLB-like domains are present in many other proteins, including Ca^{2+}-dependent protein kinase C isozymes, synaptotagmin, phospholipase C, and rabphilin-3A. Unlike the small PLA_2 isoforms, Ca^{2+} is not involved in the catalytic mechanism of $cPLA_2$.

2.2.2. REGULATION OF $cPLA_2$

$cPLA_2$ is subject to regulation by diverse agonists, including growth factors, cytokines, agonists linked to heterotrimeric G proteins, and engaged integrins. In addition, the enzyme is activated by stressful stimuli, for example, UV light, and oxidative and shear stresses. The two major mechanisms by which it is activated are a rise in cytosolic Ca^{2+} and phosphorylation. A rise in Ca^{2+} induces membrane association of the enzyme through its CaLB domain, causing a large increase in activity. Thus many agents activate $cPLA_2$ by causing the release of internal Ca^{2+} or the influx of external Ca^{2+}. The resulting increase in cytosolic Ca^{2+} induces translocation of $cPLA_2$ from the cytosol to membranes, including the endoplasmic reticulum and nuclear membranes. Phosphorylation of $cPLA_2$ is observed when cells are stimulated by many agonists. The identities of the protein kinases involved have not been fully elucidated, but mitogen-activated protein (MAP) kinases (ERKs 1 and 2) play a role. $cPLA_2$ has a consensus site for ERKs (Pro-Leu-Ser^{505}-Pro) and these kinases phosphorylate and activate $cPLA_2$ in vitro. Furthermore, phosphorylation of $cPLA_2$ correlates with agonist activation of ERKs in many cells. The enzyme can also be phosphorylated by c-Jun N-terminal kinase and p38 MAP kinase, but it has not been demonstrated that these kinases are involved in agonist regulation of the enzyme in vivo. Phosphorylation of Ser^{505} does not induce full activation of $cPLA_2$ in vivo. This is because a concurrent increase in cytosolic Ca^{2+} and membrane translocation are required for a full effect.

Growth factors and several agonists that activate heterotrimeric G proteins activate MAP kinases through a cascade of intracellular signals (Fig. 3). Thus, ligand-induced dimerization and autophosphorylation of growth factor receptors generates phosphorylated Tyr residues in their cytoplasmic tails. Cytosolic adaptor proteins such as Shc and Grb2 become associated with these residues through their SH2 domains. Grb2, in turn, binds a guanine nucleotide exchange factor that activates the small G protein Ras. Activated Ras then initiates a cascade of protein kinases that leads to the activation of MAP kinases. The mechanism(s) by which heterotrimeric G proteins activate MAP kinases is less clear, but may involve protein kinase C (PKC) and the βγ-subunits of certain G proteins. Activation of cellular PKC increases

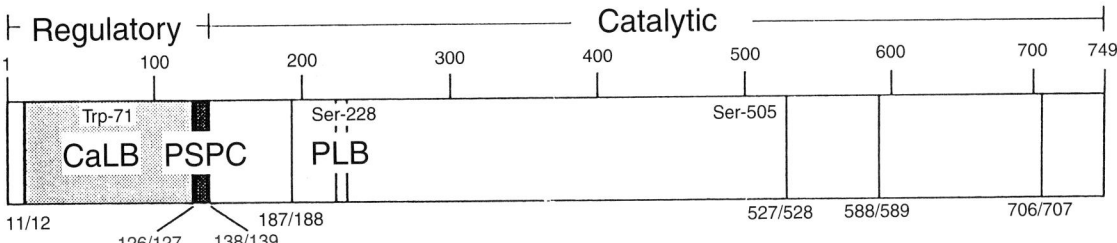

Fig. 2. Functional domains of cPLA$_2$. Shown are the Ca^{2+}-lipid-binding (CaLB), a domain identical to one in pulmonary surfactant protein C (PSPC), and residues similar to a lipase consensus sequence found in a bacterial phospholipase (PLB). This consensus sequence contains Ser^{228} which is essential for catalysis. Conserved Ser^{505}, which is the site of phosphorylation by MAP kinase, is also shown. (From Clark JD, Schievella AR, Nalefski EA, Lin L-L. Cytosolic phospholipase A$_2$. J Lipid Mediat 1995; 12:83).

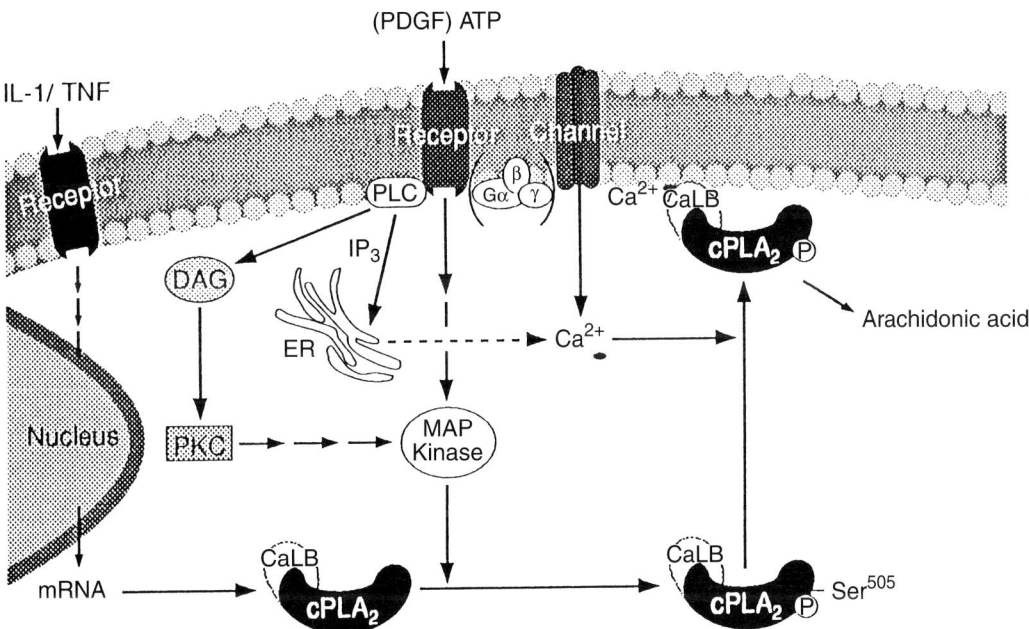

Fig. 3. Mechanisms of activation of cPLA$_2$. In the model, rapid activation of cPLA$_2$ is achieved by the synergistic actions of Ca^{2+} and phosphorylation by MAP kinase. Ligands, for example, ATP or platelet-derived growth factor (PDGF), activate receptors that lead to the activation of PLC. The mechanisms by which PLC is activated by these receptors are described in Sections 4.2–4. They involve not only G proteins (depicted) but also the tyrosine kinase activity of growth factor receptors. Hydrolysis of phosphatidylinositol 4,5-bisphosphate by PLC yields inositol 1,4,5-trisphosphate (IP$_3$), which releases Ca^{2+} from components of the endoplasmic reticulum (ER), and 1,2-diacylglycerol (DAG) which activates protein kinase C (PKC). PKC activates MAP kinase which phosphorylates and activates cPLA$_2$. Ca^{2+} ions released from the ER or entering through plasma membrane Ca^{2+} channels (*see* Section 4.6.) promote the membrane translocation of cPLA$_2$, which is required for full activation of the enzyme. Slower increases in cPLA$_2$ result from the actions of cytokines, for example, interleukin-1 (IL-1) or tumor necrosis factor (TNF), that act at both transcriptional and posttranslational levels. (From Clark JD, Schievella AR, Nalefski EA, Lin L-L. Cytosolic phospholipase A$_2$. J Lipid Mediat 1995; 12:83.)

cPLA$_2$ activity in some cells. The effect is not direct because the phosphorylation of cPLA$_2$ by PKC in vitro does not lead to activation, and there is no evidence that PKC phosphorylates the enzyme in vivo. PKC probably acts by initiating a kinase cascade leading to MAP kinase activation. Other protein kinases may be involved in the regulation of cPLA$_2$ and may act on phosphorylation sites other than Ser^{505}.

2.2.3. Lysophospholipase Activity of cPLA$_2$

cPLA$_2$ exhibits hydrolytic activity toward 1-palmitoyl-2-lyso PC in vitro that is comparable with that with 1-palmitoyl 2-arachidonoyl-PC as substrate. This lysophospholipase (LPLA) activity of cPLA$_2$ is also increased by phosphorylation of the enzyme by MAP kinase. Mutagenesis studies indicate that the same catalytic site (Ser^{228}) is involved in both the PLA$_2$ and LPLA activities. However, it is unclear that cPLA$_2$ exhibits significant LPLA activity in vivo. As noted below, Ca^{2+}-independent cytosolic PLA$_2$ also exhibits LPLA activity. In addition, several low molecular mass LPLAs have been purified and cloned from various cells and tissues. The cloned forms contain the Gly-X-Ser-X-Gly motif found in the cytosolic PLA$_2$ and other serine esterases.

2.2.4. Transcriptional Regulation of cPLA$_2$

The cytokines interleukin-1 (IL-1) and tumor necrosis factor (TNF) activate cPLA$_2$ posttranslationally and also by increasing the synthesis of the enzyme at the level of transcription in several cell types (Fig. 3). The induction of cPLA$_2$ and the concomitant increase in prostaglandin E$_1$ are inhibited by glucocorticoids. Epidermal growth factor (EGF) can also increase cPLA$_2$ mRNA in mesangial cells. The promoter region of the cPLA$_2$ gene contains several inducible elements (binding sites for AP-1 and NFκ(B), which would explain the transcriptional control.

2.2.5. Physiological Role of cPLA$_2$

The cellular function of cPLA$_2$ is to generate arachidonic acid. This serves as a precursor for various eicosanoids (prostaglandins, leukotrienes, thromboxanes, lipoxins). In addition, it may act as an intracellular messenger *per se,* for example, through effects on certain PKC isozymes. Eicosanoids are involved in the regulation of many cellular processes and physiological functions, including platelet aggregation, blood vessel contractility, cell proliferation, uterine contraction, gastrointestinal motility, lymphocyte function, hormone release, and bronchial contraction/dilation. They also play a role in mediating certain types of cell injury, for example, that induced by UV irradiation and reactive O$_2$ species.

2.2.6. Ca^{2+}-Independent Cytosolic PLA$_2$

Ca^{2+}-independent cytosolic PLA$_2$ (iPLA$_2$) is present in mammalian cells. It shows no homology to cPLA$_2$, but possess a putative Gly-X-Ser-X-Gly active site. As expected, it lacks a CaLB domain, but has eight ankyrin repeat sequences. Unlike cPLA$_2$, it does not show preference for arachidonic acid at the *sn*-2 position. It is more active toward PC, PE, and phosphatidic acid (PA) than PI, and exhibits LPLA activity. It is involved in phospholipid remodeling and thus contributes to the regulation of the arachidonic acid level in cells. However, it is uncertain if it participates in agonist-induced arachidonic acid and eicosanoid release.

3. PHOSPHOLIPASE A$_1$

Phospholipase A$_1$ (PLA$_1$) has been little studied even though its activity is higher than that of PLA$_2$ in certain tissues. It appears to be involved principally in the remodeling of membrane phospholipids. Thus phospholipids that are synthesized *de novo* in the endoplasmic reticulum and transported by phospholipid exchange proteins to other cell membranes undergo changes in fatty acid composition. This requires the release of their fatty acids by PLA$_1$ and PLA$_2$ followed by their reacylation by acyl-CoA: lysophospholipid acyltransferases. There is presently no evidence that PLA$_1$ is regulated by extracellular or intracellular signals.

A PA-preferring PLA$_1$ has been cloned from testis. It has a calculated molecular mass of 98 kDa and has a Ser-X-Ser-X-Gly putative catalytic site motif. Its mRNA is enriched in mature testis and present in high content in brain, spleen, and lung. The enzyme migrates as a 110-kDa protein on gel electrophoresis and thus resembles PLA$_1$ purified from brain. The combined action of PLA$_1$ and the LPLA activity of PLA$_2$ results in the release of significant amounts of arachidonic acid from phospholipids in vitro. It is possible that these two activities contribute significantly to arachidonic acid release in vivo.

4. PHOSPHOLIPASE C

4.1. General Properties of PLC Isozymes

The substrates of phospholipase C (PLC) in animal and plant tissues are PI and its phosphorylated derivatives, phosphatidylinositol 4-phosphate (PIP) and phosphatidylinositol 4,5-bisphosphate (PIP$_2$). PLC enzymes that act on other phospholipids are present in bacteria. A PLC activity toward PC has been described in mammalian tissues, but the enzyme has not been purified or characterized.

Mammalian PLC occurs in multiple isozymic forms that are grouped into three families (β, γ, δ) that are regulated differently (Fig. 4). They all require Ca^{2+} for activity and prefer PIP and PIP$_2$ over PI as substrate. They exhibit two conserved domains (X and Y) that are required for catalysis. They also have a pleckstrin homology (PH) domain in their N-terminus

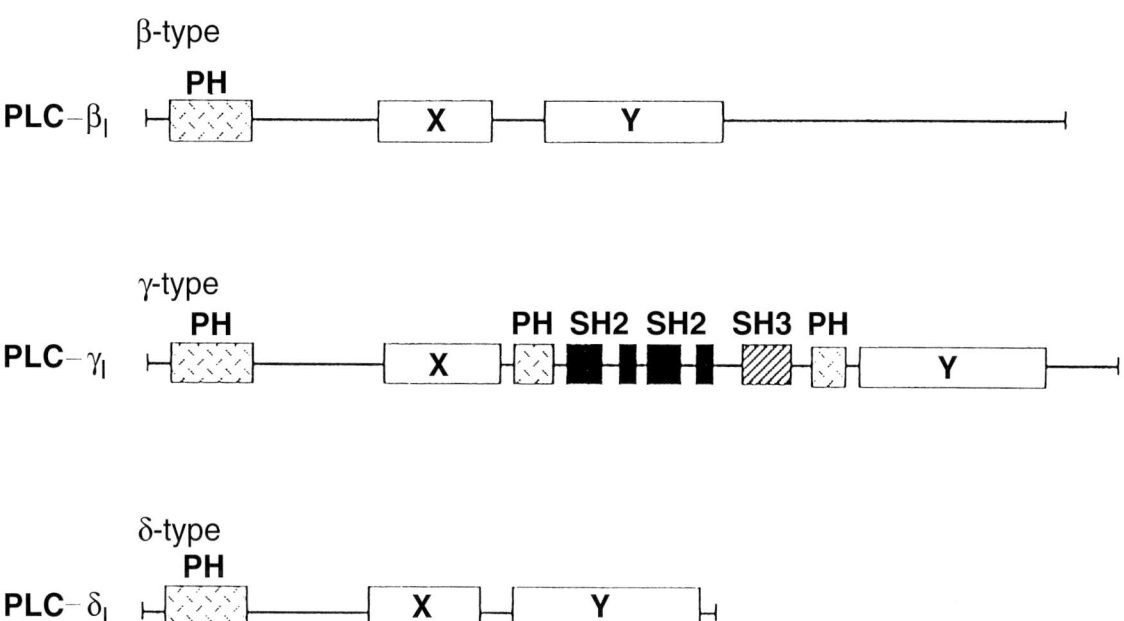

Fig. 4. Schematic structures of the β, γ, and δ isozymes of PLC. The structures show the pleckstrin homology (PH) domains and the conserved X and Y sequences. The presence of Src homology domains (SH2, SH3) in the γ isozyme is shown. (From Exton JH. Cell signalling through guanine nucleotide-binding regulatory proteins (G proteins) and phospholipases. Eur J Biochem 1997; 243:10.)

that is adjacent to four EF-hands, and their C-terminal half contains a C2 domain. PLCδ has been crystallized and the catalytic domain is in the form of an α/β or TIM barrel (first recognized in triose phosphate isomerase). The active site is located in the cleft of the α/β barrel and contains a coordinated Ca^{2+} that is required for PIP_2 hydrolysis. The hydrolysis of substrate proceeds in two sequential reactions: cleavage of inositol phospholipids to diacylglycerol and inositol 1,2-cyclic phosphates, followed by conversion of the cyclic phosphates into acyclic phosphates. The proposed general mechanism is general base/general acid catalysis. Mutagenesis studies have identified the residues involved in catalysis, stabilization of the reaction intermediate, and in the coordination of Ca^{2+}.

The PH domain of the PLC isozymes binds PIP_2 with high affinity, while the C2 domain binds Ca^{2+}, as expected. However, deletion of the Ca^{2+} binding site in C2 does not affect enzyme activity, and the putative Ca^{2+} binding sites in the EF hands also appear to be nonfunctional. This indicates that the binding site for Ca^{2+} at the catalytic site is entirely responsible for the Ca^{2+} dependence of the activity of the enzyme. Membrane binding of PLC depends on the PH domain, and it has been proposed that this tethers the enzyme to the membrane through PIP_2 binding. The role of the C2 domain in membrane binding and catalysis is unclear. It has been proposed that it fixes the catalytic domain so that it can penetrate the membrane, allowing PIP_2 hydrolysis to proceed productively.

4.2. Regulation of PLCβ Isozymes by G_q Family of G Proteins

PLCβ isozymes comprise $β_1$–$β_4$ subtypes and differ from the other PLC forms by virtue of their regulation by agonists that activate certain heterotrimeric G proteins. The first G proteins that were recognized to control PLCβ isozymes were members of the G_q family (G_q, G_{11}, G_{14}, G_{15}, and G_{16}) (Fig. 5). G_q and G_{11} are distributed widely, whereas G_{14} is present predominantly in kidney, and G_{15} and G_{16} are restricted to cells of hematopoietic lineage. Agonists that activate $G_{q/11}$ include epinephrine/norepinephrine (via $α_1$-adrenergic receptors), acetylcholine (via M1, M3, and M5 muscarinic receptors), vasopressin (via V1 receptors), bradykinin, histamine (via H1 receptors), thromboxane A_2, thyrotropin-releasing hormone and gonadotropin-releasing hormone, substance P, neurokinins, oxytocin, 5-hydroxytryptamine (via 5-HT_2 receptors), bombesin, cholecystokinin, endothelin 1, lysophosphatidic acid, glutamate (via metabotropic, mGluR, and mGluRs receptors) and purinergic agonists (via P2Y and P2U receptors). This list indicates that G_q family and PLCβ isozymes are importantly

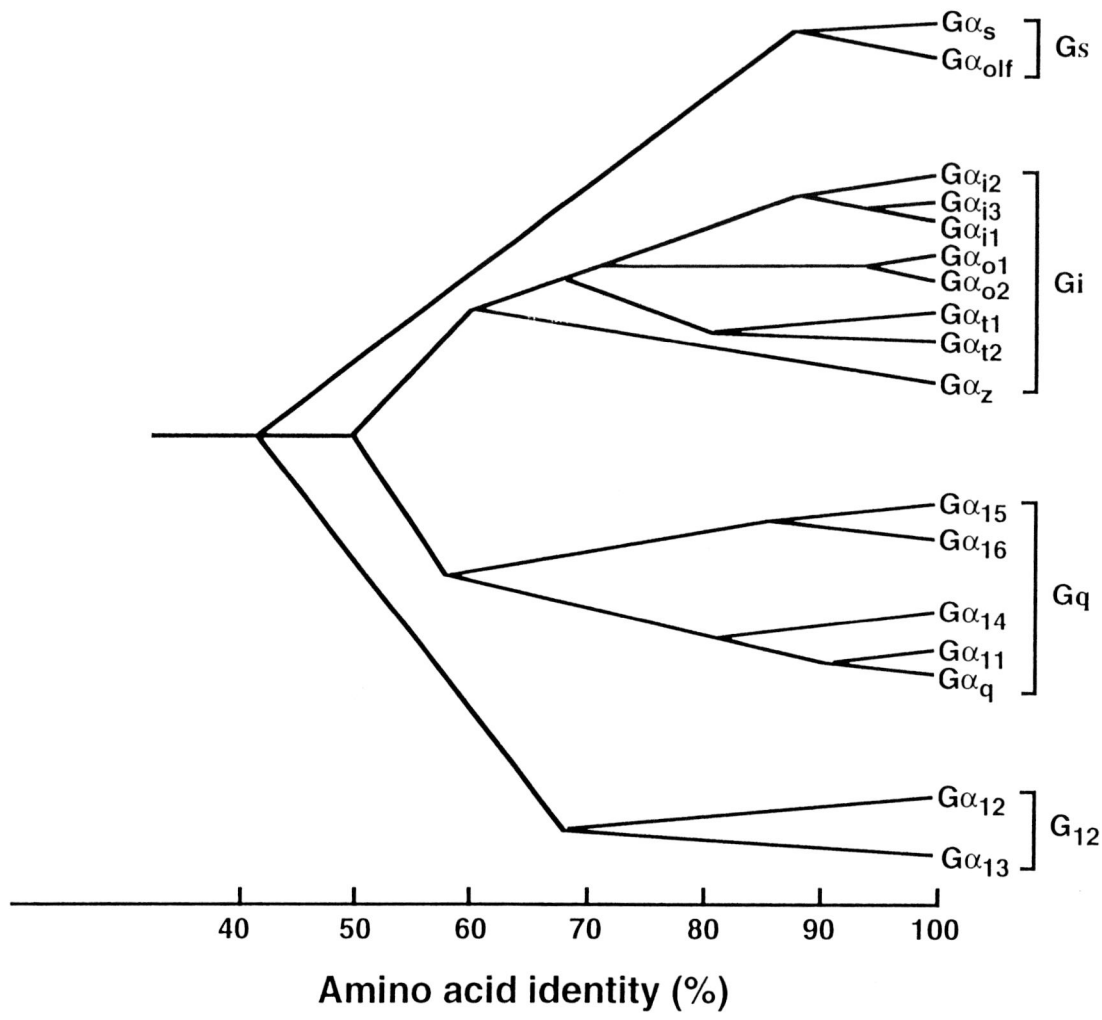

Fig. 5. Heterotrimeric G protein families. The four families of G proteins displayed in terms of the amino acid identities of their α-subunits. The G_i family is also known as the $G_{i/o}$ family. (From Simon MI, Strathman P, Gautam N. Diversity of G proteins in signal transduction. Science 1991; 252:802.)

involved in the actions of a large and diverse group of agonists.

The receptors for these agonists are located in the plasma membrane and have the seven transmembrane-spanning α-helices that are characteristic of receptors linked to heterotrimeric G proteins (Fig. 6). The receptors have extracellular N-termini of varying lengths and three extracellular loops. There are also three intracellular loops and a C-terminal tail of varying length. Agonists bind either to residues located in the transmembrane helices or in the extracellular domains. G proteins interact with the intracellular loops and the C-terminus, with sequences located in the third loop playing a major role.

Activation of the receptors by agonist binding is presumed to cause a conformational change(s) in their cytoplasmic domain(s) that promotes interaction with and activation of the G proteins. In unstimulated cells, G proteins exist as heterotrimers of α-, β-, and γ-subunits that are inactive (Fig. 7). Receptor activation involves the release of GDP from the guanine nucleotide binding site of the α-subunit and its replacement by GTP. This occurs because interaction with the receptor causes a change in the relative affinity of the α-subunit for the two nucleotides and because GTP is present at a higher concentration than GDP. The GTP-bound α-subunit then dissociates from the inhibitory complex, and the free α- and βγ-subunits interact with their effector molecules. In the case of the α-subunits of the G_q family, the effectors are PLCβ isozymes.

$G\alpha_q$ acts predominantly on the $β_1$ and $β_3$ isozymes of PLC, which are widely distributed in mammalian tissues. It also acts on the $β_4$ isozyme, but this is

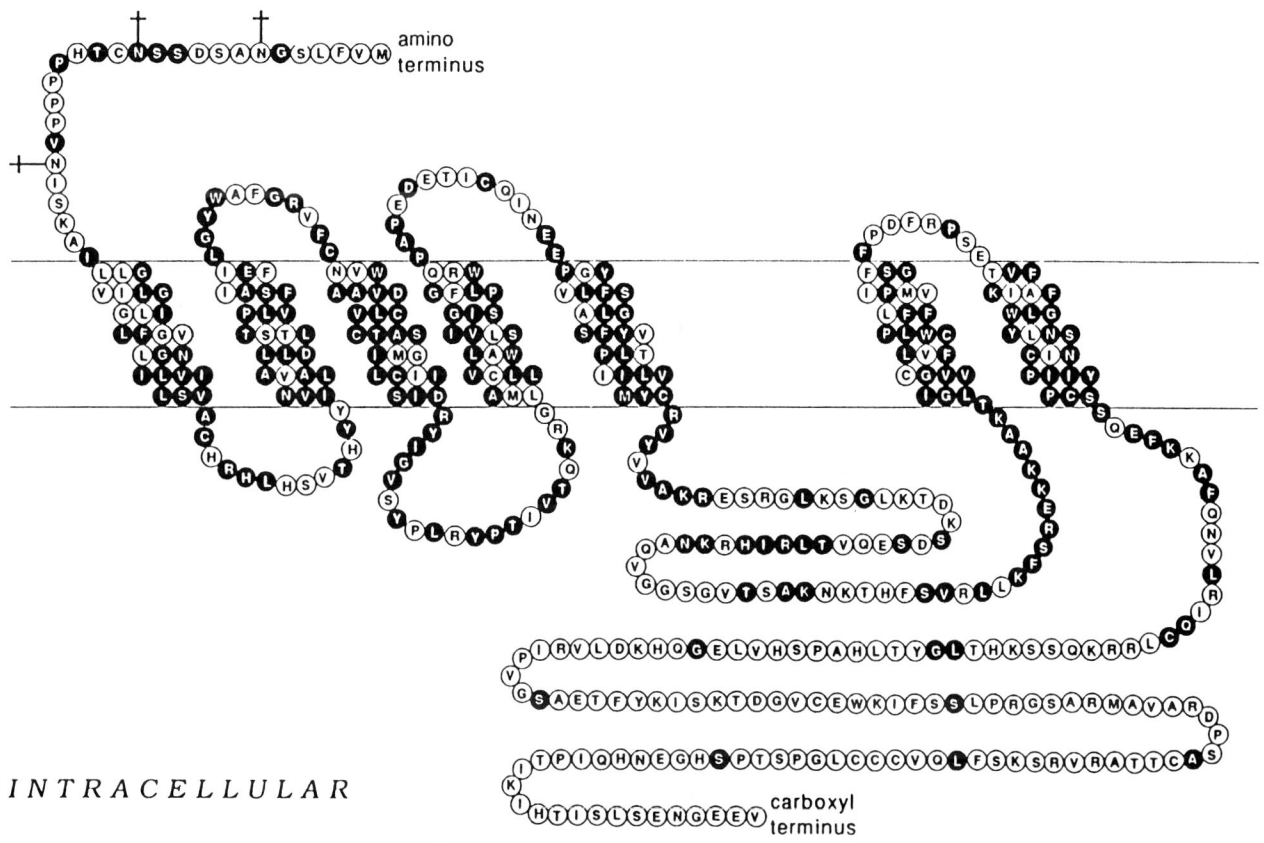

Fig. 6. Transmembrane-spanning model of the α_{iA}-adrenergic receptor. The structure is typical of receptors coupled to G proteins. Residues common to the α_{iB}-adrenergic receptor are marked in *black*. Sites of glycosylation in the extracellular N-terminus are indicated by crosses. (From Schwinn DA, Lomasney JW, Lorenz W, Szklut PJ, Fremeau RT Jr, Yang-Feng TL, Caron MG, Lefkowitz RJ, Cotecchia S. Molecular cloning and expression of the cDNA for a novel α-adrenergic receptor subtype. J Biol Chem 1990; 265:8183.)

confined to the retina. It causes little stimulation of the β_2 isozyme, which is found principally in hematopoietic cells. Its interaction domain on PLCβ$_1$ is located in the C-terminus. The domain on Gα$_q$ that interacts with PLCβ$_1$ is mainly located in a surface α-helix near a "switch" region that undergoes a marked conformational change when the α-subunit is activated.

Termination of Gα$_q$ activation of PLCβ isozymes occurs by removal of agonist and, in some cases, by desensitization of receptors. In addition, certain regulators of G protein signaling (RGS) proteins inhibit the activation of the PLC isozymes by stimulating the GTPase activity of Gα$_q$ and perhaps by competing with the isozymes for binding to the activating domain(s) on Gα$_q$. Another mechanism for termination of Gα$_q$ activation of PLCβ isozymes involves the stimulation of the GTPase activity of Gα$_q$ by these isozymes. The GTPase activating domain of PLCβ$_1$ is located in its C-terminus, that is, in the region required for Gα$_q$ activation of this isozyme.

4.3. Regulation of PLCβ Isozymes by the G_i/G_o Family of G Proteins

Other G proteins regulate PLCβ isozymes by another mechanism. These G proteins are certain members of the G_i/G_o family and the activation involves their G$_{\beta\gamma}$ subunits. Agonists that activate G_i/G_o include epinephrine/norephinephrine (via α_2-adrenergic receptors), acetylcholine (via M2 and M4 receptors), angiotensin II (via AT$_1$ receptors), adenosine (via A$_1$ and A$_3$ receptors), dopamine (via D$_2$ receptors), f-Met-Leu-Phe, 5-HT (via 5-MT$_1$ receptors), and somatostatin (via SST$_3$ and SST$_5$ receptors). However, the G_i/G_o proteins also regulate adenylyl cyclase and ion channels, and the activation of PLCβ isozymes and consequent physiological responses (see below) may not be prominent effects of these

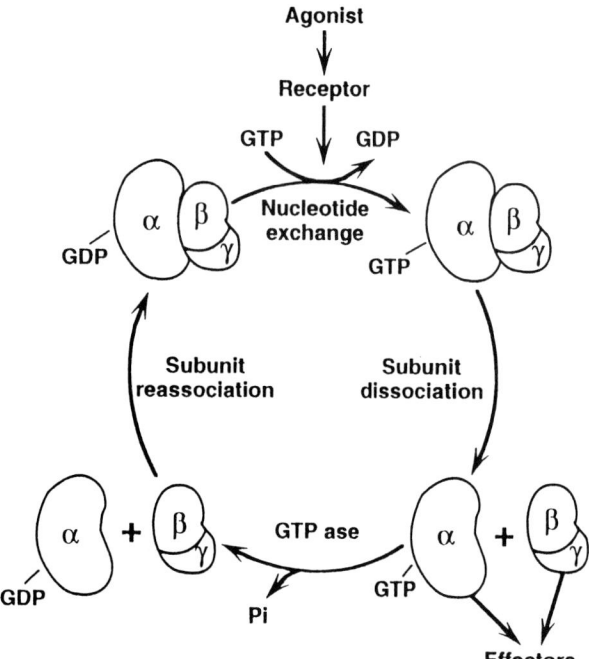

Fig. 7. Scheme of G protein-mediated signal transduction. In the basal state, G proteins exist as a heterotrimeric complex of α-, β- and γ-subunits with GDP bound to the α-subunit. Agonist binding to G protein-linked receptors induces receptor-mediated GTP–GDP exchange on the α-subunit. Binding of GTP activates the α-subunit and causes it to dissociate from the βγ complex. Both moieties are then free to interact with various effectors. Activation of the intrinsic GTPase of the α-subunit by various mechanisms results in the generation of the GDP-bound α-subunit and reassociation of the subunits to form the heterotrimeric complex. Continued activation of the receptors leads to further cycles of activation of G proteins and effectors. (From Exton JH. Phospholipid-derived second messengers. In Conn PM, Goodman HM (eds). Handbook of Physiology, Section 7. The Endocrine System, Vol. 1. Cellular Endocrinology. New York: Oxford University Press, 1998; 255.)

agonists. The absence of significant stimulation of PLCβ isozymes in some situations may be due to the requirement of high concentrations of $G_{\beta\gamma}$ subunits to elicit activation of these izymes.

G protein βγ-subunits act predominantly on the β_2 and β_3 isozymes of PLC. They have a relatively weak effect on the β_1 isozyme and do not affect the β_4 isozyme. The βγ complex apparently interacts with the β isozymes at an α-helical region on the surface of the catalytic domain. No crystal structures are available for the β-isozymes and the mechanisms by which $G\alpha_q$ and $G_{\beta\gamma}$ activate these enzymes are unknown. On the other hand, the crystal structures of several G proteins have been determined for the subunit and heterotrimeric forms. These studies indicate that the α-subunit binds to the flat top surface of the toroidal β-subunit, which is a symmetrical seven-bladed propellor structure, and also extends an α-helix along the side of blade 1 of the β-subunit. Mutational studies further indicate that the $PLC\beta_2$ and β_3 isozymes make many contacts on the top surface of the β-subunit, including specific residues in the outer strands of blades 2, 6, and 7 of the propellor. This model helps explain why the free βγ complex, but not the intact heterotrimeric G protein, is able to activate its effectors.

4.4. Regulation of PLCγ Isozymes by Tyrosine Kinases

PLCγ-isozymes comprise γ_1 and γ_2 subtypes and differ from the other PLCs by the presence of Src homology (SH2 and SH3) domains located between the X and Y regions (Fig. 4). There are two SH2 domains that target the enzyme to Tyr-phosphorylated sequences in activated growth factor receptors. The single SH3 domain is involved in binding of the enzyme to Pro-rich sequences in cytoskeletal components.

$PLC\gamma_1$ is widely distributed, whereas $PLC\gamma_2$ is expressed mainly in hematopoietic cells. Ligand binding to the receptors for epidermal growth factor, platelet-derived growth factor, fibroblast growth factor, and nerve growth factor causes dimerization of the receptors and activation of their intrinsic tyrosine kinase activity (Fig. 8). This leads to phosphorylation of the receptors and of certain cellular proteins. The phosphorylation of certain Tyr residues in the receptors creates high-affinity binding sites for cytosolic proteins containing SH2 domains. These proteins include $PLC\gamma_1$; phosphatidylinositol 3-kinase (PI 3-kinase); Ras GTPase-activating protein; protein tyrosine phosphatase 1D; and adaptor proteins such as Shc, Nck, and Grb2 (Fig. 8). Each protein binds to specific sites of autophosphorylation on the cytoplasmic tails of the growth factor receptors, with the exception of the epidermal growth factor receptor, which displays minimal site specificity. Association of $PLC\gamma_1$ with the receptors is accompanied by phosphorylation of three specific sites in the phospholipase. Phosphorylation of these sites affects the binding or activation of the enzyme.

Nonreceptor protein tyrosine kinases also phosphorylate and activate PLCγ isozymes. Thus, activation of the T-cell antigen receptor, the membrane-associated immunoglobulin M (IgM) receptor, the high-affinity IgE receptor (FcεRI), the IgG receptors (FcγRs), and the receptors for several cytokines results in the acti-

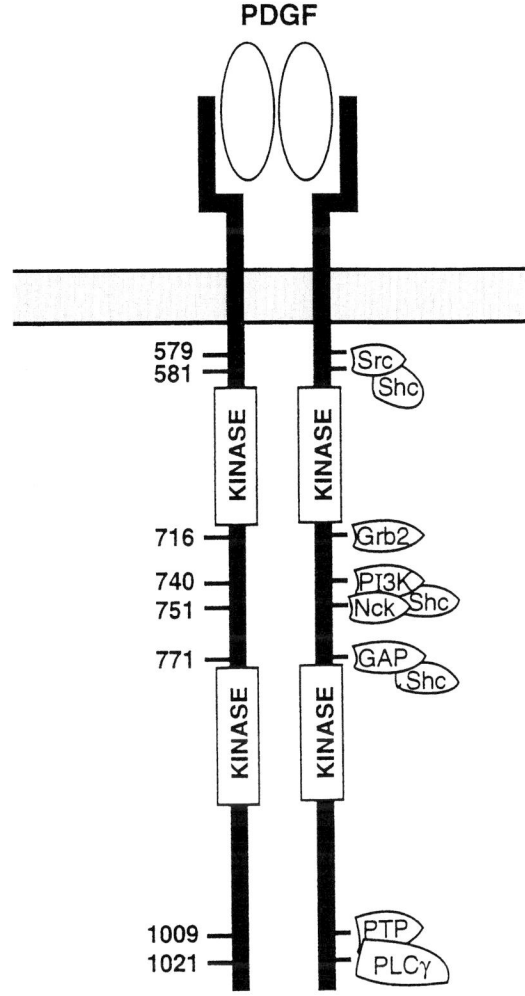

Fig. 8. Structure of the β-receptor for platelet-derived growth factor (PDGF). The figure presents schematically the binding domain for PDGF, the tyrosine kinase domains, and the numbered Tyr residues which become phosphorylated upon activation of the receptor. Cytosolic proteins that interact with specific sites of tyrosine phosphorylation through their Src-homology (SH2) domains include members of the Src family of protein tyrosine kinases; the adaptor proteins Grb2, Shc, and Nck; the regulatory subunit of PI-3 kinase (PI3K); the GTPase-activating protein of Ras (GAP); a protein tyrosine phosphatase (PTP) designated PTP1D or Syp; and the γ-isozymes of PLC (PLCγ). (From Exton JH. Phospholipid-derived second messengers. In Conn PM, Goodman HM (eds). Handbook of Physiology, Section 7. The Endocrine System, Vol. 1. Cellular Endocrinology. New York: Oxford University Press, 1998; 255.)

vation of cytosolic protein tyrosine kinases that act on PLCγ isozymes. These receptors have no intrinsic tyrosine kinase activity. The soluble tyrosine kinases that are involved are members of the Src, Syk, and Jak families.

Recent work has indicated that PLCγ can also be activated by phosphatidylinositol 3,4,5-trisphosphate (PIP_3), the product of PI 3-kinase in mammalian cells. Thus inhibition of PI 3-kinase reduces the activation of PLCγ by platelet-derived growth factor in NIH 3T3 fibroblasts and IgG receptors in platelets. The role of PIP_3 appears to be to promote the membrane binding of PLCγ, thus enhancing accessibility to its PIP_2 substrate.

4.5. Properties of PLCδ Isozymes

PLCδ comprises four isozymes ($δ_1$–$δ_4$) that differ from the β- and γ-isozymes in having a truncated C-terminal tail and hence a lower molecular mass (85 kDa). The PLCδ isozymes are not activated by heterotrimeric G proteins or by protein tyrosine kinases, but respond to changes in Ca^{2+} in the cytosolic range. $PLCδ_1$ has been crystallized, and much of our knowledge of the structure and catalytic mechanisms of the PLC isozymes derives from study of this isozyme. This information has been summarized in Section 4.1.

4.6. Functions of PLC Isozymes—IP_3 and Ca^{2+} Mobilization

The physiological substrate of all mammalian PLC isozymes is PIP_2 and its products are two signaling molecules—inositol 3,4,5-trisphosphate (IP_3) and 1,2-diacylglycerol (DAG). The function of IP_3 is to interact with specific receptors located on the cytosolic surface of certain components of the endoplasmic reticulum (ER). Activation of these receptors causes the release of Ca^{2+} from the ER and hence a rise in cytosolic Ca^{2+} (Fig. 9). The IP_3 receptors have a transmembrane domain near the C-terminus and a long N-terminal cytoplasmic region. The N-terminal segment contains the IP_3 binding domain, and the transmembrane domain has eight membrane-spanning segments that are involved in constituting a Ca^{2+} channel. IP_3 receptors are structurally and functionally similar to ryanodine receptors, and both exist as tetramers. The four C-terminal regions of the receptors cooperate to form Ca^{2+} channels that are activated by either IP_3 or ryanodine and caffeine.

Ca^{2+} released via IP_3 action promotes the further release of Ca^{2+} from adjacent ER stores through the process of Ca^{2+}-induced Ca^{2+} release. This results in the propagation of a Ca^{2+} wave in the cell. The decrease in the Ca^{2+} content of the ER stores induces the influx of Ca^{2+} through channels (CRAC) in the plasma membrane that are homologs of the *Drosophila* Trp channels (Fig. 9). The Ca^{2+} influx through

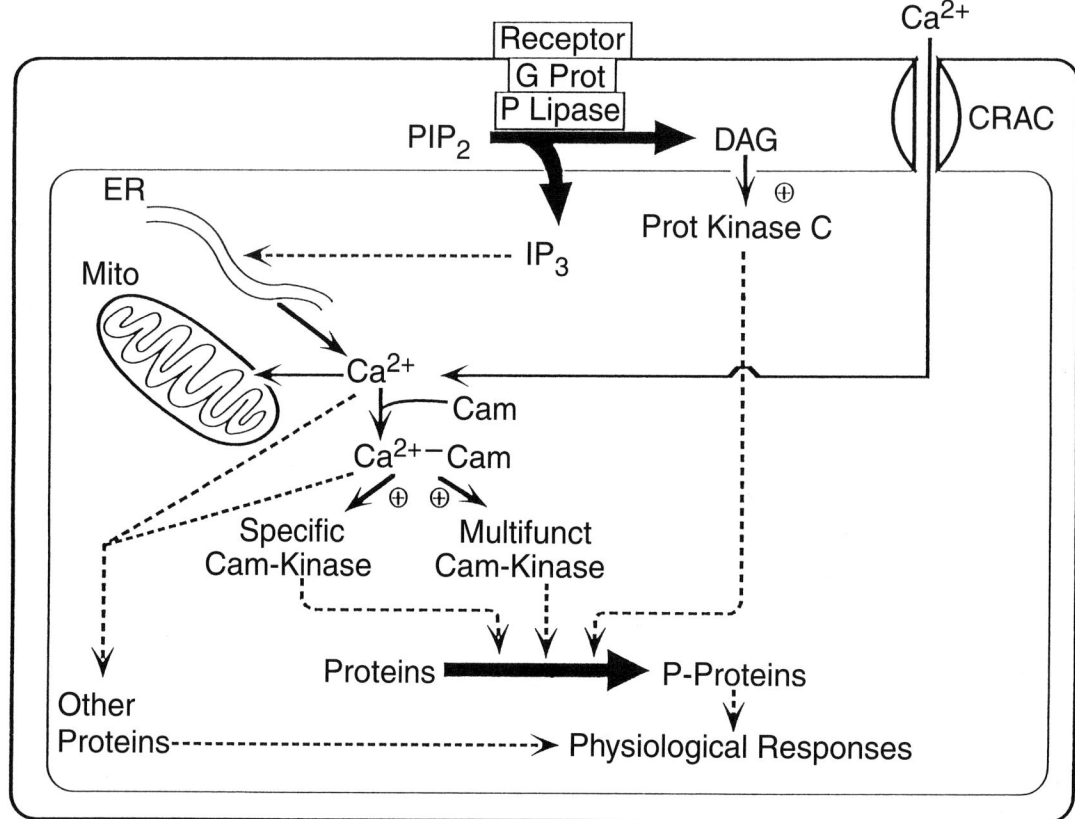

Fig. 9. Mechanisms involved in physiological responses generated by PLC activation. The figure shows the mechanisms utilized by receptors linked to G_q or $G_{i/o}$ G proteins, but similar mechanisms operate for growth factors and cytokines that activate PLC by protein tyrosine phosphorylation-dependent pathways. Ca^{2+} ions released by the action of IP_3 on components of the endoplasmic reticulum (ER) and entering through Ca^{2+}-release activated plasma membrane channels (CRAC) bind to calmodulin (Cam) and other Ca^{2+} binding proteins. The Ca^{2+}–Cam complexes act on a variety of cellular proteins including specific and multifunctional Ca^{2+}–Cam-dependent protein kinases (Cam-Kinases). DAG formation initiates a separate mechanism involving activation of protein kinase C isozymes. The activation of both types of kinases leads to the generation of physiological responses through the phosphorylation of cellular proteins as described in the text.

the channels is termed capacitative Ca^{2+} entry and is responsible for the sustained phase of cytosolic Ca^{2+} increase seen in many stimulated cells. The mechanism by which ER Ca^{2+} depletion activates Ca^{2+} entry remains unclear.

The rise in cytosolic Ca^{2+} due to its internal release and influx from the extracellular space leads to increased binding of Ca^{2+} to binding proteins such as calmodulin. The Ca^{2+}–calmodulin complex interacts with a variety of cellular proteins, altering their activities. Important targets of Ca^{2+}–calmodulin are the Ca^{2+}-ATPase that pumps Ca^{2+} out of the cell, and specific and multifunctional Ca^{2+}–calmodulin-dependent protein kinases. These Ser/Thr kinases act on myosin light chains to promote contraction of smooth muscle, phosphorylase kinase to cause glycogen breakdown, and tyrosine hydroxylase to increase catecholamine synthesis. Ca^{2+}–calmodulin-dependent protein kinases are also important in neuronal function in other ways, for example, long-term potentiation, which underlies some forms of memory.

4.7. Functions of PLC Isozymes—DAG Production and Activation of PKC

The other product of PIP_2 hydrolysis, DAG, is a major mediator of PKC activation. PKC occurs in several forms that are grouped into conventional or Ca^{2+}-dependent isozymes (α, β_I, β_{II}, γ), novel or Ca^{2+}-independent isozymes (δ, ϵ, θ), and atypical or DAG-independent isozymes (ι, γ, and ξ). Another isozyme (μ or D) has an unusual structure and does not belong in the other groups.

Schematic structures of the PKC isozymes are depicted in Fig. 10. The conventional forms have an autoinhibitory (pseudosubstrate) domain at the N-terminus followed by two regulatory membrane-

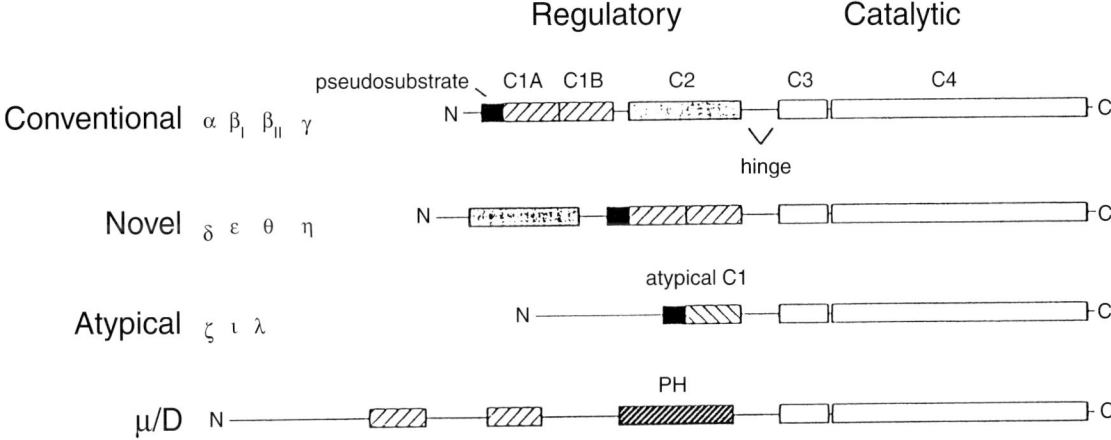

Fig. 10. Linear domain composition of PKC isozymes. C1 domains *(cross-hatched)* bind DAG or phorbol esters; C2 domains *(light gray)* bind anionic lipids and, in the case of conventional isozymes, Ca^{2+}; PH domains *(dark-hatched)* bind phosphoinositides; C3 domains bind ATP; C4 domains bind substrate; pseudosubstrate domains *(black box)* sterically block the active site. (From Newton AC, Johnson JE. Protein kinase C: a paradigm for regulation of protein function by two membrane-targeting modules. Biochim Biophys Acta 1998; 1376:155.)

targeting domains (C1 and C2). The C1 domain binds DAG, whereas the C2 domain binds acidic lipids, for example, PS. The N-terminal regulatory domain is divided by a hinge region from the C-terminal catalytic domain, which comprises C3 (ATP-binding) and C4 (substrate binding) domains. The enzyme is maintained in an inactivated state by the interaction of the autoinhibitory domain with the active site. Binding of DAG and other factors to the regulatory domain activates the enzyme by releasing the autoinhibitory domain from the catalytic site. The novel isozymes contain two C1 domains and a modified C2 domain, and the atypical isozymes contain one atypical C1 domain and no C2 domain. The µ isozyme has two C1 domains and a PH domain.

The C1 domains are globular structures with conserved Cys-rich motifs that coordinate Zn^{2+}. They bind DAG and phorbol esters, which greatly activate the enzyme. Binding of phorbol esters increases the surface hydrophobicity of the C1 domains, accounting for the fact that these esters recruit the enzyme to membranes. The C1 domain of the atypical isozymes lacks the required consensus sequence and does not bind DAG or phorbol esters. Consequently, these isozymes are not translocated or activated by these compounds.

The C2 domain of PKC resembles those of PLC and synaptotagmin. Structural studies of these domains reveal that it consists of an eight-stranded antiparallel β sandwich, with loops that form an Asp-lined mouth. This Asp-rich electronegative structure binds two Ca^{2+} ions. However, neutralization of the negative charges by Ca^{2+} does not appear to be responsible for the recruitment of C2 domain-containing proteins to membranes. Instead, the C2 domain binds membranes via residues that are on the loops that comprise the Ca^{2+} site. Novel PKC isozymes lack critical Asp residues in the Ca^{2+} binding site of the C2 domain and their membrane binding does not require Ca^{2+}. The binding of PKC isozymes to membranes also requires anionic lipids. In the absence of DAG, Ca^{2+} increases the affinity of the enzymes for these lipids, and this is believed to be mediated by the C2 domain. The kinase domain of PKC resembles that of PKA and other Ser/Thr kinases. However, all PKC isozymes have a cluster of acidic residues at the entrance to the active site. These residues are probably responsible for maintaining the basic pseudosubstrate domain at this site.

Ca^{2+} ions and DAG/phorbol esters produce synergistic activation of conventional PKC. This apparently comes about because Ca^{2+} and DAG/phorbol esters both increase the affinity of the enzyme for PS by different mechanisms. PKC can also be activated by unsaturated fatty acids, for example, arachidonic acid. This stimulation is enhanced by DAG and Ca^{2+}, suggesting that it is mediated by the C2 domain. PIP_3 also activates certain PKC isozymes, presumably by interacting with anionic lipid binding determinants in the C2 domain.

A new dimension to the understanding of the cellular role of PKC isozymes has come from the discovery of anchoring proteins for these enzymes. These are termed receptors for activated C-kinase (RACKs) or receptors for inactive C-kinase (RICKs). These anchor PKC isozymes to certain subcellular locations and may therefore determine their substrate specificities, that is, functions. In addition, there are other PKC binding proteins that are phosphorylated by the enzyme and also bind PS. These could also act as anchoring proteins.

RACKs are not usually substrates for PKC, but they enhance substrate phosphorylation by the enzyme. They show selectivity for certain PKC isozymes and are localized to certain subcellular sites. In the inactive state, PKC isozymes are postulated to be in a folded state with the RICK binding site exposed. Translocation of PKC isozymes from RICKs to RACKs involves activation of the PKCs, which is thought to cause unfolding of the enzyme and exposure of the RACK binding site. The RICK and RACK binding sites of PKC appear to be located in the C1 and C2 domains.

DAG generated by agonist activation of PLC accumulates in the plasma membrane and induces the membrane translocation and activation of PKC isozymes by the mechanism described earlier. The enzyme then associates with certain RACKs. The substrates of PKC isozymes include plasma membrane proteins such as ion channels, receptors, and other proteins involved in cell signaling. PKC stimulation also leads to activation of the MAP kinase pathway by an undefined mechanism. This leads to enhanced transcription of certain genes.

5. PHOSPHOLIPASE D

5.1. General Properties of PLD Isozymes

Phospholipase D (PLD) is widely distributed, being found in bacteria, yeast, plant, and animal cells. It acts principally on PC in mammals, but shows less substrate specificity in other organisms. It hydrolyzes phospholipids to form phosphatidic acid (PA) and the polar head group. PLD has been cloned from several mammalian, yeast, and bacterial sources. As shown in Figs. 11 and 12, the isozymes contain several conserved regions (I–IV) and within two of these are highly conserved $HXK(X)_4D$ motifs (usually abbreviated as HKD motifs). These motifs are found in other proteins (bacterial PS synthase and cardiolipin synthase, bacterial endonucleases, poxvirus envelope proteins, and a *Yersinia* murine toxin). The widespread occurrence of HKD motifs implies the existence of a PLD superfamily. Mutation of the motifs leads to complete loss of enzymatic activity, indicating that they are involved in the catalytic reaction.

The mammalian PLD isozymes that have been cloned to date are all alternatively spliced products of two genes (PLD1 and PLD2). Biochemical data suggest the existence of other mammalian isozymes, for example, isoforms that are stimulated by fatty acids. Both PLD1 and PLD2 require PIP_2 for enzymatic activity, and this is the case for plant and yeast PLDs. However, the two mammalian isoforms are regulated very differently and are located in different cellular structures. The PLD1 isozymes are regulated in vitro by many factors: they are stimulated by conventional PKC isozymes and small G proteins of the Rho and ARF families, and are inhibited by oleate. In contrast, the PLD2 isozymes exhibit high basal activity and are not regulated by the PKC isozymes, Rho proteins, or oleate, and show little or no activation by ARF proteins. The high basal activity of PLD2 is suppressed by α- and β-synucleins, which are proteins localized to presynaptic nerve terminals. It is possible that the enzyme is inhibited by other cellular proteins.

PLD1 is localized to late endosomes and lysosomes in many cell types and is present in the Golgi apparatus in some cells. In contrast, PLD2 is localized primarily to plasma membranes. The presence of PLD2 in detergent-insoluble fractions has been reported. In promonocytes, these fractions contain actin and other proteins associated with the actin-based cytoskeleton. Overexpression of PLD2 in fibroblasts causes cytoskeletal reorganization. PLD2 may also be targeted to caveolae.

As indicated earlier, both the N-terminal and C-terminal HKD motifs in mammalian PLDs are required for activity, and studies of truncated forms of the enzymes indicate that the two motifs associate to form the catalytic center. Dimerization of two HKD domains to form an active center is also seen in the crystal structure of another member of the PLD superfamily, namely the Nuc endonuclease. Studies of the catalytic mechanism for this endonuclease and the *Yersinia* murine toxin indicate that a covalent phosphohistidine intermediate is involved. It has been proposed that the reaction occurs in two steps: first the formation of a phosphatidyl–enzyme complex with the release of choline, followed by release of PA from the enzyme. Dixon and associates have suggested that one of the His residues functions as a nucleophile to attack the phosphodiester bond, whereas the other serves as a general acid and protonates the oxygen of the leaving group. It has been reported that PLD1 is posttranslationally modified by fatty acylation and that this is required for catalytic activity.

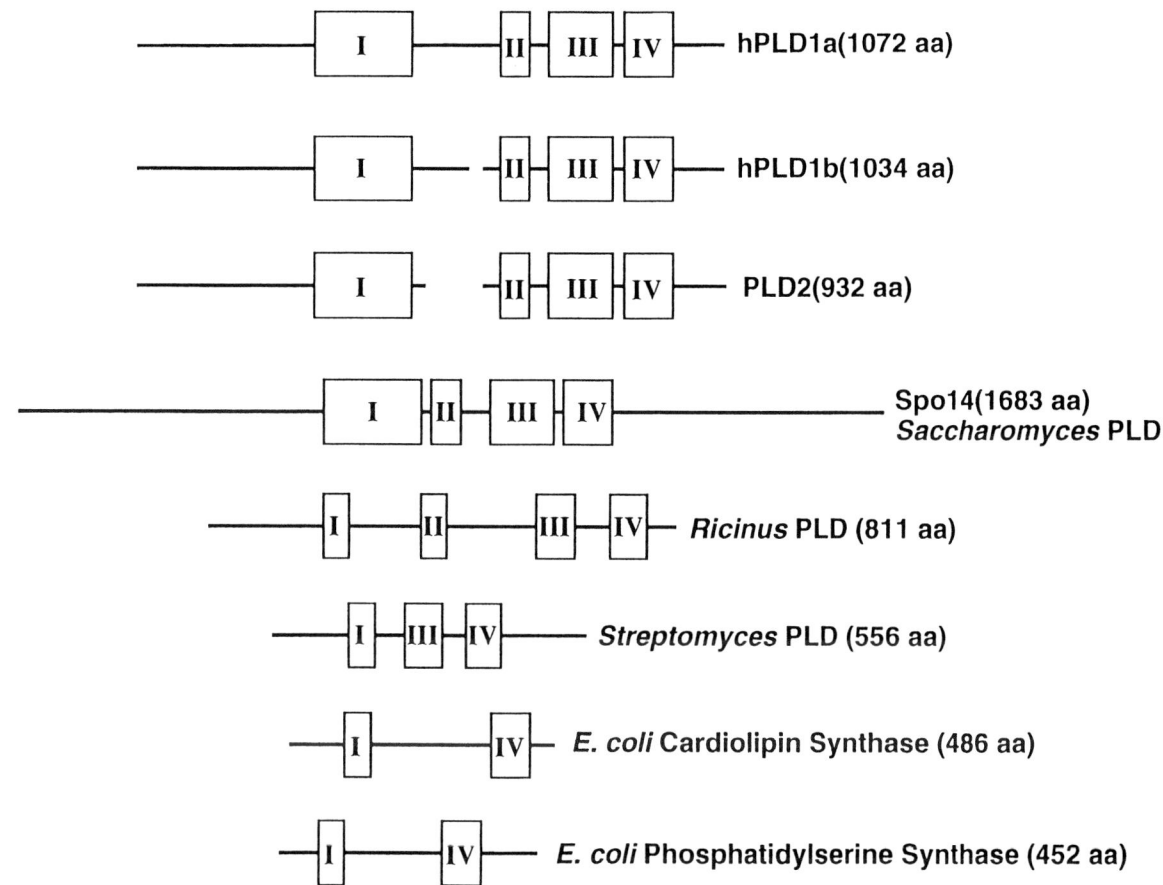

Fig. 11. Alignment of sequences of PLDs and related proteins from several species. Homologous regions are identified as boxes I–IV. (Redrawn from Morris AJ, Engebrecht JA, Frohman MA. Structural regulation of phospholipase D. Trends Pharmacol Sci 1996; 17:182.)

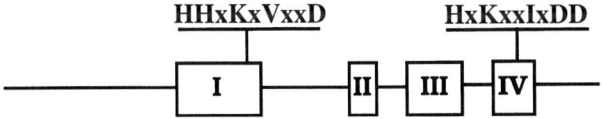

Fig. 12. Linear structure of hPLD1 showing the two highly conserved HKD motifs. There are also PH and PX (protein–protein interaction) motifs in the N-terminus.

Early studies recognized that PLD catalyzes a transphosphatidylation reaction whereby primary alcohols act as nucleophilic acceptors in place of H_2O. Because PA can be formed by other reactions in vivo or in broken cell preparations, the formation of phosphatidylalcohols is commonly utilized as an assay of PLD activity. High concentrations of primary alcohols are also used to suppress PA formation by PLD and thus explore the function of the enzyme in vivo.

5.2. Regulation of PLD1

PLD1 exists in two splice variants, one of which has a 38 amino acid deletion between conserved sequences I and II. The regulation of the two alternatively spliced forms is similar, but they are differentially expressed in various tissues. Like most PLD isozymes, PLD1 has an absolute requirement for PIP_2 or PIP_3 for activity. Other inositol phospholipids have little or no effect, indicating that the presence of phosphate groups at positions 4 and 5 on the myoinositol ring is essential. The binding site for PIP_2/PIP_3 is a highly conserved region containing many basic and hydrophobic residues between sequences II and III. The binding site is not required for membrane association, but is needed for enzymatic activity towards cellular lipid substrates.

5.2.1. Regulation of PLD1 by Protein Kinase C

Numerous studies have shown that phorbol esters activate PLD in a variety of cell types, and, as described in the previous sections, many agonists stimulate PIP_2 hydrolysis and thereby activate PKC. These findings indicate that PKC is a major regulator of PLD. Certain PKC isozymes are able to directly activate PLD1 isozymes in vitro. Surprisingly the activation does not involve ATP and phosphorylation, and is restricted to the conventional isozymes, as no other PKC isozymes produce activation under these conditions. The activation involves the regulatory domain of the PKC isozymes and is unaffected by inhibitors of PKC kinase activity. It therefore occurs by a nonphosphorylating protein–protein interaction. The PKC interaction site on PLD1 has been localized to two sequences in the N-terminus. Deletion of these sequences abolishes PKC activation of the enzyme. Deletion of the most N-terminal sequence also causes a large increase in activity. It has been proposed that this sequence contains an inhibitory domain and that PKC acts to reverse the effect of this domain. When PLD1 is simultaneously activated by PKC and Rho and/or ARF, marked synergism is observed. The mechanistic basis for this is presently unknown.

Despite the evidence that conventional PKC isozymes activate PLD by a nonphosphorylating mechanism in vitro, many studies have shown that inhibitors of PKC kinase activity decrease the activation of PLD by phorbol esters and agonists in intact cells. The magnitude of the inhibition is variable, but in many cases it is substantial. PLD can be phosphorylated by PKCα in vitro, but this does not cause activation. Therefore, direct phosphorylation of PLD by PKC cannot be a mechanism by which it is activated in vivo. Other possible mechanisms include the phosphorylation of regulatory proteins that act on the enzyme and the phosphorylation of another protein kinase(s) that phosphorylates and activates PLD. An alternative mechanism, for which there is some support, is that PKC phosphorylates a scaffolding protein and that the phosphorylation promotes the binding and productive interaction of PKC and PLD.

5.2.2. Regulation of PLD1 by Rho Family Proteins

Members of the Rho subfamily of Ras GTPases activate PLD1 in vitro. RhoA, RhoB, and RhoC are the most effective Rho proteins, but Rac1, Rac2, and Cdc42Hs have partial effects. Their potency is greatly increased by geranylgeranylation. When the Rho proteins are expressed in cells, RhoA, RhoB, and RhoC produce greater stimulation of PLD than Rac1, but Cdc42Hs is ineffective. The Rho proteins interact with PLD1 through specific residues in their Switch I (activation loop) region, which is the common effector domain for Ras GTPases. However, differences in Rho protein efficacy are determined by additional residues in or adjacent to the Switch II region. Concerning the site on PLD1 at which the Rho proteins interact, this is in the C-terminal third of the enzyme. PLD1 displays synergistic activation when concurrently stimulated by Rho proteins and PKCα. The molecular basis for the synergism is unknown.

The in vivo role of Rho proteins in agonist regulation of PLD has been explored using the C3 exoenzyme of *Clostridium botulinum,* which ADP-ribosylates and inactivates Rho proteins. The exoenzyme interacts preferentially with Rho rather than Rac and Cdc42Hs in vitro, and in vivo is selective for Rho also. Treatment of many cell types with C3 exoenzyme markedly inhibits agonist activation of PLD, although other cells may show little or no effect. These results indicate that Rho plays a major role in mediating agonist activation in PLD in some, but not all, cells. Dominant negative and dominant active forms of Rho proteins have also been utilized in a limited number of studies. These have shown that Rho and Rac play a role in the activation of PLD by growth factors and agonists linked to heterotrimeric G proteins (G_q and G_{13}) in several cell lines. In summary, there is evidence that Rho proteins are important mediators of agonist activation of PLD in many, but not all, cells.

The mechanism(s) by which Rho proteins regulate PLD in intact cells is unclear. As the proteins can directly activate the enzyme in vitro, it is logical to conclude that this is the mechanism in vivo. However, Jakobs and associates have obtained evidence suggesting that Rho acts by controlling the intracellular concentration of PIP_2. Rho activates PI 4-P 5-kinase, which is the enzyme that synthesizes PIP_2, and treatment of cells with C3 exoenzyme or tyrosine kinase inhibitors causes a decrease in PIP_2. However, it has not been shown that agonist activation of PLD is associated with an increase in PIP_2 in any cell line.

5.2.3. Regulation of PLD1 by ARF Family Proteins

Members of the ADP-ribosylation factor (ARF) subfamily of Ras GTPases activate PLD1 isozymes. There is little difference in the potency of the different ARF members, but myristoylation increases their potency markedly. The in vitro efficacy of the myristoylated ARF proteins is greater than that of the geranylgeranylated Rho proteins, but they exhibit similar concentration dependencies. The PLD1 interaction site has been localized to the N-terminus of the ARFs, but the region of PLD1 that interacts with the ARFs is unknown. When combined with Rho proteins and/

or PKCα, ARF1 produces a synergistic activation of PLD1.

As in the case of Rho, some cells exhibit a dependency on ARF for agonist activation of PLD, whereas others show no dependency. Thus brefeldin A, an inhibitor of guanine nucleotide exchange on ARF, inhibits the PLD-stimulatory effects of some agonists, but not others. There have also been reports that addition of ARF restores the activation of PLD by growth factors in permeabilized cells, and that expression of dominant negative mutants of ARF blocks the stimulation of PLD in intact cells.

Whereas there is much evidence that certain agonists activate PKC and Rho proteins in cells, there are fewer data showing that agonists activate ARF. Evidence that agonists activate ARF comes from studies showing that the agonists increase the membrane association of ARF. As this association is also increased by GTPγS in vitro, it is reasoned that the agonists promote GTP binding to ARF.

PLD1 is localized to the perinuclear region of the cell, which includes the ER, Golgi apparatus, and late endosomes. The Golgi-associated enzyme is stimulated by ARF in the presence of a nonhydrozable analog of GTP and the activation is blocked by brefeldin A. ARF-sensitive PLD is also found in the secretory vesicles of neutrophils and has been reported to be translocated to the plasma membrane by the vesicles upon stimulation by f-Met-Leu-Phe. Translocation of PLD1 from secretory granules/lysosomes to plasma membranes has also been observed in RBL-2H3 basophilic cells upon crosslinking of IgE receptors. There are also reports that ARF-stimulated PLD is present in plasma membranes as well as Golgi, endomembranes, and nuclei in HL60 promyelocytic cells and liver. The functional significance of the PLD1 activity in these intracellular organelles, its regulation by ARF, and its translocation to the plasma membrane is at present unclear.

5.3. Regulation of PLD2

PLD2 has 51% sequence identity to PLD1 and possesses the same four conserved sequences. It has high intrinsic activity, is unaffected by PKC isozymes and Rho proteins, and shows little activation by ARF. However, if the first 308 amino acids are deleted, the basal activity of the enzyme decreases markedly and it becomes very responsive to ARF, but not other stimulators. As noted in Section 5.1., the subcellular distribution of PLD2 differs from that of PLD1, that is, it localizes primarily to the plasma membrane and components of the actin cytoskeleton. Serum stimulation of fibroblasts expressing PLD2 causes redistribution of the enzyme from the plasma membrane to submembranous vesicles and the formation of filopodia-like projections indicative of reorganization of the actin cytoskeleton. PL2 is inhibited by two proteins (α- and β-synucleins) that are associated with synaptic membranes. These proteins do not affect PLD1 activity, and their action on PLD2 is not overcome by PIP_2 or activators of PLD1.

5.4. Role of Tyrosine Phosphorylation in Regulation of PLD

Agents that activate or inhibit protein tyrosine kinases (PTKs) or protein tyrosine phosphatases (PTPs) alter PLD activity in many cell types. The effects of PTK and PTP inhibitors on growth factor induced PLD activation are attributable to the fact that signal transduction from these agonists requires receptor tyrosine kinase activity. However, another explanation relates to the fact that growth factors induce the generation of H_2O_2 and other reactive O_2 species that influence PTK and PTP activity. H_2O_2 in combination with vanadate produces tyrosine phosphorylation and activation of PLD in several cell lines. This treatment causes phosphorylation of both PLD1 and PLD2, whereas activation of the receptors for epidermal growth factor induces the tyrosine phosphorylation of only PLD2. The role of tyrosine phosphorylation in the action of G protein linked agonists on PLD is less clear. There is evidence that it involves soluble PTKs of the Src family and the βγ-subunits of the G proteins. Cells transformed with vSrc show increased PLD activity and evidence has been presented that this is mediated by Ras. The effect of Ras is indirect and involves RalA, which associates with PLD1. Ras-dependent activation of RalA involves interaction of activated Ras with the guanine nucleotide exchange factors that activate Ral.

5.5. Physiological Role of PLD

The physiological role of PLD is not as well defined as that of PLA_2 or PLC. The physiologically significant product of PLD action is PA, as its cellular concentration is increased by agonist activation of the enzyme. In contrast, the cellular concentrations of choline and phosphorycholine are already high in resting cells. There are many reports of PA effects on enzymes and other proteins in vitro, but few demonstrations that these are affected in vivo. Reported targets of PA include NADPH oxidase, which is responsible for O_2^- production in phagocytes, Raf1 a Ser/Thr kinase involved in signal transudction to MAP kinase, PLCγ PI 4-kinase which synthesizes PIP_2, GTPase activating proteins (GAPs) for Ras and Rho, PKC isozymes, and an uncharacterized Ser/Thr kinase.

There is much evidence of in vivo roles for PLD and PA in O_2^- production and enzyme release in neutrophils. Other studies indicate their involvement in MAP kinase activation, mitogenesis, cytoskeletal changes, vesicle trafficking in Golgi, and exocytosis in various mammalian cells. In yeast, PLD was first recognized as SPO14, a factor required for meiosis and sporulation.

PA is metabolized to DAG through the action of phosphatidate phosphohydrolase. Although the DAG generated through PA hydrolysis would be expected to activate PKC, this is by no means clear. This is because the molecular species of fatty acids in DAG formed from PC differ from those in DAG generated by PLC action on PIP_2. Furthermore, in some cells, for example, neutrophils, HL60 cells, and MDCK (kidney) cells, the DAG produced from PC is mainly alkyl- or alkenyl-linked. The physiological role of PC-derived DAG is unclear because studies of the effects of various DAG species on different PKC isozymes have been limited and contradictory. It is possible that DAG derived from PC may be more active on novel than on conventional PKC isozymes.

Another potential product of the metabolism of PA is lysophosphatidic acid (LPA). This can be formed from PA by the action of a specific PLA_2, but it is also an intermediate in triacylglycerol synthesis and could be derived from lysophosphatidylcholine through the action of PLD. LPA is released from platelets during their activation and is found in serum, where it represents the major mitogenic activity. It may also be generated by other cells. LPA interacts with G protein linked receptors on many cell types and produces marked cellular changes including proliferation, differentiation, morphological changes, membrane depolarization, chemotaxis, aggregation, and tissue invasion. The endothelial differentiation genes Edg2 and Edg4 have been recognized to encode receptors for LPA. Activation of these receptors increases PLC activity with consequent Ca^{2+} mobilization. Both G_q and G_i appear to be involved in this response. There is also evidence that LPA activates G_{13} in some cell types and that this leads to activation of Rho proteins, thus accounting for the observed changes in the actin cytoskeleton. As the effects of LPA on mitogenesis are blocked by pertussis toxin,

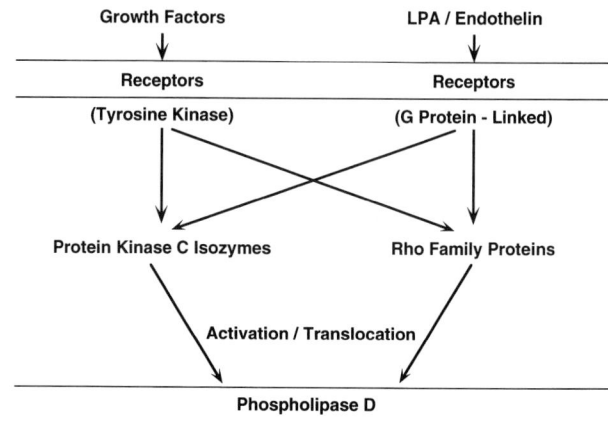

Fig. 13. Postulated mechanisms by which PLD is activated by growth factors and agonists linked to G proteins, for example, lysophosphatidic acid (LPA) and endothelin 1. For simplicity, the mechanisms by which tyrosine kinase activity and G proteins activate protein kinase C isozymes and Rho family proteins (Rho, Rac, Cdc42) are omitted. Activation of conventional and novel PKC isozymes and of certain Rho proteins is associated with their membrane translocation. Other mechanisms of PLD activation by PKC and Rho proteins involving phosphorylation may operate. The role of ARF in agonist regulation of PLD is insufficiently defined to depict. (From Exton JH. New developments in phospholipase D. J Biol Chem 1997;272:15579.)

it seems likely that G_i mediates this response through activation of the MAP kinase pathway.

An important component of the cellular actions of PLD is the changes induced in the physical properties of membranes by the decrease in PC and increases in PA and DAG. The activation of PLD appears to be restricted to the plasma membrane, late endosomes, lysosomes, and Golgi membranes, but may also occur in caveolae. A role for PLD in vesicle trafficking to and through the Golgi apparatus has been proposed. It is possible that increased activity of PLD due to ARF action could complement the effects of this small G protein on the assembly of coat proteins (COP and AP-1) that are necesssary for vesicle formation in different components of the Golgi. It could be speculated that PLD plays a similar role in vesicle budding in other organelles. In the case of late endosomes/lysosomes, the lipid involved may be lysobisphosphatidic acid which is formed from DAG and PA by the transphosphatidylation activity of PLD. The Golgi apparatus is enriched with semilysobisphosphatidic acid, which suggests that it may have a function on this organelle. The origin of this lipid is undefined, but it could arise from PLA_2 action on lysobisphosphatidic acid. It is possible that other functionally important lipids are generated through PLD action.

SELECTED READINGS

Berridge MJ. Inositol trisphosphate and diacylglycerol: two interacting second messengers. Annu Rev Biochem 1987; 56:159.

Bourne HR. How receptors talk to trimeric G proteins. Curr Opin Cell Biol 1997; 9:134.

Casey PJ, Gilman AG. G protein involvement in receptor-effector coupling. J Biol Chem 1988; 263:2577.

Clapham DE. Calcium signaling. Cell 1985; 80:259.

Clark JD, Schievella AR, Nalefski EA, Lin L-L. Cytosolic phospholipase A_2. J Lipid Mediat 1995; 12:83.

Dennis EA. Diversity of group types, regulation, and function of phospholipase A_2. J Biol Chem 1994; 269:13057.

Exton JH. Phospholipase D. Biochim Biophys Acta 1998; 1436:105.

Hamm HE. The many faces of G protein signaling. J Biol Chem 1998; 273:669.

Katan M. Families of phosphoinositide-specific phospholipase C: structure and function. Biochim Biophys Acta 1998; 1436:5.

Leslie CC. Properties and regulation of cytosolic phospholipase A_2. J Biol Chem 1997; 272:16709.

Mellor H, Parker PJ. The extended protein kinase C superfamily. Biochem J 1998; 332:281.

Newton AC, Johnson JE. Protein kinase C: a paradigm for regulation of protein function by two membrane-targeting modules. Biochim Biophys Acta 1998; 1376:155.

Nishizuka Y. Protein kinase C and lipid signaling for sustained cellular responses. FASEB J 1995; 9:484.

Rhee SG, Choi KD. Regulation of inositol phospholipid-specific phospholipase C isozymes. J Biol Chem 1992; 267:12393.

Wess J. G-protein-coupled receptors: molecular mechanisms involved in receptor activation and selectivity of G-protein recognition. FASEB J 1997; 11:346.

PART III   Cyclic AMP, Protein Kinases and Protein Phosphatases

14 Adenylyl Cyclases

Lutz Birnbaumer

CONTENTS

INTRODUCTION
REACTION CATALYZED BY ACs
FUNDAMENTAL PROPERTIES AND MOLECULAR CHARACTERISTICS OF ACs
CONCLUSION
SELECTED READINGS

1. INTRODUCTION

Cyclic 3',5'-adenosine monophosphate (cAMP) was discovered in 1958 and is historically the parent of second messengers. It has multiple well-established roles, which still keep expanding—a new effector molecule was discovered as recently as in 1998, and there may be more to come. As a second messenger, cAMP mediates the effects of innumerable hormones and neurotransmitters in their respective target cells. In addition to being a second messenger and as such the mediator for the action of extracellular stimuli, cAMP has more recently been shown to also act as a regulator of parallel pathways, playing a regulatory or gating role in which its presence gives an either red or green light to another pathway to express itself or be turned off. Regardless of its mediator or gating role, its effects come about primarily by activation of the cAMP-dependent protein kinase system (PKA) and a handful of direct effector molecules that, as of this writing, include the cyclic AMP-gated cation channels of olfactory sensory neurons and two recently discovered *rap1* nucleotide exchange factors.

In the absence of pharmacologic intervention, acute physiologic changes in intracellular cAMP levels are for the most part a reflection of changing rates of synthesis catalyzed by a family of membrane-bound enzymes, the adenylyl cyclases (ACs). While the enzymes that degrade cAMP, the cyclic nucleotide phosphodiesterases, also contribute to the steady state of cAMP levels, their activities are regulated secondary to primary effects of other signals that act by altering other signaling factors such as Ca^{2+}, tyrosine phosphorylation, and cyclic nucleotides themselves. In this chapter we cover the main structural features of ACs, their molecular diversity, and the mechanisms and complexity of their regulation. The most striking theme emerging from still ongoing studies on the properties of different ACs is that they are not only effectors of specific G-protein coupled receptors that act by generating the active forms of G-proteins, but also integrators of convergent signaling pathways that in addition to the activated arms of G-proteins (i.e., the GTP-liganded forms of stimulatory and inhibitory α-subunits and the free βγ dimers) include signals from other pathways that increase free Ca^{2+} and generate Ca–calmodulin (CaM). Among these other signaling pathways are those triggered by Ca^{2+}-permeable glutamate receptors, cyclic nucleotide gated cation channels, and receptors that as part of their response system activate phospholipase C enzymes with formation of inositol trisphosphate (IP_3) and diacylglycerol and its response system, the protein kinase C enzymes. As illustrated later, the response of ACs to secondary regulatory inputs may be conditioned on simultaneous regulation by its primary regulator, the α-subunit of G_s, $G_s\alpha$, or be set up as a response that is temporally delayed with respect an AC-independent response, so

From: *Principles of Molecular Regulation* (P. M. Conn and A. R. Means, eds.), © Humana Press Inc., Totowa, NJ.

that the AC response becomes a regulator in time of the primary signaling process.

2. REACTION CATALYZED BY ACs

ACs catalyze the intramolecular cyclization of the 5'-adenylyl moiety of ATP (hence their name) to give 3',5'-cyclic AMP plus inorganic pyrophosphate (PP_i) according to reaction I.

$$ATP \rightleftarrows 3',5'\text{-cyclic AMP (cAMP)} + PP_i \quad (I)$$

The reaction requires Mg^{2+} as cofactor, as the true substrate is Mg·ATP, and proceeds with a standard free energy change (ΔG^o) of +3.5 kcal/mol at 37° and pH 7.0. Thus, the energetics of the reaction favors synthesis of ATP rather than cAMP. In intact cells the synthesis of ATP from cAMP is prevented from occurring by the continuous removal of PP_i by nonspecific hydrolases. This pulls the equilibrium of the reaction toward formation of cAMP. Scheme II describes the kinetic path followed by the enzyme catalyzed process. It includes the formation, in the direction of cAMP synthesis, first of a ternary enzyme product complex—cAMP–E–PP—that can yield either of two enzyme–product intermediates: cAMP–E, resulting from PP_i dissociating from the ternary complex, or E–PP, resulting from cAMP dissociating before PP_i. Adenosine analogs, called P-site inhibitors, for example, 2',3'-dideoxyadenosine, are effective AC inhibitors that act by binding to the cAMP site of the E–PP and thus remove the enzyme from the system. Diffusion of a P-site inhibitor into a crystal formed of $G_s\alpha$ and the catalytic domains of AC in the presence of the forskolin allowed for identification of the catalytic site of AC.

3. FUNDAMENTAL PROPERTIES AND MOLECULAR CHARACTERISTICS OF ACs

ACs are integral membrane proteins found in the plasma membranes of cells. They colocalize with the regulatory G-proteins G_s and G_i and the receptors that activate G_s, G_i, and other regulators forming a rapid response system. They catalyze the formation of cAMP at the inner surface of the membrane. The cAMP synthesizing activity was discovered in conjunction with the discovery of cAMP itself in 1958 by Sutherland, Rall, and co-workers while investigating the mechanism by which epinephrine and glucagon activate glycogen phosphorylase in dog liver homogenates. After locating the enzymatic activity that generated the second messenger to the particulate, plasma membrane containing fraction, they noted that cAMP formation could be stimulated not only by these hormones but also by sodium fluoride (NaF), initially added to inhibit hydrolysis of ATP as it is a nonspecific inhibitor of phosphatases. Characterization of the hormone- and fluoride-sensitive AC activity in adipocyte membranes, "fat cell ghosts," led Rodbell, Birnbaumer, and coworkers to infer that hormone receptors and AC were distinct molecular entities and that hormones and fluoride shared common characteristics in their mode of action. The same group, working with liver membranes, discovered in 1970 the existence of a GTP-dependent step between hormone receptor and AC. Gilman and Ross, working with mutant T-cell lymphoma cells that had lost all measurable AC activity, S49 cyc^- cells, discovered that rather than lacking AC these cells have an AC activity measurable only if Mn^{2+} is used as cofactor rather than the commonly used Mg^{2+}. They also discovered that activity with Mg^{2+} was dependent on a "regulatory" component present in wild-type but absent in cyc^- S49 cells. More importantly, reconstitution of Mg^{2+} activity was accompanied by reconstitution of stimulation both by hormone acting through a β-adrenergic receptor and by fluoride. Catecholamine-stimulated activity required GTP. Indeed hormonal stimulation could be bypassed, and AC activity in S49 cyc^- membranes could be activated by the regulatory component plus NaF or regulatory component plus the nonhydrolyzable analog of GTP, GMP–P(MH)P. The term "stimulatory regulatory component of AC" was coined. By the mid 1980s G_s and G_i, a closely related inhibitory regulatory component of AC, substrate of pertussis toxin, had been purified and found to be heterotrimers of composition αβγ, with α's being GTP binding GTPases. $G_s\alpha$ in its GTP- or GTP analog-liganded, but not in its GDP-liganded form, stimulated the cAMP-forming activity in the G_s-deficient but catalyst-containing membranes of cyc^- S49 cells. This led to the now proven concept that $G_s\alpha$ is the active form of the G-protein and that $G_s\alpha$ acts by directly binding to AC. Although proposed early on, the direct inhibitory regulation of an AC by an activated $G_i\alpha$ subunit was not shown until 1992. It is now known that all ACs (see below) are stimulated by $G_s\alpha$, but apparently only some are inhibited by $G_i\alpha$, and that this inhibition varies not only with the affected AC type but with also presence or absence of other coregulators. As discussed in chapter 2 of this book, stimulation of AC activity by hormones (i.e., hormone-liganded receptor) is due to receptor mediated activation of G_s with attendant formation of the active GTP–$G_s\alpha$ complex. Stimulation by NaF is in fact due to aluminum fluoride, AlF_4^-, that forms with contaminating Al^{3+} and acts by binding to the GDP–$G_s\alpha$ of the inactive trimeric G_s in the site of the γ-phosphoryl group of GTP. This stabilizes the GTP conformation of the $G_s\alpha$ subunit, causing the protein to dissociate into Gβγ plus active $G_s\alpha$–GDP[AlF_4].

Purification of the catalytic component, the AC proper, was more difficult due to its lability and much reduced activity when it was separated from G_s, even when assayed in the presence of Mn^{2+}. The discovery in an extract of *Coleus forskohlii*, of an universal stimulator of ACs, the heterotricyclic diterpene derivative forskolin, led to the development of an affinity purification technique by Pfeuffer and Metzger that was applied in Gilman's laboratory to purify sufficient quantities of the bovine brain catalyst (AC) to allow for partial microsequencing and, by 1989, the molecular cloning of the full-length cDNA.

3.1. Molecular Diversity

Already prior to cloning, it had become clear that AC was likely to be a group of isoenzymes, as brain AC was stimulated by Ca^{2+} (more exactly Ca^{2+}–calmodulin, Ca–CaM), while the heart enzyme was inhibited by low concentrations of Ca^{2+}. Also, some enzymes systems showed dual positive, G_s-, and negative, G_i-mediated regulation by hormonal inputs (Fig. 1), while others did not, responding only to G_s-mediated effects of hormones. Molecular cloning confirmed the notion of that ACs are a family of proteins. The present, 1999, count is at nine (AC I through AC IX). Table 1 presents molecular properties of the cloned enzymes, chromosomal location, and their principal sites of expression.

3.2. A Cytosolic Enzyme with a Complex Membrane Anchor

ACs are membrane-bound enzymes. Not surprisingly, the secondary structure of the cloned brain enzyme cDNA predicted characteristics typical of a transmembrane enzyme (Fig. 2): a short cytosolic N-terminus is followed successively by a domain that spans the membrane six times (M1), a large cytosolic domain of ca. 40 kDa (C1), another membrane domain

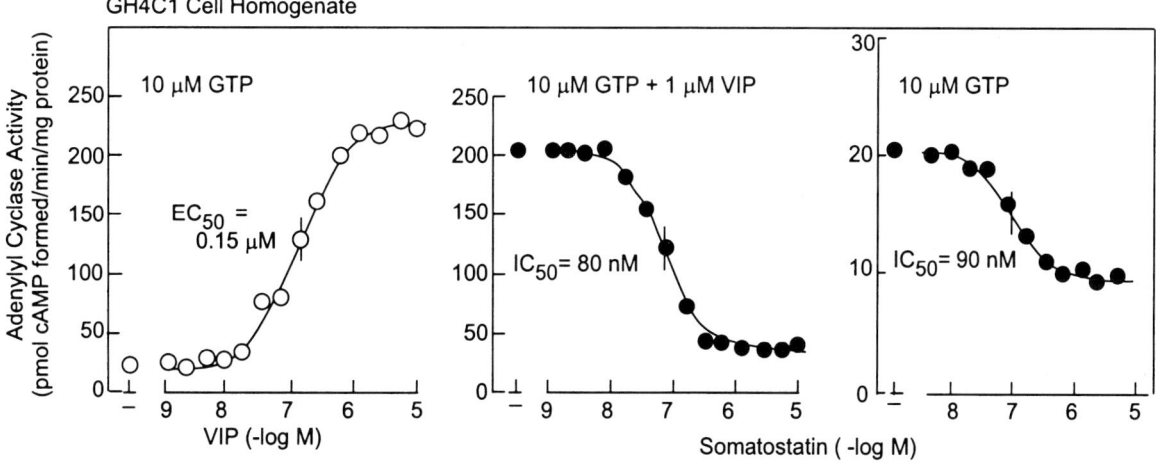

Fig. 1. Adenylyl cyclases are under dual stimulatory and inhibitory control of G-proteins operating in an uncompetitive manner. The figure shows dose–response curves for the stimulatory action of vasoactive intestinal peptide (VIP) acting via a G_s-coupled PACAP receptor and for the inhibitory action of somatostatin acting on a receptor that activates the G_i and G_o family of G-proteins. Note that activation of G_s did not interfere with the action of G_i/G_o on the ACs of the GH4C1 cells, indicating that inhibition of AC by G_i/G_o is not the result of competitive binding of Gα subunits to a common binding site.

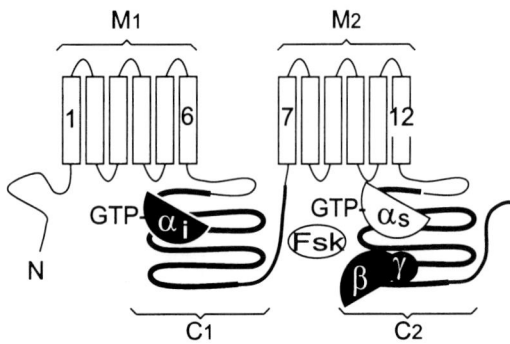

Fig. 2. Model of topological organization of adenylyl cyclase. Shown are the 12 transmembrane segments of the two membrane domains M1 and M2, and the cytosolic C1 and C2 domains that form the enzyme proper. *Thick lines* denote the sequences of C1 and C2 that form the pseudosymmetric dimer regulated by $G_i\alpha$ and $G_s\alpha$, respectively. $G\beta\gamma$ stimulates AC II and IV by binding to a sequence with a QXXER motif predicted to constitute an insert between their C2 $\beta3'$ and $\alpha3'$ structures, when compared to the AC V C2 sequence. Forskolin binding involves residues of both, C1 and C2. (Adapted from Sunuhara et al., 1996.)

similar to the first (M2), and a second cytosolic domain (C2) of ca. 25 kDA forming the C-terminus. Although this type of molecular architecture would predict a transport function for the protein as found for the P-glycoprotein or the cystic fibrosis transmembrane conductance regulator (CFTR), no transport or channel activity has so far been detected in purified or recombinant molecules. Moreover, sequence similarity searches fail to show any relationship between ACs and the transporter proteins whose topography they emulate. In contrast, significant sequence similarity was found both intramolecularly between the C1 and C2 domains and intermolecularly between the C domains of ACs and the subunits of soluble guanylyl cyclase. Soluble guanylyl cyclase is a heme-containing enzyme formed of two distinct but homologous subunits, α and β. Guanylyl cyclases are of two kinds, soluble and membrane bound, and also the membrane bound GCs, which are single-pass transmembrane enzymes, are active as dimers.

Molecular dissection and expression of fragments alone and in combination led Tang and Gilman to discover that the coexpression of M1–C1 and M2–C2 as separate molecules results in formation of a fully functional and regulated enzyme, and that the same is true for the C1 and C2 domains alone, without their respective membrane anchors. Thus, while the complete protein is M1–C1–M2–C2, the actual enzyme is a fully cytosolic C1–C2 heterodimer formed of ca. 190 amino acids each of the C1 and C2 domains and anchored to the membrane by the M1 and M2 transmembrane domains. The crystal structure of a complex of type V C1 core sequence (VC1) and type II C2 core sequence (IIC2), GTPγS-liganded $G_s\alpha$, and forskolin was solved in 1997, that of a VC1–VC2–$G_{i1}\alpha$ complex was deduced in 1998. The general structure of the C1 and C2 sequences forming the enzymatic core is a string of α-helices and β-strands arranged in the order $\beta\alpha_2\beta_2\alpha\beta_2\alpha\beta\alpha\beta\alpha_2\beta$ for C1 and $\beta'\alpha'_2\beta'_2\alpha\beta'_2\alpha'\beta'\alpha'\beta'_2$ for C2 where α-helices and β-strands are connected to each other by α–α, β–β, and α–β or β–α loops. Pooling of data allowed for refinement of a cytosolic enzyme model that is a pseudosymmetric dimer formed of the homologous C1 and C2 domains in which both subunits contribute to the formation of a catalytic site and an independent forskolin binding site. $G_s\alpha$ binds to C2

Table 1
Molecular Diversity and Expression of Mammalian Adenylyl Cyclases

Type	Sp	GenBank Accession No.	AA (No.)	Chromosomal Location		Primary Site(s) of Expression
				Human	Mouse	
I	Bov	M25579	1134	7p12–13	11A2	Brain, GH4C1
II	Rat	M80550	1090	5p15.3	13C1	Brain > olfactory bulb, lung, GH4C1
III	Rat	M55075	1144	2p22–24	12A–B	Olfact. epith., basal ganglia, GH4C1
IV	Rat	M80633	1064	14q11.2	14D3	Brain > ubiquitous, GH4C1
V	Rat	M96159	1262	3q13–21	16B5	Heart > lung,
VI	Mus	M93422	1165	12q12–13	15F	Heart >> lung, intestine, spleen, GH4C1
VII	Mus	U12919	1099	16q12–13	8C3–D	Ubiquitous, S49
VIII	Mus	U85021	1249	8q24.2	NA	Brain
IX	Mus	U30602	1353	16p13.2–3	NA	Ubiquitous, AtT20

Sp, species; Bov, bovine; Mus, mouse.

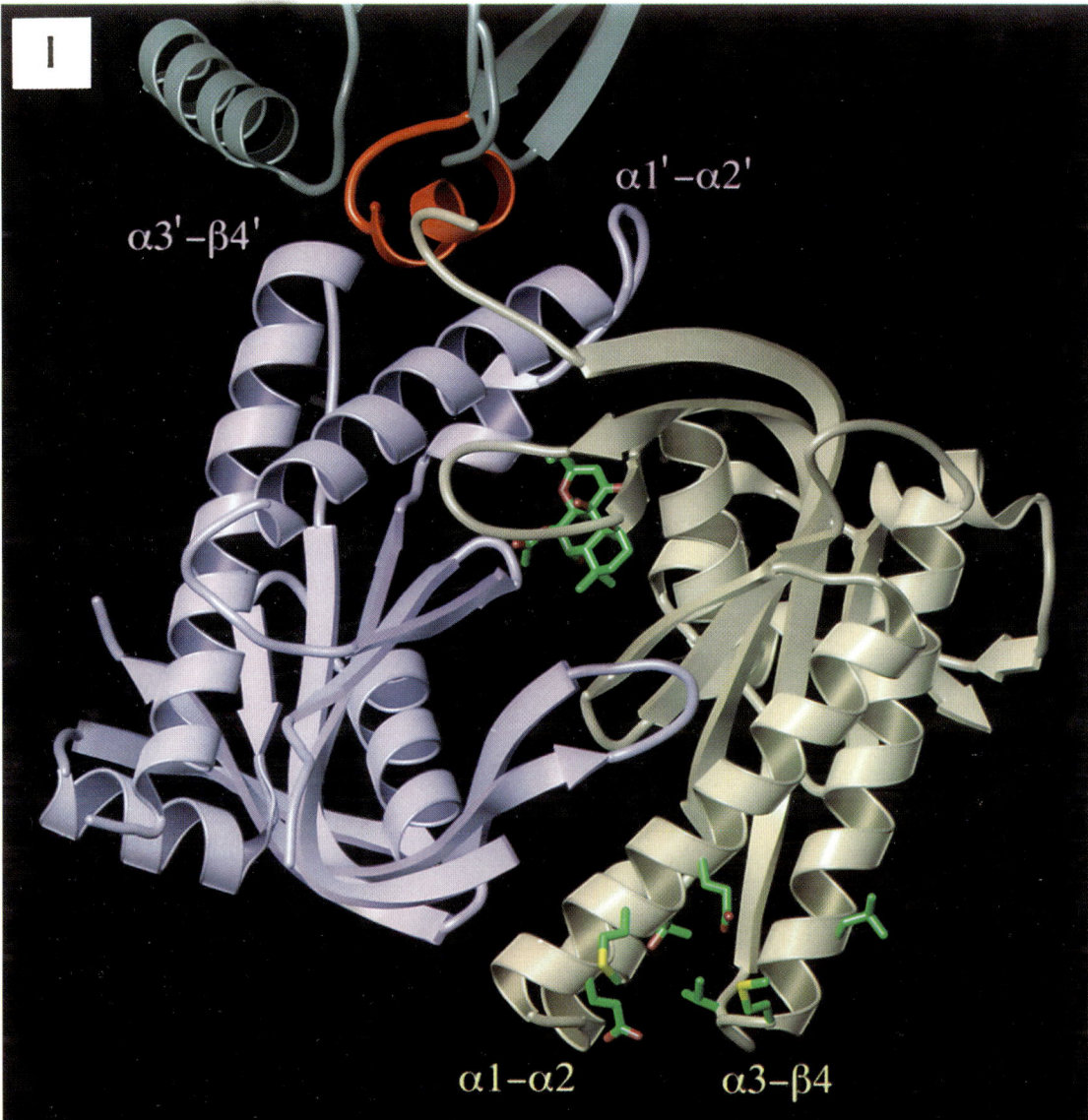

Fig. 3. Three-dimensional structure of the catalytic C1–C2 domains of AC and the Gα docking sites. **I.** Closeup of the C1–C2 dimer with forskolin bound to it (*upper center ball and stick model*). The view is the dorsal aspect from the plasma membrane towards the ventral side away from the plasma membrane (*plane of the printed page*). **Bottom right:** V-shaped $G_i\alpha$-binding region of C1: α_1–α_2 loop (held in place by α_1 and α_2), part of α_2 and α_3–β_4 loop (held in place by α_3 and β_4). **Upper right:** A portion of $G_s\alpha$ docked to the loops positioned by the V formed by backbones of $\alpha2'$ and $\alpha3'$ and $\alpha3'$ and $\beta4'$. Note that $G_s\alpha$ interacts with C2 via its switch II helix, oriented perpendicular to the plane of the membrane. Note also the pseudosymmetry of the C2 (left)–C1 (right) dimer. **II.** Two views of model of the C1–C2 enzyme complexed with $G_i\alpha$ and $G_s\alpha$ via their respective switch II helices. **(A)** lateral view where the plasma membrane would lie above the complexes, perpendicular to the plane of the printed page.   *Continued*

while $G_i\alpha$ interacts with the pseudosymmetric site of C1 (Fig. 3). Activation by forskolin or $G_s\alpha$ is presumed to come about by increasing the affinity between C1 and C2, thereby "compacting" the catalytic site and bringing the active residues of the dimeric enzyme into the optimal configuration for catalysis. Inhibition by $G_i\alpha$, in turn, is presumed to come about by disruption of the C1–C2 interaction, thus accounting for inhibitory regulation by G_i-coupled receptors with respect to stimulation by G_s-coupled receptors, which is of an uncompetitive nature as was illustrated in Fig. 1.

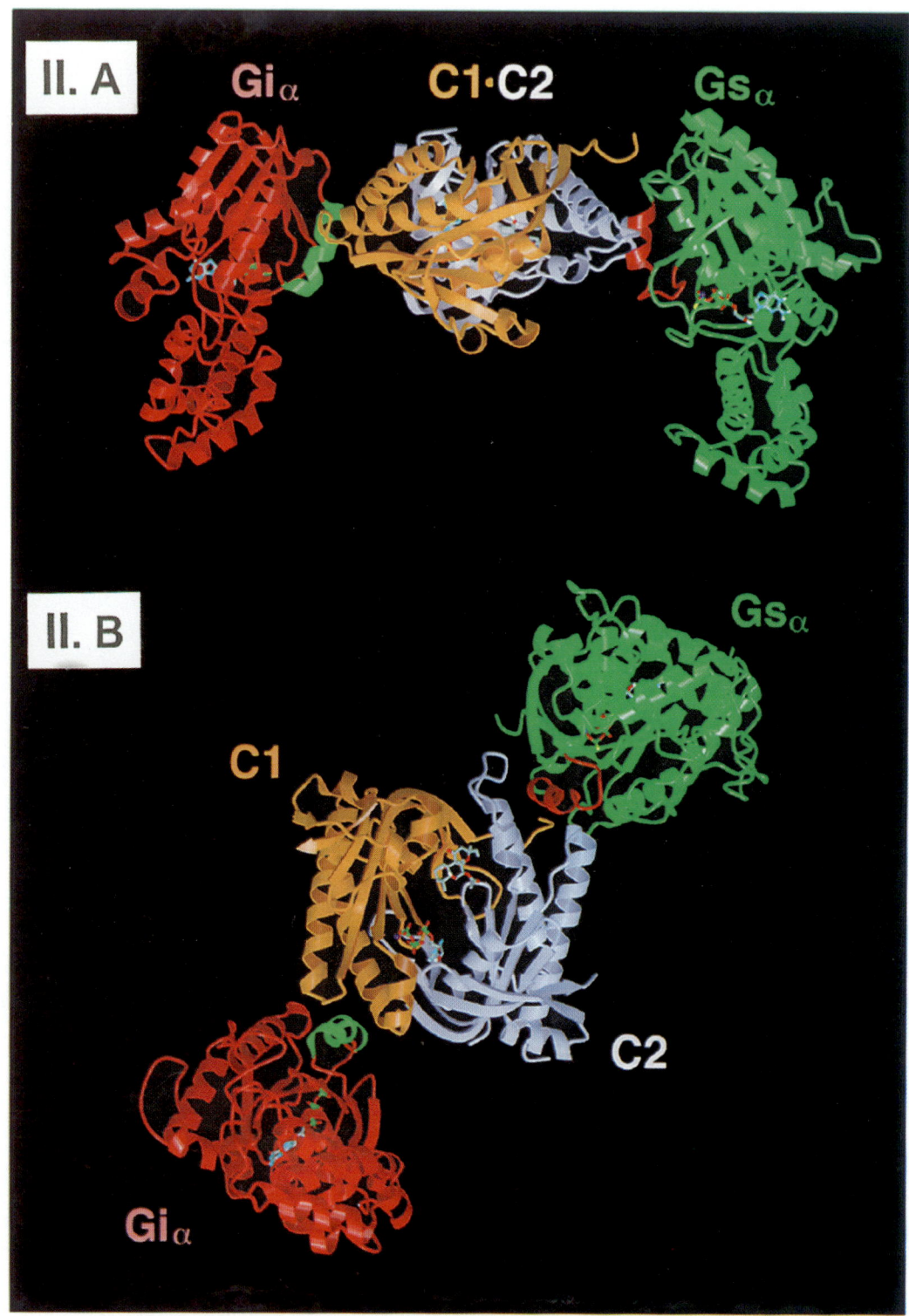

Fig. 3. *Continued* **(B)** Ventral view toward the plasma membrane. (Reprinted from J Biol Chem 1998; 273:25831–25839, with permission from the American Society for Biochemistry and Molecular Biology.)

Table 2
Summary of Regulatory Inputs of AC Subtypes

I. Stimulation
  a. All adenylyl cyclases
    i. $G_s\alpha$ subunit
       Binding site formed by the side chains of the amino acids
       forming α_1–α_2 loop, part of α_2 and the α_2–β_4 loop of the C2 domain (α_1'–α_2', α_2', and α_2'–β_4')
       α.GTP subunits of G_s and G_{olf} activated as a result of:
       — receptor catalyzed GDP/GTP exchange and formation of α-GTP plus $\beta\gamma$ dimer
       — aluminum fluoride (AlF_4^-) mimicking γ-phosphoryl group of GTP at the guanine nucleotide binding site of α
       — binding of nonhydrolyzable GTP analogs to the GTP site
         — GMP-P(NH)P
         — GTP—γS
    ii. Forskolin
        Binding pocket formed by both C1 and C2 ($\beta 5$–$\alpha 4$, $\alpha 1'$, and $\beta 2'$–$\beta 3'$)
  b. Type-dependent stimulations
    i. Ca/CaM: ACI and III synergistic with $G_s\alpha$
    ii. $G\beta\gamma$ dimer: ACII and IV, dependent on $G_s\alpha$ (coincidence detection)
    iii. PKC: types II, IV, VII

II. Inhibition
  a. All adenylyl cyclases
     P-site inhibitors: homologs of adenosine and adenylic acid (AMP) that act as end-product inhibitors that bind to the product site (cAMP site) stabilizing the E–PP_i form of the enzyme and impeding PP_i release. This blocks regeneration of the free enzyme needed to initiate a subsequent catalytic cycle.
  b. Type-dependent inhibition
    i. $G_i\alpha$: type V and VI, Ca/CaN-stimulated type I; not α_s-stimulated type I
    ii. $G_o\alpha$: only type I
    iii. Ca^{2+}: type V and VI
    iv. $G\beta\gamma$ dimer: type I
    v. PKA: types V and VI
    vi. CaMKII: type III

3.3. Type-Dependent Regulation of ACs

Successful molecular cloning of the nine subtypes of AC and the study of their differential properties as seen when they are over expression in mammalian or Sf9 and insect cells has revealed a bewildering array of regulatory features that, for the most part, had not been appreciated from studying activities in tissues or cells as cells and tissues tend to coexpress more than one AC. An illuminating case in this regard are GH4C1 cells. Analysis of their AC activity by standard methods in crude membranes reveals a "normal" AC system, stimulated by G_s-coupled receptors such as prostaglandin 1 (PGE_1) and PACAP and vasoactive intestinal polypeptides (VIP) receptors, and inhibited by G_i/G_o-coupled receptors such as somatostatin and M2-like muscarinic receptors. Molecular analysis by polymerase chain reaction (PCR) showed expression of five out of six enzymes tested for (positive for types I, II, III, IV, and VI, negative for type V; C. Day, J. Codina, and L. Birnbaumer, unpublished). Not known is of course the relative abundance of each protein. Type VI may predominate because of the marked inhibition that somatostatin receptors can elicit, and the type VI enzyme is exquisitely sensitive to inhibition by G_i (see below). On the other hand, in view of the lesser extent of inhibition seen under basal conditions, when compared to G_s-stimulated activities, it is likely that the enzyme that is seen to be inhibited under basal conditions is the type I enzyme. This enzyme is known to have a relatively high "basal" activity as compared to that of the type VI enzyme, which has low basal activity.

The nine cloned enzymes are best subdivided into subgroups the type II/IV/VII; the type V/VI; a group of neuronal Ca–CaM-stimulated enzymes types I, III, and VIII; and type IX, which although being unresponsive to Ca^{2+} or Ca–CaM, is still in need of evalua-

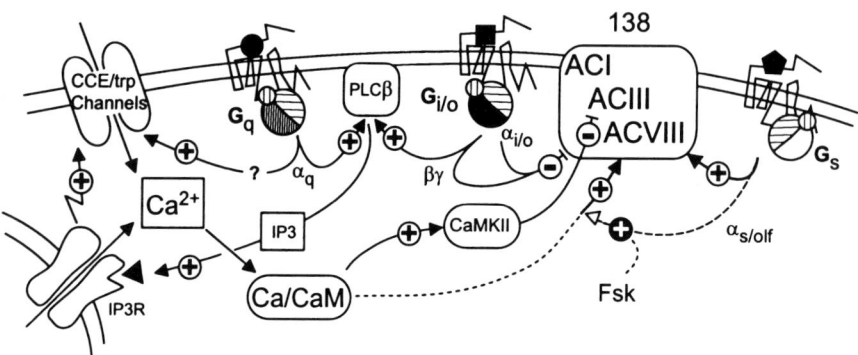

Fig. 4. Impact of G_s, $G_{i/o}$, G_q, and Ca^{2+} signaling on cAMP production by type I, III, and VIII enzymes. **Upper left:** Capacitative Ca^{2+} entry (CCE) channels activated by IP_3-liganded IP_3 receptor (IP3R) and Ca^{2+} store depletion. (Adapted from Sunahara et al., 1996.)

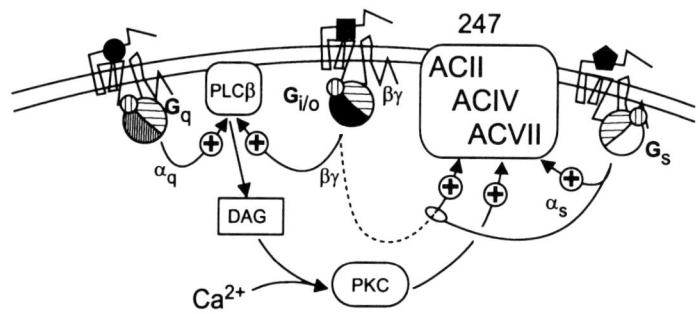

Fig. 5. Impact of G_s, $G_{i/o}$, G_q, and Ca^{2+} signaling on cAMP production by type II, IV, and VII enzymes. The inhibitory effect of PKC may apply only to the type VII enzyme. Note that stimulation by $G\beta\gamma$ is dependent on costimulation by $G_s\alpha$. (Adapted from Sunahara et al., 1996.)

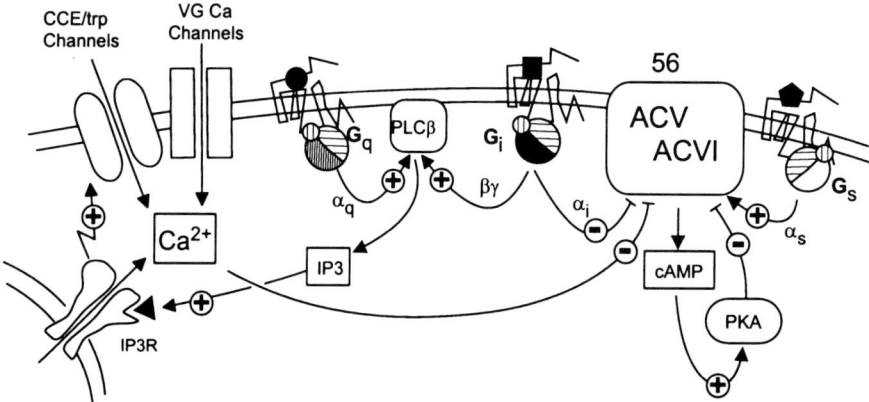

Fig. 6. Impact of G_s, $G_{i/o}$, G_q, and Ca^{2+} signaling on cAMP production by type V and VI enzymes. **Upper left:** Ca^{2+}-elevating forces: capacitative Ca^{2+} entry (CCE) channels activated by IP_3-induced IP3R activation and concomitant Ca^{2+} store depletion and voltage-gated Ca^{2+} channels. (Adapted from Sunahara et al., 1996.)

tion. All are stimulated by $G_s\alpha$ (and presumably its homolog, $G_{olf}\alpha$) and by forskolin. They differ in their response, or lack thereof, to G_i, $G\beta\gamma$, Ca^{2+}, Ca–CaM, and phosphorylation, although the last of these has been much less studied (Table 2). Figures 4–6 illustrate multiple input regulation of the I/III/VII, II/IV/VIII, and V/VI AC's on the basis of the inputs listed in Table 2.

3.4. Regulation by $G\beta\gamma$

Prior to the in vitro demonstration that purified $G_i\alpha$ had the ability to inhibit AC activity, there had been two schools of thought about the mechanism by which inhibition of AC by G_i-coupled receptors such as somatostatin, A1-adenosine, α_2-adrenergic, and M2-like muscarinic receptors comes about. One

$$\text{GDP-G}\alpha\beta\gamma + \text{GTP} \longrightarrow \text{GDP} + \text{G}\beta\gamma + \text{GTP-G}\alpha \xrightarrow[P_i]{k_{cat}} \text{GDP-G}\alpha + \text{G}\beta\gamma \longrightarrow \text{GDP-G}\alpha\beta\gamma \quad (\text{III})$$
(inactive) (active) (inactive) (inactive)

hypothesis was that activated G_i exerted it effects via $G_i\alpha$, and disregarded a role for $G\beta\gamma$. The other ascribed AC inhibition to the $G\beta\gamma$ formed on G_i activation, and disregarded an effect of $G_i\alpha$. In apparent support of the second of these hypotheses, the first cloned AC, brain type I, was found to be inhibited by $G\beta\gamma$ dimers. However, cloned type II, which was relatively poorly stimulated by $G_s\alpha$, was found to be unresponsive to $G\beta\gamma$ in its "basal" state and stimulated, rather than inhibited, by $G\beta\gamma$ provided that the enzyme was simultaneously costimulated by $G_s\alpha$. The term coincidence detection was coined to describe the fact that $G\beta\gamma$ could activate type II AC only in the context of a stimulatory $G_s\alpha$ signal. Receptors that activate other G-proteins, G_i or G_q, would therefore have no effect on AC-II unless a true G_s-coupled receptor was also activated at the sam time. The same proved to be true for the structurally related AC-IV. The site on AC II and IV interacting with $G\beta\gamma$ was located to the C2 domain, and involves a QXXER motif that is also found in other $G\beta\gamma$ effector systems.

The remainder of the ACs appear to be unresponsive to $G\beta\gamma$. Because type I AC is expressed only in neurons, $G\beta\gamma$ could not be the mediator of G_i-coupled receptors in nonneuronal cells, or in cells in which type II and IV enzymes were expressed. This function falls on $G_i\alpha$.

3.5. Inhibition by $G_i\alpha/G_o\alpha$

AC-V and -VI are now known to be the most responsive to inhibition by $G_i\alpha$ Ca–CaM-stimulated AC-I, but apparently not, or much less so, $G_s\alpha$-stimulated AC-I, is also inhibited by $G_i\alpha$. In this case $G_i\alpha$ synergizes in inhibiting the type I enzyme with the inhibition by $G\beta\gamma$ coproduced on G_i activation. Only the type I enzyme seems to be responsive to inhibition by $G_o\alpha$.

3.6. Regulation by Phosphorylation

Only limited studies have been done on the effects of protein kinases on AC activity in systems with defined AC subtypes. Inhibition of type III AC by Ca–CaM is one; inhibition of type V and VI by PKA is another. The effects of PKC are more complex. Type II and IV enzymes are stimulated by PKC-mediated phosphorylation. S49 cells AC activity is also stimulated by PKC activators, and inhibition of AC in these cells by G_i-coupled receptor appears to be suppressed by PKC activation. However, S49 cells do not appear to express type II or IV AC and the mechanism of stimulation by PKC appears to be due to the presence of AC-VII in these cells. Regulation of ACs by different kinase systems requires further investigation.

3.7. $G_s\alpha$-GAP Activity of C2

G-proteins are activated by GTP and have an intrinsic GTPase activity responsible for their deactivation, according to reaction scheme III, where the first reaction is facilitated by the receptor and the second describes the GTPase reaction and is the rate-limiting step of the cycle. As described in more detail in Chapter 2, not only the exchange and activation reaction is susceptible to control—by receptor—but also the intrinsic GTPase activity. RGS proteins (regulators of G-protein signaling) are one group of proteins that accelerate the GTP–$G\alpha$ deactivation by accelerating the k_{cat} of the GTPase reaction. They are GAPs or GTPase activating proteins. RGS proteins, of which more than 16 are known, regulate signaling of $G_{i/o}$, G_t, and G_q proteins, but apparently not G_s. G-proteins of the PLC-activating G_q family (G_q, G_{11}, G_{14}, and G_{16}) are feedback regulated by their effectors, the Cβ-type phospholipases, and therefore under control of two types of GAPs. As is the case for PLCs activated by G_q-like α-subunits, the ACs activated by GTP-$G_s\alpha$ also feed back onto $G_s\alpha$ and activate its GTPase activity. Tests of subdomains responsible for the G_s GAP activity of ACs pinpointed the C2 domain, that is, the same domain with which $G_s\alpha$ interacts to activate AC.

3.8. Effects of Ca^{2+}

AC-V and -VI are subject to direct inhibition by micromolar concentrations of Ca^{2+}. In cardiac membranes this inhibition reaches a maximum of 50% of the starting activity. In the intact cardiac myocyte, inhibition of AC-V and -VI by Ca^{2+} entering through voltage-dependent Ca^{2+} channels contributes to the initiation of the sequels that reverse sympathetic stimulation.

Types I, III, and VIII are enzymes stimulated by increases in cytosolic Ca^{2+} acting via CaM. Ca–CaM stimulation of type I and III enzymes is potentiated by either $G_s\alpha$ or forskolin, indicating a preferential action of Ca–CaM on the active form of the enzymes.

The stimulatory effect of Ca–CaM on the type I

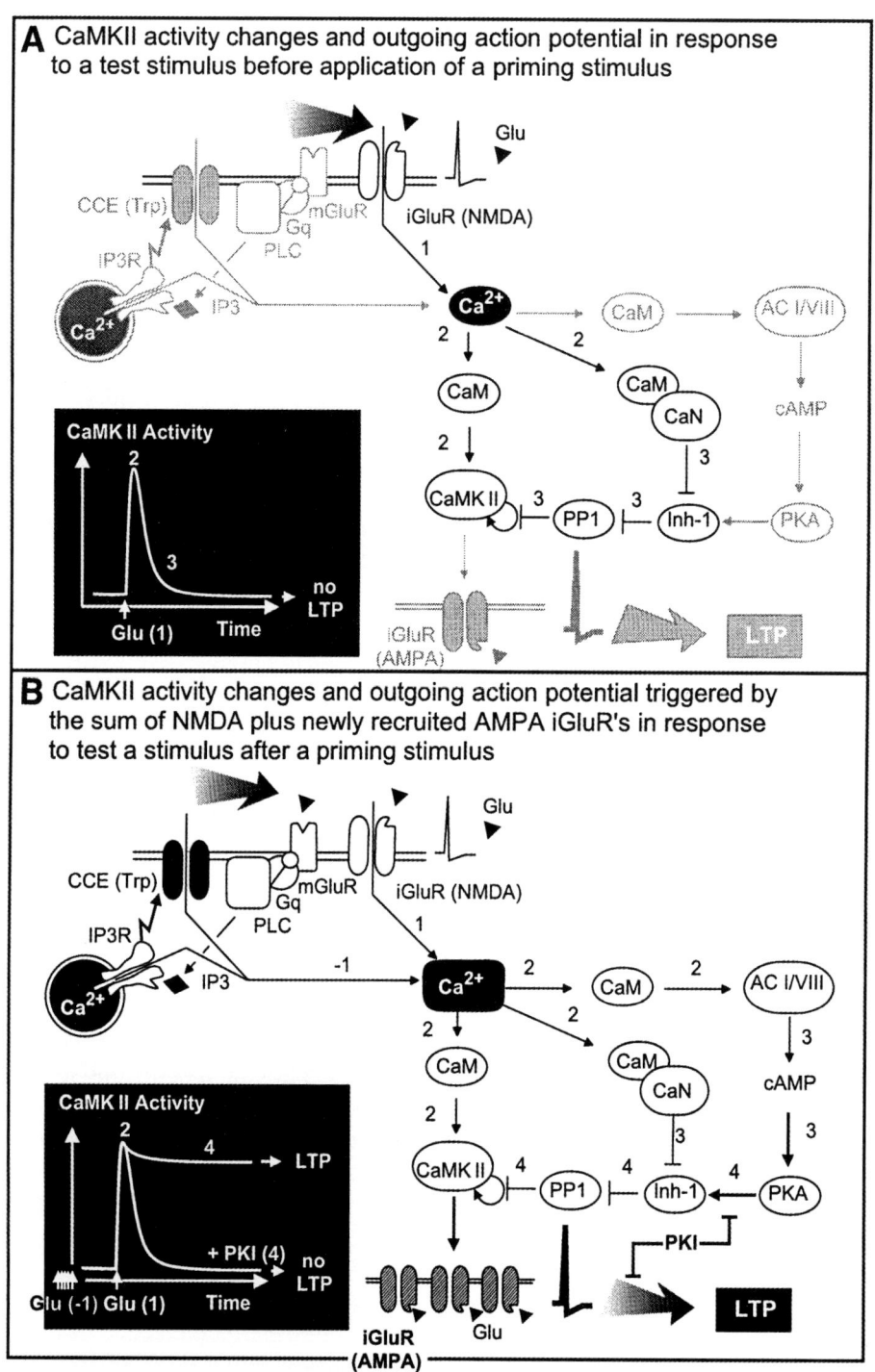

Fig. 7. Extension of life span of active Ca–Calmodulin-activated kinase II (CaMKII) by Ca^{2+}-activated feedback loops triggered by prepulse elevation of cytosolic Ca^{2+}. Note that in **A** a single pulse activated CaMKII is rapidly dephosphorylated by the activated CaN, via its effect to inhibit inhibitor 1 from inhibiting protein phosphatase 1, resulting in rapid deactivation of CAMKII and no effect of glutamate through AMPARs. In **B** the negative feedback loop is blocked by rephosphorylation of inhibitor 1 by PKA activated by cAMP generated by Ca/CaM-stimulated neuronal ACI or VIII, which results in phosphorylation and recruitment of AMPARs to the postsynaptic membrane and an increase in the depolarizing signal emerging from the postsynaptic density. (After Blitzer et al., 1998.)

enzyme in neurons is responsible for the so-called G_s-independent formation of cAMP that occurs after high-frequency stimulation of glutaminergic afferents acting on postsynaptic N-methyl-D-aspartate (NMDA) receptors. NMDA receptors are Ca^{2+} permeable postsynaptic cation channels. As diagrammed in Fig.7, cAMP formation in conjunction with activation of Ca–CaM-stimulated kinase II (CAMKII) results in stabilization of the activated state of CAMKII, which in turn is needed to elicit an increase in synaptic strength, termed early long term potentiation (LTP) in CA1 neurons of the hippocampus. Activation of CAMKII by Ca–CaM leads to autophosphorylation of the oligomeric form of the enzyme which in this manner becomes Ca–CaM independent. Thus, autophosphorylated CAMKII remains active after Ca^{2+} has fallen. Dephosphorylation (and deactivation) is catalyzed by the protein phosphatase calcineurin (CaN), also a Ca–CaM-activated enzyme. The dynamics of the system is such that the increase in CAMKII activity after a single activating pulse of glutamate is transient. The activation is stabilized, however, if cAMP levels are elevated at the time of NMDA receptor stimulation because cAMP, acting via protein kinase A, activates protein phosphatase inhibitor 1 (I-1) which prevents CaN from inactivating CAMKII. This circuit is used to facilitate CaMKII-mediated phosphorylation of AMPA-type glutamate receptors and in this way to increase synaptic strength.

A yet different response pattern to Ca^{2+}, also involving CAMKII, occurs in olfactory sensory neurons expressing the Ca–CaM-activated type III AC (Fig. 8). AC-III is both stimulated by Ca–CaM and inhibited by CAMKII. Olfactory receptors are coupled by G_{olf}, which like G_s stimulates ACs, including AC-III. The cAMP formed elicits the electrical response by activating the cyclic nucleotide gated olfactory cation channel, a Ca^{2+}-permeable channel that by allowing Na^+, K^+, and Ca^{2+} to pass causes membrane depolarization and leads to an increase in cytosolic Ca^{2+}. This Ca^{2+}, through CaM, first synergizes with $G_{olf}\alpha$, increasing cAMP still further, but subsequent to the concurrent activation of CAMKII, cAMP formation by AC-III is shut off, ensuring that the response of the sensory neuron is transient. This circuit, less the synergy between $G_s\alpha$ and Ca–CaM, is illustrated in Fig. 8.

4. CONCLUSION

ACs are a closely related family of enzymes that catalyze the formation of cAMP from ATP in response to activation by $G_s\alpha$. Their primary, secondary, and tertiary structures have been elucidated showing them to be cytosolic enzymes anchored to the plasma membrane by two hydrophobic membrane domains. In addition to being activated by $G_s\alpha$, ACs are subject to type-specific secondary regulations by inhibitory $G_i\alpha$ and $G_o\alpha$, formed upon hormonal or neurotransmitter induced activation of G_i/G_o coupled seven-transmembrane receptors, by Ca^{2+} or Ca–CaM in response to stimulation of NMDA receptors by glutamate, to activation of olfactory receptors by cAMP synthesized as a consequence of G_{olf}-mediated stimulation of AC activity, and in response to an elevation of cytosolic Ca^{2+} occurring as a consequence of the combined IP_3-induced release from internal stores and entry from the external milieu. The latter occurs secondary to IP_3-receptor activation and is an obligatory sequel to PLC activation in response to either G_q-coupled receptors or tyrosine kinase coupled receptors. The intricate, type-specific secondary regulations of cAMP formation demonstrate that ACs are not just simple elements that respond to certain extracellular stimuli, but also integrators of convergent signaling pathways.

SELECTED READINGS

Blitzer RD, Connor JH, Brown GP, Wong T, Shenmolikar S, Iyengar R, Landau EM. Gating of CaMKII by cAMP regulated protein phosphatase activity during LTP. Science 1998; 280:1940–1942.

Chen J, DeVivo M, Dingus J, Harry A, Li J, Sui J, Carty DJ, Blank JL, Exton JH, Stoffel RH, Inglese J, Lefkowitz RJ, Logothetis DE, Hildebrandt JD, Iyengar R. A region of adenylyl cyclase 2 critical for regulation by G protein βγ subunits. Science 1995; 268:1166–1169.

Dessauer CW, Tesmer JJ, Sprang SR, Gilman AG. Identification

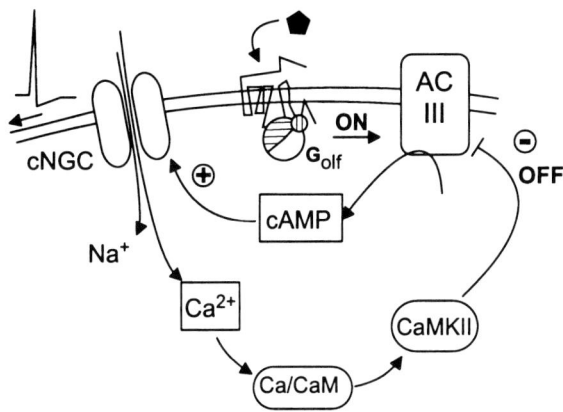

Fig. 8. On–off response of olfactory sensory neuron. (Adapted from Wei et al., 1998.)

of a G$_i\alpha$ binding site on type V adenylyl cyclase. J Biol Chem 1998; 273:25831–25839.

Hanoune J, Pouille Y, Tzavara E, Shen T, Lipskaya L, Miyamoto N, Suzuki Y, Defer N. Adenylyl cyclase structure, regulation and function in an enzyme superfamily. Mol Cell Endocrinol 1997; 128:179–194.

Pieroni JP, Jacobowitz O, Chen J, Iyengar R. Signal recognition and integration by Gs-stimulated adenylyl cyclases. Curr Opin Neurobiol 1993; 3:345–351.

Sunahara RK, Dessauer CW, Gilman AG. Complexity and diversity of mammalian adenylyl cyclases. Annu Rev Pharmacol Toxicol 1996; 36:461–480.

Tang WJ, Hurley JH. Catalytic mechanism and regulation of mammalian adenylyl cyclases. Mol Pharmacol 1998; 54:231–240.

Wei J, Zhao AZ, Chan GC, Baker LP, Impey S, Beavo JA, Storm DR. Phosphorylation and inhibition of olfactory adenylyl cyclase by CaM kinase II in neurons: a mechanism for attenuation of olfactory signals. Neuron 1998; 21:495–504.

Xia Z, Storm DR. Calmodulin-regulated adenylyl cyclases and neuromodulation. Curr Opin Neurobiol 1997; 7:391–396.

15 Cyclic Nucleotide Phosphodiesterases

Marco Conti

CONTENTS

INTRODUCTION
THE SUPERFAMILY OF PHOSPHODIESTERASES
ROLE OF PDEs IN CYCLIC NUCLEOTIDE SIGNALING
PDEs AS TARGETS FOR PHARMACOLOGICAL INTERVENTION
REFERENCES

1. INTRODUCTION

In multicellular organisms, cell functions are integrated through a large network of signals and homeostatic mechanisms that control the intracellular concentration of second messengers. This is a feature indispensable for the basal functions of the body, as well as for an efficient response to the continuous changes in the environment to which the organism is exposed. In addition to regulation of second messenger production, complex regulations of second messenger inactivation are also necessary to maintain cell homeostasis and to control cell sensitivity to extracellular cues.

For the cyclic nucleotide dependent signaling pathway, second messenger removal from the cell is accomplished by two mechanisms. Together with a poorly defined transport of cyclic nucleotide across the membrane to the extracellular space, intracellular cyclic nucleotides are inactivated via degradation by cyclic nucleotide phosphodiesterases (PDEs). These enzymes catalyze the hydrolysis of the 3′–5′ phosphate bond of the cyclic ring in cyclic nucleotides, yielding 5′-AMP or 5′-GMP (Fig. 1). Intracellular cyclic nucleotide concentration is therefore the net result of the rate of synthesis by adenylyl and guanylyl cyclases and degradation by cyclic nucleotide phosphodiesterases (Butcher and Sutherland, 1962).

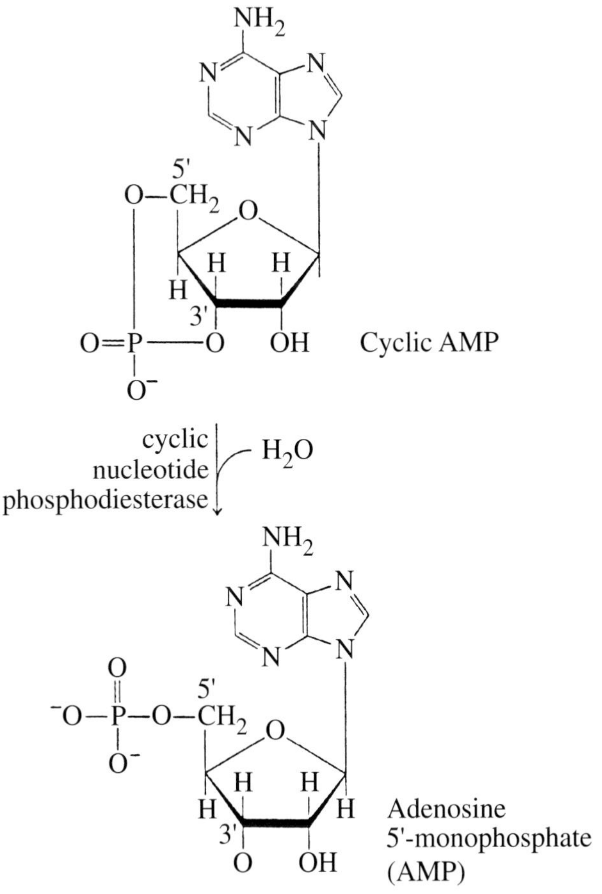

From: *Principles of Molecular Regulation* (P. M. Conn and A. R. Means, eds.), © Humana Press Inc., Totowa, NJ.

Fig. 1. Degradation of cAMP by phosphodiesterases.

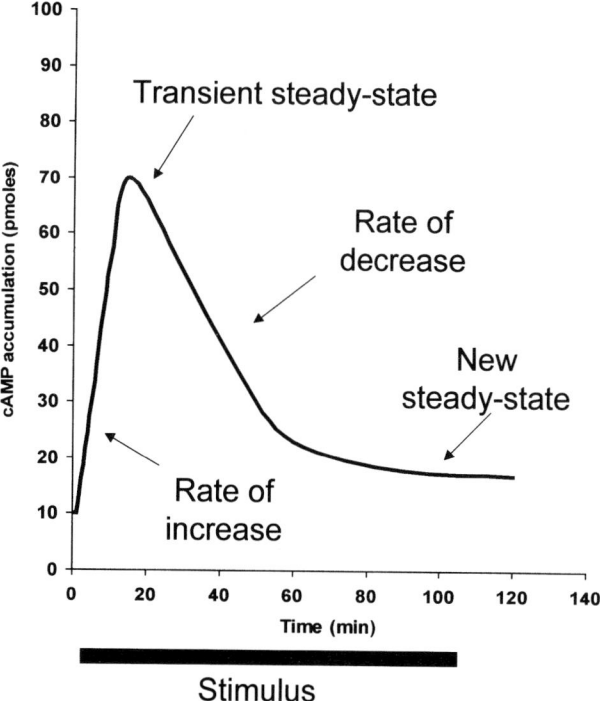

Fig. 2. The pattern of cyclic nucleotide accumulation in a cell. On exposure to the extracellular stimulus, cAMP increases at a rate that is dependent on cyclase activity. When a maximal concentration is reached, it is maintained only transiently. The cyclic nucleotide decreases at a variable rate to a new steady state. The time frame of the increase and decrease in cyclic nucleotide concentration is dependent on the cell type. In olfactory cells, cAMP accumulation reaches a maximum in 50 ms and returns to basal levels in 500 ms. In an endocrine cell, cAMP concentration reaches a maximum in 15–30 min. In neurons, cAMP accumulation reaches a maximum in seconds to minutes.

In spite of a continuous presence of hormone, neurotransmitter, or cytokine stimulus, the cell responds with only a transient increase in second messenger concentration (Fig. 2). The rapid decay of the cyclic nucleotide signal cannot be explained either by the inactivation of the ligand eliciting the response or by its dissociation from the receptor. Rather, active intracellular mechanisms control cyclic nucleotide levels during signaling (Su et al., 1976; Barber et al., 1978). Four distinct phases characterize this transient accumulation of cyclic nucleotides (Fig. 2). On activation of the membrane transduction machinery, cyclic nucleotide concentration increases from a basal to a new steady-state level that is maintained only for a short time. From this maximum, cyclic nucleotide concentration returns toward basal levels, and the final steady state that is reached depends on the intensity of the stimulus and the properties of the target cell. It may be identical to the initial steady state, or a new, slightly elevated steady state may be maintained for as long as the agonist is present. While the initial rate of cyclic nucleotide increase is dependent exclusively on the rate of synthesis (Su et al., 1976), the maximum reached, the rate of decrease, and the final steady state are all dependent on the continuous changes in the rate of degradation by PDEs. Because cyclic nucleotide degradation is mediated by a large number of PDE isoenzymes and different modes of regulation, the intensity and the duration of the cyclic nucleotide signal are to a great extent cell specific.

A second property of the cyclic nucleotide signaling shared with most other signaling pathways is that repeated stimulation of the cell rarely produces the same pattern of cyclic nucleotide concentration. The cell adjusts its response to repeated signals by modifying the duration and intensity of the cyclic nucleotide accumulation (Bunemann et al., 1998). These changes are part of the phenomenon termed cell adaptation and/or desensitization, or tachyphylaxis. Mechanisms regulating the PDE activity are essential components of this adaptation.

In addition to the temporal dimension, the spatial pattern of cyclic nucleotide accumulation specifies the intracellular response. With rare exception, cyclic nucleotides are synthesized at the level of the plasma membrane. From this site of synthesis, they diffuse until they reach their effectors, that is, protein kinases, cyclic nucleotide gated channels, or other binding proteins that are located in different compartments of the cell, often at considerable distance (Rubin, 1994; Scott and McCartney, 1994). In view of their small size, it was initially thought that cyclic nucleotides diffuse freely and instantaneously throughout the cell. However, this may not be entirely true. Numerous observations indicate that different pools of cyclic nucleotides are present in the cell and that compartmentalization of signaling may be necessary to elicit different responses. It is becoming clear that PDEs play a role in controlling cyclic nucleotide diffusion and channeling of signals (HousS\lay and Milligan, 1997).

As a final note, it should be underscored that cyclic nucleotide hydrolysis liberates the same amount of energy that is produced when ATP is converted to ADP. It is also puzzling that rapid fluxes through ATP→cAMP→AMP→ADP→ATP have been measured in the cell (Goldberg et al., 1984). This observation indicates that a large amount of energy is used for cyclic nucleotide signaling. It unclear whether the energy liberated from cyclic nucleotide degradation is solely necessary for processing information and allowing rapid responses to the ever changing environment, or whether it is utilized for any specific purpose other than signaling.

2. THE SUPERFAMILY OF PHOSPHODIESTERASES

2.1. Classification and Nomenclature

PDEs were identified soon after the discovery of the second messenger cAMP by Dr. Earl Sutherland and collaborators (Butcher and Sutherland, 1962; Sutherland and Rall, 1958). When the properties of cyclic nucleotide hydrolytic activity from different tissues were compared, considerable heterogeneity was observed, which suggested the presence of multiple isoforms (Thompson and Appleman, 1971). This complexity has been amply verified with the biochemical characterization and the molecular cloning of cDNAs and genes coding for these PDEs. In mammals, at least 35 distinct PDE proteins and 20 genes have been described (Beavo 1995; Conti and Jin, 1999). On the basis of their structure, kinetic properties, substrate specificity, and regulation, these PDE isoenzymes can be grouped into ten different families (Conti and Jin, 1999). Pharmacological agents that selectively inhibit the different families have been useful tools for the classification of the different forms. A summary of the properties of these PDE families is reported in Table 1. While members of the first seven families are relatively well characterized, little is known of the properties and regulation of the last three PDE families recently identified (Fisher et al., 1998a, b; Soderling et al., 1998). The presence of these novel PDE genes was uncovered by homology searches of sequence databases generated by the Genome Project, and it is likely that this strategy will identify additional PDE genes.

According to the most widely accepted nomenclature (Sonnenburg et al., 1995), PDEs are defined by an arabic number that indicates the family, followed by a capital letter identifying the gene within a family. The different splicing variants derived from a gene are identified by an arabic number after the capital letter. As an example, PDE1C2 is the splicing variant 2 derived from the third gene (C) composing the family 1 of PDEs. In addition to this systematic nomenclature, additional names based on the function of the PDEs have been used. For instance the PDE1s are also termed CaM-PDEs because they are regulated by calmodulin. The most commonly used names are reported in Table 1.

2.2. Overview of the Structural Properties of a PDE Protein

In general, PDEs are proteins of molecular mass ranging between 50 and 130 kDa. They aggregate as oligomeric structures composed of identical or different subunits. In some isoenzymes, as in the PDE6 present in the retinal outer segment, the subunit composition has been firmly established, whereas for other PDEs it is still an object of investigation. Although the complete tridimentional structure of PDEs are yet to be defined by protein crystallography, the analysis of the primary structure of a PDE subunit predicts the presence of several modular domains (Fig. 3).

The alignment of the primary sequence of different members of the 10 PDE families shows a region of high conservation (Charbonneau et al., 1986). This is a domain of approximately 300 amino acids characterized by several invariant residues spaced between less conserved regions (Fig. 3). Members within a family are more than 80% homologous in this region, a feature that has been used as a major criterion for classification (Conti et al., 1995). Controlled proteolysis and deletion or site-directed mutagenesis studies have demonstrated that this conserved region corresponds to the catalytic domain of a PDE. It has often been possible to express truncated PDEs of 30–40 kDa that are still active in cyclic nucleotide hydrolysis, thus indicating that the catalytic domain corresponds to the highly conserved sequence and is a modular, self-contained structure. This is the site where the cyclic nucleotide binds and is hydrolyzed, even though the exact points of contact between the substrate and protein are not known.

While comparing PDE sequences with those of several metallohydrolases, it was noticed that a conserved region within the catalytic domain contains sequences with features of a zinc binding domain (Francis et al., 1994). This has led to the proposal that zinc is required for catalysis and that the catalytic domain of PDEs may be structurally related to the catalytic center of zinc hydrolases. There may also be some distant relationship between the structure of the PDE catalytic domain and the structure of other cyclic nucleotide binding proteins. For instance, some homology with the regulatory subunits of protein kinase A (PKA) has been reported for the catalytic

Table 1
The clic Nucleotide Superfamily of Isoenzymes

Family	Common Name	Genes in the family (variants)	Substrate specificity	Extracellular signals	Intracellular regulators	Kinase	Inhibitor	Ref.
PDE1	CaM-PDE	PDE1A (2) PDE1B (1) PDE1C (5)	cGMP>cAMP cGMP>cAMP cGMP=cAMP	Acetylcholine Olfactory cues	Ca^{++} Calmodulin	PKA CaM-CK	Vimpocetine 8-Me-IBMX	Sonnenburg et al. 1995
PDE2	cGS-PDE	PDE2A (2)	cAMP/cGMP	ANF	cGMP		ENHA	Sonnenburg et al. 1991
PDE3	cGI-PDE	PDE3A (1) PDE3B (2)	cAMP=cGMP	Insulin, IGF-1 catecholamines	cGMP	PKB PKA	Milrinone Vesrarinone	Degerman et al. 1997
PDE4	cAMP-PDE	PDE4A (3) PDE4B (3) PDE4C (2) PDE4D (5)	cAMP	FSH, LH, TSH catecholamines	cAMP	PKA MAPK	Rolipram RS25344 Ariflow	Swinnen et al. 1989
PDE5	cGB-PDE	PDE5A	cGMP	Unknown	Y subunit?	PKG	Sildenafil	Francis et al. 1994
PDE6	Retina-PDE	PDE6A PDE6B	cGMP	Light	γ subunit/ transducin	PKG	Dypridamol Sildenafil	
PDE7	—	PDE7A (2) PDE7B	cAMP	TCR ligation	cAMP?	PKA?	Dypridamol	Michaeli et al. 1993
PDE8	—	PDE8A PDE8B	cAMP	Unknown	—	—	—	Soderling et al. 1998
PDE9	—	PDE9A	cGMP	Unknown	—	—	—	Fisher et al. 1998
PDE10	—	PDE10A (2)	cAMP/cGMP	Unknown	—	—	—	Soderling et al. 1999

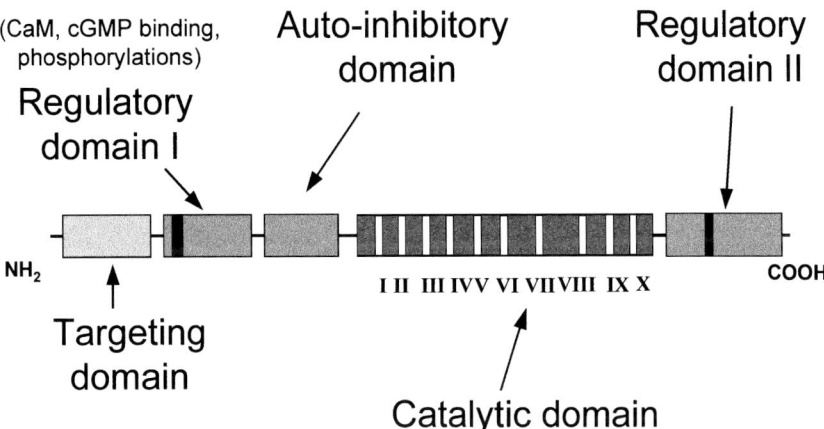

Fig. 3. Domain organization in a PDE protein. *Boxes* correspond to the different domains connected by hinge regions. The *white bars* in the catalytic domain indicated by Roman numerals correspond to the highly conserved subdomains in the catalytic center. These subdomains are characterized by residues invariant in all phosphodiesterases. The *black boxes* in the regulatory domains correspond to sites of phosphorylation by different kinases. The targeting domain at the N-terminus contains protein–protein interaction motifs. The function of the carboxy terminus regulatory domain is poorly understood.

domain (Jin et al., 1992). The significance of this homology remains unclear. Finally, several compounds have been used to probe the structure of the catalytic center of a PDE, pointing to conserved residues that may be important for substrate binding or catalysis. This is a very active field of investigation in view of its importance for drug design.

While the catalytic domain is usually located toward the C-terminus of a PDE, regulatory domains have been identified in the remaining portion of the protein, most frequently at the N-terminus. These regulatory domains are binding domains for allosteric regulators that include calmodulin in PDE1 and cGMP in PDE2, PDE5, and PDE6. Domains in close proximity to the catalytic domain within PDE6 α- and β-subunits also bind a γ inhibitory subunit, and possibly the G-protein transducin. Additional regulatory domains, mostly at the N-terminus, contain sites that are targets for phosphorylation by protein kinases. Thus, phosphorylation sites for PKA, CaM-kinase, PKG, and PKB–AKT have been identified in members of the PDE1, PDE3, PDE4, and PDE5 families (*see* Table 1). These phosphorylations usually modulate the catalytic activity of the enzyme or modify the interaction of a PDE with other modulators or with drugs.

An autoinhibitory domain exerting a negative constraint on catalysis has been mapped at the N-terminus side of the catalytic domains of PDE1, PDE2, and PDE4. The presence of this domain has been inferred by experiments of controlled proteolysis. On cleavage, several PDEs yield a constitutively active catalytic fragment with an increased V_{max} and without changes in K_m for the substrate. A commonly held view is that the phosphorylation or the binding of the allosteric modulator modifies the catalytic activity of a PDE by modulating the interaction between this autoinhibitory domain and the catalytic center (Lim et al., 1999).

Recently, it has been reported that N-terminal domains in a PDE encode signals for subcellular targeting of these proteins (Houslay and Milligan, 1997; Shakur et al., 1993). Thus, targeting domains have been identified in PDE2, PDE3, PDE4, PDE6, and PDE7. The exact mechanism of targeting to different subcellular structures is poorly understood. In some instances, posttranslational modifications such as myristylation may mediate interaction of a PDE with the membrane. In other cases, it has been suggested that PDE anchoring proteins serve to dock a PDE to different organelles. In support of this hypothesis, modular protein–protein interaction domains have been identified, mainly in PDE4. In addition to subcellular targeting, these protein–protein interactions may play an important role in modulating the activity of the enzyme.

A property common to many *PDE* genes is the presence of multiple ORFs coding for more than one PDE protein. This is due to the presence of multiple transcriptional units controlled by different promoters or to the extensive alternate splicing of the different exons present in a gene. An extreme case is the *PDE4* family of genes (Conti et al., 1995) and the orthologous *Drosophila dunce* gene (Davis and Dauwalder, 1991). The

Drosophila dunce locus spans almost 200 kb of genomic DNA, with five promoters that control the production of 10 different transcripts. Each promoter also controls the expression of the dunce mRNAs in a tissue-specific manner in different regions of the fly (Qiu and Davis, 1993). Similarly, the mammalian *PDE4D* gene codes for five different proteins that contain different N-terminal domains (Monaco et al. 1994). This property of *PDE4* underlines the modular structure of the PDE protein, where a catalytic domain is joined to several different regulatory domains.

3. ROLE OF PHOSPHODIESTERASES IN CYCLIC NUCLEOTIDE SIGNALING

By degrading and inactivating cyclic nucleotides, PDEs terminate their biological action within the cell. They also serve to control the flow of information along the cyclic nucleotide pathway. These enzymes are similar to a gate that controls cyclic nucleotide access to their intracellular targets, which include protein kinases (Beebe and Corbin, 1995), cyclic nucleotide gated cationic channels (Maeliche, 1990), and the newly discovered cyclic nucleotide binding proteins that modulate the function of the small G-proteins (de Rooij et al., 1998; Kawasaki et al., 1998). Here follows a description of the different roles that PDEs play in cell signaling.

3.1. PDEs Are Effectors in Signal Transduction Pathways: Light Activates a PDE6 Present on the Plasma Membrane of Retina Cells

In the rod and cone cells of the mammalian retina, light activates a signal transduction pathway that is very similar to that used by chemical messages such as hormones and neurotransmitters. This pathway is probably the best understood of all the signaling pathways controlling PDE activity (Yarfitz and Hurley, 1994). In the initial step, the visual pigment rhodopsin adsorbs a photon of light. Photoexcited rhodopsin collides with and activates a member of the G-protein family, transducin, which is abundantly expressed in the photoreceptor plasma membrane. This interaction with the "light receptor" promotes the exchange of GDP for GTP, yielding an activated GTP-bound α-transducin. This, in turn, causes the activation of a PDE6, with a consequent decrease in intracellular cGMP. Activation of PDE6 is terminated when the bound GTP is hydrolyzed back to GDP by the intrinsic GTPase activity of transducin. The resulting transient decrease in cGMP causes closure of the cyclic nucleo-

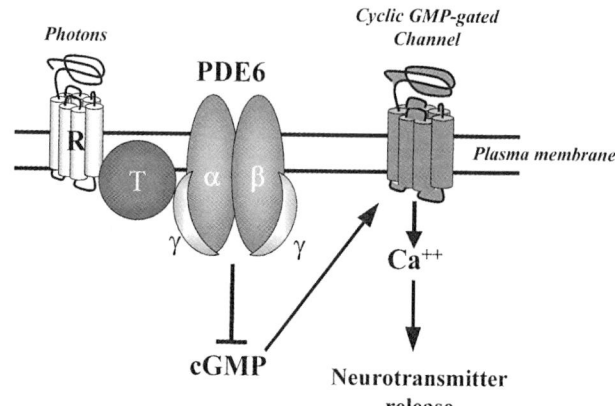

Fig. 4. Signal transduction pathway in the retina. The light receptor rhodopsin (R) is similar to other seven transmembrane domain receptors coupled to G-proteins. Rhodopsin interacts with the G-protein transducin (T). PDE6 is composed of one α and one β catalytic subunit, and two γ identical inhibitory subunits.

tide gated cationic channel, leading to a transient hyperpolarization of the photoreceptor and ultimately a decrease in neurotransmitter release (Fig. 4).

The PDE6 target for light activation is a multimeric complex composed of different subunits, depending on the cell in which it is expressed. In the retina rod cells, PDE6 is a tetramer composed of distinct α- and a β-subunits, and two identical γ-subunits (Beavo, 1995). The α- and β-subunits possess the cGMP hydrolytic activity, while the γ-subunits inhibit the α- and β-subunits. A slightly different isoenzyme with different subunit composition is found in cone cells. This is composed of two identical α'2 subunits, and one or two γ-subunits. In addition, a δ-subunit copurifes with the cone PDE6 and the soluble retina PDE6, suggesting that this additional subunit may have a role in localization of the holoenzyme (Florio et al., 1996). In addition to a catalytic domain homologous to other PDEs, each α- and β-subunit contain two noncatalytic high-affinity cGMP binding sites that function differently from the allosteric cGMP binding sites described in the cGMP-activated PDE2. In the retina PDE6, the cGMP bound to these sites may modulate the interaction with transducin, thus controlling the termination of the light stimulus.

The essential role of PDE6 in the retina function is underscored by the effect of mutations in these enzymes. Natural mutations have been described in the subunit of PDE6 and are the cause of retinal degeneration or night blindness (Farber et al., 1992). Although further studies are needed to confirm this view, a signaling pathway similar to that functioning

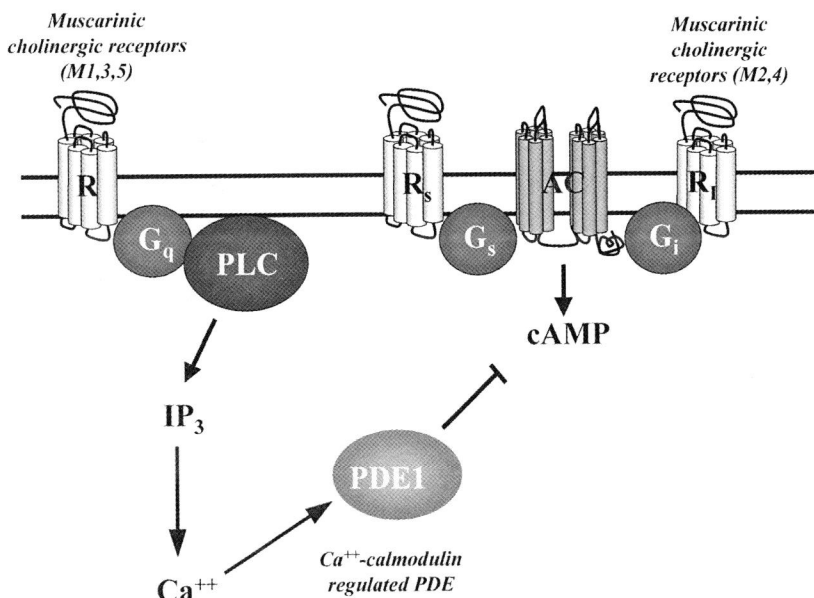

Fig. 5. Mechanism of muscarinic cholinergic agonist activation of PDE1. Muscarinic receptors M1, M3, and M5 are coupled to G_q and phospholipase C, while M2 and M4 receptors are coupled to G_i. The increased intracellular Ca^{2+} stimulates smooth muscle contractility while an increase in cAMP is inhibitory to contraction. Thus, the increase in Ca^{2+} and the decrease in cAMP produced by muscarinic cholinergic agonists have synergistic effects on contractility. The activation of PLC and the increase in Ca^{2+} stimulate PDE1 to decrease cAMP.

in the retinal outer segment has been implicated in the function of taste buds of the tongue. Over the years, several additional studies have suggested an interaction between a PDE and G-proteins (Conti et al., 1995). However, the possibility that a PDE functions as an effector of a membrane signal transduction in cells other than the retina remains to be confirmed.

3.2. PDEs Are Integrators of Different Signaling Pathways

Initially, it was thought that the different signaling pathways were organized as a linear sequence of steps functioning in parallel and that each extracellular signal would activate its own pathway. The last 10 yr of investigation of signaling have completely revolutionized this view. Signaling pathways are composed of nodes and branches, and different pathways are meshed into an intricate network of crosstalks. The concept of combinatorial signaling has been put forward to underscore the almost infinite number of second messenger combinations that control cell function. This organization of the signaling machinery accounts for the pleiotropic effects that hormones, neurotransmitters, and cytokines have on cell function. An increase in cAMP not only changes metabolic activities, but also regulates gene expression and differentiation, cell shape, and motility, as well as entry and exit from the cell cycle. These regulations are mediated by changes in virtually all the signaling pathways operating in the cell. Because PDEs are regulated by multiple second messengers, they function as a branching point in a signaling pathway and allow the integration of cyclic nucleotides pathway with other pathways. Here we provide the most relevant examples of this function.

3.2.1. PDE1 AND CA^{2+} SIGNALING

By virtue of their ability to be regulated by the Ca^{2+}–calmodulin complex, PDE1s serve to integrate the Ca^{2+} signaling pathway with the cyclic nucleotide dependent pathway. In cells where a PDE1 is expressed, an increase in cytoplasmic Ca^{2+} causes an activation of the PDE hydrolytic activity, with a consequent decrease in intracellular cyclic nucleotide concentration. This, for instance, has been observed for the muscarinic cholinergic signaling pathway (Fig. 5) (Erneaux et al., 1985; Harden et al., 1985). Muscarinic receptors are coupled either to adenylyl cyclase inhibition through a G_i-protein or to phospholipase C (PLC) activation through G_q. In spite of the fact that two signaling pathways are activated, in both cases a decrease in intracellular cAMP is produced. In those cells where cholinergic agonists stimulate PLC, phosphoinositol (PI) turnover, and inositol tri-

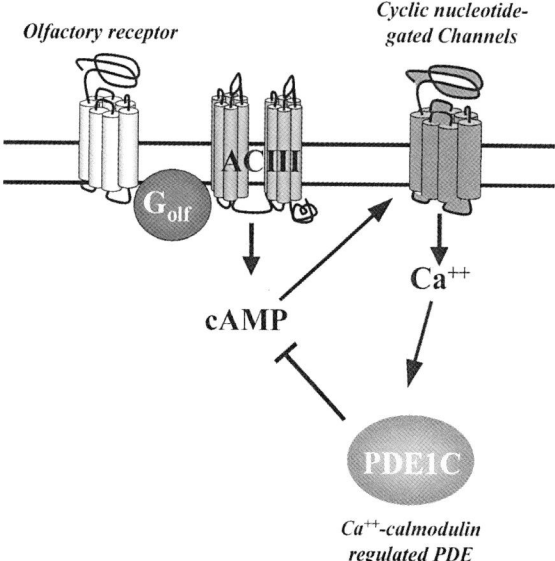

Fig. 6. Mechanism of regulation of PDE1 in the olfactory neurons. Some olfactory neurons express receptors coupled to the cyclase ACIII through a unique G protein (G_{olg}). The cyclic nucleotide gated cationic channels are sensitive to the intracellular cAMP concentration. PDE1C is localized in the cilia, where receptors and cyclases are located. The Ca^{2+} influx which follows the opening of the channel, activates PDE1C, causing the termination of the stimulus.

phosphate (IP_3) production, the resulting increase in intracellular Ca^{2+} causes the activation of a CaM–PDE. The increase in activity of this enzyme, in turn, produces a net decrease in cAMP concentration, as observed in glyoma cell lines and in the thyroid. In many instances, as in smooth muscle cells, the increase in Ca^{2+} and the decrease in cAMP have synergistic effects in stimulating contraction.

The cyclic nucleotide signaling in the olfactory system is remarkable because of the time frame in which it operates. While in endocrine cells cAMP increases minutes after peptide hormone stimulation, the cyclic nucleotide spike develops in milliseconds in the olfactory epithelium (Boekhoff and Breer, 1992). It is believed that these rapid responses are possible because of the presence on the olfactory ciliae of cyclic nucleotide gated channels and CaM PDEs in close proximity to the olfactory receptors coupled to cyclases (Fig. 6) (Beavo, 1995; Juilfs et al., 1997). The local increase in cAMP opens a cyclic nucleotide gated cation channel causing an influx in Ca^{2+}. In turn, the Ca^{2+} influx activates PDE1C, causing a rapid return of cyclic AMP to basal levels. This rapid and transient response is thought to be critical for the correct spatial and temporal perception of the olfactory cue, a discrimination particularly important in animals.

3.2.2. PDE2 AND INTEGRATION OF cGMP AND cAMP REGULATED PATHWAYS

The two classes of PDEs that are sensitive to cGMP (PDE2 and PDE3) integrate the cyclic GMP and cAMP signaling pathway. In several cells, ligands signal through an increase in intracellular cGMP concentration and also indirectly through changes in cAMP levels. This is, for instance, the case for the atrial natriuretic peptide (ANP), a peptide released by the cardiac atrium, which plays an important function in electrolyte homeostasis. It is established that the major effects of ANP on natriuresis are mediated by the inhibition of aldosterone secretion. In the glomerulosa cells of the adrenal cortex, ANP binds to and activates a membrane-bound guanylyl cyclase, thus stimulating cGMP production (Fig. 7). It has been shown that the increase in cGMP causes an activation of the cGS–PDE (PDE2) abundantly expressed in the adrenal glomerulosa cells (MacFarland et al., 1991). This cGS–PDE activation causes a decrease in cAMP, inhibiting signaling through this pathway. This type of crosstalk has been proposed to mediate the antagonistic effects of ANP on ACTH-mediated aldosterone release. A similar signaling pathway is thought to control epinephrine release by chromaffin cells of the adrenal medulla (Whalin et al., 1991).

Conversely in other cells such as platelets or smooth muscle cells, an increase in cGMP is associated with a concomitant increase in cAMP (Maurice and Haslam, 1990). In general, cells in which cGMP and cAMP have synergstic effects express a cGI–PDE (PDE3), an enzyme that hydrolyzes predominantly cAMP. Although cGMP can be hydrolyzed by PDE3, the velocity of the reaction is one order of magnitude lower than that for cAMP, so that cGMP, in essence, functions as an inhibitor. An increase in cGMP concentration in these cells therefore blocks cAMP hydrolysis, thus promoting its accumulation. This mechanism is thought to play a role in nitric oxide (NO) signaling and in the control of the vascular smooth muscle tone. In platelets, aggregation is prevented by both an increase in cGMP or cAMP.

3.2.3. PDE3 ROLE IN INSULIN AND GROWTH FACTOR ACTION

In adipocytes, lipolysis is regulated by agents that activate adenylyl cyclase and that increase intracellular cAMP (i.e., catecholamines and glucagon). The

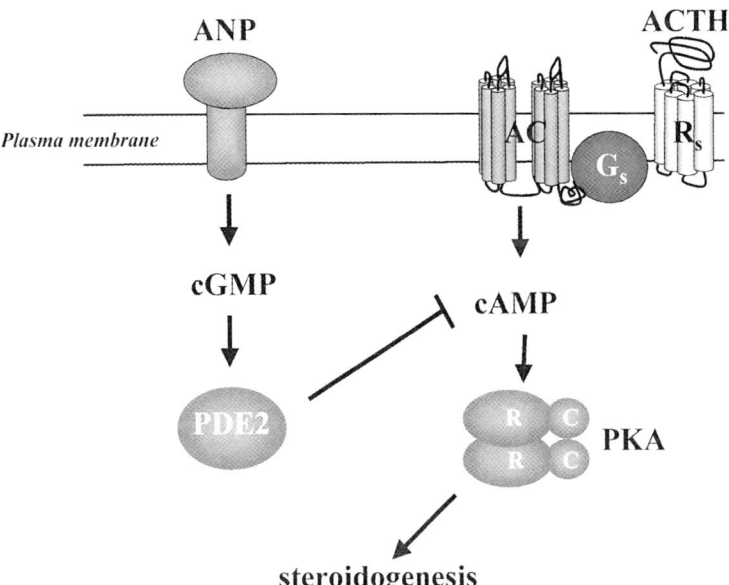

Fig. 7. Regulation of PDE2 by the atrial natriuretic peptide in the adrenal glomerulosa cells. The ANP binding guanylyl cyclase on the plasma membrane is shown on the left. ANP binding stimulates cGMP production and activation of the cGMP-stimulated PDE2. This in turn decreases cAMP levels, thus inhibiting the cAMP-stimulated steroidogenesis.

increase in cAMP activates the hormone sensitive lipase via a PKA-mediated phosphorylation. In a similar fashion, an activation of the cyclic nucleotide dependent pathway in hepatocytes stimulates glycogenolysis via a PKA-dependent activation of glycogen phosphorylase. Insulin is an important physiological regulator of both of these metabolic functions. This hormone inhibits lipolysis and glycogenolysis *via* suppression of the activity of the two enzymes. Although the exact mechanism of insulin action is still a matter of debate, it is generally accepted that an insulin-dependent decrease in cAMP levels mediates these antilipolytic and antiglycogenolytic effects on these cells. In 1970 Loten and Sneyd first reported that insulin activates cAMP hydrolysis in intact adipocytes. That this increased cAMP hydrolysis is important for insulin action has been confirmed by experiments with nonhydrolyzable cAMP analogs, as insulin does not block lipolysis activated by these analogs (Beavo, 1995). It is now clear that the activity of a PDE3 abundantly expressed in the adipose cells is the primary target for this activation. Degerman et al., 1997 have charted most of the steps of the signaling pathway impinging on this PDE3 in adipocytes (Fig. 8). The PDE3 activation is dependent on the insulin activation of phosphatidylinositol 3′-kinase (PI3K). Downstream of the PI3 kinase, a kinase activity that is activated by insulin and that phosphorylates PDE3B has been identified. On the basis of biochemical observations, as well as experiments with reconstitution systems, this kinase is most likely the recently discovered PKB–AKT. Finally, leptin, a recently discovered hormone that controls food intake, also signals through the PI3 kinases–PDE3 pathway. In β cells of the pancreas, leptin causes the activation of PDE3B, which leads to a marked inhibition of insulin secretion stimulated by the glucagon-like peptide-1.

Because the PI3K pathway is shared by many growth factor signaling pathways, it is likely that many growth related signals branch out of the tyrosine–Ras–MAP kinase cascade to regulate cAMP levels via a PDE3 phosphorylation. This is probably important for regulation of cell replication, as cAMP exerts control over several checkpoints of the cell cycle. In support of this hypothesis, a signaling pathway involving AKT and a PDE3 activation has been charted for the IGF1 activation of resumption of meiosis in the *Xenopus laevis* oocyte (Sadler, 1991; Andersen et al., 1998).

3.2.4. Regulation of PDE7 in T Lymphocytes

Antigen stimulation of inflammatory cells involves both activation of stimulatory pathways as well as removal of inhibitory constraints. For instance, activation of peripheral T cells is mediated by occupancy of both the T-cell receptor–CD3 complex and the CD28 costimulatory receptors. When both receptors are occupied, T cells are stimulated to produce inter-

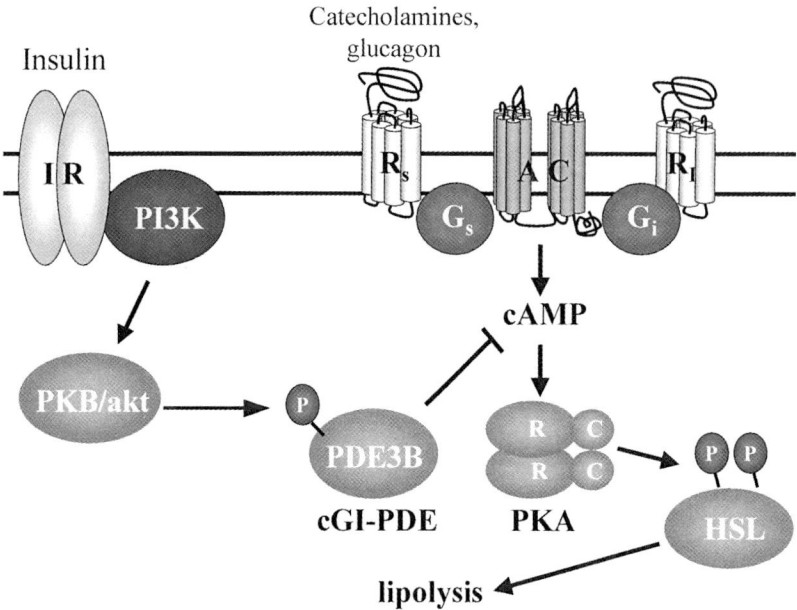

Fig. 8. Regulation of PDE3 by insulin. The insulin receptor (IR) tyrosine kinase is activated by insulin binding. The autophosphorylation of the receptor recruits the phosphatidylinositol 3′-kinase (PI3K), which increases the local accumulation of phosphatidylinositol polyphosphate. This in turn activates two lipid-dependent kinases (not shown) and recruits PKB/akt to the membrane. The phosphorylated and activated PKB/akt phosphorylates PDE3B, causing its activation. The increase in PDE activity decreases cAMP, thus blocking the hormone-sensitive lipase activation.

leukin-2 (IL-2) and to proliferate. The stimulatory limb of the pathway involves the activation of the mitogen-activated protein (MAP) kinase pathway and translocation of the nuclear factor of activated T cell (NFAT) to the nucleus. There is ample evidence that the MAP kinase pathway is inhibited by activation of the cAMP-dependent pathway and PKA. NFAT translocation is prevented by PKA phosphorylation. Furthermore, phosphorylation of *raf* by PKA prevents signaling through the MAP kinase pathway. Several reports have suggested that PDE activation may follow T-cell activation. PDE7 was recently discovered by using a yeast complementation screening. This PDE is expressed in skeletal muscle and in T cells. Recently, it has been reported that T-cell receptor activation causes an induction of PDE7 and that selectively blocking this induction with an antisense oligonucleotide prevents T-cell activation (Li et al., 1999). It is then probable that T-cell activation, among other signals, involves the relief of the cAMP inhibition through activation of PDEs.

3.3. PDEs Are Homeostatic Regulators of Cyclic Nucleotide Signaling and Cell Adaptation

The in-depth investigation of the cAMP pathway has established that signals that activate cyclase and increase cAMP synthesis are also associated with changes in PDE activity. Depending on the cell, the increase in cAMP triggers either negative feedback mechanisms activating PDEs or feed-forward mechanisms inhibiting PDE activity.

In many instances, hormones and neurotransmitters that activate cyclic AMP synthesis also cause an activation of cAMP hydrolysis. Signaling through cAMP may require minimal changes in concentration of the cyclic nucleotide within the cell. For instance, Goldberg and associates (1984) have demonstrated that in platelets where the adenine nucleotides were labeled with ^{18}O, agents that activate adenylyl cyclase also rapidly activate cAMP degradation. The net result of these contemporaneous changes in synthesis and degradation of the second messenger is a minor change in cAMP concentration but a large increase in cyclic nucleotide turnover. Since this initial observation, similar findings have been reported for many cells, pointing to the ubiquitous nature of the phenomenon. The exact physiological significance of these changes in turnover is still a matter of debate, even though there is a consensus that this is a means to increase sensitivity of the cell responsiveness and to protect the cell from excessive stimulation. It is also a mechanism that contributes to the termination of the stimulus. The transient pattern of cAMP accumulation

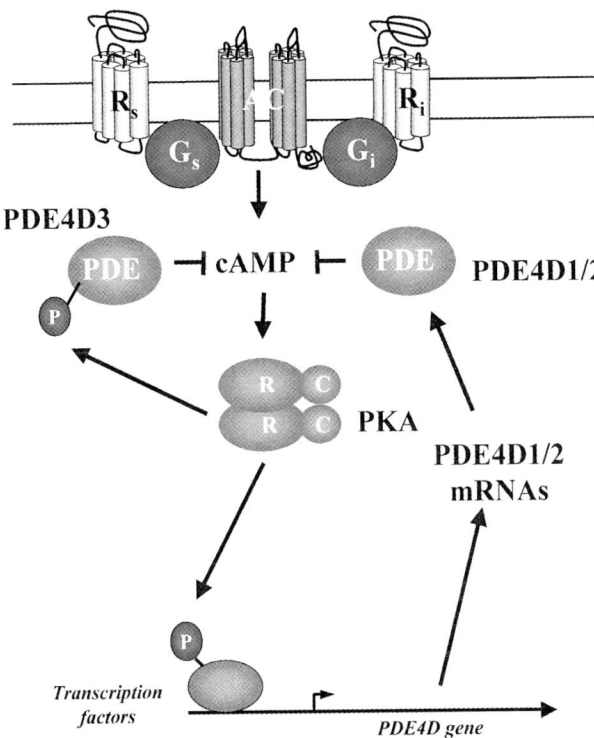

Fig. 9. Feedback regulation of PDE4 by cAMP. Short-term **(left)** and long-term **(right)** feedbacks regulate PDE4D. In the short-term feedback, an increase in cAMP activates PKA, which in turn phosphorylates and activates the PDE4D3 variant. Long-term regulation of PDE4D1/2 variants involves a PKA-mediated increase in transcription of the PDE4D gene, an increase in PDE4D mRNA, and synthesis of PDE4D1/2 variant proteins. These proteins are constitutively active in degrading cAMP.

resulting from rapid changes in synthesis and degradation may be necessary not only for quantitative but also for qualitative changes in cell function, adding specificity to the signal.

In platelets, hepatocytes, and adipocytes, it has been shown that an increase in cAMP causes activation of PKA, which phosphorylates and activates a PDE3 (Beavo, 1995; Gettys et al., 1987). More recently, a similar regulation has been reported for a PDE4 (Fig. 9) (Sette et al., 1994). A splicing variant of PDE4D is also activated via a PKA mediated phosphorylation at the N-terminus of the protein. In both cases, activation of these PDEs closely follows the increase in intracellular cAMP levels. This regulation most likely represents a dampening mechanism that limits cAMP accumulation and/or diffusion of cAMP throughout the cell. In support of this function is the observation that a PDE inhibitor specific for PDE3 or PDE4 has a major effect on cAMP accumulation. Incubation of thyroid cells with both the PDE4-specific inhibitor rolipram and thyroid-stimulating hormone (TSH) causes a 100-fold increase in cAMP levels, while TSH alone produces only a twofold increase in cAMP. Thus, PDE4s exert a major negative constraint on cAMP accumulation in these cells and oppose hormone stimulation of cyclases.

Although the above regulation involving PDE phosphorylation is short lived, there is an additional, long-term feedback mechanism that regulates PDE4 in most cells of the body. It has long been known that an increase in cAMP brings about an increase in PDE4 activity (Conti et al., 1991). This increase is sensitive to the protein synthesis inhibitor cycloheximide, indicating that PDE protein synthesis may be regulated by cAMP. Moreover, this activation requires PKA activation because cAMP does not produce an increase in PDE in a lymphoma cell line deficient in PKA. With the cloning of the PDE4 cDNAs, it has become clear that PDE4D mRNA is up-regulated by cAMP, with >100-fold increase observed in the testicular Sertoli cells after FSH stimulation (Swinnen et al., 1989). More interestingly, it has been shown that the gene that codes for the PDE forms phosphorylated by PKA also encodes the PDE variants induced by cAMP. This is due to the presence of intronic promoters that control the expression of truncated PDE forms. These small PDEs lack the N-terminus domain containing the PKA phosphorylation sites. Thus, products encoded by the same gene mediate short- and long-term feedback regulations of cAMP levels.

Since its first description, PDE4D mRNA induction has been demonstrated in a large number of cells, supporting the idea that this is a ubiquitous mechanism of regulation of cell responsiveness. Using the Sertoli cell response to FSH as a model, our laboratory has demonstrated that this regulation is a component of the desensitization process. That these two modes of PDE regulation are essential for cAMP signaling is indicated by the phenotype of PDE4D knockout mice. Decreased viability and fertility and growth retardation are several of the pleiotropic phenotypes observed in the PDE4D-null mice (Jin et al., 1999).

Although less frequently observed than the negative feedback, in some cases a positive feedback mechanism involving a PDE potentiates cAMP accumulation. This mechanism is mediated by phosphorylation of the CaM–PDE. PKA phosphorylates PDE1A in a domain in close proximity to the CaM binding domain (Sharma and Wang, 1985; Sonnenburg et al., 1995). This phosphorylation causes a major decrease in calmodulin affinity for the PDE, thus blocking the activation. The resulting inhibition of PDE activity and cAMP hydrolysis contributes to the overall increase in cAMP. This kind of regulation has been observed in Sertoli cells and in the olfactory system.

PDEs Are Involved in Signal Compartmentalization

Although this possibility has only recently received attention, it is becoming clear that PDEs are targeted to different cellular compartments and that this subcellular localization may establish distinct subcellular cyclic nucleotide pools with important consequences for cAMP signaling.

It has long been known that a considerable amount of PDE activity is recovered in the particulate fraction of the cell, indicating an interaction with different organelles. Recently, PDE subcellular localization has been further assessed by immunofluorescence and confocal microscopy (Juilfs et al., 1997). In the olfactory system, a PDE1 is located in the olfactory cilia, in close proximity to the olfactory signaling machinery, while a PDE4 is located in the body of the cell, at quite a distance from the site of cAMP synthesis (Fig. 10). In cultured cells, some PDE4 isoforms are predominantly located in the Golgi centrosomal region, while other PDE4 variants are soluble in the cytoplasm (Huston et al., 1996; Jin et al., 1998). These different localizations may have an important role in signal compartmentalization. This concept is supported by the observation that PKA regulatory subunits are targeted to different compartments through interactions with anchoring proteins (AKAPs) (Rubin, 1994; Scott and McCartney, 1994). The same may be true for PDEs. Although indirectly, observations made with PDE3 and PDE4 inhibitors support the presence of different cAMP pools in a cell. In spite of the fact that treatment of mesangial cells of the glomerolus with PDE3 and PDE4 inhibitors bring about a similar increase in cAMP and PKA activation, these inhibitors produce distinct effects on two cell functions. While PDE4 inhibitors have no effect, PDE3 inhibition blocks cell replication. Conversely, PDE4 inhibition blocks reactive oxygen metabolite production, while PDE3 inhibitors are without effect. This has been seen as an indication that the two PDEs control two different cAMP pools present in these cells.

Experiments in cardiomyocytes have further highlighted the importance of cAMP diffusion in the regulation of channels. By probing Ca^{2+} L channel permeability along the plasma membrane of these cardiomyocytes (Jurevicius and Fischmeister, 1996),

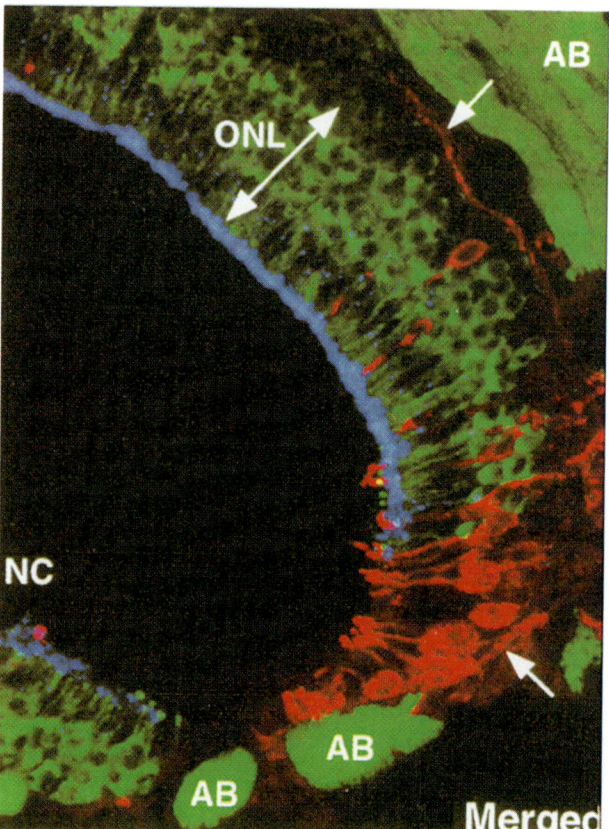

Fig. 10. Subcellular localization of PDE1 and PDE4 in the olfactory epithelium. Merged triple-labeled image demonstrating unique subsets of olfactory neurons expressing PDE1 and PDE4 and ACIII. The PDE4A staining is in *green* and is localized in the body of the olfactory neurons, while the colocalization of PDE1C2 and ACIII is in *blue* in the olfactory cilia. The *red* corresponds to a different subset of olfactory neurons expressing PDE2 (*white arrow*). ONL, olfactory neurons; AB, axon bundles; NC, nasal cavity (Courtesy of D. Juilfs).

it has been shown that β-adrenergic stimulation causes cAMP accumulation and channel openings only in the vicinity of the applied stimulus. Conversely, when PDE inhibitors are present, cAMP diffuses throughout the cell, leading to the opening of channels at a considerable distance from where the stimulus has been applied. This experiment supports the concept that PDEs play an important role in limiting the diffusion of the cyclic nucleotide.

4. PDES AS TARGETS FOR PHARMACOLOGICAL INTERVENTION

Inhibition of PDE is one of the most ancient pharmacological treatments used by mankind. It is believed that stone-age man discovered the stimulating properties of extracts of leaves and seeds from

Thea sinensis, Coffea arabica, and *Theobroma cacao,* and mention of the stimulating properties of these extracts is found in records of the most ancient civilizations. The active principles in these extracts are alkaloids called xanthines, the most potent being theophylline and caffeine. The effect of caffeine and theophylline on PDE activity was discovered soon after the discovery of cAMP. It was actually adopted as a diagnostic tool of activation of the cAMP-dependent pathway. Those cell functions that are potentiated by xanthines are most likely mediated by activation of the cyclic nucleotide pathway. More comprehensive studies have demonstrated that xanthines have additional effects on Ca^{2+} sequestration and on the adenosine pathway, but it is still accepted that a good portion of their stimulatory activity is due to the increase in cAMP that follows inhibition of the PDE activity. Treatment with xanthines produces a wide range of pharmacological effects, including stimulation of the central nervous system (CNS), increased urine production by the kidney, increased cardiac functions, and relaxation of the bronchi and vascular smooth muscle. Theophylline is still the most prescribed antiasthma drug and is used for cardiovascular disorders (Torphy, 1998). While PDE inhibition is regarded as an effective strategy to manipulate intracellular cyclic nucleotide concentration and therefore cell responses, the use of xanthines is limited by the fact that they inhibit all PDEs present in the body. Because of this lack of selectivity, xanthines produce a large array of effects, restricting the useful therapeutic margin to a narrow window. The development of a second generation of inhibitors selective for different PDE families has confirmed the concept that increased selectivity can be obtained by targeting different PDE families (Thompson, 1991). For instance, inhibitors of PDE3 have useful cardiovascular effects but produce little stimulation of the CNS, while PDE4 inhibitors are largely devoid of cardiovascular effects, while retaining the CNS and pulmonary effects of theophylline. The potency of these inhibitors has also been greatly improved. While xanthines inhibit most PDEs in the millimolar range of concentration, there are now inhibitors selective for PDE3 or PDE4 that are effective in the nanomolar range. Here we will briefly review the most relevant pharmacological application of PDE inhibitors.

4.1. PDE Inhibitors and Cardiovascular Disorders

By virtue of their inotropic and vasorelaxant effects, PDE3 inhibitors have been widely tested for cardiovascular disorders (Beavo, 1995). They are currently used in acute settings to treat patients during cardiac surgery or for patients awaiting heart transplantation. They have been also tested as a possible replacement of, or in conjunction with, digoxin in congestive heart failure. Although a large multicentric trial with milrinone was stopped prematurely because of increased death rate in patients receiving this drug, it is felt that adjustment of the dosage and the availability of new compounds will widen the use of these PDE3 inhibitors for this indication. Because of its inhibition of PDE3 activity in platelets, the PDE3 inhibitor cilostazol is currently prescribed in Japan as an antithrombotic agent and it has been recently approved for the treatment of vascular stenosis causing claudicatio intermittens (Sorkin and Markham, 1999).

With the discovery of the second messenger nitric oxide and its link to activation of guanylyl cyclase, there has been a renewed interest in inhibition of cGMP-selective PDEs as a means to produce vascular smooth muscle relaxation. While testing several compounds that inhibit the cGMP-specific PDE5, it was discovered that the corpus cavernous of the penis was particularly sensitive to these drugs. This observation has lead to the development of drugs useful in the treatment of penile erectile dysfunction (Licht, 1999). Sildenafil citrate (Viagra) is one of the most widely publicized drugs produced at the end of this century. Another area of potential use of PDE5 inhibitors is for the treatment of pulmonary hypertension (Beavo, 1995).

4.2. PDE Inhibitors and Inflammatory Disorders

The molecular probes developed during the characterization of the different PDE families have been widely used to study the expression of the different PDEs in various organs. This has led to the discovery that inflammatory cells express mostly PDE4 (Torphy, 1998). This observation, together with the notion that activation of the cAMP signaling pathway blocks activation of lymphocytes, have prompted the investigation of PDE4 inhibition as a strategy to suppress inflammation. There is now a large body of information available on the antiinflammatory effects of PDE4 inhibitors, and clinical trials are underway. The use of theophilline for the treatment of asthma is limited by the general toxicity and the cardiovascular effects of this drug. With the development of PDE4 selective inhibitors, the cardiovascular effects of theophilline should be eliminated.

4.3. PDE Inhibitors and CNS Regulation

While there are no PDE inhibitors currently on the market for the treatment of CNS dysfunction, promising studies are underway with different classes of compounds. Rolipram, a PDE4 inhibitor, was initially developed for the treatment of depression, but it was found in general no more efficacious than the more traditional drugs. However, studies are underway with new more potent PDE4 inhibitors for this indication. There is also interest in developing PDE1 and PDE4 inhibitors to improve or increase cognitive functions of the brain. It has been reported that these inhibitors have nootropic effects (memory enhancer). This property may be useful for the treatment of Alzheimer's or Parkinson patients.

BIBLIOGRAPHY

Andersen CB, Roth RA, Conti M. Protein kinase B/Akt induces resumption of meiosis in Xenopus oocytes. J Biol Chem 1998; 273:18705–18708.

Artemyev NO, Surendran R, Lee JC, Hamm HE. Subunit structure of rod cGMP-phosphodiesterase. J Biol Chem 1996; 271 (41):25382–25388.

Barber R, Clark RB, Kelly LA, Butcher RW. A model of desensitization in intact cells. Adv Cyclic Nucleotide Res 1987; 9:507–516.

Beavo JA. Cyclic nucleotide phosphodiesterases: functional implications of multiple isoforms. Physiol Rev 1995; 75:725–748.

Butcher RW, Sutherland EW. Adenosine 3′,5′-monophosphate in biological materials. J Biol Chem 1962; 237:1244–1250.

Boekhoff I, Breer H. Termination of second messenger signaling in olfaction. Proc Natl Acad Sci USA 1992; 89:471–474.

Bunemann M, Lee KB, Pals-Rylaarsdam R, Roseberry AG, Hosey MM. Desensitization of G-protein-coupled receptors in the cardiovascular system. Annu Rev Physiol 1999; 61:169–192.

Butcher RW, Sutherland EW. Adenosine 3′,5′-monophosphate in biological materials. J Biol Chem 1962; 237:1244–1250.

Charbonneau H, Beier N, Walsh KA, Beavo JA. Identification of a conserved domain among cyclic nucleotide phosphodiesterases from diverse species. Proc Natl Acad Sci USA 1986; 83:9308–9312.

Conti MJ, In SC. The molecular biology of cyclic nucleotide phosphodiesterases. Adv Nucleic Acid Res 1999; 63:1–38.

Conti M, Jin SL, Monaco L, Repaske DR, Swinnen JV. Hormonal regulation of cyclic nucleotide phosphodiesterases. Endocrinol Rev 1991; 12:218–234.

Conti M, Nemoz G, Sette C, Vicini E. Recent progress in understanding the hormonal regulation of phosphodiesterases. Endocrinol Rev 1995; 16:370–389.

Davis RL, Dauwalder B. The *Drosophila dunce* locus: learning and memory genes in the fly. Trends Genet 1991; 7:224–229.

Degerman E, Belfrage P, Manganiello V. 1997; 272:6823–6826.

de Rooij J, Zwartkruis FJ, Verheijen MH, Cool RH, Nijman SM, Wittinghofer A, Bos JL. Epac is a Rap1 guanine-nucleotide-exchange factor directly activated by cyclic AMP. Nature 1998; 396:474–477.

Erneux C, Van Sande J, Miot F, Cochaux P, Decoster C, Dumont JE. A mechanism in the control of intracellular cAMP level: the activation of a calmodulin-sensitive phosphodiesterase by a rise of intracellular free calcium. Mol Cell Endocrinol 1985; 43:123–134.

Farber DB, Danciger JS, Aguirre G. The beta subunit of cyclic GMP phosphodiesterase mRNA is deficient in canine rod-cone dysplasia 1. Neuron 1992; 9:349–356.

Fisher DA, Smith JF, Pillar JS, St. Denis SH, Cheng JB. Isolation and characterization of PDE9A, a novel human cGMP-specific phosphodiesterase. J Biol Chem 1998a; 273:15559–15564.

Fisher DA, Smith JF, Pillar JS, St. Denis SH, Cheng JB. Isolation and characterization of PDE8A, a novel human cAMP-specific phosphodiesterase. Biochem Biophys Res Commun 1998; 246:570–577.

Florio SK, Prusti RK, Beavo JA. Solubilization of membrane-bound rod phosphodiesterase by the rod phosphodiesterase recombinant delta subunit. J Biol Chem 1996; 271:24036–24047.

Francis SH, Colbran JL, McAllister-Lucas LM, Corbin JD. Zinc interactions and conserved motifs of the cGMP-binding cGMP-specific phosphodiesterase suggest that it is a zinc hydrolase. J Biol Chem 1994; 269:22477–22480.

Francis SH, Corbin JD. Structure and function of cyclic nucleotide-dependent protein kinases. Annu Rev Physiol 1994; 56:237–272.

Gettys TW, Blackmore PF, Redmon JB, Beebe SJ, Corbin JD. Short-term feedback regulation of cAMP by accelerated degradation in rat tissues. J Biol Chem 1987; 262:333–339.

Goldberg ND, Walseth TF, Eide SJ, Krick TP, Kuehn BL, Gander JE. Cyclic AMP metabolism in intact platelets determined by 18O incorporation into adenine nucleotide alpha-phosphoryls. Adv Cyclic Nucleotide Protein Phosphorylation Res 1984; 16:363–379.

Harden TK, Evans T, Hepler JR, Hughes AR, Martin MW, Meeker RB, Smith MM, Tanner LI. Regulation of cyclic AMP metabolism by muscarinic cholinergic receptors. Adv Cyclic Nucleotide Protein Phosphorylation Res 1985; 19:207–220.

Houslay MD Milligan G. Tailoring cAMP-signalling responses through isoform multiplicity. Trends Biochem Sci 1997; 22:217–224.

Huston E, Pooley L, Julien P, Scotland G, McPhee I, Sullivan M, Bolger G, Houslay MD. The human cyclic AMP-specific phosphodiesterase PDE–46 (HSPDE4A4B) expressed in transfected COS7 cells occurs as both particulate and cytosolic species that exhibit distinct kinetics of inhibition by the antidepressant rolipram. J Biol Chem 1996; 271:31334–31344.

Jin CS, Busnick T, Lan L, Conti M. Subcellular Localization of the PDE variants. J Biol Chem 1998; 273:19672–19678.

Jin S-LC, Richard F, Kuo W-P, D'Ercole AJ, Conti M. Impaired growth and fertility of cAMP-specific phosphodiesterase PDE40-deficient mice. J Biol Chem 1999; 96:11998–12003.

Jin SL, Swinnen JV, Conti M. Characterization of the structure of a low Km, rolipram-sensitive cAMP phosphodiesterase. Mapping of the catalytic domain. J Biol Chem 1992; 267:18929–18939.

Juilfs DM, Fulle HJ, Zhao AZ, Houslay MD, Garbers DL, Beavo JA. A subset of olfactory neurons that selectively express cGMP-stimulated phosphodiesterase (PDE2) and guanylyl cyclase-D define a unique olfactory signal transduction pathway. Proc Natl Acad Sci USA 1997; 94:3388–3395.

Jurevicius J, Fischmeister R. cAMP compartmentation is responsible for a local activation of cardiac Ca^{2+} channels by beta-adrenergic agonists. Proc Natl Acad Sci USA 1996; 93:295–299.

Kawasaki H, Springett GM, Mochizuki N, Toki S, Nakaya M, Matsuda M, Housman DE, Graybiel AM. A family of cAMP-binding proteins that directly activate Rap1. Science 1998; 282:2275–2279.

Li L, Yee C, Beavo JA. CD3- and CD28-dependent induction of PDE7 required for T cell activation. Science 1999; 283:848–851.

Licht MR. Use of oral sildenafil [Viagra] in the treatment of erectile dysfunction. Compr Ther 1999; 25:90–94.

Lim J, Pahlke G, Conti M. Activation of the phosphodiesterase PDE4D3 by phopshorylation: identification and function of an inhibitory domain. J Biol Chem 1999; 274:19677–19685.

Loten EG, Sneyd JG. An effect of insulin on adipose-tissue adenosine 3':5'-cyclic monophosphate phosphodiesterase. Biochem J 1970; 120:187–193.

MacFarland RT, Zelus BD, Beavo JA. High concentrations of a cGMP-stimulated phosphodiesterase mediate ANP-induced decreases in cAMP and steroidogenesis in adrenal glomerulosa cells. J Biol Chem 1991; 266:136–142.

Maelicke A. The cGMP-gated channel of the rod photoreceptor—a new type of channel structure? Trends Biochem Sci 1990; 15:39–40.

Maurice DH, Haslam RJ. Molecular basis of the synergistic inhibition of platelet function by nitrovasodilators and activators of adenylate cyclase: inhibition of cyclic AMP breakdown by cyclic GMP. Mol Pharmacol 1990; 37:671–681.

Michaeli T, Bloom TJ, Martins T, Loughney K, Ferguson K, Riggs M, Rodgers L, Beavo JA, Wigler M. Isolation and characterization of a previously undetected human cAMP phosphodiesterase by complementation of cAMP phosphodiesterase-deficient *Sacharomyces cerevisiae*. J Biol Chem 1993; 268(17):12925–12932.

Monaco L, Vicini E, Conti M. Structure of two rat genes coding for closely related rolpram-sensitive cAMP-phosphodiesterases. J Biol Chem 1994; 269:347–357.

Qiu Y, Davis RL. Genetic dissection of the learning/memory gene dunce of Drosophila melanogaster. Genes Dev 1993; 7:1447–1458.

Rubin CS. A kinase anchor proteins and the intracellular targeting of signals carried by cyclic AMP. Biochim Biophys Acta 1994; 1224:467–479.

Sadler SE. Type III phosphodiesterase plays a necessary role in the growth-promoting actions of insulin, insulin-like growth factor-I, and Ha p21ras in *Xenopus laevis* oocytes. Mol Endocrinol 1991; 5:1939–1946.

Scott JD, McCartney S. Localization of A-kinase through anchoring proteins. Mol Endocrinol 1994; 8:5–116.

Sette C, Iona S, Conti M. The short-term activation of a rolpram-sensitive, cAMP-specific phosphodiesterase by thyroid-stimulating hormone in thyroid FRTL-5 cells is mediated by a cAMP-dependent phosphorylation. J Biol Chem 1994; 269:9245–9252.

Shakur Y, Pryde JG, Houslay MD. Engineered deletion of the unique N-terminal domain of the cyclic AMP-specific phosphodiesterase RD1 prevents plasma membrane association and the attainment of enhanced thermostability without altering its sensitivity to inhibition by rolipram. Biochem J 1993; 292:677–686.

Sharma RK, Wang JH. Differential regulation of bovine brain calmodulin-dependent cyclic nucleotide phosphodiesterase isoenzymes by cyclic AMP-dependent protein kinase and calmodulin-dependent phosphatase. Proc Natl Acad Sci USA 1985; 82:2603–2607.

Soderling SH, Bayuga SJ, Beavo JA. Identification and characterization of a novel family of cyclic nucleotide phosphodiesterases. J Biol Chem 1998; 273:15553–15558.

Soderling SH, Bayuga SJ, Beavo JA. Isolation and characterization of a dual-substrate phosphodiesterase gene family: PDE10A. Proc Natl Acad Sci 1999; 96(12):7071–7076.

Sonnenburg WK, Mullaney PJ, Beavo JA. J Biol Chem 1991; 266:17655–17660.

Sonnenburg WK, Seger D, Kwak KS, Huang J, Charbonneau H, Beavo JA. Identification of inhibitory and calmodulin-binding domains of the PDE1A1 and PDE1A2 calmodulin-stimulated cyclic nucleotide phosphodiesterases. J Biol Chem 1995; 270:30989–31000.

Sorkin EM, Markham A. Cilostazol. Drugs Aging 1999; 14:63–71; discussion 72–3.

Su YF, Cubeddu L, Perkins JP. Regulation of adenosine 3':5'-monophosphate content of human astrocytoma cells: desensitization to catecholamines and prostaglandins. J Cyclic Nucleotide Res 1976; 2:257–270.

Sutherland EW, Rall TW. Fractionation and characterization of a cyclic adenine ribonucleotide formed by tissue particles. J Biol Chem 1958; 232:1077–1091.

Swinnen JV, Joseph DR, Conti M. The mRNA encoding a high-affinity cAMP phosphodiesterase is regulated by hormones and cAMP. Proc Natl Acad Sci USA 1989; 86:8197–8201.

Thompson WJ. Cyclic nucleotide phosphodiesterases: pharmacology, biochemistry and function. Pharmacol Ther 1991; 51:13–33.

Thompson WJ, Appleman MM. Multiple cyclic nuceotide phosphodiesterase activities in rat brain. Biochemistry 1971; 10:311–316.

Torphy TJ. Phosphodiesterase isozymes: molecular targets for novel antiasthma agents. Am J Respir Crit Care Med 1998; 157:351–370.

Whalin ME, Scammell JG, Strada SJ, Thompson WJ. Phosphodiesterase II, the cGMP-activatable cyclic nucleotide phosphodiesterase, regulates cyclic AMP metabolism in PC12 cells. Mol Pharmacol 1991; 39:711–717.

Yarfitz S, Hurley JB. Transduction mechanisms of vertebrate and invertebrate photoreceptors. J Biol Chem 1994; 269:14329–14332.

16 Cyclic Nucleotide-Dependent Protein Kinases

Sharron H. Francis and Jackie D. Corbin

CONTENTS

INTRODUCTION
VARIATION IN TISSUE DISTRIBUTION OF ISOZYMIC FORMS OF PKA AND PKG
SIMILAR DOMAIN STRUCTURES OF PKA AND PKG HOLOENZYMES
DISTINCT FEATURES OF PKA AND PKG THAT AFFECT THEIR
    SUBSTRATE SPECIFICITIES
MECHANISMS USED IN THE ACTIVATION OF PKA AND PKG
EXISTENCE AND USE OF SELECTIVE INHIBITORS OF PKA AND PKG
SELECTED SUBCELLULAR LOCALIZATIONS OF PKA AND PKG
INTERACTION WITH PROTEIN KINASE INHIBITOR AND
    PROTEIN KINASE INHIBITOR PEPTIDES
NEGATIVE FEEDBACK AND FEED-FORWARD CONTROL OF
    CYCLIC NUCLEOTIDE PATHWAYS
THE USE OF TRANSGENIC MICE IN PROVIDING NEW INSIGHTS INTO THE
    ROLES OF PKA AND PKG IN CYCLIC NUCLEOTIDE SIGNALING
SUMMARY
SELECTED READINGS

1. INTRODUCTION

The two cyclic nucleotides, cAMP and cGMP, are ubiquitous intracellular second messengers. The discovery of cAMP by Earl Sutherland and the demonstration that this molecule plays a key role in the actions of many hormones established the entire field of second messenger signaling and revolutionized our understanding of cellular biochemistry. cGMP was subsequently identified and has also been shown to be an important second messenger for mediating specific signaling pathways. As our understanding of the physiology of cAMP and cGMP signaling pathways continues to grow, the opportunities for interventional therapies that target the intracellular receptors for cAMP and cGMP are also rapidly expanding. The intracellular concentrations of cAMP and cGMP are modulated in response to a panoply of hormonal and chemical signals, and the resulting changes in the levels of these nucleotides provide for the key roles of these molecules in regulating myriad physiological processes.

The responsiveness of a given cell to changes in cAMP and/or cGMP is dictated by the properties of the proteins that serve as the intracellular receptors for cAMP and cGMP in that particular tissue. In most cells, there are multiple cyclic nucleotide receptors that contribute to cyclic nucleotide signaling (Fig. 1), including cAMP-dependent protein kinases (PKA) and cGMP-dependent protein kinases (PKG), the cyclic nucleotide-gated cation channels, the allosteric cyclic nucleotide-binding sites on several families of cyclic nucleotide phosphodiesterases (PDE),

From: *Principles of Molecular Regulation* (P. M. Conn and A. R. Means, eds.), © Humana Press Inc., Totowa, NJ.

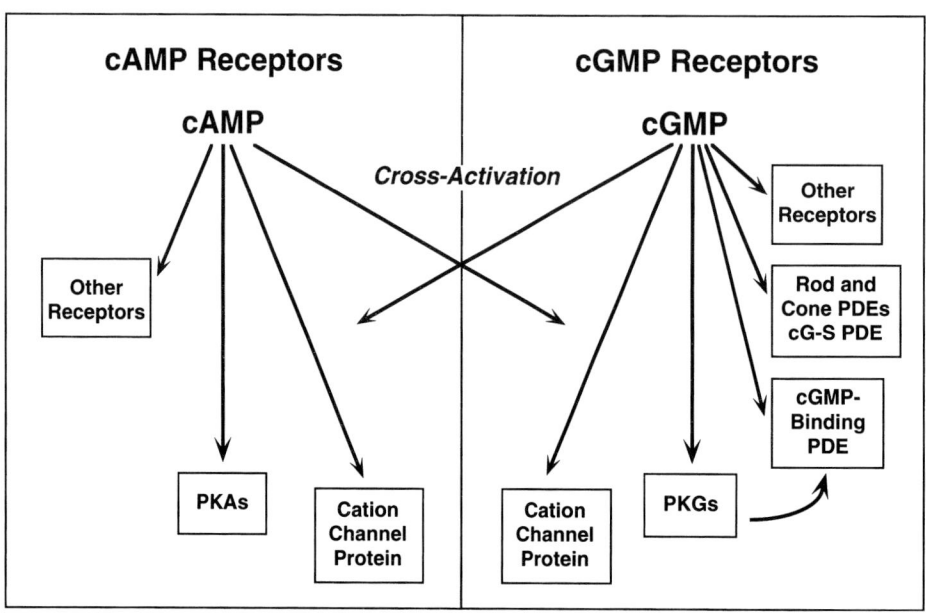

Fig. 1. Intracellular receptors for cAMP and cGMP in mammals.

the catalytic sites of certain cyclic nucleotide PDEs such as the cGMP-inhibited phosphodiesterase, and perhaps other as yet unidentified proteins. Members of a recently identified family of guanine-nucleotide-exchange factors (GEFs) bind cAMP at a single site which is homologous to the cyclic nucleotide-binding sites in PKA and PKG; in the cAMP-bound form, these proteins activate Ras in a PKA-independent manner. Although all of these receptors play important roles, the PKAs and PKGs have been shown to mediate a major portion of the physiological responses to cAMP and cGMP. As such, the PKAs and PKGs are important pharmacological targets, and our understanding of these enzymes has increased markedly in recent years. This chapter describes many of the functional properties of the PKAs and PKGs and considers approaches that continue to prove useful in studying the roles of these enzymes in intact tissues.

Both PKA and PKG are ligand-activated protein kinases, that is, cyclic nucleotide binding increases their catalytic activities, and although PKA and PKG are quite distinct in certain respects, they are also similar in a number of structural and functional features. Many of the findings derived from studies of PKA and PKG have also provided insights into the structures and functions of the superfamily of protein kinases, which may comprise 3–5% of the total protein species in mammals.

PKA was first described in 1963–1964 and was the first recognized ligand-activated protein kinase; PKG was discovered in 1970. When activated, each of these kinases catalyzes transfer of the γ-phosphate of ATP to selected serines or threonines in a consensus phosphorylation sequence in peptides/proteins that contains two basic residues (arginine or lysine), another residue C-terminal to the basic amino acids, followed by a serine or threonine residue (e.g., -RRXSX-). Although some determinants that are required to form a typical consensus phosphorylation sequence are shared by PKA and PKG, there are usually clear distinctions between the roles of these two kinases in intact tissues. PKA has been studied more exhaustively owing to the earlier emphasis on the cAMP signaling pathway. However, the importance of the cGMP signaling pathway has emerged more recently, and interest in PKGs has mushroomed.

It is assumed that alteration of either intracellular cAMP or cGMP modulates the activity of the corresponding protein kinase (Fig. 1). Cross-activation of either kinase by the other cyclic nucleotide may also exist in some cells or in certain pathological conditions. PKA and PKG activities are also modulated by autophosphorylation, subcellular localization, translocation, or change in the amount of the enzymes. It is now clear that there are hundreds of proteins that can be phosphorylated by PKA and/or PKG (a process known as *heterophosphorylation,* which is distinguished from *autophosphorylation*); the abundance

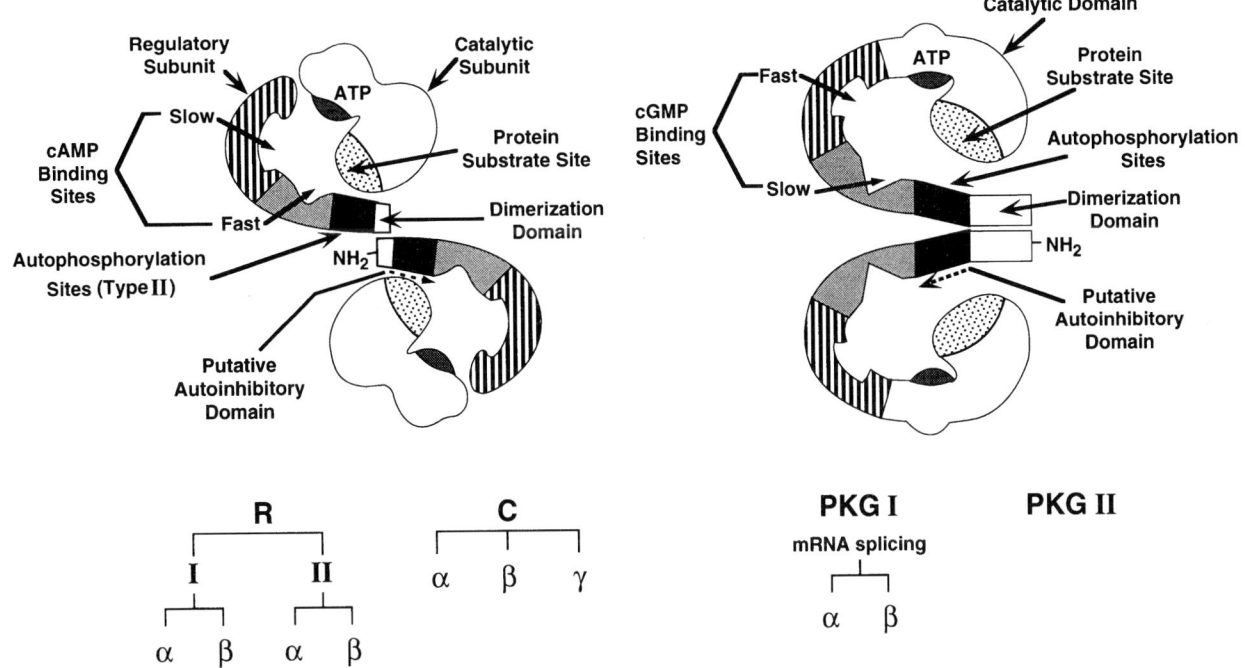

Fig. 2. Working models of PKAs and PKGs. The various functional domains that contribute to dimerization, regulation, and catalysis are depicted. Major isozymic forms of the kinases are described below the models; in addition to the PKA isoforms shown, there are numerous splice variants of both the R-subunits and certain C-subunits.

of potential PKA and/or PKG protein substrates is not surprising, as sequences that qualify as consensus phosphorylation sites for these kinases are quite common in many proteins. However, whether phosphorylation occurs or is physiologically relevant for a putative substrate must be carefully established as the consensus sequence may not be available to the particular kinase within cells. Furthermore, most kinases are known to phosphorylate a number of sites that do not conform to ideal consensus sequences for phosphorylation.

2. VARIATION IN TISSUE DISTRIBUTION OF ISOZYMIC FORMS OF PKA AND PKG VARY SIGNIFICANTLY IN THEIR TISSUE DISTRIBUTIONS

A number of isoforms of PKA and PKG have been identified; PKA is present in all mammalian tissues examined, whereas the tissue distribution of the PKG isoforms is more restricted. Substantial effort has been devoted to ascertaining the molecular basis for the differences in the various isoforms of PKA and PKG and to understanding the physiological importance, if any, of these differences. PKA is a heterotetramer comprised of two regulatory subunits (R-subunit) and 2 catalytic subunits (C-subunit). PKGs are homodimers in which the regulatory and catalytic domains are fused in a single polypeptide chain.

2.1. PKA Regulatory Subunit Isoforms Confer a Number of Specific Behavioral Properties on the PKA Holoenzyme

There are two major classes of isoforms (RI [43 kDa] and RII [45 kDa]) of the mammalian R-subunit, and there are subtypes (RIα, RIβ, RIIα, and RIIβ) within each isoform, which are products of different genes (Fig. 2). Additional forms of these R-subunits are generated by alternative mRNA splicing. To a large extent, the properties of the R-subunit isoforms confer the major chromatographic and functional differences of the various PKA holoenzymes. In the absence of cAMP, each of the RI and RII subunits can form holoenzyme complexes with each of the C-subunit isoforms with an affinity of ~0.2 nM. Both RI and RII are found in most tissues, but the relative abundance of these isoforms varies widely depending on the species and tissue; therefore, great caution must be exercised when attributing specific physiological processes to a given isoform because in a different species the tissue in question may have a different pattern of abundance of the respective R-subunit isoforms. For example, rat heart contains predominately RI, which is primarily soluble, while bovine heart contains almost entirely RII, which is largely particulate. Two subclasses of both RI and RII isoforms are denoted RIα and RIβ, and RIIα and RIIβ respectively. RIα and RIIα are expressed in many tissues, and RIβ and RIIβ appear to be more selectively expressed (central nervous system, neuroendocrine tissues, Sertoli cells, ovarian granulosa cells, and lung). A number of studies using varied approaches suggest that either the type I or type II PKA can elicit the same physiological responses in many instances. This is not surprising, because upon activation of either the type I or type II PKA holoenzyme, the same population of C-subunits would be activated; these activated C-subunits would then target the same substrates assuming equal distribution of all components within the cell. However, in certain instances, a specific role for a particular PKA is indicated. This could relate to differences in sensitivity to cAMP (type I PKA has slightly higher cAMP affinity), to localization of a particular PKA isoform near the targeted substrate(s), or to restrictions in the pattern of RI and RII expression during development.

2.2. Catalytic Subunit Isozymes Are Distinct

There are three main isoforms of the mammalian C-subunit (M_r ~ 40 kDa) (Cα; Cβ1,2,3; and Cγ) (Fig. 2), and the catalytic properties of these C isoforms differ somewhat. Cα and Cβ are ~91% identical in their primary sequences; the Cβ variants (Cβ1, Cβ2, and Cβ3) are products of alternative mRNA splicing and differ at their N-termini. The Cβ2 and Cβ3 isoforms are highly expressed in neuronal tissues. Expression of Cγ appears to be restricted to primate testis; the gene for Cγ lacks introns and apparently arose as a retroposon.

Excluding alternatively spliced forms with four R isoforms that can form homodimers, and three isoforms of C, it is possible that 24 different PKA holoenzymes are present in mammalian tissues. In addition, heterodimers of the R-subunits have been reported and would increase the number of potential isoforms. The molecular properties of these respective R- and C-subunits produce differences in the interactions that provide for PKA holoenzyme formation. As a result, differences in enzyme stability and the K_a of cAMP for these various isoforms are observed.

2.3. The Various PKG Isoforms Have Distinct Structures and Functions

Two main classes of PKG isoforms (PKGI and PKGII) occur in mammalian tissues (Fig. 2). These forms differ significantly in their tissue distributions, substrate specificities, affinities for cGMP, and subcellular localizations. In contrast to the PKA system, PKGI and PKGII have not yet been found to coexpress substantially in the same cell type. The PKGs are typically abundant in more selected tissues including smooth muscle, smooth muscle related tissues, platelets, cerebellum, intestinal epithelial cells, lung, and kidney. The concentrations of these enzymes vary significantly in different tissues. Assuming equal distribution in cytosol, it can be calculated that the typical cellular level of PKA is 0.2–2 μM, but the concentration of PKGs ranges from barely detectable levels (e.g., in adipose tissue and liver) to nearly 1 μM (e.g., in platelets). Except in certain tissues such as smooth muscle where PKGs are particularly abundant (~0.1–0.2 μM in vascular smooth muscle), the PKA concentration typically exceeds that of the PKG by 10- to 100-fold. The PKGI and PKGII also show considerable specificity in their tissue distributions, and whether these distribution patterns will hold across species lines is not known. However, evidence suggests that for the PKGs much of the selectivity in isozyme distribution is physiologically relevant.

There are two sublcasses of the PKGI isoform (PKGIα and PKGIβ) that arose by alternative mRNA splicing and to date only one form of PKGII is known (Fig. 2). The PKGIα, PKGIβ, and PKGII forms exist as homodimers (Fig. 2) containing ~75-kDa and 86-kDa subunits, respectively. The PKGI isoforms are primarily cytosolic, although PKGI(is membrane bound in platelets and is also localized to the neuromuscular junction in skeletal muscle. The PKGI isoforms more closely resemble RI than RII of PKA based on comparisons of amino acid sequence and interaction with MgATP. PKGII appears to be largely membrane bound, although it is enzymatically active when overexpressed as a soluble enzyme; among other tissues, it is abundant in the brush border of intestinal mucosa epithelial cells, selected kidney cells, and chondrocytes. The primary structures of the PKGIα and Iβ differ only in the N-terminal ~100 amino acids where they are only 36% identical. This limited difference in primary structure translates into significant enzymological differences despite the fact the remaining primary structure (which contains the cGMP binding domain and the catalytic domain) is identical in these two isoforms. PKGIα predominates in bovine lung (>90%), bovine trachealis smooth muscle (70% PKGIα vs 30% PKGIβ), cerebellum, and uterine smooth muscle. PKGIα and PKGIβ are present in approximately equal proportions in vascular smooth muscle (pig coronary arteries, bovine and human aortas), and PKGIβ predominates in rat gastrointestinal smooth muscle and rabbit corpus cavernosum.

3. SIMILAR DOMAIN STRUCTURES OF PKA AND PKG HOLOENZYMES

PKA and PKG are homologous in primary structure, but they differ significantly in many respects (Fig. 2). The inactive mammalian PKA is a tetramer comprised of two R-subunits and two C-subunits. In contrast, each of the PKGs is a homodimer in which each monomer contains a regulatory domain and a catalytic domain (Fig. 2). When cAMP binds to the two cAMP binding sites on the R-subunit of the PKA holoenzyme, the R- and C-subunits can dissociate, and the enzyme is activated. The R-subunits that dissociate from the PKA holoenzyme complex remain dimerized, but the activated C-subunits can be isolated as monomers. However, there is evidence that activation of the kinase activity can occur in the absence of physical dissociation of the holoenzyme. Both scenarios may occur in the cellular milieu with some of the C-subunit remaining juxtaposed to the R-subunits while some is free to diffuse to more distant targets. As the regulatory and catalytic components of PKGs are fused in a single polypeptide chain, they obviously cannot dissociate upon activation. However, activation of PKG by cGMP binding produces a marked conformational change in PKG as indicated by a progressive elongation of the enzyme that accompanies the cGMP-mediated activation.

In the absence of cyclic nucleotide, the catalytic activities of the PKA and PKG holoenzymes are latent owing to the interaction of the respective catalytic sites with an autoinhibitory domain that is located near the N-terminus of either R-subunit or PKG (Fig. 2). The autoinhibitory domain consists of a substrate-like sequence in the polypeptide chain as well as other structural components that are less well defined. Activation of either PKA or PKG disrupts these interactions, thereby availing the catalytic site for interaction with potential substrates.

Many of the most thorough biochemical studies of PKA have utilized either the isolated R- or C-subunit in the absence of the other component of the kinase. Similar studies in PKGs are continually influenced by interactions between the various domains. Although this feature complicates some studies, it also provides insight into interactions that may be difficult to study in PKA because this enzyme readily dissociates into R- and C-subunits. Studies of PKA and PKG frequently complement each other because the functional domains of PKG and those in the R- and C-subunits of PKA are closely related (Fig. 2). Isoforms of R-subunits and the regulatory domains of PKGs are comparable in size and have homologous amino acid sequences that contain similar functional domains. Sequences of the R isoforms are most divergent at their amino termini, as is also the case for the PKGs. This extreme N-terminal region provides important contacts for dimerization and, in some instances, also specifies subcellular localization of R-subunits and PKGs (Fig. 2).

3.1. Dimerization of the Cyclic Nucleotide Dependent Protein Kinases Involves the N-Terminus of the Respective Subunits

Although both the PKGs and the PKA R-subunits exist as stable dimers, the mechanism of dimerization appears to differ substantially between these proteins. Experimentally demonstrated differences in the dimerization characteristics account for the opposite alignment of the subunits within the working models of the holoenzymes shown in Fig. 2. In PKG, the best evidence to date supports a parallel alignment of the two monomers within the dimer, but for the R-subunit, there is strong evidence for antiparallel arrangement. Dimerization of the cyclic nucleotide-dependent protein kinases is not critical to enzymological function because the salient biochemical features of regulation and catalysis are retained in the monomeric form of PKG or in a PKA formed by the combination of a single R-subunit and a single C-subunit. However, at least for PKA, dimerization of the R-subunits is required to provide for specific subcellular localization through interactions with a multitude of proteins known as A kinase anchoring proteins (AKAPs) and to provide for cooperative interactions between the low-affinity cAMP binding sites on the two R-subunits. Selective subcellular localization of PKA is important for efficient cAMP signaling in certain instances. To date, there is no clear role for dimerization of the PKGs, and the mechanisms providing for selective anchoring of PKGs are poorly understood. Some reports of PKG anchoring proteins have recently appeared, and a G kinase anchoring protein specific for PKGIα has been identified in testis. The type I PKG was recently shown to be colocalized with dystrophin at the neuromuscular junction in skeletal muscle, but PKG is not directly anchored to dystrophin. The PKGII is anchored through a myristyl group at the N-terminus.

The R-subunits of PKAs form dimers through contacts involving the N-termini. Removal of the N-terminal 45 amino acids converts RII into the monomeric form; in the intact RII, large hydrophobic residues at positions 13 and 36 (Leu and Phe, respectively) stabilize the interaction of two RII monomers to form a dimer. A segment of sequence that is predicted to have high α-helical content in this region is strongly implicated in RII dimerization. Monomers within the RI dimer have an antiparallel arrangement and are linked by two disulfide bonds, but the disulfide linkage is not required to form a stable dimer. Dimerization in RIα is believed to be a novel motif near the N-terminus and to involve sequences with strong α-helical content comparable to that described for RII. Hydrophobic interactions are also important for RI dimerization and involve two phenylalanines (Phe[47] and Phe[52]). For both RI and RII, it cannot be ruled out that dimerization also involves contacts C-terminal to the first ~50 N-terminal residues.

PKGs also homodimerize through interactions involving the N-terminus of each monomer (Fig. 2). Although the monomers in PKGIα can be covalently linked at Cys[45], this crosslink is not required for PKGIα dimer formation, and PKGIβ lacks a cysteine in this region. Although heterodimers of the R-subunits of PKA have been demonstrated, no heterodimers of PKGIα and PKGIβ have been found. A leucine zipper has been invoked as the basis for PKG dimerization because PKGI and PKGII isoforms contain a strong leucine zipper motif near the N-termini. Furthermore, a synthetic peptide based on the PKGIα N-terminal sequence (residues 1–39) dimerizes in solution, and the modeled structure of this sequence based on biophysical analysis is consistent with that of a leucine zipper. Despite the presence of a convincing leucine zipper motif in PKGIα, PKGIβ, and PKGII, and multiple efforts to determine the mechanism providing for PKG dimerization, the arrangement of the monomers within the dimer has not been entirely resolved. Removal of the N-terminus of either PKGIα or PKGIβ converts the enzyme to a monomer. Likewise, a C-terminal truncation mutant of PKGIα that deletes only the catalytic domain is monomeric. The combined results implicate the involvement of both the N-terminal and the C-terminal domains in dimerization. An improved understanding of the molecular characteristics of the dimerization component of the PKGs will be required to determine the arrangement of the PKG monomers within the dimer with confidence.

3.2. The Autoinhibitory and Autophosphorylation Domains Overlap and Are Located Near the N-Terminus

C-terminal to the dimerization domains in the R-subunits and the PKG subunits is the autoinhibitory/autophosphorylation domain. This region provides a major portion of the autoinhibitory interactions that maintain the PKA or the PKG in a catalytically latent form in the absence of cyclic nucleotide, and modification of this region by autophosphorylation contributes important regulatory features to the enzymes.

3.2.1. AUTOINHIBITION OF CATALYSIS INVOLVES MULTIPLE STRUCTURAL COMPONENTS IN THE REGULATORY DOMAINS OF PKA AND PKG

The existence of an autoinhibitory domain within protein kinases was first proposed for PKA and was based on the hypothesis that a substrate-like sequence in the kinase structure could interact with the catalytic site to hold the enzyme in an inactive conformation. This fundamental concept of protein kinase regulation has been proven for PKA, PKG, and for many other protein kinases. In PKA, the RII subunits contain a sequence in the N-terminal portion of the protein that closely mimics that of a PKA substrate (e.g., -RRXSX-, -RKXSX-, -KRXSX-, -KKXSX-) and that undergoes autophosphorylation at the serine (P position) by the C-subunit. The RI subunits contain a pseudosubstrate sequence (-RRXG/A-) that also provides for autoinhibition but lacks a phosphorylatable serine or threonine (Table I). Most of the autoinhibitory domains of PKGs contain sequences that resemble, but do not exactly duplicate, a pseudosubstrate sequence.

A major portion of energy required for autoinhibition of PKAs or PKGs is provided by the contacts between the substrate-like sequence in the autoinhibitory domain and residues within the catalytic domain in these enzymes because deletion or modification of

**Table 1
Comparison of Substrate and Pseudosubstrate Sequences That Are Conserved in the Autoinhibitory Domains of Cyclic Nucleotide-Dependent Protein Kinases and in the Protein Kinase Inhibitor**

Prototypical PKA and PKG substrates	RRX$\overset{*}{\underline{S}}$X
PKA and PKG autoinhibitory domains	
Bovine RIα	-RRG<u>A</u>I-
Mouse RIβ	-RRG<u>G</u>V-
Bovine RIIα	-RRV$\overset{*}{\underline{S}}$V-
Rat RIIβ	-RRA$\overset{*}{\underline{S}}$V-
Protein kinase inhibitor	-RRN<u>A</u>I-
Bovine PKGIα	-RAQ<u>G</u>I-
Bovine PKGIβ	-KRQ<u>A</u>I-
Mouse PKGII	-AKA<u>G</u>V-
Drosophila PKGI	-KKQ<u>G</u>V-
Drosophila PKGII	-RAL<u>G</u>I-

Underlined residues occupy the position that is homologous to the serine ($\overset{*}{S}$) and that can be phosphorylated in the autoinhibitory domain of RII subunits.

the substrate consensus sequence dramatically reduces the inhibitory effectiveness of this region. However, results of studies using heat denaturation, limited proteolysis, and site-directed mutagenesis support the interpretation that other interactions between the R-subunit and the C-subunit also contribute to autoinhibition. Furthermore, synthetic peptides that duplicate the pseudosubstrate sequences from PKA or PKG are poor inhibitors of catalysis. An N-terminally truncated monomeric R subunit can interact with and inhibit a single C-subunit, indicating that the basic components that provide for autoinhibition of PKA are contained within a single R-subunit monomer.

Autoinhibition of PKG is provided by contacts within a single PKG monomer and involves a substrate-like sequence that is homologous to that found in the R-subunits. However, unlike the situation in PKAs, the autoinhibitory domains of most PKGs contain autoinhibitory sequences that are less than optimal motifs for substrates (Table 1). The autoinhibitory regions of PKGs are homologous in sequence to those in PKAs, but they frequently contain a single basic amino acid located at P-2 or P-3. The evolutionary necessity of preserving a canonical high-affinity pseudosubstrate site in the autoinhibitory domain of PKG may be diminished because the interactions between the regulatory and catalytic domains of this enzyme are significantly enhanced by covalent linkage of these domains. In contrast, the dissociation of PKA subunits upon activation diminishes the potential for such interactions, and the R- and C-subunits of PKA may require high-affinity docking sites to foster efficient reassociation.

The dibasic motif in the substrate-like sequence (-RRXSX-) comprising the autoinhibitory domain of PKAs is a critical component for autoinhibition. Substitution of either of the basic amino acids located at P-2 or P-3 [where the P position indicates the phosphorylatable serine/threonine (S) or the glycine/alanine that occupies this position in the pseudosubstrate site] profoundly decreases the potency with which the mutant R-subunit inactivates the C-subunit (Table 1). The autoinhibitory interactions in PKGs appear to be less rigid. Removal of 74 residues from the N-terminus of PKGIβ produces a monomer with full cGMP dependency, but only a single basic amino acid remains in the autoinhibitory pseudosubstrate site (RQAI-). Even with more extensive N-terminal truncation to remove an additional eleven residues and eliminate the entire pseudosubstrate sequence, the PKGIβ is still partially dependent on cGMP for activation. The conserved hydrophobic residue (I/V) at P+1 is also important for autoinhibition, thus, as for PKA, autoinhibition of PKG is likely to involve additional contacts in regions outside that which contains the autophosphorylation sites.

3.2.2. AUTOPHOSPHORYLATION OF PKA AND PKG ALTERS THE FUNCTIONAL PROPERTIES OF THESE ENZYMES

Upon holoenzyme formation, C-subunit rapidly phosphorylates the type II R-subunit in an intramolecular reaction at a single serine located in a consensus sequence, -RRXSX-, in the autoinhibitory domain (Table 1). This particular sequence provides an important part of the autoinhibitory potency of this region as described previously (Fig. 2). When this site is autophosphorylated, the inhibitory potency of the RII subunit is decreased ~10-fold, and the affinity of the RII for cAMP is increased. This phosphorylated type II PKA is therefore more easily activated by low levels of cAMP than is the nonphosphorylated PKA. A significant proportion of RII exists as the phosphoform in certain tissues. In the homologous sequence in type I R-subunit, there is a pseudosubstrate sequence in which either alanine or glycine replaces the serine in this sequence, and therefore it cannot be autophosphorylated. PKGIα and PKGIβ autophosphorylate multiple sites in this region of their respective structures; however, most of these sequences do not replicate a canonical PKG substrate consensus

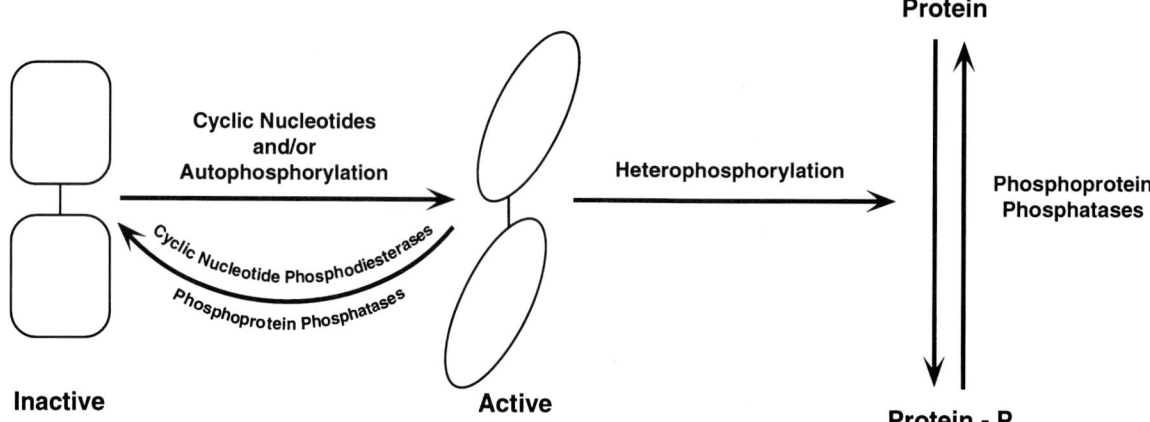

Fig. 3. Schematic depiction of the conformational changes that are associated with the activation of PKGI either by binding of cGMP or by autophosphorylation. With either cGMP binding or autophosphorylation, the type PKGI undergoes a progressive elongation that reflects an increase in the Stokes radius of the enzyme, which is associated with activation of catalysis.

sequence. The major autophosphorylation sites in PKGIα are Ser[50] (-PVPST-), Thr[58] (-PRTTR-), Ser[72] (-TYRSF-) and Thr[84] (-RKFTK-), and in PKGIβ they are Ser[63] (-QKQSA-) and Ser[79] (-QAISA-). Again, this suggests that contacts between the catalytic site of PKG and the N-terminal autoinhibitory/autophosphorylation domain are perhaps less precise than those that occur between R- and C-subunits of PKA.

The rate of autophosphorylation of PKG is increased in the presence of cyclic nucleotide, and when a high stoichiometry of autophosphorylation at the above sites is achieved, the basal (no cGMP) kinase activity and the binding affinity of PKGIα or PKGIβ for cyclic nucleotides is substantially increased. Autophosphorylation of PKGIs appears to be an intramolecular process that occurs at physiological concentrations of kinase and cyclic nucleotide (Fig. 2). Whether the PKG is substantially autophosphorylated in intact tissues is not known. High concentrations of peptide or protein substrates are required to diminish the rate or extent of autophosphorylation. Therefore, PKG autophosphorylation appears to be a preferential process compared to phosphorylation of other substrates containing prototypical PKG phosphorylation sites. Autophosphorylation by PKGII also occurs, but neither the sites modified nor the effects of this modification on enzyme function have been studied extensively.

Either cGMP binding to PKG or autophosphorylation of PKG causes a similar conformational change in the enzyme that is associated with activation (Fig. 3); high levels of cGMP and autophosphorylation do not have additive effects on PKG catalytic activity. Many protein kinases catalyze autophosphorylation, but the physiological relevance of this process in many instances is poorly understood. Type II PKA is clearly autophosphorylated in vivo, and PKGs are also likely to be autophosphorylated in the cell. The resulting sensitivity of the kinases to activation by cyclic nucleotides would be predicted to be enhanced (Fig. 3). Furthermore, the increase in the basal kinase activity of PKGs that is elicited by autophosphorylation could sustain a more prolonged PKG activation after the demise of the cyclic nucleotide signal. Return to the inactive state would depend on the rate of dephosphorylation of PKA or PKG by phosphoprotein phosphatase(s) (Fig. 3). This effect of the cyclic nucleotides to foster the activation of type II PKA and the PKGs provides a type of "feed-forward" mechanism that potentiates signaling through these pathways (*see below*).

3.3. Cyclic Nucleotide-Binding Domain

The cyclic nucleotide binding sites (~110 amino acids each) in PKAs and PKGs are homologous to the cyclic nucleotide binding domain in the cyclic nucleotide gated cation channels and to that in an evolutionarily related bacterial cAMP-binding protein, catabolite gene activating protein (CAP) (Fig. 4). As such, more is known about these cyclic nucleotide-binding sites than about other unrelated families of cyclic nucleotide-binding sites such as those occurring on cyclic nucleotide phosphodiesterases. PKAs and PKGs are distinct in the CAP family because both the R-subunit and the PKGs contain two homologous cyclic nucleotide-binding sites arranged in tandem in the primary structure (Figs. 2 and 4); the sequence of ~110 amino acids in each site is sufficient to bind

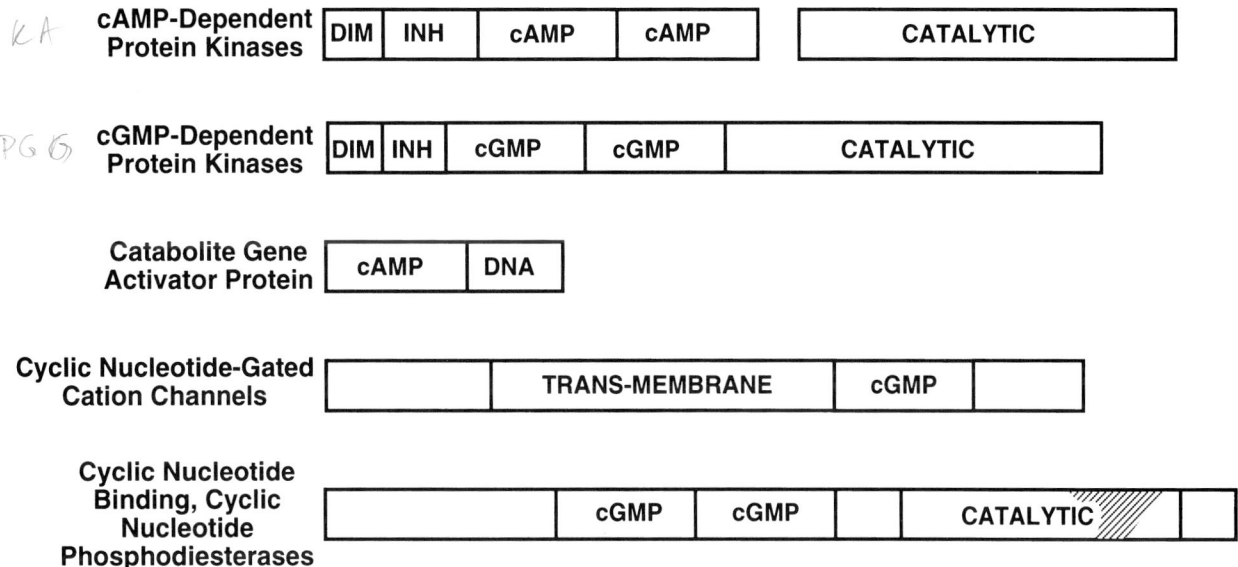

Fig. 4. Families of cyclic nucleotide-binding sites. The cyclic nucleotide-binding sites in the regulatory subunit of PKAs, PKGs, the cyclic nucleotide-gated cation channels, and the catabolite gene activator protein are derived from a common ancestral gene. The amino acid sequence homology between the respective N-terminal (A site) or C-terminal (B site) cyclic nucleotide-binding sites of PKA and PKG is greater than between the two binding sites in the same protein. The cyclic nucleotide-binding sites in the allosteric cyclic nucleotide binding sites in the PDEs are derived from another distinct family of binding sites, and the interaction of cyclic nucleotides within the catalytic site in PDEs is apparently derived from yet another family of cyclic nucleotide-binding sites. Cyclic nucleotide interaction with the catalytic sites in PDEs involves residues distributed throughout the catalytic domain which is indicated by the ill-defined boundaries of the lined areas.

the cyclic nucleotide. The mechanistic importance of two cyclic nucleotide binding sites is unclear. The cyclic nucleotide binding sites in R-subunits and PKG appear to be products of an ancient gene duplication that occurred prior to divergence of these proteins because the sequence homology among the respective sites A or sites B in these proteins is greater than that between the two binding sites in a single R-subunit or PKG. However, there are no apparent similarities in the exon organization of the PKGI gene and that of the RIα and RIβ of PKA. Specific structural features in each of the sites determine the different kinetics of cyclic nucleotide dissociation, cyclic nucleotide specificities, and cyclic nucleotide analog selectivities which are distinct for each of the respective sites in RI, RII, and PKGs.

3.3.1. THE FUNCTIONAL CHARACTERISTICS OF THE CYCLIC NUCLEOTIDE-BINDING SITES OF PKA AND PKG ARE DISTINCT

The structural elements contained in the ~100 amino acids that comprise a single cyclic nucleotide-binding site in the R subunit is sufficient to provide for cAMP binding. The two cyclic nucleotide-binding sites in cyclic nucleotide-dependent protein kinases (Figs. 2 and 4) have been dubbed as the fast and slow sites based on the relative rates of two kinetic components of the dissociation of cyclic nucleotides from these proteins. The relative binding affinities of the fast and slow cyclic nucleotide-binding sites in the cyclic nucleotide-dependent protein kinases differ by ~10-fold. In the R-subunit of PKA the more N-terminal cyclic nucleotide-binding domain (noted as site A in some instances) is the low-affinity (fast) site and the more C-terminal site (noted as site B in some instances) is the high-affinity (slow) site. However, the linear arrangement of these kinetically defined cyclic nucleotide-binding sites in the structure of type I PKGs is transposed compared with the arrangement in PKA (Figs. 2 and 4). This was unexpected because there is high sequence homology between the respective sites in PKA and PKG based on their arrangement in the sequence. The precise molecular basis for the differences in the binding affinities of these sites is not known.

Cyclic nucleotide binding to both the R-subunit and to PKGs shows positively cooperative kinetics, that is, occupation of the fast site in PKA or PKG slows dissociation from the slow site. In the R-subunit of PKA, this is mainly due to interaction of binding sites within a single subunit although there is some cooperativity between the fast sites within a PKA dimer. In PKGs it is not clear whether cooperativity results from interaction of binding domains in the same monomer or through influences between monomers. In some instances, the kinetics of cyclic nucleotide binding are also modulated by interactions with other domains within the enzymes; in both PKGIα and type I PKA, Mg^{2+}-ATP promotes cyclic nucleotide dissociation.

3.3.2. THE CYCLIC NUCLEOTIDE-BINDING SITES OF PKA AND PKG SHARE MANY STRUCTURAL FEATURES

The overall structures of the cyclic nucleotide binding sites in PKA and PKG resemble that of the single cyclic nucleotide-binding site in CAP although the kinases bind cyclic nucleotides with significantly higher affinity (~3 orders of magnitude) than does CAP; the sites contain three conserved glycines that provide critical structural features, an arginine and alanine that bind the equatorial phosphate oxygen, and a glutamic acid that interacts with the 2'OH of the ribose. In the PKA and PKG, the cyclic nucleotides are bound in the *syn* conformation, whereas CAP binds cyclic nucleotide in the *anti* conformation (Fig. 5). The X-ray crystallographic structure of both CAP and an N-terminally truncated monomeric RI have been determined. The cyclic nucleotide-binding pocket in this family of cyclic nucleotide-binding proteins is characterized by three α-helices and an eight-stranded, antiparallel β-barrel; the ribose and the cyclic phosphate lie deep in this pocket. In the R-subunit, the adenine moiety of cAMP stacks with an aromatic amino acid in each of the two binding sites. The two cyclic nucleotide binding sites are distant from each other within the monomer, which precludes direct interactions. In the R-subunit, cooperative interactions between the intrasubunit sites could occur via an extended hydrophobic surface between the sites.

Because the overall structures of the cyclic nucleotide binding sites in PKA and PKG are quite similar and the molecular features of cAMP and cGMP are also quite similar (Fig. 5), it is not surprising that each of the kinases can bind either nucleotide through a process known as cross-activation. However, the respective binding sites in these proteins exhibit ~10- to 100-fold selectivity for cAMP and cGMP, respectively, and the variation in this range results in part from the source of the enzymes and their states of autophosphorylation. A portion of the relative cAMP/cGMP selectivity is provided by a key difference in

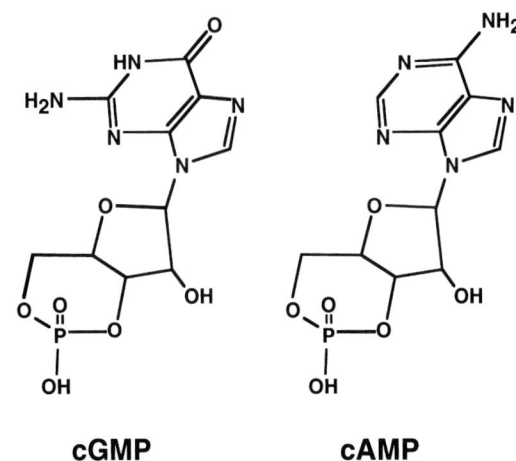

Fig. 5. Structural models of cGMP and cAMP. cGMP and cAMP have identical ribose and cyclic phosphate rings and differ only in their purine rings. Rotation around the linkage between the purine and the ribose allows the cyclic nucleotide to assume two distinct conformations, *syn* (shown) and *anti*; in the *anti* conformation the ribose and cyclic phosphate ring are rotated to the right and away from the pyrimidine portion of the purine base. The cyclic phosphate moiety is required for interaction with the cyclic nucleotide-binding sites on both proteins. Important structural features of cGMP that contribute to its specific interaction with PKG include the amino group at C-2 (which interacts with a conserved threonine in cGMP-preferring sites), the oxygen at C-6, and a protonated N-1; in contrast, cAMP has a single substitution of the purine ring, that is, an amino group at C-6 and no protonation at N-1. Cyclic nucleotide analogs that have been employed in studying the kinases include a multitude of substitutions directly on the purine ring as well as modifications of the ribose and the cyclic phosphate ring. The Rp(S) analogs that bind to PKA and PKG, but do not activate them, are modified by substitution of a sulfur at the equatorial exocyclic oxygen atom.

a single amino acid in the cyclic nucleotide binding sites in CAP-related proteins. In this family of proteins, sites that are selective for cGMP contain an invariant threonine or serine that increases the binding affinity of the site for cGMP. The threonine or serine in this position is not a negative determinant for cAMP binding, but rather contributes positive features to cGMP binding. In the cyclic nucleotide-binding sites in PKA, this position is occupied by an alanine. In cyclic nucleotide-binding sites that are modeled after CAP, the hydroxyl of this invariant threonine is proposed to form a hydrogen bond with the C-2 amino group of guanine (Fig. 5). This interpretation is supported by the results of site-directed mutagenesis of PKA, PKGs, and cyclic nucleotide gated cation channels, by molecular modeling of the sites, and by cyclic nucleotide analog-binding studies. Thus, a single conserved threonine in the cGMP-binding sites in conjunction with the conserved arginine and glutamate interacts with specific positions of cGMP (Fig. 5) to provide for its high-affinity binding. This suggests that during evolution a single mutation in these cyclic nucleotide-binding sites could have provided for much of the cAMP/cGMP selectivity differences between these two enzymes. By contrast, binding-site elements that dictate a preference for adenine vs guanine are less well understood; interaction with the C-6 amino group is important and may function indirectly by changing the electron distribution in the purine.

Cyclic nucleotide binding to both sites in either PKA or PKG contributes to the full activation of kinase activity; occupation of both the high-affinity (slow) and low-affinity (fast) sites in PKA is required for activation of the enzyme, but in PKGIα, cGMP binding to the slow site alone partially activates catalysis (Fig. 2). This quantitative difference in the requirement of site occupation for activation of PKA and PKG may relate to the interaction of the respective high-affinity sites with other domains in the kinases. In PKA, the more N-terminal low-affinity cyclic nucleotide-binding site in the R-subunit provides for critical interactions with C-subunit to form the PKA holoenzyme, but the more C-terminal high-affinity site is less important for this interaction. The low-affinity site must be occupied by cAMP in order for PKA activation to occur, and in the holoenzyme, cAMP binding to the high-affinity site provides for cAMP access to the low-affinity site. If interdomain contacts in PKG mimic those in PKA, then cGMP occupation of the more N-terminal high-affinity cyclic nucleotide-binding site in PKG might also partially disrupt interaction of this site with the catalytic domain of PKG, thereby partially activating the enzyme. Subsequent occupation of the more C-terminal low-affinity cyclic nucleotide-binding site of PKG would then complete the activation process.

3.3.3. cAMP and cGMP Can Bind To and Cross-Activate PKG and PKA, Respectively

Because PKA and PKG can bind either cAMP or cGMP, either nucleotide can activate the catalytic activities of these kinases. This characteristic also raises the possibility that cross-activation could occur under physiological conditions, that is, cAMP could cross-activate PKG and cGMP could cross-activate PKA (Fig. 1). Convincing evidence supporting cross-activation has been demonstrated in a number of systems and must be considered when interpreting physiological responses to elevation of either cyclic nucleotide. An example of this is found in studies of smooth muscle relaxation. Elevation of either cAMP or cGMP within the physiological range of these cyclic nucleotides causes vascular smooth muscle relaxation, and PKGI appears to mediate both effects. PKG has a 10- to 100-fold higher affinity for cGMP over cAMP depending on the source and modification of the enzyme. The basal cAMP level in smooth muscle tissue, and in many others, is five- to sixfold higher than that of cGMP, which allows for cross-activation of PKG when cAMP is modestly elevated. Conversely, cross-activation of PKA by cGMP has been demonstrated in an intestinal epithelial cell line (T84 cells) used to study chloride conductance. This cell line has PKA but lacks measurable PKG, and when cGMP is elevated in response to the heat-stable enterotoxin, chloride conductance across the membranes of those cells is increased due to PKA activation. Cross-activation of PKA by cGMP is also proposed to account for the increase in testosterone production when Leydig cells are exposed to atrial natriuretic peptide. This cross-activation in cyclic nucleotide signaling expands the potential participation of these kinases in the response of tissues to increases in either nucleotide and must be considered as a possible mechanism when interpreting studies relating to cyclic nucleotide effects. Cross-activation may also play an important role in signaling through cyclic nucleotide-gated channels because these proteins also have overlapping cyclic nucleotide binding specificities.

3.3.4. Cyclic Nucleotide Analog Specificities

The two homologous cyclic nucleotide-binding sites in both the PKAs and PKGs differ not only in their relative affinities for cyclic nucleotides, but they can also be distinguished by the specificity with which they bind a variety of cyclic nucleotide analogs. This property has been exploited for studying the relative contribution of PKG and PKA to various physiological functions as well as for examining the relative contribution of the PKA isoforms, as most tissues contain both type I and type II PKA.

Most cyclic nucleotide analogs interact to some extent with both the fast and slow cyclic nucleotide binding sites in PKAs and PKGs, so that studies using a single cyclic nucleotide analog cannot be interpreted unambiguously. The cyclic nucleotide analog specificities vary between the RI and RII subunits of PKA and between sites. Therefore, a combination of two analogs, each with good specificity for one of the two different sites of a particular isoform of PKA, can be used fruitfully to investigate the physiological roles of that isoform. For instance, 8-piperidino-cAMP shows fast site selectivity in RI subunit and slow site selectivity in RII subunit. Analogs derivatized at C-8 generally bind preferentially to the slow site, whereas analogs modified at C-6 or N-6 are more selective for the fast site. When slow-site selective and fast-site selective analogs that preferentially target either the type I or type II PKA are used, subsaturating concentrations of pairs of these analogs elicit a synergistic activation.

The use of cyclic nucleotide analogs in conjunction with the knowledge regarding their specific interactions with the R-subunits or the PKGs can be highly valuable in studying the physiological actions of these kinases. Cyclic nucleotide analogs have been used extensively for studies of PKA and PKG functions in a variety of intact cells and work reasonably well in some tissues because they can traverse the cell membrane in sufficient quantities to target the kinases directly. Use of either stimulatory or inhibitory analogs in "leaky" cells such as hepatocytes is usually successful and may require relatively low concentrations of analogs. However, some cell types such as adipocytes are more resistant, presumably because of poor penetration of the analogs into the cells. Therefore, it is important to use a range of analog concentrations to assess the potential effectiveness of cyclic nucleotide analogs in a given process and to determine an appropriate concentration-dependence curve. Many cyclic nucleotide analogs activate PKA and/or PKG, but others such as the (Rp)cAMP(S) are inhibitory (Fig. 5 and *see* Section 6).

There are several considerations that should accompany use of cyclic nucleotide analogs: analogs that are resistant to cyclic nucleotide phosphodiesterases are preferred; the analogs commonly show some reactivity with both PKA and PKG, so caution must be exercised in assigning function based on a limited number of analogs; and nonspecific effects such as the interaction of high concentrations of these analogs with other purine receptors (adenosine receptors, ATP-binding proteins, etc.) or effects of metabolites of these analogs (such as the "butyrate effect" demonstrated in some studies with butyryl derivatives of cyclic nucleotides) are sometimes obtained. One drawback to the use of the inhibitory analogs for intact cell studies is that they typically have relatively low affinity for PKA and PKG.

Combinations of various analogs that are specific for the type I and type II PKA isoforms, respectively, have been used in studies of intact tissues to assess the relative contributions of these PKA isoforms to a variety of physiological processes. Although the majority of studies support the interpretation that either isoform of PKA (type I or type II PKAs) can elicit a given physiological process, there appear to be exceptions. A similar approach for studying the roles of type I and type II PKGs has not been used.

The differences in analog specificities of the cyclic nucleotide-binding sites of type I and type II PKAs may relate in part to specific features in the primary sequence in the respective sites; for example, differences in the hydrophobicities of the binding pockets or spatial constraints could alter interactions. The functioning of the cyclic nucleotide-binding sites in PKAs and PKGs is also significantly influenced by other components in the structures of these enzymes. This is clearly demonstrated by the fact that PKGIα and PKGIβ have quite similar K_as for cGMP, but exhibit distinctly different analog specificities despite having identical amino acid sequences in their cyclic nucleotide-binding sites. Some analogs activate both PKGI isoforms equally well, whereas others show marked discrimination between the two. Thus, the divergent N-termini of the PKGs impart different biochemical properties to these cGMP-binding sites that are identical in amino acid sequence. The PKGII also binds cyclic nucleotide analogs with a specificity profile that differs from that of either PKGIα or PKGIβ.

3.4. The Structures of the Catalytic Moieties of PKA and PKG Are Homologous, but the Specificities of Function Exhibit Some Differences

The catalytic component of PKA resides in a separate polypeptide chain (C-subunit) from the regulatory component, the R-subunit. In contrast, the regulatory and catalytic components of PKGs reside in the same polypeptide chain (Fig. 2).

3.4.1. THE STRUCTURAL FEATURES OF THE CATALYTIC MOIETIES OF PKA AND PKG ARE QUITE SIMILAR

The catalytic components of the PKA and PKG are predicted to have similar structures; modeling of the catalytic domain of PKG using the coordinates for the bilobed C-subunit supports this prediction. Despite the strong homology between PKAs and PKGs, clear differences in the properties of the catalytic sites have been demonstrated, so that subtle features involving substrate specificity and perhaps catalysis in each may vary somewhat. Both kinases contain several functional subdomains that were defined in early studies of PKA and PKG. Some of these domains are conserved among most protein kinases; the smaller, more N-terminal lobe of the C-subunit contributes importantly to binding Mg^{2+}-ATP, and the adjacent larger lobe provides recognition sites for protein/peptide substrate binding as well as for residues involved in catalysis. The extreme C-terminal domain is not conserved among kinases, although it may have important functions in specific kinases.

The C-subunit is an asymmetric protein that exists in at least two conformational states, one that is relatively open and one that is more closed. When Mg^{2+}-ATP and a substrate analog, that is, the heat-stable protein kinase inhibitor (PKI), bind to the C-subunit, the enzyme undergoes a transition to a more closed conformation that is thought to be required for appropriate positioning of catalytic residues in the catalytic cleft. Following transfer of the γ-phosphate of ATP to the protein/peptide substrate, the catalytic subunit assumes a more open conformation that promotes release of the product. A substrate analog alone does not cause a significant change in the conformation of the C-subunit, but pronounced conformational effects are evident following addition of Mg^{2+}-ATP. Catalysis occurs in a deep cleft formed at the interface between the two lobes and involves multiple contacts from each lobe. The catalytic site is characterized by the apposition of three loops of sequence that extend into the catalytic cleft. Eight amino acids that are conserved in almost all protein kinases form interactions with residues of the peptide or protein substrate. Other conserved elements within the core kinase structure may assist in stabilizing the structure and function of the C-subunit.

The nucleotide-binding fold in the cyclic nucleotide dependent protein kinases as well as other protein kinases differs from those of other known nucleotide binding proteins, and the adenine moiety of the ATP is buried in a hydrophobic pocket sandwiched between the two lobes of the catalytic domain. The Mg^{2+}-ATP is bound through multiple contacts involving a glycine-rich segment termed the "glycine-rich loop" (Gly-X-Gly-X-X-Gly) in the small lobe of the C-subunit and through contacts with the larger lobe. Two other loops are also catalytically important. The catalytic loop contains three residues that are invariant in Ser/Thr protein kinases, and an aspartic acid (Asp^{166} in the C-subunit and Asp^{483} in PKG) in this loop may act as the catalytic base to deprotonate the substrate serine/threonine, thereby increasing the nucleophilicity of this oxygen, which allows its direct attack on the γ-phosphate of ATP. In the absence of substrate, the C-subunit has an ATPase activity in which the γ-phosphate is transferred to water. The role of segments of the C-subunit located at the extreme N-terminus (Figs. 2 and 4) or the extreme C-terminus is not fully understood. Deletion of 23 residues at the N-terminus of the C-subunit has no apparent effect on catalytic rate, substrate binding, or the ability to interact with the R-subunit. However, alterations at the C-terminus of the C-subunit adversely affect catalysis.

Both the mammalian PKAs and PKGs are covalently modified. Some forms of the C-subunit are N-terminally myristylated, but this modification does not promote membrane association; in some instances, the myristylation of the C-subunit enhances stability. The R-subunits are acetylated at the N-terminus. PKGIα is N-terminally acetylated, whereas PKGII is myristylated. The myristylation of PKGII appears to provide a critical component to the subcellular localization of PKGII to a particulate compartment. Both the C-subunit and PKGs contain phosphates; the C-subunit can be phosphorylated at Thr^{197} and Ser^{338}, and in some instances, also at Ser^{10}. Whether the phosphates in the C-subunit are introduced in cells by autocatalytic action or by the action of a protein kinase A kinase such as a phosphoinositide-dependent protein kinase is not clear. Phosphorylation at Thr^{197} is important for optimal catalytic function of the C-subunit, and this site is resistant to phosphoprotein phosphatase action. In PKG, a phosphate at the position homologous to Thr^{197}, is also implicated, but the phosphorylation site at Ser^{338} in the C-subunit is absent in PKG. Phosphorylation of the C-subunit lowers the K_m for substrates. In the crystal structure of the C-subunit, the phosphate on Thr^{197} is approximated with the P+2 position of a cocrystallized substrate analog, protein kinase inhibitor peptide (PKIP). The phosphorylated Thr^{197} and an adjacent tryptophan are also implicated in contacts between the C-subunit and the R-subunit.

3.4.2. CHARACTERISTICS OF CATALYSIS

The substrate binding properties of the C-subunit have been studied extensively. ATP in the *anti* conformation binds to a high-affinity site (K_m = 3–15 μM), and two Mg^{2+} neutralize the negative charges of the ATP phosphates. Binding the second Mg^{2+} lowers the K_m for ATP 10-fold. Under typical physiological conditions, the ATP substrate would not be rate limiting. This is relevant because Mg^{2+}-ATP binding to the free C-subunit precedes binding of protein substrate and is thought to induce the more closed conformation of the C-subunit, which then interacts with the protein substrate with higher affinity than in the apoenzyme. Because of the conservation of critical structural features in many members of the protein kinase family, it has been suggested that general changes in conformation observed with the C-subunit may also apply to other kinases such as PKGs. Peptide substrates containing the consensus RRXSX- motif assume an extended coil conformation when bound to the C-subunit and induce structural changes, perhaps contributing to the narrowing of the catalytic cleft by contraction of the two lobes around a conserved "glycine hinge," in the C-subunit.

Following formation of the enzyme–Mg^{2+}-ATP ternary complex, the γ-phosphate of ATP is rapidly transferred to the substrate, and configuration at the transferred phosphate is inverted. The phosphopeptide product rapidly exits the catalytic cleft of the C-subunit, and Mg^{2+}-ADP dissociates more slowly. The affinity of the C-subunit or PKG for Mg^{2+}-ADP is relatively high, so that at sufficiently high concentrations of Mg^{2+}-ADP, the reaction can be reversed.

4. DISTINCT FEATURES OF PKAs AND PKGs THAT AFFECT THEIR SUBSTRATE SPECIFICITIES

PKAs and PKGs can phosphorylate serine/threonine in many protein or peptide substrates (heterophosphorylation) having the consensus sequence RRXSX-, and they can also phosphorylate sites within their own sequences (autophosphorylation). The amino acids occupying the positions in the vicinity of a potentially phosphorylatable serine/threonine contribute significantly to the efficacy with which PKA or PKG phosphorylate that residue. The simple consensus sequence (-RRXSX-) is not always mandated for phosphorylation, as evidenced by some of the autophosphorylation sites in PGKIα and PKGIβ that contain either no basic residues, one basic residue, or an altered distance between the basic amino acids and the phosphorylated site; such a discrepancy is also found in the site in RI that is heterophosphorylated specifically by PKG. This site in RI is heterophosphorylated by PKGI, but not by the C-subunit of PKA and is the only known protein substrate to be absolutely selective between these kinases. In addition, the initial autophosphorylation site in PKGIα is a threonine, which occurs less frequently as the phospho-acceptor amino acid in substrates for PKG and PKA.

Even within the constraints of the shared consensus phosphorylation motif for cyclic nucleotide-dependent protein kinases, clear distinctions in substrate specificities have been made between the PKA, PKGI, and PKGII. A number of negative substrate determinants for PKA catalysis have been identified; these include threonine rather than serine at the phosphotransfer (P) site, an acidic amino acid at P-1 (i.e., immediately N-terminal to the P site), lysine in the P+2 position, or phenylalanine at P+4. A basic residue adjoining the phosphorylatable serine/threonine at P+1 is a positive determinant for PKGI compared to PKA, and there are likely to be other distinctions.

Although PKA appears to be a better catalyst than PKGI for many of the substrates that have been studied, this pattern varies widely and generalizations about catalytic efficacy of these enzymes are ill advised. Either of two serines in close proximity in histone H2b is selectively phosphorylated by PKA or PKGI. A commercially available heptapeptide based on this sequence in histone H2b (RKRSRAE) is commonly used for assaying PKGI, but it is also phosphorylated by PKA. A strong selectivity for PKGI vs PKA is also exhibited in another substrate, that is, the cGMP binding, cGMP-specific phosphodiesterase (PDE5), for which the rate of phosphorylation of a single serine (Ser^{92}) by PKGI exceeds that of PKA by ~10-fold. This preferential phosphorylation by PKG results from a phenylalanine in the P+4 position of the phosphorylation sequence of PDE5. A cytoskeletal platelet protein, vasodilator-stimulated protein (VASP), contains one site that is preferentially phosphorylated by PKA while another is preferentially phosphorylated by PKGI. The α-subunit of skeletal muscle phosphorylase b kinase is phosphorylated by PKG at a rate three times that by PKA. A substrate from cerebellum (G-substrate) can be phosphorylated at two threonines, and the affinity of PKG for this substrate is 30 times greater than is the affinity of PKA for this protein. Histone H1 also contains a site reported to be very selective toward PKG.

Studies using the PKA-specific heat-stable protein kinase inhibitor (PKI) or synthetic peptide substrate analogs based on varied amino acid sequences and including artificial residues (alcohols) have revealed distinct differences in the catalytic sites of PKA and PKGI. The relative PKG/PKA substrate selectivity for some of these peptides is quite high; the PKI is a naturally occurring potent and specific inhibitor of PKA, but it is a very weak inhibitor of PKG. Recent studies have led to the development of a reversible, selective PKG inhibitor and identification of a substrate sequence that is highly specific for PKG (TQKARKKSNA). These differences in substrate specificity are likely to reflect subtle variations in the interactions of substrate with specific contact residues in the catalytic domain of PKGI compared to those in the C-subunit.

The substrate specificity of PKGII differs markedly from that of both PKGI and the C-subunit of PKA. Although PKGI, PKA, and PKGII can phosphorylate some of the same substrates such as the cystic fibrosis transmembrane conductance regulator (CFTR) chloride channel quite well, in other instances the enzymes differ. PKGII phosphorylates the peptide (KRREILSRRPSYR) derived from CREB more efficiently than that from histone H2b (RKRSRAE), which functions well as a substrate for either PKGI or PKA. It has also recently been reported that PKGII selectively mediates increased Ca^{2+} reabsorption in response to atrial natriuretic peptide, but that PKGI cannot mediate this effect even when the catalytic moiety of the PKGI is membrane localized in a PKGII/PKGI chimer. This suggests that the substrate target for this action is a highly specific substrate for the PKGII. These differences suggest that each of these kinases is likely to preferentially phosphorylate certain substrates in vivo.

5. MECHANISMS USED IN THE ACTIVATION OF PKA AND PKG

In the absence of cAMP, the R-subunit dimer can combine with any of the C isozyme with an affinity of ~0.2 nM. When two cAMP molecules bind per R-subunit, the R undergoes a conformational change that reduces its affinity for C-subunit 10,000–100,000-fold. As a result, the tetrameric PKA can readily dissociate into dimeric R-subunits and two monomers of C, and inhibition of C by R is concomitantly relieved. Some studies using holoenzyme or heterochromatically fluorescent-labeled R-subunits and C-subunit in the PKA holoenzyme reveal that under equilibrium conditions cAMP binding to R does not cause the PKA holoenzyme to dissociate. PKA can be partially saturated with cAMP, and under basal conditions, much of the cAMP in a cell is bound to the R-subunits. However, both intrasubunit cAMP binding sites on the R-subunit must be occupied in order to activate catalysis.

Cyclic GMP binding to PKGI activates the enzyme and causes structural elongation. Activation by autophosphorylation causes a similar or identical elongation (Fig. 3). In both instances, negatively charged phosphates are added to the enzyme; these are conferred by the cyclic phosphate of the nucleotide or by the phosphate introduced by phosphorylation. The increased electronegativity could electrostatically repulse negatively charged amino acids in the catalytic domain of PKG. Increased surface electronegativity of the PKG could be provided directly by the phosphate(s) introduced by autophosphorylation or indirectly by charge neutralization of surface residues by the cyclic nucleotide phosphate because the cyclic phosphate moiety is buried in the crystal structure of the homologous cyclic nucleotide-bound CAP and R-subunit. The combined results suggest that either cGMP binding or autophosphorylation produces a perturbation to cause activation in each monomer of PKG. Although this perturbation is not necessarily the same for the two processes, within the monomer it results in a similar elongation of the dimeric structure.

The relative positions of the high-affinity and low-affinity cyclic nucleotide binding sites are transposed within the PKG and PKA monomers, which could account for the partial activation of PKG with sub-saturating cGMP because cAMP binding to the more N-terminal site in PKA is believed to be directly involved in activation of this enzyme. Cyclic nucleotide binding to the high-affinity site in each kinase could produce somewhat different interactions of the enzyme domains in the tertiary structures. Inherent differences in interactions between the respective inhibitory and catalytic domains in PKA vs PKG or the differences in the monomer alignment within the kinases could also be involved.

6. EXISTENCE AND USE OF SELECTIVE INHIBITORS OF PKA AND PKG

The search for effective and selective inhibitors of protein kinases, particularly for use in intact tissues, has been arduous. For the cyclic nucleotide dependent protein kinases, two functional domains have been targeted for the intervention by inhibitors, that is, the cyclic nucleotide binding domains and the catalytic domain of each enzyme. The Rp analogs such as RpcAMP(S), (Rp)-8-Br-cAMP(S), (Rp)-8-Cl-cAMP(S), (Rp)cGMP(S), (Rp)-8-Br-cGMP(S), and (Rp)-8-Br-PET-cGMP(S) bind to the cyclic nucleotide-binding domains of the kinases, but do not activate the enzymes (Fig. 5). These analogs block cAMP or cGMP effects because they compete for endogenous cAMP/cGMP binding to the respective cyclic nucleotide binding sites of the R-subunit or PKG, but when these analogs are bound to the kinases, they do not significantly activate these enzymes. However, these inhibitory analogs are not entirely selective and under certain conditions can be shown to inhibit both PKA and PKG. The PKA-specific heat-stable inhibitor protein (PKI) (as well as a peptide derived from PKI) is highly selective for the C-subunit of PKA; it interacts potently with components in the catalytic cleft of the PKA. However, the PKI and related peptides do not cross the cell membrane well and therefore are ineffective when added to most cells, although liver cells, which are relatively leaky, are amenable to the use of PKI added extracellularly. Recently, a synthetic inhibitory peptide that is quite selective for PKGI compared to PKA has been devised by varying the stereochemistry of an alcohol included in the peptide, but again this peptide is unlikely to be useful for intact cells. A serious problem has arisen in the protein kinase field owing to the marketing of a number of purportedly specific inhibitors for PKA or PKG. Many of these compounds are ATP analogs, and as such, they have the potential to react with many proteins including other protein kinases that utilize ATP. In fact, many of these compounds are nonspecific. Some of these "selective" inhibitors have even been shown to increase the catalytic activity of the target kinases when used in intact tissues.

7. SELECTED SUBCELLULAR LOCALIZATIONS OF PKA AND PKG

PKA and PKG are located in both the cytosol and in the particulate fractions of most cells. In many tissues significant proportions of the PKA and PKG activities are associated with the particulate fractions, and when activated, these localized kinases may selectively phosphorylate nearby substrates such as adenylyl cyclases, phosphodiesterase, protein phosphatases, ion channel proteins, or other substrate(s). Significant amounts of the type II PKA are membrane bound in many tissues, but type I PKA is also found in the membrane fraction in certain tissues. The proportion of PKA and PKG that is associated with specific subcellular compartments in most tissues has not been exhaustively analyzed.

7.1. Specific Anchoring Mechanisms Exist for Localizing PKA and PKG

The PKGII is primarily associated with the particulate fraction of cells; the PKGI isoforms are largely soluble enzymes, but in certain tissues such as platelets, PKGI is primarily membrane-bound. Either the type I or type II PKA holoenzyme or the respective dimeric R-subunits can be anchored to various subcellular compartments through interactions with A kinase anchoring proteins (AKAPs) that vary in tissue distribution, size, and primary structure. Most of the AKAPs that have been identified show a high specificity for RII, but an AKAP with dual specificity for RI or RII has been identified. The AKAPs are localized to specific subcellular structures through a targeting domain on their surfaces; these proteins interact with a short segment of sequence (contained in residues 1–79) near the N-terminus of the dimeric form of the R-subunits.

Only a few G kinase anchoring proteins (GKAP) have been reported. The association of PKGII with the particulate fraction appears to be mediated primarily by an N-terminal myristyl group. In instances where the PKGI is membrane bound, the mechanism providing for this localization is entirely clear. PKGIα has been reported to specifically bind to troponin T in skeletal and cardiac muscle cells and to a novel germ cell-specific protein, GKAP42, from testis. In both instances, these interactions appear to be largely mediated through the N-terminus of PKG and in the latter instance, cGMP binding to PKG modulates its interaction with GKAP42. The importance of colocalization of PKG with its substrate has been well demonstrated. Although both PKGI and PKGII can phosphorylate the CFTR chloride channel in vitro, when cells overexpress a PKGII that localizes to the particulate fraction, chloride secretion is increased, but not when the cytosolic PKGI is overexpressed. However, when a chimeric PKG was created with a structure that contained the N-terminus of PKGII and the catalytic component of PKGI, this enzyme localizes to the particulate fraction, where it can phosphorylate the chloride channel protein to increase chloride secretion. In this instance simple colocalization of the PKGI with the channel allows this kinase to phosphorylate this particular substrate. Likewise, overexpression of the membrane-associated PKGII in kidney cells provides for increased Ca^{2+} uptake in response to atrial natriuretic peptide, but similar overexpression of a nonmyristylated PKGII mutant does not mediate this effect, again emphasizing the importance of subcellular localization of these kinases.

7.2. PKA and PKG Can Translocate Within the Cell

Although anchoring of PKA and PKG to specific structures in the cell would tend to localize the effects of these kinases, translocation of the kinases to different subcellular compartments upon activation has also been implicated as an important aspect of regulation. In neutrophils, the PKGI has been shown to redistribute to cytoskeletal components of the cell following activation. The diffusibility of the C-subunit is widely believed to provide for the mechanism by which cAMP elicits its effect in a number of systems. Physical dissociation of PKA into its component R- and C-subunits following elevation of cyclic nucleotide provides the potential for the redistribution of these subunits within the cell. Persistent elevation of cAMP can promote translocation of the C-subunit to the nucleus while holoenzyme is excluded, and the translocated C-subunit is thought to provide for the nuclear effects of cAMP. It is widely believed that the dimeric R-subunit is excluded from the nucleus, but a limited number of studies suggests that the R-subunit can translocate to the nucleus and may regulate gene transcription. Overexpression of the C-subunit induces expression of a number of genes, an effect that is largely blocked by coexpression of the R-subunit or PKI. The mechanism by which the C-subunit translocates into the nucleus is not known, but it may involve simple diffusion along a concentration gradient. Although the C-subunit contains a nuclear translocation signal, it is not required for translocation. Upon relocation of C to the nucleus, phosphorylation of nuclear proteins, for example, transcription factors such as CREB, could alter gene expression.

Increases in nuclear C-subunit are reversible, and PKI has been implicated in the extrusion of C from the nucleus. It is possible that at least a portion of the C-subunit nonspecifically diffuses back to the cytoplasm where it can reassociate with the R-subunit. Although PKI may participate in the translocation of the C-subunit from the nucleus to the cytoplasm, the physiological signal that would regulate this interaction and modulate the removal of C-subunit from the nucleus is not clear. Nuclear PKI would presumably be free to immediately associate with and inhibit C-subunit entering the nucleus. Furthermore, the R-subunit of the PKA is considered to be the major cellular sensor for changes in cAMP levels, but it is believed to be largely excluded from the nucleus. The amount of cAMP bound to the R-subunit dictates its affinity for the C-subunit; thus, as cAMP levels diminish in the cytoplasm, the cAMP dissociation from R-subunit would favor interaction with the C-subunit, resulting in its inactivation. This dynamic effect is lost or greatly slowed when the two PKA subunits are isolated in different subcellular compartments, that is, the regulatory cAMP sensor is removed from the object to be regulated, the C-subunit. The distribution of the C-subunit between the cytosolic and nuclear compartments may result solely from the relative concentration gradient of the C-subunit which is determined by the amount of cAMP promoting dissociation of PKA. Because the diffusion rate for cAMP in the cellular milieu is quite rapid, it is also possible that the signal molecule itself could gain entry into the nucleus to activate nuclear PKA, thereby producing nuclear effects associated with cAMP elevation.

8. INTERACTION WITH PROTEIN KINASE INHIBITOR AND PROTEIN KINASE INHIBITOR PEPTIDES

PKA and PKG are distinguished by the difference in their interaction with the family of heat-stable inhibitor proteins known as cAMP-specific protein kinase inhibitor (PKI). PKIs are highly selective and potent competitive inhibitors of Cα and Cβ catalytic function (but not of Cγ or yeast C-subunit); these proteins interact only weakly with PKGs. The PKI contains a PKA pseudosubstrate site that interacts with the substrate binding site of the C-subunit (Table 1), but other high-affinity interactions also occur between PKI and the C-subunit. The intracellular concentration of PKI is thought to be less than that of the C-subunit, and the physiological role(s) for PKI are unresolved. If PKI and free C are colocalized in a particular compartment, a significant portion of the action of the activated PKA could be blocked, but physiological relevance of such a process remains obscure.

9. NEGATIVE FEEDBACK AND FEED-FORWARD CONTROL OF CYCLIC NUCLEOTIDE PATHWAYS

In most tissues, physiological responses to cyclic nucleotides typically occur within a two- to fourfold increase in either cAMP or cGMP. Counter-regulatory processes provide for maintaining the levels of these nucleotides and the resulting kinase activities within a range that allows for optimal responsiveness. Multiple mechanisms exist that either potentiate the response to elevations of cyclic nucleotide or conversely restrict the magnitude or persistence of the cyclic nucleotide signaling. Negative feedback control mechanisms involving the PKA have been established for the cAMP cascade and for the cGMP signaling

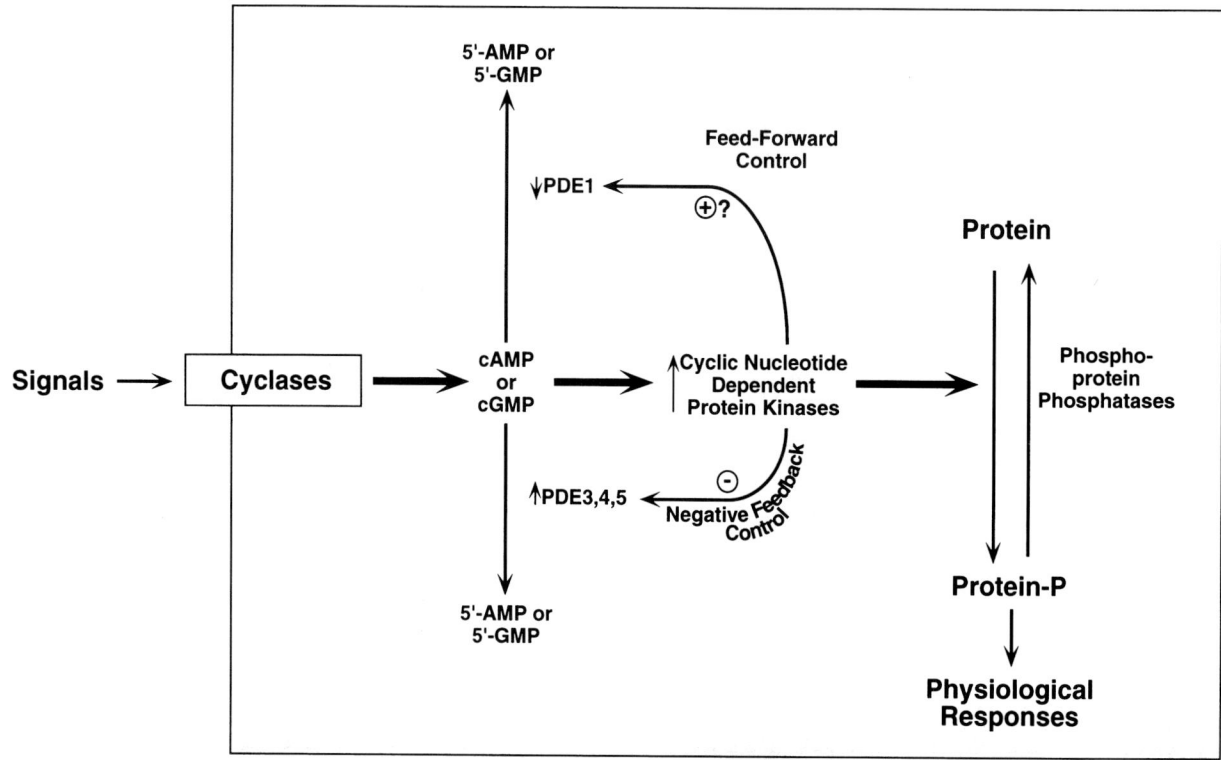

Fig. 6. Summary of mechanisms involving phosphorylation that contribute to modulation of the cAMP- and cGMP-signaling pathways. A ⊕ indicates that the effect depicted is increased as the levels of cyclic nucleotides increase; *arrows* ↑ or ↓ indicate a rapid change in the catalytic activity of the indicated enzymes that is induced by PKA/PKG phosphorylation or the state of phosphorylation of the phosphoprotein phosphatase inhibitor-1 in response to increases in the accumulation of cyclic nucleotides. Negative feedback control refers to processes that would tend to minimize the overall effectiveness of cyclic nucleotide increases whereas feed-forward control refers to events that would tend to enhance or prolong the cyclic nucleotide signal.

pathway as well (Fig. 6). In the cAMP cascade, activated PKA phosphorylates and activates two families of cyclic nucleotide phosphodiesterases (PDE3 and PDE4), which accelerates cAMP breakdown. Another mechanism that can limit cAMP accumulation involves the PDE2; cAMP binds to allosteric cyclic nucleotide binding sites in PDE2 to increase cAMP/cGMP breakdown. With persistent elevation of cAMP, the cellular concentrations of certain PDEs can also increase, thus providing an additional component of long-term negative feedback control. Phosphorylation and activation of PDE5 by PKG following elevation of cGMP provides a negative feedback control for this cyclic nucleotide pathway. However, PKA action can also foster a sustained or potentiated response to cAMP elevation resulting in a "feed-forward" type of control (Fig. 6). PKA phosphorylation of PDE1 decreases the catalytic activity of this PDE by decreasing its sensitivity to stimulation by its Ca^{2+}–calmodulin allosteric effector. In addition, PKA phosphorylation of the phosphatase inhibitor-1 produces a potent inhibition of phosphoprotein phosphatase action, thereby potentiating the effects of PKA phosphorylation of substrate proteins. Lastly, simultaneous phosphorylation of more than a single protein within a given metabolic pathway (such as glycogen synthase and glycogen phosphorylase) can potentiate the cellular response to even modest increases in cyclic nucleotide. Because phosphorylation of the synthase causes its inhibition and phosphorylation of phosphorylase causes its activation, glycogen depletion is potentiated by these two processes acting in concert.

10. THE USE OF TRANSGENIC MICE IN PROVIDING NEW INSIGHTS INTO THE ROLES OF PKA AND PKG IN CYCLIC NUCLEOTIDE SIGNALING

Since the discovery of cyclic nucleotides, a massive research effort has been devoted to discerning the signaling mechanisms and the physiological importance of these second messenger molecules. These efforts have established the importance of these nucleotides in myriad processes including regulation of carbohydrate metabolism, lipolysis, gene expression, salt and water homeostasis, muscle contraction, neurological functions, and many others. Recently, transgenic mice with null mutations for RI, RII, PKGI, or PKGII have been produced. Each of these has been phenotypically characterized, and new insights into the role of these enzymes have resulted. There have been no reports of a C-subunit null mutation in mammals, but in yeast this change is lethal.

10.1. Mice Lacking Specific Isoforms of the PKA Regulatory Subunit Have Distinct Phenotypes

Compared to normal mice, the RIIβ knockout mice are lean and have diminished white adipocyte mass, increased metabolic rate, and elevated body temperature. Studies with these mice also implicate RIIβ in the acquisition of certain motor skills and with modulation of gene expression in neural tissue. In either RIIβ or RIβ knockout mice, the level of RIα is substantially increased presumably due to its increased association with the C-subunit to form the more stable PKA holoenzyme. The potential for compensatory changes in the respective R-subunits complicates the interpretations of studies of R-subunit-null mice or studies involving overexpression of a particular R-subunit. Even in wild type animals certain tissues contain a substantial amount of free RI that is saturated with cAMP.

10.2. Mice Lacking Either the PKGI or PKGII Have Distinctive Phenotypes

Mice lacking either PKGI or PKGII have also been studied recently. PKGI-null mice have notable alterations in smooth muscle function. Motor activity in the gastrointestinal tract is highly disorganized and transit of foodstuffs through the gut is profoundly slowed. PKGI has been found in both the smooth muscle of the gastrointestinal tract as well as in some of the neurons of the enteric neural ganglia, and the absence of the PKGI from both of these tissues is likely to contribute to the dysfunction of gastrointestinal motility. Likewise, modulation of vascular smooth muscle function in these mice is impaired, consistent with the established importance of PKGI in promoting vasodilation. As a result, male PKGI-null mice have low ability to reproduce because of impaired relaxation of the smooth muscle in the vasculature of the penis, and the PKGI-null mice are somewhat hypertensive and unresponsive to nitrovasodilators. Because systemic blood pressure is controlled by a multiplicity of factors, the absence of PKGI and the cGMP signaling through this enzyme is likely to be significantly compensated by adjustments in other signaling pathways; this results in the ability of the animals to maintain a relatively normal blood pressure even in the absence of PKGI.

The PKGII-null mice have a very different phenotype from the PKGI-null mice, which is in agreement with the fact that the PKGI and PKGII are largely, if not entirely, expressed in separate tissues and involved in modulating different physiological processes. PKGII-knockout pups appear normal at birth, but defective developmental changes in the epiphyseal growth plates retard bone growth, resulting in dwarfism in adults. These mice are insensitive to heat-stable enterotoxin-induced diarrhea that typically results from the phosphorylation of the CFTR-chloride channel in response to elevated cGMP. In normal mice, cGMP accumulates in response to the enterotoxin activation of guanylyl cyclase and in turn activates PKGII to increase chloride secretion and an accompanying water loss. This suggests that in the intestinal mucosa of normal animals, the sensitivity of the intestinal epithelium to the enterotoxin is mediated by PKGII. Guanylins activate the same subset of guanylyl cyclases and also modulate fluid homeostasis, and it is likely that PKGII mediates these effects as well.

11. SUMMARY

The cyclic nucleotide-dependent protein kinases are major intracellular receptors for cyclic nucleotides. Although these proteins are highly homologous to each other and may exhibit overlapping functions in some cases, they have distinct properties that contribute to some differences in their regulatory features and physiological effects. In addition, the PKAs and PKGs can serve as models for studying mechanisms that pertain to the catalytic function and regulation of other protein kinases in the superfamily.

12. SELECTED READINGS

Butt E, Geiger J, Jarchau T, Lohmann SM, Walter U. The cGMP-dependent protein kinase-gene, protein, and function. Neurochem Res 1993; 18:27.

Doskeland SO, Maronde E, Gjertsen BT. The genetic subtypes of cAMP-dependent protein kinase—functionally different or redundant? Biochim Biophys Acta 1993; 1178:249.

Francis SH, Corbin JD. Cyclic nucleotide-dependent protein kinases: intracellular receptors for cAMP and cGUP action. Crit Rev Clin Lab Sci 1999; 52:275.

Hofmann F, Dostmann W, Keilbach A, Landgraf W, Ruth P. Structure and physiological role of cGMP-dependent protein kinase. Biochim Biophys Acta 1992; 1135:51.

Knighton DR, Zheng JH, Ten Eyck LF, et al. Crystal structure of the catalytic subunit of cAMP-dependent protein kinase. Science 1991; 253:407.

Knighton DR, Zheng JH, Ten Eyck LF, Xuong NH, Taylor SS, Sowadski JM. Structure of a peptide inhibitor bound to the catalytic subunit of cAMP-dependent protein kinase. Science 1991; 253:414.

Lincoln TM, Cornwell, TL. Intracellular cGMP receptor proteins FASEB J 1993; 7:328.

Madhusudan, Trafney EA, Xuong NH, et al. cAMP-Dependent protein kinase: crystallographic insights into substrate recognition and phosphotransfer. Prot Sci 1994; 3:176.

McKnight GS, Cummings DE, Amieux PS, Sikorski MA, Brandon EP. Cyclic AMP, PKA, and the physiological regulation of adiposity. Rec Prog Horm Res 1998; 53:139.

Pawson T, Scott JD. Signaling through scaffold, anchoring, and adaptor proteins. Science 1997; 278:2075.

Pfeifer A, Klatt P, Massberg S, et al. Defective smooth muscle regulation in cGMP kinase I-deficient mice. EMBO J 1998; 17:3045.

Rubin GS. A kinase anchor proteins and the intracellular targeting of signals carried by cAMP. Biochim Biophys Acta 1994; 1224:467

Smolenski A, Burkhardt AM, Eigenthaler M, et al. Functional analysis of cGMP-dependent protein kinases I and II as mediators of NO/cGMP effects. Naunyn-Schmied. Arch Pharmacol 1998; 358:134.

Su Y, Dostmann WR, Herberg FW, et al. Regulatory subunit of protein kinase A: structure of a deletion mutant with cAMP binding domains. Science 1995; 269: 807.

Taylor SS, Radzio-Andzelm E, Knighton DR, et al. Crystal structures of the catalytic subunit of cAMP-dependent protein kinase reveals general features of the protein kinase family. Receptor 1993; 3:165.

17 Protein Serine/Threonine Kinases

Jörg Heierhorst, Richard Pearson, James Horne, Steven Bozinovski, Bostjan Kobe, and Bruce E. Kemp

CONTENTS

INTRODUCTION
REGULATION OF PROTEINS BY SERINE/THREONINE PHOSPHORYLATION
REGULATION OF PROTEIN KINASES
EVOLUTION OF PROTEIN KINASES
SELECTED READINGS

1. INTRODUCTION

The reversible phosphorylation of proteins by protein kinases is the most extensively used molecular mechanism regulating cellular functions in eukaryotes. Essentially all major cellular processes, including cell growth and differentiation, motility, metabolism, and communication between cells, are in some way regulated by protein phosphorylation (Fig. 1). It is estimated that approx 30% of all mammalian proteins exist in different phosphorylated forms. To fine-tune the cellular response to external or internal signals toward such a diversity of substrates, protein kinases themselves must be tightly regulated and be present in a sufficiently large number of diverse forms. It is therefore not surprising that the recent genome sequencing projects revealed that a large proportion of genes code for different protein kinase catalytic subunits. The yeast *Saccharomyces cerevisiae* has 113 protein kinase genes, representing approx 2% of its genome. The number of protein kinases (435) in the nematode worm *Caenorhabditis elegans* is topped only by the 650 G-protein coupled receptor-like genes (many of which may be odorant receptors geared toward the environment rather than mediators of intercellular signaling). Likewise, it is estimated that 2% of all human genes encode protein kinases.

The protein serine/threonine kinases catalyse the transfer of the terminal γ-phosphate group from nucleotides, typically ATP but sometimes GTP, to serine/threonine hydroxyl groups on proteins to form phosphate monoesters and ADP or GDP. The crystal structure of the cAMP-dependent protein kinase (*see* Chapter 16) together with stop flow rapid kinetic studies have revealed the key features of the catalytic mechanism. The structure shows two lobes with the active site positioned in the cleft between the two lobes. The nucleotide (ATP) is bound to the small lobe of the kinase with β and γ phosphates positioned under the glycine-loop (Gly-X-Gly-X-X-Gly) and including the β-strands 1 and 2 (Fig. 2). The conserved lysine residue in β-strand 3 anchors the α–β-phosphates. Mg^{2+} plays an important role in the catalytic mechanism with an essential Mg^{2+} ion bound to a conserved aspartic acid (in the DFG sequence) in the large lobe that bridges the β and γ phosphates. A second aspartic acid residue in the large lobe is responsible for the transfer of the γ-phosphate to the hydroxyl group on serine or threonine in the protein. The protein substrate is positioned in a substrate-binding groove that

From: *Principles of Molecular Regulation* (P. M. Conn and A. R. Means, eds.), © Humana Press Inc., Totowa, NJ.

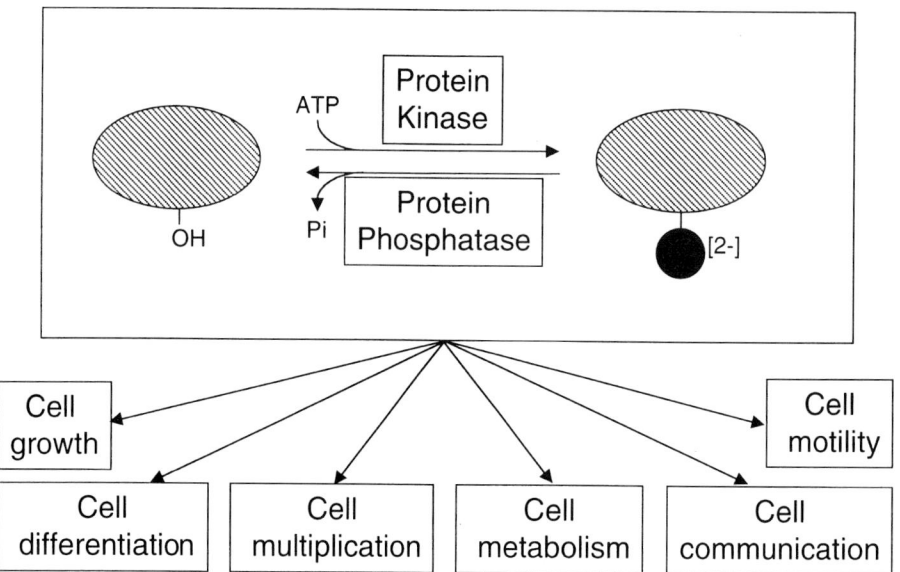

Fig. 1. Protein kinases catalyze the transfer of the γ-phosphate group of ATP to a hydroxyl side chain in proteins. Protein phosphatases reverse this reaction. Protein phosphorylation is involved in the regulation of all major functions of eukaryotic cells.

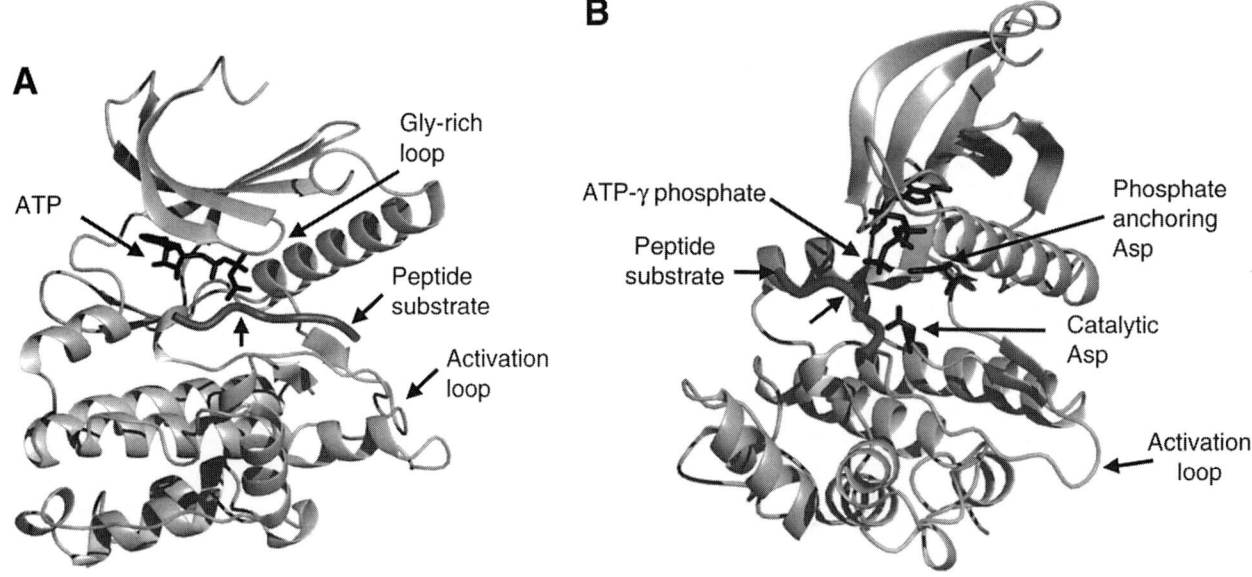

Fig. 2. (A) Protein kinase structure showing bound ATP and a peptide substrate. The ribbon diagram depicts helices and β-strands. The protein kinase structure is divided into two lobes with the upper, small lobe consisting of mainly β-strands and the lower large lobe consisting mainly of α-helices. ATP is positioned under the small lobe with the phosphate groups anchored by the Gly-rich loop. The protein substrate binding groove contains a peptide with the serine hydroxyl group positioned (indicated by an *arrow*) in-line with the γ-phosphate of ATP. The activation loop that requires phosphorylation in many protein kinases is also indicated. **(B)** Protein kinase structure showing active site residues and position of the γ-phosphate of ATP ready for transfer. Key aspartic acid residues are shown.

extends up the face of the large lobe so that the target residue is positioned within range of the γ-phosphate. Protein substrates are positioned in the active site by virtue of key residues on either side of the phosphorylatable serine or threonine that function as specificity determinants to ensure that only specific protein kinase(s) will phosphorylate these sites. Most of the key residues involved in the catalytic mechanism are present in the large lobe along with the protein substrate binding residues. Once ATP and protein sub-

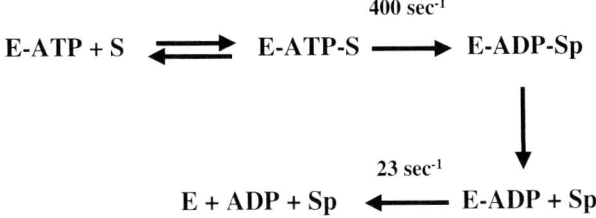

Fig. 3. The reaction scheme for protein kinase catalysis. Protein kinase (E) binds ATP and substrate (S) reversibly and forms the ternary complex E–ATP–S. There is a fast phosphoryl transfer step giving ADP and the phosphorylated substrate (Sp). ADP dissociates from the enzyme last and is the rate-limiting step.

strate is bound the transfer reaction is very rapid (in the range ~400 s^{-1}). The phosphorylated protein dissociates first, followed by ADP (in the range 23 s^{-1}) which is the rate-limiting step in the catalytic cycle (Fig. 3). The protein kinase small and large lobes are present in an open conformation prior to substrate binding and then close around a hinge region in the connecting peptide between the small and large lobes for catalysis. The protein kinase structures appear malleable, and indeed careful kinetic studies have shown that the off rate for ADP (which leaves after the phosphorylated protein) can very between 23 and 29 s^{-1} depending on the substrate that was previously bound, indicating that the substrate influences the catalytically competent structure.

Although protein serine/threonine kinases and protein tyrosine kinases are generally very similar in structure and reaction mechanism, they are considered as different subclasses of the protein kinase superfamily. This chapter therefore focuses on the regulation of proteins by serine/threonine phosphorylation and the regulation of protein serine/threonine kinases. The opposite situation, the reversal of phosphorylation by protein phophatases, is covered in another chapter of this volume.

2. REGULATION OF PROTEINS BY SERINE/THREONINE PHOSPHORYLATION

The phosphorylation of target proteins on serine/threonine residues can regulate cell function by different mechanisms with different time scales. The most immediate mode of regulation is to directly affect the catalytic activity of an enzyme target by phosphorylation, thus acting as an "on–off switch." Several non-enzymatic signaling molecules can also be regulated by phosphorylation with an immediate impact on cell function. For example, ion channels or G-protein coupled receptors at the cell membrane can be both activated and inhibited by phosphorylation. At the other end of the spectrum, protein phosphorylation can have a more delayed effect on the cell by regulating the profile of its protein composition at the transcriptional level (phosphorylation of transcription factors) or translational level (phosphorylation of ribosomal proteins). Mostly phosphorylation events affect cell function in a medium time scale. Again, this group exploits diverse mechanisms of regulation. Phosphorylation can regulate the subcellular localization of targets, or the incorporation of targets into functional effector complexes, and finally, in several cases it is known that phosphorylation can both positively or negatively regulate the lifetime of the substrates.

2.1. Regulation of the Catalytic Activity of Phosphorylation Targets

2.1.1. CELL METABOLISM

In their pioneering studies on glycogen metabolism in 1955 Fischer and Krebs showed that activation of phosphorylase by phosphorylase kinase was due to phosphorylation at Ser^{14}, and they were awarded the 1992 Nobel Prize in Medicine for this discovery. In 1968 it was found that phosphorylase kinase itself was phosphorylated and activated by the cAMP-dependent protein kinase. These discoveries formed the foundation for our understanding of many aspects of protein phosphorylation and signal transduction pathways. A number of important concepts were developed from this and subsequent studies.

- Second messengers such as cAMP, diacylglycerol, $PtdInsP_3$, and Ca^{2+} can modulate the function of intracellular proteins by controlling phosphorylation reactions.
- Kinase–kinase linked reactions can form regulatory signal transduction pathways. Signal transduction cascades linking multiple protein kinases also provides a mechanism for amplification of signals initiated at the cell surface or within.
- Specific protein inhibitors regulate protein kinases and phosphatases. These include the protein kinase inhibitor (PKI), protein phosphatases inhibitors-1 and -2 and a family of cyclin-dependent protein kinase inhibitors such as p16, p21, and p27.

When considering the regulation of metabolism several important themes have emerged. Multiple isoforms of proteins exist that differ in their regulation by phosphorylation. For example, liver pyruvate kinase is phosphorylated and inhibited by the cAMP-dependent protein kinase as a means of inhibiting hepatic glycolysis and promoting gluconeogenesis. In other tissues such as muscle the M-form of pyruvate kinase is present that does not have the corresponding site of phosphorylation and is not regulated in this way. A further example of the importance of protein substrate isoforms and the integration of protein phosphorylation with classical allosteric control is 6-phosphofructo-2-kinase. This enzyme is responsible for catalyzing the conversion of fructose 6-phosphate to fructose 2,6-phosphate, an important allosteric activator of phosphofructokinase and inhibitor of fructose 1,6-bisphosphatase. In liver, the cAMP-dependent protein kinase phosphorylates 6-phosphofructo-2-kinase at Ser^{32} near its N-terminus, thereby inhibiting it; the consequent reduction in fructose 2,6-bisphosphate serves to promote gluconeogenesis. In contrast to liver, phosphorylation of 6-phosphofructo-2-kinase in the heart promotes glycolysis; this is achieved through a different isoform of 6-phosphofructo-2-kinase that is activated by phosphorylation at several sites in its C-terminus by a variety of protein kinases. Thus, depending on the tissue specific expression of a particular isoform of 6-phosphofructo-2-kinase phosphorylation by the cAMP-dependent protein kinase can have the opposite effects on metabolism, promoting gluconeogenesis in the liver and glycolysis in the heart.

There is very tight coordination between energy demand and metabolism. Muscle contraction and glycogen metabolism are tightly coupled with calcium signaling to simultaneously initiate contraction as well as glycogen breakdown through activation of phosphorylase kinase. The AMP-activated protein kinase represents perhaps the most impressive example of how protein phosphorylation is used to coordinate metabolism. This is a metabolic stress-sensing protein kinase that is activated when ATP levels start to fall and there is a concomitant increase in the level of AMP. The role of the AMP activated protein kinase is to stimulate metabolism to restore the ATP levels (Fig. 4). As expected, all eukaryotes from fungi, yeast, plants, and nematodes to higher organisms have homologs of the AMP-activated protein kinase. In yeast it is called snf1p kinase and is responsible for adapting yeast to glucose starvation where it stimulates the secretion of invitase into the environment to break down complex sugars and scavenge glucose. In mammals, energy demand, for example, during vigorous exercise, switches on the AMP-activated

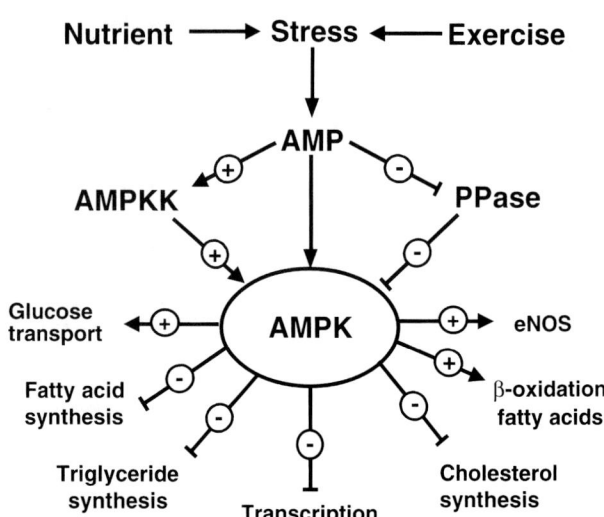

Fig. 4. Coordinated control of metabolism by the AMP-activated protein kinase. Nutritional stress or high energy demand from exercise increases AMP levels which in turn activate the AMP-activated protein kinase. This switches on energy yielding pathways including glucose transport (glycoysis) and fatty acid oxidation and simultaneously switches off energy-consuming pathways such as triglyceride and cholesterol synthesis.

protein kinase that in turn phosphorylates and inactivates acetyl-coenzyme A (acetyl-CoA) carboxylase, the enzyme responsible for producing malonyl-CoA. The reduction in malonyl-CoA stimulates fatty acid oxidation because malonyl-CoA is a negative regulator of fatty acid transport into the mitochondria via carnitine palmotyl transferase. In addition to stimulating fatty acid oxidation, the AMP-activated protein kinase plays a key role in stimulating glucose uptake in response to exercise. Activation of the AMP-activated protein kinase is complex and involves an upstream protein kinase that phosphorylates Thr^{172} in the AMP-activated protein kinase activation loop (see below). Inactivation occurs by dephosphorylation at this site catalysed by a phosphatase. AMP activates the upstream kinase and the action of the phosphatase is suppressed when AMP is bound to the AMP-activated protein kinase. The complex mechanism of AMP activation of the AMP-activated protein kinase serves to ensure that any small changes in cellular AMP levels will result in rapid activation of the AMP-activated protein kinase.

The AMP-activated protein kinase phosphorylates and activates endothelial NO synthase. Thus, during periods of energy demand and metabolic stress stimulation of NO production serves to improve the vascular supply of nutrients into tissues by relaxation of

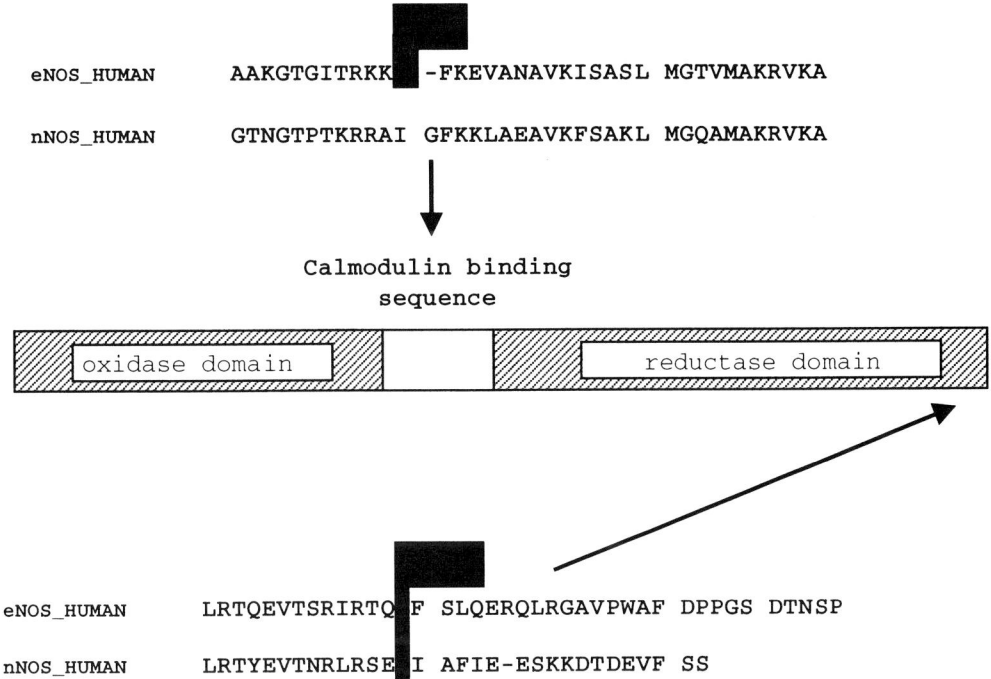

Fig. 5. Regulation of endothelial NO synthase by protein phosphorylation. Both endothelial (e) and neuronal (n) NO synthase are regulated by calmodulin that binds to a sequence between the enzyme's oxidase and reductase domains. Phosphorylation of eNOS at Thr[495] in the calmodulin binding domain inhibits NO synthase whereas phosphorylation at Ser[1177] activates the enzyme.

blood vessels. In the heart, NO also serves to dampen the mechanical activity so as to reduce energy demand. The phosphorylation of endothelial NO synthase by the AMP-activated protein kinase is also a good example of the complexity of regulation by protein phosphorylation and its integration with allosteric control. Endothelial NO synthase is regulated by Ca^{2+}–calmodulin that binds to a sequence between the oxidase and reductase domains of the protein and activates the synthase. The AMP-activated protein kinase phosphorylates endothelial NO synthase at two positions: Thr[495] in the calmodulin binding sequence only when calmodulin is not bound, and Ser[1177] in the C-terminus only when calmodulin is bound (Fig. 5). Phosphorylation at Thr[495] is inhibitory by blocking calmodulin binding whereas phosphorylation at Ser[1177] stimulates NO synthase activity. Thus, depending on the Ca^{2+}–calmodulin concentration, phosphorylation by the AMP-activated protein kinase can either stimulate or inhibit NO synthase activity. It should be noted that the regulation by phosphorylation is working in concert with the Ca^{2+}–calmodulin regulation of NO synthase and represents another layer of regulation. We expect that the neuronal and muscle forms of NO synthase (nNOS) are regulated in a similar manner except that they do not contain a phosphorylation site in the calmodulin binding sequence corresponding to Thr[495].

2.1.2. Cell Motility and Contractility

Another classic example of the direct regulation of the activity of a protein by protein kinases is the phosphorylation of myosin. Myosin II is the major actin-based molecular motor in eukaryotes. Functional myosin molecules are hexamers consisting of two larger heavy chains (MHC), each complexed with a so-called essential and a regulatory myosin light chain (MLC). The major part of the MHCs forms a long rod domain that is important for the assembly of supramolecular myosin filaments. The other end of the MHC contains the motor domain that interacts with the actin filament and has an ATP-hydrolyzing activity (ATPase). The motor domain together with the light chains forms the myosin head domain. Force and movement along the actin filament are generated by a tilt of the head as a result of hydrolysis of ATP, thereby pulling the myosin along the filament. Importantly, the myosin ATPase or motor activity is regulated by phosphorylation of the regulatory MLCs on Ser[20] (and Ser[19]). In skeletal muscle fibers, MLC phosphorylation plays a modulatory role by enhancing the force production. However, in smooth muscle

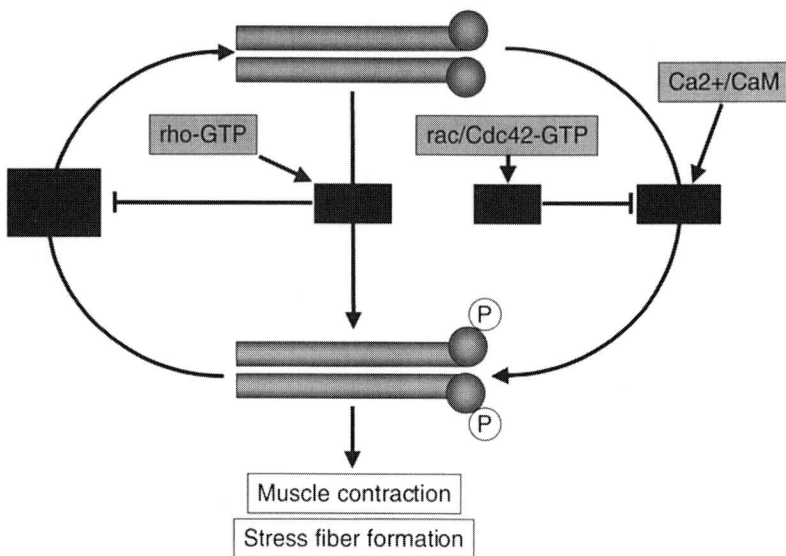

Fig. 6. Regulation of myosin by phosphorylation. Myosin is an enzyme that hydrolyzes ATP and uses the released energy for a confomational change resulting in motility. In smooth muscle cells and nonmuscle cells, phosporylation on Ser[19] and Ser[20] of the myosin regulatory light chains, regulates myosin activity, leading to contraction or stress fiber formation. Classically, myosin activation depends on myosin light chain kinases (MLCK) and Ca^{2+}–calmodulin (CaM). Recently protein kinases regulated by small GTP binding proteins, Rho-dependent kinase (rho-K) and p21-activated kinase (PAK), have been identified that contribute to the regulation of myosin activity. These G-protein dependent kinases can act on myosin directly, or indirectly by inhibiting MLCK or the myosin phosphatase (PPase).

and nonmuscle cells MLC phosphorylation is the key regulatory event that turns myosin on or off. In addition, in nonmuscle cells the phosphorylation of the MLC regulates the formation of actin stress fibers and focal adhesions at the cell membrane.

For a long time, MLC kinase (MLCK) was regarded as the major enzyme to regulate MLC phosphorylation. MLCK is an autoinhibited protein kinase that is activated by calmodulin in response to increases of the intracellular calcium concentration (*see* below). While calcium is elevated, MLCK is active and MLCs are phosphorylated, resulting in high myosin ATPase activity and smooth muscle contraction. Recently, however, a family of protein kinases with somewhat antagonistic effects on MLC phosphorylation has been discovered that may be crucial for the motility of nonmuscle cells. These protein kinases are regulated by the small GTP-binding proteins Rho, Rac, and Cdc42 (Fig. 6). Some members of this family, the Rho-dependent kinases, can phosphorylate MLCs directly and further increase MLC net phosphorylation by inactivating the myosin–phosphatase through phosphorylation of its myosin-binding subunit. Strikingly, kinases regulated by Rac and Cdc42, do the opposite—they phosphorylate and inactivate MLCK leading to less phospho-MLCs. This apparently paradoxical action of the small G-protein regulated kinases is crucial for the motility of nonmuscle cells by coordinating the cyclic assembly (phospho-MLC) and disassembly (dephospho-MLC) of stress fibers and focal adhesions.

2.2. Regulation of Protein–Protein Interactions, Subcellular Localization and Stability by Protein Phosphorylation

2.2.1. PROTEIN–PROTEIN INTERACTIONS

More and more cases are emerging where the interaction of proteins is regulated by protein phosphorylation, subsequently resulting in the regulation of cellular function. Two different mechanisms can account for phosphorylation-dependent protein interactions. First, interacting proteins could directly bind to the phosphorylation site via phosphopeptide binding sequences, or second, a "hidden" binding site could be indirectly exposed due to a conformational change resulting from phosphorylation.

Analogous to tyrosine-phosphorylated proteins, whose binding partners contain SH2 and PTB domains as specific phosphotyrosine binding modules, serine/threonine-phosphopeptide binding domains also exist. An example is the so-called 14-3-3 proteins that form dimers and contain one phosphopeptide binding site per monomer. They can therefore induce "homodimerization" by binding identical proteins to each 14-3-3 subunit. This is likely to play a role in the regulation of the protein kinase raf. Alternatively, they may form complexes of distinct phospho-proteins. As 14-3-3 proteins seem less target specific than protein kinases are substrate specific, the actual phosphorylation reaction is the main specificity determinant for 14-3-3 mediated interactions.

WW domains have recently been identified as modular phospho-serine/threonine peptide binding sites. WW domains are short regions of about 30 residues present in several proteins that were thought to bind proline-rich sequences in interacting proteins; it now appears that they may be more important for the regulated binding to phosphorylation sites. The opposite scenario seems to be the case for protein–protein interactions by the co-called PDZ domains that are abolished by serine/threonine phosphorylation of their binding sites.

The following examples illustrate how the regulation of protein–protein interactions through phosphorylation can be crucial for cells, even without affecting the catalytic activity of phosphoproteins. The first example comes from the yeast DNA damage response pathway. In response to DNA damage the noncatalytic Rad9p protein becomes phosphorylated. The major function of this phosphorylation event is to induce binding to the protein kinase Rad53p. This interaction depends on yet another modular domain, the so-called FHA domain located in the C-terminus of Rad53p. The FHA domain is essential for the activation of Rad53p by a postulated protein kinase. If this interaction is abolished, yeast are impaired in their ability to induce transcription of repair enzymes and to arrest the cell cycle until the damage is repaired, resulting in a severe loss of viability. Phosphorylation of Rad9p is crucial for the survival of the cell. A similar example is found in the differentiation of striated muscle. Here, the small noncatalytic protein telethonin is phosphorylated by the protein kinase domain located in the C-terminus of the large structural protein titin (>30,000 residues) and telethonin can then associates with the N-terminal region of titin as a crucial event in the formation of myofibrils.

2.2.2. SUBCELLULAR LOCALIZATION OF PROTEINS

As the subcellular localization of soluble proteins is generally controlled via a set sequence of protein–protein interactions, the regulation of the subcellular localization of proteins by phosphorylation is in some way just another facet of the regulation of protein interactions by phosphorylation. Most examples of

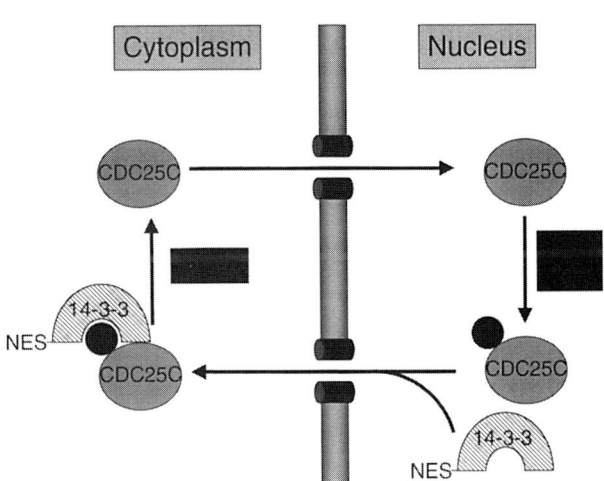

Fig. 7. Subcellular localization of Cdc25C regulated by phosphorylation on Ser^{216}. Cdc25C is phosphorylated by the Chk1 or Cds2 protein kinases, and binds a 14-3-3 protein that serves as an attachable nuclear export signal (NES) to shuttle it out of the nucleus.

phosphorylation-regulated subcellular localization relate to the regulation of the nuclear localization of proteins. Phosphorylation can affect the nuclear localization of proteins two ways. Some proteins are preferentially localized to the nucleus following phosphorylation, whereas other proteins are excluded from the nucleus if they are phosphorylated. The classical mechanism for proteins to be imported into the nucleus—unless they are small enough to freely traffic through nuclear pores—is to bind to the soluble import factor importin-α via a basic nuclear localization signal (NLS). In many cases the interaction with importin-α can be abolished by phosphorylation in, or in close proximity to, the NLS by various protein kinases, resulting in their cytoplasmic retention. Phosphorylation can also prevent the nuclear export of proteins, for example, in the case of cyclin B1 that accumulates in the nucleus after it is phosphorylated at the G_2/M transition of the cell cycle. An interesting variation of the regulation of nuclear localization by phosphorylation has recently been identified for the Cdc25 phosphatase that is crucial for the initiation of mitosis by dephosphorylating the inhibitory Tyr^{15} in the mitotic Cdc2 cyclin dependent protein kinase. Cdc25 phosphorylated on Ser^{216} binds to 14-3-3 proteins that serve in this case as "attachable" nuclear export signals. Cdc25 is therefore retained in the cytoplasm during interphase, preventing the access to its substrate (Fig. 7). There are several protein kinases that can phosphorylate Cdc25 on this residue, and the checkpoint kinases Chk1 and Chk2/Cds1 may be the most relevant physiologically for this process. When the cell senses that all requirements for an accurate mitosis are fulfilled, activity of these kinases is reduced, and the unphosphorylated Cdc25 can enter the nucleus to signal the onset of mitosis.

2.2.3. Protein Stability

Phosphorylation on serine or threonine residues can be crucial for regulating a protein's fate, and again, this can work two ways. A prominent example of a protein whose stability is enhanced in response to phosphorylation is the tumor suppressor p53. One function of p53 is to act as a transcription factor for p27, an inhibitor of mitotic cyclin-dependent kinases. Phospho-p53 accumulates in the cell, whereas dephospho-p53 is more rapidly destroyed. A candidate protein kinase regulating p53 stability is the ATM protein that functions as an effector in mammalian DNA damage and oxidative stress pathways. As a result, cells lacking ATM have reduced levels of p53, and humans lacking ATM suffer from a syndrome called ataxia-telangiectasia that comprises increased susceptibility to tumor development, increased sensitivity to ionizing radiation, and neuronal degeneration. It is currently not known if there is a specific mechanism for the destruction of unphosphorylated p53, or if the latter is simply less stable.

Mechanisms for a plethora of proteins that are degraded in a phosphorylation-dependent manner have recently been elucidated. Such proteins mostly function in the cell cycle and are recognized by socalled F-box proteins only when they are phosphorylated. The F-box proteins, of which more than 400 are currently known, associate with their targets via leucine-rich repeat or WD40 modular domains, and interact through the F-box motif with Skp1 in the socalled SCF ubiquitin ligase complex. In this way, the phosphorylated proteins become ubiquinated, earmarking them for degradation via the 26S proteasome. In some sense, this phosphorylation-dependent protein degradation adds a component of irreversibility to the primarily reversible nature of protein phosphorylation.

3. REGULATION OF PROTEIN KINASES

The crucial role protein kinases play in eukaryotic signaling pathways underscores the need for tight regulation of the kinases themselves. As in the case of the regulation of substrates by protein phosphorylation, there are also a variety of mechanisms to regulate the activity of the kinases, some examples of which are described in detail below. Noncatalytic regulators control protein kinases and this regulation can either be positive or negative. For example, the cAMP-dependent protein kinase catalytic subunit is constitutively active, but inhibited by a regulatory subunit that dissociates only in response to increased cellular cAMP levels. Conversely, other kinases, such as the CaM kinases, are autoinhibited and activated by association with calmodulin as an activating subunit in response to increased intracellular Ca^{2+}. Probably the most widespread mechanism of protein kinase regulation is the regulation of protein kinase activity through phosphorylation by "upstream" protein kinases within a kinase cascade and it is amazing that the sole purpose of a large number of protein kinases is to regulate other kinases. In the case of the mitogen- or stress-activated protein kinases it is common to have three or four kinases activating the next member of the series before another cellular effector is phosphorylated (see the yeast mating response described later)! The regulation of several protein kinases apparently even employs a two-tiered system by combining different means of activation, for example, in the case of the CaM kinases I and IV that require an upstream kinase kinase, as well as Ca^{2+}–calmodulin for full activity. Finally, the recent demonstration of scaffold proteins that organize protein kinases into functional complexes may be seen as a similar backup system.

3.1. Intrasteric Regulation of Protein Kinases

Early on it was found that protein kinases could often be activated by proteolysis. This included the classic examples of phosphorylase kinase, protein kinase C, and the myosin light chain kinase. Subsequent studies, using protein crystallography of twitchin kinase, titin kinase, and CaM kinase I, showed that this was due to the mechanism of autoinhibition that involved part of the protein's structure binding to the active site and inhibiting enzyme activity. Activation by proteolysis involves the removal of the autoinhibitory sequence. Protein kinase activation occurs by the binding of an activator such as diacyl glycerol for protein kinase C or the calcium binding protein calmodulin for the myosin light chain kinase. Before the crystal structures of several autoinhibited protein kinases were obtained, some aspects of the mechanism were inferred by indirect studies. In the case of the myosin light chain kinase it was observed from the amino acid sequence that a region overlap-

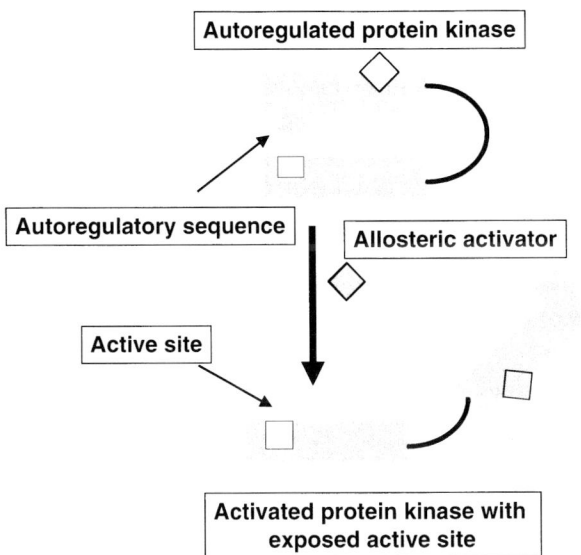

Fig. 8. Intrasteric regulation model for an autoregulated protein kinase. Part of the protein kinase structure contains an autoregulatory sequence that folds into the active site and blocks enzyme activity. The protein kinase is switched on by the binding of an allosteric regulator that removes the autoregulatory sequence from the active site.

ping the calmodulin binding sequence beyond the catalytic domain contained a juxtaposition of basic residues resembling a substrate recognition sequence. This autoregulatory sequence was termed a pseudosubstrate sequence. Protein kinase C was also found to contain a pseudosubstrate sequence near its N-terminus. Synthetic peptides corresponding to the pseudosubstrate sequences were found to act as potent inhibitors and truncation mutagenesis of the pseudosubstrate sequence rendered the kinases constitutively active. In the case of protein kinase C an antibody raised against the pseudosubstrate sequence was found to activate the enzyme.

The term intrasteric was introduced to describe pseudosubstrate autoregulation emphasizing that the autoregulatory sequence acted directly at the active site (Fig. 8). This contrasts to the classical allosteric regulation where a ligand binds to a regulatory site distinct from the active site. In general, allosteric effectors bear no structural resemblance to the target protein substrate. Intrasteric regulation is not restricted to protein kinases but occurs in protein phosphatases (e.g., calcineurin) and numerous other protein examples are now emerging. The structure of twitchin kinase revealed that the mechanism of autoinhibition was much more sophisticated than the simple pseudosubstrate model initially proposed. In Fig. 9 the autoregulatory sequence of twitchin kinase extends up the face of the large lobe along the substrate binding grove. Several basic residues in the autoregulatory sequence make similar pseudosubstrate-like contacts with the large lobe similar to the protein kinase inhibitor peptide (PKI; see Chapter 16). This segment is terminated with Trp[348] that is largely exposed and makes contacts across to the glycine-rich loop involved in ATP binding. The remarkable feature about the intrasteric inhibition of twitchin kinase is that the autoregulatory sequence makes an extensive number of contacts and essentially all residues that have been identified as critical for catalysis make contacts with the autoregulatory sequence (Fig. 10). For example, the conserved lysine that would normally neutralize the charges on the α–β phosphates of ATP contacts the backbone carbonyls of the inhibitory sequence residues 348 and 351. In the autoinhibited twitchin kinase structure a Pro-Ala-Pro-Gln-Pro sequence directs the inhibitory sequence deep in the active site with positioning of a 3_{10}-helix to block ATP binding. The catalytic aspartic acid forms an electrostatic contact with Arg[355] in the autoinhibitory sequence and the aspartic acid that would otherwise bind the metal ion Mg^{2+} (in the DFG sequence) in the large lobe that bridges the β- and γ-phosphates binds Phe[351] in the autoinhibitory sequence. The interface between the autoregulatory sequence and the catalytic core is extensive, with 40 hydrogen bonds, 388 van der Waals contacts, and 18 water-mediated interactions. There are 47 residues in the autoregulatory sequence comprising 179 atoms that contact 77 residues comprising 193 atoms within the catalytic small and large lobes. In terms of surface area of interaction 21% of the catalytic core and 47% of the surface of the autoregulatory sequence are buried. As expected, there is a high degree of complementarity between the buried surface of the autoregulatory sequence and the corresponding surface of the catalytic core. Indeed, the degree of shape complementarity between these surfaces is comparable to other protein–protein interactions including antibody–protein antigen interactions. The extensive number of contacts between the autoregulatory sequence and the kinase active site in twitchin kinase is much greater than for the cAMP-dependent protein kinase and its inhibitor peptide that is simply a potent protein substrate competitor (see Chapter 16).

This comprehensive mechanism of autoregulation is also conserved in the human titin kinase, and to a somewhat lesser degree in CaM kinase I. So far, no

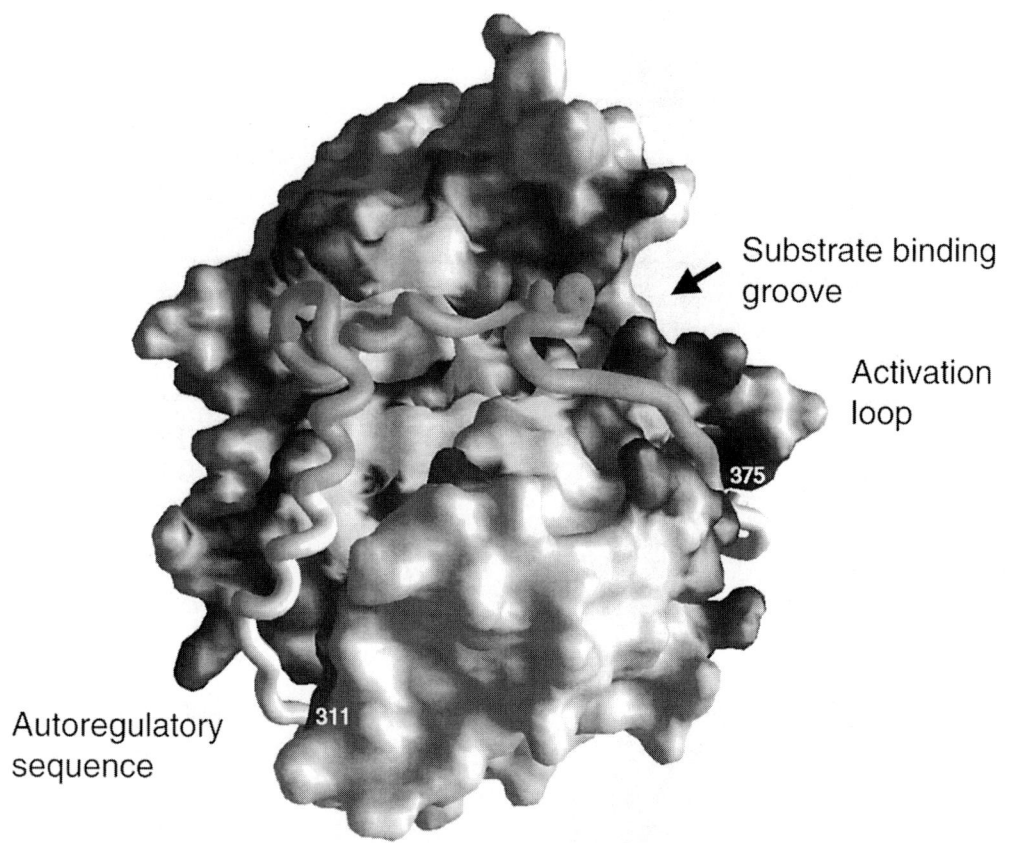

Fig. 9. Intrasteric control of twitchin kinase. Shown is a surface illustration of the structure of twitchin kinase with its autoregulatory sequence (shown as a worm from residue 311 to 375). It extends up the face of the kinase along the protein substrate binding grove into the active site. (For detailed contacts between the autoregulatory sequence and the kinase catalytic domain, see Fig. 10.)

structures of activated complexes have been solved. However, it can be anticipated that a dramatic rearrangement of the autoinhibitory contacts must be achieved by calmodulin, to render these kinases active.

3.2. Regulation of Protein Kinase Activity by Phosphorylation

Many serine/threonine protein kinases are regulated by changes in their phosphorylation state. Most of these protein kinases require phosphorylation of a conserved threonine residue within the so-called activation loop of the catalytic domain, which is located in the proximity of the kinase active site cleft, and in the linear sequence found between the DFG and APE motifs (Fig. 2A). This phosphorylated residue then interacts with multiple residues within the catalytic domain to modify the active site of the enzyme to a conformation suitable for substrate binding and catalysis. This is the simplest form of regulation by phosphorylation and can occur via autophosphorylation in the case of the cAMP-dependent protein kinase, or via the activity of an upstream kinase, discussed later for the activation of p70^{s6k} or protein kinase B (PKB) by phosphatidylinositol-dependent protein kinase 1 (PDK1).

Additional levels of regulation by phosphorylation are also commonly employed. In contrast, the cyclin-dependent protein kinases are phosphorylated at an inhibitory site (Tyr^{15}) in the glycine-rich loop that must be dephosphorylated by specific phosphatases (Cdc25) to permit activation. Many other protein kinases are reversibly phosphorylated on serine, threonine, and tyrosine residues outside the catalytic domain and in a large number of cases these are regulatory events that modulate the activation process. Phosphorylation thus provides one mechanism for exquisite control of these enzymes.

The related protein kinases, p70^{s6k} and PKB, provide examples of the extremely complex and intricate

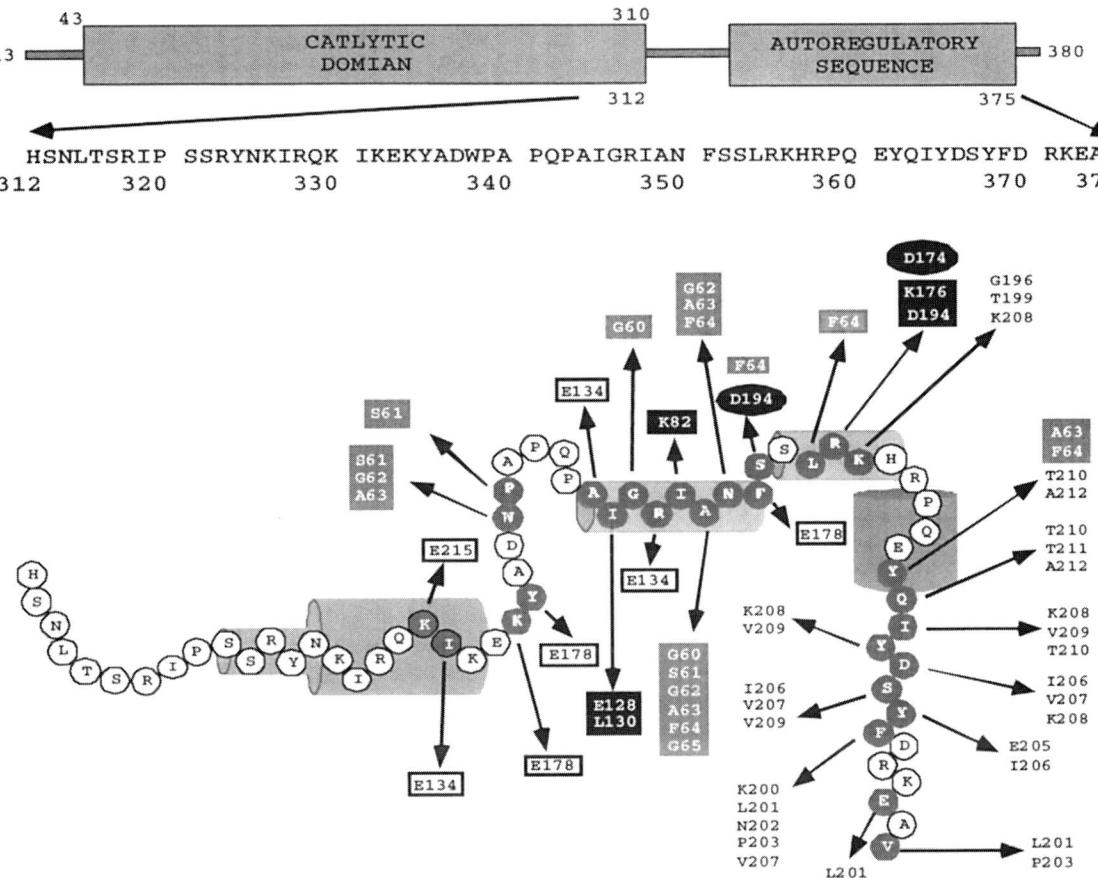

Fig. 10. Intramolecular contacts between the twitchin kinase autoregulatory sequence and the catalytic core. Four major types of contacts occur between the autoregulatory sequence and the catalytic core. Protein-substrate-like contacts are shown in *lined boxes*. Contacts to residues involved in ATP binding are shown in *gray boxes*. Contacts to key catalytic residues are shown in *black boxes* or *black ovals* for the active site Asp residues. Contacts to residues in the activation loop region are *unboxed*.

control of kinase activity by phosphorylation. p70^{s6k} plays a major role in regulating mRNA translation by phosphorylating the ribosomal protein S6, while PKB plays more diverse roles, including the regulation of glycogen metabolism and cellular survival. Despite their distinct roles, these two enzymes share many common features. Each is activated in response to a wide range of mitogenic signals including platelet-derived growth factor (PDGF), epidermal growth factor (EGF), insulin, thrombin, and nerve growth factor (NGF). The activation of each kinase is dependent on the activity of phosphatidylinositol 3′-kinase (PI-3 kinase). Importantly, the activation of both p70^{s6k} and PKB requires phosphorylation on equivalent sites within the catalytic domain (Thr^{229} and Thr^{308} respectively) and the C-terminus (Thr^{389} and Ser^{473}, respectively). Each is also phosphorylated on another equivalent site (Ser^{371} and Thr^{451}, respectively) although the significance of this phosphorylation in regulating the enzymes is unclear (*see* Fig. 11).

3.2.1. p70^{s6k} Phosphorylation and Structure

The activation of the p70^{s6k} by numerous stimuli is associated with its phosphorylation at multiple sites, and can be visualized by polyacrylamide gel electrophoresis that generates multiple, slower migrating bands. Treatment of p70^{s6k} with phosphatase removes the slower migrating bands. A total of 10 phosphorylation sites has been identified within the kinase sequence (Fig. 11A). These can be divided into functional groups, and evidence is accumulating of interdependence between phosphorylation sites. Ser^{411}, Ser^{418}, Thr^{421}, and Ser^{424} that are all followed by proline residues are clustered within a 14-residue sequence that is an autoinhibitory sequence apparently acting via its interaction with the N-terminal Module I (*see*

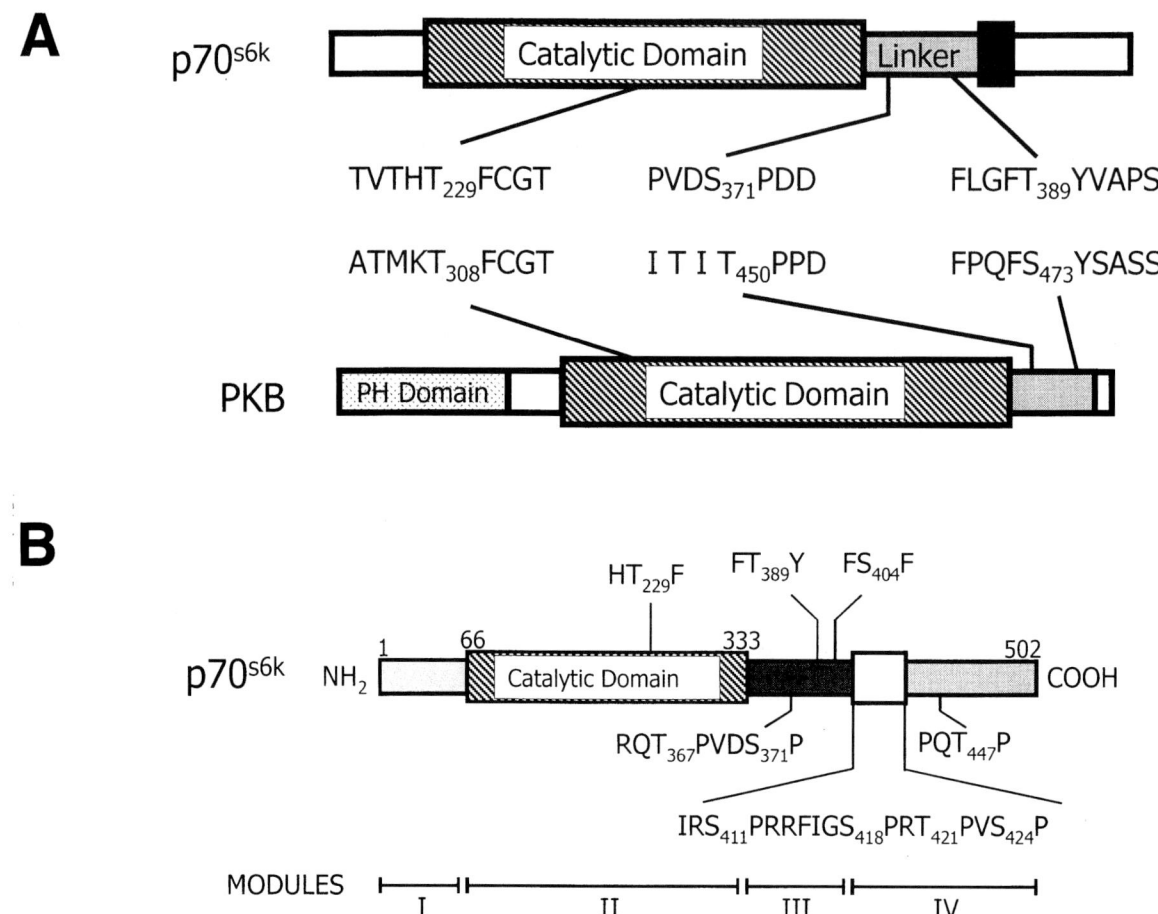

Fig. 11. Regulation of p70^s6k and PKB. **(A)** Homologous regulatory phosphorylation sites in p70^s6k and PKB. A schematic representation of the domain structure of each kinase is shown with the p70^s6k autoinhibitory sequence in *black*. The homologous sequences surrounding the essential regulatory phosphorylation sites are aligned. **(B)** p70^s6k phosphorylation sites. A schematic representation of p70^s6k showing the location of all the known phosphorylation sites. The conserved catalytic domain is shown in *black* and the autoinhibitory domain is *white*. The linker region between these two domains contains two of the essential regulatory phosphorylation sites, Ser^{371} and Thr^{389}. The locations of the proposed regulatory modules are indicated.

Fig. 11B). Phosphorylation of Thr^{229}, Thr^{389}, and Ser^{404} was identified on the basis of the sensitivity of these sites to the drug rapamycin. These sites are flanked by large hydrophobic residues suggesting they are phosphorylated by a common upstream kinase. Phosphorylation of Thr^{389} and Thr^{229} is essential for enzyme activity (*see* below). Very recently, three additional sites were identified (Thr^{367}, Ser^{371}, and Thr^{447}), also with adjacent C-terminal proline residues. Like Thr^{229} and Thr^{389}, Ser^{371} is homologous to residues conserved in many members of the protein kinase C family as well as Akt.

Structure–function analysis with individual sites substituted with acidic residues to mimic phosphorylation, or with alanine revealed that Thr^{389}, Thr^{229}, and Ser^{371} are essential for enzyme activity, while the autoinhibitory sites played a modulatory role. The kinase has been subdivided into four functional domains: an N-terminal domain (module I), the catalytic domain containing Thr^{229} (module II), a linker region containing Ser^{371} and Thr^{389} (module III), and a C-terminal domain containing the "autoinhibitory sites" (module IV). It appears that in the quiescent state, modules I and IV interact to inhibit phosphorylation of Thr^{389}. Binding of a postulated regulator breaks this interaction, allowing phosphorylation of the inhibitory sites in module IV that in turn allows phosphorylation of Thr^{389} (by a postulated protein kinase) and subsequently the final activating phosphorylation at Thr^{229} by PDK1.

3.2.2. PKB Phosphorylation and Structure

As noted earlier, PKB also lies downstream of PI-3 kinase and the role of the PI-3 kinase products, $PI(3,4,5)P_3$ and $PI(4,5)P_2$, is better defined than for $p70^{s6k}$. $PI(3,4,5)P_3$ and $PI(4,5)P_2$ play a dual role in the activation of PKB. Phospholipid binding of the PH domain of the kinase results in its translocation to the plasma membrane and releases an autoinhibitory function of the PH domain. Activation of the kinase is subsequently achieved via its phosphorylation at Thr^{308} by PDK1 and at Ser^{473} by PDK2, a postulated kinase. Mutational analysis indicates that constitutive phosphorylation at Ser^{124} and Thr^{450} by unknown upstream kinases may be required to prime the enzyme for activation PDK1.

There are thus similarities between the regulation of $p70^{s6k}$ and PKB. The context and location of the essential phosphorylation sites is conserved and each is kept in an inactive conformation, unable to be phosphorylated within the active site until secondary coactivating signals are received by the kinase. In the case of $p70^{s6k}$, binding of a postulated regulator disrupts the interaction between the N- and C-termini, permitting an ordered set of phosphorylation reactions requiring Thr^{389} phosphorylation before the final activation by PDK-1. For PKB, the costimulatory binding molecules are $PI(3,4,5)P_3$ and $PI(4,5)P_2$ and this binding allows PDK-1 phosphorylation of Thr^{308}. Phosphorylation of Ser^{473} is apparently a separate reaction required for maximum enzyme activity. The conserved nature of the hydrophobic motif surrounding Thr^{389} and Ser^{473} implies a common upstream kinase (PDK2) may phosphorylate these sites.

In summary, phosphorylation of common sites within the kinase structure is required for activity, and the access of essential upstream kinases is regulated individually by the requirements for binding modulatory proteins and/or auxiliary phosphorylation reactions.

3.3. Regulation of Protein Kinase Cascades by Scaffold Proteins

Whether protein kinases phosphorylate their substrates or not is not only determined by their activation, but also by their access to their targets. It has become apparent that several protein kinases are organized into functional signalling complexes by stable association with so-called scaffold proteins (*see* Fig. 12). This was first noted in the yeast *Saccharomyces cerevisiae*, where Ste11p acts as an "upstream" pro-

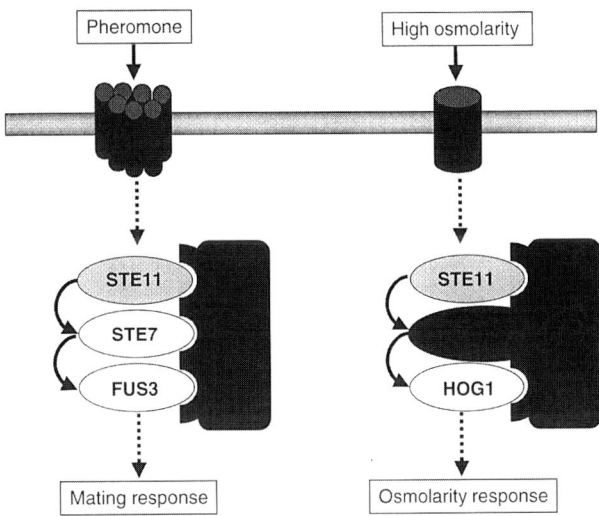

Fig. 12. Regulation of protein kinase cascades by scaffold proteins. In yeast, Ste11p, Ste7p, and Fus3p form a protein kinase cascade leading to serial activation in the response to pheromones. They are linked into a complex by the Ste5p scaffold protein. Ste11p is also part of a protein kinase cascade activated in response to high extracellular osmolarity, together with Pbs2p and Hog1p kinases. In this pathway, Pbs2p serves a dual function as kinase and scaffold protein. Assembly of protein kinases into complexes through scaffold proteins prevents Ste11p activation by pheromone to cross-activate the osmolarity response, and conversely, Ste11p activation by high osmolarity will not activate the mating response.

tein kinase that regulates the activity of other kinases further down the line, in both the mating response and osmolarity response pathways. However, Ste11p activated by mating pheromones will not activate the osmolarity response, and vice versa. This was due to its association with the scaffold protein Ste5p in the mating response that aligns Ste11p with its target Ste7p, another kinase, and Fus3p, a third protein kinase that is regulated by Ste7p. Conversely, in the osmolarity response pathway, the protein kinase Pbs2p seems to double up as a scaffold by linking its activator Ste11p with its substrate Hog1p.

In mammals, JIP1 may play a similar role to Ste5p in linking three protein kinases—MLK3, MKK7, and JNK—that are sequentially activated by cytokines or stress. Another protein, MP1, seems to play a similar role by bundling two other protein kinases, MEK1 and ERK1.

The conservation of scaffolds in the pathways mentioned here is not too surprising as all kinases in question are related members of the MAP kinase family. The principle of scaffolding kinases is not restricted to MAP kinase pathways, but includes the scaffold protein IKAP that organizes three protein kinases (NIK, IKKα, and IKKβ) required for activation of the NFκB transcription factor as well as AKAPS for the cAMP-dependent protein kinase. It seems likely that further scaffold proteins will be identified in other protein kinase signaling pathways in the future.

4. EVOLUTION OF PROTEIN KINASES

It is clear protein serine/threonine kinases and protein tyrosine kinases play a major role in eukaryotic signal transduction pathways. Despite this, there are no such kinases in prokaryotes, as these only contain histidine kinases. So where do they come from? One possibility is that protein kinases evolved from bacterial defense enzymes that confer antibiotic resistance. This possibility was noted when the crystal structure of a bacterial aminoglycoside antibiotic metabolizing enzyme was solved. Surprisingly, this enzyme had a three-dimensional fold very similar to eukaryotic protein kinases. In addition, this enzyme has a very similar reaction mechanism to protein kinases, in that it transfers the γ-phosphate group from ATP to hydroxyls in the antibiotic. Interestingly, bacterial aminoglycoside kinases can be inhibited with some protein kinase inhibitors that act as competitive ATP antagonists. Therefore, there is a possibility that the study of protein kinases may contribute to solving a seemingly unrelated problem—the increasing resistance of bacteria to clinically used antibiotics.

SELECTED READINGS

Downward J. Lipid-regulated kinases: some common themes at last. Science 1998; 279: 673.

Johnson LN, Noble NEM, Owen DJ. Active and inactive protein kinase: structural basis for regulation. Cell 1996; 85:149.

Kemp BE, Parker MW, Hu S-H, Tiganis T, House C. Substrate and pseudosubstrate interactions with protein kinases: determinants of specificity. Trends Biochem Sci 1994; 19:440.

Kemp BE, Mitchelhill KI, Stapleton D, Michell BJ, Chen ZP, Witters LA. Dealing with energy demand: the AMP-activated protein kinase. Trends Biochem Sci 1999; 24:22.

18 Protein Serine/Threonine Phosphatases

Shirish Shenolikar

CONTENTS

HISTORICAL PERSPECTIVE
EMERGING COMPLEXITY OF PROTEIN SERINE/THREONINE PHOSPHATASES
REGULATORY PARADIGMS FOR PROTEIN PHOSPHATASES
PATHOPHYSIOLOGY OF PROTEIN PHOSPHATASES
FUTURE OF PROTEIN PHOSPHATASES
SELECTED READINGS

1. HISTORICAL PERSPECTIVE

Reversible protein phosphorylation now seems so firmly established as a mechanism for regulating physiological processes in eukaryotic cells that the early work that led to its discovery is often forgotten. The first evidence that covalent modification controls the functions of cellular proteins came from studies of glycogen metabolism nearly 60 yr ago. By 1940, adrenaline was known to promote glycogen breakdown in skeletal muscle by increasing phosphorylase activity. Moreover, two forms of this enzyme had been isolated, one that required the allosteric activator, 5′-AMP, termed phosphorylase *b*, and another, phosphorylase *a*, that was fully active in the absence of 5′-AMP. In 1943, phosphorylase *a*, for the first time, was converted in vitro to phosphorylase *b,* using a cellular activity called the PR enzyme. During the next 10 yr, several mechanisms were proposed to explain the activity of the PR enzyme. At first, as the name implied, the PR or *p*rosthetic group-*r*emoving enzyme, was thought to remove pyridoxal phosphate, the necessary cofactor or prosthetic group that gives phosphorylase *a* its characteristic yellow color. Subsequently, it was noted that conversion of phosphorylase *a* to *b* was accompanied by a reduction in its apparent molecular weight, leading to the suggestion that the PR enzyme was a protease. In 1955, coincident with the discovery of phosphorylase kinase, the enzyme that converted phosphorylase *b* to *a*, the PR enzyme was firmly established as a protein phosphatase. As the first protein serine/threonine phosphatase to be discovered, the skeletal muscle PR enzyme is appropriately known today as protein phosphatase-1 (PP1). As the best characterized serine/threonine phosphatase, PP1 serves as the prototype for other eukaryotic protein phosphatases.

Studies in the early 1980s demonstrated the presence of phosphorylase phosphatase activity in tissues that did not undertake glycogen metabolism. This suggested that this protein phosphatase served many other functions. However, purification of this enzyme by different tissues failed to identify these functions. In fact, purified preparations of phosphorylase phosphatase showed an alarming lack of substrate specificity in vitro, dephosphorylating many phosphoproteins. The broad substrate specificity of the early phosphatase preparations fostered the longstanding misconception that protein phosphatases were nonspecific or pleotropic, and therefore could not transduce specific physiological signals. Subsequently, it was found that two different serine/threonine phosphatases, PP1 and PP2A, accounted for the phosphorylase phosphatase activity in mammalian tissues. PP1

From: *Principles of Molecular Regulation* (P. M. Conn and A. R. Means, eds.), © Humana Press Inc., Totowa, NJ.

and PP2A were distinguished by their remarkable selectivity for the dephosphorylation of two subunits, α and β, of phosphorylase kinase.

More recent studies show that PP1, PP2A, and other serine/threonine phosphatases possess much greater substrate selectivity than at first realized, both in terms of the phosphoproteins and the specific phosphoserines/threonines that they recognized. Specificity of these enzymes is only in part encoded in the catalytic subunits but this can be greatly enhanced by the association of the catalytic subunits with various regulatory proteins. PP1, PP2A, and other phosphatases are now known to exist as multisubunit complexes that display unique cellular functions. Interestingly, the wide range of molecular weights reported for phosphorylase phosphatases, presumably reflecting different PP1 complexes, had frustrated the early efforts to define the molecular components of this enzyme. It is now clear that it is through this association with regulatory subunits that PP1 controls many different physiological processes including carbohydrate, lipid, and protein biosynthesis and includes among its substrates many enzymes, receptors, ion channels, cytoskeletal proteins, and transcription factors.

Finally, there persists a misconception that protein phosphatases are largely unregulated, and therefore protein kinases represent the primary routes for signal transduction. This view prevailed for more than 30 yr after the discovery of the first protein phosphatase. However, work over the last 10 yr has identified many regulatory subunits that not only define the subcellular localization of the protein serine/threonine phosphatases but also dictate substrate recognition by these enzymes. What is even more remarkable is that these proteins participate in elaborate regulatory mechanisms that modulate phosphatase activity in response to physiological stimuli. Analysis of these regulators has emphasized the close communication that exists between protein kinases and phosphatases to ensure the appropriate physiological response and hints at errors in kinase–phosphatase crosstalk as contributing factors to human disease.

2. EMERGING COMPLEXITY OF PROTEIN SERINE/THREONINE PHOSPHATASES

Once it was realized that cells contained many different phosphatase complexes sharing common catalytic subunits, better criteria had to be developed to characterize cellular protein serine/threonine phosphatases (Cohen, 1989). However, the lack of well-defined phosphoprotein substrates slowed early efforts to characterize these enzymes on the basis of their substrate profile. For instance, not all the serine/threonine phosphatases dephosphorylate phosphorylase *a*. This substrate is primarily useful for analyzing PP1 and PP2A, which together account for >95% of phosphorylase phosphatase activity in eukaryotic cells. Today, molecular cloning has provided considerable new information on the primary structure of protein phosphatases but without appropriate substrate(s) little progress can be made in defining their function and regulation. Fortunately, many of the newly discovered serine/threonine phosphatases dephosphorylate the enzyme, phosphorylase kinase, showing a preference for either the α- or β-subunit. However, PP1 remains the only phosphatase that significantly favors the dephosphorylation of the β-subunit. Most other cellular phosphatases preferentially dephosphorylate the α-subunit of phosphorylase kinase and are thus collectively referred to as type-2 enzymes. Other criteria, such as the dependence on divalent cations for activity, did separate some type-2 phosphatases. For instance, calcineurin, or PP2B, is a calcium–calmodulin-dependent phosphatase and currently the only serine/threonine phosphatase directly activated by a second messenger (*see* Chapter 9). This enzyme has drawn much attention from the scientific community as it is the target of cyclosporin and FK506, the two most effective immunosuppressive drugs in use today. PP2C was identified by its unique requirement for millimolar concentrations of magnesium. All the major serine/threonine phosphatases—PP1, PP2A, PP2B, and PP2C—can be readily activated in vitro by manganese ions, consistent with the fact that they are all metalloenzymes. X-ray crystallography of PP1, PP2B, and PP2C has established the conserved architecture of the metal-containing catalytic sites (Fig. 1). This is interesting as there is little primary sequence homology between PP2C and the other phosphatases. Yet, the three-dimensional structure of PP2C, specifically the catalytic site, is remarkably similar to that in PP1 and PP2B. Finally, endogenous protein inhibitors, such as inhibitor-1 (I-1) and inhibitor-2 (I-2), that selectively inhibit PP1 and environmental toxins that inhibit PP1 and PP2A to varying degrees, have been extremely useful in characterizing these enzymes from many species and in delineating their functions in living cells. Nevertheless, there remains a need for new and improved experimental tools to investigate the cellular functions of phosphatases.

The real progress in the identification of protein

Chapter 18 / Protein Serine/Threonine Phosphatases

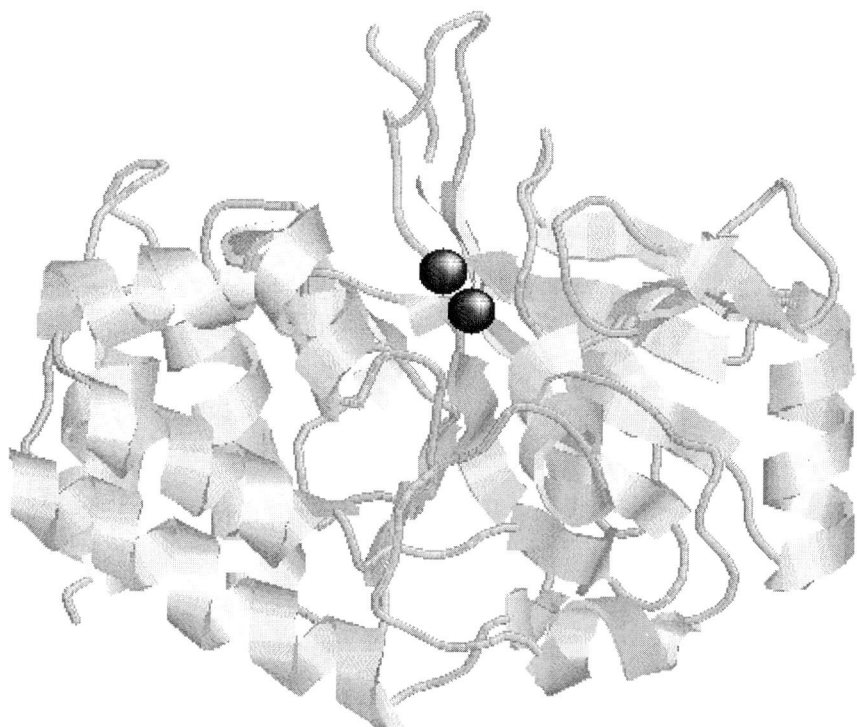

Fig. 1. Three-dimensional structure of the PP1 catalytic subunit. X-ray crystallography of the PP1 catalytic subunit was undertaken with and without the toxin inhibitor, microcystin-LR. This highlighted the relatively rigid structure of the enzyme and suggested that natural compounds that fit into the bimetallic catalytic site will be effective phosphatase inhibitors.

serine/threonine phosphatases came in the mid-1980s from the molecular cloning of the two major cellular phosphatases, PP1 and PP2A. Comparison of the primary structure of the two enzymes highlighted six regions of sequence homology that became the hallmarks for identifying other members of this enzyme family (Barton et al., 1995). Polymerase chain reaction (PCR) cloning, based on the conserved sequences, has identified structurally related phosphatases from many different species (Table 1). The data suggest that the three major eukaryotic phosphatases—PP1, PP2A, and PP2C—are expressed in all animals and plants and their primary structures have been remarkably conserved through evolution. A calcium–calmodulin-activated phosphatase activity has been detected in plants but its structural similarity to calcineurin remains unknown. Amazingly, PP1 and PP2A show nearly 90% sequence identity between fungi and humans. This raises the very likely possibility that these protein serine/threonine phosphatases serve crucial functions in eukaryotic cells so that few changes in their structure have been tolerated over millions of years of evolution. Molecular cloning has also identified several new enzymes that are structurally related to PP1 and PP2A, but form their own PPP family (Cohen, 1997). Comparison of their primary structures has segregated these enzymes into seven families. However, sequencing of the *C. elegans* genome identified several new protein serine/threonine phosphatase genes that do not readily fall into the already defined PPP families. This suggests the existence of additional eukaryotic PPP families.

Any efforts to elucidate the physiological functions of a protein phosphatase must face the challenge of accommodating multiple genes that encode an individual catalytic subunit. For example, there are three human *PP1* genes—α, β (previously called δ), and γ. Moreover, two isoforms are derived from the *PP1γ* gene through alternate splicing. The fruit fly, *Drosophila melanogaster*, has four *PP1* genes and some plants have more than 10 genes encoding the PP1 catalytic subunit. Isolation of mammalian PP1 complexes containing a single PP1 isoenzyme has prompted the speculation that the individual isoenzymes can distinguish between different regulatory subunits and form unique enzyme complexes. However, in vitro studies with recombinant PP1 catalytic subunits have not provided particularly compelling evidence for such specificity in protein–protein interactions. On the other hand, genetic studies in lower

**Table 1
Protein Serine/Threonine Phosphatase Catalytic
Subunits—The PPP Family**

Enzyme	Other Names	Organisms
PPP1	PP1α	Mammals
	PP1β or δ	[Plants
	PP1γ	Drosophila
		Fungi]
	Ppq1	S. cerevisiae
	PPY	Drosophila
	Ppz1	S. cerevisiae
	Ppz2	S. cerevisiae
PPP2	PP2A Cα	Mammals
	PP2A Cβ	[Plants
		Drosophila
		Fungi]
	Ppg1	S. cerevisiae
PPP3	PP2B Aα	Mammals
	PP2B Aβ	[Drosophila
		Fungi]
PPP4	PPX	Mammals
		Drosophila
	Pph3	S. cerevisiae
PPP5	PP5	Mammals
		Drosophila
	Ppt1	S. cerevisiae
PPP6	PPV	Mammals
		Drosophila
	Sit4	S. cerevisiae
	Ppe1	S. pombe
PPP7	PP7	Mammals
	RdgC	Drosophila
PPPM	PP2Cα	Mammals
	PP2Cβ	[Plants
	PP2Cγ	Fungi
		C. elegans]

eukaryotes, in which individual phosphatase catalytic subunits were mutated, argue for unique as well as overlapping functions for different phosphatase isoenzymes (Shenolikar, 1994). The facile genetics of fungi and fruit fly provide us with one of the most exciting experimental approaches for resolving the complexity of phosphatase catalytic subunits. By linking specific phenotypes with mutations or disruption of individual phosphatase genes, these studies have defined the multiple roles anticipated for these protein serine/threonine phosphatases. In addition, they have yielded experimental systems to analyze structurally related enzymes from other species. For example, mammalian, plant and fly *PP1* genes successfully replace the defective *PP1* genes in fungi and rescue their aberrant phenotypes. This emphasizes the conservation in function of protein serine/threonine phosphatases across species. However, not all mammalian or plant *PP1* genes are equally effective in rescuing the fungal phenotype, pointing to functional differences in these isoenzymes.

As mutations in a phosphatase catalytic subunit can be linked to many different phenotypes in fungi, this provides further evidence that a single phosphatase can control many physiological processes. The data suggest that the diversity in phosphatase function comes from the presence of different regulatory subunits. For instance, the number of potential PP1 regulatory subunits now exceeds forty. PP2A, a heterotrimeric enzyme consisting of a catalytic subunit bound to two regulatory subunits, may form as many as 100 different complexes (Wera and Hemmings, 1995). With the increasing number of phosphatase catalytic and regulatory subunits being identified, there is growing support for the suggestion that up to 20% of the human genome may be dedicated to genes that control protein phosphorylation. So, the challenge for future research is to develop experimental strategies that will help to unravel this complexity of protein kinases, phosphatases and their regulators.

A number of reviews have been written in recent years that describe the biochemical properties of protein serine/threonine phosphatases (Cohen, 1989; Shenolikar and Nairn, 1991). The reader is referred to these for more information on the structure-function of the major cellular phosphatases. In keeping with the theme of this book on molecular regulation, this chapter focuses on the emerging paradigms for regulating the activity of protein serine/threonine phosphatases. Understanding the mechanisms that control protein phosphatases may help to establish their importance in cell signaling and provide new insights into the physiology and pathophysiology of the serine/threonine phosphatases.

3. REGULATORY PARADIGMS FOR PROTEIN PHOSPHATASES

Hormones, neurotransmitters and growth factors act through cell surface receptors to stimulate protein kinases and initiate cascades of protein phosphorylation that dictate physiological responses. If constantly opposed by cellular phosphatases, which may sometimes display higher specific activity or turnover rate than kinases, the cellular response mediated by protein phosphorylation would be both blunted and brief. This would in essence be like driving one's car with its brakes on, resulting in both slow progress of the vehicle and waste of fuel. Cellular mechanisms that suppress protein serine/threonine phosphatase activity, would remove the "brakes" on signal transduction and thus amplify and prolong the cell's response to hormones. Consistent with this notion, one of the first phosphatase regulators identified, a phosphoprotein known as inhibitor-1, facilitates hormone signaling in many mammalian tissues.

Some hormones, such as adrenaline, activate protein kinases and stimulate protein phosphorylation in the target tissues. These hormones are counteracted by other hormones, such as insulin, that reverse the phosphorylation events in part by activating protein phosphatases. A simple mechanism by which hormones and neurotransmitters can activate a phosphatase is by increasing intracellular calcium and stimulate the calcium–calmodulin-activated protein phosphatase, calcineurin (PP2B). However, most phosphatases do not respond directly to intracellular second messengers and more elaborate mechanisms are needed for the hormonal control of these enzymes. For example, insulin, epidermal growth factor and platelet-derived growth factor all increase PP1 activity. This might be achieved by negating endogenous PP1 inhibitors or modulating other PP1 regulators to increase phosphatase activity. Physiological studies suggest that insulin exploits both of these mechanisms to activate PP1.

Finally, there are mechanisms for crosstalk between protein kinases and phosphatases that result in the delayed activation of the phosphatases. In this setting, the hormone initiates intracellular protein phosphorylation by kinases and among the proteins phosphorylated is a phosphatase regulator. Phosphorylation of this protein enhances phosphatase activity, and consequently the physiological response is downregulated. This kind of crosstalk mechanism functions as a "timer" to establish the duration of the cell's response to hormone. In this chapter, we discuss examples of phosphatase regulatory mechanisms that function as amplifiers and timers to control cell signaling. This should illustrate that the days of considering serine/threonine phosphatases as unregulated enzymes are over. It is now clear that protein kinases and phosphatases function in a close partnership to control cell physiology.

3.1. Phosphatase Inhibitors Coordinate Hormone Signals

The first endogenous phosphatase inhibitors to be identified were two thermostable proteins, termed inhibitor-1 (I-1) and inhibitor-2 (I-2). These proteins were isolated from skeletal muscle extracts and shown to inhibit the skeletal muscle phosphorylase phosphatase (PP1). Nanomolar concentrations of these proteins inhibited PP1 activity but they had little effect on PP2A or other type-2 phosphatases, even at 1000-fold higher concentration. What makes these proteins particularly interesting is that their inhibitory activity is modulated by protein phosphorylation. Thus these proteins play a key role in cell signaling. Like the prototypic phosphatase that they inhibit, I-1 and I-2 led the way for the discovery of other cellular phosphatase inhibitors. For instance, recent studies have identified highly specific and potent endogenous inhibitors for PP2A and PP2B, although the precise role of these proteins in modulating cell signals remains to be determined. Whether cells contain specific inhibitors for all serine/threonine phosphatases is uncertain but the insights provided by PP1 inhibitors suggest that such proteins will greatly facilitate hormone signals.

The kinetics of I-1 phosphorylation put it among the best known in vitro substrates of protein kinase A (PKA). Thus, I-1 is likely to be one of the first proteins to be phosphorylated in cells in response to hormones and neurotransmitters that elevate intracellular cAMP. The phosphorylation of I-1 is absolutely required for PP1 inhibition. Thus, I-1 suppresses PP1 activity in response to hormones and facilitates the phosphorylation of proteins that are substrates for PKA as well as other protein kinases (Fig. 2). Physiological studies have provided evidence that PP1 inhibition by I-1 enhances GSK-3 phosphorylation of glycogen synthase in skeletal muscle and propagates calcium–calmodulin-dependent protein kinase II (CaMKII) signals in hippocampal neurons. The inhibition of PP1 by I-1 as a mechanism may also impose cAMP control over the phosphorylation of proteins that are not direct substrates of PKA. In this regard, the disruption of the mouse gene encoding DARPP-32 (dopamine and cAMP-regulated phosphoprotein of apparent M_r 32,000), a PP1 inhibitor closely related to I-1, significantly impaired dopamine signaling mediated by D1 receptors. This emphasized the importance of this phosphatase regulatory mechanism for hormone signaling through the second messenger,

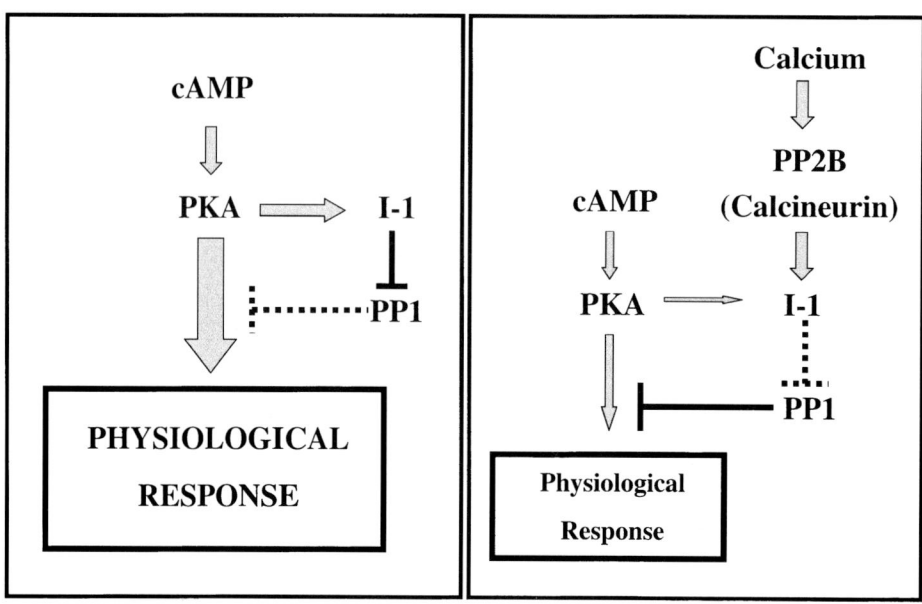

Fig. 2. Phosphatase regulation by the second messengers, cAMP and calcium. **(Left)** Hormone-induced increases in intracellular cAMP activate PKA and promotes phosphorylation of proteins that mediate the hormone response. However, PP1 often antagonizes PKA actions in mammalian cells. Thus the phosphorylation and activation of the PP1 inhibitor, I-1, promotes cAMP signaling in these cells. **(Right)** Calcium entry activates the calcium–calmodulin-activated phosphatase, calcineurin or PP2B, and promotes the dephosphorylation of I-1, one of the best known substrates for this enzyme. The inactivation of I-1 leads to increased PP1 activity and down-regulation of the physiological response mediated by PKA and other kinases.

cAMP. Protein kinase C (PKC) mediated phosphorylation of a protein known as CPI17 also produces a potent PP1 inhibitor. Hence, there may be a parallel mechanism by which CPI17 amplifies and prolongs signals initiated by lipid second messengers that activate PKC.

Having phosphoproteins such as I-1 and CPI17 inhibit a phosphatase raises interesting questions about how the enzyme distinguishes between these proteins and phosphoprotein substrates. Although the molecular basis for phosphatase inhibition by I-1 and other inhibitors is still not fully understood, this regulatory mechanism makes the following prediction. As target phosphatases do not recognize these proteins as substrates, other protein serine/threonine phosphatases are required to reverse the activity of these inhibitors. For example, I-1 is dephosphorylated in vitro and in vivo by PP2A and PP2B. Cell-permeable compounds that inhibit these serine/threonine phosphatases enhance I-1 phosphorylation in cells. In neurons and skeletal muscle, PP2B is the predominant I-1 phosphatase, providing another point of antagonism between two second messengers, calcium and cAMP, in these tissues. In any case, this form of regulation represents a novel paradigm for signal transduction, a protein phosphatase cascade in which one phosphatase increases the activity of another. With the increasing awareness that many phosphatase regulators undergo reversible phosphorylation, we are likely to see many more protein phosphatase cascades.

In contrast to I-1 and CPI17, I-2 and NIPP-1 do not require phosphorylation to inhibit PP1 activity. In fact, these two phosphatase inhibitors are inactivated by protein phosphorylation. The PP1/I-2 complexes isolated from cytosol of many mammalian tissues is inactive. The phosphatase activity of this complex is increased when I-2 is phosphorylated in a hierarchical manner by two different kinases, casein kinase II and GSK-3. Phosphorylation of I-2 by casein kinase II does not increase phosphatase activity. However, once phosphorylated by casein kinase II, I-2 becomes an excellent substrate for another kinase, GSK-3. This phosphorylation triggers PP1 activity. The phosphorylated PP1/I-2 complex then autodephosphorylates and through a slow conformation change returns to its latent or inactive state. The physiological importance of this transient increase in PP1 activity is poorly understood but a number of interesting proposals have been put forward. One of these suggests that I-2 plays the role of a PP1 chaperone and that its association with PP1 and subsequent phosphorylation aids in refolding newly synthesized catalytic subunits into active phosphatases. It has also been speculated that the PP1–I-2 complex delivers the enzyme to various subcellular locations where PP1 is exchanged to form new complexes with other PP1 regulators or targeting subunits.

Another latent phosphatase complex, found in nuclei, contains the PP1 catalytic subunit bound NIPP-1, a *n*uclear *i*nhibitor of *PP1*. The PP1–NIPP-1 complex is activated by two protein kinases, casein kinase II and PKA. Biochemical studies suggest that the two kinases act synergistically to activate the PP1–NIPP-1 complex as a protein phosphatase. Although dephosphorylation of NIPP-1 is required to return the PP1–NIPP-1 complex to its inert state, whether this occurs by autodephosphorylation or by the action of another nuclear phosphatase remains unknown. NIPP-1 contains a C-terminal RNA-binding motif found in a bacterial endoribonuclease and NIPP-1 binds artificial RNA templates in vitro. Immunodepletion of PP1–NIPP-1 complexes from nuclear extracts suggests that this complex controls mRNA splicing in mammalian cells. Thus, NIPP-1 and I-2 appear to be multifunctional proteins and their precise role in the control of cellular PP1 activity remains to be determined.

Following in the footsteps of the PP1 inhibitors, two endogenous proteins, I_1PP2A and I_2PP2A, have been isolated. These proteins selectively inhibit PP2A activity. Moreover, two proteins, CAIN and CABIN, have been identified in mammalian cells that inhibit calcineurin or PP2B activity. These phosphatase inhibitors are similar to I-2 and NIPP-1 in that they do not appear to require phosphorylation to inhibit enzyme activity but preliminary evidence suggests that one or both PP2A inhibitors are phosphoproteins. Whether reversible phosphorylation of these inhibitors modulates PP2A activity remains to be seen but the immediate challenge is to define the role of these proteins in the physiological control of their target phosphatase.

3.2. Subcellular Targeting Defines Protein Phosphatase Functions

The first evidence that phosphatase functions may be regulated by subcellular targeting came from analysis of the glycogen-bound phosphatase from rabbit skeletal muscle. This enzyme differed from the isolated catalytic subunit in that the glycogen-bound phosphatase was a better glycogen synthase phosphatase. Following digestion of the glycogen particle with amylases, the glycogen-bound phosphatase was isolated and shown to be a complex of PP1 catalytic subunit bound to a 160-kDa regulatory subunit. The 160-kDa subunit, called G_M, also binds glycogen.

Addition of G_M enhanced the activity of the PP1 catalytic subunit as a glycogen synthase phosphatase. Subsequent studies identified several different glycogen-targeting subunits from liver, adipose tissue, and smooth muscle. In adipocytes, a homologous protein, PTG (*p*rotein *t*argeting to *g*lycogen) binds PP1 but also functions as a scaffold to bind its substrates, phosphorylase kinase and glycogen synthase. G_M also transduces hormone signals to regulate PP1 function and control glycogen metabolism (Fig. 3). Biochemical studies suggest that on phosphorylation by PKA, G_M releases the PP1 catalytic subunit into the cytosol and reduces its glycogen synthase phosphatase activity. PKA also phosphorylates the cytosolic inhibitor, I-1, which inhibits the free PP1 catalytic subunit. The coordinate regulation of G_M and I-1 ensures that the PP1 catalytic subunit, released from glycogen, does not dephosphorylate other phosphoproteins that may be important for the physiological response. As G_M and I-1 share a common PP1 binding domain, PP1 binding to the phosphorylated I-1 in the cytosol also prevents its return to the glycogen granule where it could dephosphorylate glycogen synthase and turn off the hormone response. Thus, adrenaline is very efficient at promoting the phosphorylation of glycogen-bound enzymes and eliciting glycogen breakdown. Insulin, on the other hand, promotes dephosphorylation of these same proteins. This occurs by the phosphorylation of G_M on a different serine. This modification maintains the association of PP1 with glycogen but greatly enhances glycogen synthase phosphatase activity and thus promotes glycogen deposition. As G_M also mediates PP1 binding to the sarcoplasmic reticulum (SR), adrenaline and insulin may utilize similar mechanisms to regulate the phosphorylation–dephosphorylation of SR proteins implicated in skeletal muscle contraction. G_M established a novel paradigm for phosphatase regulation in which the targeting subunit mediates the subcellular localization of PP1, determines substrate recognition, and communicates hormonal signals to control PP1 activity.

Another phosphatase targeting mechanism was identified in smooth muscle. Preparations of smooth muscle myosin contain a tightly associated protein phosphatase that dephosphorylated myosin light chain and promoted smooth muscle relaxation. The myosin-bound phosphatase was shown to be a complex of three polypeptides, PP1 catalytic subunit, a 130-kDa myosin binding subunit (also known as MBS) and a 20-kDa regulatory subunit. The 130-kDa polypeptide is capable of binding both PP1 and myosin. The 20-

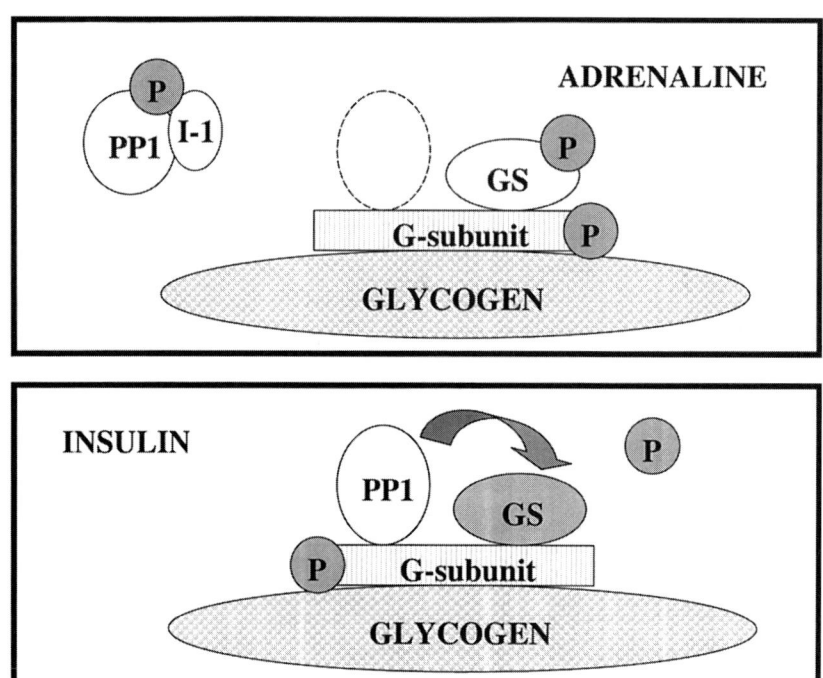

Fig. 3. Hormonal regulation of the glycogen-bound phosphatase. PP1 localization to the glycogen granule is mediated by the glycogen-targeting or G-subunit. Adrenaline promotes the phosphorylation of the skeletal muscle G-subunit, GM, to release PP1, which then associates with the PKA-activated cytosolic inhibitor, I-1. Thus adrenaline facilitates the phosphorylation of glcogen synthase, GS, and inhibits glycogen synthesis (**top**). Insulin promotes the phosphorylation of GM at a different serine and enhances PP1 activity. The dephosphorylation of GS by PP1 enhances glycogen synthesis (**bottom**).

kDa subunit did not bind PP1 but shares sequence homology with the C-terminus of the 130-kDa subunit and binds myosin. Thus, the two regulatory subunits create a high-affinity myosin binding complex. PP1 binding to the 130-kDa subunit enhances its activity as a myosin phosphatase. Interestingly, the PP1–MBS complex is very poor at dephosphorylating other PP1 substrates such as the glycogen enzymes. Thus, MBS directs PP1's localization to myosin and restricts its substrate specificity to favor the dephosphorylation of myosin light chains. MBS also binds a protein kinase, p160 ROCK. ROCK is activated by the binding of a small GTP binding protein, Rho, which transduces signals for growth factors and hormones (Chapter 2). Following its activation ROCK phosphorylates both MBS and myosin light chain. Phosphorylation of MBS inhibits the associated myosin phosphatase activity (Fig. 4). Thus, ROCK coordinates the phosphorylation of myosin with the inhibition of the myosin phosphatase to promote smooth muscle contractility in the absence of calcium. This kinase–phosphatase crosstalk accounts for the calcium sensitization reported in many physiological studies in smooth muscle. The ROCK/MBS/PP1 complex also binds to and regulates two other structural proteins, the spectrin binding protein, adducin, as well as moesin and other members of the ERM family of the actin binding proteins. Thus, the regulatory complex identified in smooth muscle appears to control an array of structural proteins implicated in cell shape and motility in other nonmuscle cells.

Although G_M and MBS represent two targeting–regulation paradigms for PP1, there is considerable evidence that other protein serine/threonine phosphatases also form signaling complexes. For instance, the *A-kinase anchoring protein*, AKAP79, binds not only PKA but also calcineurin or PP2B, calmodulin, and PKC, and thus may represent one of the most elaborate signaling complexes found in mammalian cells (Chapter 11). The newly discovered AKAP220 binds PKA and PP1. However, the precise physiological role of these kinase–phosphatase complexes and the mechanisms that modulate their functions remain unknown. Recent studies have shown that the heterotrimeric forms of PP2A associate with several different protein kinases including CaMKIV, p70rsk, and

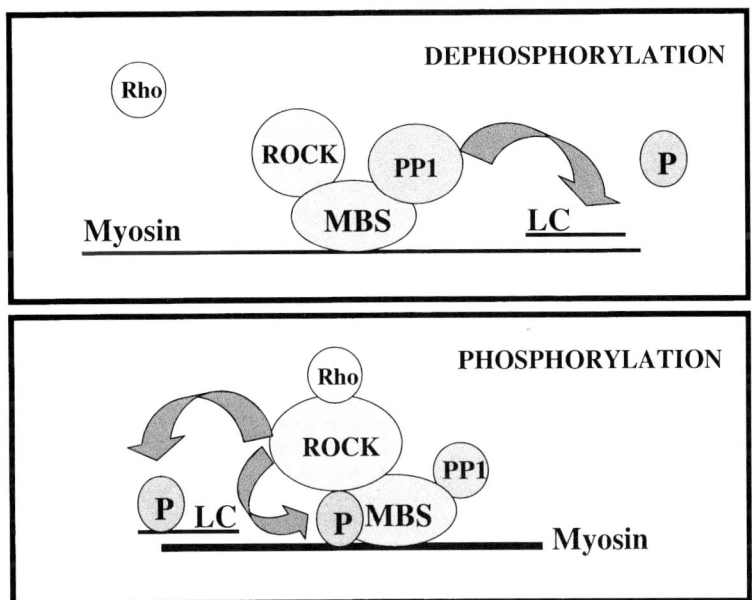

Fig. 4. Regulation of the myosin-bound phosphatase. PP1 is bound to smooth muscle myosin via the myosin-binding subunit (MBS). This leads to the dephosphorylation of myosin light chain (MLC) and smooth muscle relaxation (**top**). MBS also binds the Rho-activated protein kinase, ROCK. Rho binding enhances ROCK activity which phosphorylates MLC and MBS. Phosphorylation of MBS inhibits the activity of the myosin-bound phosphatase and facilitates myosin contraction (**bottom**).

the p21-activated protein kinase, PAK-1. The association of PP2A controls the phosphorylation state and activity of these kinases. On the other hand, casein kinase II association enhances PP2A activity. Thus, signaling complexes containing kinases and phosphatases that constantly monitor each other's function may act as rheostats or volume controls that determine the throughput or flow of information through these signaling units and thus define the size or strength of the physiological response.

Imunohistochemical experiments have provided important clues to cellular functions of protein serine/threonine phosphatases. For example, *PP4* was localized to spindle pole bodies. Although the disruption of the *Drosophila PP4* gene was embryonic lethal, analysis of dying embryos suggested that *PP4* regulates nucleation of microtubules at the centrosomes to form the mitotic spindle. Thus, cells lacking *PP4* showed aberrant spindles and experienced difficulties in chromosome segregation and cell division. Biochemical studies, on the other hand, showed that *PP4* associates with the transcription factor, c-Rel, and controls gene expression. Thus, different experimental approaches provide different clues to the physiological functions of a protein phosphatase. Nevertheless, the notion that cellular functions of protein serine/threonine phosphatases are dictated by their subcellular localization seems firmly established.

6.3. Covalent Modifications of Protein Phosphatase Subunits

Like the phosphatase regulatory subunits discussed previously, the catalytic subunits of several protein serine/threonine phosphatases also undergo reversible protein phosphorylation. The phosphorylation of PP1 catalytic subunit on a conserved C-terminal threonine is mediated by cell cycle regulated or cyclin dependent protein kinases and inhibits enzyme activity. This event is thought to be critical for cells to enter into mitosis. However, it is uncertain whether the phosphorylated phosphatases undergo auto-dephosphorylation as seen in vitro or other enzymes are required to reset the PP1 activity when cells exit out of mitosis. PP2A activity may be regulated by both serine and tyrosine phosphorylation. Both the catalytic and regulatory subunits of PP2A undergo protein phosphorylation in vitro. The limited in vivo studies suggest that the phosphorylation of PP2A inhibits enzyme activity and propagates signals mediated by the mitogen-activated protein kinases, ERK1 and ERK2. The phosphorylation of calcineurin by calcium–calmodulin-dependent protein kinase II has been shown in vitro to modulate its sensitivity to calcium–calmodulin but the physiological significance of this finding remains unclear. Similarly, PP2C is an excellent substrate for PKC but this phosphorylation appears to have no effect on enzyme activity. Future studies will determine whether phosphorylation of phosphatase catalytic subunits occurs in cells and is coordinated with the changes in other regulators, such as inhibitors and targeting subunits, to control protein dephosphorylation.

Finally, the PP2A catalytic subunit is the major methylated protein in mammalian cells. Indeed, C-terminal methylation may extend to other members of the PP2A-like protein serine/threonine phosphatases. The precise role of methylation remains unclear. It has been suggested that it may modulate enzyme activity or determine the types of heterotrimeric complexes formed by the PP2A catalytic subunit. These in turn would modify the substrate specificity of the enzyme and reversible methylation of PP2A may be involved in signal transduction in mammalian cells.

4. PATHOPHYSIOLOGY OF PROTEIN PHOSPHATASES

Given the partnership between protein kinases and phosphatases to regulate cell physiology, there are a growing number of situations when deregulation of protein phosphatases leads to disease. For instance, a chromosomal translocation in many acute undifferentiated leukemias results in the fusion of I_2PP2A, also known as SET, with a nucleoporin. This may deregulate the nuclear pool of PP2A to induce leukemia. Yet other data show that I_2PP2A is overexpressed in Wilm's tumor. The chronic inhibition of PP2A in these cells may enhance mitogenic signaling mediated by mitogen-activated protein (MAP) kinases and other mediators to promote cell growth and transformation. Additional evidence for PP2A deregulation as the cause of human cancers comes from studies of cell transformation by DNA tumor viruses. Polyoma middle T antigen and simian virus small T antigen associate with PP2A occupying the site for the B regulatory subunit. This diverts the normal functions of this phosphatase to enhance host cell growth and increase viral replication. Yet other viral proteins, such as the EBNA2 nuclear antigen derived from the Epstein–Barr virus, associate with PP1 and control phosphoproteins involved in the transcription of ribosomal RNA genes. Finally, the product of the *HOX11* gene, isolated from a chromosomal breakpoint in human T-cell leukemia, has been implicated in onco-genesis. This protein binds to both PP1 and PP2A and disrupts the G_2–M cell cycle checkpoint to promote cell transformation.

A similar strategy is utilized by a growing number of toxins produced by soil bacteria, freshwater and marine microorganisms, and insects, as well as some commonly used herbicides to inhibit PP1 and/or PP2A. Structure–activity studies have linked the inhibition of these two major cellular phosphatases to tumor promotion by the cell-permeable compounds. Occasionally, therapeutically useful compounds have been obtained from natural products that inhibit protein serine/threonine phosphatases. For instance, the most effective immunosuppressive drugs in use today, FK506 and cyclosporin, bind to intracellular receptors, known collectively as immunophilins, to inhibit calcineurin/PP2B activity. Inhibition of calcineurin prevents the dephosphorylation and nuclear translocation of a T-cell transcription factor, NFAT, that controls the expression of cytokine genes. The remarkable efficacy of cyclosporin and FK506 as calcineurin inhibitors has made these the most effective drugs to prevent tissue rejection after organ transplant surgery. Similar benefits may be derived from compounds that inhibit PP1 and PP2A to facilitate signaling in diseases associated with hormone resistance. Recent studies suggest that the anticancer drug fostreicin may inhibit PP1, PP2A, and PP4 to suppress tumor cell growth.

Non-insulin-dependent or type II diabetes shows a hereditary component. Analysis of mutations in genes encoding signal transduction molecules that could account for insulin resistance in skeletal muscle of type II diabetics identified a point mutation in G_M. Physiological studies suggest that this mutation combines with obesity to induce hyperinsulinema in these individuals, leading ultimately to insulin resistance and diabetes. It is likely that future studies will identify structural changes in other subunits of protein serine/threonine phosphatases as contributing factors to human disease.

5. FUTURE OF PROTEIN PHOSPHATASES

More than 60 different catalytic and regulatory subunits of protein serine/threonine phosphatases have been identified. Given the permutations and combinations of catalytic and regulatory subunits, as many as 400 different phosphatase enzymes may exist in eukaryotic cells. Some estimates suggest that we may be seeing only the tip of the iceberg with >90% of the phosphatase genes still to be discovered. So the goal of future research will undoubtedly be to unravel the structure and function of these new enzymes. These studies will most likely begin by defining subcellular localization of these proteins as a guide to their function. Another approach will be to employ biochemical and genetic strategies to isolate interacting proteins that may represent either regulators or substrates. The third approach is to disrupt or mutate the genes encoding specific phosphatases to unravel their biological functions. Finally, efforts are being made to develop selective inhibitors, cell-permeable compounds that will lower or eliminate enzyme activity, and thus reveal the functions of the phosphatases. In the absence of specific inhibitors, antisense reagents will be used to reduce enzyme levels and decipher their cellular functions. By far the biggest challenge for future researchers is to determine how physiological stimuli regulate the newly discovered phosphatases and the consequences of phosphatase regulation to cell function. In this regard, the regulatory paradigms discussed in this chapter may act as guideposts to set the course for future studies to establish the importance of protein serine/threonine phosphatases in cell physiology.

ACKNOWLEDGMENTS

Work in the author's laboratory is supported by Grants DK52054 and DK55881 from the National Institutes of Health and a feasibility grant from the American Diabetes Association. The author wishes to thank the members of his laboratory, specifically John H. Connor, Carey Oliver, and Ryan Terry-Lorenzo, for helpful comments on the chapter.

SELECTED READINGS

Barton GJ, Cohen PTW, Barford. Conservation analysis and structure prediction of protein serine/threonine phosphatases. Sequence similarity with diadenosine tetraphosphatase from *Escherichia coli* suggests homology to protein phosphatases. Eur J Biochem 1994; 220:225.

Cohen P. The structure and regulation of protein phosphatases. Annu Rev Biochem 1989; 58:453.

Cohen PTW. Novel protein serine/threonine phosphatases: variety is the spice of life. Trends Biochem Sci 1997; 22:245.

Shenolikar S. Protein serine/threonine phosphatases—new avenues for cell regulation. Annu Rev Cell Biol 1994; 10:55.

Shenolikar S, Nairn AC. Protein phosphatases: recent progress. Adv Second Messenger Phosphoprotein Res 1991; 23:1.

Wera S, Hemmings BA. Serine/threonine protein phosphatases. Biochem J 1995; 311:17.

19 Protein Tyrosine Phosphatases

Cynthia V. Stauffacher and Harry Charbonneau

Contents

OVERVIEW
CLASSIFICATION OF PROTEIN TYROSINE PHOSPHATASES
STRUCTURE AND CATALYTIC MECHANISM OF PROTEIN TYROSINE PHOSPHATASES
IDENTIFICATION OF SUBSTRATES FOR PROTEIN TYROSINE PHOSPHATASES
REGULATION OF PROTEIN TYROSINE PHOSPHATASES
PHYSIOLOGICAL ROLES OF PROTEIN TYROSINE PHOSPHATASES
POTENTIAL ROLE OF PROTEIN TYROSINE PHOSPHATASES IN HEALTH AND DISEASE
SELECTED READINGS

1. OVERVIEW

The importance of tyrosine phosphorylation in diverse cellular processes such as cell growth, gene expression, metabolism, differentiation, cytoskeletal dynamics, and cell motility was discovered primarily through studies of the protein tyrosine kinases (PTKs). The discovery that defective PTKs are encoded by oncogenes that induce cell transformation sustained the early focus on these enzymes. This work established the concept that defects in signaling pathways dependent on tyrosine phosphorylation can lead to diseases including cancer. For many years, research on tyrosine phosphorylation centered around the kinases and very little was known about the phosphatases that remove phosphate from tyrosine residues. Because it is a dynamic and reversible process, protein tyrosine phosphorylation is necessarily governed by the combined action of protein tyrosine phosphatases and PTKs. Consequently, an understanding of protein tyrosine phosphatases and their action in opposing PTKs is essential for understanding any signaling pathway involving tyrosine phosphorylation.

In the mid-1980s, Nick Tonks and Ed Fischer at the University of Washington initiated studies to identify and isolate protein tyrosine phosphatases. Prior to this, protein tyrosine phosphatase activity had been described but the enzymes responsible for this activity had not been well characterized. In 1988, Tonks and Fischer reported the purification and characterization of PTP1B, a protein tyrosine phosphatase from human placenta. Shortly thereafter, the amino acid sequence of PTP1B was determined. This sequence established that protein tyrosine phosphatases were not related to the protein Ser/Thr phosphatases and were a distinct family of enzymes. The sequence similarity between PTP1B and CD45, a cell surface protein of lymphocytes, provided the first indication that, like PTKs, the protein tyrosine phosphatases exist as transmembrane, receptor-like forms. This work established the protein tyrosine phosphatase field and generated a great deal of interest. Over the past decade, we have learned that there is a large and surprisingly diverse group of enzymes with protein tyrosine phosphatase activity. All of them use the same catalytic mechanism and active site design although many exhibit little sequence and structural similarity. We now know that a subset of these enzymes not only dephosphorylate Tyr residues but also Ser and Thr residues as well. Surprisingly, several enzymes that were originally thought to have protein substrates actually dephosphorylate nucleic acids and lipids instead. Consider-

From: *Principles of Molecular Regulation* (P. M. Conn and A. R. Means, eds.), © Humana Press Inc., Totowa, NJ.

able progress has been made in understanding the catalytic mechanism employed by these enzymes and there is at least one crystal structure representing each of the four major subgroups of tyrosine phosphatases. Protein tyrosine phosphatases are involved in regulating many aspects of cell physiology ranging from cell cycle control to cell motility. Although the mechanisms regulating tyrosine phosphatases are not yet completely defined, it is clear that these enzymes are subjected to exquisite controls. Like the PTKs, defects in tyrosine phosphatase function and regulation are likely to result in disease. In this chapter, we present highlights of what has been learned during the past decade about protein tyrosine phosphatase structure, function, and regulation.

2. CLASSIFICATION OF PROTEIN TYROSINE PHOSPHATASES

Enzymes customarily referred to as protein tyrosine phosphatases employ a similar catalytic mechanism involving formation of a thiophosphate intermediate at an essential Cys residue located within a hallmark active site motif (Cys-X_5-Arg). These tyrosine phosphatases bear no resemblance to the protein Ser/Thr phosphatases and have distinct sequences, structures, catalytic mechanisms, and evolutionary ancestry. The protein tyrosine phosphatases are commonly divided into four groups: tyrosine-specific, dual-specific, low molecular weight, and Cdc25 phosphatases. The structural organization of representative enzymes from all four groups are depicted in Fig. 1. Aside from their signature active site motifs, these four groups exhibit little sequence similarity to one another. However, all four groups are generally considered part of the same family, although primary sequences and three-dimensional structures suggest that they do not have a common evolutionary origin. In addition, several of these enzymes do not attack proteins; instead, they remove phosphate from nonproteinaceous substrates. Protein tyrosine phosphatases have been identified in a variety of organisms including plants and animals and are probably expressed in all eukaryotes. A recent survey of genome sequences reveals that orthologs of tyrosine phosphatases from all four groups are encoded in some but not all prokaryotic species. It is not known whether all these prokaryotic counterparts act on protein substrates. In this chapter, the phrase CX_5R phosphatases is used collectively to refer to all four groups of enzymes containing this motif.

The first protein tyrosine phosphatases discovered exhibited a high degree of specificity for phosphotyrosine residues. The tyrosine-specific enzymes, designated herein as PTPs, contain a highly conserved catalytic domain of about 240 residues (see Fig. 1). The PTPs are the most abundant of the groups, with more than 70 different enzymes that exist as both membrane receptors and nontransmembrane forms. Receptor PTPs (RPTPs) have an extracellular segment, a single transmembrane domain, and in most cases, two tandem catalytic domains within their cytoplasmic segments. The diversity among RPTPs arises primarily from differences in the size and sequence of their extracellular domains. The nonreceptor PTPs are distinguished primarily by the noncatalytic sequences that are linked to the conserved catalytic domain. Like other signaling enzymes, the noncatalytic sequences typically have regulatory functions.

The dual specificity phosphatases (DSPs) are distinguished by their ability to dephosphorylate Ser/Thr as well as Tyr residues. These DSPs exhibit little or no sequence similarity to the much larger group of PTPs and contain a significantly smaller catalytic core (see Fig. 1). To date, no receptor forms of DSPs have been identified. Despite the lack of sequence similarity, the crystal structure of VHR, a human DSP, shows that its topology is very similar to that of several PTP structures, not only at the active site but also in surrounding regions. The overall similarity in structure is consistent with the evolution of DSPs and PTPs from a common ancestor. A subset of DSPs including MKP1, PAC1, MKP3, hVH5, VHR, and others display sequence similarity with one another and share the ability to specifically dephosphorylate and inactivate mitogen-activated protein (MAP) kinases (see below). A characteristic feature of the MAP kinase phosphatases is the presence of two short CH2 domains ($\approx$ 25 residues) in their N-terminal noncatalytic sequences. These CH2 domains have limited but recognizable similarity to two regions flanking, but not including, the active site of Cdc25 phosphatases. Other DSPs such as Cdc14 or KAP are involved in cell cycle regulation.

The Cdc25 phosphatases have dual specificity and control cell division by dephosphorylating regulatory sites within cyclin-dependent protein kinases (see below). The active site motif (Cys-X_5-Arg) is located within a C-terminal catalytic domain ($\approx$ 180 residues) that is conserved among Cdc25 orthologs (see Fig. 1). The larger N-terminal region is thought to have a regulatory function. With the exception of limited similarity to the MAP kinase phosphatases noted previously, Cdc25 has no significant sequence similarity

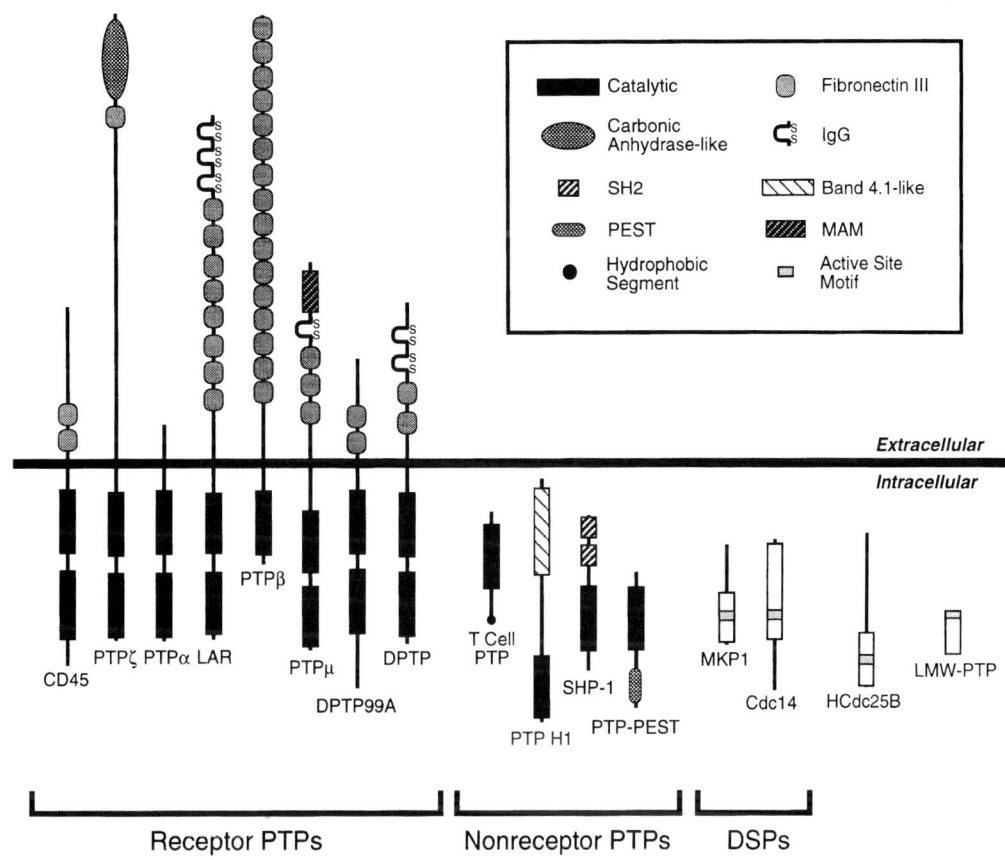

Fig. 1. Structural relationship among protein tyrosine phosphatases. The structural organization of representative enzymes from the four major groups of protein tyrosine phosphatases are depicted. Both nonreceptor and receptor PTPs are shown on the left as indicated, with the *solid black boxes* denoting the location of the conserved 240-residue domain common to all of these enzymes. Two members of the dual specificity enzymes are shown with the location of the Cys-X_5-Arg active site motif delineated by the *small shaded box*. The two structurally unrelated Cdc25 and LMW-PTPs are shown on the right of the figure with the *shaded box* depicting the active site motif that is the only common structural feature of these enzymes. The panel in the upper right identifies all domains that are represented in the proteins shown here.

with the PTPs, DSPs, or the low molecular weight protein tyrosine phosphatases outside of the active site motif. Despite this lack of similarity, Cdc25 was considered to be related to the DSPs until its crystal structure was determined. Except for the active site, the architecture of the Cdc25A catalytic domain does not match that of any PTP or DSP, but instead is very similar to Rhodanese. This structure suggests that Cdc25 evolved separately from the other CX_5R phosphatases, and implies that the structural similarity at the active site and nearby neighboring regions has resulted from convergent evolution.

The low molecular weight protein tyrosine phosphatases (LMW-PTPs) are 18-kDa enzymes that are widely expressed among eukaryotes and have been implicated in regulating certain mitogenic signaling pathways. The LMW-PTPs contain a Cys-X_5-Arg active site motif and employ a catalytic mechanism similar to that of the PTPs, DSPs, and Cdc25. The active site in the LMW-PTPs is located near the N-terminus (*see* Fig. 1), whereas it is positioned near the C-terminus of catalytic domains from the PTPs and DSPs. In regions outside of their active sites, the LMW-PTPs and the three other tyrosine phosphatase groups have no significant sequence similarity and are unlikely to have a common evolutionary ancestor. Thus, it can be argued that the extant tyrosine phosphatases have evolved separately from three distinct progenitors, but during evolution have converged and arrived at the same active site topology and chemistry for cleaving phosphomonoester bonds.

A unique class of proteins has been identified that contains sequences exhibiting a significant degree of similarity to PTPs and DSPs except that they lack one or two essential residues required for catalysis. The prototype for this group of proteins, STYX, contains a Gly in place of the invariant active site Cys residue of the Cys-X_5-Arg motif. Replacement of the Gly to Cys by site-directed mutagenesis converts STYX into an active phosphatase. STYX and the other STYX-like proteins are thought to bind and recognize phosphoproteins, which are most likely the substrates for the enzymes from which they diverged. With this binding property, the STYX-like proteins could have a regulatory function. The biological properties of the STYX-like protein known as Sbf1 suggests that it has such a function and prevents dephosphorylation of substrates by myotubularin, the enzyme it most closely resembles.

An important development in this field is the discovery that several CX_5R phosphatases do not act on phosphoprotein substrates. These enzymes appear to be most closely related to the DSPs but the similarities are relatively low in regions outside of the active site. The first evidence for nonproteinaceous substrates came from the discovery that the N-terminal domain of mRNA capping enzymes contains the PTP-like signature motif and that the RNA triphosphatase activity of this enzyme was dependent on the Cys residue of the Cys-X_5-Arg motif. This RNA triphosphatase activity is an essential reaction required to form the 5' cap structure present on all eukaryotic mRNAs. Subsequently, two enzymes previously assumed to be DSPs, a baculoviral phosphatase (BVP) and a human phosphatase (PIR1), were shown to have both RNA 5'-triphosphatase and diphosphatase activities. The diphosphatase activity distinguishes BVP/PIR1 from the mRNA capping enzymes, suggesting that these represent two distinct subsets of enzymes with activity toward nucleic acids.

PTEN (phosphatase and tensin homolog deleted on chromosome ten) was first thought to be a DSP, but subsequent studies have shown that this enzyme has lipid phosphatase activity and dephosphorylates the 3-position of phosphatidylinositol (3,4)-bisphosphate and phosphatidylinositol (3,4,5)-trisphosphate. PTEN regulates a phosphoinositide-dependent signaling pathway that results in activation of the AKT/PKB protein Ser/Thr kinase, which in turn stimulates cell survival and proliferation. PTEN is a tumor suppressor gene and loss of function mutations are found in many human tumors. Many researchers in the field speculated that given their capacity to antagonize oncogenic tyrosine kinases, the PTPs might be tumor suppressors. It is ironic that the first tumor suppressor among the CX_5R phosphatases would prove to be a lipid phosphatase. The possibility that PTEN and the BVP/PIR1 enzymes could also target phosphoproteins has not been excluded, but the physiologic role of the PTP-like domain of the mRNA capping enzymes is indisputable.

Arsenate reductase from budding yeast contains the Cys-X_5-Arg motif and exhibits significant sequence similarity with the catalytic domain of yeast Cdc25, indicating that arsenate reductase and Cdc25 might have diverged from the same progenitor. Certain prokaryotes express an arsenate reductase that differs from the eukaryotic reductase. Although the sequence similarity to LMW-PTPs is relatively low, the prokaryotic reductase may have diverged from the LMW-PTPs.

3. STRUCTURE AND CATALYTIC MECHANISM OF PROTEIN TYROSINE PHOSPHATASES

The canonical PTP domain structure, represented here by the X-ray crystallographic structure of PTP1B, is shown in Fig. 2A. PTP1B is a compact globular α/β protein, with a central nine-strand mixed parallel/antiparallel β-sheet at its core. This highly twisted sheet is surrounded on both sides by α-helices, with five α-helices on one side and two on the other. At the center of this structure, a β-strand–loop–α-helix contains the $CysX_5Arg$ active site motif. The Cys residue is located at the C-terminus of the β-strand and the Arg lies at the N-terminus of the α-helix so that they flank the loop that connects these secondary structure elements. The intervening loop, known as the P loop, forms the bottom of the active site, whose walls are formed by two additional loops of the protein. The phosphotyrosine recognition loop (below left of the active site in Fig. 2) presents one wall with a Tyr against which the phosphotyrosine substrate can pack. Residues in this recognition loop also interact with side chain and backbone atoms in polypeptide phosphosubstrates and govern substrate specificity. The WPD loop (above right of the active site in Fig. 2) contains the catalytic Asp and upon substrate binding undergoes a large conformational change that closes a second hydrophobic wall against the substrate. These two loops form a deep pocket that selects for phosphotyrosine substrates and excludes phosphoserine and phosphothreonine residues.

The crystal structures of catalytic domains from both nonreceptor and receptor PTPs are remarkably similar. Surprisingly, the VHR dual specificity phosphatase shows the same basic structure, despite having very little sequence identity with the PTPs. The VHR

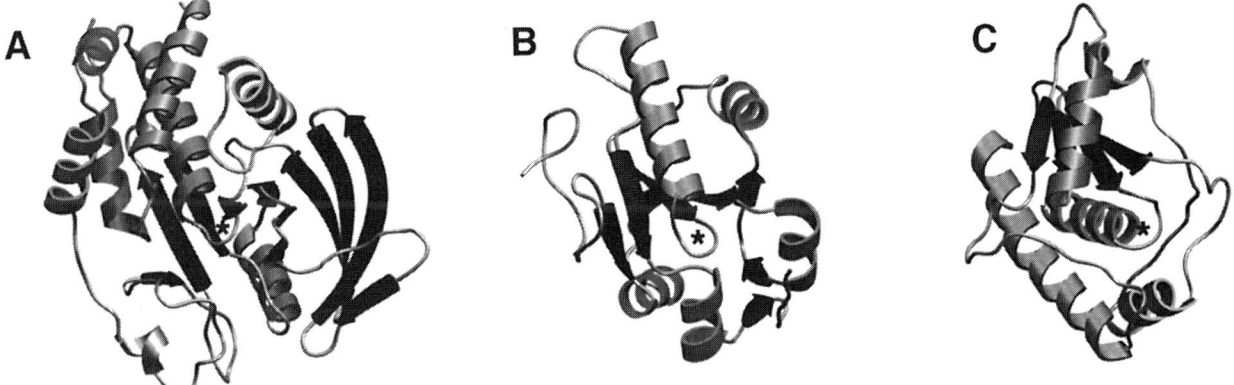

Fig. 2. Ribbon drawing representations of the structures of protein tyrosine phosphatases. Structures shown are (A) the canonical catalytic domain of the PTPs, represented by the structure of the catalytic domain of PTP1B; (B) the cell cycle control phosphatase Cdc25; and (C) the LMW-PTP, BPTP. α-Helices are shown as curled ribbons; β-strands as flat ribbons with *arrowheads* indicating the C-terminus of the strand. The core unit of secondary structure that is common to all three representatives of the protein tyrosine phosphatases is a four-strand parallel β structure with flanking helices containing a β–loop–α unit that encompasses the Cys-X_5-Arg active site motif. All three molecules are oriented such that the active site P loop, denoted with an *asterisk*, lies at the center of the structure and the four parallel strands are viewed end on.

structure is somewhat smaller, with a five-stranded central sheet, and the three flanking antiparallel β-strands are replaced by a helical structure, but the positions of the core catalytic residues are identical to those of the PTPs. One major difference is the lack of a phosphotyrosine recognition loop, which results in a shallower active site for VHR that can accommodate both phosphotyrosine and phosphoserine/threonine substrates.

The core structure of Cdc25, shown in Fig. 2B, has a five-stranded parallel central β-sheet, surrounded on both sides by α-helices. Central to this sheet is the β–loop–α structure containing the active site motif of the P loop. Although the secondary structure elements appear to be the same, the topology of this protein is completely different from the central core of the PTPs and the DSPs. The P loop forms the base of a very shallow active site with no surrounding loops that could confer selectivity for substrates. This implies that Cdc25 recognizes a larger substrate surface instead of an isolated loop on the phosphoprotein substrate.

The structure of the LMWPTP, shown in Fig. 2C, has a four-stranded parallel β-sheet surrounded by five helices. At the center of this molecule is the same β–loop–α structure with the P loop forming the base of the active site. Although there is no sequence identity between the LMW-PTPs and PTPs, the core structure (secondary structure elements and P-loop configuration) surrounding their active sites are remarkably similar. Two extended loops on each side of the active site in the LMW-PTP form hydrophobic walls, against which the phosphotyrosine substrate can stack. On the right of the structure (Fig. 2C) is a long loop that functions like the phosphotyrosine recognition loop of the PTPs and contains a Tyr or Trp followed by variable residues that determine the substrate selectivity of the enzyme. On the left is a flexible loop containing two adjacent Tyr or Trp residues that pack against the substrate and the catalytic Asp. This loop, which is functionally equivalent to the WPD loop of the PTPs, forms a deep active site pocket that selects for phosphotyrosine residues.

Figure 3 shows details of the active site from PTP1B and a LMW-PTP. The two structures are overlaid here to demonstrate the features that define a CX_5R phosphatase active site and to illustrate their structural similarity. The P loop forms the circular structure at the bottom of the figure, starting with the invariant Cys of the active site motif. Seven P-loop residues have main chain nitrogens that face the inside of the loop and bind three of the four anionic oxygens of the phosphate moiety (shown at the center of the loop). This phosphatebinding cradle is completed by the invariant Arg residue of the signature motif, which hydrogen bonds to two of the phosphate oxygens, providing a flexible side of this circular structure that accommodates the substrate and stabilizes the transition state intermediate of the reaction. The side chain of the Cys sits at the bottom of the loop, centered under the phosphate. The side chains of the P-loop residues extend back into the structure, where they

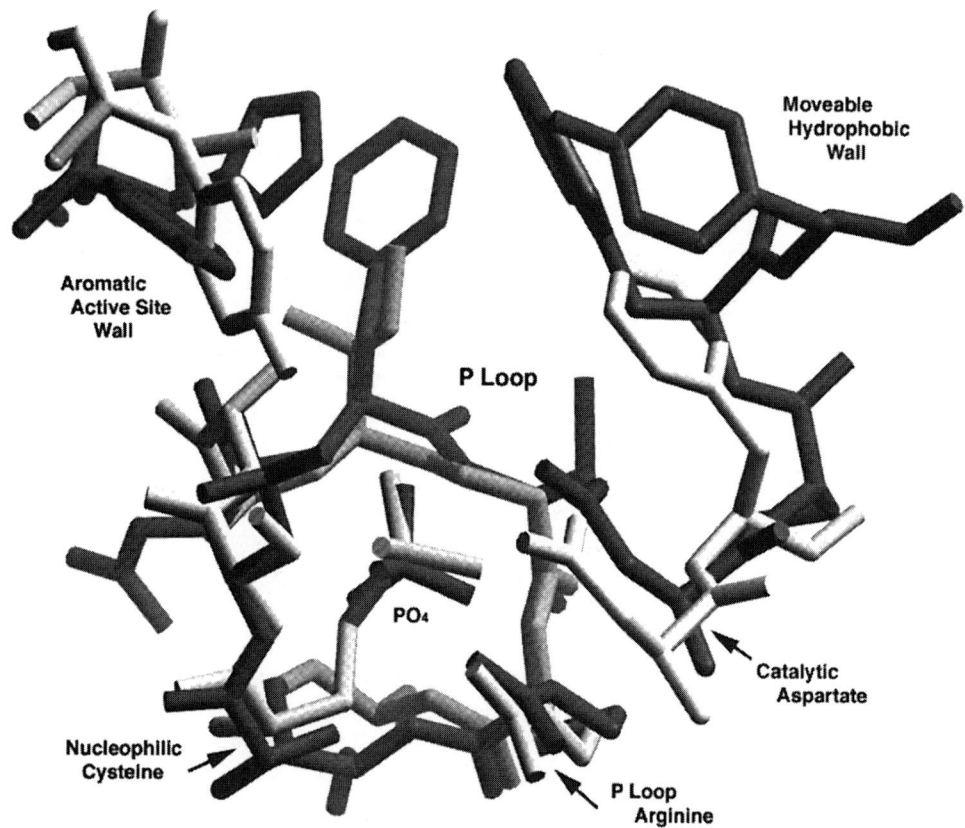

Fig. 3. Close up view of the active site of protein tyrosine phosphatases. The active site residues of PTP1B (nonreceptor PTP) with the phosphoenzyme intermediate (*light gray*) are superimposed with those of LTP1 (LMW-PTP) containing a bound phosphate (*dark gray*). The three critical residues of the active site structure are indicated by *arrows*. Primary residues on both walls of the active site that contact the bound phosphotyrosine substrate are shown for both structures. The depth of the active site, which excludes phosphoserine/threonine substrates, can readily be seen.

form hydrogen-bond networks stabilizing their unusual backbone structure. The aromatic residues of the phosphotyrosine recognition loop that forms one wall of the active site are shown on the left of Fig. 3. Residues of the moveable wall, including aromatics that can interact with the Tyr ring of the substrate, are on the right (Fig. 3). At the bottom of this flexible structure is the catalytic Asp.

All CX_5R phosphatases employ a two-step mechanism that requires no metal ions or cofactors and depends only on the essential catalytic Cys and Arg residues and the conserved Asp residue that functions as a general acid catalyst (*see* Fig. 4). The phosphorylated substrate binds to the enzyme via the phosphorylated Tyr group (for PTPs or LMW-PTPs) which interacts with the P-loop residues in the active site as shown in Fig. 3. As the anionic oxygens of the phosphate group bind to the backbone amides of the P-loop residues and the conserved Arg, the phosphorus atom is fixed in close proximity to the nucleophilic side chain of the catalytic Cys. In the PTPs, the WPD

$$E + R\text{-}OPO_3^{2-} \rightleftharpoons E \cdot R\text{-}OPO_3^{2-} \xrightarrow{\text{Step 1}} E\text{-}P \xrightarrow{\text{Step 2}} E + P_i$$
$$\phantom{E + R\text{-}OPO_3^{2-} \rightleftharpoons E \cdot R\text{-}OPO_3^{2-} \xrightarrow{\text{Step 1}}}\downarrow$$
$$\phantom{E + R\text{-}OPO_3^{2-} \rightleftharpoons E \cdot R\text{-}OPO_3^{2-} \xrightarrow{\text{Step 1}}} R\text{-}OH$$

Fig. 4. General reaction scheme for the two-step reaction catalyzed by CX_5R phosphatases. The enzyme (E) first binds phosphosubstrate ($ROPO_3^{2-}$) to form the enzyme substrate complex. In step 1, the thiophosphate intermediate (E–P) is formed as the dephosphorylated product is released from the enzyme as described in the text. The phosphoenzyme intermediate is then hydrolyzed by water in step 2, resulting in the release of inorganic phosphate. In most cases, step 2 is rate limiting.

loop then closes on the substrate and is stabilized both by hydrophobic interactions between the ring of the phosphotyrosine substrate and the aromatic residues on the loop and by hydrogen bond contacts with the P-loop Arg. Similar movements are thought to occur with the LMW-PTPs.

In the first step of the reaction, the Cys residue, which has an anomalously low pK_a, makes a nucleophilic in-line attack on the phosphate group cleaving the scissile P–O bond, releasing the dephosphorylated substrate, and simultaneously generating a thiophosphate intermediate (see Fig. 4). A proton donated by the catalytic Asp residue assists P–O bond cleavage by protonating the phenolic oxygen of the tyrosine leaving group. The phosphate of the phosphoenzyme intermediate interacts with the active site P loop in a position similar to that shown in Fig. 3.

In the second step (see Fig. 4), which is rate limiting, a water molecule makes an in-line attack on the phosphoenzyme intermediate, releasing the phosphate product from the active site Cys and returning the phosphatase to its native form. In PTPs, this water molecule is positioned for nucleophilic attack by the catalytic Asp and a conserved Gln residue. In this step, there is no residue in the LMW-PTPs that functions as the Gln of the PTPs. However, activation of LMW-PTPs by exogenous molecules, such as purines, has been shown in structural studies to be due to the positioning of a water molecule in the active site for nucleophilic attack, in this case by the purine and the catalytic Asp.

These structural mechanistic principles do not apply directly to Cdc25, whose active site structure is very different from the canonical PTP and LMW-PTPs. Although the nucleophilic Cys is present, the Arg extends out above the P loop in a very different configuration than in the other CX_5R phosphatases. The two deep walls around the active site are also missing, along with the catalytic Asp that resides on the moveable wall. The proton-donating group for the first step of the reaction might be the Glu in the second position of the signature sequence of Cdc25, but this has not yet been demonstrated. This Glu extends out above the P loop in the structure.

4. IDENTIFICATION OF SUBSTRATES FOR PROTEIN TYROSINE PHOSPHATASES

Identification of substrates is crucial for fully understanding CX_5R phosphatase functions. However, linking a phosphatase with a particular physiologic phosphoprotein substrate is a challenging endeavor, requiring multiple approaches. With the exception of certain DSPs, many of the tyrosine phosphatases exhibit rather broad substrate specificity in vitro. Even when candidate phosphoprotein substrates are available in sufficient quantities, information garnered from in vitro assays is often of limited use in assigning in vivo substrates. Perhaps the greatest strides toward identification of tyrosine phosphatase functions have come from genetic studies (see below).

Although genetic studies provide invaluable clues regarding potential substrates, establishing a direct effect of a phosphatase on the state of phosphorylation of specific protein substrates is not usually possible with this approach alone.

A powerful biochemical approach to substrate identification was introduced about 3 yr ago. This procedure, known as substrate trapping, is based on the fact that many CX_5R phosphatases can be mutated and converted into proteins having little or no enzymatic activity but retaining the ability to bind substrates with selectivity and affinity comparable to that of the wild-type enzyme. Replacement of the essential active site Cys residue with Ser or Ala completely inactivates these enzymes, whereas mutating the invariant catalytic Asp residue to Ala produces a phosphatase with detectable but extremely low activity and unchanged K_m. The Asp mutation appears to form a superior trapping agent having higher affinity for substrates. Replacement of the Cys residue may reduce affinity for substrate because the thiol contributes significantly to the initial binding of the phosphatase, whereas the Asp residue is not involved in binding. These substrate trapping agents can be immobilized on a solid support and used as an affinity reagent to purify substrates from cell extracts.

In an early example of this strategy, a trapping mutant of PTP-PEST, a tyrosine-specific enzyme containing C-terminal PEST (Pro-Glu-Ser-Thr) sequences, was used to identify p130[Cas] as a physiologic substrate. The adapter molecule, p130[Cas], is thought to be involved in cell adhesion. PTP-PEST, like many other PTPs, has robust activity against numerous substrates in vitro, yet p130[Cas] was the only substrate identified for PTP-PEST in multiple cell types. This and similar observations with other PTPs strongly suggest that some PTPs exhibit considerable selectivity for substrates in vivo and may not be as promiscuous as indicated by in vitro studies. Subtrate trapping has been used successfully with a number of other PTPs and DSPs.

5. REGULATION OF PROTEIN TYROSINE PHOSPHATASES

If we are to understand fully the role of tyrosine phosphatases in signal transduction, it is necessary to know how they are regulated. Because PTPs exhibit robust activity in vitro that is comparable to that of fully active tyrosine kinases, these phosphatases have the potential to suppress or abrogate the propagation of signals dependent on tyrosine phosphorylation. Consequently, there must be regulatory mechanisms

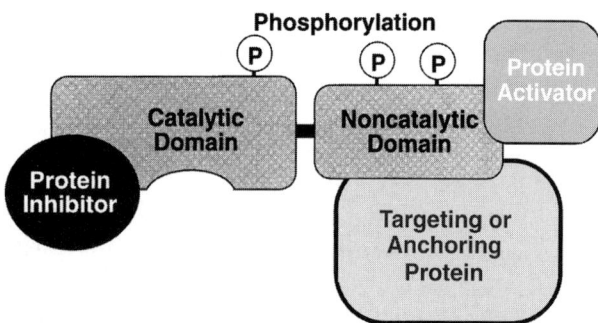

Fig. 5. Modes of regulation for nonreceptor protein tyrosine phosphatases. This diagram depicts the modular structure of many nonreceptor enzymes and illustrates several of the major regulatory mechanisms that have been found to control their activity and/or function as discussed in the text.

that inhibit tyrosine phosphatases at appropriate times and subcellular locations following the activation of tyrosine phosphorylation-dependent signaling pathways. In contrast, some PTPs and DSPs act in a positive fashion to promote signal propagation in response to mitogens. In these cases, regulatory mechanisms must exist to activate or guide these phosphatases to their targets with the appropriate timing. To have these phosphatases acting on their targets without suitable controls could lead to the constitutive activation of pathways controlling proliferation or differentiation.

Like many other enzymes involved in signal transduction, most protein tyrosine phosphatases have a modular structure in which catalytic domains are attached within the same polypeptide chain to noncatalytic sequences having specific functional properties of their own (see Fig. 1). Noncatalytic sequences may contain modules that mediate the anchoring or targeting of tyrosine phosphatases to their substrates or to subcellular locations where their targets reside. Noncatalytic sequences also provide sites for binding inhibitors or activators and for phosphorylation as illustrated in Fig. 5. In the case of RPTPs, their noncatalytic sequences reside outside of the cell and provide sites for binding of ligands that could potentially control their activity and/or location on the cell surface. Certain regulatory mechanisms may directly involve the catalytic domain itself. Although advances in defining regulatory mechanisms for CX_5R phosphatases have been limited, some basic themes are emerging; these are described in the following sections.

5.1. Regulation by Targeting

Protein–protein and protein–phospholipid interactions are essential for the propagation of signals from the plasma membrane to the nucleus and various other sites within the cell. A classic example are the receptor tyrosine kinases, which in response to ligand binding, recruit multiple enzymes and adaptor proteins to their phosphorylated cytoplasmic tails to assemble complexes capable of initiating multiple downstream signaling events. Intracellular signaling enzymes and proteins are often bound together into large multiprotein signaling complexes. In this way, the specific protein–protein interactions needed to propagate signals are not necessarily limited by diffusion. The organization of signaling molecules into complexes contributes to specificity and provides a means for both spatial and temporal control of signals. In some cases, anchoring or scaffolding proteins also modulate the activity of the enzymes they bind. Targeting or anchoring of CX_5R phosphatases to different proteins and sites within the cell is an important aspect of their regulation. Restricting these phosphatases to specific protein partners or to specific regions of the cell influences their access to substrates and also serves to define the signaling pathways they regulate. One function of the noncatalytic regions of PTPs and DSPs is to recognize and bind other proteins (see Fig. 5).

Perhaps the best examples of targeted CX_5R phosphatases are SHP-1 and SHP-2, PTPs that have two, tandem Src-homology 2 (SH2) domains linked to the N-terminus of their catalytic domains. SH2 domains bind phosphotyrosine residues with specificities governed by flanking sequences. SHPs are recruited to proteins when Tyr residues in sites recognized by their SH2 domains are phosphorylated by PTKs. This recruitment process brings SHPs into close proximity to their substrates and restricts the range of phosphoproteins that they encounter. The SH2 domains of these SHPs have dual functions because they not only control access to substrates but also regulate activity as discussed later. The T-cell PTP (TCPTP) is also regulated in part by a targeting mechanism. Two splice variants of TCPTP differing only in noncatalytic sequences located at their C-termini are expressed in humans and rodents. A 48-kDa splice variant contains a 34-residue hydrophobic C-terminus and is localized to the endoplasmic reticulum. In contrast, a smaller 45-kDa splice variant, in which the hydrophobic tail is replaced by a short hydrophilic segment, is localized to the nucleus. Thus, alternative splicing provides a mechanism to control the subcellular localization of TCPTP. There is evidence that the 45-kDa variant is translocated out of the nucleus in response to appropriate mitogenic signals (see later). This observation highlights the fact that the localiza-

Chapter 19 / Protein Tyrosine Phosphatases

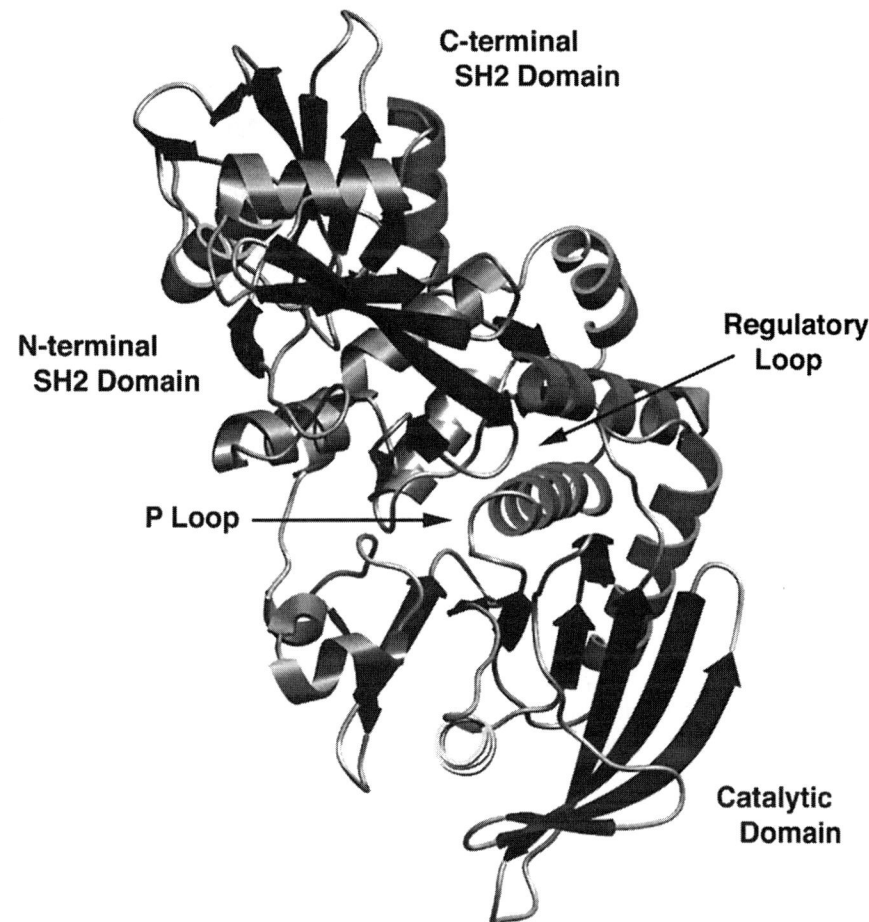

Fig. 6. Ribbon drawing of the structure of SHP-2. The molecule is oriented in approximately the same view used in Fig. 2 with the β-loop–α structure that contains the active site motif at the center of the catalytic domain. The two SH2 (Src Homology 2) domains curl over the top of the catalytic domain, with their binding clefts at approx 90° relative to each other. The N-terminal SH2 domain extends toward the active site, with a loop from this domain protruding into the P loop to block the access of substrate to the active site. Residues on this loop interact either directly or through water-mediated hydrogen bonds to the active site P loop residues, and to the catalytic Cys and Asp residues.

tion of tyrosine phosphatases need not be static because there are mechanisms to move them from one site to another.

Through interactions mediated by their noncatalytic sequences other PTPs and DSPs are specifically associated with their substrates. For example, a polyproline region in the noncatalytic C-terminus of PTP-PEST binds its substrate, p130^{Cas}. The DSP, MKP3, specifically interacts with its MAP kinase substrates through its N-terminal noncatalytic domain. Binding of its kinase substrate activates MKP3 up to 30-fold. In both of these cases, the noncatalytic domains have a direct role in governing substrate specificity. It is not known whether these phosphatase–substrate complexes also include the kinase that phosphorylates the substrates. Such complexes might provide a fast and efficient mechanism for potentially integrating two input signals, one acting on the kinase and another acting through the phosphatase.

5.2. Regulation by Inhibitors or Activators

The two SH2 domains of SHP-1 and SHP-2 not only target these enzymes to specific proteins throughout the cell, but are also involved in regulating activity. The binding of phosphotyrosine-containing peptides to SH2 domains of SHPs results in up to a 100-fold increase in activity. Biochemical studies suggested that the SHPs might be regulated by an autoinhibitory mechanism similar to that of Ca^{2+}–calmodulin-dependent protein kinases and protein Ser/Thr phosphatases. The crystal structure of SHP-2, shown in Fig. 6, confirms its regulation via a similar autoinhibitory mechanism. In the inactive state observed in the crystal, the N-terminal SH2 domain interacts with the catalytic domain so that a loop is inserted into the active site cleft. Residues from this loop interact with active site residues in a manner resembling that of the phosphosubstrate and sterically occlude the active site. When interacting with the active site region, the N-terminal SH2 domain assumes a conformation incapable of binding a phosphotyrosine-containing peptide or protein. The structure suggests that when the N-terminal SH2 domain engages ligand the resulting conformational change releases it from contact with the active site. Thus, the N-terminal SH2 domain can assume two conformations; one binds the active site and is inhibitory and the other binds ligand but not the active site. The C-terminal SH2 domain has minimal contacts with the catalytic domain and has a conformation capable of interacting with its ligand. The C-terminal SH2 domain may bind one phosphotyrosine within a protein target and facilitate the binding of a second nearby phosphotyrosine residue to the N-terminal SH2 domain. A diagram illustrating the regulation of SHP-2 by the autoinhibitory mechanism outlined here is shown in Fig. 7. Several other PTPs such as TCPTP and PTP-H1 might be regulated in an analogous manner, as suggested by the fact that removal of their noncatalytic sequences results in significant stimulation of their activity.

Protein inhibitors provide an important means of regulating Ser/Thr phosphatases and protein kinases. Recent studies have revealed the first example of a protein inhibitor for a CX_5R phosphatase. Cdc14, a DSP from budding yeast, that controls cell cycle progression, is both sequestered in the nucleolus and inhibited through its interaction with the anchoring protein Net1. It is likely that regulation by protein inhibitors will be employed by other CX_5R phosphatases.

5.3. Regulation by Phosphorylation

Many of the CX_5R phosphatases can be phosphorylated in vitro and in vivo on Ser, Thr, and Tyr residues. Phosphorylation, which can occur in both catalytic and noncatalytic regions, has the potential to influence activity or to control protein–protein interactions. As we have discussed with reference to SH2 domains, phosphorylated Tyr residues can create specific binding sites for regulatory or signaling proteins. In addition, sites of Ser phosphorylation are recognized and bound by 14-3-3 proteins. Perhaps one of the best examples of regulation by phosphorylation is the Cdc25 phosphatase that controls when cells enter

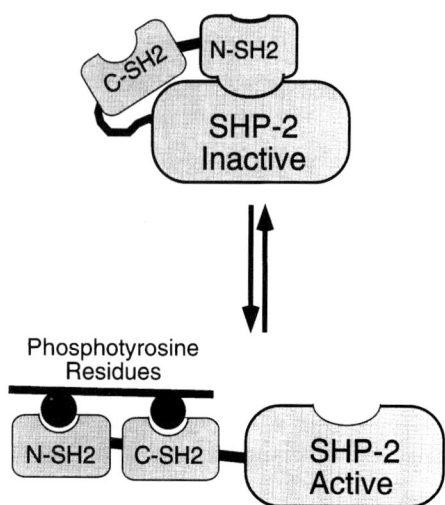

Fig. 7. Model for the autoinhibitory regulation of SHP-2. This diagram illustrates how SHP-2 is regulated through an autoinhibitory mechanism. The upper cartoon portrays the structure of autoinhibited, inactive SHP-2 that exists in the absence of phosphotyrosine-containing ligand and corresponds to the crystal structure of Fig. 6. The activated form of SHP-2 depicted on the bottom is based on biochemical studies; there is no corresponding crystal structure. The N- and C-terminal SH2 domains are labeled as N-SH2 and C-SH2, respectively. As depicted in the inactive protein, the peptide-binding site in the N-SH2 domain is in a conformation that does not permit binding of phosphotyrosine as observed in the structure shown in Fig. 6.

mitosis. The activity of Cdc25 is stimulated upon phosphorylation by the Cdc2 cyclin-dependent protein kinase and by Polo/Plx kinase, but is inhibited when phosphorylated by the Chk1 kinase (*see* below). There are other examples, but in many cases the effects of phosphorylation on CX_5R phosphatase activity or function are not understood and the kinase(s) responsible have not yet been identified.

5.4. Regulation of Receptor Protein Tyrosine Phosphatases

The structural organization of the RPTPs led researchers to postulate that these transmembrane enzymes might initiate signaling in response to ligands that bind to their extracellular segments. The concept that the RPTP external domains act as receptors has now been confirmed with the identification of ligands for several RPTPs. However, little is known about the effects of ligand binding on phosphatase activity or other functional properties of the RPTPs. Surprisingly, the external domain of the CD45 RPTP is not required for its essential role in T-cell activation, suggesting that the external domain might be dispensable for certain functions of other RPTPs.

The crystal structure of the membrane-proximal catalytic domain (D1) from RPTPα has inspired a model for how ligand binding could modulate the activity of RPTPs. The structure of the D1 domain of RPTPα is very similar to that for the catalytic domains of the nonreceptor PTPs. However, the D1 domain of RPTPα exists as a dimer in the crystal and in solution. In this dimer, each monomer contains a helix–turn–helix segment that occludes the active site of its partner. This observation led to the suggestion that oligomerization in response to ligand binding might suppress or completely inhibit RPTP activity (see Fig. 8A). Although some evidence supports this model, it should be emphasized that the membrane-distal (D2) domain was not present in this structure and its potential effect on dimerization of D1 was not assessed. In addition, the activity of the D1 dimer in solution has not been directly measured in vitro and there is no direct evidence for the existence of dimers in vivo. Moreover, this is unlikely to be a general model because the inhibitory segment from RPTPα is not conserved in all RPTPs.

Two crystal structures of catalytic domains from RPTPs indicate that this "inactivation by oligomerization" model might not be applicable to all PTPs. In the structure of the D1 domain of RPTPμ, dimers were observed, but the two monomers did not interact in a manner resembling those from RPTPα and consequently the active sites were not blocked. Recently, the structure of the complete intracellular segment (D1 and D2) of the LAR RPTP has been determined. In this structure the LAR domains are monomers and the active sites of both domains are predicted to be accessible to substrates. The structure predicts that dimerization of LAR D1 in the inhibitory mode observed with the RPTPα D1 domain cannot occur owing to steric interference from segments of D2. A number of alternative models have been proposed to explain the effects of ligands on RPTP function. One attractive possibility is that ligands induce or mediate the clustering of RPTPs with other transmembrane proteins which could bring them into contact with their substrates or regulators and thereby modulate their activity and/or function (see Fig. 8B). The putative substrates in this case would be the cytoplasmic segment of the co-receptor molecule or any interacting intracellular protein(s). The ligands in this model could be on an apposing cell or located within the membrane of the cell bearing the RPTP. With the

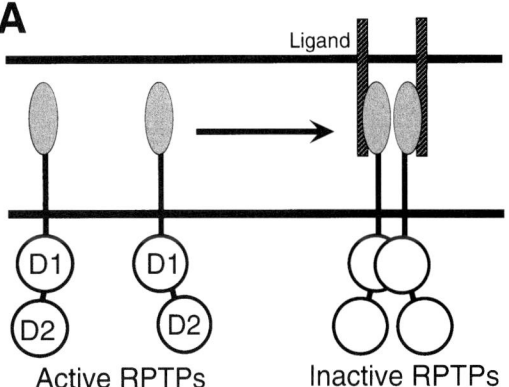

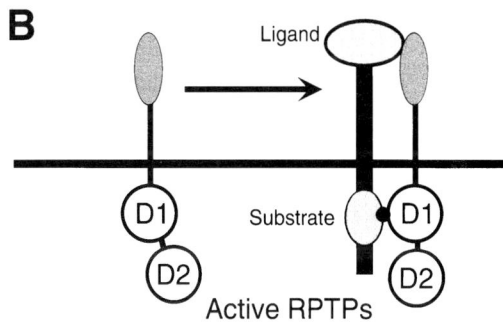

Fig. 8. Models for the regulation of receptor PTPs by their cognate ligands. Two potential effects of ligand binding on RPTP activity and function are shown. (A) In this model, ligand binding induces receptor oligomerization as illustrated. The ligand-induced RPTP oligomer (dimer shown here) is inactive because the RPTP active sites in the D1 domain are occluded and substrate cannot bind. In this depiction, the ligand is a cell surface molecule on an apposing cell, but the ligand could reside in the same membrane as the RPTP. (B) This diagram illustrates one of several models for the positive regulation of RPTPs by ligand. In this case, the ligand is a transmembrane protein on the same cell. The extracellular domain of this cell surface protein binds the RPTP and its cytoplasmic tail interacts with an RPTP substrate. Recruitment of the ligand to the RPTP brings the associated substrate into a position where it is accessible to the RPTP catalytic domain. A multivalent ligand on an apposing cell could mediate substrate recruitment in a similar manner.

possible exception of RPTPα, the effect of ligand binding on the activity or function of RPTPs is not understood and remains a major question in RPTP research.

The role of the D2 domain in the function and/or regulation of RPTPs is unclear owing in part to contradictory data regarding their catalytic potential. The D2 domain from RPTPα has low but detectable activity as measured in vitro. Other studies failed to detect activity for D2 domains from several RPTPs. This result is expected for certain RPTPs having D2 domains that should possess no intrinsic activity because they are missing residues (e.g., Cys, Arg, or Asp) that are known to be essential for catalysis. In other cases, predictions about activity cannot be made from sequence alone. As described later, two amino acid side chains in the LAR D2 domain were not oriented in positions favoring catalysis. However, the structure suggested that steric hindrance would not prevent their rearrangement to positions favoring catalysis in response to conformational changes induced by a suitable substrate. Mutational analyses have shown that inactivating the D2 domain from several RPTPs by replacement of the essential Cys does not abrogate biological functions.

For those RPTPs in which it lacks activity, the D2 domain may have other functions. For example, despite having no activity, the D2 domain of the CD45 RPTP is required for some of its cellular functions. In addition, the activity of certain D1 domains measured in vitro is influenced by the presence of the D2 domain. The crystal structure of the two tandem catalytic domains of LAR reveals that there are significant interactions between the D1 and D2 domains, suggesting that conformational changes in the D2 domain could affect the D1 domain. The structure of the LAR catalytic domains indicated that the D2 domain is not active. However, the overall topology of the D2 active site region resembled that of other active PTPs. Hence, it was proposed that D2 might bind phosphoproteins but not hydrolyze them and thereby serve as a phosphotyrosine binding domain. Several proteins bind the D2 domain of LAR but the mode of binding and functional consequences are not completely understood. Thus, the D2 domains of RPTPs may bind phosphosubstrates and/or regulatory proteins that could influence the activity of the D1 domain and/or control the location of RPTP on the cell surface.

6. PHYSIOLOGICAL ROLES OF PROTEIN TYROSINE PHOSPHATASES

6.1. Role of CD45 in Lymphocyte Activation

CD45 is an RPTP with a glycosylated extracellular domain, a single transmembrane segment, and two tandem PTP catalytic domains. The extracellular domain is thought to have a ligand binding function; however, the ligand(s) for CD45 have not yet been identified. The expression of CD45 is limited to nucleated cells of hematopoietic origin, where it occupies about 10% of the cell surface. Relative to many other proteins with signaling functions CD45 is surprisingly abundant. The size of the extracellular domain varies in length from 400 to 550 residues because of alternative mRNA splicing. Splice variants differ at their N-termini and are expressed in a manner that depends on the cell type, developmental stage, and antigenic stimulation.

To carry out their immune function peripheral T and B lymphocytes must be activated to a fully functional state upon encountering antigen. Quiescent B lymphocytes that recognize antigen are activated to proliferate and differentiate into antibody secreting plasma cells or long-lived memory cells that persist after antigen is eliminated. Activation of B cells requires lymphokines produced by helper T cells. When T cells bind an antigen presented by the major histocompatibility complex (MHC) on a macrophage or target cell, they are also activated to secrete lymphokines, proliferate, and differentiate into mature cells equipped to carry out their helper or cytolytic functions. Both T and B lymphocytes have multiprotein receptors on their surface that bind antigen and initiate a complex signaling cascade that culminates in their differentiation and proliferation. Antigen receptor signal transduction is also important for certain stages of lymphocyte development.

Genetic and biochemical analyses have demonstrated that CD45 is required for the development and antigen-dependent activation of T lymphocytes. This discussion focuses on the role of CD45 in T-cell activation. To describe how CD45 participates in this process, we will consider the basic structural organization of the T-cell antigen receptor (TCR) and the early signaling events that are triggered by antigen binding. The TCR is a large complex of several distinct integral membrane proteins that is not only responsible for antigen recognition but must also initiate signaling events leading to T-cell activation. Figure 9 shows the subunits and organization of the TCR. The disulfide-bonded αβ heterodimers of the TCR are responsible for binding the MHC–antigen complex on an adjacent antigen-presenting cell. The subunits of the αβ dimer are members of the immunoglobulin (IgG) superfamily and have large extracellular domains with very short cytosolic segments. The αβ chains are tightly associated with a CD3 complex that is composed of γ, δ, and ε subunits, which have IgG-like extracellular domains and form noncovalently

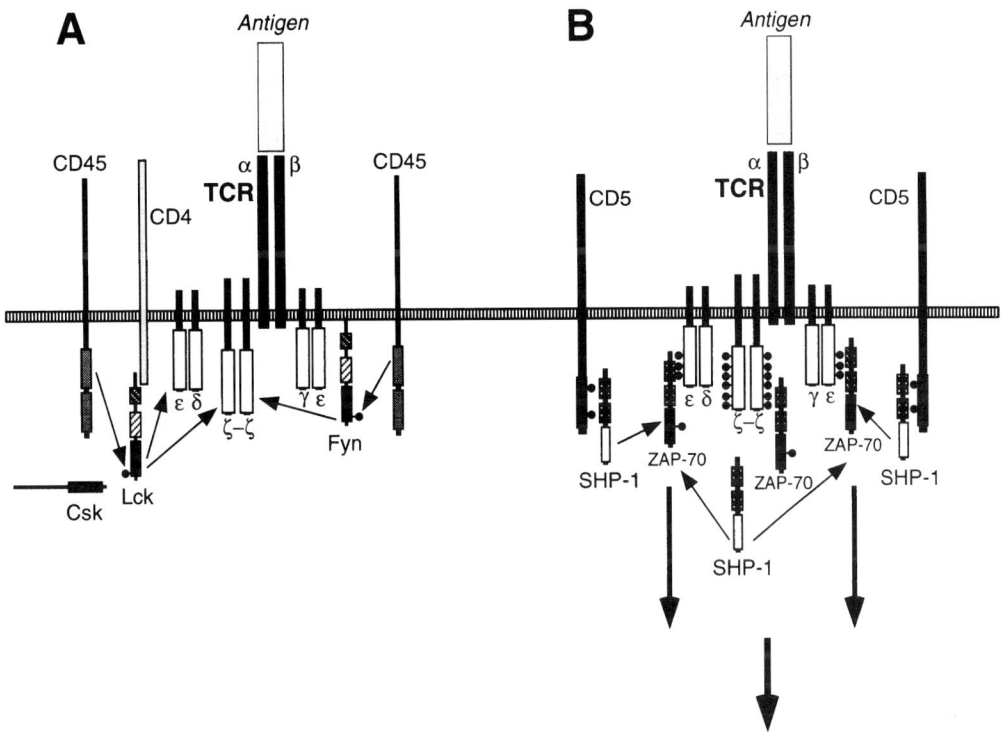

Fig. 9. The role of CD45 and SHP-1 in T-cell activation. This schematic illustrates a model of the major signaling events that follow engagement of the TCR with antigen with an emphasis on highlighting the positive action of CD45 and the negative functions of SHP1. This is not a comprehensive depiction of TCR signaling and many components and steps have been omitted. (A) This diagram illustrates the components of the TCR and the associated Src family kinases. As the arrows show, CD45 targets a negative regulatory site in the C-termini of the Src family kinases, Lck and Fyn, and thereby activates them. Thus, CD45 opposes the action of the Csk kinase, which inhibits Src kinase activity by phosphorylating a C-terminal Tyr residue. Activated Lck and possibly Fyn then phosphorylate the ITAMs (*open boxes*) of the TCR subunits. (B) Results of ITAM phosphorylation on signal propagation as well as the action of SHP-1 in suppressing or negatively regulating this signaling pathway. Once phosphorylated, the ZAP-70 kinase is recruited to ITAMs of the TCR subunits as illustrated and activated via phosphorylation by the Src kinases. The CD4 coreceptor and the Src family kinases are not shown in (B) for clarity. As the *large arrows* indicate, activation of ZAP-70 leads to activation of multiple downstream pathways. There may be at least two mechanisms by which SHP1 down-regulates TCR signaling. SHP1 may be recruited directly to ZAP-70, as the *arrows* in the figure indicate, and directly dephosphorylate and inhibit its kinase activity. SHP1 may also be recruited to ITIM sequences in the cell surface receptor CD5. When recruited to CD5, ZAP-70 is one of the major targets for SHP1 as illustrated by the *arrows*. This scheme may not illustrate all the substrates or functions of CD45 and SHP-1 in T-cell activation.

linked ε–γ and ε–δ heterodimers. The third component of the TCR is the disulfide-linked ζ–ζ homodimer. It is the CD3 complex and the ζ–ζ homodimer that are involved in transmitting a signal across the membrane upon antigen engagement despite the fact that their cytoplasmic segments all lack intrinsic enzymatic activity.

One of the earliest signaling events observed following antigen binding is the activation of PTKs and a concomitant increase in the tyrosine phosphorylation of multiple intracellular proteins. Subsequently, phospholipase Cγ is activated, releasing inositol trisphosphate and diacylglycerol leading to elevated Ca^{2+} and activation of protein kinase C. Other distal signaling events include the activation of Ras and the MAP kinase signaling cascade. Through mechanisms that are not yet completely understood these signaling events ultimately lead to T-cell proliferation and the changes in gene expression that result in T-cell differentiation.

The initial tyrosine phosphorylation that occurs upon antigen binding to the TCR results from the rapid activation of Lck and Fyn, two PTKs of the Src family (*see* Fig. 9). The Fyn kinase is thought to be associated with the ε-chains of the CD3 complex, whereas Lck associates with either the CD4 or CD8 coreceptors. The primary targets of these PTKs are the immunoreceptor tyrosine-based activation motifs (ITAMs) located in the intracellular domains of the γ, δ, and ε CD3 subunits and the ζ–ζ homodimer.

Tyrosine phosphorylation of ITAMs in the TCR results in the recruitment of ZAP-70, a protein kinase of the Syk PTK family. Once recruited to the TCR, ZAP-70 is phosphorylated and activated by Lck or Fyn. Activation of the Src family PTKs and ZAP-70 is essential for normal T-cell activation and development. The downstream targets of ZAP-70 include SLP-76 and other adaptor proteins. Phosphorylation by ZAP-70 leads to activation of phospholipase Cγ and other distal signaling pathways that ultimately induce differentiation and proliferation. Effective TCR-mediated lymphocyte activation also requires signals initiated by the CD28 cell surface protein and may also involve signaling pathways activated by adhesion molecules that engage their ligands when T cells adhere to antigen presenting cells.

Numerous studies have shown that CD45 is required for TCR-mediated signal transduction. The strongest evidence comes from the analysis of CD45⁻ T cells, which fail to hydrolyze phosphatidylinositol, elevate intracellular Ca^{2+}, secrete cytokines, or proliferate in response to antigen ligation. More importantly, CD45⁻ cells do not respond to antigen binding with the rapid increase in tyrosine phosphorylation observed in normal cells. Because tyrosine phosphorylation is one of the earliest signaling events following antigen stimulation, this finding suggested that CD45 was involved in steps that were immediately downstream of the TCR. Studies in CD45-deficient mice revealed that CD45 was required for normal T-cell development, a process requiring TCR signal transduction. Thus, the results from knockout mice also indicate that CD45 has a crucial role in TCR signaling.

The lack of tyrosine phosphorylation in CD45⁻ cells is a paradoxical finding, as PTPs reverse the action of PTKs. However, the regulatory properties of the Src family kinases such as Lck can explain how dephosphorylation results in enhanced tyrosine phosphorylation. A PTK, known as Csk, inhibits Src family kinases by phosphorylating a C-terminal regulatory tyrosine (Tyr^{505} in Lck). Thus, CD45 is postulated to oppose the action of Csk and up-regulate Lck and Fyn activity during TCR signaling (*see* Fig. 9). This notion is supported by data showing that the regulatory site in Lck is hyperphosphorylated in CD45⁻ cell lines and CD45 effectively dephosphorylates this site in vitro. Moreover, Lck has been found to physically interact with CD45. This model predicts that Lck activity in CD45-deficient cells should be suppressed. In many studies, decreased Lck activity has been observed in CD45⁻ cells. However, some researchers have found that Lck activity is actually increased in the CD45 mutant cells. CD45 is also capable of negatively regulating Src kinases by dephosphorylating a Tyr residue located in the activation loop of the catalytic domain.

CD45-deficient mice have been employed to better understand the discrepancies in Src family kinase activity measured in CD45⁻ cells. As described previously, T-cell development, which requires TCR-mediated signaling, is blocked in CD45 knockout mice. In recent studies, an Lck transgene in which Tyr^{505} was replaced by Phe was introduced into CD45⁻ mice. T cells developed normally in these transgenic mice, indicating that the Lck Y505F mutant rescues the defect in TCR signal transduction. This study provides strong support for a model in which CD45 couples antigen stimulation to downstream signaling pathways by dephosphorylating the negative regulatory site of Lck and/or Fyn (*see* Fig. 9). Once activated, these Src kinases are able to phosphorylate ITAMs when antigen is bound, initiating the signaling pathways that lead to T-cell activation. How antigen stimulation influences CD45 activity, if at all, is not known. It has been suggested that by opposing the Csk kinase CD45 dephosphorylates the negative regulatory site in Src family kinases prior to antigen ligation. In this model, CD45 primes the Src kinases and sets the threshold for TCR signaling. The preactivated Src kinases are then recruited to the TCR when antigen is bound resulting in the rapid phosphorylation of ITAMs. An attractive feature of this model is that the Src kinases and their substrates are physically separated from CD45, which is not recruited to the TCR upon antigenic stimulation. This physical separation prevents the immediate dephosphorylation of ITAMs and other targets by CD45. Besides the Src family PTKs, little is known about other potential CD45 substrates. There is evidence that CD45 can dephosphorylate the ζ-chains of the TCR. It is possible that CD45 targets the ζ-subunits and other substrates as part of a negative feedback mechanism that terminates or limits TCR signaling.

The role of CD45 in the activation of B cells resembles its function in T-cell signaling. Antigen binding to the B-cell antigen receptor (BCR) also triggers an initial increase in PTK activity that ultimately leads to differentiation and proliferation. In the absence of CD45 expression, BCR-mediated signaling in response to antigen binding is much less efficient but not abrogated. In CD45⁻ B cells, the antigen-induced activation of Lyn kinase and Ca^{2+} mobilization are significantly reduced, indicating that CD45 acts as positive regulator of B-cell activation just as it does in T cells. Little is known about the function of CD45 in other cells of hematopoietic origin such as mast cells or macrophages. Its presence in a variety of immune cells and its extraordinary abundance on the cell surface suggests that it has multiple functions.

The function of the extracellular domain of CD45 has not been defined. The external domain is believed to be the binding site for ligands, but thus far no physiologic ligands have been identified. Such ligands may be cell surface proteins residing on interacting cells or in the same membrane; however, a circulating macromolecule or low molecular weight compound cannot be excluded. Surprisingly, several studies have shown that the external segment of CD45 is not essential for lymphocyte activation and that it is only necessary for the internal catalytic segment to be located in proximity to the membrane. Although the external domain is not essential, there is evidence that the splice variants of CD45, which possess distinct N-termini and patterns of glycosylation, have differential effects on the intensity of TCR signaling. It is possible that the external domains of CD45 influence TCR-mediated signal transduction in ways that cannot be detected with the current techniques used to monitor lymphocyte activation. It is likely that ligand binding to the external domain will prove to be crucial for yet undiscovered functions.

Relatively little is known about how CD45 is regulated during lymphocyte activation or other immune cell functions. Phosphorylation appears to be an important mode of regulation, as CD45 is a phosphoprotein in vivo and contains both phosphoserine and phosphotyrosine in its intracellular domains. Mitogenic stimulation results in phosphorylation of both Ser and Tyr residues, suggesting that phosphorylation is a potential mechanism for regulating CD45 during lymphocyte activation. Treatment of T cells with ionomycin or phorbol esters has been reported to elicit changes in Ser phosphorylation and the activity of CD45. Whether the modest alterations in CD45 activity are physiologically relevant and a direct result of changes in phosphorylation is not known. There are additional reports that phosphorylation of CD45 in vitro by both tyrosine and Ser/Thr kinases can enhance activity; again, it is not clear if these effects are of physiologic significance. An acidic insert in the distal PTP domain (D2) of CD45 is multiply phosphorylated in vivo by casein kinase 2, but the effects of this phosphorylation on CD45 function are not known. Much remains to be learned about how phosphorylation modulates CD45 function(s).

Protein–protein interactions also appear to be an important means for regulating CD45. A 30-kDa protein termed CD45-AP associates with CD45 from lymphocytes. CD45-AP is an integral membrane protein with a single transmembrane segment, a 145-residue cytosolic tail, and a very short external segment. CD45-AP and CD45 interact directly with one another through their transmembrane segments. T and B cells of mice lacking CD45-AP exhibited reduced proliferation on antigen stimulation, indicating that CD45-AP was not essential for antigen receptor signaling but was necessary for achieving full lymphocyte activation. Recent studies have suggested that CD45-AP might link Lck and either the CD8 or CD4 coreceptors together with CD45. Such interactions may facilitate the activation of Lck when antigen receptors are stimulated.

6.2. SHP-1 Negatively Regulates Hematopoietic Cell Signaling

The expression of SHP-1 is restricted primarily, but not exclusively, to cells of hematopoietic origin, whereas the closely related enzyme, SHP-2, is found in most cells and tissues. Substantial evidence has established that the primary function of SHP-1 is to negatively regulate signaling through cell surface receptors of hematopoietic cells. SHP-1 opposes the PTKs that are activated when multiprotein receptors such as the TCR bind their ligands. In addition to the antigen receptors of T and B cells, SHP-1 also downregulates FcγRIII receptors of natural killer cells and certain cytokine receptors that initiate signaling upon ligand binding by activating PTKs of the Src and Janus families, respectively. By opposing these receptor-associated PTKs, SHP-1 terminates or suppresses downstream signals and is involved in setting the threshold for activation of these immune receptors. Perturbations in the balance between PTKs and PTPs such as SHP-1 might contribute to autoimmune and inflammatory diseases.

The impact of imbalances between the PTKs and PTPs on immune function is evident in the phenotypes of *motheaten* (*me*) and *motheaten* viable (*me^v*) mice. These mice carry naturally occurring point mutations in the *SHP-1* gene that result in improper splicing of RNA transcripts. Homozygous *me/me* mice are null mutants because the splicing defect results in a complete absence of *SHP-1* expression, whereas the *me^v/me^v* mice express a form of the enzyme that has 10–20% of normal activity. Hence, the *me^v/me^v* mice have a less severe phenotype and a longer life span (approx 9 wk after birth) than the *me/me* mice (2–3 wk after birth).

The *motheaten* phenotype is complex and these animals suffer from severe immune deficiency, autoimmune, and chronic inflammatory disease. Their death results primarily from pneumonitis due to accumulation of large numbers of macrophages and neutrophils in the lung. The amassing of macrophages and neutrophils in the skin causes the hair and pigment loss that is the basis for their *motheaten* designation. Other defects in homozygous *motheaten* mice include hyperresponsiveness of BCR- and TCR-mediated signaling, elevated numbers of a subset of B cells that is associated with autoimmunity, and high concentrations of serum immunoglobulins. The *motheaten* mice have been invaluable for investigating the function and regulation of *SHP-1* in leukocytes. Despite the pleiotropic effects of these mutations on immune function, most studies with *motheaten* mice have consistently shown that *SHP-1* negatively regulates a variety of signal transduction pathways in cells of hematopoietic origin.

We have discussed the positive role of CD45 in promoting B-cell activation when antigen binds to the B-cell antigen receptor. The negative role of *SHP-1* in B-cell activation is highlighted by the finding that the me^v mutation restores normal B-cell development and BCR-mediated signaling in CD45-deficient mice. Hence, the absence of *SHP-1* rescues the deficiency in B cells resulting from lack of CD45. This confirms that CD45 acts to stimulate PTK activity during lymphocyte activation, whereas *SHP-1*, when appropriately activated, opposes these PTKs.

Most hematopoietic cells have cell surface molecules or receptors that participate in terminating or negatively regulating their response to foreign antigens. Many of these cell surface proteins are involved in returning the immune system to its basal state after antigen elimination and/or suppressing or preventing a response to self-antigens. For example, the FcγRIIB receptor recognizes the constant region of IgG complexed with antigens and inhibits B cell activation via the BCR, thus helping to terminate the immune response. The CD22 cell surface lectin can also negatively regulate B-cell signal transduction pathways. The killer cell inhibitory receptor (KIR) binds MHC class I molecules and suppresses the activation of natural killer cells and a subset of T cells, preventing their attack on host cells.

These and other negative regulatory cell surface molecules have in common the ability to suppress or down-regulate signals emanating from the multisubunit receptors that contain ITAMs (*see* earlier) in their cytoplasmic segments. Recall that when phosphorylated by Src-family kinases, these ITAMs provide docking sites to recruit additional PTKs or other signaling molecules that contain SH2 domains and initiate downstream signaling events. Another important shared feature of the negative regulatory receptors is the presence of a Tyr residue within a sequence motif (V/I/LXYXXL) termed ITIM for immunoreceptor tyrosine based inhibitory motif. When the Tyr of ITIMs is phosphorylated, it recruits SH2 domain-containing enzymes such as SHP-1. As noted previously, SHP-1 has well-documented roles in the negative regulation of certain cytokine receptors such as the erythropoietin receptor of red blood cell progenitors and interleukin-3 (IL-3) receptors. These cytokine receptors also have ITIM-like motifs that recruit SHP-1.

Numerous studies suggest a paradigm by which SHP-1 functions to negatively regulate leukocyte signaling. Cell surface molecules or receptors (e.g., KIR, CD22, etc.) with negative regulatory functions colocalize or associate with receptors (e.g., TCR, BCR, FcγRIII) that initiate signaling through the rapid PTK-dependent phosphorylation of their ITAM-containing subunits. When sequestered together with the positive receptors, the ITIMs of inhibitory receptors are phosphorylated on their tyrosines, which then triggers the recruitment of SHP-1 via its SH2 domains. When bound to ITIMs, SHP-1 activity is increased up to 100-fold and because of its location near ITAM-containing receptors, it is capable of dephosphorylating and thus inactivating signaling proteins including the activated PTKs and/or their targets. The substrates of SHP-1 probably depend on the ITIM-containing receptor molecules with which it is associated. The details of this inhibitory mechanism are not yet understood. There are reports describing the direct association of SHP-1 with cytoplasmic signaling proteins such as Grb2, Vav, SLP-76, and ZAP-70. It is not known whether these interactions are mediated via ITIMs or, with the exception of ZAP-70, if any of these proteins are SHP-1 substrates.

The cell surface glycoprotein, CD5, is a negative regulator of TCR-mediated lymphocyte activation (*see* Fig. 9B). Although it is not a perfect match to the consensus sequence, CD5 contains an ITIM motif in its cytoplasmic tail. Interaction with CD72 located on adjacent B cells or other ligands is thought to induce the association of CD5 with the ζ-subunits of the TCR leading to the phosphorylation of ITIMs within CD5 and the subsequent recruitment of SHP-1 (*see* Fig. 9). The TCR ζ-chains, the ZAP-70 and Syk protein kinases, and phospholipase C_γ are dephosphorylated when CD5 is engaged presumably through the action of SHP-1. The Src-family kinases that are activated at a very early stage of TCR signaling do not appear to be substrates for SHP-1. Dephosphorylation of ZAP-70 or the ζ-chains is sufficient to attenuate downstream signaling pathways. Some functions of SHP-1 may not be mediated through ITIM-containing receptors. SHP-1 has been reported to be activated through a direct interaction with the ZAP-70 kinase and to dephosphorylate it (see Fig. 9), Thus, SHP-1 may act in two ways to abrogate T-cell signaling.

It is important to point out that at least two other enzymes can be sequestered to phosphorylated ITIM motifs. These include the related protein phosphatase SHP-2 and SHIP, an inositol polyphosphate 5′-phosphatase that hydrolyzes inositol 1,3,4,5-tetrakisphosphate (IP_4) and/or phosphatidylinositol (3,4,5)-trisphosphate (PIP_3). It is thought that SHIP may act to oppose the elevation of intracellular Ca^{2+} that is induced upon ligation of many lymphocyte receptors. ITIMs within the FcγRIIB receptor, a negative regulator of BCR signaling, bind SHIP but not the SHP-1 protein phosphatase. The cell-surface molecule CTLA-4, a negative regulator of T-cell activation, associates with SHP-2, which also contributes to suppressing TCR signal transduction. This is one of the few cases where SHP-2 has a negative role; as described in Section 6.3.1., SHP-2 has a positive role in many signaling pathways.

6.3. Role of Protein Tyrosine Phosphatases in Growth Factor and Cytokine Signaling

The binding of growth factors, cytokines, and certain hormones to their cell surface receptors results in the rapid activation of PTKs. These initial tyrosine phosphorylation events result in diverse changes in cell physiology and metabolism, cell growth and differentiation, and cell motility and shape. The receptors for many growth factors and certain hormones have intrinsic tyrosine kinase activity within their intracellular domains. Ligand-induced oligomerization activates these receptor tyrosine kinases (RTKs) and facilitates transphosphorylation of multiple Tyr residues within their cytoplasmic domains. These phosphotyrosine residues serve as sites for the docking of adaptor molecules that can activate diverse downstream signaling mechanisms. Although devoid of intrinsic kinase activity, cytokine receptors trigger tyrosine phosphorylation of their cytoplasmic domains through the action of associated JAK family kinases that are activated when ligands trigger receptor oligomerization. The signals emanating from both the growth factor RTKs or cytokine receptors are propagated by signaling pathways, most of which include further activation of downstream protein kinases such as those of the MAP kinase cascade.

Many CX_5R phosphatases have been implicated in regulating mitogenic signaling both at the level of the receptor and further downstream. Much of the evidence linking PTPs to these signaling pathways remains indirect and circumstantial, primarily because of a lack of knowledge regarding substrates. In this section, we focus on the tyrosine-specific enzyme SHP-2 and a subset of the DSPs known as the MAP kinase phosphatases (MKPs). Both types of phosphatases have been intensively studied and they will serve to illustrate how enzymes of the PTP family regulate mitogenic signal transduction.

6.3.1. SHP-2/Csw Function Positively in Mitogenic Signaling

Both genetic and biochemical studies have shown that SHP-2 from vertebrates and its orthologs encoded by the *corkscrew* (*csw*) and *ptp-2* genes from *Drosophila* and *C. elegans*, respectively, are required for signaling through multiple RTKs and cytokine receptors. SHP-2 has been shown to function downstream of receptors for epidermal growth factor (EGF), insulin, fibroblast growth factor (FGF), platelet-derived growth factor (PDGF), and numerous cytokines such as IL-3, IL-2, IL-5, and growth hormone.

Many developmental processes in both vertebrate and invertebrate organisms depend on RTK signaling pathways. In genetic studies, Csw was identified first as a component of the Torso RTK signaling pathway that determines cell fate in terminal regions of the *Drosophila* embryo. Subsequent studies in *Drosophila* revealed a role for the Csw phosphatase in multiple RTK signaling systems including differentiation of the R7 photoreceptor of the ommatidium in the compound *Drosophila* eye, which is controlled through the Sevenless (Sev) RTK. The *C. elegans ptp2* phosphatase is essential for oogenesis and also functions downstream of the EGF receptor homolog. Mice, in which the SH2 domains of SHP-2 were deleted, exhibited serious developmental defects. It has been suggested that SHP-2/Csw and their orthologs may function in most if not all RTK signaling pathways.

The phosphatase activity of SHP-2, Csw, and their orthologs is required for their positive role in promoting RTK and cytokine signal transduction. With their ability to oppose protein tyrosine kinases, the positive role for SHP-2/Csw phosphatases in these pathways might appear paradoxical. Although many of the molecular details are not yet understood, there appear to be at least three distinct mechanisms whereby SHP-2/Csw carry out their positive function in mitogenic signal transduction. SHP-2 and Csw function downstream of the receptors but act both upstream and in parallel to the Ras/MAP kinase cascade (*see* below) that mediates many of the effects of growth factors and cytokines.

Signals arising from RTKs, cytokine receptors, and many other cell surface receptors are propagated through one or more MAP kinase cascades that deliver signals to the nucleus to trigger gene expression and cell proliferation as shown in Fig. 10. There are many MAP kinase signaling modules but they all are comprised of three protein kinases that act on one another in series. The terminal protein kinase in the series is the MAP kinase (MAPK), which is a Ser/Thr protein kinase that is activated when phosphorylated on Thr and Tyr residues within the "activation loop" of the kinase catalytic domain. MAPK is downstream of and phosphorylated by the dual specificity protein kinase, MAP kinase kinase (MAPKK), which is, in turn, activated upon phosphorylation by a MAP kinase kinase kinase (MAPKKK). Raf kinase functions as a MAPKKK that is linked to RTKs and many other receptors through the Ras GTP binding protein. Ras is regulated by the guanine nucleotide exchange factor known as Sos. When the cytoplasmic domain of an RTK is phosphorylated in response to ligand, the SH2 domain containing adaptor protein Grb2 translocates to the receptor carrying its binding partner Sos. Once recruited near the plasma membrane via the interaction of Grb2 with an RTK, Sos stimulates the formation of an activated Ras–GTP complex, which in turn activates the Raf kinase and initiates the MAP kinase cascade (*see* Fig. 10).

A signaling mechanism for Csw and SHP-2 has been identified in studies of the *Drosophila* Torso RTK and several mammalian growth factor receptors such as the PDGF receptor (*see* Fig. 10). Ligand binding and the subsequent tyrosine phosphorylation of Torso provide a docking site for the recruitment of Csw via its SH2 domain. On association with Torso, Csw is phosphorylated on a Tyr residue located near its C-terminus. Once this site is phosphorylated, Csw binds the SH2 domain of Drk (the *Drosophila* homolog of Grb2). Thus, Drk and its constitutively associated partner Sos are linked to the Torso RTK through Csw. When translocated to a site on Torso near the membrane, Sos activates Ras and triggers the MAP kinase cascade as outlined previously. Acting in this capacity, Csw is not only a phosphatase, but also

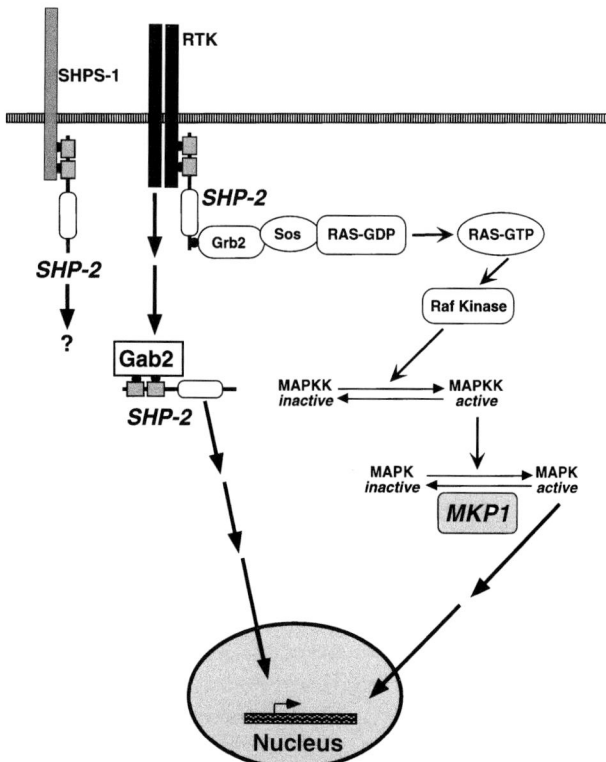

Fig. 10. Role of protein tyrosine phosphatases in mitogenic signaling. Three major roles of SHP-2 in mediating signaling from RTKs are illustrated in this generalized scheme. SHP-2 is recruited to RTKs when ligand binding leads to autophosphorylation of Tyr residues recognized by the SH2 domains SHP-2. Other adaptor proteins and signaling enzymes that are typically recruited to RTKs are not shown. When SHP-2 is recruited to the RTK, it is phosphorylated on a C-terminal Tyr, creating a binding site for SH2 domains of Grb2. SHP-2 then couples Grb2 to the activated RTK, thus triggering the "classic MAP kinase" signaling cascade as shown. SHP-2 is also recruited to the Gab2 adaptor protein, which may be its substrate. The association of SHP-2 and Gab2 activates gene expression in a pathway that acts in parallel with the MAP kinase cascade as shown. Finally, SHP-2 is recruited to SHPS-1 and other members of a family of membrane proteins that are phosphorylated upon growth factor activation. Recruitment of SHP-2 to this complex may lead to MAP kinase activation but this mechanism is not well understood. The role of MKP1 in negatively regulating the MAP kinase cascade is also illustrated. As shown, MKP1 opposes the MAP kinase kinase that phosphorylates and activates MAP kinase. This scheme represents a generic model for three positive roles of SHP-2 in mitogenic signaling, but it is not meant to suggest that all three modes of signaling are used by a single receptor or that all three modes would operate simultaneously in a single cell.

an adaptor molecule linking other signaling proteins to Torso. When Csw is bound to Torso, its phosphatase activity is stimulated on engagement of its SH2 domain and it dephosphorylates a Tyr residue within the cytoplasmic domain of the receptor. In the phosphorylated state, this Tyr residue serves as a site for binding Ras–GAP, which is a negative regulator of Ras. Thus, the displacement of Ras–GAP from Torso by Csw also promotes Ras activation. With certain mammalian growth factor receptors, SHP-2 also acts as an adaptor protein linking RTKs to Grb2 and Sos, which then leads to activation of the Ras–MAP kinase cascade. It is not known whether SHP-2 also dephosphorylates and regulates RTKs in a manner analogous to Csw.

Csw also binds and dephosphorylates the *Drosophila* adaptor protein Dos that is required for signaling by Sev and other RTKs with developmental roles. Dos has an N-terminal pleckstrin homology (PH) domain with sequence similarity to a segment of Gab1, IRS-1, and IRS-2, which are mammalian growth factor receptor substrates. Moreover, the C-terminus of Dos has a polyproline region and tyrosines that are potential binding sites for SH3 and SH2 domains. Genetic analyses indicate that Csw and Dos either act in parallel or upstream of Ras in Sev signaling.

A recent study has identified Gab2 as an important adaptor protein required for signaling downstream of cytokine and RTK receptors (*see* Fig. 10). Upon cytokine or growth factor stimulation, Gab2 is phosphorylated and binds SHP-2 as well as other signaling proteins such as the adaptor protein Shc or the p85 subunit of phosphoinositide-3 kinase (PI-3 kinase). The available evidence suggests that Gab2 is a substrate of SHP-2. As mentioned previously, Gab2 and Dos have limited sequence similarity and related structural organization. The Gab2/SHP2 complex is thought to activate cytokineinduced gene expression in a pathway that is parallel to the Ras–MAP kinase cascade (*see* Fig. 10). These studies suggest that SHP-2 and Csw have a second, conserved mode of action in mitogenic signaling in which they act together with the adaptor proteins Grb2 and Dos, respectively, to propagate signals that activate gene expression.

SHP-2 may function through yet a third mechanism in growth factor and cytokine signaling (*see* Fig. 10). In response to insulin and other mitogens, a transmembrane glycoprotein known as SHPS-1 is phosphorylated permitting the subsequent recruitment of SHP-2. Although the precise role of SHPS-1 and its homologs is not fully understood, it is possible that recruitment of SHP-2 to this membrane protein also results in activation of the Ras–MAP kinase signaling pathway. It is not yet known whether SHP-2 must interact with all three potential targets (the receptor, Gab2, or SHPS-1) to fully execute its positive role in mediating signaling from a single RTK or cytokine receptor. Further studies including identification of substrates will be required to clarify the precise function of SHP-2 and Csw in mitogenic signal transduction.

6.3.2. Dual Specificity Phosphatases Negatively Regulate Mitogenic Signaling by Dephosphorylating MAP Kinases

A subset of the DSPs suppresses growth factor and cytokine signaling pathways that are mediated by the MAP kinase cascade. The prototype for this group of DSPs is MKP1 (MAP kinase phosphatase 1) which specifically dephosphorylates both the Thr and Tyr residues located within the activation loop of the MAP kinase catalytic domain. Because both residues must be phosphorylated for MAP kinase to be fully activated, MKP1 has the capacity to inactivate MAP kinase and thereby terminate signaling by this pathway as illustrated in Fig. 10. Expression of MKP1 is induced in response to growth factors and stress (e.g., heat shock, exposure to oxidants, UV irradiation, etc.). The delayed expression of MKP1 following mitogenic stimulation or stress demonstrates that MKP1 acts as a feedback inhibitor of MAP kinase signaling. This function is consistent with the genetics of the *Drosophila puckered* gene that encodes an MKP1-like enzyme that regulates a MAP kinase homolog in a similar manner.

Besides MKP1, at least nine other related DSPs with activity toward MAP kinases have been identified. More than a dozen MAP kinase signaling modules are present in mammalian cells including the ERK1/2, JNK/SAPK, and p38/RK protein kinases and the existence of multiple MKP1-like phosphatases may reflect the diversity among these MAP kinases. Several of the MAP kinase phosphatases such as MKP3/PYST1 and PYST2 are expressed constitutively and are not induced by stress or growth factors. They are also found in the cytosol instead of the nucleus. Some of the MAP kinase phosphatases are selective and target a subset of the MAP kinases. MKP3/PYST1 exhibits specificity for the ERK kinases, whereas the hVH5/M3–6 phosphatases selectively target the JNK/SAPK kinases that mediate the response of cells to heat, osmotic shock, or UV radiation.

MKP1 and related DSPs are not the exclusive regulators of MAP kinase. In yeast, the tyrosine-specific enzymes, PTP1 and PTP2, are involved in regulating MAP kinases and in some cases both a MAP kinase phosphatase and a PTP may act together to regulate the same kinase. In addition, Ser/Thr protein phosphatases may also act on MAP kinases. These findings suggest that the physiologic regulation of MAP kinases by dephosphorylation is complex and may involve interplay among several types of phosphatases.

6.3.3. Role of Nonreceptor PTPs as Potential Negative Regulators of RTKs

Many other tyrosine-specific PTPs have been implicated in regulating RTKs, cytokine receptors, or the pathways that are downstream of them. For example, biochemical studies have implicated PTP1B, a nontransmembrane PTP, as a negative regulator of insulin receptor signaling. This function has been corroborated by studies of mice in which the *PTP1B* gene is disrupted. In response to insulin treatment, homozygous *PTP1B*[-/-] mice exhibit increased insulin sensitivity and enhanced phosphorylation of the insulin receptor and its substrate, IRS-1. These knockout mice and biochemical data indicate that the insulin receptor itself may be the PTP1B target, but other targets in the pathway may exist as well. The *PTP1B*[-/-] mice have no other detectable developmental defects and other abnormalities, suggesting that there may be other PTPs that can act as functional replacements or that PTP1B has a narrow role in regulating RTKs.

The T cell PTP (TCPTP), the closest relative of PTP1B, has been implicated in regulating the epidermal growth factor receptor (EGFR). A 45-kDa splice variant of the TCPTP resides in the nucleus, but is translocated to the cytoplasm in response to EGF treatment. Substrate trapping experiments indicate that the EGFR itself and the associated Shc adaptor protein are physiologic substrates. The 45-kDa TCPTP variant inhibited the EGF-induced association of the EGFR with Shc and Grb2, but had no effect on MAP kinase activation. This suggests that this form of TCPTP could act to inhibit or suppress a discrete subset of pathways downstream of the receptor. A distinct 48-kDa variant of the TCPTP resides in the endoplasmic reticulum and may also act on the EGFR, but the physiologic role of this enzyme is not understood.

6.3.4. LMW-PTPs Regulate Receptor Tyrosine Kinases

The LMWPTP is involved in regulating signal transduction through the platelet-derived growth factor (PDGF) RTK. Interestingly, LMWPTP in unstimulated NIH3T3 cells is found in the cytoplasm and in association with the cytoskeleton. On PDGF stimulation, soluble LMWPTP transiently associates with the PDGF receptor. Studies suggest that once bound LMWPTP dephosphorylates the receptor and represses downstream signaling by Src kinase and STAT (signal transducer and activator of transcription) pathways, but not by phospholipase Cγ, PI3 kinase, and MAP kinase. In contrast, the cytoskeleton-associated fraction appears to be transiently phosphorylated and activated by Src kinase, which leads to dephosphorylation of cytoskeletal proteins. LMWPTP in the two subcellular locations responds to PDGF with different kinetics. The cytoskeleton-associated fraction exhibits a maximum response at 40 min, whereas the soluble enzyme responds more rapidly with a maximum response at 5 min. Potential effects of LMWPTP dephosphorylation on the cytoskeleton are not known. It will be interesting to learn whether there is any overlap between LMWPTP and the classic PTPs in modulating mitogenic signaling.

6.4. Protein Tyrosine Phosphatases and Cell Adhesion

Cell–cell interactions and contacts that cells make with the surrounding extracellular matrix (ECM) affect cell shape and motility and induce growth and differentiation. In development, these cell contacts are particularly crucial because they are needed to organize cells into tissues and organs. Cell–cell communication also provides cues to neurons as they make connections to one another and innervate muscle or other tissues. Alterations or defects in normal cell adhesion are thought to underlie the events that allow malignant tumor cells to become invasive and metastasize.

6.4.1. Role of Receptor PTPs in Cell Adhesion

Many different cell surface proteins participate in cell–cell and cell–ECM adhesion. The neural cell adhesion molecule (NCAM) represents a group of membrane glycoproteins from the immunoglobulin (IgG) superfamily that mediate cell adhesion by both homophilic and heterophilic mechanisms and have a crucial role in the development of the central nervous system. The NCAM extracellular domain contains five tandem IgG-like domains and two tandem fibronectin type III (FNIII) repeats. The striking similarity between the extracellular segments of LAR, one of the first RPTPs identified, and NCAM provided the initial indication that RPTPs might function in cell adhesion (*see* Fig. 11). Indeed, the majority of RPTPs that have been identified to date possess extracellular segments that resemble cell adhesion molecules. As

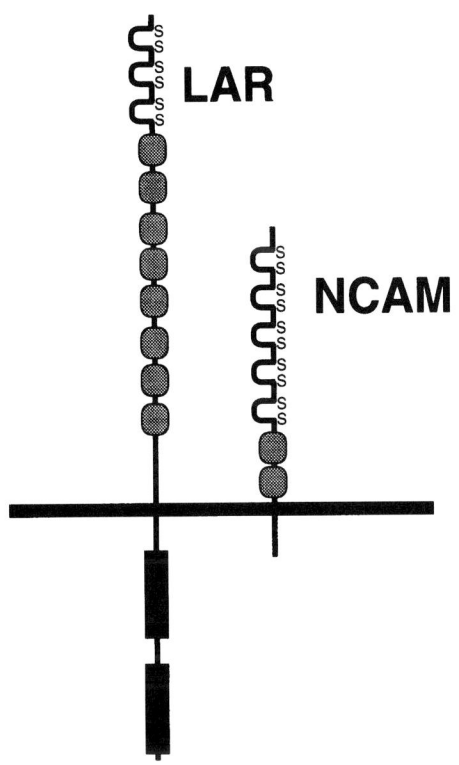

Fig. 11. Similarity between the extracellular domains of receptor PTPs and cell adhesion molecules. The structural organization of the LAR RPTP is shown along with that for the 140-kDa form of the neural cell adhesion molecule (NCAM). The domains of these proteins are labeled as in Fig. 1.

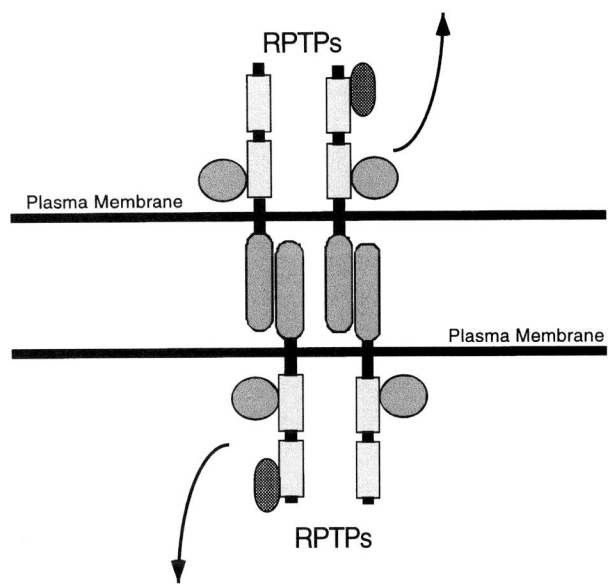

Fig. 12. Receptor PTPs can mediate cell adhesion and have the potential to initiate signaling pathways by dephosphorylating proteins residing at the plasma membrane. Homophilic binding between RPTPs on apposing cells is shown schematically. This mode of binding is observed between PTPκ and PTPμ. The interaction between the cytoplasmic tails of the RPTPs and intracellular signaling proteins, some of which may be substrates or regulators, is depicted. The *arrows* indicate hypothetical signaling pathways from the membrane to the interior of the cell.

outlined later, a growing body of evidence using both genetic and biochemical approaches supports a role for RPTPs in cell adhesion and in development. The presence of tandem phosphatase domains in the cytoplasmic domain of RPTPs distinguishes them from most other cell adhesion molecules, which have no obvious intracellular binding motifs or enzymatic activities that could directly initiate signaling in response to ligand binding.

The first evidence supporting a role for RPTPs in cell adhesion was obtained in experiments showing that PTPμ and PTPκ could induce cell–cell aggregation through a homophilic mechanism in which an RPTP on one cell binds the same protein on an adjacent cell as shown in Fig. 12. Both of these RPTPs possess similar external segments having an N-terminal MAM (meprin/A5/μ) domain, an IgG-like domain and three FNIII repeats (*see* Fig. 1). Despite the similarity in their extracellular domains, PTPμ cannot bind to PTPκ. Interestingly, both PTPμ and PTPκ interact with cell adhesion molecules known as cadherins. Using a Ca^{2+}-dependent, homophilic mode of binding, cadherins support cell–cell adhesion and are constit-

uents of adherens junctions. PTPμ and PTPκ might regulate cadherins and adhesion junctions by controlling their state of phosphorylation.

Genetic studies have revealed that three RPTPs from *Drosophila* (DPTP99A, DPTP69D, and DLAR) are required for certain motor neurons to connect with their appropriate muscles and to follow appropriate paths to their target sites in the developing nervous system of embryos. These studies establish a function for these three RTPs in axon guidance and pathfinding and suggest that they might act as cell adhesion molecules that recognize ligands on other cells or in the ECM. RPTPβ (also termed PTPζ) is expressed on glial cells and astrocytes within the nervous system of mammals and may have a role in directing neuronal growth but through a distinct mechanism. The external domain of RPTPβ/ζ is unusual in that it contains an N-terminal carbonic anhydrase-like domain (CAH) that has no enzymatic activity, a single FNIII repeat followed by a "spacer" segment having no obvious protein motifs. The CAH domain of RPTPβ/ζ binds contactin, a 140-kDa glycosylphosphatidylinositol (GPI)-anchored cell surface protein expressed by neurons. The interaction between the extracellular domain of RPTPβ/ζ expressed on a glial cell and contactin associated with Nr-CAM on an adjacent neuron leads to differentiation of the neuron and outgrowth of neurites. The effect, if any, of contactin binding on RPTPβ activity or the glial cell is not known.

The extracellular domain of a specific LAR splice variant binds the laminin/nidogen complex, a major component of the ECM. Through its interaction with cells, the laminin–nidogen complex has been implicated in controlling their growth, motility, and differentiation. LAR is localized to focal adhesions, specialized sites where cells in culture anchor their cytoskeleton to the ECM. LAR has also been found at other sites of cell–cell contact. LAR-deficient mice have defects in mammary gland development. The cytoplasmic D2 domain of LAR interacts with a multifunctional signaling protein termed TRIO. Two guanine nucleotide exchange domains within TRIO act on the Rho and Rac GTPases to influence the actin cytoskeleton and modulate cell motility and growth. A family of related proteins, termed α-liprins, also interacts with the D2 domains of LAR. The α-liprins and the associated β-liprins might function to localize or direct LAR to specific regions on the cell surface. These properties of LAR indicate that it is involved in certain developmental pathways and may be involved in regulating or generating signals in response to the interaction of cells with the ECM.

Although these exciting findings clearly point to crucial roles for RPTPs in cell adhesion and development, many questions remained unanswered about their role in these processes. Do ligands on adjoining cells or within the matrix affect PTPase activity and if so, how? Is phosphatase activity required for the developmental processes involving RPTPs? For example, it is possible that only the extracellular domains of DLAR and the other Drosophila PTPs would be sufficient to drive motor axon guidance. What signaling pathways are regulated by RPTPs? Genetic studies of the CLR-1 RPTP of C. elegans have provided answers to the latter two questions.

Like other RPTPs, the external segment of CLR-1 resembles cell adhesion molecules with IgG-like and FNIII repeats. Animals homozygous for clr-1 mutant alleles have a characteristic translucence due to the accumulation of fluids and are infertile and immobile because of developmental defects. Genetic analysis demonstrates that the phosphatase activity of the D1 but not the D2 domain of CLR-1 is required for negative regulation of signaling through EGL-15, the C. elegans homolog of the fibroblast growth factor receptor kinase. EGL-15 is required for proper migration and guidance of sex myoblasts during development. The precise substrate of CLR-1 is unknown, but EGL-15 itself or phosphorylated forms of adaptor proteins and other signaling molecules that are downstream of the receptor are good candidates. The identity of the ligand(s) for CLR-1 is also unknown. With the exception of CD45 and RPTPα, which act on Src family kinases, this important work is the first to define how an RPTP might carry out its function in cell adhesion. This promising system, which is amenable to further genetic studies, should provide new insight into the roles of RPTPs in development and cell adhesion. It will be important to determine whether modulation of development through negative regulation of RTK signaling will be a general paradigm for other RPTPs.

6.4.2. Role of CD45 in Cell Adhesion

A new direction in CD45 research is the analysis of its role in regulating cell adhesion. Cell–cell interactions are an important aspect of lymphocyte activation and the adhesion of leukocytes to other cells, to one another, and/or to the extracellular matrix is crucial for their immune function. At sites of infection or inflammation, leukocytes use specific adhesive proteins to bind endothelial cells and migrate to underlying tissues where they mediate the immune response.

In addition to its crucial role in lymphocyte activation, new studies have shown that CD45 is involved in modulating the adhesive properties of hematopoietic cells. In macrophages, CD45 is localized to sites of focal adhesion along with Lyn kinase and β_2 integrin. It has been suggested that CD45 might associate with the cytoskeleton at these sites. Macrophages from $CD45^-$ mice initially adhered more readily than normal cells; however, they were unable to sustain β_2 integrin-mediated adhesion to tissue culture dishes. At sites of adhesion, CD45 negatively regulates Src family kinases by dephosphorylating an activating Tyr residue located in the activation loop of their catalytic domains. CD45 also limits or negatively regulates $\alpha_5\beta_1$ integrin-mediated T-cell adhesion to fibronectin. Both the transmembrane and cytosolic PTP domains of CD45, but not its extracellular segment, were required for its role in regulating adhesion to fibronectin. Only the cytoplasmic PTP domains are needed for lymphocyte activation. The requirement for the transmembrane segment suggests that CD45-AP, which requires this region for interaction with CD45, might mediate the action of CD45 in controlling adhesion. Analysis of the role of CD45 in regulat-

6.4.3. ROLE OF NONRECEPTOR PTPS AND LMW-PTPS IN CELL ADHESION

Regulation of cell adhesion is not the sole province of RPTPs. New findings have implicated nonreceptor PTPs in regulating integrin signaling and modulating cell motility. Integrins are a class of heterodimeric transmembrane receptors that bind fibronectin, laminin, collagen, and other ECM proteins and are a major component of focal adhesions. The cytoplasmic domains of integrins are linked to actin and thereby facilitate bidirectional communication between the cytoskeleton and the ECM. Just as cells, primarily via tyrosine phosphorylation, can signal to integrins to regulate interactions with the ECM, it is now clear that in response to the ECM, integrins initiate intracellular signaling pathways that influence cell shape, motility, and differentiation.

Several studies suggest that SHP-2 is localized to focal adhesions. By dephosphorylating focal adhesion kinase, paxillin, and other components at focal contacts, SHP-2 is thought to modulate integrin signaling, thus influencing cell motility and other cellular processes. Most data suggest a positive role for SHP-2 in propagating signals arising from the ECM via integrins. New evidence suggest that the nonreceptor enzyme, PTP-PEST, is also involved in regulating cell motility in fibroblasts. As noted earlier, PTP-PEST exhibits a high degree of substrate specificity for p130^{Cas}, an adaptor protein implicated in controlling the actin cytoskeleton. Dephosphorylation of p130^{Cas} is required for integrin-mediated cell migration and by controlling tyrosine phosphorylation of p130^{Cas}, PTP-PEST influences fibroblast motility.

LMW-PTPs are involved in regulating cell adhesion through their ability to modulate signaling events downstream of EphB1 RTKs. The Eph family of RTKs direct cell–cell interaction and cell migration and are involved in controlling developmental processes such as axon pathfinding and angiogenesis. Eph RTKs engage ephrins, their membrane-bound ligands, through cell–cell contacts. On cell–cell contact, multimeric ephrin-B1 activates EphB1 RTKs and triggers the recruitment of LMW-PTP to a specific phosphotyrosine residue in its cytoplasmic domain. Although its targets are not known, the data indicate that recruitment of LMW-PTP to the EphB1 RTK is required to mediate downstream adhesion events such as increased attachment of cells to fibronectin. It will be important to determine how LMW-PTP binds the RTK and to identify the substrates of this phosphatase that mediate its effects on cell adhesion.

6.5. Role of Dual Specificity Phosphatases in Cell Cycle Progression

Two DSPs are directly involved in regulating cell division. The Cdc25 phosphatase activates a protein kinase that triggers mitosis, whereas the Cdc14 phosphatase is required for terminating mitosis. Given their critical roles in governing the timing of mitosis, it is not surprising that the activity of both phosphatases is tightly controlled.

As all eukaryotic cells divide, they proceed through an ordered and precisely timed sequence of events known as the cell cycle. Two crucial phases of the cell cycle are interphase, in which chromosomes are duplicated, and mitosis or M phase, in which the chromosomes are condensed, attached to microtubule-containing spindles, and segregated to opposite poles. At the end of mitosis, the cell undergoes cytokinesis, a mechanical process by which the cell is cleaved into two daughter cells. Progression through the cell cycle involves the coordination of many complex processes, some occurring at distinct sites within the cell. Failure to properly coordinate this complex series of events, execute them in the proper sequence, and faithfully duplicate chromosomes is catastrophic. Thus, an intrinsic regulatory mechanism or "clock" exists for ensuring the order and timing of cell cycle events.

Genetic analyses of temperature-sensitive cell division cycle (*CDC*) mutants of budding and fission yeast together with biochemical studies in *Xenopus* oocytes were instrumental in deciphering the mechanisms of cell cycle control, the essential features of which are universal among eukaryotes. These studies have shown that reversible protein phosphorylation is a prominent mechanism for regulating cell cycle progression. The precisely timed activation of cyclin-dependent Ser/Thr protein kinases (Cdks) triggers key cell cycle transitions such as those that occur at the onset of mitosis and initiation of DNA replication. Cdks are inactive in the absence of a regulatory cyclin subunit and inhibited when specific sites are phosphorylated. Thus, Cdk activity is modulated throughout the cell cycle by periodic oscillations in cyclin levels as well as changes in their state of phosphorylation. Cdk activity is also controlled by protein inhibitors that are synthesized and degraded at precise points in the cell cycle. Higher eukaryotes have multiple Cdks; the Cdk involved in starting mitosis is Cdc2 or Cdk1. Yeast have one major Cdk for cell cycle

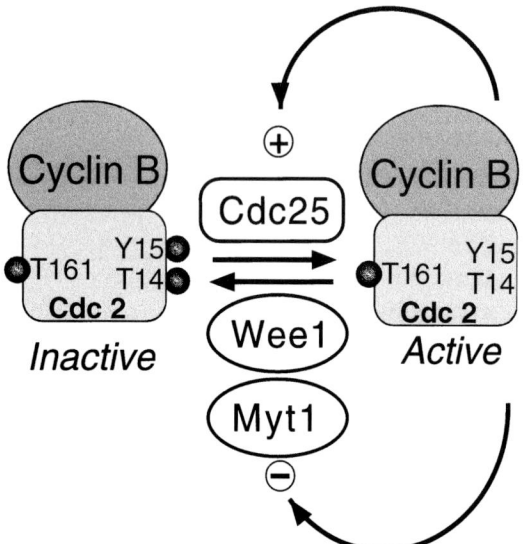

Fig. 13. Cdc25 controls entry into mitosis by dephosphorylating and activating the Cdc2 cyclin-dependent kinase. Cdc25 activates Cdc2 by dephosphorylating Thr^{14} and Tyr^{15} and opposes the Wee1 and Myt1 kinases. Cdc2 is inactive when these two sites are phosphorylated thus preventing premature entry into mitosis. A small amount of Cdc2 activity accelerates the activation in a positive feedback mechanism because phosphorylation of Cdc25 by Cdc2 activates the phosphatase, whereas phosphorylation of Wee1 inhibits its activity.

regulation termed Cdc28 and Cdc2 in budding and fission yeast, respectively.

6.5.1. CDC25 CONTROLS ENTRY INTO MITOSIS

An active heterodimer consisting of Cdc2 and a B-type cyclin initiates mitosis. To achieve full activity, Cdc2 not only needs to interact with the cyclin B subunit, but must also be phosphorylated on Thr^{161}, a conserved residue in its activation loop. When the complex with B cyclin is formed in interphase, Cdc2 is readily phosphorylated on Thr^{161} by CAK (Cdk activating kinase), Tyr^{15} by the Wee1 kinase, and Thr^{14} by Myt1. Phosphorylation of Tyr^{15} and Thr^{14} inhibits Cdc2 preventing its premature activation during interphase. Cdc25 dephosphorylates Tyr^{15} and Thr^{14} and opposes the action of the Wee1 and Myt1 kinases (*see* Fig. 13). Mitosis is initiated when Cdc2 kinase activity increases in response to the concurrent activation of Cdc25 and suppression of Wee1. Because Cdc25 is activated and Wee1 inhibited when they are phosphorylated by Cdc2/cyclin B, a positive feedback mechanism contributes to the activation of the phosphatase (*see* Fig. 13). Exactly how this feedback loop is initiated is not known, but it may involve other kinases that phosphorylate and activate Cdc25 such as the Polo kinases. In mammals there are three forms of Cdc25. It is thought that Cdc25C and Cdc25B are involved in entry into mitosis, whereas Cdc25A is involved in regulating the Cdk2, which starts DNA replication.

Cdc25 is also involved in blocking mitosis when DNA is damaged or is not fully replicated. In response to DNA damage, the Chk1 checkpoint protein kinase phosphorylates human Cdc25C on Ser^{216}, creating a binding site for 14-3-3 proteins thus preventing activation of Cdc2 and delaying the onset of mitosis. The 14-3-3 proteins bind to phosphoserine residues, but the mechanism by which they prevent the activation of Cdc2 by Cdc25C is not fully understood. In fission yeast, the binding of the 1433 protein (Rad24), which contains a nuclear export signal, drives the export of Cdc25 from the nucleus thus preventing Cdc2 activation. In *Xenopus*, binding of the 14-3-3 proteins also controls the intracellular localization of Cdc25 by blocking nuclear import of Cdc25 and permitting the net nuclear export of Cdc25, which contains a nuclear export sequence in its N-terminus. In contrast, a recent report suggests that phosphorylation of human Cdc25 by Chk1 may have a direct effect on its activity.

6.5.2. CDC14 IS REQUIRED FOR EXIT FROM MITOSIS

In the budding yeast, the Cdc14 phosphatase terminates mitosis so that a new round of cell division can begin. This transition, known as exit from mitosis, requires inactivation of Cdc28, the major Cdk in this organism. Inactivation of Cdc28 occurs by two mechanisms, the ubiquitination and subsequent proteolytic degradation of B-type cyclins and the synthesis of the Cdc28 inhibitor Sic1. The role of Cdc14 in these two processes is just beginning to be understood. Cdc14 dephosphorylates Hct1, a protein that promotes the ubiquitination of B-cyclins when dephosphorylated. There is evidence that Cdc14 also promotes the synthesis of the Sic1 inhibitor by targeting Swi5, a transcription factor for the *SIC1* gene that must be dephosphorylated in order to enter the nucleus. Cdc14 may also target Sic1 itself, which is protected from premature degradation by dephosphorylation. Little is known about Cdc14 in mammals or humans, but there are at least two Cdc14 forms, each of which have multiple splice variants.

The activity of Cdc14 is tightly regulated during the cell cycle. Throughout interphase and until the end of mitosis (anaphase), Cdc14 is sequestered to the nucleolus and maintained in an inactive form by the multifunctional Net1 anchoring protein. Net1 not only sequesters Cdc14 and other proteins to the nucleolus, but is also a potent inhibitor of Cdc14 activity. In many aspects, this mode of regulation resembles that employed by the protein inhibitors of the Ser/Thr phosphatases.

7. POTENTIAL ROLE OF PROTEIN TYROSINE PHOSPHATASES IN HEALTH AND DISEASE

The protein tyrosine phosphatase field is relatively new; about one decade has passed since the isolation and characterization of the first of these enzymes. Many PTPs have not been studied in detail and there is much to be learned about their function and regulation. Yet, what has been accomplished so far links these enzymes to many different aspects of cell regulation and gives every indication that they will be meaningful to human health and disease.

Many investigators in this field have viewed protein tyrosine phosphatases as good candidates for tumor suppressors because of their ability to oppose oncogenic protein tyrosine kinases. The only one to turn up so far is PTEN, which targets lipids instead of proteins. However, it is too early to exclude the possibility of discovering other CX_5R phosphatases that are tumor suppressors. The extensive involvement of these phosphatases in mitogenic signal transduction attests to the likelihood of their involvement in tumorigenesis. Many of the mutations leading to cancer occur with high frequency in proteins with cell cycle regulatory functions, and accordingly there is a great deal of interest in targeting these and related proteins in the development of potential anticancer drugs. Hence, further analyses of human Cdc25 and other tyrosine phosphatases involved in regulating cell division will be an important area of research with relevance to cancer. Cell adhesion and cell motility, phenomena that clearly involve both receptor and nonreceptor tyrosine phosphatases, are likely to be significantly impaired in cells that are metastasizing. Determining how tyrosine phosphatases are involved in these processes should spawn new ideas about how invasive tumor cells might be impeded.

As we have discussed, CD45 and SHP-1 execute crucial roles in lymphocyte development, lymphocyte activation, and leukocyte adhesion. The phenotypes of mice that fail to express functional forms of these enzymes dramatically illustrate the importance of these PTPs for normal immune function. Hence, these and other protein tyrosine phosphatases are proving to be important for investigating the molecular basis of immunodeficiencies, leukemias, and autoimmune and chronic inflammatory diseases. The potential of PTPs in managing or controlling the immune response is highlighted by studies showing that antibodies to a specific isoform of CD45 are able to prevent and reverse renal allograft rejection in mice.

Exciting findings directly demonstrate that PTP1B is a potential therapeutic target in the treatment of type 2 diabetes and suppressing obesity. PTP1B knockout mice exhibit increased sensitivity to insulin treatment and resistance to weight gain when given high-fat diets. Perhaps most importantly, these mice appeared to be normal and healthy, indicating that specific PTP1B inhibitors might be free of side effects. The protein tyrosine phosphatases may have potential connections to cardiovascular diseases, although this possibility has received little notice thus far. Some of the early events in atherogenesis are postulated to involve an inflammatory response requiring cell adhesion, migration, and proliferation. The available evidence suggests a role for protein tyrosine phosphatases in all of these processes. It is reasonable to surmise that we are just seeing the tip of the iceberg when it comes to applications of protein tyrosine phosphatases in medicine. The future holds much promise for both basic and clinical scientists with interests these enzymes.

SELECTED READINGS

Barford D, Das AK, Egloff M-P. The structure and mechanism of protein tyrosine phosphatases: insights into catalysis and regulation. Annu Rev Biomol Struct 1998; 27:133.

Denu JM, Stuckey JA, Saper MA, Dixon JE. Form and function in protein dephosphorylation. Cell 1996; 87:361.

Flint AJ, Tiganis T, Barford D, Tonks NK. Development of "substrate-trapping" mutants to identify physiologic substrates of protein tyrosine phosphatases. Proc Natl Acad Sci USA 1997; 94:1680.

Keyse SM. Protein phosphatases and the regulation of MAP kinase activity. Semin Cell Dev Biol 1998; 9:143.

Neel BG. Role of phosphatases in lymphocyte activation. Curr Opin Immunol 1997; 9:405.

Neel BG, Tonks NK. Protein tyrosine phosphatases in signal transduction. Curr Opin Cell Biol 1997; 9:193.

Peles E, Schlessinger J, Grumet M. Multi-ligand interactions with receptor-like protein tyrosine phosphatase β: implications for intercellular signaling. Trends Biochem Sci 1998; 23:121.

Plas DR, Thomas ML. Negative regulation of antigen receptor signaling in lymphocytes. J Mol Med 1998; 76:589.

Tonks NK, Neel BG. From form to function: signaling by protein tyrosine phosphatases. Cell 1996; 87:365.

Unkeless JC, Jin J. Inhibitory receptors, ITIM sequences and phosphatases. Curr Opin Immunol 1997; 9:338.

Van Vactor D. Protein tyrosine phosphatases in the developing nervous system. Curr Opin Cell Biol 1998; 10:174.

Van Vactor D, O'Reilly AM, Neel BG. Genetic analysis of protein tyrosine phosphatases. Curr Opin Genet Dev 1998; 8:112.

PART IV SIGNALING MECHANISMS INITIATED BY NUCLEAR RECEPTORS

20 The Mechanism of Action of Steroid Hormone Receptors

Donald P. McDonnell

Contents

Introduction
The Steroid Hormone Receptors
The Classical Model of Steroid Hormone Action
Steroid Receptor Isoforms and Subtypes
The Role of Ligand in Steroid Receptor Mediated Signal Transduction
Receptor-Associated Proteins and Steroid Receptor Pharmacology
An Updated Model of Steroid Hormone Action
Selected Readings

1. INTRODUCTION

Estrogens, progestins, androgens, glucocorticoids, and mineralocorticoids are a group of structurally similar lipophilic steroids responsible for the regulation of a wide variety of cellular processes in different target organs. The glucocorticoids and mineralocorticoids, for instance, are primarily involved in the regulation of cellular homeostasis and metabolism whereas estrogens, progestins, and androgens, the sex steroids, are responsible for the development and maintenance of reproductive function. Despite their different roles, it has become apparent that the steroid hormones are mechanistically similar and that insights gleaned from the study of each hormone has advanced our understanding of the class of molecules as a whole. The objective of this chapter is to provide an updated model of steroid hormone action and to discuss the impact of recent insights on our understanding of the pharmacology of steroid hormones.

From: *Principles of Molecular Regulation* (P. M. Conn and A. R. Means, eds.), © Humana Press Inc., Totowa, NJ.

2. THE STEROID HORMONE RECEPTORS

The cDNAs for each of the steroid receptors (SRs) have been cloned and used to develop specific ligand responsive transcription systems in heterologous cells. This has permitted the application of reverse genetic approaches to define the functional domains within each of the receptors. A schematic that outlines the organization of the major functional domains within the steroid receptors is shown in Fig. 1.

The ligand binding domain (LBD), encompassing approx 300 amino acids, is located at the C-terminus of each receptor. Crystallographic analysis of the agonist bound forms of the progesterone and estrogen receptors have indicated that this domain consists of 12 short α-helical structures that fold to provide a complex ligand binding pocket. The LBD also contains sequences that facilitate receptor homodimerization, and that permit the interaction of apo-receptors with inhibitory heat-shock proteins. An activation function (AF-2) required for receptor transcriptional activity is also contained within the LBD.

The DNA binding domain (DBD) is a short, 66–70 amino acid region located in the center of the receptor. This module binds as a dimer to target genes. Within

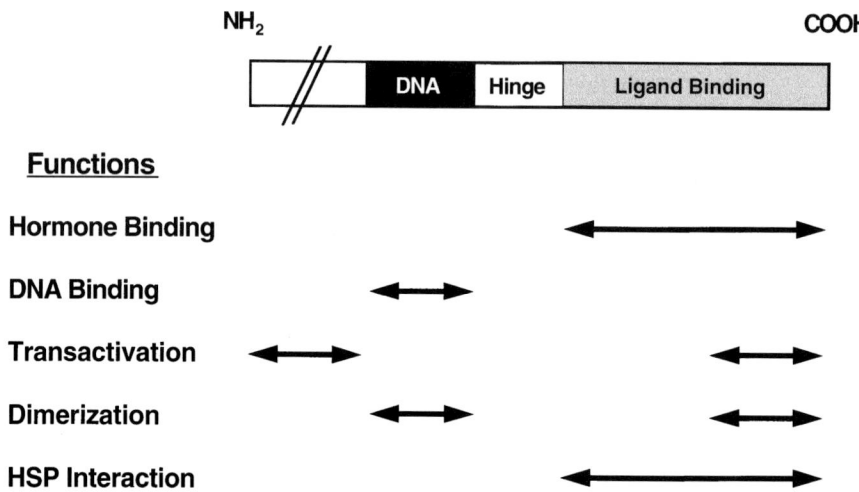

Fig. 1. The domain structure of the steroid hormone receptors is similar.

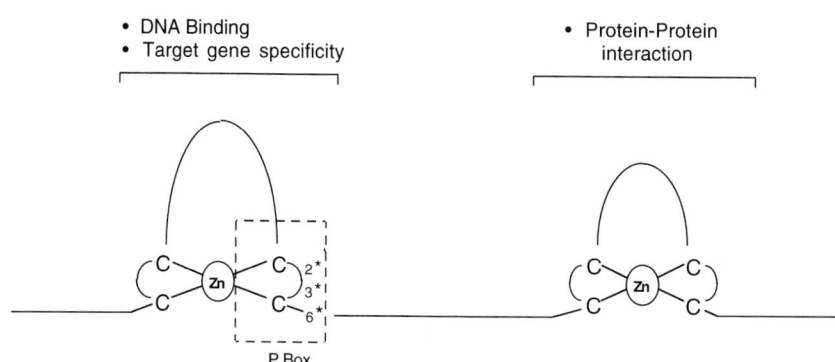

Fig. 2. The DNA binding domain of the steroid hormone receptors. Schematic of the DNA binding domain of the steroid hormone receptors. The DNA binding domain contains two distinct finger structures that permit the interaction of the steroid hormone receptors with their cognate DNA response elements. The ability of individual receptors to discriminate between similar SR DNA response elements is determined by the P-box domain within the first finger. Those receptors that have a similar P-box (GR, PR, MR, and AR) recognize a similar DNA response element.

the DBD there are nine cysteine residues, eight of which chelate two zinc atoms, forming two fingerlike structures that allow the receptor to interact with DNA (Fig. 2). Structural analysis has revealed that the first finger of each monomer binds directly into the DNA major groove whereas the second finger mediates protein–protein interactions between each monomer.

The modular nature of these fingers was demonstrated by "finger-swapping" experiments that showed that the N-terminal zinc finger specifies DNA target site selectivity. Further dissection of this first finger led to the identification of three amino acids within a short "P-box" domain that were shown to be critical for target site selectivity. Interestingly, several receptors, the glucocorticoid (GR), progesterone (PR), mineralocorticoid (MR), and androgen (AR) all have the same P-box sequence and interact with the same DNA response element. The estrogen receptor (ER), which has a unique P-box sequence, recognizes a different response element. These rules have been shown to hold up very well in vitro. However, because GR, MR, PR and AR do not regulate the same genes it is implied that target site selection in vivo is more complex.

The DNA and hormone binding domains are separated from each other by a hinge region of variable length. A specific role for this domain has not yet been defined, although recent evidence suggests that it may permit the interaction of the steroid receptors with corepressor proteins (*see* below) and may contain a minor transcriptional activation domain.

The entire N-terminus of each steroid receptor also appears to constitute a functional domain. This hinge region, also called the "hypervariable domain," is the most divergent among the SRs and has been the most difficult to dissect. It has been shown that this domain contains one of the two major transcriptional activation domains contained within each receptor. The N-terminal activation domain (AF-1) is complex and as yet the mechanism by which it contributes to the overall transcriptional activity of SRs is not known. It has been shown recently, however, that the ER, AR, and PR N-termini can interact directly with their respective C-termini and regulate the interaction of these receptors with the transcription apparatus. This provides a potential molecular mechanism with which to explain the observed synergistic actions of the AF-1 and AF-2 domains within these receptors. A complete appreciation of the role of the N-terminus in SR action will require crystallization of a full-length form of at least one of the receptors.

3. THE CLASSICAL MODEL OF STEROID HORMONE ACTION

It is generally believed that steroid hormones enter cells from the bloodstream by simple passive diffusion. Thus, although they have the ability to enter all cells they exert activity in cells only where they encounter a specific receptor protein. The inactive GR resides in the cytoplasm whereas ER and PR are predominantly located in the nucleus. Consensus as to the precise localization of the inactive AR and MR has not yet emerged. In the absence of ligand, these receptor proteins are maintained in an inactive form by their association with a large heat-shock protein complex. On binding ligand however, the receptors

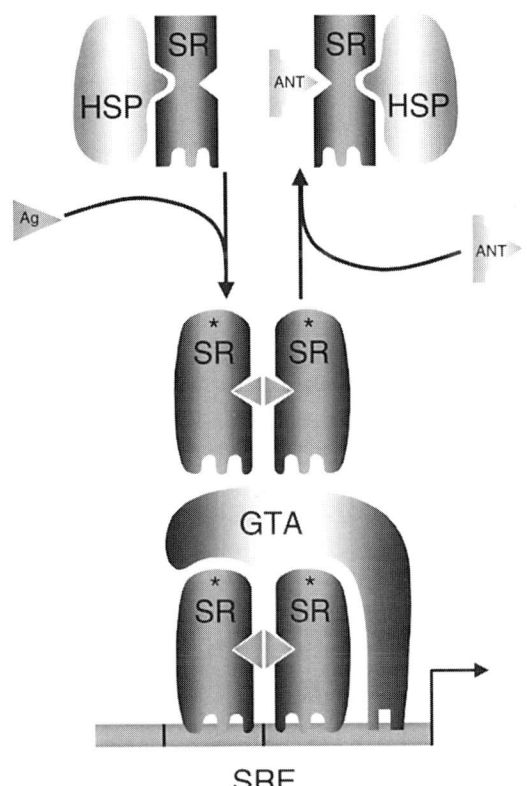

Fig. 3. The established models of steroid hormone action. The classical models of steroid hormone action suggested that in the absence of ligand, the steroid receptor (SR) existed in the nuclei of target cells in an inactive form. On binding an agonist, the SR underwent an activating transformation event which displaced inhibitory heat shock proteins and facilitated the interaction of the receptor with specific DNA response elements within target gene promoters. The activated receptor dimer could then interact with the general transcription machinery and positively or negatively regulate target gene transcription. In this model the role of the agonist was that of a "switch" that merely converted SR from an inactive to an active form. Thus, when corrected for affinity, all agonists were qualitatively the same and would evoke the same phenotypic response. By inference, SR antagonists, compounds that oppose the actions of agonists, were considered to competitively bind to SRs and freeze the receptor in an inactive form. As with agonists, this model predicted that all antagonists were qualitatively the same.

undergo an activating conformational change that promotes the dissociation of inhibitory proteins. This event permits the formation of receptor homodimers that are capable of interacting with specific high-affinity DNA response elements located within the regulatory regions of target genes (Fig. 3). The DNA-bound receptor can then exert a positive or negative effect on target gene transcription.

In the classical models of SR action, it is proposed

that ligands function merely as switches that bind to their cognate receptor and permit its conversion from an inactive to an active state in an all or nothing manner. This implies that steroid hormone receptor pharmacology is very simple and that for a given receptor, when corrected for affinity, all agonists are qualitatively the same. Furthermore, it suggests that antihormones function simply as competitive inhibitors of agonist binding, freezing the target receptor in an inactive state within the cell. Under most experimental conditions, this simple model is sufficient to explain the observed biology of the known SR agonists and antagonists. However, analysis of the systems where this model does not hold indicated that SR pharmacology is more complex than originally anticipated. In this regard, the studies which have probed the complex pharmacology of the antiestrogen tamoxifen have been very informative. This compound is widely used as a breast cancer therapeutic and has recently been approved for use as a breast cancer chemopreventative in high-risk patients. In ER-positive breast cancers, tamoxifen opposes the mitogenic action of estrogen(s) by binding to the receptor and competitively blocking agonist access. However, it has become clear in recent years that tamoxifen is not a pure antagonist and that it can exhibit estrogen-like activities in some target organs. This is most apparent in the skeletal system where tamoxifen, like estrogen, increases lumbar spine bone mineral density and in the cardiovascular system, where both compounds have been shown to decrease low-density lipoprotein (LDL)-cholesterol. The ability of tamoxifen to mimic estrogen in some cells and manifest antagonist activities in others have led to it being reclassified as a selective estrogen receptor modulator (SERM) as opposed to antagonist. The observation that tamoxifen was a SERM indicated that the simple models of ER action were incorrect, as it was now apparent that different ER–ligand complexes were not recognized in the same manner in all cells. This was an important clinical finding as it suggested for the first time that it was possible to develop compounds that, acting through their cognate receptor, could manifest different activities in different cells. The unexpected pharmacological activity exhibited by tamoxifen begged a reevaluation of the classical models of ER action and initiated the search for the cellular systems that enabled different ER–ligand complexes to manifest different biologies in different cells. These ongoing investigations have provided significant insight into ER action and have helped to update the established models of SR action.

4. STEROID RECEPTOR ISOFORMS AND SUBTYPES

One mechanism to explain the cell-selective action of SR ligands is that they may activate different receptor isoforms (derived from the same gene) or subtypes (derived from similar genes). This concept has been well established for the α- and β-adrenergic systems, where it has been shown that different receptor subtypes have distinct ligand preferences and that selectivity can be explained by differences in the expression of these subtypes. Until recently, the parallel between this system and that of the nuclear receptors was not obvious. However the identification and characterization of novel steroid receptor isoforms and subtypes has shed new light on this issue (Fig. 4).

4.1. Progesterone Receptor Isoforms

The progesterone receptor was the first for which *bona fide* receptor isoforms were shown to exist. Human PR exists in one of two distinct forms, hPR-A (94 kDa) and hPR-B (114 kDa), within target cells. These proteins are produced from distinct mRNAs that are derived from different promoters within the same gene. Thus, these receptors differ only in that the hPR-B isoform contains an additional 164 amino acid extension at its N-terminus. In most progesterone responsive tissues these two receptor isoforms are expressed in equimolar amounts. This apparent 1:1 relationship is so widespread that until about 10 yr ago it was considered that the hPR-A isoform was merely an artifact and that it was derived from hPR-B by proteolysis during biochemical fractionation. It has now been established that these two proteins are produced in a deliberate manner by the cell and that they are not functionally equivalent. The first evidence in support of this hypothesis came following the cloning and subsequent functional analysis of the chicken progesterone receptor (cPR) cDNA. Specifically, upon expression in heterologous cells, it was found that although the A- and B-forms of cPR displayed identical ligand binding preferences, they activated different target genes. It was subsequently shown that the N-terminal sequences, which distinguished cPR-B from cPR-A, were important in determining target gene selectivity. This concept was reaffirmed when the cloned hPR-B and hPR-A were analyzed in a similar manner. With few exceptions in the systems examined thus far, it has been observed that hPR-B alone functions as a transcriptional activator in response to progesterone whereas hPR-A displays minimal or no activity. Further analysis revealed that

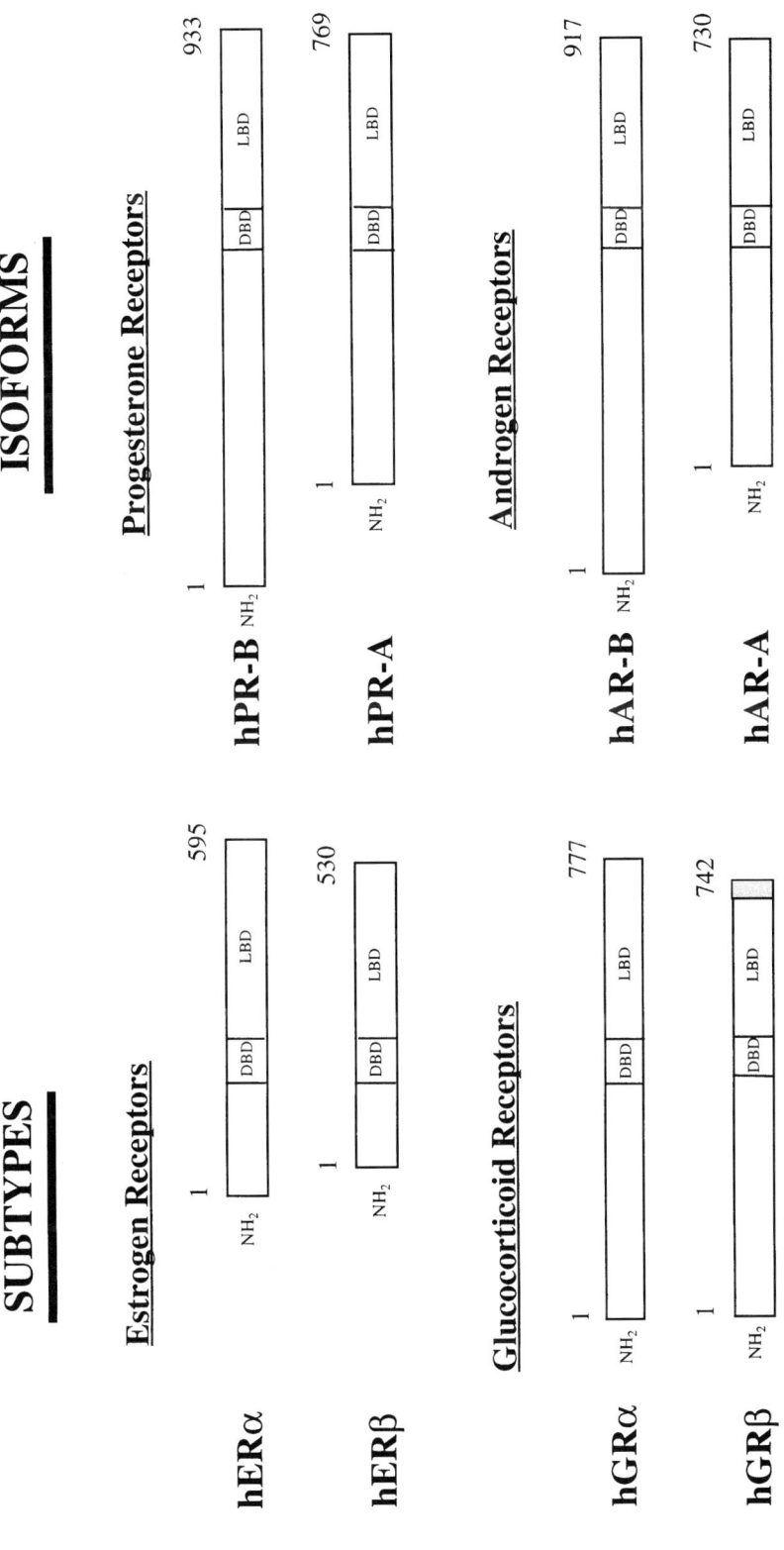

Fig. 4. Steroid hormone receptor isoforms and subtypes.

hPR-A functioned primarily as a ligand-dependent transdominant modulator of the transcriptional activity of hPR-B. Thus, the ability of hPR-B to activate target gene transcription was influenced by the cellular concentration of hPR-A. Surprisingly, it was also determined that ligand activated hPR-A could also inhibit the transcriptional activity of agonist activated ER, AR, and MR. Thus, by virtue of having two functionally different receptor isoforms, a single hormone, such as progesterone, can have completely different functions in target cells.

4.2. Androgen Receptor Isoforms

To date two distinct isoforms of the androgen receptor (AR-A and AR-B) have been identified and shown to be expressed in many androgen responsive cells. These isoforms appear to be produced by alternate initiation of translation from two distinct ATGs within the coding region of a single message giving rise to proteins of 110 kDa and 87 kDa, respectively. The relative expression level of the two isoforms appears to vary considerably from cell to cell. In general, the larger AR-B predominates by fivefold over AR-A. In contrast to other systems where SR isoforms have been identified, the two AR isoforms appear to be identical in terms of their response to agonists and antagonists. No remarkable consequence of their coexpression in target cells has been observed. It is possible that as additional androgen responsive systems and pathways are characterized specific roles for AR-A and AR-B will be identified.

4.3. Estrogen Receptor Subtypes

The identification of functionally distinct PR and AR isoforms introduced a new dimension to progesterone action though it was not until a second estrogen receptor was cloned in 1995 that the general significance of isoforms (or subtypes) in SR signaling was established. ERα and ERβ are encoded by different genes, and although they share significant amino acid homology in their ligand binding domains they are not pharmacologically equivalent. Both receptors bind the endogenous estrogen 17β-estradiol with equivalent affinity. However, when binding analyses were extended to additional compounds, significant differences in ligand preferences were noted. The biological and pharmacological consequences of these differences remain to be determined.

Although ERβ was discovered relatively recently, significant progress has been made in elucidating its role in estrogen signaling. It has been determined for instance that the expression pattern of ERβ does not exactly mirror that of ERα. It has been found that there are tissues where both receptors are expressed whereas in others such as lung, the urogenital tract, and the colon, ERβ alone has been detected. The recent generation of mice in which ERβ has been genetically disrupted and the demonstration that the phenotype of these animals is dissimilar to that observed in ERα knockout mice reaffirmed the distinct roles of these two receptors in the endocrinology of estrogen.

A specific role for ERβ in tissues where it is the only estrogen receptor expressed has not yet been elucidated. The high amino acid homology between ERα and ERβ within the DNA binding domain suggests but does not prove that these receptors may regulate the same genes. It is also possible that ERβ may interact with target genes in a manner that does not require direct contact with DNA regulatory elements within target genes. It has been shown in transfected cells that both ERα and ERβ can interact with the AP-1 transcription factor and that this interaction permits these receptors to activate transcription through an AP-1 response element. The physiological significance of the ER–AP-1 interaction remains to be determined, however, it has been shown that 17β–estradiol functions as an ERα agonist and an ERβ antagonist when examined in this system. In addition, ICI182,780, which functions as a pure antiestrogen in most systems, is an ERβ agonist on AP-1 responsive promoters. Clearly, ERα and ERβ are functionally distinct, providing target cells with two distinct mechanisms to respond to receptor agonists and antagonists.

4.4. Glucocorticoid Receptor Subtypes

One of the reiterating themes that has evolved in steroid hormone action is that processes or principles that are initially defined in one receptor system are usually found to be important for multiple receptors in this family. Not surprisingly therefore, two glucocorticoid receptor (GR) subtypes isoforms have been identified and characterized. The biological activity of the two forms of GR has been characterized extensively. The predominant form, GRα, has an intact ligand binding domain and in reconstituted glucocorticoid responsive systems in vitro it functions as a ligand-dependent transcription factor. A second isoform, GRβ, arises as a consequence of alternate splicing which leads to the production of a protein with a unique C-terminus. The protein produced from this mRNA does not bind hormone; however, its expression in vivo has been confirmed using antibodies that are directed against the unique C-terminus of this protein. These studies have also indicated that unlike GRα, the GRβ protein is located primarily in the nuclei of target cells. There has been much debate as to the function of the GRβ protein. Several groups have found that when GRβ is overexpressed in target cells it can function as a dominant negative inhibitor of GRα action. The significance of this finding was enhanced upon the identification of specific cells in which GRβ is expressed at levels greater or equivalent to that of GRα. It appears that GRβ is a key regulator of GRα activity and that the cellular response to glucocorticoids and antiglucocorticoids will be determined in part by the relative expression level of these two receptors. Finally, it is worth mentioning that MR and GR could also be considered to be GR-subtypes as these proteins exhibit similar ligand binding specificity and are nearly as closely related to each other as the ERα and ERβ isoforms.

The identification of SR isoforms and the definition of specific functions that they modulate has introduced a new dimension in SR signaling. Using these isoforms, mechanisms have evolved by which the same hormone can have different biological activities in different cells. Understanding the regulatory mechanisms that control the expression levels of the individual forms of each receptor is likely to provide novel targets for pharmaceutical intervention.

5. THE ROLE OF LIGAND IN STEROID RECEPTOR MEDIATED SIGNAL TRANSDUCTION

The finding that SRs exist in multiple forms within target cells suggests that some of the tissue selective actions of SR agonists and antagonists can be explained by their ability to differentially regulate the action of one specific receptor isoform or subtype. Whereas this is likely to be true, a specific example of a receptor subtype-selective steroid receptor ligand has not yet emerged. However, the ability to generate subtype selective ligands for the related retinoic acid receptors and the demonstration that these compounds can function as cell-selective retinoids suggests that receptor subtype selective SR ligands will be identified. Regardless, it has become apparent from the study of antiestrogens that the identical ligand operating through the same receptor can manifest different biological activities in different target cells. In breast, for instance, where ERα predominates, all of the known antiestrogens oppose the mitogenic actions of estrogen. In the endometrium, however, where ERα also predominates, it has been found that tamoxifen functions as a partial estrogen mimetic whereas other compounds, such as raloxifene, GW5638, and ICI182,780, function as pure antiestrogens in vivo. Thus, the same compounds, acting through ERα, manifest different biological activities in the breast and the endometrium. This finding was not in agreement with the classical models of ER action that indicated that ligands basically fell into two classes, agonists and antagonists. This paradox has been the subject of much investigation leading to the observation that different compounds can induce different alterations in ER structure and that not all structures are functionally identical. It is implied, therefore (discussed in more detail later), that the cell possesses the cellular machinery to distinguish between these dissimilar complexes and that the identification and characterization of the specific components of these systems is the key to the development of the next generation of tissue-selective SR modulators.

Much of what we know about the effect of ligands on SR structure is based on studies of different ER–ligand complexes. It was demonstrated, initially, using differential sensitivity to proteases, that the hormone binding domain within ER adopted different shapes upon binding estradiol and tamoxifen, and that these structures were dissimilar to that of the apo-receptor. Thus, receptor conformation was affected by the nature of the bound ligand. This relationship between structure and function was later confirmed by the observation that agonists and antagonists induce different alterations in PR structure. Further analysis revealed that the majority of the structural changes that occurred in PR were at the extreme C-tail of the receptor and that removal of the C-terminal 42 amino acids of hPR-B permitted the antagonist RU486 to function as an agonist. Interestingly, a similarly positioned domain enables ER to discriminate between different compounds and, not surprisingly, removal of 35 amino acids from the C-terminal tail of ER abolishes its ability to distinguish between agonists and antagonists. A similar role for the carboxyl tail in AR pharmacology has also been proposed.

The crystal structures of the ER–estradiol and ER–tamoxifen complexes have recently been determined, confirming the important role of the C-tail in determining the pharmacology of the steroid receptors. This new structural information has revealed that agonist activation of ER permits the formation of a unique pocket within the receptor that allows it to interact with the general transcription machinery through the mediation of receptor associated proteins called adaptors or coactivators (*see* below). In the presence of the SERM tamoxifen, however, the C-tail of ER is positioned in a manner that it occludes this coactivator binding pocket so that a productive association with the cellular transcription apparatus is not possible. In addition to tamoxifen, there are several additional SERMs that manifest distinct activities in vivo. One of these compounds, raloxifene, has recently been approved as a SERM for the treatment of osteoporosis. This compound is different from tamoxifen in that it does not exhibit estrogenic action in the postmenopausal endometrium. Although clearly biologically different, tamoxifen and raloxifene were found to have very similar profiles in the protease digestion assay. In addition, the crystal structures of the ER–tamoxifen and ER–raloxifene complexes were shown to be virtually indistinguishable. Initially, these results appeared to be at odds with the hypothesis that linked receptor structure to function. However, some recent data from our group have reconciled these potential discrepancies.

The protease digestion assay is a crude technique to assess the changes that occur on the surface of a receptor in the presence of a specific ligand. Because of the nature of the assay, however, it is unlikely to be useful in the detection of small but biologically relevant changes in receptor structure. To circumvent this problem we have used phage display technology to identify small peptides whose ability to bind ER was affected differentially by the nature of the ligand bound to the receptor. The rationale behind this approach was that in the vast complexity of the peptides available in these libraries it may be possible to find peptides that have the ability to distinguish between two very similar receptor–ligand complexes. Using this approach, a series of high-affinity peptide probes were identified that, in addition to being able to distinguish between ER–estradiol and ER–tamoxifen complexes, were able to distinguish between several different ER–SERM complexes (Fig. 5). Recently, this approach has been extended to PR and it has been observed in a similar fashion that different PR ligands manifest different biology in different cells and that peptide probes can be identified whose interaction with the receptor is influenced by the nature of the ligand bound to PR. These findings establish a firm relationship between SR structure and function and it is likely that when extended to GR, AR, and MR, a similar relationship between the structure of specific ligand–receptor complex and function will be observed. The next frontier is to identify the components of the cellular systems that transmit the information contained within a specific SR–ligand complex to the transcription apparatus.

6. RECEPTOR-ASSOCIATED PROTEINS AND STEROID RECEPTOR PHARMACOLOGY

The steroid hormone receptors are ligand-dependent transcription factors that on activation by ligands associate with specific DNA response elements located within the regulatory regions of target genes. The DNA-bound receptor can then positively or negatively influence gene transcription by altering RNA polymerase II activity. However, because RNA polymerase does not appear to interact directly with the SRs, it is implied that there must be additional factors that allow these two proteins to communicate. It has recently become clear that there are at least two functional classes of proteins that are involved in recognizing the activated receptor. One class, the general transcription factors, are components of the basic transcription machinery whose expression levels are generally invariant from cell to cell. The second class of proteins, the cofactors, are not components of the general transcription machinery and can exert either a positive or a negative effect on SR transcriptional activity. Those cofactors that interact with agonist activated SRs have been called coactivators whereas those that interact with apo-receptors or antagonist activated receptors have been called corepressors. Importantly, the expression level of the nuclear receptor coactivators and corepressors varies from cell to cell. Not surprisingly, it has become apparent that differences in the relative expression levels of coactivators and corepressors has a profound influence on the pharmacology of SR ligands.

6.1. Steroid Receptor Coactivators

One of the most well characterized coactivator proteins, steroid receptor coactivator 1 (SRC-1), was identified in a yeast two-hybrid screen as a protein that interacted with agonist-activated PR. Subsequently, this protein has been shown to interact with agonist-activated ER, GR, and PR. It appears that SRC-1 increases target gene transcription by linking the hormone-activated SR with the general transcription machinery, stabilizing the transcription preinitiation complex, and nucleating a large complex of proteins that together have the ability to acetylate histones and facilitate chromatin decondensation. In addition to SRC-1, more than 30 additional coactivators have been identified and characterized. Cumulatively, these

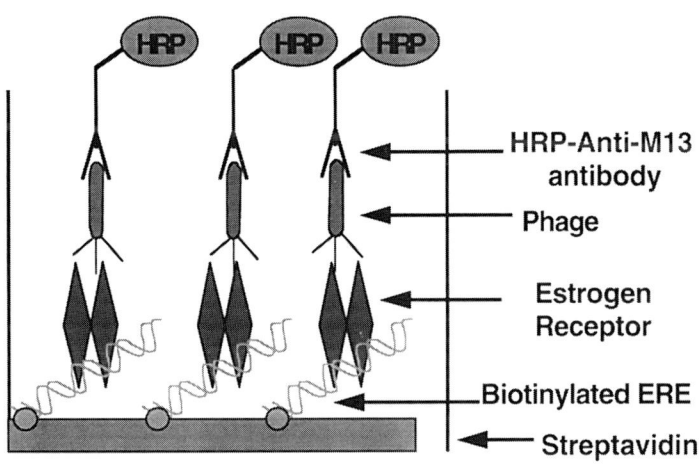

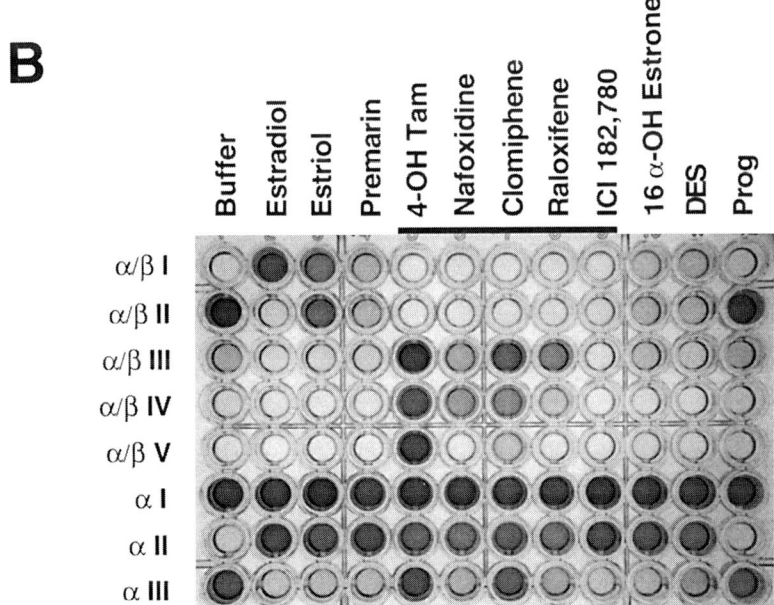

Fig. 5. Fingerprinting the surface of different ER–ligand complexes using conformation-sensitive peptide probes. (A) Random peptide libraries were constructed in an M13 bacteriophage. Each of the resulting bacteriophage expresses a unique random peptide on their surface pilus. Screens were subsequently performed to identify specific peptides whose interaction with ER was influenced by the nature of the bound ligand. The bacteriophage identified in this manner were used to develop an enzyme-linked immunosorbent assay (ELISA) to monitor changes that occur in ER upon its interaction with different ligands. Specifically, a biotinylated ERE was used to immobilize recombinant ER on streptavidin-coated plates. This complex was then incubated with the ligand to be tested. An aliquot of a different class of ER-interacting bacteriophage was added to each well. Binding of the bacteriophage was assessed enzymatically using an anti-M13 antibody coupled to horseradish peroxidase. (B) Fingerprint analysis of ER conformation in the presence of different ER ligands. Immobilized ER was incubated in the presence of saturating concentrations of the indicated ligands and the resulting complexes were incubated with aliquots of bacteriophage expressing eight different peptides. This figure has been published previously in a similar form (Proc Natl Acad Sci USA, 96:3999–4004, 1999) and is reproduced and presented here with permission (Copyright 1999, National Academy of Sciences, Washington, DC).

analyses have revealed that (a) the expression level of these coactivators varies from cell to cell, (b) coactivators demonstrate specific receptor preferences, (c) a given receptor can interact with more than one type of coactivator, and (d) the conformation of the receptor adopted in the presence of a specific ligand can determine which coactivators are engaged. These findings strongly support the hypothesis that differential cofactor expression is the most important determinant of SR pharmacology.

With the discovery of the nuclear receptor coactivators and the characterization of their biochemical properties has come a new understanding of the mechanism by which differently conformed receptor–ligand complexes are recognized in the cell. The studies that have been performed with ER are the most informative. As mentioned previously it has been shown that ER, in the presence of estradiol, undergoes a conformational change that allows the formation of pockets on the receptor that permit it to interact with coactivators. Thus, because estradiol induces an identical conformational change within ER, the phenotypic consequence of a cells exposure to estradiol will depend on the properties of the coactivators expressed in each target cell and their relative expression levels (Fig. 6). The situation becomes more complicated when the role of coactivators in mediating the cell selective action of SERMs, such as tamoxifen, are considered. It has been shown that the tamoxifen-induced conformational changes within ER do not allow the coactivator binding pocket to form properly. These alterations prevent, or hinder, the interaction of those coactivators that require the coactivator binding pocket in order to interact with ER. Thus, in cells where this type of coactivator is important, tamoxifen can function as an antagonist. It is becoming clear, however, that not all coactivators rely on the coactivator binding pocket to the same extent. Thus, the relative agonist/antagonist activity of tamoxifen depends on the ability of the tamoxifen–ER complex to engage a coactivator in target cells. As the repertoire of cofactors increases we are likely to find that targeting specific cofactor–receptor complexes will yield pharmaceuticals that manifest their activities in a cell- or tissue-restricted manner.

6.2. Nuclear Receptor Corepressors

The discovery of SRC-1 and other coactivators that positively affect ER transcriptional activity has helped us to understand how agonist-activated ER communicates with the general transcription machinery. However, as is typical in biology, each positive acting

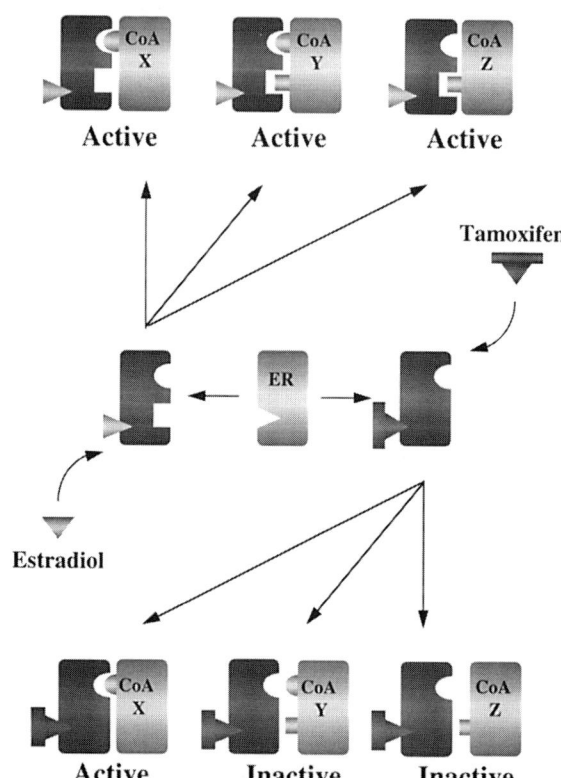

Fig. 6. A molecular explanation for the tissue-selective agonist/antagonist activity of the SERM tamoxifen. The estrogen receptor undergoes different conformational changes on binding the full agonist estrogen or the SERM tamoxifen. The estradiol induced conformation allows ER to interact with any coactivator protein expressed in target cells, and thus it can activate transcription. The tamoxifen-induced conformational change, conversely, is more restrictive and allows the interaction of ER with only a subset of available coactivators. In those cells where the tamoxifen–ER complex can engage a coactivator, this compound can manifest agonist activity. In other contexts, tamoxifen functions as an antagonist.

factor or process is opposed by some system that operates in the opposite manner. Thus, it was not surprising that transcriptional corepressors, cofactors that bind to SRs and reduce their transcriptional activity, were discovered. To date, two nuclear corepressor proteins have been identified that appear to be important in SR signaling. These proteins, NCoR and SMRT, were initially found as proteins which interacted with DNA bound thyroid hormone or retinoid X receptors and permitted these proteins to repress basal transcription in the absence of hormone. However, it has now been shown that these proteins can interact with either PR or ER in the absence of hormone or in the presence of antagonists. Under these conditions, the corepressors nucleate a large protein complex that represses target gene transcription by deacetylating histones and facilitating chromatin condensation. The physiological importance of the corepressors in ER pharmacology was suggested by recent studies that demonstrated that passage of breast tumors in mice from a state of tamoxifen sensitivity to insensitivity was accompanied by a decrease in the expression level of the corepressor NCoR. A similar process, if occurring in humans, could explain how cells become resistant to the antiestrogenic actions of tamoxifen.

7. AN UPDATED MODEL OF STEROID HORMONE ACTION

On ligand binding the SRs can interact as dimers with specific DNA response elements within target genes. It is now apparent that the conformation of the receptor is influenced by the nature of the bound ligand and that the shape of the SR–ligand complex is a critical determinant of whether or not it can activate transcription. In the presence of a full agonist, the conformation adopted by an SR facilitates a displacement of the corepressor proteins and a recruitment of coactivator proteins. This permits the assembly of a histone acetylating complex on DNA with a concomitant increase in target gene transcription. In this model, pure antagonists drive SRs into a conformation that favors corepressor interaction and assembly of a histone deactylating complex, resulting in a decrease in gene transcription. Importantly, the activity of compounds that display mixed agonist/antagonist activities appear to relate to their ability to differentially alter SR conformation and the ability of the corepressor and coactivator proteins within a given target cell to recognize these complexes. Finally, SR signal transduction, complex though it is, does not exist in a vacuum. It has been shown that signaling events initiated at the cell membrane, such as those generating increased levels of cAMP, also impinge on ligand activated SRs. These signaling pathways may function by affecting receptor phosphorylation or they may operate by modifying the interaction of the SRs with nuclear receptor coactivators and corepressors. Clearly, the classical models of SR action need to be updated to accommodate the insights that have emerged from the study of the genetic and molecular pharmacology of the steroid receptor family of receptors.

SELECTED READINGS

Evans RM. The steroid and thyroid hormone receptor superfamily. Science 1988; 340:889.

Horwitz KB, Jackson TA, Bain DL, et al. Nuclear receptor coactivators and corepressors. Mol Endocrinol 1996; 10:1167.

McDonnell DP, Clemm DL, Hermann T, Goldman ME, Pike JW. Analysis of estrogen receptor function in vitro reveals three distinct classes of antiestrogens. Mol Endocrinol 1995; 9:659.

Norris JD, Fan D, Stallcup MR, McDonnell DP. Enhancement of estrogen receptor transcriptionl activity by the coactivator GRIP-1 highlights the role of activation function 2 in determining estrogen receptor pharmacology. J Biol Chem 1998; 273:6679.

Oakley RH, Webster JC, Sar M, Parker CR, Cidlowski JA. Expression and subcellular distribution of the β-isoform of the human glucocorticoid receptor. Endocrinology 1997; 138:5028.

Onate SA, Tsai S, Tsai M-J, O'Malley BW. Sequence and characterization of a coactivator for the steroid hormone receptor superfamily. Science 1995; 270:1354.

Perlmann T, Evans RM. Nuclear receptors in Sicily: all in the famiglia. Cell 1997; 90:391.

Shibata H, Spencer TE, Onate SA, et al. Role of co-activators and co-repressors in the mechanism of steroid/thyroid receptor action. Rec Prog Horm Res 1997; 52:141.

Vegeto E, Allan GF, Schrader WT, et al. The mechanism of RU486 antagonism is dependent on the conformation of the carboxy-terminal tail of the human progesterone receptor. Cell 1992; 69:703.

Wagner BL, Norris JD, Knotts TA, Weigel NL, McDonnell DP. The nuclear corepressors NCoR and SMRT are key regulators of both ligand- and 8-bromo-cyclic AMP-dependent transcriptional activity of the human progesterone receptor. Mol Cell Biol 1998; 18:1369.

Weinberger C, Hollenberg SM, Ong ES, et al. Identification of human glucocorticoid receptor complementary DNA clones by epitope selection. Science 1985; 228:740.

Wen DX, Xu Y-F, Mais DE, Goldman ME, McDonnell DP. The A and B isoforms of the human progesterone receptor operate through distinct signaling pathways within target cells. Mol Cell Biol 1994; 14:8356.

21 Orphan Nuclear Receptors

Deepak S. Lala and Richard A. Heyman

CONTENTS

INTRODUCTION
STRUCTURAL AND FUNCTIONAL DOMAINS OF NUCLEAR RECEPTORS
CLASSIFICATION AND BIOLOGY OF ORPHAN RECEPTORS
IN SEARCH OF A LIGAND
NOVEL HORMONE SIGNALING PATHWAYS
ORPHAN RECEPTORS AS DRUG TARGETS
SUMMARY AND PERSPECTIVES
SELECTED READINGS

1. INTRODUCTION

The nuclear receptor superfamily consists of a class of transcription factors comprising more than 100 different proteins. In contrast to membrane-bound receptors, the nuclear receptors are intracellular and act by controlling the activity of genes directly. Most members of this family bind directly to small lipid-soluble signaling molecules, or ligands, which owing to their lipophilic nature can easily enter the target cell. This superfamily includes known receptors for steroid hormones, such as the glucocorticoid receptor (GR), estrogen receptor (ER), progesterone receptor (PR) (*see* Chapter 20) and nonsteroid hormone receptors such as the vitamin D receptor (VDR), thyroid hormone receptor (TR), retinoic acid receptor (RAR), peroxisome-proliferator activated receptor (PPAR), and retinoid X receptor (RXR), all of which play key roles in animal development, physiology and human disease. The steroid hormones are derived from cholesterol, share a common chemical structural motif, and act in a classic endocrine manner. In contrast, ligands for the nonsteroid receptors are chemically diverse, including vitamin D, thyroid hormone, retinoids, and prostanoids. Furthermore, the source of the ligands for this class of receptors may be either endocrine, or generated intracellularly and not secreted (intracrine), or may be modified within the cell from an apohormone (Fig. 1).

Sequence analysis and functional studies, based on the original members of this superfamily (i.e., steroid, vitamin D, thyroid, and retinoic acid receptors), revealed that both steroid and nonsteroid receptors share common structural motifs consisting of highly conserved DNA and ligand binding domains (*see* below). Based on these sequence similarities, a large number of proteins have been isolated within the last decade. Many members of this class of the intracellular receptor (IR) superfamily have been termed "orphan receptors," which, owing to their conserved amino acid sequence, are presumed to be functional members of the IR family but whose physiological ligands (if any) are unknown. This term, coined in the late 1980s, was used to describe the first of what has now become the largest group of nuclear receptors. It is now realized that in some cases these proteins may be ligand-independent transcription factors and may not really fulfill the more classical definition of a nuclear receptor. However, the term orphan receptor is still used, for more or less historical reasons, to refer to any protein that is related by sequence identity to other members of the IR. These receptors

From: *Principles of Molecular Regulation* (P. M. Conn and A. R. Means, eds.), © Humana Press Inc., Totowa, NJ.

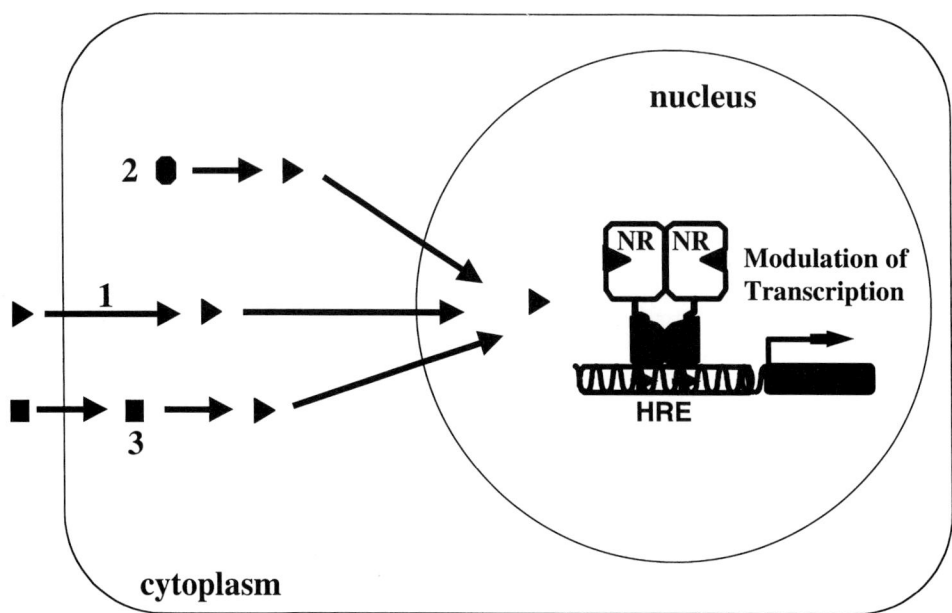

▶ **Ligand**

■ **Ligand precursor or apohormone generated outside the cell**

● **Ligand precursor synthesized within the cell**

NR is nuclear receptor; HRE is hormone response element

Fig. 1. A schematic diagram illustrating ligand activation of nuclear receptors acting through their specific hormone response element (HRE). Note that the ligand may be generated extracellulary (1, endocrine); modified within the cell from an apohormone (2); or generated intracellularly (3, intracrine). Also, an example of a nuclear receptor binding to its HRE as a homodimer is shown; receptors may also interact with their respective HREs as heterodimers or as monomers (*see* Section 2.).

are phylogenetically diverse and are found in virtually every animal species examined including mammals, insects, nematodes, and nonvertebrates. To date, more than 30 different subfamilies of orphan receptors, many with multiple members, have been identified. Our emerging knowledge of orphans has provided unique insight into novel biological pathways. The study of these fascinating receptors in recent years has already resulted in discoveries with major impact on various areas of biology including normal development and physiology as well as human disease and drug discovery.

In this chapter, we first discuss the common structural features that allow for DNA binding and transcriptional activation of nuclear receptors. Next, we focus on specific orphan nuclear receptors and summarize the recent progress made in elucidating their biological function. This includes receptor gene disruption and expression studies, and target gene and ligand identification. Finally, we discuss methods for identifying orphan receptor ligands/signaling pathways and their implications for discovery of therapeutic drugs for humans. The text is meant to be factual and concise, and the information has been limited primarily to selected mammalian orphan receptors. For further information on specific or additional nuclear receptors the reader is directed to the list of selected readings.

2. STRUCTURAL AND FUNCTIONAL DOMAINS OF NUCLEAR RECEPTORS

Amino acid sequence analysis and functional studies reveal that nuclear receptors share a high degree of structural similarity in their DNA and ligand binding domains (Fig. 2). The N-terminal A/B region is not well conserved, and in most receptors, contains a transcriptional activation function 1 (AF-1) domain that functions independently of ligand binding. The hallmark feature of the nuclear receptor family is a

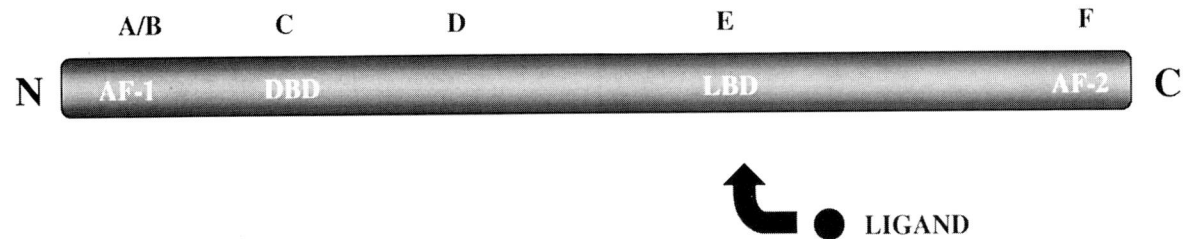

Fig. 2. Basic nuclear receptor structure and functional domains (**A–F**). Region **A/B** is the transactivation function-1 (AF-1) domain. Regions **C** and **E** contain the highly conserved DNA binding and ligand binding domains. These two regions are connected through a flexible hinge region (**D**). The F region comprises the highly conserved AF-2 domain.

highly conserved DNA binding domain (DBD, region C) containing two zinc finger motifs that mediate specific interaction with DNA. The C-terminal domain of the receptor, also reasonably well conserved, consists of a rather complex region that participates in ligand binding and dimerization (D, E, and F regions). In many receptors, this domain also contains a second highly conserved activation function domain 2 (AF-2 or τ_c) that is ligand dependent.

2.1. DNA Binding Domain

Nuclear receptors regulate transcription primarily through direct interaction of the DBD (region C), consisting of approximately 70 amino acids present within the amino terminal region of the receptor, with specific nucleotide target gene sequences termed hormone response elements or HREs (Fig. 2). Many studies, including recent X-ray crystallographic studies, have been performed on DBDs of various nuclear receptors providing detailed information about how they interact with HREs in a sequence-specific manner (see below).

Most steroid receptors, such as GR and ER, bind as homodimers to inverted repeats (IR) of the consensus sequence AGAACA separated by a specific number of nucleotides (n, where $n = 3$) (Fig. 3A). On the other hand, nonsteroid receptors and the majority of orphan receptors bind as hetero- or homodimers to direct repeats (DR) of the sequence AGGTCA also separated by a specific number of nucleotides (n, where $n = 0–5$) (Fig. 3B). Certain orphan receptors heterodimerize with RXR, and these include liver X receptor (LXR), which binds to DR4 motifs; and constitutive androstane receptor (CAR), which binds to DR5 sequences. Other receptors, such as hepatic nuclear factor 4 (HNF-4), bind as homodimers to a DR1 motif. In addition, some orphan receptors bind as monomers to the sequence AGGTCA, such as steroidogenic factor 1 (SF-1) and retinoic acid-related orphan receptor (ROR). The binding of receptor dimers is often cooperative and requires multiple independent dimerization determinants located within the DBD and the ligand binding domain (LBD). How are these related receptors able to gain specificity for their unique target HREs? This is determined in large part by the number of nucleotides between each repeat sequence for the receptors that bind as hetero- or homodimers. For example, the X-ray structure of a RXR–TR heterodimer bound to its cognate HRE (DR4) has provided evidence that a DNA-supported asymmetric dimerization interface located within the DBDs of RXR and TR is the molecular basis for receptor heterodimers to distinguish between closely related HREs with different spacers. In addition, the 5′-flanking sequence also plays an important role in determining specificity. This becomes especially important in the case of receptors that bind as monomers. For example, RORα, a member of the ROR family of orphan receptors, binds to HREs consisting of a single copy of the core recognition sequence AGGTCA preceded by a six-basepair (bp) A/T rich sequence. Structure–function analysis of different members of the monomeric DNA binding receptors suggest that they may utilize distinct subdomains present within the C-terminal extension of their DBDs for precise recognition of the 5′-flanking sequence preceding the AGGTCA core element (Fig. 3B).

2.2. Ligand Binding and Dimerization Domain

The C-terminal region of the receptor consists of approximately 225 amino acids that is capable of autonomous ligand binding. This LBD (region E) is connected to the DBD through a flexible hinge region (D). In addition to functioning as an LBD, this region also possesses a homo- and heterodimerization interface, and a hormone-dependent AF-2, or τ_c domain that can interact with receptor cofactors. A dimerization interface referred to as the I-box, present within the C-terminal LBD, has been identified as a region

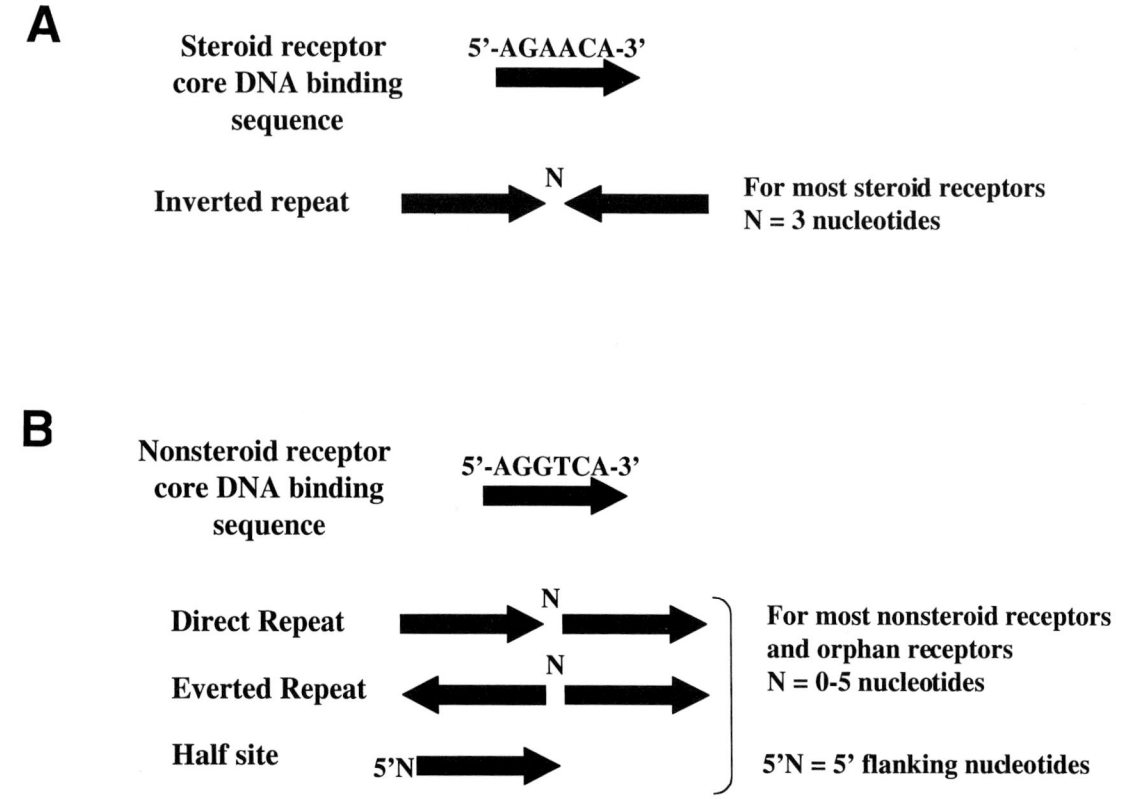

Fig. 3. (A) Most steroid receptors bind as homodimers to inverted repeats (IR) of the core sequence AGAACA spaced by three nucleotides. (B) Most nonsteroid and orphan receptors bind either as heterodimers (with RXR) or homodimers to direct repeats of the core sequence AGGTCA spaced by zero to five nucleotides (DR0–DR5). Some orphan receptors also bind to everted repeats. In addition, several orphan receptors bind as monomers to a half-site containing a single AGGTCA motif. For these receptors binding specificity appears to be determined by the nature of the 5′-flanking sequence.

that mediates cooperative binding to DNA. Structural studies indicate that ligands can regulate AF-2 activity by binding to and altering the structure of the LBD. An agonist-induced conformational change involves repositioning of helix 12, the most C-terminal helix of the LBD. This change is thought to lead to a productive interaction with coactivator proteins and an activated receptor complex. Antagonist binding leads to an alternative orientation of helix 12, partially burying residues required for coactivator recognition and AF-2 function, leading to inhibition. Thus, agonist and antagonist binding to receptors can lead to different conformational changes within the LBD (for more information see Chapter 23).

2.3. RXR Dimerization and Heterodimer Response to RXR Ligands

As mentioned previously, RXR acts as an essential heterodimer partner for a number of nonsteroid receptors, including many orphan receptors. The nonsteroid members of the nuclear receptor superfamily that have ligands are all RXR heterodimers (e.g., VDR, TR, RAR). In addition, RXR itself is ligand responsive and binds 9-cis-retinoic acid with high affinity (see Section 6.). It is thus important to understand the responsiveness of different RXR heterodimers to RXR ligands.

In certain instances, RXR fails to bind its ligand and acts as a "silent or nonpermissive" partner (e.g., RXR–VDR; RXR–TR). These dimers do not respond to RXR ligands in the presence or absence of the RXR partner's ligand. Sometimes, RXR can acquire the ability to respond to its specific ligand if the partner is occupied with its own ligand (e.g., RXR–RAR). Here, RXR is considered conditionally silent. In contrast, when RXR forms a heterodimer with several orphan receptors, it is responsive to its ligand even in the absence of the partner's ligand. These heterodimers (e.g., RXR–LXR; RXR–NGFI-B; RXR–PPAR) are RXR-ligand responsive and RXR is "active or permissive" (Fig. 4). Finally, on many RXR-heterodimers, simultaneous addition of ligands

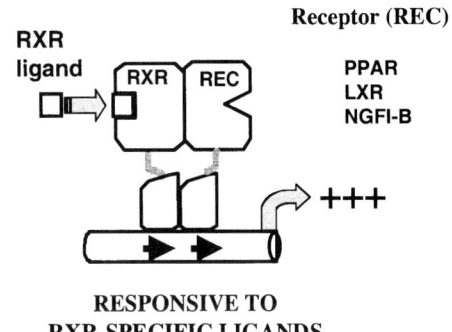

Fig. 4. Nuclear orphan receptor classification based on DNA binding preference. Shown are the four classes of receptors and their preferred mode of binding to HREs. Also, some examples in each class are provided (*see* text for details).

for both receptors leads to synergistic activation of the heterodimer (e.g., RXR–RAR, RXR–PPAR). Therefore, the ability of RXR heterodimers to respond to RXR ligands is primarily dependent on the specific partner. Thus, not only does RXR play a crucial role in the binding of several orphan receptors to specific DNA sequences, but heterodimerization with RXR can also lead to the ability of the orphan receptor–RXR complex to respond to RXR ligands.

Therefore, in addition to being an obligate heterodimer partner for several orphan receptors, RXR plays a critical role in generating diversity of hormonal responses.

3. CLASSIFICATION AND BIOLOGY OF ORPHAN RECEPTORS

It is important to note that the classification of orphan receptors is rather complex and can be done using different criteria. We have chosen a simple and broad method of classification based on the ability of the receptors to bind HREs either as heterodimers with RXR, as homodimers, or as monomers. In some cases, receptors may have more than one mode of binding (e.g., as monomers and heterodimers); however, we have classified them based on their preferred or primary mode of binding. Furthermore, certain receptors that lack a typical nuclear receptor DBD and exhibit unusual DNA specificities, or whose HREs remain unidentified, have been classified as atypical receptors.

Based on these criteria, orphan nuclear receptors can be classified into four groups (Fig. 5):

1. Orphan receptors that bind DNA as heterodimers with RXR
2. Orphan receptors that bind DNA as homodimers
3. Orphan receptors that bind DNA as monomers
4. Atypical orphan receptors

Within each category, representative examples are provided including a brief summary of our current knowledge of their biological activity and for the purpose of this chapter only vertebrate receptors are considered. When an orphan receptor is deemed "adopted" and is no longer considered an orphan is debatable. Therefore, in recent cases where putative ligands have been identified, we still classify them as orphan receptors. In addition, in many cases, while these putative ligands/activators modulate transcriptional in a cell-based transfection assay (*see* Section 4.), it is not entirely clear if these compounds are in fact *bona fide* physiological ligands that directly bind receptors. Finally, the RXR and PPAR families of receptors, which are in an advanced state of investigation and whose physiological ligands have been identified, are discussed in Section 6. Table 1 lists the orphan receptors, grouped in their gene families, along with a brief description of their mode of DNA binding, tissue distribution, chromosomal localization, and whether or not targeted disruption of their gene has been accomplished. A summary of the phenotypes observed as a consequence of receptor gene disruption or natural mutations is provided in Table 2.

3.1. Orphan Receptors that Heterodimerize with RXR

3.1.1. CAR

Murine constitutive androstane receptor (mCAR) or CARβ is closely related to the human orphan

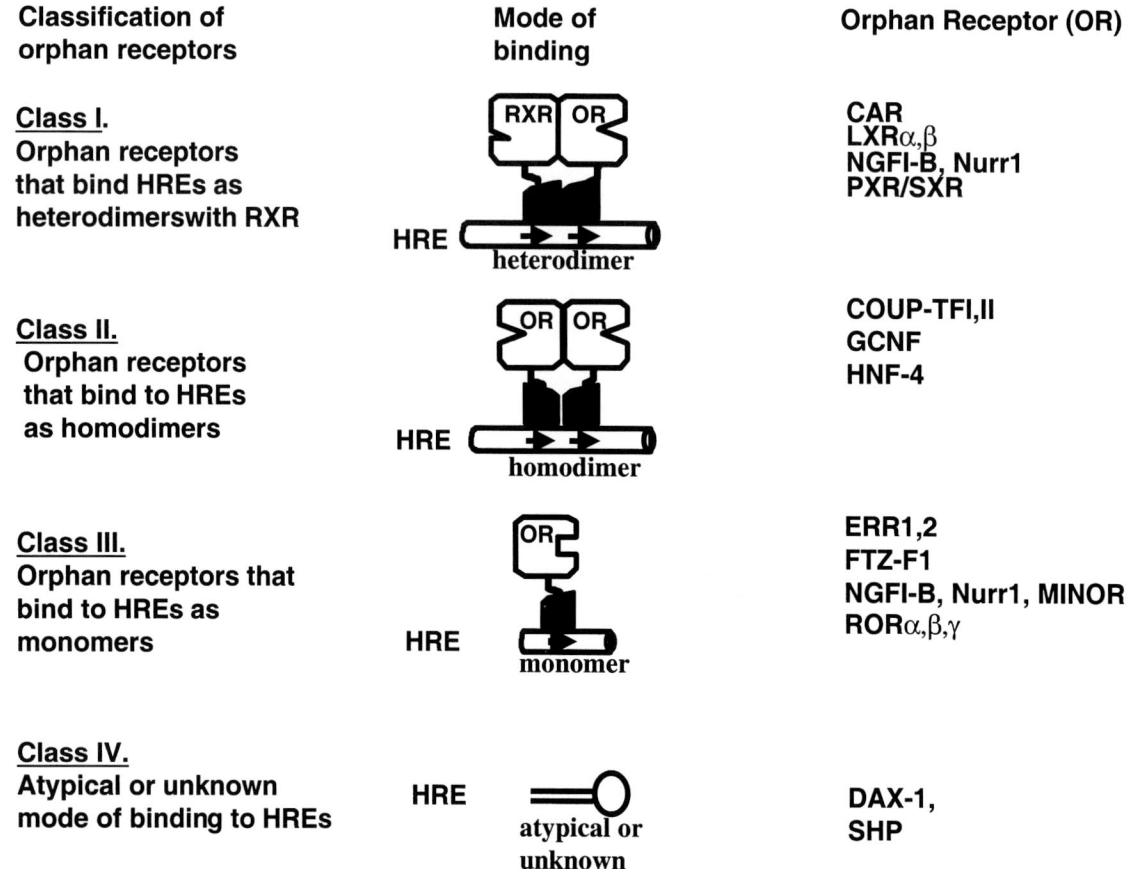

Fig. 5. Silent and active roles of RXR in transcriptional activation. The ability of RXR to respond to its ligand appears to be determined by its heterodimer partner. RXR heterodimers such as RXR–VDR and RXR–TR are not responsive to RXR ligands and RXR is referred to as a silent partner. RXR heterodimers with the orphan receptors LXR, PPAR, NGFI-B, and FXR represent another class where RXR responds to its ligand even in the absence of its partners ligand. In this case RXR is considered active. *See* text for details.

MB67, also called hCAR. Both receptors are highly expressed in the liver. In addition to the most abundant mCAR1 isoform, the *mCAR* gene expresses a truncated mCAR2 variant that is missing the C-terminal portion of the ligand binding/dimerization domain. CAR forms a heterodimer with the RXR and binds to the retinoic acid response element from the promoter of the RAR β2 isoform (βRE or a DR5). Consistent with its lack of a critical heterodimerization interface, the mCAR2 variant does not bind this site. The cytochrome P450 gene *CYP2b10* has been identified as a potential target gene for CAR–RXR heterodimers.

Both mCAR1 and hCAR act as constitutive transcriptional activators. This activity is dependent on the presence of the conserved C-terminal AF2 transcriptional activation motif. Steroids such as androstanol and androstenol have been reported to inhibit the constitutive activity of CARβ. This ligand-dependent inhibition of CARβ is unique, because in most cases ligand binding leads to receptor activation.

3.1.2. FXR AND RIP14

Farnesoid X receptor (FXR) was originally isolated from rat liver and also interacts with RXR. FXR bears strong homology to the insect ecdysone receptor (EcR) within its DBD (~80%) and FXR–RXR heterodimers interact with the EcR response element (EcRE) that consists of an inverted repeat of the consensus receptor binding hexamer separated by one base pair (IR1). It is highly expressed in the liver, kidney, and gut of the developing embryo. In the adult rat kidney, FXR is expressed in areas rich in renal tubules: the medullary rays and stripe. Additional expression is also observed in the adrenal cortex. FXR is activated by the insect juvenile growth hormone III (JH III) that is derived from farnesyl pyrophosphate, a key precursor in the mevalonate biosynthetic pathway. A

Table 1
Vertebrate Orphan Receptors

Receptor Gene Family	DNA Binding Mode	Sub-types	Names[a]	Species[b]	Drosophila Homolog	Tissue Distribution[c]	Chromosomal Localization[d]	K/O[e]
COUP-TF	Homodimer/ heterodimer	α	COUP-TFI, ear-3	h,r,a,f,i	SVP	Widespread, developing organs, CNS	h 5q14; m 13	+
		β	COUP-TFII, ARP-1, ear-2	h,r,a,c		Widespread, developing organs, CNS	h 15q26; m 7	+
		γ		h,r		Widespread, fetal liver	h 19	−
DAX-1	Unknown[f]		DAX-1	h,r		Adrenals, gonads, pituitary, hypothalamus	h Xp21; m X	*
ERR	Monomer/ homodimer	α	ERRα, ERR1	h,r		Widespread, esp. CNS	h 11q12−13	+
		β	ERRβ, ERR2	h,r		Kidney, heart	h 14q24.3	+
FTZ-F1	Monomer	α	Two isoforms: SF-1, Ad4BP; ELP	h,r,b,i	FTZ-F1	Adrenals, gonads, brain, hypothalamus, pituitary	h 9q33; m 2	+
		β	FTF, LRH-1, PHR-1, FF1rA	h,r,a,i		Liver, pancreas		−
FXR	Heterodimer		FXR, RIP14	r		Kidney, liver, gut, adrenals		−
GCNF	Homodimer		GCNF, RTR	h,r		Germ cells	m 10	−
HNF-4	Homodimer		HNF-4	h,r,i	dHNF4	Liver, kidney, intestine, pancreas	h 20q12−q13.1	+
LXR	Heterodimer	α	LXRα, RLD-1	h,r		Liver, kidney, spleen, fat, intestine, pituitary, adrenals	h 11q23.3	−
		β	UR, NER, OR-1, RIP15	h,r		Widespread	h 19q13.3	−
MB67	Heterodimer		MB67, CAR	h,r		Liver		−
NGFI-B	Monomer/ heterodimer[g]	α	NGFI-B, Nur77, N10, NAK1, TR3, TIS1	h,r,i	DHR38	Thymus, brain, adrenal gland, muscle, testis	h 12q13; m 15	+
		β	NURR1, NOT, RNR-1, TINUR	h,r		Brain, regenerating liver	h 2q22−q23	−
		γ	NOR-1, MINOR, TEC	h,r		Fetal brain, lung; adult heart, skeletal muscle	h 9q	−
ONR1	Heterodimer		ONR1	a		Xenopus embryos		−
PPAR	Heterodimer	α	PPARα	h,r,a		Liver, heart, kidney, brown fat	h 22q12−q13.1; m 15	+
		β	PPARβ, PPARδ, NUC-1, FAAR	h,r,a		Widespread	h 6p21.1−p21.2; m 17	−
		γ	PPARγ	r,a		Adipose	h 3p25; m 6	−
PXR	Heterodimer		PXR, SXR, PAR	r,h		Liver, small intestine		−
REV-ERB	Monomer/ homodimer	α	Rev-ErbA-α, ear-1	h,r,i		Widespread, esp. skeletal muscle, brown fat	h 17q21	−
		β	Rev-Erbβ, RVR, BD73	h,r		Widespread	m 14	−
ROR	Monomer/ homodimer	α	RORα, RZRα	h,i		Widespread, esp. peripheral blood leukocytes	h 15q21−q22; m 9	+,*
		β	RZRβ	r		Brain, retina	m 19	−
		γ	RORγ, TOR	h,r		Skeletal muscle, thymus	h 1q22−23; m 3	−

Continued

Table 1
Continued

Receptor Gene Family	DNA Binding Mode	Sub-types	Names[a]	Species[b]	Drosophila Homolog	Tissue Distribution[c]	Chromosomal Localization[d]	K/O[e]
RXR	Homodimer/ heterodimer	α	RXRα	h,r,a,c,f,i	USP	Widespread, esp. skin, liver, kidney, lung, muscle	h 9q34; m 2	+
		β	RXRβ, H2RIIBP	h,r,a,f		Widespread	h 6p21.3; m 17	+
		γ	RXRγ	h,r,a,c,f		Heart, brain, lung, adrenal, kidney, muscle, liver	h 1q22–q23; m 1	+
SHP	Unknown[h]		SHP	h,r		Liver, pancreas, heart		–
TLX	Monomer/ homodimer		Tlx	r,c,f,i	Tll	Developing brain, eye		–
TR2	Homodimer	α	TR2	h,r		Testis, prostate, seminal vesical	h 12q22; m 10	–
		β	TR4, TAK1, TR2R1	h,r		Widespread, esp. testis, brain, kidney, skeletal muscle	h 3p25	–

Reproduced with minor modifications from Nuclear orphan receptors: the search for novel ligands and signaling pathways. Hormones and Signaling, pp. 307–358, Academic Press, 1998.

[a]Indicates alternate names for the receptor.
[b]Species indicates homologs or related receptors in other species: a = amphibian; b = bovine; c = chicken; f = fish; h = human; i = invertebrate; r = rodent.
[c]Tissue distribution refers to major sites of mRNA expression.
[d]m, mouse; h, human.
[e]K/O refers to targeted gene knockouts that have (plus) or have not (minus) been reported. An *asterisk* indicates a natural gene mutation or disruption.
[f]DAX-1 lacks a conventional DNA binding domain, but has been reported to bind to a DR5 type sequence and hairpin loop structures (*see* text for details).
[g]NGFI-B binds to DNA as a monomer; however, on specific response elements, NGFI-B can also heterodimerize with RXR (*see* text for details).
[h]SHP lacks a conventional DNA binding domain and although it can heterodimerize with RXR, DNA binding has not been demonstrated thus far.
*Indicates naturally occurring mutations.

Table 2
Summary of Orphan Nuclear Receptor Mutants

Orphan Nuclear Receptor	Phenotype of Mutant Mice and/or Human Patients
COUP-TFI	Mutant mice die perinatally exhibiting abnormal development of a subset of neurons in the peripheral nervous system.
DAX-1	Humans with mutant DAX-1 exhibit adrenal hypoplasia congenita (AHC) with hypogonadotropic hypogonadism (HGG).
ERRβ	Severely impaired placental development and failure, characterized by abnormal chorion formation followed by a failure of diploid trophoblast self-renewal and an increased formation of giant cells.
HNF-4	MODY patients have normal insulin sensitivity but suffer from a defect in glucose-stimulated insulin secretion, suggesting a defect in pancreatic β-cell function, rather than insulin resistance, as the primary defect.
	Mice deficient in HNF-4 die *in utero* and mutant embryo development is severely disrupted.
LXR α	Mutant mice are normal at first, are viable and fertile; however, upon challenge with a high-cholesterol diet lose their ability to catabolize dietary cholesterol. This is accompanied by a rapid accumulation of large amounts of cholesterol in the liver that eventually leads to impaired hepatic function.
NGFI-B	Mice do not exhibit grossly abnormal functions.
Nurr1	Knockout mice demonstrate a lack of normal development of midbrain dopaminergic neurons and die soon after birth.
RORα	Staggerer *(sg)* mice (natural *RORα* mutant) show tremor, body imbalance, hypotonia, as well as small size and die shortly after weaning. The cerebral cortex of these mice exhibits a cell-autonomous defect of the Purkinje cells.
	RORα defective mice exhibit a phenotype similar to the *sg* mice.
SF-1	Mutant mice exhibit adrenal and gonadal agenesis, impaired gonadotrope function, ablation of a specific region of the hypothalamus, and male-to-female sex reversal of the internal and external genitalia. Animals die soon after birth.

Summarized above are the resulting phenotypes observed upon nuclear receptor mutations created in mice through gene disruption and/or naturally occurring mutations in mice and humans.

number of metabolites within the mammalian mevalonic acid pathway, such as farnesol, farnesal, farnesyl acetate, and geranylgeraniol also activate FXR, farnesol being the most efficacious. None of these compounds, however, have been shown to bind directly to FXR, and its *bona fide* natural ligand remains unknown. The FXR–RXR heterodimer is a permissive dimer and responds to RXR ligands. While additional studies are required to elucidate the function of FXR, the FXR activators, all-*trans*-farnesol and juvenile hormone III, have been shown to accelerate the formation of a mature stratum corneum (SC), the outermost layer of the epidermis and functional epidermal permeability barrier, suggesting a potential role for FXR in epidermal barrier development.

RIP14 (RXR interacting receptor 14) is the mouse homolog of FXR, also binds to EcREs, and like FXR, is highly expressed in liver and kidney with some expression in the adrenal and gut. Unlike FXR, however, RIP14 is not activated by farnesol. Instead, it is activated by metabolites of the retinoid, all-*trans*-retinoic acid (ATRA), that binds to the retinoic acid receptor (RAR). The synthetic retinoid TTNPB ([*E*]-4-[2- (5, 6, 7, 8-tetrahydro-5, 5, 8, 8-tetramethyl-2-naphthalenyl)propen-1-yl] benzoic acid) also activates RIP14. It is thus possible that RIP14 represents an additional member of the retinoid receptor family playing an important role in retinoid signaling in RIP14 expressing tissues. Confirmation of this requires the identification of its physiological ligand.

3.1.3. THE LXR FAMILY

Liver X receptor alpha (LXRα, also called RLD-1) and LXRβ (also called UR, OR1, RIP15, and NER) are related members of the orphan nuclear receptor superfamily that require RXR as an obligate heterodimer partner to bind their HRE (e.g., DR4). Like the FXR–RXR heterodimer, LXR–RXR heterodimers respond to RXR ligands. LXRα is highly expressed in several tissues where cholesterol metabolism occurs, including liver, kidney, intestine, adrenals, adipose tissue, and spleen. LXRβ is ubiquitously expressed and has been detected in most tissues tested. Detailed studies have suggested that while it is broadly expressed in the fetal brain, LXRβ expression is restricted in the postnatal and adult brain. Both LXRs are activated by a specific set of hydroxylated cholesterol molecules, such as 22(*R*)-hydroxycholesterol and 24(*S*)-hydroxycholesterol. Many of these molecules serve as key intermediates in the rate-limiting steps of bile acid and steroid hormone production (*see* section 5. for details). Also, the cytochrome *P450* gene encoding CYP7a, or 7α-hydroxylase, the rate-limiting enzyme in bile acid production, contains an LXR response element and is an important target for LXRα. Thus, the LXRs appear to act as sensors of cholesterol. This has been confirmed by studies using gene manipulation to create mice that lack functional LXRα. These animals appear normal at first and are viable and fertile; however, when challenged with a high cholesterol diet lose their ability to catabolize dietary cholesterol, being unable to tolerate any amount of excess cholesterol over that which they synthesize *de novo*. This is accompanied by a rapid accumulation of large amounts of cholesterol in the liver that eventually leads to impaired hepatic function. The regulation of several other crucial lipid metabolizing genes is also altered in LXRα knockout mice when given a high cholesterol diet. These studies have established LXRα as a key regulator of cholesterol homeostasis in the liver.

3.1.4. PXR

Pregnane X receptor (PXR) is an orphan receptor that also requires RXR as an obligate heterodimer partner. Two full-length clones, mPXR.1 and mPXR.2, have been isolated from mice. Northern blot analysis has identified at least three transcripts in the liver and intestine where it is expressed at high levels; low level expression is also seen in the kidney and stomach. Similar sites of expression were also observed using *in situ* hybridization techniques with mouse embryonic tissues. Thus, mPXR expression appears to be restricted to specific tissues including the liver and intestine, both in the embryo and the adult. The cytochrome P450 3A (*CYP3A*) family of genes appear to be targets for PXR. These genes are expressed in PXR expressing tissues and are involved in the hydroxylation of steroid hormones and a large variety of xenobiotics. Promoters within the *CYP3A1* and *-A2* genes contain DR3 motifs that have been shown to bind PXR–RXR heterodimers.

In transient transfection studies, PXR is activated by a large number of steroids including pregnenolone and progesterone and their 17-hydroxylated derivatives, as well as synthetic glucocorticoids and antiglucocorticoids. Interestingly, PXR combines features of both steroid and nonsteroid receptors, because, unlike steroid receptors, it forms heterodimers with RXR; however, similar to steroid receptors, it is activated by steroids.

One full-length clone (hPXR) from humans, also called SXR (steroid and xenobiotic receptor), corresponding to mPXR.1 has been isolated and may be its human homolog. hPXR or SXR is also expressed in the liver and intestine, however, unlike mPXR, trancripts were undetectable in the stomach and kidney as determined by Northern blotting. An everted repeat of two nuclear receptor half sites separated by six nucleotides (ER6) serves as an hPXR response element. The human *CYP3A4* monooxygenase gene that plays a major role in the biotransformation of drugs contains ER6 motifs within its promoter and is a likely target for hPXR. *CYP3A4* is expressed in the liver and intestine at high levels, has a broad substrate specificity, and is capable of metabolizing more than 60% of all clinically used drugs. This includes contraceptive steroids, various antibiotics, and immunosupressive agents. hPXR is activated by a large number of compounds including the antibiotic rifampicin; the calcium channel antagonist nifedipine; steroid antagonists such as tamoxifen; and synthetic steroids such as dexamethasone, estradiol, corticosterone, and phytoestrogens. Although direct binding to the receptor has not been demonstrated, some of these compounds promote hPXR interactions with coactivators in biochemical experiments, suggesting they serve as PXR ligands. Thus, hPXR–SXR appears to act as a receptor with broad specificity that may be involved in the detoxification and catabolism of such compounds. In addition, its ability to regulate *CYP3A4* suggests an important role in drug metabolism and drug–drug interactions.

3.2. ORPHANS THAT BIND AS HOMODIMERS

3.2.1. THE COUP–TF FAMILY

The COUP-TF (chicken ovalbumin upstream promoter-transcription factor) family of orphan receptors consists of three members, COUP-TFI (also called ear-3), COUP-TFII (also called ARP-1), and ear-2. Of the three, COUP-TFI and -II are more closely related to each other than to ear-2. COUP-TF homologs have been cloned in several species, ranging from *Drosophila* to human, and share a high degree of sequence homology suggesting a conservation of function through evolution. COUP-TFs bind as homodimers to AGGTCA direct repeats or palindromes with various spacings. The highest affinity for these proteins is a DR1 sequence on which they also form heterodimers with RXR. COUP-TF response elements have been identified in the promoters of genes involved in lipid transport, muscle differentiation, and steroid hormone biosynthesis. In these cases, the COUP-TFs act as negative regulators of transcription both in vitro and in vivo. The mechanisms for this negative regulation include competitive DNA binding, competition for RXR with other heterodimeric partners, or protein–protein interactions with transcriptional coregulator molecules. Thus, COUP-TFs modulate VDR, TR, and RAR pathways via competition for RXR. The ratio of the different receptors within a cell, their affinities for RXR and/or specific HREs is also likely to play a role in these types of competitive inhibition. In some cases, COUP-TFs can also act as transcriptional activators depending on the promoter context and/or availability of coactivator molecules in a cell.

The COUP-TFs are highly expressed in the developing nervous system indicating a possible involvement in neuronal development and differentiation. In the mouse, the COUP-TF genes (I and II) exhibit expression patterns that overlap extensively. Gene knockout studies demonstrate that mCOUP-TFI null animals die perinatally and mutant embryos show abnormal development of a subset of neurons in the peripheral nervous system. Interestingly, although COUP-TFI expression is widespread, the knockout phenotype is very specific, suggesting that in certain tissues COUP-TFI activity is compensated by COUP-TFII.

3.2.2. GCNF

The orphan receptor germ cell nuclear factor (GCNF) demonstrates specific and high-affinity binding to the direct repeat DNA element AGGTCAAGGTCA (DR0) as a homodimer. It can also bind to extended half-sites, such as TCAAGGTCA, that serve as binding sites for the nuclear receptor SF-1. However, in contrast to SF-1, which binds as a monomer, GCNF appears to bind as a homodimer to these sequences. GCNF is unable to transactivate through either of its binding sites, although it is able to confer repression of basal transcriptional activity through the DR0 sequence.

During development, GCNF exhibits a restricted brain-specific expression pattern, whereas GCNF expression in the adult is germ cell specific. *In situ* hybridization studies demonstrate that GCNF mRNA is expressed specifically within germ cells in the adult male and is most abundant in stage VII round spermatids. GCNF expression first occurs in the testis of 20-d-old mice, when round spermatids first emerge. Thus, in the male, GCNF expression occurs after meiosis and may play a role in the morphological changes of the maturing spermatids. In contrast, female expression of GCNF occurs in growing oocytes that have yet to complete the first meiotic division. The receptor has been shown to bind to DR0 sequences present within the promoters of protamine genes *Prm1* and *Prm2*, suggesting these may serve as GCNF target genes. Based on these data, GCNF appears to be important for certain aspects of gamete development and neurogenesis.

3.2.3. HNF-4

Hepatocyte nuclear factor-4 (HNF-4) is an orphan receptor that is expressed in the liver, pancreas, intestine, and kidney. During mouse development, HNF-4 expression is observed early, even before organogenesis occurs. It is first seen in the primary endoderm of implanting blastocysts and later in the extraembryonic visceral endoderm cells. Homozygous loss of functional HNF-4 in mice is an embryonic lethal and mutant embryo development is severely disrupted.

HNF-4 binds DNA as a homodimer to direct repeats of the nuclear receptor core sequence AGGTCA separated by one or two nucleotides (DR1, 2) with high affinity. It can also bind to DR1 and DR2 sequences that differ from the canonical DR, although with slightly lower affinity. In transient transfection experiments, HNF-4 demonstrates strong constitutive activity acting through both kinds of HREs. Interesingly, long-chain fatty acids have been reported to modulate the transcriptional activity of HNF-4 by directly binding the receptor as their acyl-CoA thioesters. Altering the chain length and the degree of saturation of these fatty acyl-CoA ligands can lead to activation or inhibition of HNF-4 transcriptional activity.

HNF-4 interacts with regulatory sequences present within promoters of several genes whose products are involved in diverse functions such as, cholesterol, fatty acid, glucose, and amino acid metabolism as well as liver development and differentiation. Studies using embryonic cells showed that loss of functional HNF-4 leads to impaired expression of several genes involved in glucose transport and metabolism. Thus, HNF-4 appears to be critical for regulating these processes and for maintaining glucose homeostasis.

Genetic evidence directly linking HNF-4 to a role in controlling glucose homeostasis in humans came from clinical studies using subjects with maturity-onset diabetes of the young (MODY). This is a genetically heterogenous monogenic form of non-insulin-dependent (type 2) diabetes mellitus (NIDDM) characterized by early onset and by autosomal dominant inheritence. Prediabetic MODY patients have normal insulin sensitivity but suffer from a defect in glucose-stimulated insulin secretion suggesting a defect in pancreatic β-cell function, rather than insulin resistance, as the primary defect. Linkage analysis have localized the genes that are mutated in MODY on human chromosomes 20, 7 and 12 (called *MODY1, 2,* and *3,* respectively). *MODY1* is the gene encoding HNF-4, and, HNF-1α a target for HNF-4, has been associated with *MODY3*. These important discoveries suggest this form of NIDDM is primarily a result of aberrant gene expression and ligands modulating HNF-4 activity may lead the way to novel treatments for diabetes.

3.3. Orphan Receptors that Bind as Monomers

3.3.1. THE ERR FAMILY

ERR α and -β (estrogen-related receptor α and β) (originally called ERR-1 and -2) represent a family of orphan receptors that bear strong homology to the DBD of the estrogen receptor. The receptors do not bind to estrogens in vitro, nor do they activate transcription in their presence; however, the ERRα gene itself is estrogen responsive. ERRα is highly expressed in the kidney, heart, and brown adipocytes, tissues that metabolize fatty acids. It binds as a monomer to an ERRE (ERRα response element) consisting of an extended half-site with a core sequence consisting of AGGTCA. An ERRE is present within the promoter of the medium-chain acyl coenzyme A dehydrogenase (MCAD) gene that encodes an enzyme involved in the β-oxidation of fat. ERRα has been shown to regulate MCAD and may play an important role in regulating cellular energy balance in vivo. Other ERRα target genes include the estrogen-regulated lactoferrin gene where it can accentuate its estrogen-dependent activity. ERRα has also been shown to act as a transcriptional repressor of the simian virus 40 (SV40) late promoter and may act as a key regulator of the early-to-late switch of *SV40* gene expression.

ERRβ can also bind to ERREs and has been identified as a cell-type-specific repressor of hormone pathways including those controlled by glucocorticoids, retinoids and estrogen. This may occur through titration of receptor cofactors and/or direct competition for DNA. ERRβ also binds as a homodimer to estrogen response elements (EREs), suggesting possible crosstalk between the two receptor signaling pathways. ERRβ exhibits a highly specific pattern of expression during mouse development. During early embryonic development, ERRβ gene expression is restricted to extraembryonic tissues. ERRβ transcripts are seen in the extraembryonic ectoderm at 5.5 d postcoitum (d.p.c.) and more prominently at 6.0 d.p.c. Later, ERRβ is expressed by ectodermally derived regions of the amniotic fold that form the chorion (7.5 d.p.c.). As the chorion fuses with the ectoplacental cone, ERRβ expression is decreased and by 8.5 d.p.c. it is seen only at the free margin of the chorion, and is not detected at subsequent developmental stages. ERRβ mutant animals die at 10.5 d.p.c. and have severely impaired placental development and failure, characterized by abnormal chorion formation followed by a failure of diploid trophoblast self-renewal and an increased formation of giant cells. Thus, ERRβ is essential for normal placental formation.

3.3.2. THE FTZ-F1 FAMILY

Steroidogenic factor 1 (SF-1, also called Ad4BP), and LRH-1 (or PHR-1 or FTF) are the mammalian homologs of the *Drosophila* transcription factor FTZ-F1, which regulates transcription of the *fushi tarazu* homeobox gene in fly embryos. Four distinct transcripts appear to be encoded by the SF-1 gene, embryonal long terminal repeat binding protein (ELP)1, ELP2, ELP3, and SF-1. These are formed through alternative promoter usage and/or differential splicing and differ in their expression profiles possibly serving distinct functional roles. SF-1 binds as a monomer to its response element (SF-1RE) which has been identified within promoters of many cytochrome P450 steroid hydroxylases. It acts primarily as an activator and plays an important role in steroidogenic gene regulation. The gene encoding steroidogenic acute regulatory protein (StAR), an essential factor required for cholesterol mobilization from the outer to the inner mitochondrial membrane during steroidogenesis, is also induced by SF-1. In addition, the scavenger receptor BI (SR-BI), a receptor for high-density lipoprotein (HDL) cholesterol, is activated by SF-1. SF-1 is expressed in all steroidogenic tissues such as the adrenal, gonads, and at relatively low levels in the placenta. It is also expressed in the spleen, where cholesterol metabolism occurs. Thus, SF-1 plays an important role in cholesterol metabolism. SF-1REs have also been identified within promoters of other genes, whose products act as pituitary-derived mediators of gonadal steroidogenesis, such as luteinizing hormone (LH) and follicle stimulating hormone (FSH). SF-1 is expressed within pituitary gonadotropes and the ventro-medial hypothalamus (VMH) and is essential for the regulation of the hypothalamic–pituitary–steroidogenic axis.

Detailed *in situ* studies showed SF-1 expression occurs early during development of the gonads and the adrenal primordium, suggesting a role in steroidogenic organ development. This was confirmed by genetic studies in mice where SF-1 was inactivated using targeted gene disruption and these mice exhibited adrenal and gonadal agenesis. In addition, these mice had impaired gonadotrope function, ablation of a specific region of the hypothalamus and male-to-female sex reversal of the internal and external genitalia. Furthermore, SF-1 regulates the müllerian inhibiting substance (MIS) gene, a key mediator of male sexual differentiation. However, it remains unclear whether SF-1 activates or represses MIS promoter activity. These results show that SF-1 plays essential roles in the regulation of the hypothalamic–pituitary–steroidogenic organ axis at various levels, and in the complex processes of endocrine differentiation.

Transfection studies have shown that constitutive SF-1 activity is generally higher in cells that are steroidogenic in comparison to nonsteroidogenic cells, indicating SF-1 activity is cell type specific. Its activity can be influenced by secondary messengers including cAMP and phorbol esters, suggesting phosphorylation may play a role in SF-1 activation. Also, several proteins including receptor coactivators, and other transcription factors, have been reported to regulate SF-1 function. All these activities occur in a cell- and promoter-specific manner.

In addition, specific oxysterols such as 25-, 26-, or 27-hydroxycholesterol, known suppressors of cholesterol biosynthesis, also enhance SF-1-dependent transcriptional activity via specific HREs. These compounds have also been shown to increase SF-1 dependent transactivation of the human StAR promoter as well as other promoters of human steroidogenic P450 genes. Again, these responses are cell type specific. Although it is clear that certain oxysterols can activate SF-1, further studies are needed to determine whether they act as *bona fide* physiological ligands.

LRH-1 is encoded by a different gene and is expressed in the liver and pancreas, tissues that do not express SF-1. Both receptors share similar amino acid sequences, especially within the DNA binding domain and, like SF-1, LRH-1 also appears to bind to its target sequence as a monomer. The α-fetoprotein gene has been identified as a potential target for LRH-1.

3.3.3. THE NGFI-B FAMILY

Nerve growth factor induced gene B (NGFI-B, also called *Nurr77, TR3/NAK1, TIS1, N10*) was originally isolated as a gene induced by nerve growth factor (NGF) in the rat pheochromocytoma cell line PC12. Other family members include Nurr1 (Nur-related factor 1, also called RNR-1, NOT,TINUR, RNR-1) isolated from a mouse brain cDNA library and NOR-1 (neuron derived orphan receptor, also called MINOR, TEC, CHN), identified from cultured rat fetal forebrain cells. Their genes are rapidly and transiently induced by a wide variety of stimuli and are classified as immediate-early genes. The inducers include chemical stimuli such as cAMP, phorbol esters, peptide hormones, growth factors, neurotransmitters, and physical stimuli such as membrane depolarization, mechanical agitation, and a magnetic field. Furthermore, activation of T cells via the T-cell receptor leads to rapid induction of *NGFI-B* gene expression. NGFI-B and Nurr1 bind to nuclear receptor half-sites as monomers. Receptor activity on the NGFI-B response element (NBRE) can be regulated through phosphorylation of a serine residue within its DBD. Both of these receptors can also dimerize with RXR (via NBRE and DR5 sequences) and the heterodimers respond to RXR ligands. In addition, NGFI-B can bind as a homodimer. In contrast, NOR-1 binds as a monomer and has not been reported to heterodimerize with RXR. The *NGFI-B* subfamily of genes is expressed in several tissues, notably in the pituitary, thymus, and adrenal gland. The three receptor isoforms are differentially expressed in the brain at different stages during development. Nurr1 is highly expressed early during midgestation, whereas NOR-1 expression is observed at later times and *NGFI-B* is expressed in the adult brain. In rat brain cell cultures, NOR-1 antisense oligonucleotides induce cell migration and aggregation with extension of neuritelike processes. Nurr1-deficient mice demonstrate a lack of normal development of midbrain dopaminergic neurons and die soon after birth. These findings implicate the *NGFI-B* family members as critical players in neuronal differentiation and function; however, the exact roles need to be elucidated. *NGFI-B* and *Nurr1* play a role in regulation of the hypothalamic–pituitary–adrenal (HPA) axis through the regulation of corticotropin releasing factor (CRF) and POMC (proopiomelanocorticotropin), a precursor for several neuropeptides including adrenocorticotropic hormone (ACTH). *NGFI-B* also regulates *CYP21*, a steroid hydroxylase gene encoding an enzyme involved in glucocorticoid and mineralocorticoid biosynthesis. In addition, *NGFI-B* plays a role in programmed cell or apoptosis of T cells following T-cell antigen receptor (TCR) stimulation. However, NGFI-B knock-out mice do not exhibit grossly abnormal functions of the HPA axis or TCR-mediated apoptosis, suggesting a possible redundancy among *NGFI-B* family members. These receptors may play a role in cancer based on the identification of fusion genes, between a gene involved in Ewing's sarcoma (EWS), and NGFI-B. In addition, treatment of breast cancer cells with retinoids leads to an alteration in the expression of *NGFI-B*, *Nurr1*, and *NOR-1* genes. Retinoids exhibit antiproliferative and differentiating effects in a large number of tissues and cells and are one of the most promising compounds for cancer treatment and prevention. Thus, together with their ability to be regulated by retinoids and through heterodimerization with RXR, these receptors may play important roles in retinoid signaling in cancer cells.

3.3.4. The ROR Family

The retinoic acid-related orphan receptor (ROR) family, also called retinoid Z receptor (RZR), consists of three subtypes, α, β, and γ. Their DBDs are about 70% homologous to the RAR; however, they do not appear to be activated by retinoids. The *RORα* gene gives rise to multiple transcripts (*RORα1–4*) that share a common DBD and LBD but contain distinct N-terminal domains that can modulate their DNA binding activity. RORα isoforms bind as monomers to HREs (ROREs) and function as constitutive activators in transient transfection experiments. The γF-crystallin gene is a target for this receptor. The gene encoding RORα resides on mouse chromosome 9 in proximity to the *staggerer* gene (*sg*). Mice homozygous for a mutation in the *sg* gene show tremor, body imbalance, hypotonia, as well as a small size and die shortly after weaning. The cerebral cortex of these mice exhibits a cell-autonomous defect of Purkinje cells that are small and abnormal in morphology, and are reduced by 60–90% in numbers. RORα is involved in Purkinje cell development and the gene encoding RORα in *sg* mice harbors a 6.5-kb genomic deletion that removes part of its LBD. RORα knock-out mice generated by disruption of its DBD using homologous recombination also show defects similar to the *sg* mice. Thus, the absence of functional RORα is responsible for the *sg* defect. Other defects in the *sg* mouse and RORα knockout animals, particularly in the immune system have also been identified.

RORβ is expressed in the brain and retina and functions as a cell-type-specific activator with constitutive activity. RORβ is activated by melatonin, the pineal gland hormone, and a class of thiazolidinediones having antiarthritic activity. However, these observations are controversial because they have not been reproduced in other laboratories.

The mouse ROR gamma (*mRORγ*) gene has a complex structure consisting of 11 exons separated by 10 introns spanning more than 21 kb of genomic DNA. *RORγ* is localized to chromosome 3 in mice, a region that corresponds to band 3F2.1–2.2, and is expressed as two mRNAs, 2.3 and 3.0 kb in size, that are derived by the use of alternative polyadenylation signals. Human *RORγ* has mapped to chromosome region 1q21. *RORγ* is highly expressed in the skeletal muscle and thymus.

3.4. Atypical Nuclear Receptors

3.4.1. DAX-1

DAX-1 (Dosage-sensitive sex reversal-adrenal hypoplasia congenita critical region on the X chromosome) is an unusual member of the nuclear receptor superfamily of transcription factors that contains no canonical zinc finger or any other known DNA binding motif. Mutations of *DAX-1* cause a complex endocrine phenotype that includes adrenal hypoplasia congenita (AHC) with hypogonadotropic hypogonadism. In humans with AHC, the adrenal cortex is disorganized, cytomegalic, fetal in appearance, and the definitive adult cortex does not develop, indicating *DAX-1* is essential for normal adrenal development. *DAX-1* maps to Xp21, a region associated with dosage-sensitive sex reversal wherein males with an extra copy of Xp21 have impaired testicular development and consequent sex reversal. Thus, *DAX-1* is also important for sex determination in humans.

Although *DAX-1* lacks the DBD present within nuclear receptors, it is capable of binding to unique DNA hairpin structures in vitro where binding is efficient for stems composed of 10–24 nucleotides but less efficient with shorter stems. Although a loop is required for this binding there is no strong sequence specificity. Such hairpin structures are also formed in vivo and found in the promoters of the *DAX-1* and *StAR* (an essential factor required for steroidogenesis) genes. As a consequence of binding to these structures, *DAX-1* acts as a powerful inhibitor of *StAR* expression and blocks steroid hormone synthesis in Y-1 adrenocortical tumor cells. This repression occurs through a mechanism of allosteric inhibition by which *DAX-1* impairs the binding of another nuclear receptor, *SF-1*, that acts as a transcriptional activator of *StAR* expression.

Expression of *DAX-1* and *SF-1* exhibit a striking colocalization within the adult adrenal cortex, gonads, hypothalamus, and pituitary and demonstrate striking similarities in their mutant phenotypes. These data suggest both receptors regulate common endocrine pathways. This may occur completely independently of each other, or the two receptors may either act sequentially or interact directly, to regulate target genes crucial for endocrine development. Interestingly, a putative *SF-1* binding site has been identified within the *DAX-1* gene promoter through which *SF-1* has been postulated to regulate *DAX-1* expression. However, *SF-1* knockout mice continue to express *DAX-1* in the embryonic gonad and hypothalamus, and *DAX-1* expression in the fetal testis is independent of *SF-1*, suggesting that *DAX-1* expression does not require *SF-1* in these cases. It is unlikely that *DAX-1* regulates *SF-1* expression, as *SF-1* expression either precedes or coincides with that of *DAX-1*, and the *SF-1* knockout phenotype is more severe than that seen in patients with *DAX-1* mutations. Another possibility is that the two receptors interact to form heterodimers. Although both proteins have been shown to physically interact in vitro, it is unclear whether this occurs in vivo. Thus, the precise mechanisms by which *DAX-1* and *SF-1* coregulate common pathways leading to endocrine differentiation remain undefined.

3.4.2. SHP

Small heterodimer partner (SHP) is an orphan member of the nuclear hormone receptor superfamily similar to *DAX-1*, in that it contains a putative dimerization and ligand-binding domain but lacks the conserved DNA binding domain. However, unlike DAX-1, no putative DNA binding sequences have been identified for SHP. SHP interacts with several conventional and orphan members of the receptor superfamily, including RXR, TR, ER, and CAR. It is highly expressed in the spleen and small intestine of adult human tissues. It is also expressed in the fetal liver and adrenal gland. In mammalian cells, SHP specifically inhibits DNA binding and transactivation by the superfamily members with which it interacts. It therefore appears to function as a negative regulator of receptor-dependent signaling pathways. At least in the case of ER, competition for coactivator binding appears to be one mechanism by which SHP inhibits receptor function.

4. IN SEARCH OF A LIGAND

One of the most exciting aspects of orphan receptor research is the search and identification of receptor ligands. The identification of natural ligands can, in some instances, lead to unexpected and unique insights into physiological signaling and potentially novel approaches to treat human disease. Besides other biological properties described previously, elucidation of its ligand is key to understanding orphan receptor function. Over the past few years, a number of different laboratories have identified novel ligands/activators for several orphan receptors. Here, we discuss some approaches to identify ligands.

The most common and successful approach in the search for new ligands or activators has been the cell-based cotransfection assay. In this assay, plasmid DNA encoding the receptor of interest, along with another plasmid encoding a reporter gene capable of expressing an easily measurable product, are introduced into cultured cells. This can be done via several different means including DNA precipitation using calcium phosphate, trapping the DNA in lipid vesicles, or electroporation. These transient transfection techniques allow the DNA to enter the cell and be expressed under appropriate conditions. The reporter gene is driven by a promoter containing multiple response elements specific for the receptor, such that the expression of the reporter gene is dependent on regulation of orphan receptor activity by a potential ligand. For many receptors, in the absence of ligand, no transcriptional activity is observed. The presence of ligand leads to transcriptional activation which is measured by monitoring reporter activity (Fig. 6). With receptors that exhibit high constitutive activity, ligands that either repress or enhance their basal activity may be identified. This assay can be carried out in a 96- or 384-well plate, and, using this approach it is possible to screen a large number of potential ligands individually, or in pools of 5–10 compounds. The source for these compounds can be natural or

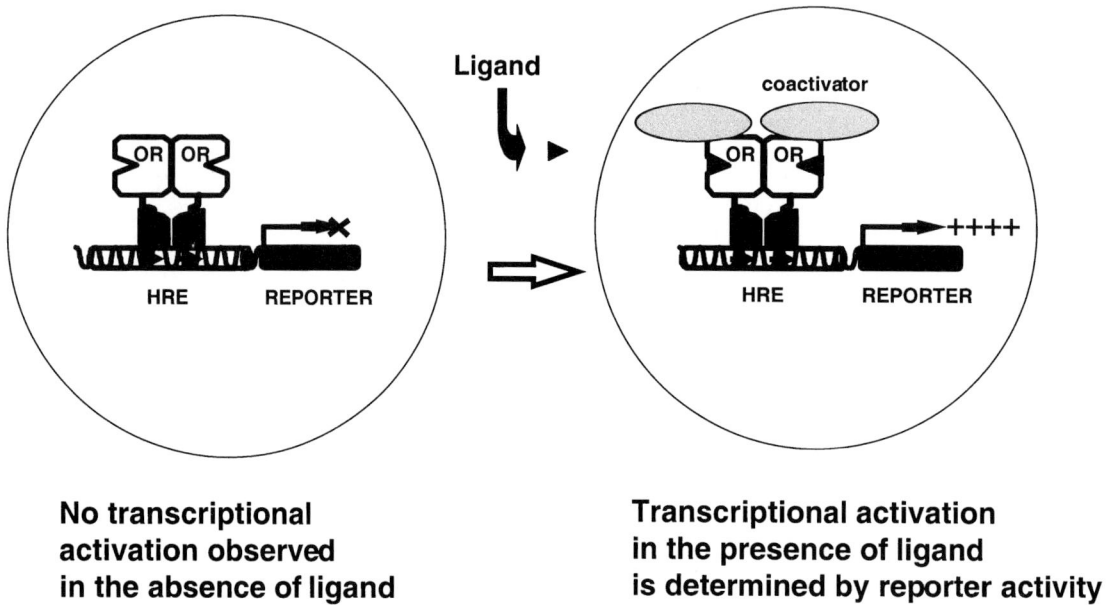

Fig. 6. Cell-based assay for measuring ligand-dependent transactivation. Shown above is a depiction of the transient assay. Plasmid DNAs encoding the receptor of interest and a reporter linked to the receptor HRE are introduced into a cell. In the absence of ligand no transcriptional activation is observed. In the presence of ligand, ligand-bound receptor interacts with coactivators leading to an increase in transcriptional activity. This is determined by measuring the expression of the reporter gene (*see* text for more detail).

synthetic, chemical libraries, tissue or cell extracts, whole embryos, or serum. Compounds may modulate receptor function by interacting directly with the receptor, or through metabolic conversion within a cell to the real ligand. Furthermore, direct binding can be determined through biochemical based receptor binding experiments using labeled compounds.

Recent developments in our understanding of the mechanism of ligand-dependent nuclear receptor function have led to the development of novel biochemical assays to discover receptor ligands. As mentioned earlier, a conformational change occurs within the LBD of a receptor on ligand binding, and leads to altered physical interactions between receptor LBD and cofactors. These interactions can be measured using various biochemically based technologies. An example of this is a method called fluorescence resonance energy transfer (FRET). This technology depends on the unique property of a lanthanide (e.g., europium [Eu]) to display a pronounced difference in its excitation and emission maxima and a long fluorescence time, on binding its chelate. Glutathione S-transferase (GST)-receptor fusion proteins are indirectly labeled with the lanthanide, using Eu-labeled GST antibodies. A second fluorophore is attached to the cofactor of interest which mediates the FRET.

The cofactor–fluorophore complex can be added to microtiter plates along with fluorogenic receptor fusion proteins (labeled with Eu–GST antibodies), and ligand-dependent receptor–cofactor interactions can be studied by FRET in the presence or absence of ligand Excitation of europium at 337 nm results in emission at 620 nm; the second fluorophore can absorb this if the receptor and cofactor are in close proximity and subsequently emits at 665 nm. Thus, this transfer of energy between the two fluorophores is a direct measure of the proximity of the two proteins (Fig. 7). A major advantage of this type of approach is the *high-throughput* nature, as several hundreds of thousands of compounds can be screened in a very short period of time. For this reason, this approach is attractive to pharmaceutical companies involved in identifying synthetic ligands from large compound libraries. A major disadvantage, however, is compounds that may be precursors to the real ligand, and would have to be metabolized within a cell to generate the true ligand, would not be detected.

Owing to various advantages and disadvantages associated with any given approach it is thus wise to employ different methods, cell-based and biochemical, to identify receptor ligands. Although in many instances this can be considered analogous to looking

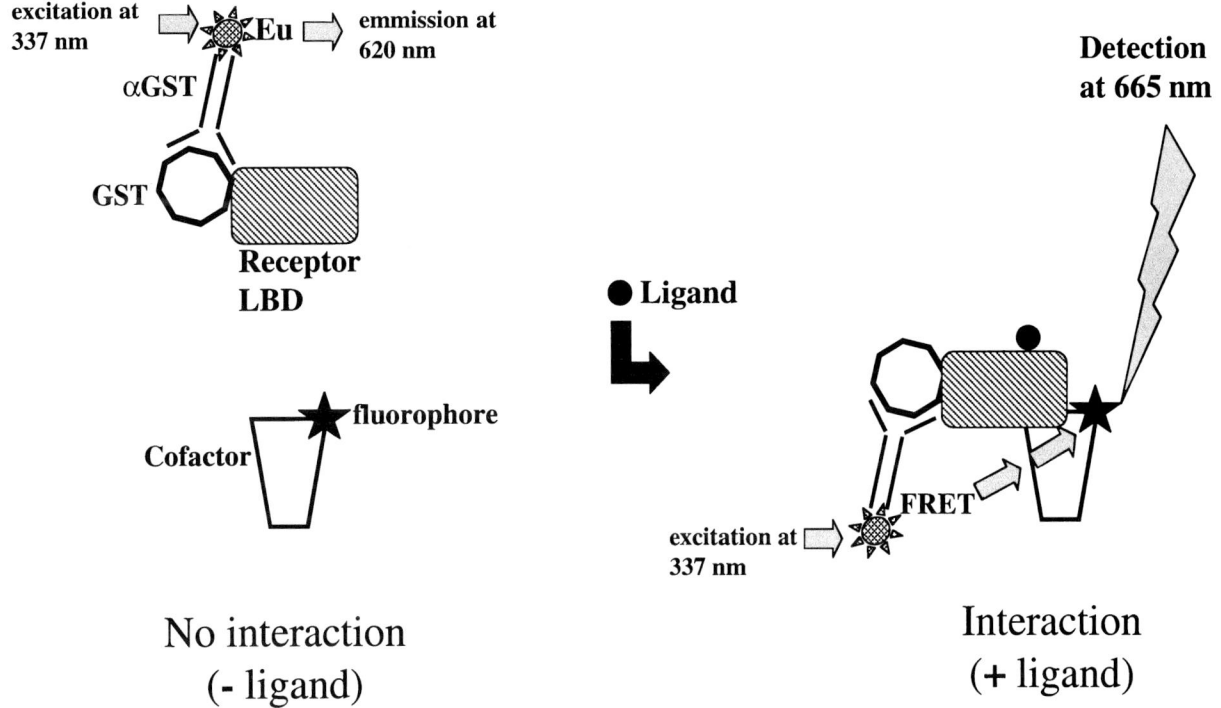

Fig. 7. Ligand-dependent receptor–cofactor interactions in the presence or absence of ligand. As described in the text, a GST-receptor fusion protein is indirectly labeled with a lanthanide (e.g., Eu) bound to an anti-GST antibody (αGST). A cofactor of interest is bound to another fluorophore (an acceptor molecule) that mediates the fluorescence energy resonance transfer (FRET). If the receptor interacts with the cofactor in the presence of ligand, FRET occurs upon excitation of Eu at 337 nm, resulting in a specific emission at 665 nm. This assay may also be modified using receptors bound to their specific HREs.

for a needle in a haystack, the promise of discovering new biological pathways and ligands has provided the motivation for various laboratories to embark upon the enticing search for receptor ligands. Some of these searches have led to the discovery of important hormone signaling pathways and are discussed in the next section.

5. NOVEL HORMONE SIGNALING PATHWAYS

About 8–10 yr ago it was predicted by leaders in the field of nuclear receptors that the orphan receptors would provide much excitement in the future. The study of these receptors holds many promises and opportunities, not only for discovering new biology, but also for the identification of novel and sometimes previously unsuspected classes of ligands. By identifying new ligands, potentially new principles of physiology can be determined. These predictions have come true, and in fact, some recent discoveries of unique and unanticipated nuclear receptor signaling pathways have created a fervor in the field. Some of these novel pathways are described below and summarized in Fig. 8.

5.1. FXR/RIP14: Farnesoid and Retinoid Signaling Pathways

The mevalonic acid biosynthetic pathway serves a critical role in higher eukaryotes leading to the synthesis of cholesterol, bile acids, retinoids, vitamin D, steroid hormones, carotenoids, porphyrin, ubiquinone, dolichol, and farnesylated proteins. The last precursor common to all branches of this pathway is farnesyl pyrophosphate, which serves as a key metabolic intermediate. The activation of FXR by farnesol, an isoprene intermediate in this pathway, was the first example of a receptor regulated by intracellular metabolites, suggesting the existence of a potentially novel signaling system in vertebrates. Activation of FXR by isoprenoids and its expression in isoprenoidogenic tissues suggests it plays a critical role in the mevalonate biosynthetic pathway.

Natural retinoids such as ATRA activate RIP14, the mouse ortholog of FXR. The synthetic RAR agonist

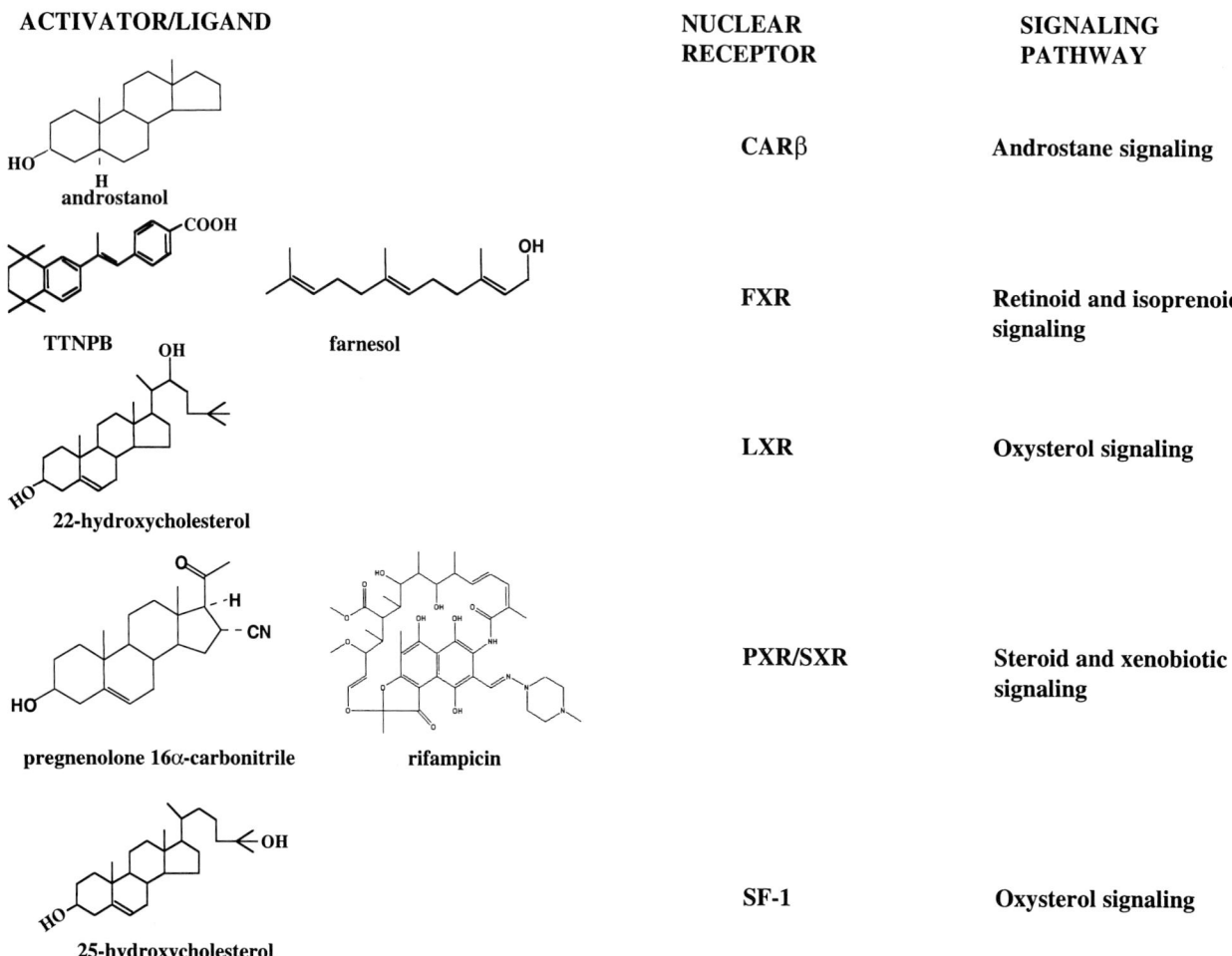

Fig. 8. Structures for orphan nuclear receptor activators and ligands and their implication in mediating various novel signaling pathways.)

TTNPB is also an activator, although both compounds activate at supraphysiological concentrations and in a time-dependent manner, suggesting that a retinoid metabolite or a novel terpenoid may be a ligand for FXR–RIP14. Thus, FXR may play a role in retinoid signaling acting as a novel retinoic acid receptor.

5.2. LXR and SF-1: Oxysterol Signaling Pathways

The LXRs have been shown to be activated by a specific class of naturally occurring oxidized derivatives of cholesterol (oxysterols). These include 22(R)-hydroxycholesterol (OHC), 24(S)-OHC, and 24,25(S)-epoxycholesterol, which act as intermediates in a variety of biologically important pathways, including bile acid production and steroid hormone biosynthesis. Activation occurs in the micromolar range and is compatible with the concentrations reported for these compounds in vivo. Activation of LXRα by these molecules would lead to induction of its target gene *CYP7a*, that encodes the rate-limiting enzyme in bile acid production. Hence, LXRα appears to play an important role in the conversion of cholesterol to bile acids. Interestingly, oxysterols such as 25-, 26-, and 27-(OHC), potent suppressors of cholesterol biosynthesis and cellular uptake, do not activate LXR significantly. Thus, while one class of oxysterols regulates cholesterol biosynthesis and cell entry, the other acts to promote cholesterol catabolism.

SF-1 also plays an important role in cholesterol metabolism acting at multiple control points of the steroidogenic pathway. In its case, 25-, 26-, and 27-(OHC) function as SF-1 activators at concentrations comparable to those previously shown to inhibit cholesterol biosynthesis. The promoters of specific genes (e.g., *StAR*) whose products act at critical steps in cholesterol metabolism are activated by these oxysterols via *SF-1*. StAR also increases P450c27 activity, which converts cholesterol to 26- or 27-(OHC), and is expressed in steroidogenic tissues. Activation of *SF-1* by oxysterols would amplify the transcription of *SF-1* target genes and "lock" the cell in a steroidogenic mode. These results suggest a novel mechanism for increased steroidogenesis.

Interestingly, the different oxysterols differ in their rank order potency for LXR and *SF-1*, with 25-(OHC) preferentially activating *SF-1* and 22(*R*)-OHC and 24-(OHC) preferentially activating LXR. Thus, there appears to be a group of nuclear receptors that responds to different classes of oxysterols.

The above studies have led to the important concept of intracellular oxysterols as "feed-forward activators" of nuclear receptors that act as "sensors" and control crucial steps within key metabolic pathways.

5.3. PXR–SXR: Steroid and Xenobiotic Signaling Pathways

Murine PXR (mPXR) activity is modulated by a number of natural and synthetic steroids including naturally occurring pregnanes and synthetic glucocorticoids and antiglucocorticoids. Some of these steroids, called *catatoxic* steroids, are known for their ability to induce cytochrome P450 gene expression which confers nonspecific protection or resistance against a number of drugs and xenobiotic compounds by accelerating their catabolism. An interesting feature of these catatoxic compounds is that although many are steroid receptor agonists (e.g., dexamethasone) a large number are antagonists (e.g., pregnenolone-16α-carbonitrile, PCN). Both these compounds can activate mPXR, indicating a novel pathway that is independent of other steroid receptors (e.g., GR). Activation of mPXR by these steroids may lead to their clearance in the liver through the actions of *CYP3A1*, an mPXR target gene. Besides activating *CYP450* genes, PCN is also known to affect cholesterol homeostasis by regulating its biosynthesis, storage, and secretion. Thus, PXR activators may regulate cholesterol and steroid homeostasis through feedback and feed-forward loops respectively.

Human PXR (hPXR) or SXR is also activated by a large number of steroid agonists and antagonists such as estrogen and tamoxifen, drugs such as rifampicin and nifedipine, xenobiotic compounds, and dietary bioactive compounds such as phytoestrogens. Many of these compounds induce the catabolic CYP450 enzymes including CYP3A4, which metabolize a number of steroids and xenobiotics including 60% of clinically relevant drugs in humans. Rifampicin is involved in drug–drug interactions by inducing CYP3A4 and altering the metabolism of other drugs. Long-term administration of rifampicin to patients with tuberculosis leads to increased steroid clearance. This can, in some cases, lead to misdiagnosis of Cushing's syndrome owing to an increase in urinary steroid levels. Steroid levels return to normal when the drug is withdrawn. In patients with Addison's disease, who mainly lack the ability to synthesize adrenal steroids, administration of rifampicin leads to a rapid depletion of endogenous steroids. The activation of PXR–SXR by rifampicin implicates this receptor in drug–drug interactions and drug metabolism. Also, activation by a large number of steroids and xenobiotic compounds suggests it acts to clear circulating receptor ligands and other compounds in the liver in humans as well. Clearance of steroid hormones is important in maintaining hormonal homeostasis where the ligand can be cleared from the blood through the liver, after mediating its effect on target tissues through specific receptors.

Thus, PXR–SXR is a novel steroid and xenobiotic sensor that acts as a broad-specificity receptor involved in the clearance and detoxification of a number of naturally occurring as well as ingested bioactive compounds. The aforementioned studies indicate the existence of another novel steroid hormone signaling pathway, via a unique receptor involved in sterol and hormonal homeostasis as well as drug metabolism and interactions.

5.4. CARβ: Androstane Signaling Pathway

CARβ is a novel steroid receptor that appears to participate in a unique mode of ligand-dependent deactivation in response to steroids. It exhibits significant constitutive activity when combined with RXR, but in response to the androstane metabolites, androstenol and androstanol, this activity is switched off. These ligands are naturally occurring, and are thought to function as inverse agonists owing to their ability to reverse transcriptional activation by nuclear receptors. This effect is stereospecific; only 3α-hydroxy, 5α-reduced androstanes are inhibitors and appear to act by promoting coactivator release from the LBD. CARβ therefore defines an unexpected androstane signaling pathway that functions in a manner opposite to that of the conventional nuclear receptor pathways.

These discoveries have given rise to several important concepts in nuclear receptor signaling. First, the activation of a number of these receptors by intracellular metabolites indicates the presence of an intracrine signaling pathway in addition to conventional endo-

Table 3
Nuclear Receptor Based Drugs

Nuclear Receptor	Drug	Activity on Receptor	Indication
Androgen receptor	Casodex	Antagonist	Metastatic prostate cancer
Estrogen receptor	Nolvadex	Antagonist	Treatment and prevention of breast cancer
Glucocorticoid receptor	Decadron	Agonist	Endocrine disorders and inflammation
Mineralocorticoid receptor	Aldactone	Antagonist	Treatment of hyperaldosteronism
Peroxisome proliferator activated receptor γ	Rezulin	Agonist	Diabetes
Progesterone receptor	Mifepristone	Antagonist	Contraception
Retinoic acid receptor	Accutane	Agonist	Acne
Retinoid X receptor	Targretin	Agonist	Certain types of cancer
Thyroid hormone receptor	Synthroid	Agonist	Treatment of certain types of hypothyroidism
Vitamin D receptor	Rocaltrol	Agonist	Treatment of osteoporosis

Representative drugs that act via members of the intracellular receptor superfamily are listed above (only trade names are given). Their mode of action may be either through activating (agonist) or inhibiting (antagonist) receptor function.

crine pathways. Second, also in contrast to traditional endocrine systems, activation for some of the receptors requires concentrations in the high nanomolar or mid- to high micromolar range (e.g., PXR, FXR, SF-1, LXR) rather than the low nanomolar concentrations required for activation of conventional receptors (e.g., RAR, VDR, TR). Third, oxysterols that had been previously demonstrated to function as repressors have now been shown to function as activators through a unique class of nuclear receptors (e.g., LXRs and SF-1). Fourth, the activation of PXR–SXR by steroid metabolites and xenobiotics suggests a mechanistic explanation for the paradoxical features of catatoxic compounds. This also led to the "steroid sensor hypothesis" that proposes that detoxification and clearance of various endogenous hormones, dietary steroids, drugs, and xenobiotics are regulated through the action of a few broad specificity sensing receptors rather than a large number of specific ones. Finally, the ligand-mediated inhibition, rather than activation, of CARβ by androstane metabolites suggests the possibility of naturally occurring inverse agonists or antagonists defining a unique mode of nuclear receptor signaling.

6. ORPHAN RECEPTORS AS DRUG TARGETS

The nuclear receptor superfamily has attracted particular interest not only in academia but also in the pharmaceutical industry, because members of this family are the targets of a large numbers of drugs. In fact, all ligand-dependent receptors serve as novel therapeutic targets (Table 3). Thus by analogy, the orphan receptors serve as promising new molecular targets for drug discovery.

6.1. Orphan Receptors That Have Set a Precedent for Drug Discovery

During the past few years, the identification of therapeutic agents acting through two different orphan receptors, RXRs and PPARs, has provided an important proof-of-principle for orphan receptors as targets for future drug discovery efforts. Research on both receptors has led to the development of new drugs that are currently being used to treat diseases such as lymphomas, leukemias, breast cancer, HIV-related Kaposi's sarcoma, and type II diabetes. These two receptors provide examples illustrating the potential for orphan receptors as *bona fide* molecular targets for drug discovery.

6.1.1. RXR

The RXR family of receptors contains three subtypes called RXR α, β, and γ. RXRα was originally cloned from the liver and after a search for activators, it was found to be transcriptionally responsive all-*trans*-retinoic acid. However, it was hypothesized that a metabolite of all-*trans*-retinoic acid was the active ligand when it was determined that all-*trans*-retinoic acid did not bind to RXR. This led to the first successful screening for a novel orphan receptor ligand resulting in the identification of the novel vitamin A metabolite, 9-*cis*-retinoic acid, as a high-affinity ligand for the retinoid X receptor (RXR). Interestingly, 9-*cis*-retinoic acid is also a high-affinity ligand for the RARs and thus is a pan-agonist for the retinoid receptors. The ability of retinoids such as 9-

cis-retinoic acid to function as potent differentiation and anticancer agents suggested that RXRs play an important role in prevention and treatment of a variety of neoplasms, including leukemia, lung cancer, breast cancer, and Kaposi's sarcoma. This has led to the clinical development of this compound for cancer therapy and it was recently approved for the treatment of AIDS-related Kaposi's sarcoma.

During the past few years numerous laboratories have focused on the design and synthesis of RXR-selective ligands. These, synthetic RXR ligands, termed rexinoids, have been characterized in a variety of endocrine signaling pathways. For example, in preclinical models of cancer prevention and therapy, the RXR selective compound LGD1069, also referred to as Targretin, is a highly efficacious cancer agent. Furthermore, the combination of Targretin and tamoxifen is more effective than treatment with either agent alone. Finally, in a recent series of studies examining mammary tumors that failed to respond to tamoxifen, Targretin caused complete or partial regression in the majority of the breast tumors. Thus, RXR compounds show promise in preclinical trials of cancer and Targretin has entered human clinical trials for cancer therapy.

Because RXR serves as a common heterodimeric partner for many nuclear receptors, an important question that has been further explored is whether rexinoids (e.g., RXR-selective agonists) may replicate the activity of ligands for several of these receptors. Because PPARγ is a critical transcription factor that mediates adipocyte differentiation and is a target for antidiabetic agents, a series of investigations has studied whether RXR ligands would alter insulin and glucose signaling. The results of these studies have shown that RXR agonists induce adipocyte differentiation, and in mouse models of non-insulin-dependent diabetes mellitus (NIDDM) and obesity these compounds can function as insulin sensitizers decreasing hyperglycemia and hyperinsulinemia. This antidiabetic activity can be further enhanced by combination treatment with a PPARγ agonist, such as the thiazolidinediones. These data suggest that activation of RXR:PPARγ dimer with rexinoids may provide a novel and effective treatment for NIDDM.

6.1.2. PPAR

Members of the PPAR subfamily are one of the many nuclear receptors that form heterodimers with the RXRs. There are three PPAR subtypes referred to as PPAR α, β, and γ and these receptors differ in their tissue distribution as well as in their relative ability to be activated by a variety of different ligands. PPARα is highly expressed in tissues such as liver and kidney and is activated by certain long-chain fatty acids, including arachidonic acid, linoleic acid, and certain eicosanoids including 8(S)-hydroxyeicosatetraenoic acid (8[S]-HETE). In addition, the fibrates, which are used in the treatment of hyperlipidemia, activate PPARα. Activation of PPARα has been shown to regulate the transcription of several key enzymes involved in fatty acid metabolism including β-fatty acid oxidation and bifunctional enzyme. Thus, PPARα specific drugs such as gemfibrozil and fenofibrate function by regulating fatty acid metabolism leading to the lowering of lipids in humans.

In contrast to the role of PPARα in lipid catabolism, PPARγ is predominantly expressed in adipocytes and is thought to stimulate adipogenesis through regulation of genes involved in adipocyte differentiation. Several naturally occurring ligands for PPARγ have recently been identified, including the prostaglandin metabolite, 15-deoxy-Δ12,14 prostaglandin J2 and components of the lipid coat low-density lipoprotein, including 9- and 13-hydroxyoctadecadienoic acid (9- and 13-HODE). In addition, the antidiabetic class of drugs such as the thiazolidinediones (TZDs), which function in humans as insulin sensitizers are potent activators of PPARγ. During the past few years numerous thiazolidinediones have undergone human clinical trials and are effective as insulin sensitivers. Interestingly, there is a good correlation in preclinical models of NIDDM between the potency of PPARγ agonists and their ability to lower glucose in vivo. These data strongly support the hypothesis that PPARγ is the molecular target for the insulin sensitization of these new class of therapeutic agents. The discovery that PPARγ is a key regulator of these processes has led to the development of new pharmacological approaches to combat diseases such as diabetes and Syndrome X (obesity, hyperlipidemia, hypertension, and adult-onset diabetes).

The RXRs and the PPARs demonstrate that orphan receptors, once "adopted," have the future promise to serve as targets for new therapeutic agents. Although quite early in our understanding of the biological activity of the many orphans, emerging biology of the LXRs, HNF4, and SF-1 suggest that modulation of their activity with agonists or antagonists may regulate key pathways in cholesterol, bile acid metabolism, and other important metabolic pathways. Ligands for receptors such as members of the NGFI-B family may modulate central nervous system (CNS) function. Based on the receptors whose ligands are well characterized, it is tempting to speculate that many orphan receptors for which ligands are discovered will serve as therapeutic targets for the treatment of distinct diseases.

7. SUMMARY AND PERSPECTIVES

The nuclear receptors, including the orphan receptors, currently represent the largest family of transcription factors. The study of these proteins during the past 10 yr has made tremendous impact on our knowledge of physiology and metabolic regulation. Indeed new ideas in endocrinology have emerged through the identification of novel signaling molecules acting via nuclear receptors that at one time were considered orphans. The use of genetic techniques has aided immensely in our understanding the biological role of these receptors through gene knockout and knock-in studies. Since many nuclear receptors play important roles in development and in the adult, the ability to generate adult- specific and/or tissue-specific knockouts will become increasingly important. In molecular genetics, mutations in the genes for HNF-4, DAX-1, and RORα have provided direct links between receptors and diseases such as diabetes and adrenal hypoplasia in humans and the *staggerer* phenotype in mice. Structural studies with X-ray crystallography and other methods have provided a wealth of information regarding the mechanism of nuclear receptor function in general. These studies have led to the identification of conserved structural determinants of the LBD, suggesting it may be possible to predict or design ligands for orphan receptors in the future. Although remarkable progress has been made, many questions remain unanswered. Which orphans are ligand dependent and which ones are ligand independent? How many more orphans and signaling pathways remain undiscovered? Which of these receptor-dependent pathways are going to lead the way as novel therapeutic targets for the treatment of human diseases? While one can speculate on some of these questions, one thing is clear. Based on discoveries made in the past decade, the study of this family of transcription factors promises to provide continued excitement well into the next millennium.

SELECTED READINGS

Mangelsdorf DJ, Evans RM. The RXR heterodimers and orphan receptors. Cell 1995; 83:841–850.

Willy PJ, Mangelsdorf DJ. Nuclear orphan receptors: the search for novel ligands and signaling pathways. Horm Signal 1998; 307–358.

Blumberg B, Evans RM. Orphan nuclear receptors—new ligands and new possibilities. Genes Dev 1998; 12:3149–3155.

Enmark E, Gustafsson J-A. Orphan nuclear receptors—the first eight years. Mol Endocrinol 1996; 10:1293–1307.

Maruyama K, Tsukada T, Ohkura SB, Hosono T, Yamaguchi K. The NGFI-B subfamily of the nuclear receptor superfamily (Review). Int J Oncol 1998; 12:1237–1243.

Zanaria E, Muscatelli F, Bardoni B. An unusual member of the nuclear hormone receptor superfamily responsible for X-linked adrenal hypoplasia congenita. Nature 1994; 372: 635–641.

Qiu Y, Pereira FA, DeMayo FJ. Null mutation of the COUP-TFI results in defects in morphogenesis of the glossopharyngeal ganglion, axonal projection, and arborization. Genes Dev 1997; 11:1925–1937.

Parker KL, Schimmer BP. Steroidogenic factor 1: a key determinant of endocrine development and function. Endocrin Rev 1997; 18:361–377.

Lala DS, Syka PM, Lazarchik SB. Activation of the orphan nuclear receptor SF-1 by oxysterols. Proc Natl Acad Sci 1997; 94:4895–4900.

Peet DJ, Turley SD, Ma W. Cholesterol and bile acid metabolism are impaired in mice lacking the nuclear oxysterol receptor LXRα. Cell 1998; 93:693–704.

Rastinejad F, Perlmann T, Evans RM, Sigler PB. Structural determinants of nuclear receptor assembly on DNA direct repeats. Nature 1995; 375:203–211.

Dussault I, Fawcett D, Matthyssen A, Bader J-A, Giguere V. Orphan nuclear receptor RORα-deficient mice display the cerebellar defects of staggerer. Mech Dev 1998; 70:147–153.

Yamagata K, Furuta H, Oda N. Mutations in the hepatocyte nuclear factor-4α gene in maturity-onset diabetes of the young (MODY1). Nature 1996; 384:458–460.

Hertz R, Magenheim J, Berman I, Bar-Tana J. Fatty acyl-CoA thioesters are ligands of hepatic nuclear factor-4α. Nature 1998; 392:512–516.

Zhou G, Cummings R, Li Y. Nuclear receptors have distinct affinities for coactivators: characterization by fluorescence resonance energy transfer. Mol Endocrinol 1998; 12:1594–1604.

22 Coactivators and Corepressors

Neil J. McKenna, Zafar Nawaz, Sophia Y. Tsai, and Ming-Jer Tsai

CONTENTS
   COACTIVATORS
   COREPRESSORS
   SELECTED READINGS

1. COACTIVATORS

1.1. Eukaryotic Transcription

At the vast majority of eukaryotic promoters, transcription is mediated by RNA polymerase II (RNA pol II). Activated transcription at these promoters is believed to require assembly of a transcriptional preinitiation complex, comprising general transcription factors (GTFs) such as TFIIB, TFIID, a multicomponent complex composed of TATA binding protein (TBP) and the TBP-associated factors (TAF$_{II}$s), as well as RNA pol II itself. In many cases, the net rate of assembly of such complexes is enhanced by transcription factors that bind specific recognition sequences within promoters. Functional interactions between activated transcription factors and general transcription factors are thought to be mediated by *coactivators*, which may be broadly defined as molecules recruited by transcription factors to enhance their transcriptional activity. Within this definition lies a wide range of functions that together are required for efficient activation by transcription factors.

This section briefly describes classes of molecules that act as coactivators for members of the nuclear receptor superfamily. For a thorough review of these nuclear receptor coactivators the reader is referred to

From: *Principles of Molecular Regulation* (P. M. Conn and A. R. Means, eds.), © Humana Press Inc., Totowa, NJ.

recent reviews (Horwitz et al., 1996 and McKenna et al., 1999). The nuclear receptor superfamily comprises functionally similar ligand-inducible transcription factors that regulate transcription of target genes in response to hormonal ligands. These molecules have a common structure, particularly in the case of conserved activation functions (AFs) in their C-termini (AF-2) and N-termini (AF-1). Upon binding of the receptor to specific DNA sequences (response elements) in the promoter of their target genes, these AFs are believed to initiate a sequence of interactions that results in assembly and stabilization of GTFs at the target promoter.

1.2. Evidence for the Existence of Coactivators

Although a number of direct interactions between nuclear receptors and GTFs have been reported, such as those between TFIIB and thyroid hormone receptor (TR), TBP and retinoid X receptor (RXR) AF-2, estrogen receptor (ER) and TBP, and TFIIB and vitamin D receptor (VDR), abundant historical evidence has suggested that receptors recruit specific non-GTF targets after ligand binding. Squelching studies, which identified interference between receptors in transient cotransfection assays, implied that receptors competed for essential cofactors, the levels of which limited the transcriptional potential of the competing pools of receptor in these assays. The identification

Table 1
Selected Nuclear Receptor Coactivators

| Cofactor | Alternative Names | Function | | | | Reference |
		Activation Domains	Acetylation	Ubiquitin Pathway	ATPase	
SRC-1 family (Section 1.3.)						
SRC-1	NCoA-1, ERAP-160	√	√			Onate et al., 1995; Xu et al. 1998 and references therein
SRC-2	TIF2 NCoA-2	√				Voegel et al., 1998 and references therein
SRC-3	p/CIP, ACTR, AIB-1, TRAM-1	√	√			Li and Chen, 1998 and references therein
Cointegrators						
CPB		√	√			
p300		√	√			See Chapter by _____, this volume
Others (Section 1.4.)						
TAF$_{II}$s		√				
TRAPs/DRIPs			√			Rachez et al., 1998 and references therein
Trip-1	Sug-1			√		See references in McKenna et al., 1999
E6-AP		√		√		McKenna et al., 1998 and references therein;
RPF-1				√		Nawaz et al., 1999 and references therein
BRG-1	SWI2/SNF2				√	See references in McKenna et al., 1999

of liganded estrogen receptor-interacting proteins such as the estrogen receptor associated proteins (ERAPs) and receptor interacting proteins (RIPs) reinforced the supposition that specific downstream targets, distinct from GTFs, are contacted by activated receptors.

1.3. The SRC Family

1.3.1. SRC-1/NCoA-1

In 1995, our laboratory described the first cloning and characterization of a common transcriptional coactivator for nuclear receptors, steroid receptor coactivator-1 (SRC-1), a prototype for an emerging family of nuclear receptor coactivators, the SRC family (Onate et al., 1995; Table 1). This structurally and functionally redundant family consists of three members: SRC-1 (NCoA-1), SRC-2 (GRIP-1/TIF2; *see* Section 1.3.2.), and SRC-3 (p/CIP/ACTR/RAC3/AIB-1/TRAM-1; *see* Section 1.3.3.).

SRC-1 interacts with and enhances the ligand-dependent transactivation of a broad range of nuclear receptors, including PR, GR, ER, TR, and RXR. Importantly, SRC-1 partially reverses the squelching of PR transactivation by cotransfected ER, indicating that it can relieve the titration of a common, limiting factor recruited in vivo by the AF-2s of ER and PR for efficient transactivation. In addition, SRC-1 contains two autonomous, transferable activation domains that are capable of stimulating transcription when fused to the DNA binding domain of the yeast transcription factor GAL4. Sequence analysis of the amino terminal region of mSRC-1/NCoA-1 has identified bHLH (for basic helix–loop–helix) and PAS (for Per/Arnt/Sim) domains. The conservation in the SRC family of these domains, known to mediate functional interactions between proteins containing these domains, indicates that SRC family members might extend the regulatory compass of nuclear receptors to signaling pathways mediated by other bHLH/PAS factors. In addition, sequence analysis of SRC-1 and other SRC family members has indicated their conservation of short LXXLL motifs, or NR boxes, which appear to contribute to the specificity of coactivator–receptor interactions (Heery et al., 1997). Targeted deletion of SRC-1 in mice confirms the in vivo function of SRC-1 as a coactivator for nuclear receptors, as well as the biological role of redundancy within the SRC family. Although both sexes are viable and fertile, the response of steroid target tissues to hormonal stimulation is significantly reduced in the SRC-1 null mutant, although overexpression of SRC-2/TIF2 (*see* Section 1.3.2.) appears to be a compensatory mechanism in certain tissues of the mutant (Xu et al., 1998).

1.3.2. SRC-2: GRIP-1/TIF2

Another member of the SRC family, transcription intermediary factor-2 (TIF2 or hSRC-2) and GR-interacting protein (GRIP-1 or mSRC-2) associates in a ligand-dependent manner with receptor hormone binding domains and enhances their transcriptional potential. Like SRC-1, TIF2/GRIP-1 contains defined activation domains capable of stimulating transcription when tethered to a heterologous DNA binding domain. In addition, TIF2, like SRC-1, is capable of relieving transcriptional squelching by ER (Voegel et al., 1998 and references therein).

1.3.3. SRC-3: p/CIP/RAC3/ACTR/AIB-1/TRAM-1

In terms of sequence identity, the most highly variable member of the SRC family is a polymorphic protein referred to as p/CIP (p300/CBP cointegrator-associated protein), ACTR, RAC-3 (receptor-associated coactivator 3), AIB-1 (amplified in breast cancer-1), TRAM-1 (thyroid receptor activator molecule), and SRC-3 (Li and Chen, 1998 and references therein). Li and Chen (1998) have proposed that the inclusive name of SRC-3 be adopted for this protein. In parallel with SRC-1 and TIF2/GRIP1, SRC-3 interacts with and coactivates a wide variety of nuclear receptors in a ligand-dependent manner, including RAR, TR, RXR, GR, ER, and PR. The p/CIP isoform, however, exhibits markedly different specificity from other SRC family members. It fails to coactivate RAR-mediated transcriptional activation, but enhances the transcriptional activity of a number of different activators, including STAT5 and cAMP response element binding protein (CREB), factors previously shown to be primarily dependent upon the transcriptional cointegrator CREB-binding protein (CBP) for efficient activation (*see* Chapter 23). This variable specificity might reflect the considerable sequence divergence between p/CIP and other SRC-3 isoforms, which exhibit only minor sequence variations (Li and Chen, 1998 and references therein).

1.4. Other Coactivators

1.4.1. E6-AP/RPF-1

The identification by our laboratory and others of the E3 ubiquitin–protein ligases RPF-1 and E6-associated protein as coactivators for nuclear receptors has implicated protein degradation as an important component of transcriptional activation. E3 ubiquitin–protein ligases target proteins for degradation by the ubiquitin pathway. E6-AP is recruited by and enhances activation by receptors such as PR, ER, and AR in a ligand-dependent manner. In addition, E6-AP partially reverses squelching between ER and PR and contains an intrinsic activation function in its N-terminal domain. Interestingly, our laboratory has recently shown that E6-AP and RPF-1 synergistically enhance PR transactivation, and that they may exist in a common complex in vivo (McKenna et al., 1998; Nawaz et al., 1999 and references therein).

1.4.2. Trip-1/Sug1

The requirement of targeted protein degradation for efficient activation by nuclear receptors is further implied by the identification of Trip-1, a protein that interacts with TR and RXR baits in a yeast two-hybrid assay in a ligand-dependent manner. A member of the CAD (conserved ATPase domains) family of proteins, Trip-1 exhibits considerable sequence similarity with the yeast transcriptional mediator Sug1, a factor required for suppression of a mutation in the transcriptional activation domain of GAL4. Although originally suggested to be a component of the Pol II holoenzyme complex, the existence of Sug-1 in the 2MDa yeast 26S proteosome complex has been reported, and has been correlated with reduced ubiquitin-dependent proteolysis in *sug1* mutants. While the precise role of protein degradation in nuclear receptor transactivation is currently unclear, it is widely thought that activated receptors undergo functionally distinct interactions with multiple protein complexes. It may well be that timely recruitment of protein-degrading pathways to the promoter processes one complex(es), allowing the receptor to interact with another complex(es) during the next phase of transcription. Alternatively, ubiquitin pathways may be required to clear liganded receptor itself from the promoter (*see* references in McKenna et al., 1999).

1.4.3. TRAPs/DRIPs

By adopting a biochemical approach to the identification of nuclear receptor coactivators, two groups have identified multiprotein complexes that interact with liganded TR (TRAPs) and VDR (DRIPs). When added to TR- and VDR-dependent in vitro transcription systems, purified TRAPs and DRIPs enhanced the transcriptional activity of their respective receptors. Although the exact nature and function of these complexes is unknown, they have been shown to contain no previously identified coactivators such as SRC family members, CBP, ERAP140, RIP140, and GTFs. A model has been proposed that such complexes may mediate repetitive rounds of transcription mediated by TRα and VDR. Initial recruitment of complexes containing the cointegrator CBP and members of the SRC family (see Section 1.3.1.) would be followed by displacement of these complexes (a step conceivably mediated by ubiquitin pathway-linked proteases) and interaction of the receptor with TRAP/DRIP-like complexes (Rachez et al., 1998 and references therein).

1.5. Chromatin Modification by Coactivators

The default state of many genes is thought to be one in which transcriptional activity is repressed by local chromatin structure, which acts to block access of *trans*-acting factors to the promoter. Core histones are thought to be central to the cohesion of this repressive state, and their abundance of positively charged lysine side chains have been suggested to interact with the negatively charged phosphate backbone of the DNA double helix. Acetylation of core histones is known to reduce their affinity for DNA by reducing their net positive charge and weakening their interaction with DNA. Since the identification of the yeast transcriptional adaptor GCN5 and its human homolog PCAF as proteins that catalyze the transfer of acetyl groups to histone lysine side chains (histone acetyltransferases or HATs), it has become clear that many nuclear receptor coactivators possess this activity. The SRC family members SRC-1 and SRC-3/ACTR, the cointegrators p300 and CBP, the p300/CBP-associated factor PCAF, and at least one member of the DRIP complex contains HAT activity. It seems, however, that acetylation by coactivators is not limited to histones: DNA binding by the transcription factors p53 and GATA-1 is stimulated after their acetylation by p300, and the DNA binding domain of the PR appears to be a target for acetylation by PCAF (M. Burcin, *personal communication*).

In addition to targeting HATs to DNA, evidence suggests that nuclear receptors also recruit protein complexes involved in the manipulation of chromatin domains to favor transcriptional activation. Members of the SWI/SNF complex, which couples ATP hydrolysis to noncovalent chromatin remodeling, historically have been associated with transactivation by nuclear receptors, and recent data have elucidated the molecular basis of this association. Mutations in SWI protein-encoding genes prevent transactivation in yeast of a GR-responsive reporter gene in the presence of cotransfected GR, whereas a wild-type yeast strain was able to support GR-dependent transactivation. Further, a human homolog of the SWI2/SNF2 proteins, BRG-1, interacts with ER in a ligand-dependent manner in a yeast two-hybrid assay, and the nucleosomal remodeling activity of the SWI/SNF complex is required for GR function in yeast. To substantiate this in a mammalian context, we have shown that GR regulation of a stably integrated MMTV promoter is dependent upon recruitment of complexes containing BRG-1 (see references in McKenna et al., 1999).

The nuclear receptor-interacting proteins TIF-1α and TIF-1β have been implicated by association in processes of chromatin remodeling. TIF-1α interacts with the heterochromatin-associated proteins mHP1α, MOD1 (HP1β), and MOD2 (HP1γ) which in turn interact with mSNF2-β, a member of the SWI/SNF chromatin-remodeling complex. A model has been suggested for TIF-1s in transcriptional regulation, in which formation of transcriptionally inactive heterochromatin by TIF-1s effects repression, and ligand-dependent association of TIF-1s with receptors mediates euchromatin formation (Le Douarin et al., 1998 and references therein) (*see also* Section 2.2.1.3.).

1.6. Transcriptional Mediation by Coactivators

Although chromatin modification is the most thoroughly characterized process in the cascade of events that likely accompany transcriptional coactivation by nuclear receptor coactivators, another fundamental component is their mediator function, which refers to their ability to direct interactions between activated receptor, GTFs and, ultimately, RNA polII. Their possession of multiple autonomous activation domains and receptor interaction motifs suggests that coactivators occupy a pivotal position in the assembly of the final preinitiation complexes at transcriptionally active promoters. To support this assertion, SRC-1 has been shown to interact directly with a number of basal transcription factors including TBP and TFIIB.

Our laboratory has shown that on naked DNA, a dominant-negative form of SRC-1 specifically titrates PR-mediated reporter gene activity in an in vitro transcription assay, suggesting that chromatin disassembly is insufficient *per se* for PR target gene activation. A two-step model for transcriptional induction by steroid receptors, suggests that chromatin modification by SRC-1, PCAF, and other coactivators (step 1) accompanies their mediation of functional interactions between activated receptor and GTFs (step 2) during efficient transcriptional activation (Jenster et al., 1997 and references therein).

1.7. Coactivators in Disease States

Since their initial characterization, several nuclear receptor coactivators have been implicated in disease states. AIB-1/hSRC-3 is consistently overexpressed in a high percentage of primary breast tumors and cultured breast cancer cell lines, often against a background of relatively low expression levels of SRC-1 and SRC-2/TIF2. These results indicate that AIB-1/hSRC-3 is involved in breast tumorigenesis, and that the selective advantage that this overexpression evidently affords is not constrained by the comparatively low levels in these tumors of SRC-1 and TIF2. There are also instances where levels of coactivator are normal, but their affinity for a specific receptor is reduced, as is the case with the TR in patients with generalized resistance to thyroid hormone (GRTH). A variety of TR mutations in GRTH patients are now known to affect the interaction of these TR mutants with coactivators such as SRC-1. A TR AF-2 mutant, E457D, which bound hormone normally and recruited NCoR in the absence of hormone, failed to bind SRC-1 in the presence of hormone, and was a strong dominant negative inhibitor of wild-type TR transactivation. In addition, it has been shown that although a GRTH TR mutant, T277A, containing a mutation in the ligand binding cavity bound ligand normally, it exhibits impaired recruitment of SRC-1 and SRC-3/ACTR in comparison with wild-type TR.

1.8. Nuclear Receptor Coactivators: A Working Model

The plethora of nuclear receptor coactivators identified in the literature to date should not necessarily be taken to imply that they act simultaneously with the activated receptor in a single large complex. Chromatin modification and contact with GTFs (mediation) (Sections 1.5. and 1.6.) are only two of many functions that coactivators undertake for efficient transcriptional regulation by nuclear receptors. In addition, evidence suggests that promoter-specific patterns of coactivator recruitment govern expression of specific gene networks in vivo. To this end, coactivators appear to exist, in the steady state at least, in smaller subcomplexes, which likely undergo combinatorial association into higher order, possibly promoter specific comformations. Indeed, liganded receptor has been shown to interact with complexes that contain none of the more familiar coactivators. Within this model, there may be hierarchical interactions, and evidence suggests that the interactions of liganded receptor with members of the SRC-1 family may be relatively stable (McKenna et al., 1998), implying that such complexes may be important intermediates during transcriptional activation. The role of protein-processing complexes recruited by coactivators such as E6-AP conceivably involves enabling a single receptor dimer to interact with multiple protein complexes over a given time period (Fig. 1). The recruitment of these and other coactivator enzyme activities by nuclear receptors during transcriptional activation highlights the multifunctional nature of this process (Korzus et al., 1998; McKenna et al., 1998 and references therein). Such complexity is reiterated in the functions of another class of proteins recruited by nuclear receptors for efficient regulation of transcription, the corepressors.

2. COREPRESSORS

2.1. Introduction

2.1.1. Transcriptional Repression

Like activation, transcriptional repression also plays an important role in normal cellular processes. Members of a nonsteroid subfamily of the nuclear receptor superfamily, such as thyroid hormone receptor (TR), retinoid acid receptor (RAR), and vitamin D receptor (VDR) as well as many orphan receptors are known to be involved in transcriptional repression. These receptors have the ability to repress the transcription of their target genes in the absence of hormones. However, in the presence of hormone, they activate the transcription of their target genes.

Gene repression by nuclear hormone receptors appears to involve different mechanisms. In the first case, steric hindrance, the simplest mechanism of repression, involves blocking the access of transcription factors to the promoter by unliganded receptors. Second, interference and/or sequestration involves specific protein–protein contacts between an unliganded receptor and the components of the transcription initiation complex. Third, unliganded receptor

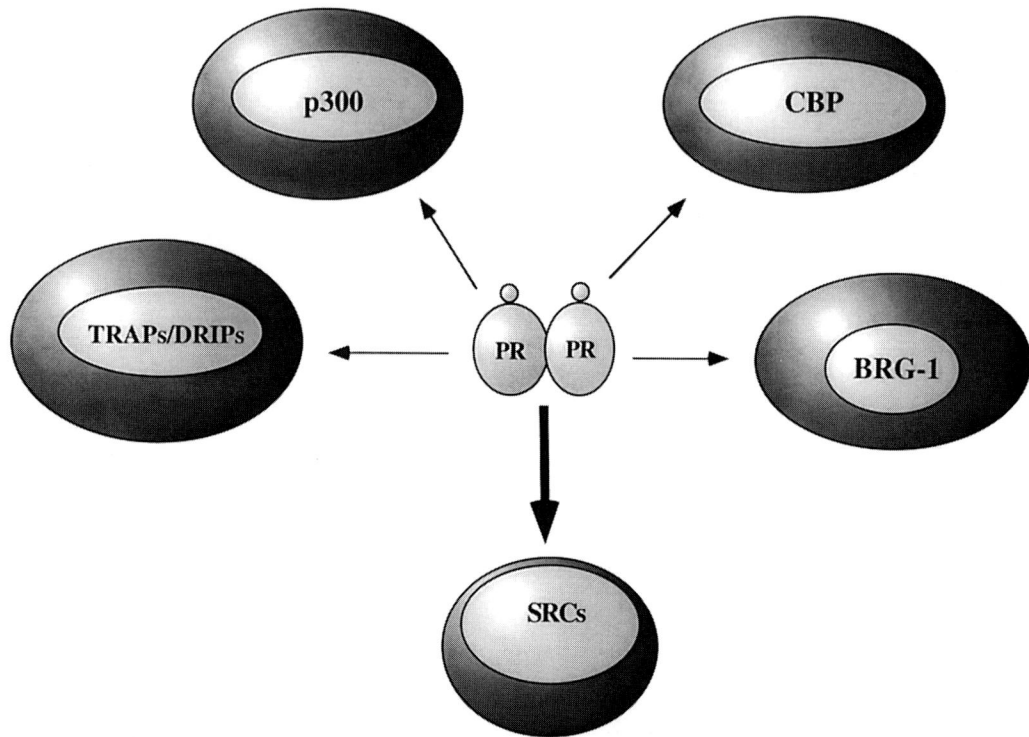

Fig. 1. Interaction of activated receptor with multiple coactivator complexes during transcriptional regulation. The association of SRC family members with liganded PR appears to be relatively strong and may be an important intermediate in PR transactivation. (Adapted from McKenna et al., 1998.)

recruits corepressors, enzymes and other proteins to the promoter region of target genes, thereby altering the acetylation status of histones and establishing transcriptionally inert chromatin.

2.1.2. Evidence for the Existence of Corepressors

The repressor function of nuclear hormone receptors has been localized to their ligand binding domains (LBD), and this function of the receptors is transferable to a heterologous DNA binding domain. This suggests that these receptors contain active repressor domains. Nuclear receptors may mediate their repressor function in part through the basal transcription factor, TFIIB. The supporting evidence for this comes from the fact that the LBD of TR interacts with the N-terminal region of TFIIB (Baniahmad et al., 1995). This interaction is hormone sensitive, suggesting that it is important for TR repressor function. The LBD of TR can be divided into two halves, each of which has no repressor activity by itself. However, when expressed together, they can work in *trans* to elicit full repressor activity, suggesting that at least two target interactions are required for repressor function.

The repressor function of TR can be reduced by overexpression of either the LBD of TR or the v-*erb*A, a mutant form of TR that has lost the ability to bind to hormone and exhibits a strong constitutive repressor activity. Furthermore, a chimeric receptor containing amino acid residues 120–392 of chicken TR fused to the yeast GAL4 DNA binding domain and the transactivation domain of the herpes simplex virus VP16 does not activate transcription when expressed alone. However, this chimeric receptor is able to activate transcription when coexpressed with unliganded TR and RAR. These findings imply the existence of cellular corepressors that are necessary for efficient repression by nuclear receptors. The existence of corepressors is substantiated further by the identification of amino acids in receptors that are critical for both repressor function and corepressor interaction. Like nonsteroid receptors, evidence also indicates the existence of corepressors for steroid receptors such as progesterone (PR) and estrogen (ER) receptors. These receptors recruit corepressors when bound to antihormones, but it is unclear at present whether these receptors are associated with corepressors in the absence of hormone (Smith et al., 1997 and references therein).

In recent years, experiments using the yeast two-hybrid and biochemical screening assays have identi-

Table 2
Nuclear Receptor Corepressors and Other Corepressors

Corepressor	Homologs	Interacting Protein(s)	kDa	Section (Accession No.)
Nuclear receptor corepressors				
N-CoR/RIP13	TRAC1, SMRT/TRAC2	TR, RAR, VDR, PR, ReV-Erbs, RORα, COUP-TFs, DAX-1, PLZF, SAP30, SIN3, TAFII32, TAFII70, TFIIB, mSiah2	270	2.2.1.1. (U35312)
SMRT/TRAC2	N-CoR/TRAC1	TR, RAR, RXR, PPARα, PR, ER, RORα, v-erbA, COUP-TFs, PLZF, SIN3, LAZ3/BCL6, HDAC-2	168	2.2.1.2. (U37146)
TIF1α		RXR, RARα, VDR, PR, ER, HP1, MOD1	112	2.2.1.3. (X99644)
TIF1β	KAP-1	KOX1/ZNF10, HP1, MOD1, KRAB	92	2.2.1.3. (X97548)
SUN-CoR		TR, RevErb, SMRT, N-CoR	16	2.2.1.4. (AF031426)
NSD1		RAR, TR, RXR, ER	285	2.2.1.5. (AF064553)
SSN6/TUP1		yα-2MCM1	106	2.2.1.6. (283218)
SIN3		MaD-MaX, SAP30, RPD3/HDAC-1, PLZF	105	2.2.1.7. (2137222)
KRAB family corepressors				
KAP-1	TIF1β	KRAB, KOX1/ZNF10, HP1, MOD1	89	2.2.2.1. (013263)
KRIP-1	TIF1	KRAB	92	2.2.2.2. (U67303)
Universal corepressors				
DR1		E4BP4, TBP	19	2.2.3. (1928868)
DRAP1		DR1	21	2.2.3. (1244714)
Other corepressors				
NAB1		NGFIA/Egr, KROX20	63	2.2.4.1. (1197669)
NAB2		NGFIA/Egr, KROX20	58	2.2.4.1. (1206027)
Gro		Hairy, Engraled, Tcf/Lef	79	2.2.4.2. (121620)
HIPKs		NK-Homeoproteins	130–133	2.2.4.3. (AF077658-60)

fied and characterized several different corepressor proteins for a diverse group of transcription factors. Based on the class of transcription factors with which these corepressors interact, they can be classified into four different groups as shown in Table 2: nuclear hormone receptor corepressors, KRAB family corepressors, universal corepressors, and other corepressors.

2.2. Corepressor Classes

2.2.1. Nuclear Hormone Receptor Corepressors

Several different corepressor proteins for nuclear hormone receptors have been cloned and characterized, a list of which is shown in Table 2. For a more thorough review of nuclear receptor corepressors, see Horwitz et al. (1996) and McKenna et al. (1999).

2.2.1.1. N-CoR/RIP 13. Nuclear receptor corepressor (N-CoR)/RXR interacting protein 13 (RIP 13), a 270-kDa protein, was originally identified as a retinoid X receptor (RXR), RAR, and TR interacting protein. N-CoR/RIP 13 interacts with the hinge region (CoR box) of the nuclear hormone receptors only in the absence of hormone. Addition of hormone results in the release of N-CoR/RIP 13 from receptors. NCoR also interacts with antihormone-bound steroid hormone receptors, orphan receptors, and several other diverse groups of proteins such as histone deacetylases and a protein, Siah2, that targets NCoR for degradation by proteosomal degradation pathway enzymes. Deletion analysis of N-CoR/RIP 13 has identified two receptor interacting domains (RIDs) in the C-terminus and three silencing domains (SDs) in the N-terminus region of the protein (Horlein et al., 1995).

2.2.1.2. SMRT/TRAC2. The *s*ilencing *m*ediator for the *r*etinoid and *t*hyroid hormone receptor (SMRT)/*t*hyroid *r*eceptor *a*ssociated *c*ofactor 2 (TRAC2) was isolated from a human lymphocyte cDNA library. SMRT, a 168-kDa protein, shares a high degree of homology with N-CoR: the C-terminus of SMRT has 48% similarity with the RID of N-

CoR and the N-termini of these proteins share 44% similarity. Analogous to N-CoR/RIP13, SMRT/TRAC2 binds to the hinge region of nuclear hormone receptors in the absence of ligand. It also interacts with antihormone-bound steroid hormone receptors, orphan receptors and histone deacetylases (Chen and Evans, 1995; Smith et al., 1997; Nagy et al., 1997 and references therein).

2.2.1.3. TIF-1s. *T*ranscription *i*ntermediary *f*actor 1 (TIF1α), a 112-kDa protein, is a RING finger protein that contains a Cys/His cluster (PHD finger), a bromo-related domain, and a coiled-coil domain, a domain responsible for mediating protein–protein interactions. TIF1 interacts with nuclear hormone receptors in a hormone-dependent manner and down-regulates RXRα-, RAR-, and ER-mediated transcriptional activities. Based on the observation that TIF1α interacts with mHP1 and mMOD1, the mouse homologs of the *Drosophila* heterochromatin protein-1, it has been suggested that TIF1 affects the transcriptional activity of nuclear hormone receptors by inducing chromatin remodeling. A second member of the TIF-1 family, TIF1β, also known as KAP-1 (*K*RAB-*a*ssociated *p*rotein-1), is a 92-kDa protein that is highly homologous to TIF1α. LIke TIF1α, TIF1β recruits mHP1 and mMOD1. In addition, it also interacts with the KRAB domain of human zinc finger factor KOX1/ZNF10. TIF1β represses transcription when fused to a heterologous DNA binding domain, but conversely activates transcription by GR and C/EBPα at the acid glycoprotein promoter. Given these data, it may be that TIF1β represses or activates transcription in a context specific manner through its interaction with chromatin-remodeling proteins, as has been proposed for TIF1α (Le Douarin et al., 1998 and references therein).

2.2.1.4. SUN-CoR. *S*mall *u*nique *n*uclear receptor *c*orepressor (SUN-CoR), is a highly basic 16-kDa nuclear protein. It enhances the transcriptional repression function of TR and the orphan receptor RevErb and contains an intrinsic silencing domain. In addition to interacting with receptor, SUN-CoR recruits SMRT and NCoR, suggesting that these three corepressors may be part of a final corepressor complex devoted to transcriptional repression by nuclear hormone receptors.

2.2.1.5. NSD1. The 285-kDa *n*uclear receptor-binding *SET-d*omain containing protein 1 (NSD1), differs from other corepressors in several respects. It possesses both repression and activation functions, and contains a SET domain, a motif found in proteins that differentially modulate chromatin structure depending on developmental context. NSD1 contains two nuclear receptor interacting domains, NID^{-L} and NID^{+L}. Whereas the NID^{+L} domain of NSD1 interacts with receptors when they are bound to hormone, NID^{-L} interacts with the hinge region of RAR and TR in the absence of hormone. The NID^{+L} domain of NSD1 interacts with the helix 12 of the receptor, a region known to be required for binding of AF-2 coactivators such as SRC family members.

2.2.1.6. SSN6/TUP1. SSN6/TUP1 is a general corepressor in yeast which is required for glucose-dependent repression of several genes and also acts as a corepressor for nuclear hormone receptors. The evidence for this comes from mutant yeast strains that lack SSN6/TUP1. The activities of ER and PR are enhanced in these SSN6 null mutant yeast over their activities in wild-type yeast. Similarly, the activity of the antihormone-bound receptor is also potentiated in mutant yeast, suggesting a possible role of SSN6/TUP1 as a corepressor of nuclear hormone receptors.

2.2.1.7. SIN3. SIN3 (RPD-1) is a 105-kDa protein that contains four paired amphipathic helices (PAH) motifs known to be important for protein–protein interaction. Disruption of the *sin3* locus in yeast suggests that it negatively regulates the yeast HO (mating type switching) gene as well as the activity of PR in yeast (Nawaz et al., 1994). Cloning of the mammalian homologs of SIN3, mSIN3a and mSIN3b, has confirmed the role of SIN3 in gene repression. Mammalian SIN3 interacts with a complex containing the Mad and Max transcription factors and represses transcription of their target genes. In addition, a number of recent studies suggest that SIN3 represses the transcriptional activity of nuclear hormone receptor and other transcription factors by recruiting histone deacetylase activity to the promoter of target genes.

2.2.2. KRAB Corepressors

The *K*ruppel *a*ssociated *b*ox (KRAB) domain is found in many human zinc finger proteins. This domain consists of ~75 amino acids and can be divided into two subdomains, an A-box and a B-box. The A-box domain is present in every KRAB domain and is required for transcriptional repression by KRAB domain-containing proteins. The KRAB domain represses transcription of target genes by associating with the corepressor proteins KAP-1 and KRIP-1.

2.2.2.1. KAP-1. *K*RAB *a*ssociated *p*rotein 1

(KAP-1) is an 89-kDa protein that interacts with the KRAB domain of human zinc finger proteins and represses transcription. The KRAB interaction domain of KAP-1 contains coiled-coil and B-box motifs, suggesting that these motifs may mediate the KRAB–KAP-1 interaction. Like nuclear hormone receptor corepressors, KAP-1 also contains a silencing domain that is separable from the KRAB interacting domain.

2.2.2.2. KRIP-1. KRAB-A interacting protein 1 (KRIP-1) is an 92-kDa protein that interacts with the KRAB-A region of human zinc finger proteins and modulates the transcriptional repression activity of this region. KRIP-1 is highly homologous to the TIF1s and is a member of the B-boxes coiled-coil subfamily of the RING finger proteins. In addition, KRIP-1 contains an intrinsic silencing domain.

2.2.3. Universal Corepressors

This group contains corepressors that repress general transcription rather than promoter-specific transcription. An example is DR1, a 19-kDa protein that interacts with TBP and recruits a complex containing the DR1-associated protein DRAP-1 to the promoter region of the target gene. DR1 also interacts with the transcription repressor E4BP4 protein and modulates its repressor function. E4BP4 mutants that fail to interact with DR1 are also deficient in their transcription repression function.

2.2.4. Other Corepressors

2.2.4.1. NABs. A 63-kDa protein, NAB1, specifically interacts with and represses transcription of the zinc finger protein NGFIA and the closely related proteins Krox20 and NGFIC. However, NAB1 has no effect on the transcriptional activity of Egr3/NGFIG. The silencing domain of NAB1 is localized to a NAB conserved domain 2 (NCD2), a region found in the C-terminal region of all NAB proteins. NAB2 is another member of the evolutionarily conserved family of NAB corepressors that repress transcription of NGFIA and Krox20. NAB2 is a 58-kDa protein that, as with NAB1, contains a silencing domain localized within the NCD2 domain of the protein.

2.2.4.2. Groucho. Groucho (Gro) is a 79-kDa protein that interacts with and mediates repression by the Hairy bHLH proteins, the Engrailed homeodomain proteins, and high mobility group proteins such as TCF/LEF-1. Gro contains multiple tandem repeats of the WD-40 repeat motif, known to mediate protein–protein interaction. In *Drosophila*, Gro converts Dorsal from an activator to a repressor and plays a critical role in the dorsal/ventral patterning system. Gro also represses Wingless/Wnt signaling activity by interacting with both TCF/LEF-1 in both *Xenopus* and *Drosophila* systems.

2.2.4.3. HIPKs. Homeodomain interacting protein kinases (HIPKs) are proteins that interact with NK-homeodomain transcription factors and modulate their repressor function. HIPKs contain a conserved protein kinase domain, a homeodomain interaction domain, and an N-terminal silencing domain. At present, the role of protein kinase activity of the HIPKs in gene repression is not clear.

2.3. Molecular Mechanism of Action of Corepressors

2.3.1. Interference and/or Sequestration

This mechanism involves specific protein–protein interactions between corepressor proteins and the proteins of the transcription initiation complex. The interaction of the corepressor with the components of the transcription initiation complex may inhibit or destabilize preinitiation complex assembly by either sequestering target protein(s) from the PIC or by physically blocking the assembly of the complex. Recruitment of TFIIB is a critical step in the assembly of the PIC and is required for the recruitment of RNA polymerase II into this complex. N-CoR has been shown to interact with the basal transcription factor TFIIB and may act to sequester TFIIB from the preinitiation complex, thereby potentially blocking a critical step in its assembly.

2.3.2. Covalent Modification

This mechanism involves recruitment of histone deacetylases (chromatin remodeling enzymes) by corepressors to the promoter region of the target genes, thus compacting nucleosomes into transcriptionally inert chromatin by promoting the deacetylation of histones. The supporting evidence for the involvement of histone deacetylation in gene repression is provided by numerous recent studies. These studies suggest that nuclear corepressors, SMRT/N-CoR, SIN3, and the histone deacetylases form a multiprotein complex that is essential for gene repression. Furthermore, inhibitors of histone deacetylases, such as trichostatin A (TSA), and antibodies against N-CoR, SIN3, or histone deacetylases relieve the repression, further confirming the role of these proteins in recruiting histone deacetylases to effect gene repression (Heinzel et al. 1997; Nagy et al., 1997). Interestingly, CBP, a HAT previously characterized as a general coactivator, targets TCF for acetylation, thereby uncoupling its interaction with its coactivator, Armadillo, and repressing the Wnt/Wingless signaling pathway. The clear inference in this case is that acetylation and deacetylation are not intrinsically positive or negative stimuli for transcription, but rather that their effects are context specific.

3. SELECTED READINGS

Baniahmad A, Leng X, Burris TP, Tsai SY, Tsai MJ, O'Malley BW. The tau 4 activation domain of the thyroid hormone receptor is required for release of a putative corepressor(s) necessary for transcriptional silencing. Mol Cell Biol 1995; 15:76.

Chen JD, Evans RM. A transcriptional co-repressor that interacts with nuclear hormone receptors. Nature 1995; 377:454.

Heery DM, Kalkhoven E, Hoare S, Parker MG. A signature motif in transcriptional coactivators mediates binding to nuclear receptors. Nature 1997; 387:733.

Heinzel T, Lavinsky RM, Mullen TM, et al. A complex containing N-CoR, mSin3 and histone deacetylase mediates transcriptional repression. Nature 1997; 387:43.

Horlein AJ, Naar AM, Heinzel T, et al. Ligand-independent repression by the thyroid hormone receptor mediated by a nuclear receptor co-repressor. Nature 1995; 377:397.

Horwitz KB, Jackson TA, Bain DL, Richer JK, Takimoto GS, Tung L. Nuclear receptor coactivators and corepressors. Mol Endocrinol 1996; 10:1167.

Jenster G, Spencer T, Burcin M, Tsai SY, Tsai M-J, O'Malley BW. Steroid receptor induction of gene transcription—a two-step model. Proc Natl Acad Sci USA 1997; 94:7879.

Korzus E, Torchia J, Rose DW, et al. Transcription factor-specific requirements for coactivators and their acetyltransferase functions. Science 1998; 279:703.

Le Douarin B, You J, Nielsen AL, Chambon P, Losson R. TIF1 alpha: a possible link between KRAB zinc finger proteins and nuclear receptors. J Steroid Biochem Mol Biol 1998; 65:43.

Li H, Chen JD. The receptor-associated coactivator 3 activates transcription through CREB-binding protein recruitment and autoregulation. J Biol Chem 1998; 273:5948.

McKenna NJ, Lanz RB, O'Malley BW. Nuclear receptor coregulators: molecular and cellular biology. Endocr Rev 1999, 20:321–344.

McKenna NJ, Nawaz Z, Tsai SY, Tsai M-J, O'Malley BW. Distinct steady state nuclear hormone receptor coregulator complexes exist in vivo. Proc Natl Acad Sci USA 1998; 95:11697.

Nagy L, Kao HY, Chakravarti D, et al. Nuclear receptor repression mediated by a complex containing SMRT, mSin3A, and histone deacetylase. Cell 1997; 89:373.

Nawaz Z, Baniahmad C, Burris TP, Stillman DJ, O'Malley BW, Tsai MJ. The yeast SIN3 gene product negatively regulates the activity of the human progesterone receptor and positively regulates the activities of GAL4 and the HAP1 activator. Mol Gen Genet 1994; 245:724.

Nawaz Z, Lonard DM, Lehman EL, et al. The Angelman Syndrome-associated gene, E6-Ap, is also a coactivator for the nuclear hormone receptor family. Mol Cell Biol 1999; 19:1182.

Onate SA, Tsai SY, Tsai MJ, O'Malley BW. Sequence and characterization of a coactivator for the steroid hormone receptor superfamily. Science 1995; 270:1354.

Rachez C, Suldan Z, Ward J, et al. A novel protein complex that interacts with the vitamin D3 receptor in a ligand-dependent manner and enhances VDR transactivation in a cell-free system. Genes Dev 1998; 12:1787.

Smith CL, Nawaz Z, O'Malley BW. Coactivator and corepressor regulation of the agonist/antagonist activity of the mixed antiestrogen, 4-hydroxytamoxifen. Mol Endocrinol 1997; 11:657.

Voegel JJ, Heine MJS, Tini M, Vivat V, Chambon P, Gronemeyer H. The coactivator TIF2 contains three nuclear receptor-binding motifs and mediates transactivation through CBP binding-dependent and -independent pathways. EMBO J 1998; 17:507.

Xu J, Qiu Y, DeMayo FJ, Tsai SY, Tsai MJ, O'Malley BW. Partial hormone resistance in mice with disruption of the steroid receptor coactivator-1 (SRC-1) gene. Science 1998; 279:1922.

23 CREB Binding Protein–Coactivator Complexes

Riki Kurokawa and Christopher K. Glass

CONTENTS

INTRODUCTION
RECRUITMENT OF CBP AND P300 BY CREB
RECRUITMENT OF CBP AND P300 BY NUCLEAR RECEPTORS
MECHANISMS OF TRANSCRIPTIONAL COACTIVATION BY CBP AND P300
ROLES OF CBP AND P300 IN DIFFERENTIATION AND DEVELOPMENT
CBP AND P300 AS INTEGRATORS OF NUCLEAR SIGNALING
SELECTED READINGS

1. INTRODUCTION

CREB binding protein (CBP) and p300 are structurally related coactivator proteins that are required for the function of several classes of regulated transcription factors (Fig. 1). CBP and p300 function in part by serving as molecular scaffolds that direct the assembly of multiprotein coactivator complexes (Fig. 2). In addition to interactions with sequence-specific transcription factors, CBP and p300 also interact with accessory coactivators, components of the RNA polymerase II holoenzyme, and proteins with kinase and histone acetyltransferase (HAT) activities. Many of these factors appear to be phosphorylated or subject to other types of posttranslational modifications, suggesting that CBP–p300 coactivator complexes are themselves targets of signal transduction pathways. CBP and p300 also possess intrinsic acetyltransferase activity, raising intriguing questions as to why coactivator complexes might contain redundant enzymatic functions.

CBP was initially identified as a protein that interacts with the phosphorylated form of the cAMP-responsive transcription factor CREB, while p300 was identified on the basis of its interaction with the adenovirus E1A oncoprotein. Subsequent studies have indicated that these proteins serve as essential coactivators for a large number of signal-dependent transcription factors. Biochemical and cell-based assays have demonstrated a high degree of functional conservation, with CBP and p300 often exhibiting nearly indistinguishable properties. These observations are consistent with the highly conserved domain structures of the two proteins, illustrated in Fig. 2. Genetic studies generally support the idea that CBP and p300 exert similar functional roles in vivo, although each protein appears to have specific functions as well.

Information concerning the diverse biological roles of CBP and p300 and the mechanisms by which they function as transcriptional coactivators is accumulating extremely rapidly. In this chapter, we begin with a consideration of mechanisms by which CBP and p300 are recruited to promoter and/or enhancer elements by sequence-specific transcription factors, using CREB and the retinoic acid receptor as well-studied examples. We next review current concepts concerning the molecular mechanisms by which

From: *Principles of Molecular Regulation* (P. M. Conn and A. R. Means, eds.), © Humana Press Inc., Totowa, NJ.

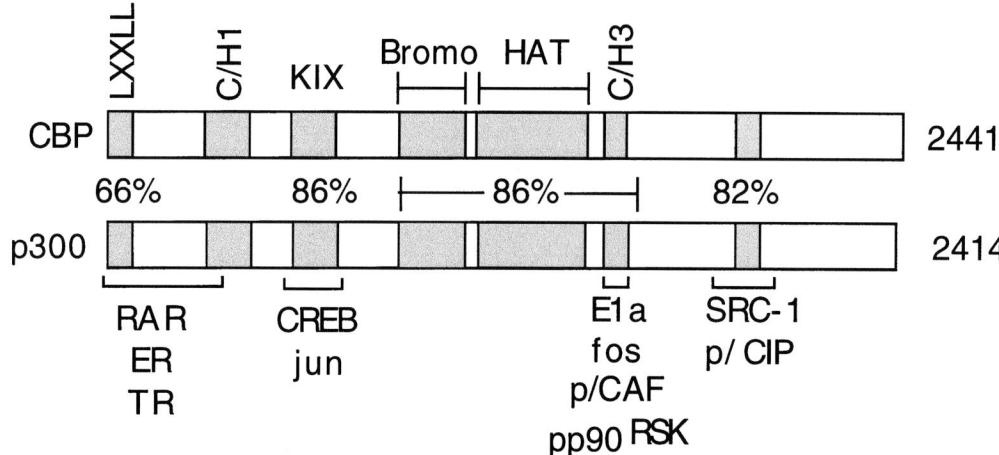

Fig. 1. Domain structures of CBP and p300. Regions of interaction with representative transcription factors are indicated. For a more extensive listing, *see* Table 1. Percentages indicate percent amino acid identity. The bromo domain has been suggested to facilitate interactions with chromatin.

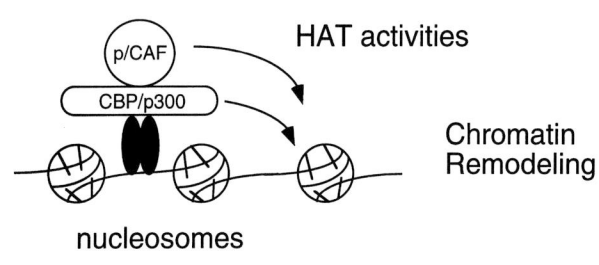

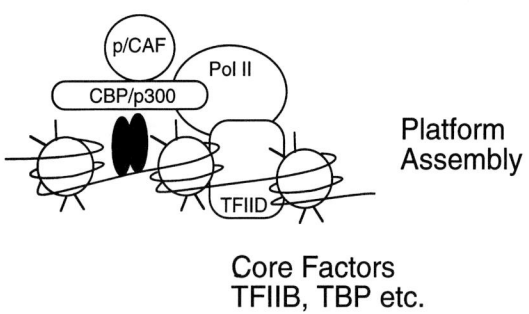

Fig. 2. HAT and platform assembly functions of CBP and p300. **(Top)** Recruitment of CBP–p300 complexes to a promoter region allows the intrinsic and associated HAT activities of CBP–p300 and p/CAF to modify N-terminal lysines of nucleosomal histones. Other factors, such as HMG proteins, may also be modified (*not shown*). **(Bottom)** As a consequence of histone acetylation, the local chromatin environment is thought to be remodeled, facilitating access of additional transcription factors. CBP–p300 may participate directly in the recruitment and assembly of these factors by serving as molecular scaffolds.

CBP–p300 stimulate transcription once recruited to a promoter or enhancer element. This chapter concludes with a brief overview of recent findings concerning the biological roles of CBP and p300 that have resulted from genetic studies and discuss the potential role of CBP–p300 coactivator complexes as integrators of signal-dependent transcription.

2. RECRUITMENT OF CBP AND P300 BY CREB

To function as coactivators, CBP and p300 must be recruited to promoter and/or enhancer elements by sequence-specific transcription factors. A representative list of sequence-specific factors that utilize CBP–p300 is provided in Table 1. In many cases, these factors have been demonstrated to participate in direct interactions between CBP–p300, while in other cases the recruitment of CBP–p300 appears to be indirect.

CREB represents the best-studied example of a protein that directly recruits CBP–p300 in a signal-dependent manner. CREB is a leucine-zipper transcription factor that mediates transcriptional responses to signaling events that result in synthesis of cAMP and activation of protein kinase A. The transactivation domain of CREB is bipartite, consisting of a kinase-inducible domain (KID) and a constitutive activation domain termed Q2. These domains function synergistically to activate transcription of CREB target genes in response to hormonal stimulation. Transcriptional activation by the KID domain requires that it interact with CBP and/or p300. This

Table 1
Sequence-Specific Transcription Factors Demonstrated to Utilize CBP and/or p300

Sequence-Specific Factor	Function
AML1 (CBF α_2, PEBP2α)	Myeloid differentiation
AP-1 c-Fos	AP-1 transcription
c-Jun	
JunB	
ATF-2	CREB/ATF transcription
β2/NeuroD	Central nervous system and gastric lineage differentiation
c-Myb, A-Myb	Myb transduction
C/EBPβ	C/EBP transcription
CREB/CREM/ATF-4	cAMP signaling
Cubitus interruptus	Hedgehog signaling
Dorsal	*Drosophila* development
Elk-1	MAP kinase-dependent signaling
Ets-1	Ets transcription
GATA-1	Erythroid differentiation
HIF-1a	Cellular response to hypoxia
IRF-3	Response to viral infection
Mi	Melanocyte development
Myo-D/MEF2C/bHLH proteins	Myogenic differentiation
NF-κB component p65 (RelA)	Inflammatory response
NFAT	Immune response
Nuclear receptors:	Nuclear hormone receptor signaling
Androgen	
ER	
Glucocorticoid	
HNF4	
Peroxisome proliferator-activated receptor	
RAR	
RXR	
SF-1	
TR	
p45/NF-E2	Hematopoiesis
p53	Tumor suppressor
Sap-1a	MAP kinase-dependent signaling
Smad2, Smad3, Smad4	TFG β signaling
SREBP-1 and -2	Cholesterol metabolism
Stat 1	Interferon-α and -γ signaling
Stat 2	Interferon-α signaling
YY1	Transcription up- and down-regulation

interaction is highly dependent on the phosphorylation of Ser^{133} within the KID domain by protein kinase A. Once phosphorylated, the KID domain interacts with a highly conserved region of CBP and p300 referred to as KIX (Fig. 3). Intriguingly, the structure of the phosphorylated KID domain undergoes a random coil to α-helical folding transition on binding to KIX, resulting in the formation of two α-helices. One of these helices is amphipathic and interacts with a hydrophobic groove defined by two helical motifs of the KIX domain. The second helix of the KID domain contacts a different face of one of the KIX helical motifs, with the phosphate group of Ser^{133} of the KID domain forming a hydrogen bond to the side chain of Tyr^{658} of KIX. These hydrogen bonds provide a structural explanation for the phosphorylation dependence of the CREB–CBP interaction.

These studies may prove to be prototypic for interactions of other proteins with CBP and p300. In addition to CREB, the KIX domains of CBP and p300 recognize the transcriptional activation domains of a number of other transcription factors, including Myb,

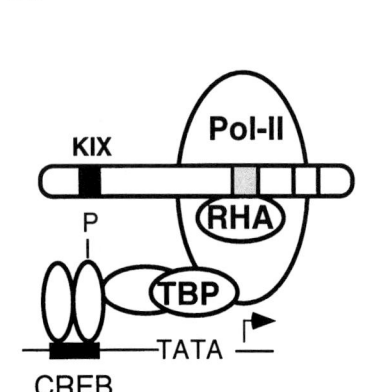

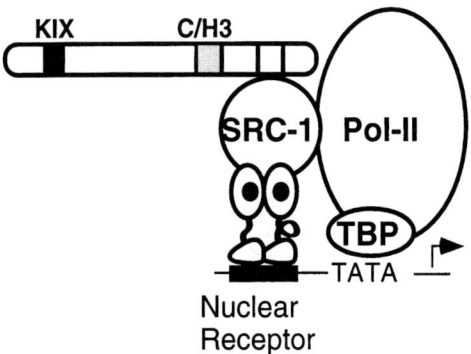

Fig. 3. Different mechanisms of CBP–p300 recruitment by CREB and nuclear receptors. (A) CREB recruits CBP in a phosphorylation-dependent manner by interacting with the KIX domain. Transactivation also requires interaction between a second region of CREB and the TAFII130 component of TFIIB. The CREB–CBP complex has been proposed to recruit RNA polymerase II through an interaction involving the C/H3 domain of CBP and RNA helicase A (RHA). (B) Nuclear receptors recruit CBP–p300 complexes through ligand-dependent interactions with p160 coactivators such as SRC-1. In addition, at least some nuclear receptors may be able to interact directly with an N-terminal region of CBP and p300 that contains an LXXLL recognition motif.

Jun, Cubitus Interuptus, STAT1, the HTLV-1 virally encoded Tax protein, and SREBP. In these cases interactions appear to be weaker than those of CREBP but constitutive. There is no clear consensus KIX-recognition motif that can be gleaned by sequence alignments of these factors. However, structural algorithms and biochemical studies suggest that many of these activation domains may also undergo a random coil to α-helical transformation on binding to CBP and p300.

3. RECRUITMENT OF CBP AND P300 BY NUCLEAR RECEPTORS

CBP and p300 interact with and are required for ligand-dependent transcription by a number of nuclear receptors, including the retinoic acid receptor (RAR) and the estrogen receptor (ER). These factors appear to use an alternative mechanism for the recruitment of CBP and p300. Members of the nuclear receptor superfamily of ligand-dependent transcription factors are defined by a highly conserved central DNA binding domain and a somewhat less conserved C-terminal domain that mediates ligand binding, dimerization, and ligand-dependent transcriptional activation. Most nuclear receptors bind to their DNA target sequences as either dimers, as in the case of the ER, or as heterodimers with retinoid X receptors (RXRs), as in the case of RAR.

Ligand-dependent transcriptional activation by nuclear receptors depends on a short conserved motif in the C-terminal end of the ligand binding domain, termed AF-2. This region was initially identified in the ER by point mutations that abolished ligand-dependent transcription without influencing ligand binding or transcriptional activation. Subsequent studies defined a similar functional requirement for this motif in other members of the nuclear receptor family, including the thyroid hormone and retinoic acid receptors. These observations, in concert with the finding that nuclear receptors could antagonize each other's function, led to the idea that the AF-2 domain undergoes a conformational change on the binding of ligand that results in the recruitment of a functionally limiting set of coactivator proteins that are used in common by several nuclear receptors.

Crystal structures of the unliganded RXR ligand binding domain and agonist-bound RARα, TRα, and ERα ligand binding domains have generally confirmed the hypothesis that the AF-2 region undergoes a ligand-dependent conformational change. The overall structures of the RXR, TR, RAR, and ER ligand binding domains are very similar. The ligand binding domain is folded into a three-layered antiparallel α-helical sandwich comprising a central core layer of three helices packed between two additional layers of helices. This helical arrangement creates a "wedge-shaped" molecular scaffold that establishes a ligand binding cavity at the narrower end of the domain. The remaining secondary structural elements, a small two-stranded antiparallel β-sheet and the C-terminal helix containing the AF-2 activation helix, are located at the ligand binding end of the molecule. In the unliganded RXR structure, the AF-2 helix extends away from the ligand binding domain, while in the agonist bound RARγ, TRα, and ER ligand binding domain structures, the AF-2 helix is tightly packed against the body of the ligand binding domain and makes direct contacts with ligand. Intriguingly, the structure of the ER ligand binding domain bound to the antagonists raloxifene or tamoxifen demonstrates a distorted conformation of the AF-2 helix compared to the agonist-bound structure. In concert, these studies are consistent with the idea that ligand-dependent changes in the conformation of the AF2 helix result in the formation of a surface that facilitates coactivator interactions.

Although the RAR can interact weakly with the N-terminus of CBP–p300, several lines of evidence indicate that ligand-dependent recruitment requires an intermediate set of coactivator proteins. A biochemical search for potential coactivators of the ER led to the identification of proteins of approx 160 kDa as predominant cellular proteins able to interact with the ligand binding domain in a ligand and AF-2-dependent manner. A similar profile of proteins was subsequently observed to interact with the retinoic acid and retinoid X receptors. The first of these proteins to be molecularly cloned was steroid receptor coactivator-1 (SRC-1). Using a variety of approaches, cDNAs encoding additional p160 proteins have been isolated form murine and human libraries by several laboratories. These proteins comprise a gene family consisting of three members, referred to as SRC-1/NcoA-1, GRIP1/TIF-2, and p/CIP/AIB1/ACTR. As illustrated in Fig. 4, members of the p160 family of coactivators share common structures. These each contain a conserved N-terminal bHLH, PAS A domain, although the function of this domain remains unclear. The central conserved region (~200 amino acids) is essential for its interaction with nuclear receptors and is referred to as the nuclear receptor interaction domain, while the C-terminal region contains a CBP interaction domain.

Inspection of the amino acid sequences of the nuclear receptor interaction domains of p160 factors revealed the presence of leucine-rich motifs of consensus sequence LXXLL, where L represents leucine and X any amino acid. These motifs were also found to be present in a number of other proteins demonstrated to interact with nuclear receptors in a ligand-dependent manner. Three of these motifs are found in the nuclear interaction domains of the p160 factors, while two related motifs are found in the CBP–p300 interaction domain (Fig. 4). Biochemical experiments have demonstrated that these LXXLL motifs play a crucial role mediating in the interaction of p160 coactivator proteins with liganded nuclear receptors. Recent crystal structures demonstrate that the LXXLL motif forms a short α-helix that is held at one end by an acidic residue within the AF-2 domain. The other end of the LXXLL helix is held by a conserved basic residue in helix 3, with the leucine residues packing into an intervening hydrophobic cavity (Fig. 5). These structures suggest that the binding of a ligand causes the AF-2 domain to close upon the LXXLL helix and hold it tightly against the ligand binding domain. Amino acids N-terminal and C-terminal to the LXXLL motif appear to make additional contacts with the ligand binding domain that may play roles in determining specificity of nuclear receptor–coactivator interaction, possibly specifying which coactivators will bind to a particular nuclear receptor dimer or heterodimer with highest affinity. Intriguingly, two LXXLL motifs from a single p160 protein can interact with the both subunits of an RAR–RXR heterodimer, suggesting that multiple interaction motifs may mediate cooperative interactions.

The docking of p160 proteins to RAR–RXR heterodimers appears to be essential for effective recruitment of CBP–p300 in response to the binding of ligands. Biochemical studies indicate that the association of CBP with RAR–RXR heterodimers requires the presence of both the nuclear receptor interaction domain and the CBP interaction domain of these proteins. Consistent with these findings, experiments in cultured cells indicate that the p160 protein SRC-1 is unable to function as a coactivator of the RAR when the CBP interaction domain is deleted. Conversely, CBP is unable to function as a coactivator of the RAR when the p160 interaction domain is deleted. In concert, these findings suggest a model in which the assembly of a CBP–nuclear receptor complex requires the presence of p160 proteins as structural adapter molecules (Fig. 3). Thus, different classes of transcription factors appear to use distinct mechanisms to recruit CBP–p300 complexes.

4. MECHANISMS OF TRANSCRIPTIONAL COACTIVATION BY CBP AND P300

CBP and p300 are thought to contribute to transcriptional activation by acting as molecular scaffolds and by catalyzing acetylation of diverse substrates that

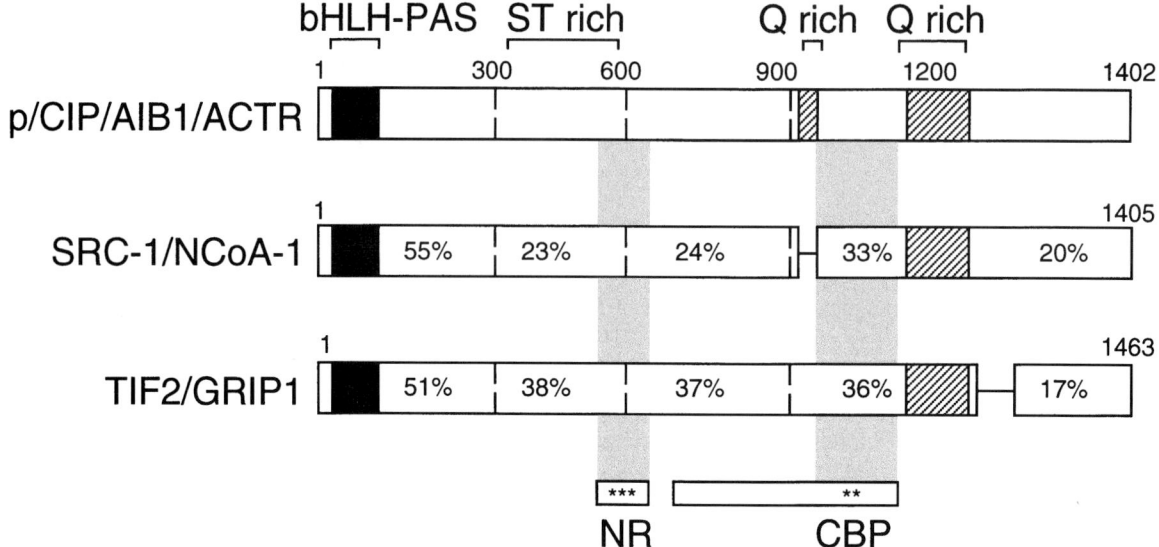

Fig. 4. Domain structure of members of the p160 family of nuclear receptor coactivators. Percentages represent percent amino acid identity with pCIP. The bHLH-PAS domain is related to a protein–protein interaction domain found in a subset of bHLH proteins that include Per, Arnt, and Sim. The nuclear receptor interaction domain contains three repeat motifs of consensus sequence LXXLL. The CBP interaction contains two similar motifs.

include histones and sequence specific transcription factors (Fig. 1). The acetyltransferase activities of CBP and p300 were initially demonstrated using histones as substrates. Although there is as yet no direct evidence that histones are acetylated by CBP and p300 in vivo, substantial circumstantial evidence makes this highly likely. The N-terminal tails of histones H2, H3, and H4 are rich in lysine, and are thought to influence nucleosome–DNA interactions that determine accessibility of core transcription factors to promoter regions. In vitro, CBP and p300 not only acetylate free histones, but also histones assembled into nucleosome complexes, suggesting that nucleosomes can be similarly modified in vivo. Rates of gene transcription roughly correlate with the degree of histone acetylation, with hyperacetylated regions of the genome being more actively transcribed than hypoacetylated regions. Thus, the specific recruitment of a complex with HAT activity to a promoter may play a critical role in overcoming repressive effects of chromatin structure on transcription.

In vitro transcription assays of ER activity are consistent with a role of p300 in overcoming chromatin-mediated repression. When ER activity is assessed on naked DNA templates, high rates of transcription are observed that are only modestly stimulated by estrogen. When assessed on chromatinized templates in the presence of defined core transcription factors, both basal and ligand-dependent activities are markedly repressed. Addition of p300 does not significantly influence basal transcription, but markedly stimulates ligand-dependent activity. This activity is required for the initial, but not subsequent, rounds of transcriptional initiation, raising the possibility that p300 plays a role in establishing a promoter architecture that is permissive to transcription.

An intriguing question concerning the roles of acetyltransferase activities arises from the observation that a number of complexes implicated to play roles in transcriptional harbor acetyltransferase activities. Why should this activity be so redundant? HAT activities have also been documented in the 250-kDa subunit of TFIID, p/CAF, GCN5, and weak activities described for members of the p160 family of coactivators. One possible explanation is that different sequence specific coactivators require different HAT activities,

p/CAF is a homolog of the yeast coactivator protein GCN5, which is a component of the ADA2 coactivator complex. Like GCN5, p/CAF contains a C-terminal region that contains protein–protein interaction

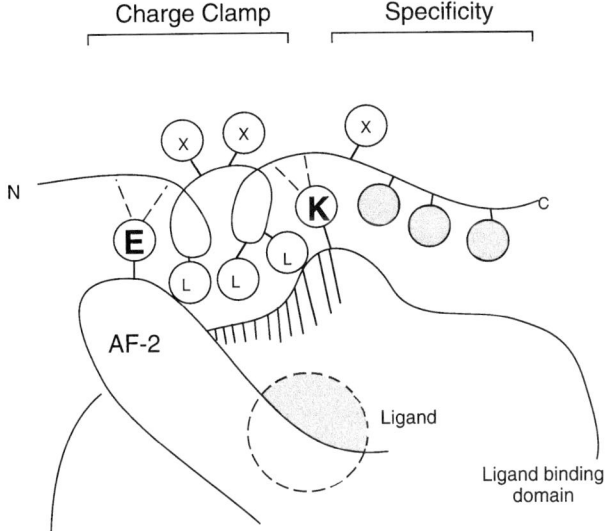

Fig. 5. Mechanisms of interaction of nuclear receptors with LXXLL recognition motifs. The ligand binding domain of a nuclear receptor is represented with the AF-2 motif contacting ligand and in the active configuration. The LXXLL motif of a p160 coactivator is shown as a helix that is clamped between a conserved glutamic acid residue (E) in the AF-2 motif and a lysine residue (K) in the body of the ligand binding domain. The leucine residues of the LXXLL motif pack tightly into a hydrophobic pocket between the charge-clamp. Residues labeled as X are projecting into the solvent and are not expected to contribute to binding. Shaded residues C-terminal to the LXXLL motif can potentially interact with additional residues in the LBD and contribute to specificity of binding.

Table 2
Cofactors and Enzymes Demonstrated to Interact with CBP and/or p300

Factor	Function
p/CAF	HAT, adaptor
p160 factors	Nuclear receptor coactivators
SRC-1/NCoA-1	
TIF2/GRIP1/NCoA2	
p/CIP/ACTR/AIB1	
pp^{90}RSK	Protein kinase; MAP kinase-dependent signaling
TBP	Basal transcription
TFIIB	Basal transcription
E1A	Adenoviral transformation
SV40 large T antigen	Viral transformation

motifs. In addition, p/CAF contains an N-terminal extension not conserved in yeast GCN5 that also appears to mediate additional protein–protein interactions. p/CAF was initially identified as a factor that interacts with the C/H3 domains of p300 and CBP. Subsequent studies have demonstrated that p/CAF is capable of interacting with the N-terminus of CBP, as well as with SRC-1/NcoA-1 and p/CIP. The specific roles that p/CAF and p/CIP play in mediating transcriptional activation by CREB function remain to be determined, but it is likely that they relate at least in part to their ability to serve as adaptor molecules. The HAT activity of p/CAF does not appear to be required for CREB activity, as a p/CAF mutant containing point mutations that abolish HAT activity can rescue the effects of knocking out endogenous p/CAF by antibody microinjection. In contrast, the HAT activity of CBP was required for transcriptional activation by CREB. Intriguingly, purification of a p/CAF complex from cells resulted in the copurification of a number of histone-like TAFs, including hADA2, hTAF$_{II}$30, hTAF$_{II}$31, hTAF$_{II}$20 and hTAF$_{II}$15, but not CBP or p300. These findings raise the possibility that p/CAF may be a component of several different complexes with distinct functional roles.

It is clear that the acetyltransferase activities of CBP and p300 exert important effects on other substrates, particularly sequence-specific transcription factors. For example, the C-terminus of p53 is efficiently acetylated by CBP and p300. The C-terminus of unmodified p53 inhibits its ability to bind to DNA. On acetylation, this inhibitory effect is relieved, allowing high-affinity DNA binding. Additional non-histone proteins that have been identified as substrates of p300 and CBP include HMG I(Y). Intriguingly, acetylation of HMG 1(Y) has been proposed to decrease its DNA binding activity and destabilize the interferon-β (INFβ) enhansosome. This activity could potentially represent a built-in switch that enables transcription to be activated very transiently, as exhibited by the IFNβ gene.

5. ROLES OF CBP AND P300 IN DIFFERENTIATION AND DEVELOPMENT

Although homologs of CBP and p300 are not present in yeast, highly related proteins have been found in *C. elegans* and *Drosophila*. Genetic studies indicate that CBP–p300 homologs are essential for normal development in each of these organisms. As predicted by cell-based and biochemical studies, mutations in CBP and p300 have resulted in striking defects in the function of many classes of signal-dependent transcription factors. In *Drosophila*, disruption of one allele of the dCBP gene resulted in embryonic lethality, indicating that the levels of CBP are functionally limiting. The most severely affected embryos exhibit a twisted morphology, suggesting defects in mesodermal differentiation. This phenotype could be accounted for in part by a failure of mutant embryos to express the basic helix–loop–helix transcription factor twist. Twist expression is dependent on dorsal, which is homologous to the NFκB family of transcription factors that require CBP and p300 for activity. Mutant embryos also exhibited defects in hedgehog signaling due to an essential role of dCBP as a coactivator of the transcription factor cubitus interuptus. In *C. elegans,* inhibition of the expression of the CBP–p300 homolog, cbp-1, caused developmental arrest of embryos, with a failure of mesodermal, endodermal, and hypodermal cells to develop. Intriguingly, all of the embryos exhibited evidence of neuronal differentiation.

Evidence that CBP and p300 are essential for normal development in humans and are functionally limiting has been suggested by studies of Rubenstein–Taybi syndrome, which is characterized by a spectrum of disorders that include mental retardation and craniofacial abnormalities. Individuals with this syndrome have been found to have deletions and point mutations affecting one allele of CBP, suggesting that developmental abnormalities result from haploinsufficiency of CBP. In mice, disruption of one allele of the CBP gene resulted significant fetal mortality and in a constellation of craniofacial and skeletal abnormalities in the surviving heterozygotes. These findings support the possibility that haploinsufficiency of CBP fully accounts for the developmental abnormalities observed in the Rubenstein–Taybi syndrome in humans. Intriguingly, the severity of abnormalities in CBP-deficient mice was influenced significantly by genetic background, which may help to explain the variability and severity of the abnormalities seen in Rubenstein–Taybi syndrome patients.

Mice heterozygous for a null allele of p300 also exhibited a significant degree of fetal lethality that was strain dependent, but did not exhibit skeletal abnormalities observed in the CBP-deficient mice. Furthermore, crosses of CBP$^{+/-}$ mice with p300$^{+/-}$ mice resulted in no births of double heterozygous animals, indicating an exquisite sensitivity to the overall gene dosage of CBP and p300. Animals nullizygous for CBP died at 8–10 d of gestation, while p300-null animals died between d 9 and 11.5 of gestation. The causes of embryonic lethality in CBP-null mice have not yet been reported. However, embryos lacking p300 exhibited defects in neurulation, cell proliferation, and heart development. In addition, cells derived from p300-deficient embryos displayed specific transcriptional defects and proliferated poorly. Intriguingly, CREB-dependent transcription in p300$^{-/-}$ fibroblasts was nearly equivalent to that in wild-type fibroblasts, while retinoic acid receptor function was markedly impaired. In concert with the difference in the skeletal phenotypes of CBP$^{+/-}$ and p300$^{+/-}$ mice, these findings indicate that CBP and p300 have at least some nonredundant functions. Analysis of the expression of bone morphogenetic proteins in CBP-heterozygous mice supported that notion that CBP/p300-dependent signaling pathways are conserved from insects to vertebrates. Bone morphogenetic protein-7 (BMP-7) mRNA levels were decreased in CBP-heterozygous mice. Similarly, the expression of *decapentapleigic (dpp),* the *Drosophila* homolog of bone morphogenetic protein, was significantly reduced in dCBP$^{+/-}$ flies. In *Drosophila, dpp* expression is controlled by cubitus interuptus, which as noted earlier requires CBP as a coactivator. Intriguingly, CBP apparently is involved in repression of the transcription factor TCF to antagonize wingless signaling.

Intriguingly, the translocation t(18;16)(p11;p3) that occurs in a subset of acute myeloid leukemia patients results in an in-frame fusion of the protein MOZ to a largely intact CBP protein. MOZ is a 2004 amino acid protein that is characterized by two C4HC3 zinc fingers and a single C2HC zinc finger in conjunction with a putative acetyltransferase signature. The presence of an acetyltransferase function in MOZ raises the possibility that a dominant MOZ–CBP fusion could mediate leukemogenesis via aberrant chromatin acetylation. In addition, by placing CBP under the transcriptional control of the MOZ promoter, altered levels of CBP expression might also influence cellular growth.

6. CBP AND P300 AS INTEGRATORS OF NUCLEAR SIGNALING

The observation that CBP and p300 are required for the activities of a large number of signal-dependent transcription factors and appear to be functionally limiting in many cells raises the intriguing possibility that these coactivators may play critical roles in the integration of complex developmental and homeostatic events (Fig. 6). Competition for CBP complexes could potentially serve as the basis, at least in part, for many examples of mutual antagonism between transcription factors. For example, activation of the glucocorticoid receptor has been demonstrated to inhibit the activity of AP-1-dependent promoters, while activation of AP-1 inhibits glucocorticoid responses. While crosstalk between these pathways

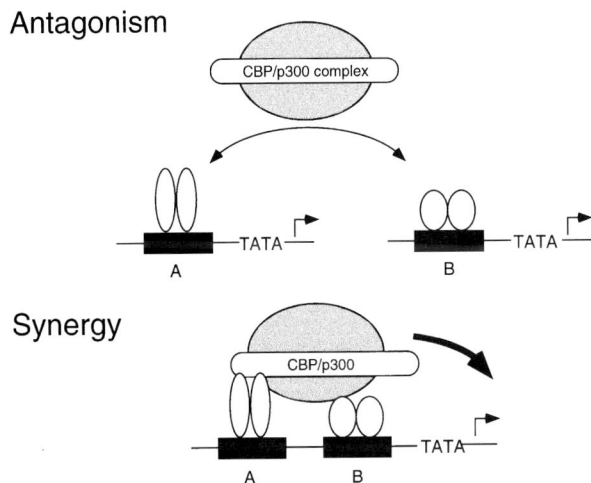

Fig. 6. Integration of transcriptional responses by CBP–p300. Transcription factors A and B both require CBP–p300 complexes to activate transcription. If CBP is limiting in the cell, simultaneous activation of A and B leads to mutual antagonism of their activities on promoters that contain response elements for B but not A, or A but not B. In contrast, promoters that contain response elements for both A and B will respond synergistically to activation of both factors due to cooperative binding of CBP complexes. These interactions can be further modulated by modifications (e.g., phosphorylation, acetylation) of either the coactivator complexes or their target transcription factors.

can conceivably occur at many levels, mutual antagonism can be reversed in some cases by overexpression of CBP, consistent with the idea that simultaneous recruitment of CBP–p300 complexes by AP-1 factors and glucocorticoid receptors limits their transcriptional responses. Such a mechanism has been suggested to account for antagonistic interactions between many classes of transcription factors, including other nuclear receptors, STAT factors, and members of the NF-κB family. In addition to antagonistic interactions in trans, CBP–p300 complexes may also play a role in synergistic interactions that occur between many of these same classes of transcription factors in *cis*. Because CBP and p300 are multivalent proteins and interact with other multivalent adapter molecules, they may be able to form simultaneous contacts with closely spaced transcription factors bound to promoter or enhancer elements. Indeed, CBP–p300 complexes have been suggested to play an architectural role in establishing the INFβ enhansosome. Finally, posttranslational modifications of CBP and p300, for example, by phosphorylation, may influence cellular programs of transcription by altering interactions with specific classes of transcription factors.

SELECTED READINGS

Akimaru H, Chen Y, Dai P. *Drosophila* CBP is a co-activator of cubitus interruptus in hedgehop signalling. Nature 1997; 386:735.

Bannister AJ, Kouzarides T. The CBP co-activator is a histone acetyltransferase. Nature 1996; 384:641.

Chrivia JC, Kwok RP, Lamb N, Hagiwara M, Montminy MR, Goodman RH. Phosphorylated CREB binds specifically to the nuclear protein CBP. Nature 1993; 365:855.

Danielian PS, White R, Lees JA, Parker MG. Identification of a conserved region required for hormone-dependent transcriptional activation by steroid hormone receptors. EMBO J 1992; 11:1025.

Eckner R, Ewen ME, Newsome D. Molecular cloning and functional analysis of the adenovirus E1A-associated 300-kD protein (p300) reveals a protein with properties of a transcriptional adaptor. Genes Dev 1994; 8:869.

Glass CK, Rose DW, Rosenfeld MG. Nuclear receptor coactivators. Curr Opin Cell Biol 1997; 9:222.

Kamei Y, Xu L, Heinzel T. A CBP integrator complex mediates transcriptional activation and AP-1 inhibition by nuclear receptors. Cell 1996; 85:403.

Kraus WL, Kadonaga JT. p300 and estrogen receptor cooperatively activate transcription via differential enhancement of initiation and reinitiation. Genes Dev 1998; 12:331.

Nakajima T, Uchida C, Anderson SF, Parvin JD, Montminy M. Analysis of a cAMP-responsive activator reveals a two-component mechanism for transcriptional induction via signal-dependent factors. Genes Dev 1997; 11:738.

Nolte RT, Wisely GB, Westin S. Ligand binding and co-activator assembly of the peroxisome proliferator-activated receptor-γ. Nature 1998; 395:137.

Ogryzko VV, Schiltz RL, Russanova V, Howard BH, Nakatani Y. The transcriptional coactivators p300 and CBP are histone acetyltransferases. Cell 1996; 87:953.

Petrij F, GIles RH, Dauwerse HG. Rubinstein–Taybi syndrome caused by mutations in the transcriptional co-activator CBP. Nature 1995; 376:348.

Radhakrishnan I, Pérez-Alvarado GC, Parker D, Dyson HJ, Montminy MR, Wright PE. Solution structure of the KIX domain of CBP bound to the transactivation domain of CREB: a model for activator:coactivator interactions. Cell 1997; 91:741.

Shiau AK, Barstad D, Loria PM. The structural basis of estrogen receptor/coactivator recognition and the antagonism of this interaction by tamoxifen. Cell 1998; 95:927.

Tanaka Y, Naruse I, Maekawa T, Masuya H, Shiroishi T, Ishii S. Abnormal skeletal patterning in embryos lacking a single Cbp allele: a partial similarity with Rubinstein–Taybi syndrome. Proc Natl Acad Sci USA 1997; 94:10215.

Wurtz JM, Bourguet W, Renaud JP. A canonical structure for the ligand-binding domain of nuclear receptors. Nature Struct Biol 1996; 3:206.

Yao T-P, Oh SP, Fuchs M. Gene dosage-dependent embryonic development and proliferation defects in mice lacking the transcriptional integrator p300. Cell 1998; 93:361.

PART V MOLECULAR REGULATION OF CELL PROLIFERATION AND DEATH

24 Cell Cycle Checkpoints

Sally Kornbluth

CONTENTS

INTRODUCTION
CELL CYCLE CHECKPOINTS: SIGNAL TRANSDUCTION PATHWAYS REGULATING CELL CYCLE PROGRESSION
CONCLUSIONS
SELECTED READINGS

1. INTRODUCTION

1.1. The Cell Cycle Clock

In eukaryotic cells, cell division proceeds through an orderly series of events in which the cellular contents are duplicated, divided, and distributed to daughter cells. This progression through the cell cycle is governed by the sequential activation and inactivation of members of a family of serine/threonine kinases known as the Cdks (cyclin-dependent kinases). Inactive as monomers, the Cdks require associated positive regulatory subunits, the cyclins, for their catalytic activity. Once active, distinct cdk–cyclin pairs phosphorylate substrates responsible for executing the various events of the cell cycle.

Cyclins derive their name from their characteristic oscillaiton in abundance during the cell cycle; the transient activation of each Cdk at its appropriate time of action in the cycle is, in part, controlled by the availability of its cognate cyclin. However, Cdk–cyclin complexes are also regulated both positively and negatively by phosphorylation of the Cdk subunit, by binding of stoichiometric inhibitors, and by factors controlling their subcellular localization. The cyclical changes in Cdk activity are commonly referred to as the "cell cycle clock," while processes contingent on the ticking of the clock, such as DNA replication, chromosome condensation, spindle assembly, and chromosome segregation are referred to as "downstream events" of the cell cycle.

1.2. Ordering the Cell Cycle

Cell cycle events must occur in a defined order. If, for example, cells attempted to segregate their chromosomes before completing DNA replication, daughter cells would receive incomplete copies of the genome—a clearly disastrous outcome. In addition, the cell cycle clock must be prepared to stop ticking, at least temporarily, to allow rectification of errors or repair of any damage that occurs during cell duplication. For example, cells must postpone initiation of replication and entry into mitosis until DNA damage lesions are repaired. In theory, several possible mechanisms could guarantee the correct ordering and fidelity of cell cycle processes. Dependent events could be temporally coordinated if the activation of, or marshalling of resources for, the later (dependent) event required more time than it took to complete the earlier, prerequisite event. For example, if the time taken to assemble the mitotic spindle consistently exceeded that required for DNA replication, then spindle formation and chromosome segregation would never precede the completion of replication. In the absence of any perturbations, cell cycles organized in this way could, in principle, function effectively. However, if, for example, replication were stalled for any reason,

From: *Principles of Molecular Regulation* (P. M. Conn and A. R. Means, eds.), © Humana Press Inc., Totowa, NJ.

the time required for spindle assembly might elapse before replication were completed.

An alternative organization of the cell cycle, which takes into account this problem, is one in which later events of the cell cycle depend upon "factors" (i.e., molecular byproducts or physical structures) produced by earlier events. For example, if a chromosomal structure produced at the completion of DNA replication were required to nucleate spindle assembly, then spindle assembly would never occur until the end of replication. Several lines of evidence, both genetic and pharmacological, suggest that this "substrate–product" mechanism is not, in general, used for coordinating the eukaryotic cell cycle. Chief among these is the observation that cellular mutations can be obtained that will allow later events to occur even when earlier events have not been properly completed. For example, while cells treated with the microtubule depolymerizing drug nocodazole normally halt prior to exit from mitosis, several different genetic mutations can override this cell cycle arrest, leading to premature inactivation of mitotically active enzymes and loss of metaphase sister chromatid adhesion. Similarly, DNA polymerase inhibitors that prevent progression of the replication fork concomitantly stop entry into mitosis; this cell cycle block can be short-circuited in many cell types by treatment with caffeine. If substrate–product mechanisms enforced cell cycle ordering, it would not be possible to circumvent a prior event to execute a later one because the required "product" of the earlier event would not yet exist.

Collectively, these and other compelling data gave rise to the idea of cell cycle checkpoints—signaling pathways that monitor events dependent on the ticking of the cell cycle clock and stop the clock (and hence dependent later events) until each monitored event is completed. For example, if DNA replication is blocked, a signaling pathway prevents entry into mitosis. Similarly, if DNA is damaged, it is the role of the checkpoint to delay cell cycle progression until the damage can be repaired. Mutational or pharmacological inactivation of checkpoint signaling pathway components allows cell cycle events to proceed without properly completing earlier events, accounting for the effects of drugs such as caffeine.

2. CELL CYCLE CHECKPOINTS: SIGNAL TRANSDUCTION PATHWAYS REGULATING CELL CYCLE PROGRESSION

Cell cycle checkpoints function not only to ensure the orderly occurrence of intrinsic cell cycle events, but also to accommodate insults from extrinsic events that might interfere with the proper duplication and division of cellular contents. It would be disadvantageous for a cell to initiate either DNA replication or mitosis with damaged DNA. Therefore, checkpoint pathways also serve to stop the Cdk–cyclin-mediated activation of these processes. Hence, cell cycle checkpoints can be thought of as signal transduction pathways leading from the process or structure being monitored (e.g., DNA replication/DNA integrity, spindle integrity) to components of the cell cycle clock. In this regard, cell cycle checkpoint pathways can be considered to have three classes of components: sensors, signal transducers, and signal receivers. The sensors detect "problems" such as DNA damage and convey this information to signal transducers, which in turn relay the signal to signal receivers, which are often integral components or direct regulators of the cell cycle clock.

To understand the functioning of cell cycle checkpoints, it is useful to examine several checkpoints in molecular detail. For this purpose, we consider the spindle assembly checkpoint, which detects errors in the processes required for proper mitotic chromosome segregation and the DNA damage checkpoint acting at G_2/M to prevent entry into mitosis with damaged chromosomes. Additional checkpoints preventing S phase entry in the presence of damaged DNA and preventing nuclear division in yeast cells that fail to bud properly are also discussed briefly.

2.1. The Spindle Assembly Checkpoint

A series of carefully orchestrated events ensures that replicated DNA will be equally distributed to daughter cells at the time of cell division. After mitotic condensation of the replicated chromosomes, the kinetochores of the sister chromatids attach to microtubules leading to opposite poles of the mitotic spindle. All of the attached chromosomes then congress to align on the metaphase plate. Once this has occurred, sister chromatids separate simultaneously; it is this event that defines the metaphase/anaphase transition (Fig. 1). It is the job of the spindle assembly checkpoint to prevent execution of the metaphase/anaphase transition if bioriented attachment of the chromosomes on the spindle is not properly acheived. Indeed, in animal cells, exit from mitosis (or progression through anaphase) is delayed or prevented by the spontaneous occurrence of chromosomes lagging in spindle attachment or by experimental manipulations that result in a disruption of even a single chromosomal attachment. Unchecked errors in the orderly

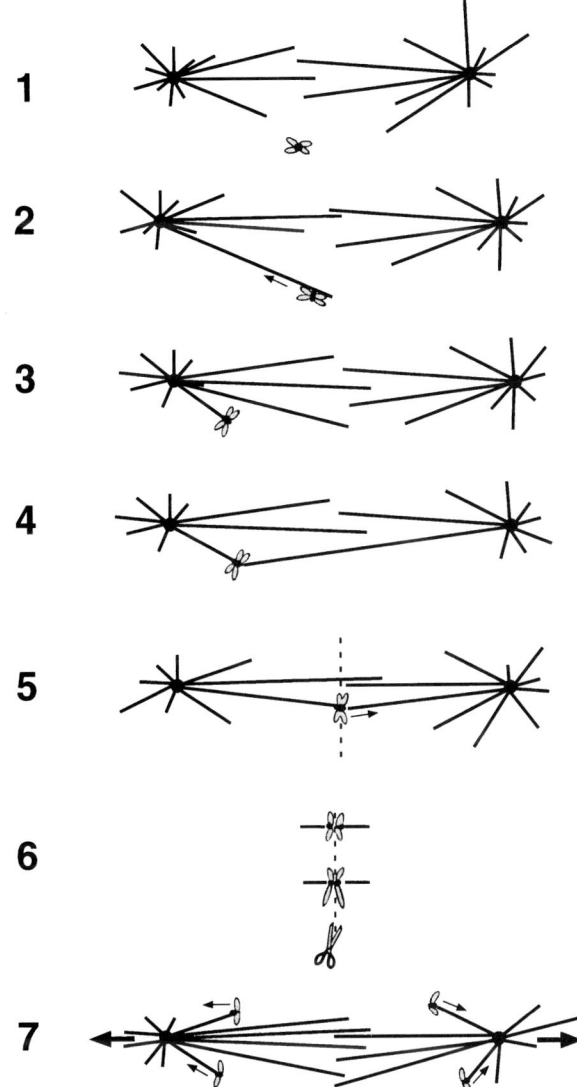

Fig. 1. The mitotic spindle and progression through mitosis. Shown are various stages in the attachment of chromosomes to the mitotic spindle and progression from metaphase to anaphase. For simplicity, only one representative chromosome is shown to illustrate the behavior of the entire chromosomal complement. (1) Prior to spindle attachment. (2) Chromosome attaches to one pole. (3) Kinetochore attaches end-on. (4) Microtubule from opposite pole is captured by distal kinetochore. (5) Chromosomes congress to the metaphase plate. (6) The metaphase/anaphase transition. (7) Anaphase; chromosomes move to the poles and poles move apart.

attachment and segregation of chromosomes may lead to aneuploidy, with daughter cells receiving either too few or too many chromosomes; cancer and developmental defects are among the possible consequences for the organism.

While there has been considerable debate about precisely which steps in chromosome segregation are monitored by the spindle assembly checkpoint, the leading candidates appear to be (1) proper attachment of spindle microtubules to chromosomal kinetochores and/or (2) the level of tension generated at the kinetochore by proper attachment of the chromosomes to both ends of the mitotic spindle. Although all of the chromosomes eventually attach to the mitotic spindle in a bipolar fashion, attachment of all of the chromosomes does not occur absolutely synchronously. In animal cells at mitosis, one can observe the process of chromosomal attachment, waiting until only one kinetochore of one chromsome from the entire complement remains detached. If the remaining unattached kinetochore is destroyed with a laser, anaphase onset is no longer delayed, despite the fact that the chromosome is not properly attached to the spindle, and hence lacks the proper tension. This suggests that some element at the kinetochore is monitored independently of tension. If affinity purified antisera against a particular family of kinetochore proteins is microinjected into G_2-arrested cultured cells, the cells will arrest in mitosis prior to the onset of anaphase, even if there do not appear to be any grossly observable defects in spindle assembly or chromosomal attachement. Similar effects are observed in yeast cells when kinetochore proteins are mutant.

Perhaps the most compelling evidence for monitoring of spindle tension across the kinetochore comes from studies of meiotic metaphase in praying mantis cells. These organisms have an unusual sex chromosome arrangement such that in 10% of meiotic divisions, one of the X chromosomes is spontaneously monoattached, residing close to one of the poles. Although these cells will arrest indefinitely in metaphase of meiosis I, progression through anaphase can be restored if a glass needle is used to pull the monoattached chromosome toward the other pole, thereby restoring tension across the kinetchore.

Although unattached kinetochores appear to generate a checkpoint signal, either through lack of tension or simple lack of chromosome/microtubule interactions, how these defects are sensed to generate a signal is still a bit obscure. A monoclonal antibody generated for a different purpose has serendipitously been found to recognize a phosphorylated epitope present only on unattached kinetochores. The staining of monoattached chromosomes by this monoclonal (called 3F3) is decreased when micromanipulator needles are used to exert tension across the kinetochore, suggesting that this epitope may play a role in sensing the tension across the kinetochore.

2.1.1. SIGNAL TRANSDUCTION IN THE SPINDLE ASSEMBLY CHECKPOINT

Once an unattached chromosome is sensed, how is this information relayed to receivers responsible for stopping the cell cycle? A series of clever genetic screens in the budding yeast, *Saccharomyces cerevisiae* have served to uncover key participants in the spindle assembly checkpoint signaling pathway. In this yeast, the cell cycle state of the cell can be determined by examining the size ratio of the mother cell/bud. At mitosis, there is no nuclear envelope breakdown, as in higher eukaryotes, and the spindle microtubules, which are nucleated by spindle pole bodies, span the interior of the nucleus. In this organism, spindle formation can start even as S phase is occurring. If spindle assembly is very slow, it is necessary to activate the spindle assembly checkpoint, to prevent premature anaphase progression. Therefore, one group used low concentrations of benomyl, an inhibitor of microtubule polymerization in yeast to slow spindle assembly and thereby to create an artificial situation in which the spindle assembly checkpoint would be required for cell viability. Cells bearing mutations in some element of the checkpoint system would be expected to exit mitosis with incompletely assembled spindles and die when plated on benomyl, but would survive unperturbed in the absence of benomyl. It would be expected that benomyl-sensitive mutants would include tubulin mutants, other proteins sensitizing the spindle to disruption by benomyl and true spindle assembly checkpoint mutants. However, unlike the first two types of mutants, which would quickly cease cell division in the presence of benomyl, cells mutant in checkpoint elements would initially continue to divide rapidly (without a checkpoint to restrain their cell cycle progression), but would then lose viability due to aneuploidy. Using such a screen, a series of mutants were selected and named *Mad* mutants (mitotic arrest defective). In another yeast screen, mutants defective in the spindle assembly checkpoint were selected by subjecting yeast to transient high doses of benomyl; it was anticipated that cells defective in the checkpoint would be unable to pause their cell cycle after complete depolymerization of the spindle, resulting ultimately in lethality. Mutants selected in this screen were termed *bub* mutants ("budding uninhibited by benzimidazole"). Consistent with their roles in the spindle assembly checkpoint, one of each of the above mutants (*Mad2* and *Bub1*) was detected on *unattached* kinetochores, potentially acting as part of the sensing/signaling mechanism at the kinetochore. Using the twist of yeast biology that allows spindle assembly to begin during S phase, Murray and colleagues showed that *Mad* mutants could survive after benomyl treatment if their S phase was sufficiently prolonged (by artificially slowing DNA synthesis with the drug hydroxyurea) to allow for completion of spindle assembly. This is consistent with the notion that the primary role of a checkpoint is to allow time for errors to be rectified before the cell cycle proceeds.

That the signaling pathway from the kinetochores to the cell cycle machinery may involve protein phosphorylation is suggested by the fact that two of the proteins identified in the yeast screens are protein kinases. Another one of these proteins, MAD1, is a phosphoprotein whose phosphorylation depends on BUB1, BUB3, and MAD2. Moreover, as alluded to previously, the monoclonal 3F3 detects a phosphoepitope at the kinetochore in a manner contingent on the spindle attachment status of the chromosome.

2.1.2. RECEIPT OF THE SIGNAL TO STOP THE METAPHASE/ANAPHASE TRANSITION

The ultimate aim of spindle assembly checkpoint signaling is to stop anaphase onset until problems with chromosome attachment to the spindle are rectified. Regulatory events governing anaphase sister chromatid separation have been studied most intensively in the two yeasts, *S. cerevisiae* and *S. pombe*. In these organisms, anaphase is initiated when specific inhibitors of anaphase (Pds1 in *cerevisiae* and Cut 2 in *pombe*) are ligated to the proteosome-targeting moeity, ubiquitin, and then degraded by proteosomes. The covalent attachment of ubiquitin is accomplished by an E3 ubiquitin ligase complex termed the cyclosome, or anaphase promoting complex (APC). When the spindle assembly checkpoint is activated by errors in chromosome/spindle dynamics, Pds1 escapes degradation by the APC/proteasome pathway, thereby preventing anaphase onset. How is this accomplished? A central element in this pathway was defined when it was found that Mad2 (recall that this protein sits at unattached kinetochores) could physically associate with a protein (called slp1 in *S. pombe* and Cdc20 in *S. cerevisiae*) which conferred on the APC the ability to target Pds1 for degradation. Indeed mutants of slp-1 and Cdc20 which no longer could bind Mad2 could override the spindle assembly checkpoint, thereby allowing anaphase onset with a defective spindle. Moreover, simple overproduction of Mad2 could lead to preanaphase M phase arrest, suggesting that Mad2 might act to titrate Slp1 away from the APC. Collectively, these data suggest a model whereby Mad2, acting in some as-yet-undefined way with other Mad/bub gene products, can signal the presence of kinetochore/microtubule attachement disturbances by physically preventing slp/APC-mediated degradation of pds1.

2.2. The G_2/M DNA Damage Checkpoint

When DNA is damaged, one of the physiological responses of the cell is to arrest in G_2, prior to mitotic entry. The presumed purpose of this arrest, or pause, in the cell cycle is to allow the cell time to repair the damage before proceeding through mitosis. Were this not the case, damaged DNA and/or broken chromosomes would be passed on to daughter cells at the time of division. The molecular pathway inhibiting entry into mitosis until DNA damage is repaired is referred to as the G_2 DNA damage checkpoint. While DNA damage induces many cellular responses, including an induction of genes encoding proteins responsible for damage repair, we will restrict our discussion here to direct effects on cell cycle progression.

Entry into mitosis in all eukaryotic cells is controlled primarily by the action of the Ser/Thr kinase Cdc2 (also called Cdk1) complexed to its activating subunit, cyclin B. To understand the molecular action of the DNA damage checkpoint, it is necessary to consider the levels at which Cdc2–cyclin B can be regulated. As mentioned earlier, cyclins oscillate in abundance throughout the cell cycle; the B-type cyclins are synthesized steadily throughout interphase and are degraded abruptly at the time of exit from mitosis. Although Cdc2–cyclin B complexes form in step with the continuous accumulation of cyclin B, these complexes are held inactive until the G_2/M transition, at which time the kinase activity of this complex spikes precipitously. The inactivity of these complexes throughout interphase is attributable in large part to phosphorylation of the Cdc2 subunit on two regulatory sites, Thr[14] and Tyr[15]. Phosphorylation of Tyr[15] in animal cells is catalyzed by two enzymes, Wee1, located predominantly in the nucleus, and Myt1, a membrane-bound kinase believed to act on cytoplasmic pools of Cdc2. Myt1 is a dual-specificity enzyme that also catalyzes phosphorylation of Thr[14]. At the time of entry into mitosis, both Thr[14] and Tyr[15] are abruptly dephosphorylated by a dual-specificity phosphatase, Cdc25. Hence, Wee1, Myt1, and Cdc25 all provide potential loci of action for regulatory pathways aimed at controlling activation of Cdc2/cyclin B. That these are, indeed, key targets for the DNA damage checkpoint is suggested by the following observations: Mutation of Tyr[15] of Cdc2 to phenylalanine greatly reduces the efficacy of the checkpoint, allowing cells to enter mitosis in the presence of damaged DNA. Similarly, overproduction of the phosphatase, Cdc25 can abrogate delay of G_2/M.

Much of our understanding of the DNA damage checkpoint comes from yeast genetic screens designed to identify mutants that were particularly sensitive to radiation. As it turns out, mutation in many of these *"rad"* genes led to reduced cell viability after irradiation because the cells bearing them did not properly arrest the cell cycle to allow for damage repair. How are these genes connected to checkpoint control? Interestingly, it has been shown that a number of rad gene products function to "process" DNA damage lesions into a form which will be recognizeable by the checkpoint "sensors." There is circumstantial evidence that the lesion that is ultimately recognized is single-stranded DNA, and a number of rad proteins have been proposed to interact directly with these single stranded regions to generate a signal; the exact nature of such a signal is unclear.

At the stage of signal transmission, the molecular role of some of these proteins in the operation of the DNA damage checkpoint becomes much more clear. A number of proteins (including rad proteins) that are essential for the damage checkpoint have clearly defined biochemical functions that link them in a series of defined reactions to modulation of Cdc2 tyrosine phosphorylation. Several of these proteins encode protein kinases (rad3 and Chk1), while others (rad24 and rad25) encode 14-3-3 proteins, a class of small signaling molecules that bind their targets in a phosphorylation-dependent manner. When cells incur DNA damage, Chk1 becomes phosphorylated and its activity as a kinase is increased. In *S. pombe*, this activation of Chk1 is dependent on one of the rad proteins, rad3, though it is not clear whether the phosphorylation of Chk1 is directly catalyzed by rad3. Intriguingly, rad3 is similar at the sequence level to ATM, mutations in which lead to development of the disease ataxia telangiectasia in humans. These patients are particularly radiation sensitive, displaying an increased cancer susceptibility.

The link between all of these proteins and Cdc2 tyrosine phosphorylation came from a series of biochemical experiments using recombinant proteins and cell lysates of both mammalian and yeast origin. First, it was found in coimmunprecipitation experiments using either yeast proteins or their mammalian counterparts that Chk1 and Cdc25 could interact physi-

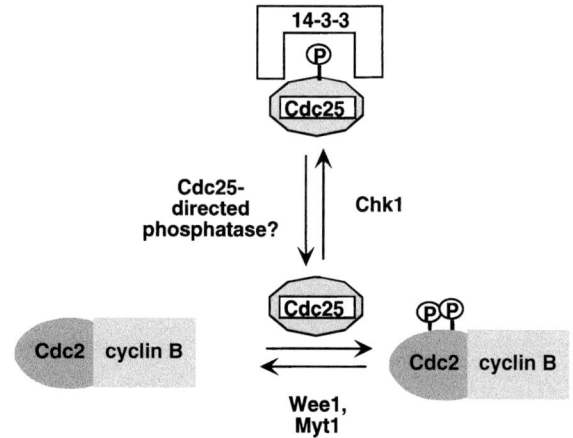

Fig. 2. The DNA damage checkpoint regulates Cdc2–cyclin B through modulation of Cdc25. Cdc2 is phosphorylated on its negative regulatory residues, Thr^{14} and Tyr^{15}, by Wee1 and Myt1. The Cdc25 phosphatase dephosphorylates these sites, leading to activation of Cdc2–cyclin B kinase activity and entry into mitosis. DNA damage triggers activation of the Chk1 kinase, which phosphorylates Cdc25. On phosphorylation, Cdc25 is bound by 14-3-3 proteins, preventing Cdc2 dephosphorylation. Presumably, the phosphorylation of Cdc25 can be reversed by a Cdc25-directed phosphatase.

cally. Moreover, Chk1 could phosphorylate a site on Cdc25, Ser^{216}, that looked to be part of a consensus binding site for 14-3-3 proteins (e.g., rad 24). Indeed, it was shown that phosphorylation by Chk1 enabled Cdc25 to bind to 14-3-3 protein in extracts from cultured cells and a mutant Cdc25 molecule in which Ser^{216} had been changed to the nonphosphorylatable Ala could no longer bind 14-3-3. Perhaps most striking is the observation that expression of the S216A mutant of Cdc25 in cells could seriously impair operation of the DNA damage checkpoint, suggesting that Chk1-mediated phosphorylation of Cdc25 is one of the key events in preventing entry into mitosis in the presence of damaged DNA (Fig. 2).

What then is the effect of phosphorylating Cdc25 on S216? Convincing evidence from a number of groups suggests that 14-3-3 binding does not alter the enzymatic activity of Cdc25. Rather, the most likely possibility is that 14-3-3 serves either to sequester Cdc25 in a particular subcellular locale or target it to a particular location, thereby preventing Cdc25-mediated dephosphorylation of Cdc2.

Although this story is appealingly elegant, there are still holes to be filled. How is phosphorylation/activation of Chk1 phosphorylation achieved? It is attractive to speculate that rad3/ATM accomplishes this task, but there is currently no evidence to support the hypothesis that this is phosphorylation is direct. Even if this were the case, it is not clear how rad3/ATM is activated by DNA damage.

It remains possible that the DNA damage checkpoint pathway modulates not only Cdc25, but also the kinases that phosphorylate cdc2. Indeed, Wee1 can be phosphrorylated by Chk1 in vitro, although the consequences of this phosphorylation are not clear. In addition, the possible involvement of the membrane-bound Myt1 kinase has not been fully explored.

While regulation through tyrosine phosphorylation may account for much of the DNA damage checkpoint functioning, the nonphosphorylatable mutant of Cdc2 that interferes with checkpoint function does not abrogate the checkpoint entirely. This suggests that there may be other levels of regulation of Cdc2–cyclin B involved in checkpoint operation. For example, there is some evidence that stoichiometric inhibitors may bind to and inhibit Cdc2–cyclin B kinase activity. Moreover, Cdc2–cyclin B complexes that are normally localized to the cytoplasm during interphase can be made to override checkpoints if they are inappropriately localized to the nucleus, suggesting that some aspect of checkpoint function may modulate the subcellular localization of Cdc2–cyclin B. All of these possibilities offer avenues for future experimentation. Additional checkpoints are described in brief in the following sections.

2.2.1. THE G1/S DAMAGE CHECKPOINT

The point of cell cycle arrest in response to DNA damage depends on the cell cycle phase at which the damage was incurred. While the G_2 damage response was discussed in detail previously, DNA damage occurring before S phase results in a block to progression from G_1 to S phase. In mammalian cells, this response is targeted toward two key signal receivers in the cell cycle clock machinery, the cyclin-dependent kinases, Cdk2 and Cdk4, required for the G_1/S transition. The mechanism of this arrest involves one of the most frequently mutated genes in human cancers, the transcriptional regulator, *p53*. Like *pds1* in the spindle assembly checkpoint, *p53* is stabilized in response to checkpoint activation. Indeed, phosphorylation of the inherently unstable *p53* in response to DNA damage (potentially by the ATM gene product), prevents its ubiquitination by the E3 ubiquitin ligase, MDM2. The stabilization of *p53* leads to increased transcription of a number of *p53*-responsive genes, including, *p21cip/waf1*, a direct, stoichiometric inhibitor of Cdk2–cyclin and Cdk4–cyclin kinases. Therefore, DNA damage-induced stabilization of *p53* leads to synthesis of *p21* which is, at least in part, responsible for a block to G_1/S progression.

2.2.2. THE DNA REPLICATION CHECKPOINT

The DNA replication checkpoint prevents the onset of mitosis when DNA replication is inhibited. Experimentally, this checkpoint can be triggered when cells are treated with DNA replication inhibitors such as hydroxyurea or aphidicolin; under these conditions, cells will arrest prior to entry into mitosis. Like the DNA damage checkpoint, this arrest can be overriden by caffeine. This checkpoint shares many features with the G_2/M DNA damage checkpoint, including prevention of Cdc2 Tyr^{15} dephosphorylation. Indeed, phosphorylation of S216 on cdc25 is prolonged when replication is blocked, although the kinase(s) responsible (human Chk2/*S. pombe* Cds1–*S. cerevisiae* Rad53) appear to be distinct from Chk1. It has also been shown recently that inhibition of DNA replication will stabilize the wee1 protein, suggesting that Cdc2 tyrosine phosphorylation may be partially increased through this mechanism upon triggering of the DNA replication checkpoint.

It is likely that different sensors will be involved in detecting stalled replication forks and damaged DNA. In *S. cerevisiae* and *S. pombe*, proteins that are themselves critical for DNA replication (e.g., polε, polα) appear to be essential for operation of the replication checkpoint; mutation of genes encoding these proteins not only abrogates DNA replication, but also allows progression into M phase in the presence of unreplicated DNA. This suggests that some feature of replication may be monitored by proteins participating in fork movement.

2.2.3. THE MORPHOGENESIS CHECKPOINT

In the yeast *S. cerevisiae*, environmental stresses such as alterations in termperature or media osmolarity lead to a delay in bud formation, resulting from transient depolarization of the actin cytoskeleton. In response to such delays, yeast cells activate a morphogenesis checkpoint, which delays nuclear division to wait for proper bud formation. In this checkpoint, it appears that some element of the actin cytoskeleton is monitored, although the precise mechanism behind this is unclear. Interestingly, the signal receiver in this pathway appears to be the Cdc2 homolog Cdc28. Mutating the negative regulatory tyrosine (in this case, Tyr^{19}, rather than Tyr^{15}) to phenylalanine overrides the checkpoint delay. Moreover, it has been convincingly demonstrated that swe1, the *S. cerevisiae* homolog of wee1, a protein normally stable only in G_1/early S phase, is stabilized throughout S phases and G_2 upon activation of the morphogenesis checkpoint. This stabilization leads to continued phosphorylation of Cdc28 on Tyr^{19}, thereby preventing cell cycle progression.

CONCLUSIONS

The molecular details of checkpoint operation are beginning to emerge as a result of recent intense research in this field. Significant questions remain; it will be particularly important to unravel the "sensing" mechanisms that are used to detect errors in the processes being monitored. On a practical level, knowledge of checkpoint regulators may offer the potential for novel therapeutic strategies in battling cancer. Checkpoint operation often goes awry during carcinogenesis; forcing tumor cells to override checkpoints and "divide themselves to death" is a potential therapeutic strategy that will almost certainly be explored in the coming years.

SELECTED READINGS

Boddy MW, Furnari B, Mondesert O, Russell P. Replication checkpoint enforced by kinases Cds1 and Chk1. Science 1998; 280:909–912.

Chen RH, Waters JC, Salmon ED, Murray AW. Association of spindle assembly checkpoint component XMAD2 with unattached kinetochores. Science 1996; 274:242–246.

Hoyt MA, Totis L, Roberts T. *S. cerevisiae* genes required for cell cycle arrest in response to loss of microtubule function. Cell 1991; 66:507–517.

Lew DJ, Reed SI. A cell cycle checkpoint monitors cell morphogenesis in budding yeast. J Cell Biol 1995; 129:739–749.

Li R, Murray AW. Feedback control of mitosis in budding yeast. Cell 1991; 66:519–531.

Li X, Nicklas RB. Mitotic forces control a cell cycle checkpoint. Nature 1995; 373:630–632.

Lydall D, Weinert T. Yeast checkpoint genes in DNA damage processing: implications for repair and arrest. Science 1995; 270:1488–1491.

Nicklas RB, Ward SC, Gorbsky GJ. Kinetochore chemistry is sensitive to tension and may link mitotic forces to a cell cycle checkpoint. J Cell Biol 1995; 130:929–939.

Peng CY, Graves PR, Thoma RS, Wu Z, Shaw AS, Piwnica-Worms H. Mitotic and G2 checkpoint control: regulation of 14-3-3 protein binding by phosphorylation of Cdc25c on serine-216. Science 1997; 277:1501–1505.

Sanchez Y, Wong C, Thoma RS, Richman R, Wu Z, Piwnica-Worms H, Elledge SJ. Conservation of the Chk1 checkpoint pathway in mammls: linkage of DNA damage to Cdk regulation through Cdc25. Science 1997; 277:1497–1501.

Weinert TA, Hartwell LH. The *Rad9* gene controls the cell cycle response to DNA damage in *Saccharomyces cerevisiae*. Science 1988; 241:317–322.

25 Molecular Regulation of Apoptosis

Rosemary B. Evans-Storms and John A. Cidlowski

Contents
Introduction
Activation of Apoptosis
Commitment to and Execution of Cell Death
Disassembly and Phagocytosis of the Cell
Conclusions
Selected Readings

1. INTRODUCTION

Regulated cell death is involved in such fundamental physiological processes as the resolution of structure during metamorphosis and embryogenesis, the regulation of immune cell maturation, target cell death induced by cytotoxic T lymphocytes and natural killer (NK) cells, and the response of hormone- and growth-factor-dependent tissues to the addition or withdrawal of ligands. Such death is also observed during normal tissue turnover, where it is usually counterbalanced by cell division. Disruption of this equilibrium by either a decrease in cell death or an increase in cell proliferation causes hyperplasia with a concomitant increase in the risk of tumor formation. This physiologic cell death is known as apoptosis (or programmed cell death), and contrasts sharply with necrosis, which tends to be caused by profound physical damage to a cell and often results in an inflammatory response.

Cells undergoing apoptosis are characterized by distinct morphological alterations including cell shrinkage, nuclear condensation, and the blebbing of apoptotic vesicles containing intracellular components. These vesicles are specifically recognized and engulfed by phagocytic cells; thus, an inflammatory response is not elicited because intracellular contents are never released into the extracellular environment.

From: *Principles of Molecular Regulation* (P. M. Conn and A. R. Means, eds.), © Humana Press Inc., Totowa, NJ.

Characteristic biochemical alterations that occur during apoptosis include the cleavage of DNA between nucleosomes to form fragments that are multiples of approximately 180 basepairs and/or into 50- to 300-kb fragments, the reorientation of phosphatidylserine from the cytoplasmic to the extracellular face of the plasma membrane, and the loss of the mitochondrial membrane potential. The regulation of the signal transduction pathways that lead to cellular disruption during apoptosis has been the focus of intense study in recent years.

Cells are exquisitely sensitive to small changes in their environment, and can adapt quickly to new conditions. In many instances, a rapid response is facilitated by the constitutive expression of key components of signal transduction or effector pathways. Many cells are able to undergo apoptosis in the presence of inhibitors of macromolecular synthesis, suggesting that the distal components of the apoptotic pathway are constitutively expressed in these cells. Such inhibitors can actually induce apoptosis in some cell lines independent of additional stimuli, implying that the constitutively expressed effectors of apoptosis are normally inhibited by the presence of suppressive molecules. Protein synthesis is required for the induction of apoptosis in some circumstances such as glucocorticoid treatment of cultured intact murine and human thymocytes. However, chromatin in isolated nuclei from these same cells can be induced to

Table 1
Proteins Known to be Cleaved by Caspases During Apoptosis

Substrate	Function and Effect of Cleavage on Activity
DNA protein kinase catalytic subunit (DNA-PK$_{cs}$)	DNA repair (−)
Poly (ADP-ribose) polymerase	DNA repair (−)
Cytosolic phospholipase A$_2$	Arachidonic acid metabolism (+)
Focal adhesion kinase (FAK)	Cell attachment and survival (−)
Mitogen-activated protein kinase kinase kinase 1 (MEKK1)	Activation of JNK (+)
p21-activated kinase 2 (PAK-2)	Activation of JNK (+)
PITSLRE-kinase α2-1	Signal transduction during apoptosis (+)
Protein kinase C—delta isoform (PKC Δ)	Signal transduction during apoptosis (+)
Mdm-2	Regulation of p53 degradation (−)
Retinoblastoma (Rb) protein	Regulation of cell cycle (−)
Sterol regulating element-binding protein (SRBEP)	Transcription factor (+)
Actin	Cytoskeletal protein (−)
Fodrin	Membrane-associated cytoskeletal protein (−)
Gelsolin	Actin-regulating protein (+)
Keratins 18 and 19	Intermediate-filament proteins (−)
Nuclear mitotic apparatus protein (NuMa)	Nuclear matrix protein (−)
DFF 45/ICAD	Inhibition of DFF 40/CAD activity (−)

undergo condensation and internucleosomal DNA degradation by incubation with Ca^{2+} and Mg^{2+}. These data suggest that although synthesis of some components of the signal transduction cascade are required for apoptosis to occur in thymocytes, terminal effectors of apoptosis that cause nuclear alterations are always present in these cells.

Much of our insight into the regulation of apoptosis has been derived from elegant genetic studies performed in the nematode *Caenorhabditis elegans*. The expression of the genes *ced*-3 and *ced*-4 is sufficient for apoptosis to occur during normal development of this organism, while the expression of *ced*-9 is able to inhibit this cell death. The *ced*-3 gene shares homology with a family of cysteine proteases in higher animals designated caspases (for *C*ys protease cleaving after *Asp* residues), many of which are involved in the induction or execution of apoptosis. The human *ced*-4 homologue *apaf*-1 is required for the activation of a specific caspase whose function is required for most examples of apoptosis in human cells, and the human *ced*-9 homologue, *bcl*-2, is able to suppress apoptosis in a variety of systems. Homologs of these three genes comprise the minimal central apoptotic pathway in all organisms in which apoptosis occurs.

The evolutionary development of additional caspases (more than 14 have been identified so far) has added regulatory complexity to the pathways controlling apoptosis in higher animals. These caspases have similar amino acid sequences and substrate specificities, each cleaves after an aspartic acid residue, and each recognizes a motif of at least four amino acids proximal to the cleavage site. However, other structural elements must be required for substrate recognition, as not all proteins with a particular motif are cleaved. Caspases are expressed as proenzymes containing three domains, and are activated by proteolytic processing followed by association of two resulting subunits to form a heterodimer. This processing occurs at caspase recognition sites; thus, all appear to be activated either by autodigestion or by other caspases, and are therefore subject to regulation at several steps during activation.

Caspases known to be involved in apoptosis can be divided into two groups: those involved in the initiation of the apoptotic signal (initiator caspases), and those involved in the disassembly of the cell (effector caspases). The initiator caspases include caspase-2 (which may also have effector functions), -8, -9, and -10, and the effector caspases include -3, -6, and -7. Activation of the initiator caspases starts a cascade in which one caspase acts upon another, ultimately causing a large amplification of effector caspase activity. Caspase substrates that are known to play a role in apoptosis are described in Table 1. These include proteins that are activated by caspase cleavage (releasing an active subunit) such as p21-activated kinase 2 and cytosolic phospholipase A2; proteins that are inactivated by cleavage including Mdm-2, the catalytic subunit of DNA protein kinase

Chapter 25 / Molecular Regulation of Apoptosis

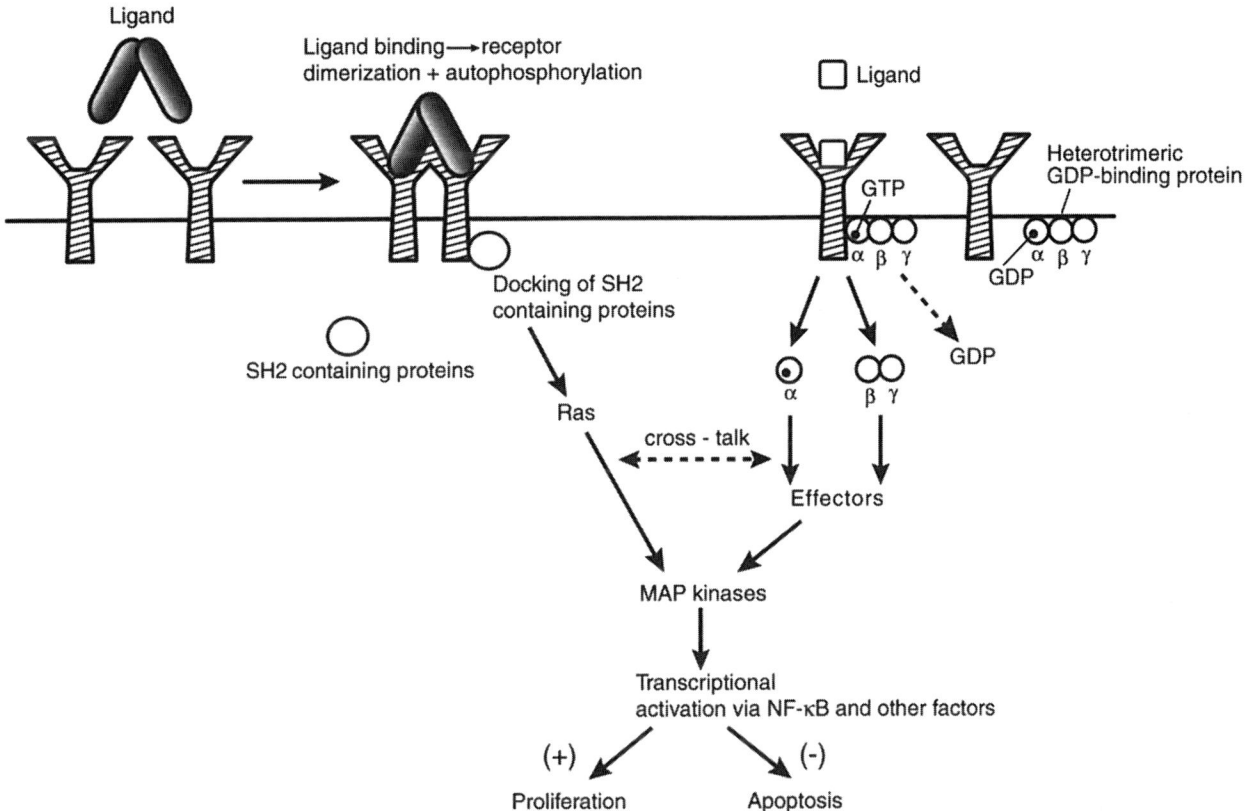

Fig. 1. Modulation of apoptosis by receptor tyrosine kinases and G-protein coupled receptors. Ligand binding to some receptor tyrosine kinases and G-protein coupled receptors activates one or more of three MAP kinase cascades. These kinases then regulate the activity of various transcription factors. Subsequent alteration of gene transcription leads to suppression of apoptosis in certain cell types.

and poly (ADP-ribose) polymerase; as well as several nuclear and cytoskeletal structural proteins. In addition, a number of caspase substrates have been identified that do not appear to be involved in apoptosis including the 70-kDa protein in the U1 SnRNP, HnRNP proteins C1 and C2, and huntingtin.

2. ACTIVATION OF APOPTOSIS

A delicate balance exists in each cell between proliferative/differentiation signals and antiproliferative signals. This allows cells to carry out their predestined function at the appropriate time and place in an organism. When this function has been fulfilled, or when too many mutations have been accrued by a cell to allow it to function normally, the cell is programmed to die by apoptosis. The stimulus required to induce apoptosis may be transduced by activation or deactivation of certain receptors (cell surface or nuclear),

or by physical stress such as radiation, hyperthermia, hypoxia, or chemically induced damage.

2.1. Receptor Tyrosine Kinases and G-Protein Coupled Receptors

The binding of growth factors and certain other molecules such as prostaglandins and the bioactive peptide endothelin to their cell surface receptors is a survival signal and therefore suppresses the apoptotic pathway. Activation of these receptors generally stimulates one or more of three mitogen-activated protein (MAP) kinase cascades: the extracellular receptor kinases (ERK); Jun N-terminal kinase, also known as the stress associated protein kinase (SAPK/JNK); and/or p38. The ERK pathway is commonly activated by receptors whose cytoplasmic domains are tyrosine kinases such as those for epidermal growth factor and platelet-derived growth factor, as well as by some G-protein coupled receptors such as that for prostaglandin F$_{2\alpha}$ (Fig. 1). Both SAPK/JNK and p38 may also be activated by stressors such as γ-irradiation, as well as inflammatory cytokines including interleukin (IL)-1β, tumor necrosis factor (TNF)-α, and Fas ligand. MAP kinases exert their effects by regulating the activity of various transcription factors, although the target genes of these transcription factors may vary with cell type.

The withdrawal of required growth factors from a cell induces apoptosis. Because these growth factors regulate MAP kinases, it might be predicted that inactivation of these kinases correlates with cell death. Surprisingly, induction of apoptosis by withdrawal of nerve growth factor (NGF) from pheochromocytoma cells is associated with activation of both SAPK/JNK and p38, and inhibition of the ERKs. In contrast, in other cell types, apoptosis does not involve JNK or p38 activation. Activation of the ERK's does protect some cell types from apoptosis induced by various stimuli including serum deprivation and ultraviolet irradiation.

The transcription factor NF-κB is activated via many receptor tyrosine kinases and G-protein coupled receptors. NF-κB has been implicated in suppression of apoptosis in several cell types, and in at least one case this is due to increased expression of specific apoptosis inhibitory molecules. In contrast, this transcription factor has a proapoptotic effect in some T cells. Therefore, as will be seen with many signal transduction pathways implicated in apoptosis, the cell type and apoptotic stimulus determine the final outcome.

2.2. "Death" Receptors

In contrast to the effect of growth factors, a number of cytokines that bind to cell surface receptors that are members of the TNF receptor superfamily can induce apoptosis. These death receptors possess similar cysteine-rich extracellular domains but lack substantial homology in their cytoplasmic regions except in the "death domain," which spans approximately 80 amino acids and which is required for transduction of the death signal. Members of this family include TRAMP (also known as DR3, Apo3, WSL-1, or LARD), the TRAIL receptors 1 and 2, avian CAR1, DR4, the Fas receptor (also known as CD95 or Apo1), and the TNF receptors (TNFR1 and TNFR2). TRAMP is expressed at highest levels in lymphocytes and thymocytes and is predicted to function in lymphocyte maturation. The only function currently attributed to TRAIL receptors is the induction of apoptosis in tumor cells. Two proteins related to the TRAIL receptors are decoys (DcR1 and DcR2) that are primarily expressed in normal cells, do not have a functional death domain, and confer resistance to TRAIL action. CAR1 and DR4 are poorly characterized. The most extensively studied members of this group are the Fas receptor (Fas R) and the TNF-1 receptor (TNFR1). The regulation of apoptosis by these receptors is a useful model for predicting the behavior of other members of this superfamily.

The Fas receptor is the only member of this family that appears to act primarily in the apoptotic pathway. This receptor is expressed in diverse tissues, but is chiefly involved in the modulation of the immune response. Binding of Fas ligand (Fas L) to this receptor mediates cell killing by cytotoxic T and NK cells, and induces apoptosis in activated T cells. Expression of Fas L also confers immune privilege on the eye (and probably other tissues) and on some tumors by inducing apoptosis in infiltrating cytotoxic T cells. The importance of Fas signaling in the immune system is emphasized by the phenotypes of several strains of mice in which the Fas signal transduction pathway has been interrupted by the alteration of either the Fas receptor or Fas ligand. The negative selection of T cells occurs normally in both strains of animal but they all develop autoimmune disorders and accumulate immature CD4$^-$CD8$^-$ T cells. Production of autoantibodies in these mice also suggests that the Fas receptor may be involved in appropriate B-cell response.

Proteins that contain death domains can associate with one another by interaction of these regions. Binding of Fas ligand to the Fas receptor causes receptor oligomerization via the death domains and subsequent recruitment of the adapter protein FADD/Mort 1 (which also contains a death domain) to the clustered receptor. This adapter protein then recruits caspase-8 to form the death-inducing signaling complex (DISC). This initiator caspase then undergoes activation by autodigestion and acts on other caspases in an amplifying cascade that culminates in the final stages of apoptosis. In some instances, the serine/threonine kinase RIP associates with FADD (instead of procaspase-8) and induces apoptosis by an unknown mechanism. Fas receptor can alternatively associate with Daxx (instead of FADD), which induces apoptosis by activation of the JNK pathway. The interaction of the Fas receptor with these proteins is illustrated in Fig. 2. Other proteins that can interact with the Fas receptor include sentrin, FAF1, which is able to potentiate apoptosis in L cells, and FAP-1, a protein tyrosine phosphatase that inhibits Fas-mediated apoptosis of T cells.

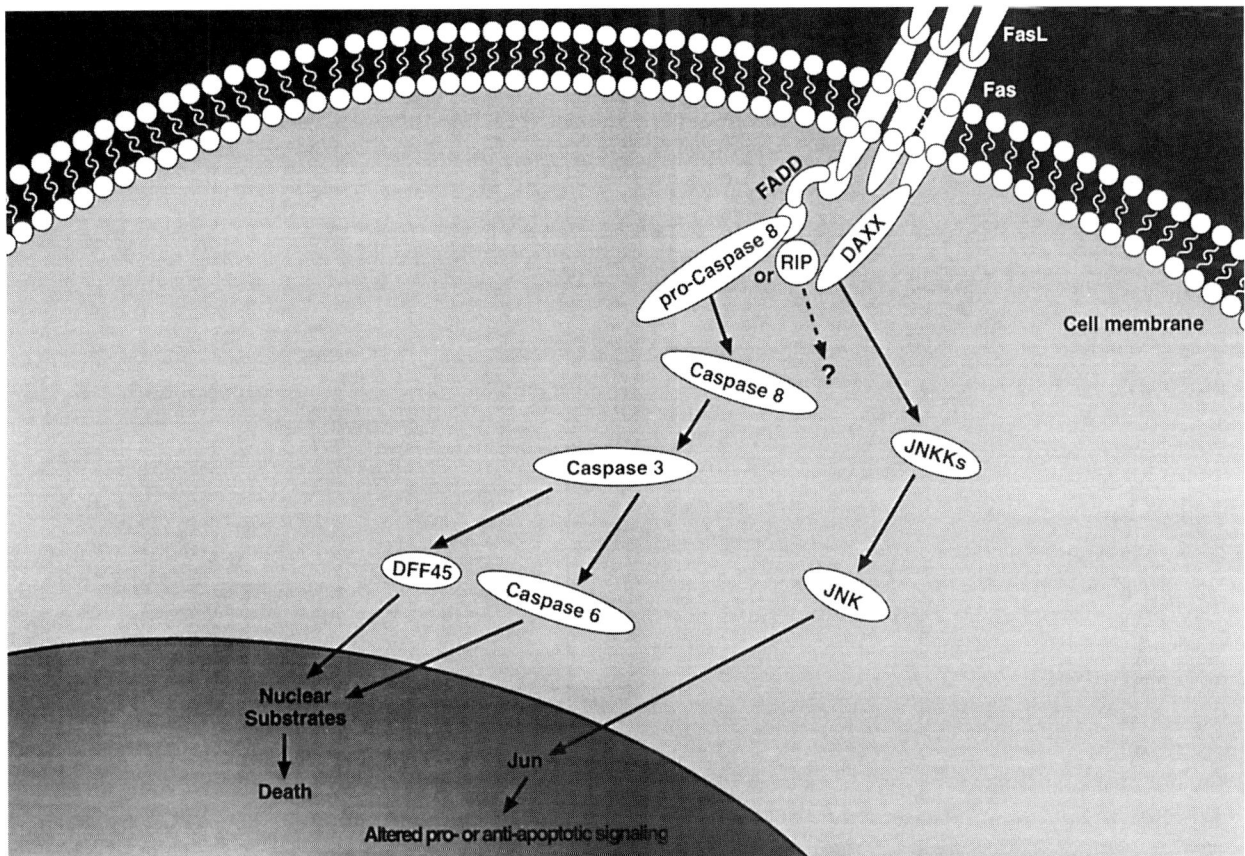

Fig. 2. Fas-mediated death signalling. Binding of Fas ligand to the Fas receptor causes receptor oligomerization followed by recruitment of adapter proteins to the cytoplasmic domain. The best characterized of these adapter proteins are FADD and DAXX. Binding of FADD causes recruitment of procaspase-8 (which is then activated) to form the death inducing signaling complex. In some cases, proteins other than procaspase-8 (such as RIP) may associate with FADD and regulate cell death via poorly characterized pathways. Association of DAXX with the Fas receptor instead of FADD causes activation of Jun Kinase, followed by altered transcription of genes involved in regulation of apoptosis.

Engagement of the TNFR1 can cause proliferation, differentiation, or apoptosis (the latter generally occurs only in the presence of protein synthesis inhibitors). These pleiotropic effects are determined by which proteins are recruited to the adapter protein TRADD, which binds to TNFR1 after ligand binding.

One protein that may associate with TRADD is FADD. As is the case when FADD associates with the Fas receptor, procaspase-8 is then recruited to FADD and activated, and the downstream events of apoptosis commence. Binding of TRAF2 to TRADD causes activation of the transcription factor NF-κB, which is associated with suppression of apoptosis in most cell types. Another protein that may associate with TRADD is RIP, which may also mediate NF-κB activation. In other instances, RIP recruits RAIDD/CRADD, leading to activation of caspase-2 and therefore apoptosis. Additional proteins that can associate with the TNFR1 include 55.11, sentrin, TRAP-1, MADD, and FAN. The functions of the first three are unknown; however, association with MADD leads to activation of ERK and SAPK/JNK, and association with FAN is able to activate neutral sphingomyelinase (NSMase) and thereby induce generation of ceramide, which has been implicated in induction of apoptosis.

The role of ceramide in apoptosis has been the subject of intense investigation as well as some controversy. Ceramide is generated by the activity of ceramide synthetase or by the hydrolysis of sphingomyelin by neutral or acidic sphingomyelinases. Ceramide may initiate differentiation, proliferation, or apoptosis depending on the cell type. Synthetic analogs of this sphingolipid can induce apoptosis in many cells; however, ceramide signaling is not universally required for induction of apoptosis, and the importance of this messenger in signal transduction via physiological inducers of programmed cell death is still a matter of debate. Although the Fas receptor and TNFR1 can both activate acidic and neutral sphingomyelinases, studies using various inhibitors of ceramide signaling suggest that they play no role in apoptosis induced by these receptors. Supporting this conclusion are experiments performed with cells from patients with Neimann–Pick disease, which lack functional acidic sphingomyelinase. These cells are resistant to apoptosis induced by ionizing radiation, but not to apoptosis induced by Fas ligand or TNF-α. Thus, it appears that ceramide may be important in some apoptotic signal transduction pathways but not others, or perhaps only in certain cell types or in combination with other modulatory signals.

2.3. Nuclear Receptors

Many members of the steroid/thyroid superfamily of nuclear receptors can regulate apoptosis in certain cell types after ligand binding. For example, engagement of the vitamin D receptor induces apoptosis in breast cancer cells and keratinocytes, and binding of ligands to the peroxisome proliferator activated receptor α (PPARα) or progesterone receptor can suppress apoptosis in rat hepatocytes and endometrial cells, respectively. Some of these hormones may also have opposing effects depending on the cell type. Examples include retinoids, which can induce apoptosis of cervical carcinoma cells but suppress cell death in activated thymocytes; and thyroid hormone, which is responsible for induction of the apoptosis that resolves structure during amphibian metamorphosis but inhibits programmed cell death of mammalian cerebellar granule neurons. The regulation of apoptosis by the glucocorticoid receptor has been the most extensively studied in this superfamily, making it a useful prototype to describe the regulation of apoptosis by all nuclear receptors.

Glucocorticoid hormones cause massive involution of the thymus gland due to apoptosis of immature thymic lymphocytes. Certain subpopulations of mature lymphocytes can also undergo glucocorticoid induced apoptosis including NK cells and cytotoxic T cells. Gene transcription seems to be necessary for induction of apoptosis by glucocorticoids in some cell types, while in other cells suppression of transcription by these hormones may be required. These effects are known to be mediated through the glucocorticoid receptor, which is localized to the cytosol in the absence of ligand, but which translocates to the nucleus after binding hormone. The receptor is then able to interact with DNA at glucocorticoid-responsive elements and affect gene transcription in either a positive or negative fashion. In addition, glucocorticoid receptor can act by antagonizing the function of the transcription factors NF-κB, AP-1, Oct-1, and CREB. This antagonism may occur by protein–protein interactions between the glucocorticoid receptor and these factors, or in some cases by steric hindrance of binding at overlapping recognition sites in DNA. The differing ways in which glucocorticoids may modulate gene transcription are illustrated in Fig. 3.

Subtractive DNA hybridization, differential display polymerase chain reaction (PCR), and two-dimensional electrophoresis of proteins have been utilized in repeated attempts to identify genes whose transcription must be activated or repressed in different cells for induction of apoptosis by glucocorticoids. More than 30 transcripts from several different cell types have been detected in this manner. Unfortunately, neither an identity for most of these genes nor a specific role for them during apoptosis has definitively been determined. Induction of apoptosis in the CEM human leukemia cell line by glucocorticoids correlates with suppression of c-*myc* expression and induction of c-*jun*, which encodes one of the subunits of the transcription factor AP-1. Similarly, apoptosis in a murine thymoma cell line has also been associated with suppression of c-*myc* expression in addition to repression of c-*myb*. In contrast, c-*myc* expression increases in rat thymocytes induced to undergo apoptosis with glucocorticoids. However, as in CEM cells, the transcription of c-*jun* is increased, as is expression of c-*fos* (which is the other subunit of AP-1). Although the effect of glucocorticoids on c-*myc* differs in these cell types, all show increased expression of one or more of the subunits of AP-1. T lymphocytes from mice in which c-*fos* has been inactivated by homologous recombination have reduced susceptibility to apoptosis induced by glucocorticoids, further implicating AP-1 in induction of programmed cell death. Other genes that have been associated with regulation of apoptosis by glucocorticoids include NUC18 (a cyclophilin-related nuclease), calmodulin, β-galactoside binding protein, and a recently described gene called *srg*-3. This latter gene is highly expressed in thymocytes, testis, and brain, and appears to be a transcriptional activator. Expression of antisense transcripts of SRG3 reduced apoptosis in mouse thymoma cells after glucocorticoid treatment. Regrettably, it is still unclear if any of the same genes must be regulated by glucocorticoids for induction of apoptosis in all cell types or with all stimuli.

Although glucocorticoids are well known for their ability to induce apoptosis in the thymus, they actually

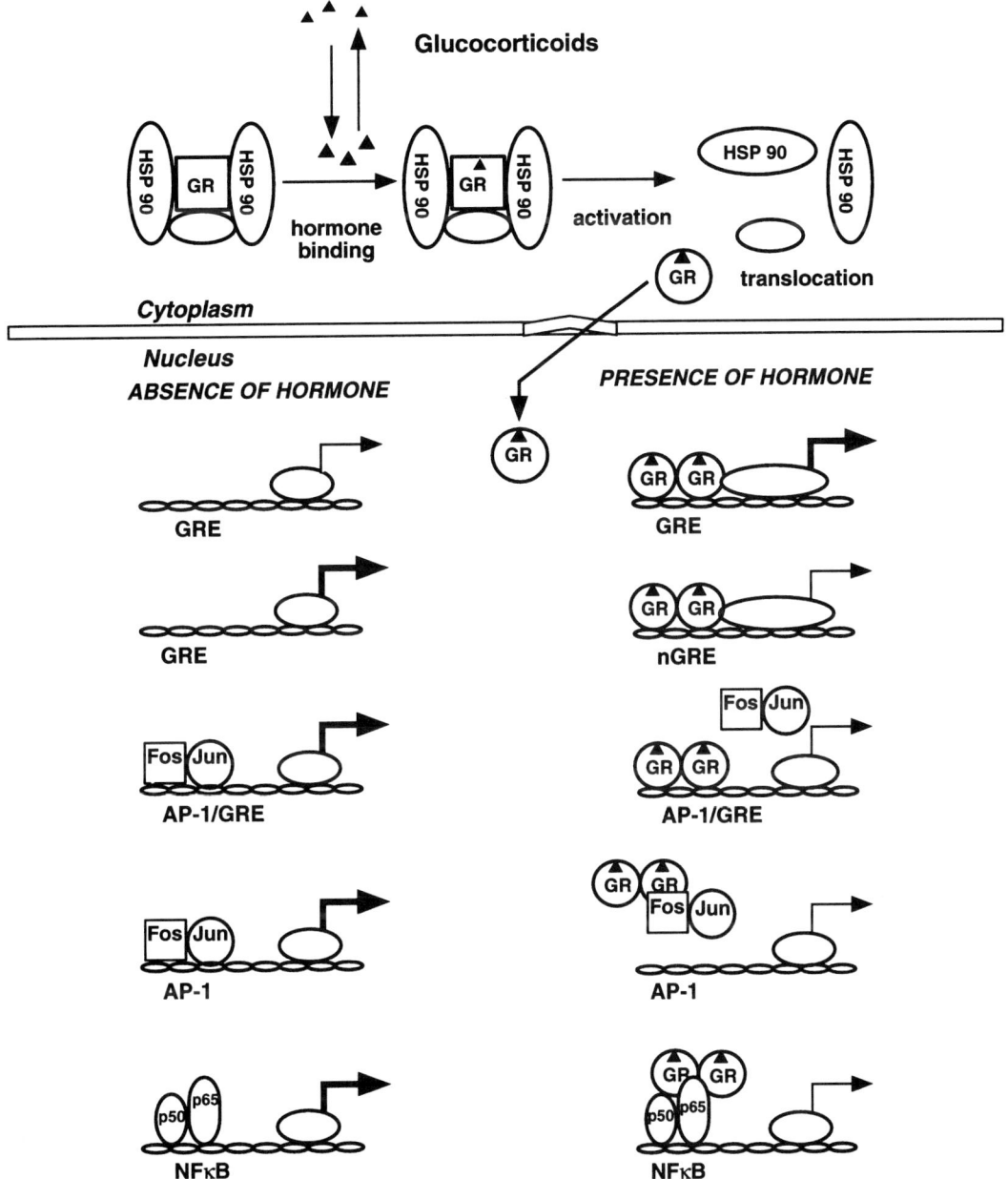

Fig. 3. Potential mechanisms for glucocorticoid mediated regulation of apoptosis. Glucocorticoids can regulate apoptosis in a positive or negative fashion (depending on the cell type) by altering transcription of genes involved in this process. Binding of hormone to the glucocorticoid receptor can affect transcription of such genes by inducing association of the receptor with positive or negative glucocorticoid regulatory elements (GRE's) in DNA. This receptor can also alter transcription by inhibiting the activity of other transcription factors either by direct interaction with these factors, or by steric hindrance of the binding of these factors to DNA in overlapping regulatory elements.

protect uterine and mammary epithelial cells, human neutrophils, prostate cells, mouse fibroblasts, and rat hepatoma cells from programmed cell death. This effect requires protein synthesis, implying that these hormones suppress apoptosis by inducing transcription of specific genes rather than by repressing the activity of other transcription factors with which the glucocorticoid receptor is known to interact. The expression of various coactivators or corepressors that bind with and affect the function of the glucocorticoid receptor may also partially explain the differential sensitivity of various cell types to glucocorticoids.

An intriguing member of the steroid/thyroid hormone receptor superfamily is *Nur77*. The expression of this orphan receptor (which has been reported to transactivate in the absence of ligand) is induced by growth factors in many cell types, and is associated with apoptosis in some cells of the immune system. However, studies in transgenic mice have produced conflicting data. One suggested that a dominant negative mutation of *Nur77* causes abnormal antigen-induced negative selection of thymocytes, while another suggested that a loss-of-function of this gene has no effect on apoptosis in the thymus or in peripheral lymphocytes. This might be explained by differences in the T-cell receptors in these mice, or in expression of homologous receptors in these strains of animal that may have functions that are redundant for those of *Nur77*. Of great interest are recent observations that *Nur77* can heterodimerize with the retinoid X receptor and possibly with the glucocorticoid receptor, suggesting a complex level of regulation between receptors of the steroid/thyroid superfamily.

2.4. Physical Damage

Apoptosis caused by hypoxia due to decreased blood flow is an important component of the pathology of myocardial infarction and stroke. Other physical stressors that can induce apoptosis include radiation (ultraviolet and ionizing), and hyperthermia. Some portions of the apoptotic signal transduction pathways are shared by these inducers. All increase AP-1 DNA binding activity and lead to generation of ceramide and reactive oxygen species. Hyperthermia and both types of radiation induce JNK and p38 kinase activity. However, γ-irradiation is unique in this group because it can induce oxidative DNA damage and/or double-stranded breaks in chromatin, both of which cause p53 to accumulate, followed by either cell cycle arrest or apoptosis depending on the cell type and other factors. How p53 induces apoptosis is still a matter of debate, but may include induction of other proapoptotic genes, transrepression of antiapoptotic genes, and/or nontranscriptional mechanisms.

3. COMMITMENT TO AND EXECUTION OF CELL DEATH

3.1. Energy Metabolism

Energy in the form of ATP is required in the early stages of apoptosis. This can be supplied by oxidative phosphorylation and/or glycolysis. Oxidative phosphorylation occurs primarily in the inner membrane of mitochondria but also in membranes of the endoplasmic reticulum and nucleus. This process invariably causes generation of reactive oxygen species (ROS) such as superoxide and hydrogen peroxide. Left unchecked, these ROS can cause lipid peroxidation with concomitant damage to membranes, proteins, and nucleic acids. Cells are generally able to control ROS-mediated damage by enzymatic conversion of these compounds to less toxic forms or by reaction with antioxidants such as glutathione and thioredoxin. Damage caused by ROS is implicated in apoptosis induced by most stimuli, as the levels of ROS rise during practically all examples of programmed cell death, addition of ROS or depletion of antioxidants is able to induce apoptosis in some cell types, and introduction of antioxidants can attenuate apoptosis induced by all stimuli except activation of the glucocorticoid and Fas receptors.

The stage at which cells are irreversibly committed to apoptosis is still a matter of debate. All apoptotic cells shrink prior to DNA fragmentation, and this may mark such a commitment step for many cells. This loss of cell volume may be required for alteration of cell structure and subsequent release of apoptotic enzymes, concentration of proapoptotic factors in the cell, or alteration of gene transcription. Presumably, energy is required to power ion pumps that keep apoptotic cells shrunken. If these pumps shut down, a net movement of water into the cell would be predicted, causing swelling (as occurs in necrosis).

Other energy-requiring processes that occur during apoptosis include protein synthesis, which is needed for induction of apoptosis in some cell types, and active transport of large molecules into the nucleus. Inhibitors of such active transport block Fas-induced apoptosis. Energy could also be necessary for the characteristic membrane blebbing observed during apoptosis. Such blebbing after induction of programmed cell death with hydrogen peroxide has been associated with ATP-dependent actin polymerization. Inhibition of actin polymerization with low doses of cytochalasin D in a different model system has been shown to inhibit blebbing but not other manifestations of apoptosis. In addition, interaction of myosin light chain (which is expressed in most cell types) with actin has been associated with energy-dependent alterations of the cytoskeleton.

3.2. Mitochondria

A mitochondrial membrane permeability transition (MPT or $\Delta\Psi_m$) occurs during both apoptosis and necrosis. This $\Delta\Psi_m$ occurs before DNA fragmentation, and is thought to be caused by the opening of pores in the inner membrane of the mitochondria which allows equilibration of ions with the cytosol, including the protons that drive the chemiosmotic generation of ATP. The levels of ATP should decrease precipitously after these pores open. Intracellular ATP content appears to determine whether apoptosis or necrosis will occur after a particular stimulus. The probability that necrosis will occur rather than apoptosis increases as ATP levels fall. Why is necrosis not induced after the opening of the mitochondrial pores? It has been suggested that these pores can open and close intermittently. Thus, perhaps the longer the pores are open, the more likely it is that necrosis will occur. Alternatively, glycolysis or oxidative phosphorylation in other organelles may supply enough ATP for apoptosis to occur in some cell types even after these pores open in the mitochondria.

The release of various intermitochondrial proteins (probably through the pores described earlier) including cytochrome c, a protease called AIF, and procaspase-2 and -9 has been associated with the onset of apoptosis. Cytochrome c is an electron carrier that is normally localized to the intermembrane space of the mitochondrian. This protein is required in combination with ATP to activate Apaf-1, the mammalian homolog of ced-4, which seems to be required for all apoptosis that involves mitochondria. Apaf-1 subsequently activates procaspase-9 which leads to downstream apoptotic events. AIF has been shown to activate caspase-3 and to directly induce nuclear apoptosis. The interaction of the mitochondria with the apoptotic pathway is illustrated in Fig. 4.

Mitochondria have been proposed to be global regulatory components of apoptosis. However, this assumption is now being questioned because two apoptotic signaling pathways have been identified for the Fas receptor, only one of which involves these organelles. In some cell types, engagement of this receptor leads to rapid activation of caspase-8 with subsequent activation of caspase-3. In other cell types, activation of these caspases is delayed owing to a reduced level of formation of the death inducing signalling complex. Although a loss of mitochondrial membrane potential is observed in both types of cells, activation of caspases precedes this in the first type, but occurs after the $\Delta\Psi_m$ in the second type. Thus, the former apoptotic pathway is not dependent on the mitochondria. Supporting a diminished role for mitochondria in some cases of Fas-induced apoptosis are studies showing that Apaf-1 is not involved in Fas-mediated apoptotic pathways in thymocytes and activated T cells. In addition, thymocytes from Apaf-1 null mice mature normally, and undergo normal activation of caspase-8 after engagement of the Fas receptor. However, activation of caspase-8 is reduced in these cells after treatment with glucocorticoids or etoposide. Embryonic stem cells derived from these null mice are resistant to many inducers of apoptosis but embryonic fibroblasts are not, suggesting that there is a differential requirement for Apaf-1 in different cell types, or alternatively that there are other proteins expressed in some cells that may substitute for Apaf-1.

3.3. Calcium

Calcium is an important second messenger that is released from the endoplasmic reticulum after activation of many receptors. This ion is involved in diverse physiologic responses including cell proliferation, muscle contraction, and apoptosis. The importance of Ca^{2+} in apoptosis is supported by data demonstrating that intracellular Ca^{2+} buffers inhibited the death of human leukemia cells induced by etoposide, and of thymocytes after irradiation or treatment with glucocorticoids. In addition, specific Ca^{2+} channel blockers inhibited the death of a human T cell line induced by cadmium. In contrast, calcium is able to promote survival of neutrophils and some types of neurons. As with many of the components of the apoptotic pathway, the requirement for calcium appears to be cell type and/or stimulus specific.

Calcium-dependent enzymes that could be involved in regulation of programmed cell death include kinases, phosphatases, proteases, and nucleases (which are discussed at length in Section 3.5.). Activation of protein kinase C (PKC) has been observed during apoptosis of T-cell hybridomas, and in PC12 cells after withdrawal of nerve growth factor; and DAP kinase (which is associated with the cytoskeleton) is a positive modulator of interferon-γ-induced apoptosis. In contrast, PKC activation by phorbol esters inhibits Fas-induced apoptosis of Jurkat cells. The protein phosphatase calcineurin potentiates apoptosis of T cells induced by glucocorticoids, but inhibits apoptosis of mouse thymocytes and a human B-cell line induced by thapsigargin or anti-immunoglobulin M (IgM) antibodies, respectively. Activation of the protease calpain is associated with apoptosis of muscle satellite cells starved for serum; thymocytes treated with glucocorticoids, irradiation, calcium ionophores, and forskolin; and β-amyloid induced death of cultured rat hippocampal neurons. However, calpain is not involved in thymocyte death after heat shock or treatment with valinomycin, nor is it implicated in Ca^{2+}-induced DNA fragmentation in isolated thymocyte nuclei. The calpain inhibitory

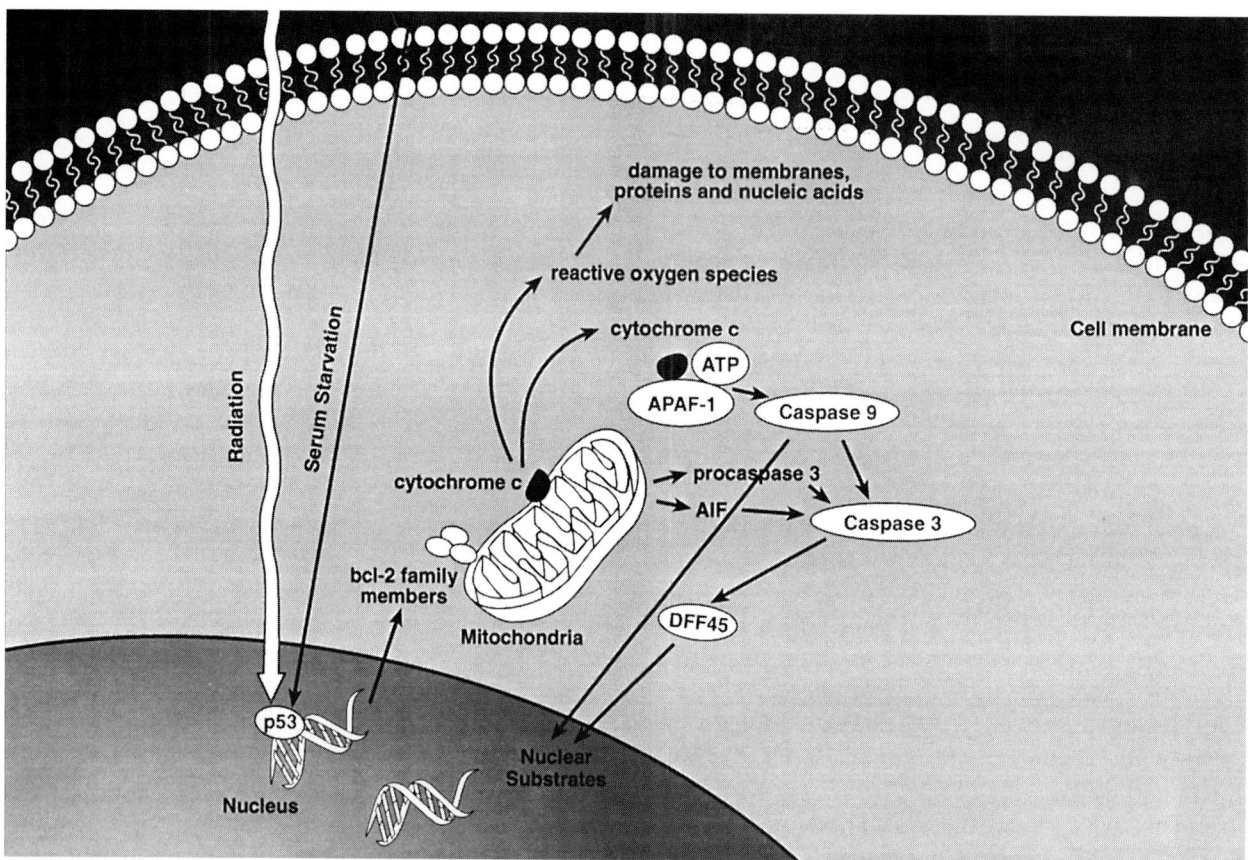

Fig. 4. Mitochondria dependent pathways for apoptosis. During the early stages of programmed cell death, cytochrome c is released from the intermembrane space of mitochondria, and in combination with ATP leads to activation of APAF-1. This leads to activation of downstream caspases that cleave many cellular substrates. The release of other components (such as AIF) from mitochondria has also been implicated in regulation of apoptosis, as has the liberation of reactive oxygen species that can cause damage to membranes, proteins, and nucleic acids.

protein calpastatin is cleaved by caspases during anti-Fas and staurosporine-induced apoptosis of Jurkat cells, providing additional evidence of a role for calpain during apoptosis. Finally, an inhibitor of nuclear scaffold (NS) protease suppressed lamin B1 degradation and chromatin cleavage in rat thymocyte nuclei incubated with Ca^{2+} and Mg^{2+}; however, additional proteases have also been associated with degradation of lamins during programmed cell death.

Although these data certainly implicate calcium in induction of apoptosis in some instances, other studies complicate the interpretation of some of these findings. Thapsigargin is an inhibitor of the Ca^{2+} ATPase of the endoplasmic reticulum. Inhibition of this pump causes release of Ca^{2+} from this organelle followed by capacitative calcium entry through the plasma membrane. Various experiments using this compound suggest that depletion of the calcium stores in the endoplasmic reticulum triggers apoptosis, rather than the increase in cytosolic calcium due to capacitative influx. For example, pancreatic β-cells induced to undergo apoptosis by thapsigargin still died when the Ca^{2+} influx through the plasma membrane was blocked or when cytosolic Ca^{2+} was buffered. These investigators suggested that apoptosis in this model was due to the liberation of intermediates of arachidonic acid metabolism. Release of unidentified apoptosis regulatory molecules or enzymes such as DNase I or NS protease from this organelle could also be involved in regulation of programmed cell death after treatment with thapsigargin.

3.4. Bcl-2 Family Members

The discovery of the *bcl*-2 gene was of pivotal importance for apoptosis research. This gene was identified by virtue of its localization near the t(14;18) chromosomal breakpoint in many human lymphomas, which causes increased transcription of the gene. Subsequently, the Bcl-2 protein was shown to suppress apoptosis when overexpressed in a lymphocyte cell line. Since that time, 26 *bcl*-2-related genes have been cloned and their function during apoptosis partially analyzed. Some of these proteins are antiapoptotic and some are proapoptotic. All of these proteins contain at least one of four conserved Bcl-2 homology (BH) motifs. Every antiapoptotic member except E1B-19K of adenovirus contains both BH1 and BH2. These include *bcl*-2, *bcl*-x_L, *bcl*-w, *mcl*-1, A1, NR-13, BHRF1, LMW5-HL, ORF16, KS-*bcl*-2, *boo*, and *ced*-9. In addition, there are two subclasses of proapoptotic Bcl-2 family members. The Bax subfamily consists of *bax*, *bak*, *bok*, and Diva; the BH3 subfamily includes *bik*, *blk*, *hrk*, BNIP3, Bim, *bad*, *bid*, and *egl*-1, *bcl*-x_S, and *bod*. The expression of these genes is described in Table 2.

The activity of many of these Bcl-2 family members is regulated by heterodimerization, and in some cases by phosphorylation. These proteins also vary in their tissue and/or differentiation specific expression, and in their affinity for different partners. Antiapoptotic family members and those of the proapoptotic Bax subfamily can act in the absence of a partner, but members of the BH3 subfamily apparently must heterodimerize with antiapoptotic family members to exert their effect. Therefore, both the balance of pro- vs antiapoptotic proteins in a cell and their ability to bind to one another may regulate the ability of the cell to undergo programmed cell death.

Potential ways in which Bcl-2 and Bcl-x_L may protect cells from apoptosis include regulating cellular redox potential and/or membrane permeability, preventing cytochrome *c* efflux from mitochondria, regulating ion homeostasis, anchoring proapoptotic proteins to mitochondrial membranes, or inhibiting association of Apaf-1 with procaspase-9. Structural similarity of the other antiapoptotic family members with Bcl-2 and Bcl-x_L suggest they may act in a similar manner. Many of the proapoptotic Bcl-2 family members act by binding to and thereby abrogating the effects of the antiapoptotic proteins. However, the proapoptotic protein Diva functions by binding to Apaf-1 and preventing association of this protein with Bcl-x_L. Members of the Bax subfamily also seem to be able to induce apoptosis by damaging organelles by insertion into membranes. These proteins are able to directly induce release of cytochrome *c* from mitochondria, with subsequent activation of downstream caspases. The proapoptotic protein Bid is normally cytosolic, and is cleaved by caspase-8 after engagement of the Fas receptor or TNFR1 in certain cells. Truncated Bid then translocates to the mitochondria, induces clustering of these organelles, and causes release of cytochrome *c*. It is inevitable that as each of these proteins is studied in depth, unique modes of interaction with the apoptotic machinery will be identified.

3.5. DNA Degradation

Internucleosomal DNA degradation is a hallmark of apoptosis. Why DNA is cleaved in such a unique manner rather than being cleaved randomly is unknown. Previous speculation that genes being actively transcribed were somehow being specifically targeted for cleavage has not been substantiated. Chromatin is also cleaved into fragments of 30–500 kb during apoptosis. These large fragments may correspond to cleavage of the loops and rosettes of DNA that comprise higher order chromatin structure. Although it is tempting to speculate that these larger fragments act as substrates for subsequent internucleosomal cleavage, this hypothesis has not been proven.

Internucleosomal DNA degradation can be induced in isolated nuclei of thymocytes, human leukemia cells, and mouse liver cells simply by incubating them in the presence of Ca^{2+} and Mg^{2+}. This suggests that these nuclei contain a constitutively expressed nuclease responsible for this specific fragmentation. However, the enzyme(s) responsible for either internucleosomal DNA degradation or the cleavage of DNA into 30- to 500-kb fragments observed during apoptosis have not been definitively identified. Although one nuclease could cause both types of cleavage, the activity responsible for cleavage into large fragments is active in the presence of Mg^{2+} alone, but internucleosomal cleavage has been shown to require the presence of both Ca^{2+} and Mg^{2+}. Additional regulatory proteins could conceivably lead to alternate substrate specificity of a single enzyme. Many investigators have searched for enzymes that can recapitulate the specific internucleosomal DNA cleavage observed during apoptosis and that possess characteristics previously described for apoptotic nucleases including formation of termini with 3′-hydroxyl groups, a dependence on Ca^{2+} and Mg^{2+} for optimal activity, and inhibition by Zn^{2+} and aurintricarboxylic acid. A number of enzymes have been identified that fulfill the ion and inhibitor sensitivity requirements but that either do not form correct termini, or do not cause internucleosomal cleavage, including NUC18, a 22–26-kDa nuclease, and a 36-kDa nuclease. DNases I and II are able to induce internucleosomal DNA cleavage, but neither is inhibited by aurintricarboxylic acid, and DNase II is not

Table 2
Expression of BCL-2 Family Members

Antiapoptotic Proteins	Expression
A1	Primarily hematopoietic cells
Bcl-2[a]	Widespread
Bcl-x_L[a]	Widespread
Bcl-w	Widespread, but null mice only exhibit failed spermatogenesis
Boo	Ovary and epididymis
Ced-9	*C. elegans* cells that survive during development
Mcl-1[a]	Widespread
NR-13	Chicken-expression unknown
BHRF1	Epstein–Barr virus
E1B-19K	Adenovirus
KS-bcl-2	Human herpesvirus 8 (Kaposi sarcoma-assoc. virus)
LMW5-HL	African swine fever virus
ORF16	Equine herpesvirus type 1

Proapoptotic Proteins	Expression
Bad	Widespread-highest in testes, breast, colon, spleen
Bak	Widespread-highest in certain keratinocytes
Bax	Widespread
Bcl-x_S[a]	Not done
Bid	Not done
Bik	Not done
Bim	Not done
Blk	Not done
BNIP3	Not done
Bod	Ovary
Egl-1	*C. elegans*
Hrk	Not done

[a]Expression of Bcl-2, Bcl-x_L, and Mcl-1 is widespread; however, Bcl-x_L is essentially absent in peripheral blood cells and certain neurons, and Bcl-2 is absent in cortical thymocytes and muscle cells. Bcl-2 expression tends to overlap expression of Bcl-x_L and Mcl-1, but in an inverse gradient (e.g., Bcl-2 tends to be present in less differentiated cells in tissues such as the epidermis while Bcl-x_L and Mcl-1 tend to be present in more differentiated cells of this same tissue).

Bcl-x_L and Bcl-x_S are formed by alternate splicing of the same message. No studies have been performed in which expression of these two proteins in various tissues could be distinguished.

inhibited by Zn^{2+}, and forms termini with 5′-hydroxyl groups. In addition, DNase I is primarily localized to secretory vesicles, and DNase II is most active at an acidic pH rarely experienced by cells, further arguing against a role for either enzyme in apoptosis. Perhaps some of these nucleases that cannot cleave DNA between nucleosomes are responsible for cleavage of DNA into the larger fragments.

What appeared to be a breakthrough in the identification of an apoptotic nuclease responsible for internucleosomal DNA degradation was the purification of a heterodimeric protein called DNA fragmentation factor (DFF) from human cells. This nuclear factor is composed of a 40-kDa subunit and a 45-kDa inhibitory subunit that is a substrate for caspase-3. Caspase cleavage of DFF45 in the nucleus causes release of DFF45 from DFF40. DFF40 has been shown to have internucleosomal DNase activity that is stimulated by the presence of histone H1 and high mobility group (HMG) proteins, which are normally associated with chromatin, and which may act to target the nuclease to the internucleosomal regions. DFF40 also triggers chromatin condensation when incubated with isolated nuclei. Thus, it appeared that the apoptotic nuclease responsible for internucleosomal DNA degradation had been identified at last. On further examination, however, it was found that a DFF45 mutant that could inhibit alterations in isolated nuclei induced by addition of caspase-3-activated cytosolic extracts could not do so when the cytosolic extract had been activated by staurosporine, suggesting that a nuclease other than DFF40 was responsible for DNA degradation in these nuclei.

4. DISASSEMBLY AND PHAGOCYTOSIS OF THE CELL

Blebbing of apoptotic bodies occurs during the terminal stages of apoptosis. Cellular components must be degraded into smaller pieces to allow for packaging and formation of these apoptotic vesicles. Therefore, dissolution of the nuclear and cytosolic matrix must occur before or at the time of membrane blebbing. Very little is known about disassembly of the cell or formation of apoptotic vesicles. Several proteins of the nuclear and cytoplasmic matrix are caspase substrates including actin, fodrin, and several nuclear lamins. However, because actin reorganization regulates membrane blebbing, it is unlikely that the entire cytoskeleton is degraded prior to vesicle formation.

An inflammatory response will be elicited if apoptotic vesicles are allowed to undergo further deterioration in the extracellular space. Therefore, they must be specifically recognized and engulfed by phagocytic cells. Phosphatidylserine reorients to the outer face of the plasma membrane during apoptosis, and may be recognized by macrophages via a phosphatidylserine receptor. However, recognition of vesicles by other populations of macrophages appears to be mediated by the vitronectin receptor. Dendritic cells preferentially utilize the vitronectin receptor for recognition of apoptotic bodies, but also may use the phosphatidylserine receptor under certain circumstances. Finally, hepatocytes may use their asialoglycoprotein and galactose-specific receptors for the recognition of the remnants of apoptotic cells. Because the mechanisms of cellular dissolution and the subsequent phagocytosis of apoptotic cells are poorly understood, they represent important targets for future research.

5. CONCLUSIONS

The importance of apoptosis in regulation of normal cellular processes should be apparent. However, the significance of aberrant regulation of programmed cell death in various diseases should not be overlooked. The suppression of apoptosis by endogenous hormones, environmental compounds, and the mutation of regulatory genes may allow for the expansion of cells that would normally be targeted for death. This hyperplasia may then increase the probability of tumor development. The pathology of autoimmune disorders and neurological diseases such as Alzheimer's also appears to be partially due to an abnormal regulation of apoptosis. Programmed cell death is also involved in numerous other disorders including cellular damage after viral infection. Apoptosis is a tremendously important mechanism involved in regulation of normal and abnormal cellular processes. As such, it promises to remain a popular research topic for years to come.

SELECTED READINGS

Anderson P. Kinase cascades regulating entry into apoptosis. Microbiol Mol Biol Rev 1997; 61:33.

Cidlowski JA, King KL. The biochemistry and molecular biology of glucocorticoid-induced apoptosis in the immune system. Rec Prog Horm Res 1996; 51:457.

Cory S. Regulation of lymphocyte survival by the *bcl-2* gene family. Annu Rev Immunol 1995; 13:513.

Cohen GM. Caspases: the executioners of apoptosis. Biochem J 1997; 326:1.

Hale AJ, Smith CA. Apoptosis: molecular regulation of cell death. Eur J Biochem 1996; 236:1.

Jehn BM, Osborne BA. Gene regulation associated with apoptosis. Crit Rev Eukaryot Gene Express 1997; 71:179.

Kiess W, Gallaher B. Hormonal control of programmed cell death/apoptosis. Eur J Endocrinol 1998; 138:482.

McConkey DJ, Orrenius S. The role of calcium in the regulation of apoptosis. Biochem Biophys Res Commun 1997; 239:357.

Mignotte B, Vayssiere J-L. Mitochondria and apoptosis. Eur J Biochem 1998; 252:1.

Nagata S. Apoptosis by death factor. Cell 1997; 88:355.

Nicholson DW, Thornberry NA. Caspases: killer proteases. Trends Biochem Sci 1997; 22:299.

Papa S, Skulachev VP. Reactive oxygen species, mitochondria, apoptosis, and aging. Mol Cell Biochem 1997; 174:305.

Raffray M, Cohen GM. Apoptosis and necrosis in toxicology: a continuum or distinct modes of cell death? Pharmacol Ther 1997; 75:153.

Rowan S, Fisher DE. Mechanisms of apoptotic cell death. Leukemia 1997; 11:457.

Schulze-Osthoff K, Ferrari D. Apoptosis signaling by death receptors. Eur J Biochem 1998; 254:439.

26 T Cells and Immunosuppression

Andrew W. Taylor

Contents
Introduction
Mechanisms of Immunosuppression
Conclusions
Selected Readings

1. INTRODUCTION

The most effective host defensive mechanism to eliminate infectious pathogens is adaptive immunity. Adaptive immunity provides for a specific, coordinated, and long-lasting immunity. This defense involves the cognizant interaction between antigen presenting cells and T cells. Through this interaction the activated T cells mediate an immune response that is specific and tailored for defense against a specific infectious pathogen or noxious substance. Also, through adaptive immunity there is memory that affords the host a rapid and vigorous response against reencounter. This immune response is focused on infecting pathogens and noxious substances; however, the resulting inflammatory response is not. Immunogenic inflammation mediates changes in the vasculature, necrosis, and scarring of surrounding tissues. Occasionally immunity turns against the host's own tissues in autoimmune disease. It also complicates and threatens survival of graft transplants. Fortunately, there are physiological mechanisms that terminate T-cell activation to maintain immune homeostasis and prevent autoimmunity.

Several tissues and organs that are structurally and functionally devastated by immunogenic inflammation have adapted several mechanisms to prevent the activation of effector T cells that mediate inflammation. These tissues and organs are collectively called immune privileged. The mechanisms by which they regulate immunity within their immune privileged microenvironment have suggested the means by which immunity can be made to terminate inflammatory T-cell responses and also activate noninflammatory effector T-cell responses. This chapter discusses the mechanisms that terminate T-cell activity. The main focus is on the mechanisms that terminate and suppress the activation of mature type 1 T helper (Th1) cells. It is activated Th1 cells that are responsible for mediating and initiating the immunogenic inflammatory response, delayed type hypersensitivity (DTH), associated with pathogen resistance, autoimmune disease, and transplant graft rejection.

1.1. Mechanisms of T-Cell Activation

The activation of T cells involves two signals, one through antigen recognition and the second through accessory activation receptors (Table 1). The antigen-associated signal is the binding of the T cell receptor (TCR) to specific antigen peptides presented in the cleft of major histocompatibility complex (MHC) antigens on the surface of antigen presenting cells (APC). There are two classes of MHC molecules, both of which present processed antigen. On APC class I MHC presents processed intracellular antigens, while class II MHC presents processed exogenous antigen.

The TCR proteins are encoded by genes that have

From: *Principles of Molecular Regulation* (P. M. Conn and A. R. Means, eds.), © Humana Press Inc., Totowa, NJ.

Table 1
Interactions Between APC and T Cells

APC	T cell
Antigen Binding (Signal 1)	
MHC class I + processed antigen	TCR + CD8
MHC class II + processed antigen	TCR + CD4
Accessory Activation Signals (Signal 2)	
B7	CD28
ICAM-1	LFA-1

Table 2
Types of Effector Th Cells

Th Cell Type	Characteristic Lymphokines Produced	Mediated Immune Functions
Th1	INF-γ, TNF, IL-2	DTH Tumor rejection Transplant graft rejection Mediate B cell switching to complement-fixing IgG
Th2	IL-4, IL-10, IL-5	Allergic responses B-cell development
Th3	TGF-β, IL-4, IL-10	Suppression of Th1 and Th2 activity

undergone genomic recombination in the region coding for the TCR antigen binding site. This chromosomal rearrangement accounts for the vast array of antigens that TCR can bind. However, TCR binding of antigen is restricted to the antigen presented in the context of MHC encoded in the same host genome as the T cells. T-cell activation to MHC of other individuals in the same species is an allogeneic immune response and is a cause of chronic transplant graft rejection. The mechanisms that restrict T cells to endogenous MHC molecules are beyond the scope of this chapter; however, they are part of the mechanisms that mediate central tolerance preventing development of most autoantigen reactive T cells. There are also minor histocompatibility antigens. The minor histocompatibility antigens are a poorly defined group of antigen presenting proteins. Several of them have class 1 like properties, suggesting they are associated with activation of CD8+ T cells.

Along with the antigen-binding TCR there are CD4 or CD8 molecules on the surface of the T cells. The CD4 and CD8 proteins bind to specific classes of MHC molecules on the APC. CD4 binds MHC class II molecules and CD8 binds MHC class I molecules. Mature T cells express either CD4 or CD8. Engagement of the CD4 or CD8 molecule to their respective MHC mediates the intracellular phosphorylation of the antigen–MHC engaged TCR proteins. In the aggregate TCR–CD4 or TCR–CD8 complexes engaged to antigen–MHC impart signal one initiating T-cell activation.

The second signal that T cells need to receive for activation is mediated through the interaction of CD28 or LFA-1 on the T cells binding B7 and ICAM-1 respectively on the APC. These second signals amplify the intracellular signals initiated by the antigen–MHC-bound TCR. The need for these accessory signals is less in primed T cells than in naïve T cells. At this point the T cells now efficiently produce interleukin-2 (IL-2) and up-regulate IL-2 receptors. IL-2 as an autocrine binds to its receptors on activated T cells and promotes T-cell progression through proliferation. The result is clonal expansion of antigen-specific T cells.

1.2. The Different Types of T-Cell Responses

The differentiation of effector Th cell is separated into three general types according to the lymphokines produced and the effector immune response elicited (Table 2). The DTH-mediating effector T cells are the type 1 Th (Th1) cells. Th1 cells are characterized by their secretion of interferon-γ (INF-γ) and tumor necrosis factor (TNF). The type 2 Th (Th2) cells mediate B-cell growth and allergic immune responses. The characteristic lymphokines produced by Th2 cells are IL-4 and IL-10. A third type of Th (Th3) cell has only recently been characterized. They are the result of oral tolerance and possibly the effect of aqueous humor (fluid from the anterior chamber of the eye) factors on activated T cells. These Th3 cells produce transforming growth factor-β (TGF-β) and IL-4 or IL-10. They have the ability to suppress autoimmune diseases mediated by Th1 cells with a potential to also suppress Th2-mediated responses. The path of differentiation by the activated T cells is associated with the manner that antigen is presented, with the type of APC, and the elements within the microenvironment where the T cells are activated. It is also at this level that the innate immune response can influence the induction of adaptive immunity.

The intracellular signals that emanate from the TCR regulate T-cell proliferation, lymphokine production, and expression of surface receptors. The affinity between the TCR and processed antigen-bound MHC regulates the intensity and fidelity of tyrosine phosphorylation of CD3 proteins, TCR-associated molecules. Changes in the affinity due to amino acid substitution of the processed antigen or to presentation of antigen by unfavorable MHC binding mediate differential TCR signals within activated T cells. High-affinity binding of TCR to antigen presented MHC leads to induction of Th1 responses, whereas low-affinity binding leads to Th2 responses and possibly Th3 development.

Presentation of antigen by macrophages and dendritic cells usually leads to inducing Th1 cells. This may be due to the role of macrophages and dendritic cells in innate immunity. Potential pathogens express pathogen-associated molecular patterns (PAMP), which are molecules that are intrinsic to the pathogen's structure or are necessary for its survival. The PAMP induce IFN-γ, IL-12, and TNF production, antigen processing, and class II MHC expression on macrophages and dendritic cells. Such activated macrophages and dendritic cells mediate activation of Th1 cells because they can present antigen in a microenvironment rich in Th1 cell growth factors (IFN-γ, and IL-12). On the other hand, B cells when activated by binding antigen to their surface immunoglobulin, process the internalized antigen and present it on class II MHC to T cells in a manner the promotes Th2 development. This promotion of Th2 responses has been speculated to be caused by B cell over expression of accessory activating protein B7.2 and production of IL-1. Another possibility has been suggested that Th2 development is a default T-cell differentiation pathway. Activation of Th cells in the absence of IL-12 and IFN-γ leads to development of Th2 cells; however, this is some what questionable because there exist strains of mice that appear to default to Th1.

The innate immune response to PAMP is not limited to inducing a Th1-mediated adaptive immune response. Activation of NKT cells mediates induction of Th2 cells. These NKT cells respond to antigen presented in the cleft of CD1, a MHC class-I like molecule expressed on cells such as intestinal epithelial cells, and secrete IL-4 influencing development of Th2 cells. Other cells that can influence the Th cell activation microenvironment are natural killer (NK) cells that secrete IFN-γ when activated by cells not expressing MHC class I. Under these conditions NK cells promote development of Th1 cells. Microenvironments rich in TGF-β, such as the anterior chamber of the eye, have the potential to promote development of Th3 cells. In addition, lymphokines produced by activated T cells can also influence the differentiation of other activated T cells. The lymphokines IFN-γ and IL-4 both suppress the production of each other and thus promote development of Th1 cells or Th2 cells, respectively. Activated Th3 cells through TGF-β, with IL-4 or IL-10 production, can suppress the activation and possibly the development of both Th1 and Th2 cells.

Since Th cell activation and differentiation is a coordinated sequence of antigen recognition, accessory protein binding, and cytokine activity, it can be regulated at the level of antigen presentation and by the microenvironment in which the Th cells are activated. It is considered that within immune privileged sites such as the eye and brain the parenchymal cells that make up the microenvironment contribute factors that locally regulate immune responses. More conventional tissues sites such as the skin use similar regulatory mechanisms as immune privileged sites in a limited manner to maintain immune homeostasis.

1.3. The Need for Terminating T-cell Activity

It is reasonable to think that survival of the host relies on the activation of an effective immune response; however, some immune responses can also jeopardize host survival. The need to terminate a T-cell response is necessary to maintain homeostasis. Here T-cell responses are terminated to prevent activation when there is no noxious pathogen, or when a noxious pathogen has been cleared. Termination of T-cell responses is necessary when there is a hypersensitivity response that permanently damages tissues and organs such as in the eye. There is also the desire to therapeutically terminate T-cell activity to treat autoimmune disease and transplant graft rejection. The mechanisms of terminating Th-cell activity are the means by which the adaptive immune response establishes homeostasis, self-tolerance, and immune privilege.

2. MECHANISMS OF IMMUNOSUPPRESSION

The mechanisms for terminating Th-cell activity, generally called immunosuppression, can be grouped into four categories (Table 3). Suppression of Th-cell activity can be through anergy, active cell death, passive cell death, and cytokine mediated. Each of these suppressive mechanisms is linked to some aspect of Th-cell activation and differentiation.

Table 3
Mechanisms of Immunosuppression

Type of Suppression	Inducer
Anergy	Signal 1 without signal 2
	Signal 2 without signal 1
Active death	Fas ligand
Passive death	Lack of presented antigen
Cytokine mediated	TGF-β, IL-10, IL-4

2.1. Suppression of T-Cell Activity by Anergy

Activation of Th cells requires that the TCR and the accessory molecule CD28 are engaged to their counter molecules antigen/MHC and B7 respectively, on the surface of APC. A Th cell that receives only one of the signals becomes functionally unresponsive in failing to produce IL-2. The functional unresponsiveness can be mediated by either the Th cell receiving an adequate TCR signal, but lacking the accessory CD28 signal. The reverse is also true; a CD28 signal with a weak TCR signal due to altered presented antigen will also mediate functional unresponsiveness in the Th cell. Such states of antigen presentation may occur in tissues in which innate immunity was sufficient for elimination of a potential pathogen. Here there is no need for a Th-cell response. It is also possible that once a pathogen has been successfully cleared the reduction of PAMPs causes the APC to reduce expression of accessory molecules while still presenting antigen, thus terminating the no longer needed effector Th cell activity.

There is also the potential for unresponsiveness to be actively induced in Th cells. Activated T cells reduce their expression of CD28 in exchange for expressing CTLA-4 which also binds to B7. Unlike CD28, engagement of CTLA-4 mediates suppressive signals within the activated Th cell. The result of the CTLA-4–B7 interaction is induction of unresponsiveness on the part of the Th cell. Because CTLA-4 is expressed only on activated Th cells it could be considered part of a Th cell self-regulating mechanism.

In conventional immune sites, anergy can result from the presentation of foreign antigen in the absence of inflammation. The lack of PAMPs with the foreign antigen results in APC not expressing accessory molecules along with reduced levels of MHC. This may be very important in preventing Th-cell responses to presented autoantigens. Because some anergic Th cells will proceed into apoptosis, induction of anergy to autoantigens has been considered a mechanism of peripheral tolerance. In immune privileged tissues the expression of MHC is suppressed. This lack of MHC expression most likely results in Th cells receiving inadequate activation signals and therefore having the potential to induce anergy.

2.2. Suppression of T-Cell Activity by Active Cell Death

Active cell death is mediated through Fas antigen expressed on the surface of activated Th cells. The Fas antigen binds to its ligand (FasL) that is expressed on the surface of other cells, including other activated Th cells. Apoptotic death is characterized by membrane rearrangement, enzymatic degradation of DNA, and the formation of cellular blebs that are absorbed by surrounding cells. Unlike apoptosis in other cells the constitutive intracellular expression of Bcl-2 and Bcl-x does not protect the Th cells from Fas-mediated apoptosis. This Fas-mediated apoptosis is promoted by repeated stimulation of the Th cells. In addition, IL-2 instead of delivering a stimulatory signal, potentiates the Fas-mediated apoptosis.

The importance of Fas-mediated Th-cell death has been shown in mice that are defective in Fas or FasL. These mice suffer from autoimmune diseases, but display normal adaptive immune defenses to infectious pathogens and to immunizations with foreign antigens. Peripheral tolerance to autoantigens must be mediated in part through Fas-mediated apoptosis. The potential for repeated activation of autoreactive Th cells in the periphery could promote Fas-mediated apoptosis. The Fas-mediated apoptosis is also important in mediating immune homeostasis. Because activated Th cells up-regulate both Fas and FasL, it is possible that excessive numbers of activated Th cells could regulate the extent of their activity by mediating apoptosis among themselves. This may also be one of the mechanisms that terminates Th cell activity once a noxious pathogen has been cleared. Th cells that escape this regulation may go on to be memory Th cells.

The cells of immune privileged tissues extensively express FasL. It has been speculated that it is the constitutive expression of FasL in these tissues that promotes graft survival and suppression of Th1 cell activity. However, recently it has been found that the aqueous humor from normal eyes suppresses Fas–FasL-mediated apoptosis and that TGF-β is the anti-apoptotic factor. Because most immune privileged tissues are sites of constitutive TGF-β production, the role of Fas-mediated apoptosis in immune privilege is still questionable. However, this does demonstrate the importance of locally produced cytokines, that is, TGF-β production, in regulating T-cell activity, which is discussed in detail later. In addition, FasL expressing tissues are susceptible to neutrophil-mediated cytotoxicity mediated through Fas–FasL-induced neutrophil degranulation, which is also suppressed by the factors of the ocular immune privileged microenvironment.

2.3. Suppression of T-Cell Activity by Passive Cell Death

The passive death of Th cells is also known as death by neglect. This is based on the phenomena that antigen-specific Th cells that do not encounter their antigen drop in frequency and may even be depleted. Both naïve and memory Th cells are sensitive to passive cell death. The need for stimulation to maintain immunological memory has suggested that long-term memory is a result of antigen remaining in the host following immunization or pathogenic infection. Passive death of T cells is Fas-independent apoptosis. The activation of the Th cells up-regulates intracellular expression of Bcl-2 and Bcl-x, which unlike in Fas-mediated apoptosis, counter the apoptotic signals in the stimulated Th cells. Also, IL-2 acts as a rescue factor and prevents the Fas-independent apoptosis. Through the mechanisms of passive cell death, mature Th cells that have not or no longer encounter their antigen are eliminated.

2.4. Cytokine Mediated Immunosuppression

2.4.1. Regulation of Th-Cell Development

The types of Th cells are defined by the predominance of specific cytokines they produce (Table 2). Th1 cells are defined by the production of IL-2, IFN-γ, and TNF, whereas Th2 cells are defined by the production of IL-4 and IL-10. Th3 cells are defined by the production of TGF-β along with IL-4 and IL-10. The cytokines produced also describe the effector immune response mediated by each type of Th cell. Th1 cells mediate immunogenic inflammation, activation of cytolytic CD8+ T cells, and promote B-cell class switching to complement-fixing antibody production. Th2 cells mediate allergic responses, B-cell growth, immunoglobulin class switching to noncomplement fixing antibodies, and suppress macrophage inflammatory activity. There is not a full understanding of effector Th3 cell activity, but it is clear that they suppress the activity of autoreactive Th1 cells, Th2-mediated allergic disorders, and since they produce

Table 4
Cytokine Regulation of Th Cell Development

T-Cell Type	Cytokines That Promote Development	Cytokines That Suppress Development
Th1	IL-12, INF-γ	IL-4, IL-10, TGF-β
Th2	IL-4, IL-1	IL-12, INF-γ, TGF-β
Th3	TGF-β with IL-4 or IL-10[a], α-MSH	IFN-γ, TNF?

[a]IL-4 and IL-10 can, with TGF-β, act as growth factors for Th3 development.

TGF-β they may mediate B-cell immunoglobulin class switching to IgA.

The cytokines that each Th cell type produce not only promotes their own development, but also suppress the development of the other types (Table 4). For example IFN-γ promotes Th1 cell development but inhibits Th2 development. The IL-10 from Th2 cells suppress APC activation of Th1 cells. Cytokines produced by the APC and other cells in the microenvironment of T-cell activation can also regulate the development of specific types of T cells. APC producing IL-12 and NK cells producing IFN-γ promote Th1 cell development. NKT cells produce IL-4 and promote Th2 cell responses, and so do APC suppressed in IL-12 and IFN-γ production. Th cells activated in TGF-β-rich microenvironments such as in the eye can develop into Th3 cells. The presence of specific cytokines in a tissue microenvironment can promote the activation of certain immune responses while suppressing others. Mechanisms that select specific immune response have been found within the normal microenvironment of the eye and the brain.

2.4.2. Oral Tolerance

The most profound example of cytokine-mediated immunosuppression is the suppression mediated by Th3 cells induced in oral tolerance models. Here a host undergoing autoimmune disease mediated by autoantigen-specific Th1 cells are fed autoantigen. The processing of autoantigen through the gut-associated lymphoid tissues (GALT) mediates development of TGF-β, IL-4, and IL-10 producing Th cells. The importance of these Th cells is in mediating immunoglobulin class switching of B cells to IgA production while not mediating hypersensitivity to ingested substances. This is the mechanism by which GALT protects gut tissues from infection and prevents induction of allergies to nutrients. The oral tolerance models take advantage of GALT-mediated induction of Th3 cells to ingested autoantigens. These Th3 cells suppress autoimmune disease in an antigen specific manner. Even though IL-4 and IL-10 are immunosuppressive cytokines, TGF-β has been found to be the common immunosuppressive cytokine produced by the regulatory Th cells induced through oral tolerance models against Th1-mediated experimental autoimmune uveitis and encephalomyelitis, and against Th2-mediated bronchial asthma. The difficulty in mediating Th3 development but not activity in IL-4 or IL-10-deficient mice has suggested that along with TGF-β, IL-4, and IL-10 are potential Th3 growth factors. Oral tolerance may not be the only means to induce Th3 cells. The immune privileged ocular microenvironment may be another physiological site that promotes induction of Th3 cells. Recently it has been found that aqueous humor through its constitutive immunosuppressive factors TGF-$β_2$ and α-melanocyte-stimulating hormone (α-MSH) induce activation of TGF-β producing T cells with regulatory activity.

2.4.3. IMMUNE PRIVILEGED TISSUES

Immune privilege is the description that was given by Medawar and his colleges in the 1940 to identify tissue sites such as the eye and brain where allografts are afforded prolonged survival (Table 5). The mechanisms that mediate prolonged allograft survival were thought to be passive mechanisms of immunological ignorance. This was based on anatomical observations that immune privileged sites have no direct lymphatic drainage and are contained within a blood–tissue barrier. Even though these are descriptions of immune privileged tissues, the immune system does respond to antigen placed and expressed within the immune privileged tissue microenvironment. However, the immune response is highly regulated, affecting systemic as well as regional immune responses to antigen within the immune privileged tissue microenvironment. Immune privileged tissues actively manipulate the immune system through the collective mechanisms of immunosuppression.

The most studied immune privileged tissue is the eye; however, some generalities can be made about other immune privileged tissues including the brain. Central to mediating immune privilege in the eye is the constitutive production of immunomodulating factors. These factors include cytokines, growth factors, and neuropeptides. The factors are found in the aqueous humor, the fluid that fills anterior chamber of the eye. Similar immunosuppressive factors are found in the cerebrospinal fluid.

Table 5
Features of Immune Privileged Tissue Sites

1. Prolonged survival of allografts
2. No direct lymphatic drainage
3. Blood–tissue barrier
4. Low expression of MHC molecules
5. High expression of Fas ligand
6. Constitutive production of immunosuppressive and antiinflammatory cytokines and neuropeptides
7. Induction of APC-mediated immune deviation
8. Induction of regulatory T cells

Macrophages treated with aqueous humor or with one of the aqueous humor immunosuppressive factors, TGF-$β_2$, change their manner of antigen processing, produce TGF-β, and fail to produce IL-12. The result of these changes is that they stimulate $CD8^+$ T cells that suppress subsequent antigen exposures from inducing effector Th1-cell activities. In addition, such macrophages mediate Th2 responses that induce non-complement-fixing antibody production by B cells. If it is an antigen that normally mediates a Th1 response, then the factors of the immune privileged eye have mediated through APC a systemic immune deviation away from the expected Th1 response. This deviation is seen when immunization is done via the ocular anterior chamber. The subsequent systemic response to immunizing through the anterior chamber is called anterior chamber associated immune deviation (ACAID). The mechanisms of ACAID prevent induction of immunogenic inflammation to ocular antigens by preventing the development of T cells that may mediate inflammation to antigen presented in the eye. In addition, because the ocular factors change the manner of antigen presentation by the APC, a more modern view of immunological ignorance can be proposed in that Th1 cells remain ignorant of antigen presented by aqueous humor treated APC.

Besides TGF-$β_2$, migration-inhibition factor, the neuropeptides α-MSH, vasoactive intestinal peptide, calcitonin gene related peptide, and glucocorticoids are also constitutively present in aqueous humor. Each of these factors in various degrees suppresses the production of IFN-γ by Th1 cells. The factors TGF-$β_2$ and α-MSH can also directly suppress activation of Th1 cells. In addition, α-MSH promotes induction of TGF-β producing T cells that have Th3-like activity. The TGF-β produced by the Th cells can further contribute to the immunosuppressive and antiinflammatory microenvironment of the immune privileged eye through the production of TGF-β, IL-4, and IL-10. Because such cells can efficiently suppress autoimmune diseases, induction of Th3-like cells may be important in maintaining tolerance to autoantigens.

Because several of the aqueous humor factors have immunomodulating and anti-inflammation activity, especially α-MSH, the ocular microenvironment has adapted several mechanisms to protect itself from the induction of inflammation. To prevent the induction of immune-mediated inflammation within the eye cytokine-mediated immunosuppression is employed. Along with cytokine-mediated immunosuppression other T cell terminating mechanisms such as anergy and Fas-mediated death are also present in the normal ocular microenvironment. Because many of these Th-cell terminating mechanisms have been observed in the brain it is possible that the central nervous system may also actively manipulate Th-cell activity just as the ocular microenvironment.

3. CONCLUSIONS

3.1. Implications for Understanding the Mechanisms Inducing Autoimmunity

The mechanisms of terminating T-cell activation are important in preventing unwanted and unnecessary T-cell activity. Normally the majority of autoreactive T cells are selected out and deleted through mechanisms of central tolerance in the thymus. Induction of autoimmunity must suggest that a protective mechanism has failed. Immune privileged tissues are not only regional immune sites that in the extreme terminate T-cell activity, but are also targets of autoimmune responses. Animal models of experimental autoimmune uveitis and encephalomyelitis have shown that immune privileged sites can also give insight into the regional tissue conditions that fail to terminate T-cell activity leading to autoimmunity.

3.1.1. Loss of Regulating Factor Activity

The leakage of blood-borne proteins through breaks in the blood–tissue barrier due to infection or trauma is considered one of the leading causes abolishing immune privilege. The introduction of serum factors that are not normally found in the immune privileged microenvironment neutralize TGF-β_2 and accelerate degradation of the immunosuppressive neuropeptides. Blood-derived growth factors have the potential to induce the cells of the microenvironment to make factors that enhance inflammatory and immune responses. Changes in the composition of immunomodulating factors from normal can lead to up-regulating antigen processing and presentation of autoantigens along with a failure to suppress Th1-cell activation and function. Therefore, changes in cytokines and growth factors within a regional tissue site have the potential to create a microenvironment with high levels of antigen processing and presentation, which would include enhanced processing and presentation of autoantigens, along with enhanced levels of proinflammatory cytokines and neutralized immunosuppressive factors, all of which promote destructive autoimmune responses.

3.1.2. Loss of Mediating Selective Lymphokine Production by T Cells

If the normal immune privileged microenvironment mediates the production of specific T-cell lymphokines, then changes in the microenvironment also mean losing the ability to regulate the production of specific lymphokines by T cells activated in the ocular microenvironment. The TGF-β producing T cells, induced by ocular immunomodulating factors, can contribute to immune privilege in two ways. First, their production of TGF-β can contribute to and reinforce the normal immunosuppressive and anti-inflammatory microenvironment of the immune privileged tissue. Second, because the TGF-β producing cells are antigen specific in their activation but not in their suppressive activity, their activation anywhere to antigen would result in suppressing surrounding T-cell activity, such as the regional activation of autoreactive Th1 and Th2 cells. Therefore, changes in the normal tissue microenvironment can deviate the normal regional immune response to an unregulated or inappropriate T-cell response. If this happens to a foreign antigen, then a pathogen may escape immune defenses; and if it is to an autoantigen, then autoimmune disease could result.

3.1.3. Innate Immunity

The ability to induce experimental autoimmune uveitis and encephalomyelitis relies on immunizing with autoantigen in adjuvant. The mycobacteria in the adjuvant delivers PAMP to the immune system. As discussed previously, PAMP activate NF-κB-regulated inflammatory cytokines. In addition, some bacterial antigens are cross-reactive with autoantigens. This adjuvant effect is so potent that it overrides the normal mechanisms that suppress induction of inflammatory autoimmunity. It also antagonizes the influence of the immune privileged microenvironment to suppress innate and adaptive inflammatory immune responses.

Another animal model of experimental uveitis is

induced by systemic injections of nonlethal amounts of endotoxin (a potent PAMP). Here the blood–tissue barrier of the anterior chamber breaks down in response to endotoxin in the blood. Macrophages within the eye up-regulate production of nitric oxide, oxygen reactive intermediates, and inflammatory cytokines. The eyes with endotoxin-induced uveitis no longer mediate the expected immunosuppression associated with immune privilege. Even though endotoxin-induced uveitis is a limited disease, it has been suggested that this period of lost immune privilege may permanently alter the ocular microenvironment, making it susceptible to a subsequent autoimmune responses. Therefore, the induction of autoimmune disease may involve pathogen-associated factors that activate extreme innate responses that lead to activating unregulated and autoreactive immunity.

3.2. Implications for Therapeutic Uses of the Mechanisms of Immunosuppression

Our increasing understanding of the mechanisms that terminate T-cell activity has led to potential therapeutic approaches to prevent and cure autoimmune disease and to treat other immune-mediated events such as transplant graft rejection (Table 6). The mechanisms of oral tolerance brings forth the potential to cure and prevent autoimmune disease through induction of Th3 cells (antigen-specific, TGF-β-producing, regulatory T cell). Oral tolerance can be considered as an immunization to autoantigen under conditions that favor development of Th3 cells over Th1 and Th2 development. To prevent transplant graft rejection, induction of active T-cell death through transfection of Fas ligand genes into transplanted tissues has been shown to promote allograft survival. Unfortunately, some FasL-expressing grafts undergo acute graft rejection because of Fas-mediated neutrophil degranulation and cytotoxicity. Other therapeutic approaches will most likely be developed to utilize the molecular biology of terminating T-cell activities. These may even involve direct treatment of inflamed tissues or rejecting tissue transplants.

The objective of this chapter was to present the various host specific mechanism that terminate inflammatory Th-cell activities within unique regional immune privileged tissue sites. As there is a need to terminate T-cell activity when a pathogen is cleared and to prevent T-cell activation when there is no pathogen, there are specific physiological mechanisms that suppress T-cell activation and activity. The suppression is mediated through anergy, active cell death (Fas-mediated), passive cell death (death by neglect), and cytokine-mediated immunosuppression including induction of TGF-β-producing regulatory T cells. By defining these mechanisms of immunosuppression, it is possible to develop successfully therapies to treat autoimmune diseases, promote survival of transplanted allografts, and even tailor an immunization to promote the most effective host defense against a potential pathogen.

Table 6
Therapies Using the Mechanisms of Immunosuppression

Disease	Therapy	Mechanism
Autoimmune	Oral tolerance	Induction of Th3 cells
Allograft rejection	Viral gene transfection	Fas-mediated T cell death

SELECTED READINGS

Alberola-Ila J, Takaki S, Kerner JD, Perlmutter RM. Differential signaling by lymphocyte antigen receptors. Annu Rev Immunol 1997; 15:125.

Fearon DT, Locksely RM. The instructive role of innate immunity in the acquired immune response. Science 1996; 272:50.

Miller JFAP, Morahan G. Peripheral T cell tolerance. Annu Rev Immunol 1992; 10:51.

Schwartz RH. Immunological tolerance. In Paul W (ed). Fundamental Immunology, 4th ed. Philadelphia: Lippincott-Raven, 1999, p. 701.

Streilein JW. Immunoregulatory mechanisms of the eye. Prog Retinal Eye Res 1999; 18:357.

Taylor AW. Ocular immunosuppressive microenvironment. Chem Immunol 1999; 73:72.

VanParijs L, Abbas AK. Homeostasis and self-tolerance in the immune system: turning lymphocytes off. Science 1998; 280:243.

Weiner HL. Oral tolerance: immune mechanisms and treatment of autoimmune diseases. Immunol Today 1997; 18:335.

Weiss A. T-lymphocyte activation. In Paul W (ed). Fundamental Immunology, 4th ed. Philadelphia: Lippincott-Raven, 1999, p. 411.

Part VI Rational Drug Discovery

27 Rational Drug Discovery and the Impact of New, Advanced Technologies

Jonathan Greer

CONTENTS

INTRODUCTION
DRUG DISCOVERY IS A MULTIDISCIPLINARY PROCESS
CHOOSING A DISEASE AND A TARGET
LEAD IDENTIFICATION
LEAD OPTIMIZATION
THE END GAME
CONCLUSIONS
SELECTED READINGS

1. INTRODUCTION

Administration of effective pharmaceutical agents is among the most efficacious and cost-effective ways of treating disease. Therefore, the discovery of new pharmaceutical agents to improve existing therapy and to introduce treatments for previously untreatable conditions is of paramount importance, especially in the current environment of managed health care and the concerted drive to decrease the ever increasing cost of health care. To achieve this goal of continued and even accelerated discovery of new and effective drugs, the drug discovery paradigm has had to evolve in major ways over the past decade or two and is continuing to do so with the discovery and implementation of a broad range of new and advanced technologies.

From: *Principles of Molecular Regulation* (P. M. Conn and A. R. Means, eds.), © Humana Press Inc., Totowa, NJ.

1.1. Historical Perspective of Drug Design

Discovery of novel drugs requires the execution of several critical steps summarized in Fig. 1. The first requirement is to identify a disease to be targeted. Traditionally, the first step was to develop a suitable animal model for this disease. For example, an in vivo model frequently used for hypertension was the spontaneously hypertensive rat. Compounds, generally hundreds, were tested in this model to search for a "hit," which would reduce blood pressure. When such a compound was found, further in vivo studies, such as pharmacokinetics, toxicity studies and the like, were carried out to determine whether the compound was suitable for clinical development. Medicinal chemistry was performed to optimize the lead compound, when necessary, and testing the new compounds for improvement was carried out entirely in the respective in vivo animal models.

Although the major steps in the discovery of novel pharmaceutical agents (Fig. 1) remain largely the same, the methods used to execute them have changed

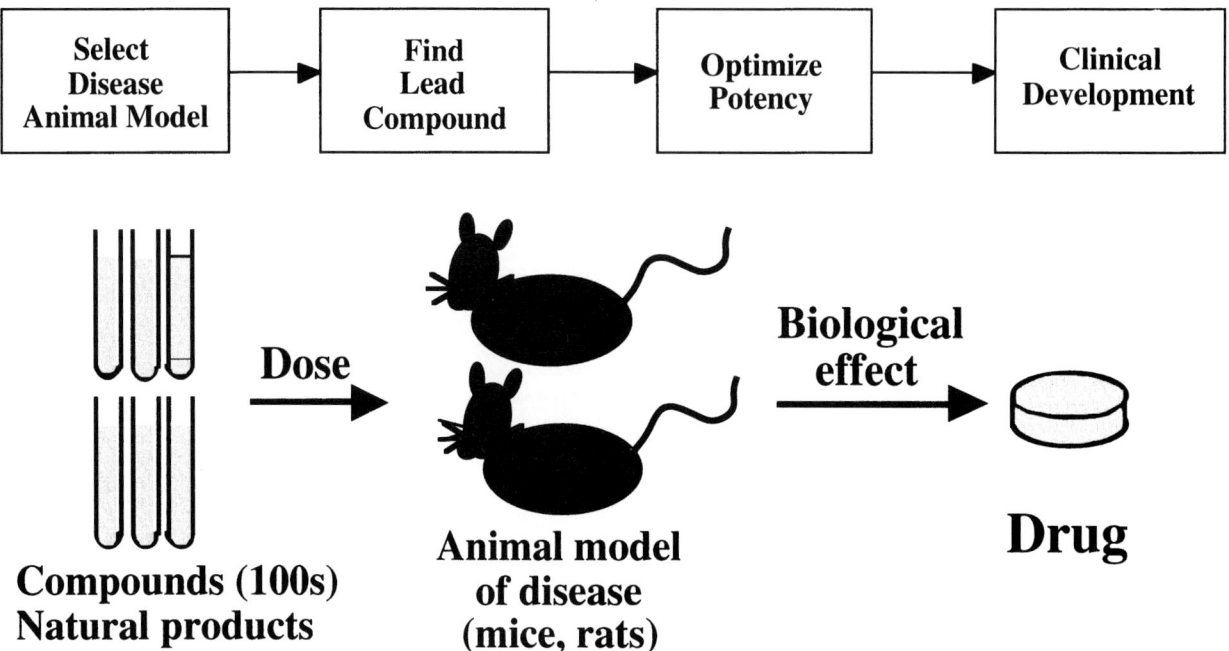

Fig. 1. Historic method of discovering drugs using animal disease models to screen and find lead compounds for optimization and eventual clinical development.

dramatically over the past several decades. These changes were necessitated by inadequacies in virtually every step of the aforementioned process, including:

- Many diseases under study today are quite complex and a suitable animal model may not be available.
- The difficulty and high cost of testing compounds in vivo make the chances of success too low for today's drug discovery process. Thousands of compounds, indeed hundreds of thousands of compounds, have to be tested, which would be prohibitive in terms of the huge number of animals and the long times required to perform these tests.
- Relatively large amounts of compound are needed for in vivo tests, tens to hundreds of milligrams, which can be expensive and time consuming to synthesize.
- The compound hits found are rarely potent enough to be drugs themselves and require significant improvement in potency, as well as other critical properties, such as pharmacokinetics, bioavailability, safety, etc. Frequently, thousands of compounds are synthesized during the optimization stage. These compounds would have to be tested in vivo in the appropriate animal models which also adds tremendously to the cost and decreases the feasibility of the project.
- When an animal model is used to screen for a hit, we do not know the mechanistic basis of the drug's action, information that may be quite important in the proper treatment of the patient.
- The use of an animal model limits the discovered hits to compounds that are both active and also have good bioavailability and pharmacokinetics. Compounds that are active but not bioavailable would be missed.
- An in vivo testing strategy does not take advantage of the large body of biochemical and physiological knowledge developed painstakingly in the past twenty-five years, which allows us to identify and pursue known molecular targets for many diseases.

In response to these deficiencies, new, more advanced technologies had to be developed to make the drug discovery process more efficient, effective, and powerful. The recent economic pressures that have been added to the health care environment have only accelerated the need to evolve the drug discovery process. This chapter reviews the new important methods that have emerged to improve this process.

Chapter 27 / Rational Drug Discovery and the Impact of New, Advanced Technologies

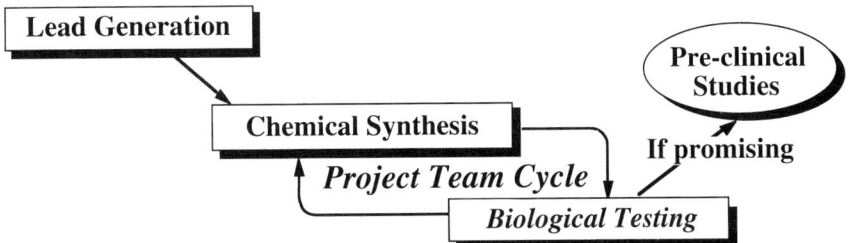

Fig. 2. Method of discovering drugs: the traditional medicinal chemistry cycle, which involves repeated rounds of chemical synthesis and biological testing. The experimentally produced biological testing results drive the compound design process. When a suitable compound is found, it is taken for preclinical and then clinical testing for safety and efficacy.

2. DRUG DISCOVERY IS A MULTIDISCIPLINARY PROCESS

Drug discovery is best carried out by a project team dedicated to the discovery of a small molecule agent to modulate the effects of a particular molecular target for the treatment of a specific disease. A wide variety of scientists from different disciplines must come together to achieve this difficult goal, including medicinal chemists and pharmacologists, but also frequently molecular biologists, biochemists, and structural biologists. These scientists must work closely to achieve the goal of discovering the drug. The basic process, which is iterative, is summarized in Fig. 2. A lead compound is found by some means (*see* Section 4.). The chemists synthesize the lead and the biologists test it in the primary assay for potency. Based on the potency measured, the chemists decide what to synthesize next. This compound is then tested in the assay and the resultant structure–activity relationships are used to direct further rounds of synthesis and testing. As the compounds achieve improvement in potency, additional assays are added to test for the other requirements that are essential for developing a drug. Many of these important properties are summarized in Fig. 3.

3. HOW DO WE CHOOSE A DISEASE AND A TARGET?

3.1. Rational Target Selection

Perhaps the major change in the drug discovery process is due to the huge accession of information about the biochemical pathways of the human body. If one contrasts the typical metabolic pathway chart of 25 yr ago with that of today, the difference is orders of magnitude in the number of proteins/enzymes/receptors that have been identified. In addition, many multiprotein highly complex pathways,

A compound must have:
- Potency
- Novel, patentable structure
- Specificity
- Metabolic stability
- Oral activity
- Good oral absorption
- Formulation in a solid dosage form
- Good pharmacokinetics
- No toxicity

Fig. 3. The critical properties necessary for a efficacious and safe drug. The order approximates the sequence in which the various properties are considered by the project team.

such as intracellular signaling, antibody production, or complement activation have been elucidated and, indeed, continue to be revealed today through the combination of sophisticated molecular biological and biochemical experiments.

Emerging from this vast body of biological knowledge are many of the molecular bases for the currently prevalent human disease conditions. This knowledge base permits the drug discovery team to identify one or more critical molecular targets, typically proteins, in the pathway that controls the biological process that has malfunctioned. The activity of these targets can then, in principle, be modulated by either inhibition or stimulation in the case of an enzyme, or agonism or antagonism for a receptor, as the respective pathway may require. Of course, in many cases the details of the pathway are known only from molecular biological studies and/or from studies in animals. Whether the mechanism will actually work in humans will sometimes not be determined until efficacy studies are actually carried out in humans.

It is useful to consider examples of how this sort of rational molecular target selection occurs. To carry

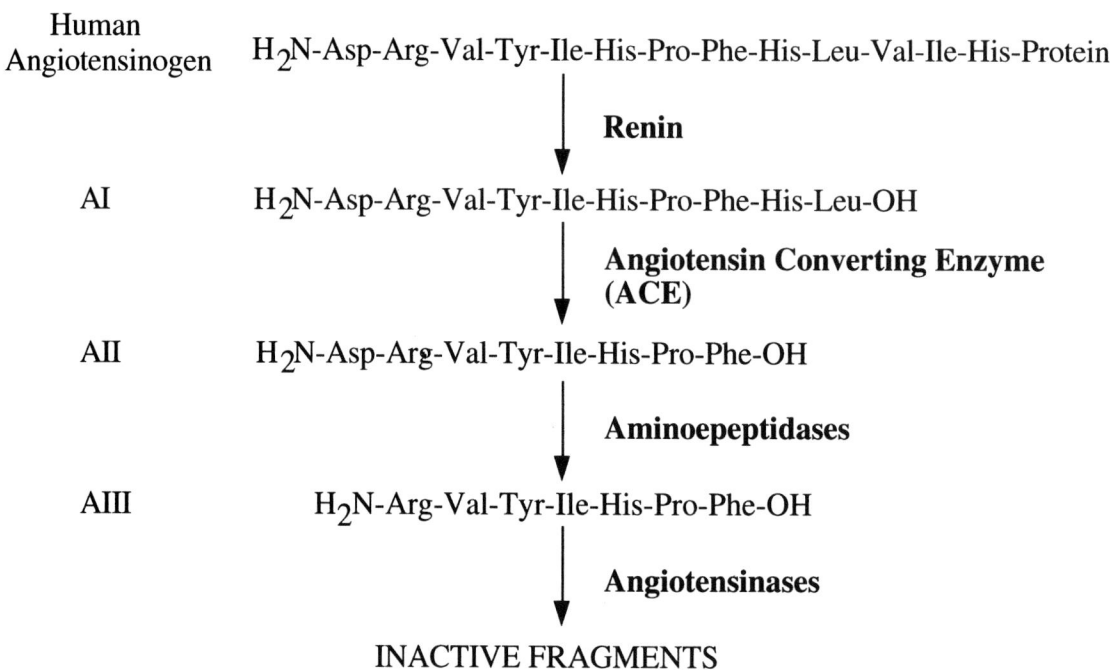

Fig. 4. The renin–angiotensin system. In the first step in this pathway, the protein angiotensinogen is cleaved by the enzyme renin to give angiotensin I, a decapeptide, that is relatively inactive. Next angiotensin converting enzyme (ACE) cleaves the C-terminal dipeptide to form angiotensin II, a very potent pressor molecule in the body. This octapeptide is subsequently inactivated by further cleavage and digestion. Inhibition of the formation of angiotensin II should function as an antihypertensive agent.

our example of hypertension above forward, the target for disease treatment is no longer the spontaneously hypertensive rat. Instead, one begins with knowledge of the pathways in the body that affect blood pressure. In particular, we consider the renin–angiotensin system, a pathway (Fig. 4) that was discovered to be critical for maintenance of blood pressure level and attempt to inhibit it at appropriate stages to produce the desired hypertensive effect. In this pathway, angiotensin converting enzyme (ACE) is essential for the production of angiotensin II, one of the most potent pressor molecules in the body. Inhibition of this enzyme has led to the production of several compounds, such as captopril and enalopril, which are highly effective and highly successful drugs for the treatment of hypertension. Another target for hypertension, using a very different pathway, has been antagonism of the α_1-adrenergic receptor by compounds such as prazosin or terazosin, which are also very efficacious. Knowledge of the mechanism of action of terazosin allowed the expansion of the use of this drug to the effective treatment of the completely different, widespread condition of benign prostatic hyperplasia (BPH) when it became clear that this disorder was also due to overactive α_1-adrenergic receptor action.

Thus, the first step in the new drug discovery process is the identification of an appropriate molecular target that is critical for management of the disease of interest. Typically this is done by careful study of the biological literature, especially the most recent discoveries. When a potential molecular target has been identified, the next step is to verify the utility or validity of the target for disease control. Its utility may be in question for a variety of reasons. First, knowledge of the actual pathway may be only rudimentary. This is especially true for the most recent discoveries. Second, the interaction of multiple components in a complex biological pathway, such as intracellular signaling, for example, may be such that the chosen molecule may be easily bypassed and turn out to be not at all essential in the homeostatic environment of the fully constituted cell or organism. To test these questions and validate the target may not be simple. It may require intensive research on the pathway, something that may not be possible in the economic environment of a pharmaceutical company. In addition, it may need the identification of an as yet nonexistent, sufficiently potent, and specific inhibitor, agonist, or antagonist as a probe ligand for the system to be able to actually test how such a compound affects the system.

3.2. Target Subtype Specificity

One of the factors that has led to the reexamination of many "old" molecular targets and reinitiation of projects is the discovery of enzyme or receptor subtypes. New discoveries indicate that what was originally thought to be a single receptor for a particular natural ligand or a single enzyme for a particular substrate turns out to exist in several different forms. These forms often differ in the physiological role they perform. The compounds/drugs that have previously been discovered and also marketed for a disease may cause side effects because of the unwanted consequence of drug action on a particular subtype that is different from the one actually responsible for the therapeutic effect. The discovery of a more selective compound would therefore produce a better, safer drug.

Discovery of subtypes has traditionally been found pharmacologically by the chance synthesis of a compound that is selective and therefore distinguishes the different subtypes in the particular in vitro assays being used in the project. More recently, subtypes are found readily using molecular biological methods that search for closely homologous sequences once the first sequence for a receptor or an enzyme is found. When such sequences are found, the next step is to determine the tissue distribution and other characteristic properties of the respective forms that can often give an indication of what their individual distinctive functions might be. Of course, when the source is molecular biological, the challenge remains to find a compound that is selective in order to take advantage of the selectivity from a drug design perspective. It is possible that three-dimensional structure determination and/or molecular modeling (*see* Section 5.2.) may suggest how to produce subtype-selective compounds.

Examples of subtype-selective projects are very common today. To give several illustrations, prazosin and terazosin were cited earlier as antihypertensive agents. It was also pointed out that terazosin has been used for BPH. If we want a drug for the treatment of BPH, the antihypertensive effects of these compounds may be considered side effects. The antihypertensive effects are due to the α_{1B} receptor subtype while the effects on BPH are due to α_{1A} (Fig. 5). Thus, an α_{1A}-specific compound would be a better drug for the treatment of BPH and several of these are currently in preclinical and clinical development. Another very current example is in the field of analgesics. A large number of analgesics, especially nonsteroidal anti-inflammatory drugs (NSAIDs) are widely used today including aspirin, ibuprofen, naproxen, etc. All have the side effect of causing stomach and intestinal bleeding because they work by inhibiting the enzyme cyclooxygenase (COX). COX we now know is essential for the maintenance of a healthy stomach and intestinal lining. When this enzyme is inhibited by an NSAID, it not only acts as an antiinflammatory, but also causes stomach and intestinal lining deterioration which can lead to bleeding and ulcers when the drugs are taken chronically. Several years ago, it was discovered that intestinal lining maintenance is due to a constitutive form of the enzyme, now called COX-1, while analgesia can be affected by inhibiting an induced form of the enzyme called COX-2. Two drugs specific for COX-2 have just been approved by the FDA and are revolutionizing the antiinflammatory analgesic treatment of patients.

3.3. Specificity of Approach, for Example, Cytotoxic Agent for Cancer

The vast knowledge base about human biological function that exists and is constantly increasing has opened up whole new areas of drug discovery that did not exist before. This is true in all the fields of drug discovery, including antiinfectives, cardiovascular, metabolic, oncologic, immunological, urinary, and even the more complex neurological diseases.

Thus, for example, cancer therapy has been dominated for decades by cytotoxic agents that depend for their efficacy upon the fact that tumor cells proliferate while most somatic cells of the body do not. However, this class of therapy has been greatly limited in that some critically important cells of the body, such as bone marrow cells, do continue to reproduce, and thus these agents are all too often ineffective because of dose-limiting toxicities. Recent discoveries about the molecular mechanism of cellular transformation and tumor growth have led to a range of drug targets for cancer that are cytostatic and tumor specific because they are aimed at specific tumor mechanisms. Among targets currently in investigation and even in clinical testing are:

- Matrix metalloprotease inhibitors to limit the ability of tumor cells to invade the extracellular space
- Antiangiogenesis agents to prevent tumor vascularization which is necessary for tumor growth
- Farnesyl transferase inhibitors because tumor cells are much more heavily dependent for survival on farnesylation

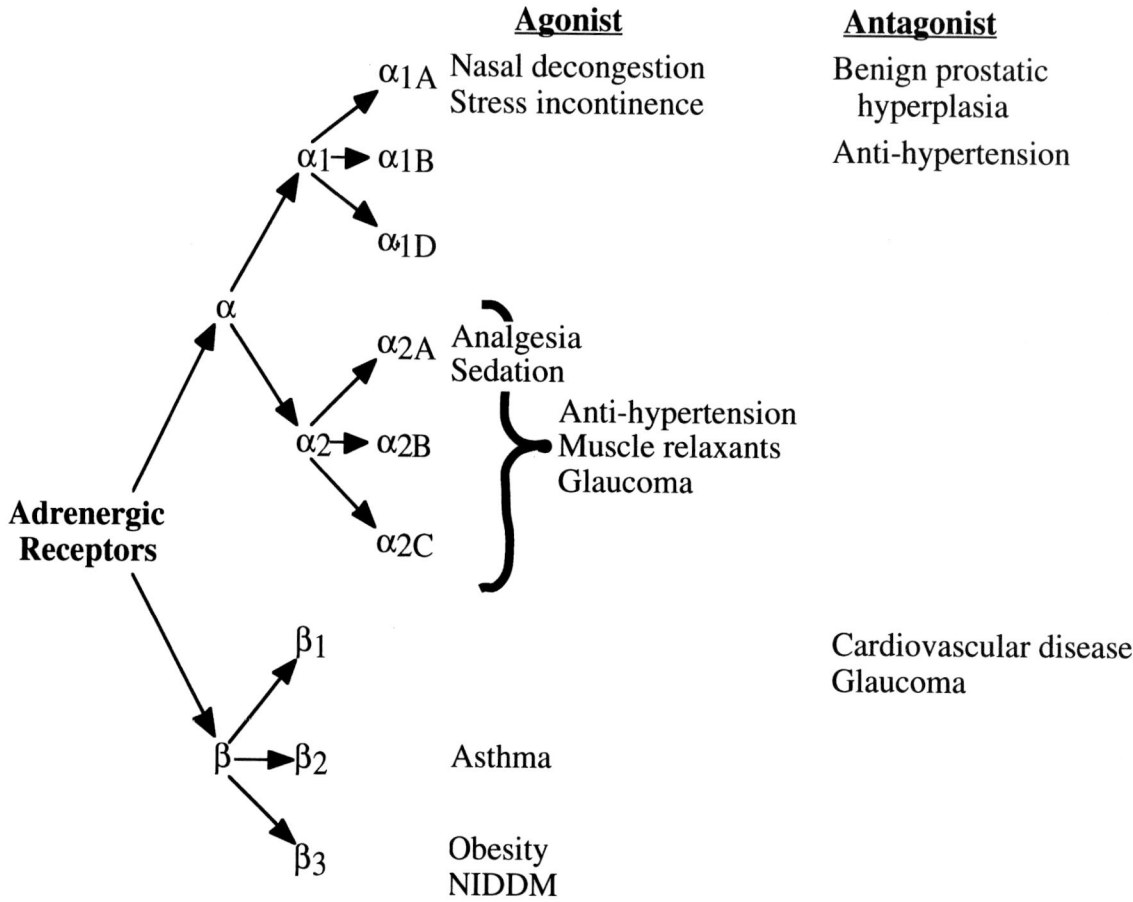

Fig. 5. Adrenergic receptor subtypes and their potential clinical indications as agonists and antagonists.

- Telomerase inhibitors to prevent tumor cells from reproducing indefinitely

Another illustration of new areas of drug discovery opening up is development of antivirals. This is a field that has been largely barren over the past years with very few effective drugs available. The emergence of AIDS and the intense study that the HIV virus has undergone because of the severity of the disease and the challenge it has presented to health care has brought about a revolution in our knowledge base in the field of virus research. Thus, for HIV itself, there are several targets that have been successfully pursued to produce drugs, including reverse transcriptase and HIV protease, the mainstays of RNA virus drug discovery, but also integrase, Rev, and tat. Two drugs are on the market for the influenza virus targeting the enzyme neuraminidase and efforts are ongoing for the common cold rhinovirus based on the unique structure of the virus capsid.

As our knowledge and understanding of the mechanism of disease conditions continues to be revealed, it is expected that many more new projects, pursuing novel targets, will be initiated over the years to come.

3.4. Genomics, the Human Genome Project and Other Genomes (Bacterial)

In addition to considerable research in basic biochemistry and molecular biology that is adding to our understanding of human diseases, the advent of the Human Genome Project (Hudson, 1998) is expected to increase our knowledge base of biological processes enormously and should, in principle, provide the molecular source for *all* human-based biological targets. The determination of the sequence of all proteins and biological macromolecules from the human genome opens this possibility. However, this research is a classic example of nonhypothesis driven research. That is, one is generating exhaustively the sequence of all human proteins with little or no accompanying knowledge of what these biomolecules do or how they function alone or in concert in the cell or the body. Therefore, realizing the value of this sequence information will require a large investment in basic research. Clearly, it will take decades to reveal the full import of the biological function of all the open reading frames that will be identified as the human genome project progresses to completion over the next 3–4 yr. However, by using bioinformatics judiciously, one can begin to identify most likely candidates for immediate follow-up. Nevertheless, each of these candidate target molecules requires detailed and even intensive study to confirm its general biological role and, in particular, its suitability as a drug discovery candidate for a specific human disease.

Target validation is a very time consuming and labor intensive biological exercise. To speed up this crucial effort, a number of new methodologies, called collectively functional genomics, have been identified. These include array technologies in which a large number of diverse oligonucleotides, representing different gene sequences, are attached to a wafer and used to recognize cDNA/mRNA libraries for particular disease conditions. Other techniques involve knocking out genes to see the effect of their loss. These include antisense and ribozyme methods, as well transgenic animals.

Although the major effort underway is to determine the full human genomic sequence, a significant effort has also begun to sequence the genomes of critical bacteria that are involved in infection. The search for new antibacterial and antiviral targets using genomics is currently an extremely active area of research (Schmid, 1998). It is a remarkable fact that all antibiotics of the past two decades have been aimed at the same very small number of targets, the D-Ala-D-Ala ligases (penicillins and cephalosporins), the ribosome (macrolides), and DNA gyrase (floxacins). Virtually no new targets have emerged. It is anticipated that this situation will change rapidly in the next few years as the results of bacterial genomics are successfully translated into new antibiotic drug discovery targets.

Several examples of the use of genomics to discover new drug targets are beginning to emerge. One such example is the discovery of cathepsin K, a new member of the cysteine protease family that was identified by analyzing new gene sequences from a cDNA library produced from osteoclast cells, cells involved in bone resorption. Subsequent studies indicated that inhibition of this protease may provide a novel therapy for osteoporosis. New chemical entities to inhibit this enzyme are being produced using combinatorial chemistry and structure-based drug design, new technologies that are described later in this chapter (*see* Section 5.).

4. LEAD IDENTIFICATION

4.1. Assay development

Absolutely essential to the pursuit of the selected molecular target, whether for concept validation or for initiation of a drug discovery effort, is development of a suitable primary in vitro biological assay for the target (*see* Fig. 2). For an enzyme, this usually is an enzyme assay; for a receptor, one generally chooses receptor binding. The primary assay may take several different forms, depending on the purpose for which it is designed, such as lead identification and/or project structure–activity determination. The primary assay is usually followed by secondary assays to confirm the function of the compounds such as a functional assay to accompany receptor binding to distinguish agonists from antagonists.

The choice of what assay to use often depends on the throughput necessary for the testing. A typical assay used by a project team will need to be able to test 10–30 compounds a week, depending on the number of chemists working on the project and how easy the compounds are to synthesize. The assay must be robust, accurate, and reproducible, as the chemists depend critically upon the assay results to decide what compounds to prepare next.

One of the first uses of an assay in the lifeline of a project is for the identification of a suitable lead structure for the chemists to pursue. There are many potential sources for such a lead compound. These are discussed in more detail in the following sections.

4.2. Medicinal Chemistry-Based Methods for Lead Identification

For many interesting targets, compounds are already known that could be a potential start point for drug design.

4.2.1. NATURAL SUBSTRATE OR LIGAND

The most obvious lead candidate is the natural ligand for the molecular target. For an enzyme target it would usually be a substrate analog. The substrate itself is rarely a good lead because it is, of course, turned over rapidly by the enzyme and thus labile. However, in many cases, a simple modification is known that converts the substrate to an inhibitor. For example, peptide substrates of aspartic proteinases can be easily converted to inhibitors by changing the scissile peptide bond to a reduced amide or an hydroxy isostere (Fig. 6). This modification has been used extensively for such targets as renin inhibition for

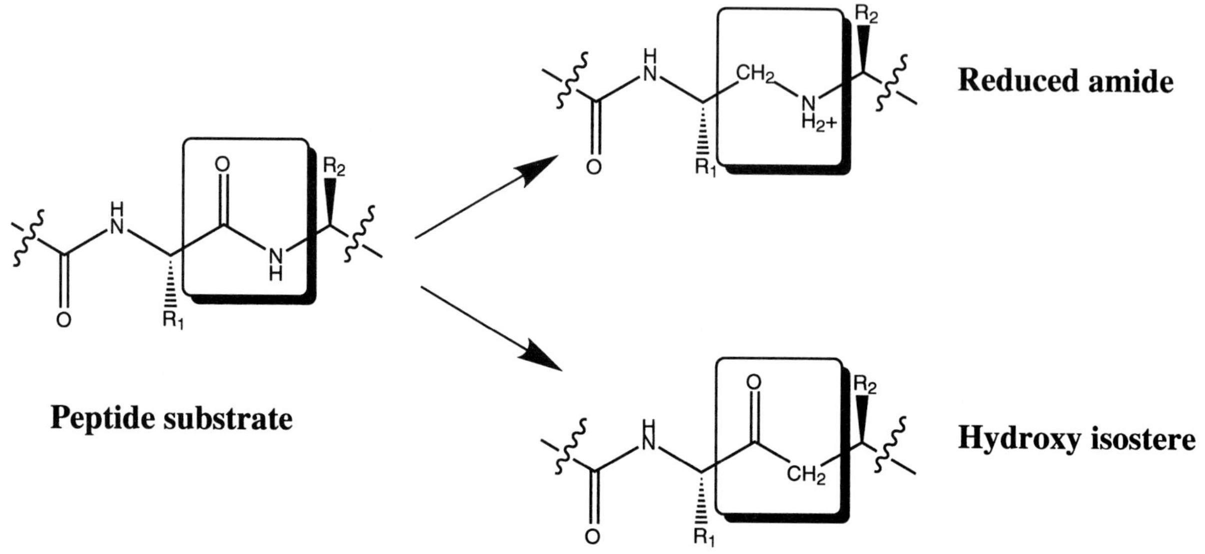

Fig. 6. Conversion of a substrate of an aspartic proteinase into an inhibitor by forming either the reduced amide or the hydroxy isostere are just two of many possibilities that have been employed in studies described in the literature.

hypertension and HIV protease inhibition for the treatment of AIDS.

Another approach that can be used with enzymes that require cofactors, such as ATP, NADH, tetrahydrofolic acid, *S*-adenosyl-methionine, etc., is to start with the cofactor itself or an analog as the lead. This has the apparent disadvantage that many enzymes usually use the same cofactor which could, in principle, create problems of specificity. However, while many enzymes do share cofactors, the binding sites on the respective proteins for the cofactor are generally very different and it is usually easy to achieve specificity.

When the molecular target is a receptor, many of which are membrane bound, the natural ligand may provide a reasonable starting ligand. However, frequently there are issues that need to addressed for the natural ligand to be useful in drug design. Typically, the natural ligand may be an agonist. If an antagonist is desired, some modification will have to be found that changes the compound's function to an antagonist. This has been successfully achieved in a number of cases. Another problem with the natural ligand is that it usually binds with similar affinity to all the receptor subtypes. When a compound that is specific for a particular subtype is desired, some chemical modification will need to be introduced to achieve this specificity. Discovering that change may be quite difficult. Finally, some natural ligands are proteins, for example, erythropoietin or insulin, both of which are used as drugs; or peptides, such as gonadotropin releasing hormone or endothelin. However, the desired drug is usually intended to be a small molecule in order to achieve oral activity. Starting with the natural peptide or protein ligand then requires reduction to a small molecule which is often very difficult to achieve. This approach, called peptidomimetics, will be reviewed separately in a later section (Section 4.4.).

4.2.2. Literature or Competitors' Compounds

A second source of leads for beginning a medicinal chemistry effort is often compounds that have been reported in the scientific or patent literature. These compounds have usually been discovered either when the molecular target was being characterized or during a previous or existing drug design project on this target. When the latter is the case, then one is beginning a new project on this target either because the original series had some flaw that needs to be improved upon or because of new knowledge as to subtype specificity and more specific compounds are desired. In both these cases, while the published compounds may present a starting point, clearly a strategy for departure from this series to achieve the desired goal of improved properties or specificity has to be developed. In addition, one must be careful in this strategy to produce a new series that is not covered or even clouded by the previously published or patented literature. As noted in Fig. 3, a patentable compound is almost always essential for clinical development and successful commercialization of a compound.

4.2.3. HIGH-THROUGHPUT SCREENING

The most common technique used to find lead compounds is screening. The nature of screening has changed dramatically over the past several decades. As noted in the introduction, it began by the screening of tens to hundreds of compounds in in vivo models. It then progressed to the use of in vitro assays and the screening of thousands of compounds in a time frame of 6 mo to a year. Most recently, with the expanded use of 96-well or even 384-well technology, the use of mixtures of compounds, and the introduction of automation, parallelization, and robotics, rates of several hundred thousand compounds over 3–6 mo can be achieved routinely. High-throughput screening methodology translates directly into more hits found, more rapidly, for more targets.

The ability to find hits using high-throughput screening depends critically on the quality of the screening assay. Typically, special effort is needed to transform a routine assay into one that can be run in high-throughput mode. An assay that is used for high-throughput screening has more stringent requirements in terms of signal-to-noise ratio and robustness in order to achieve the very high throughput. It must be well designed and carefully address the molecular target or mechanism of action that is desired by the project to detect the correct "hits."

Many different types of assays are used for different types of targets. They can be divided into two basic classes: functional assays and affinity assays. The functional assays comprise many different measurements of activity including enzyme assays, receptor functional assays, cell-based assays, gene expression, and many others. Methods of detection include optical density, fluorescence, radioactivity, etc. Success rates differ for different types of targets. The highest success rate is found for enzyme inhibitors and receptors that have defined binding pockets. The poorest success rates result when the compound must interfere with a protein–protein interaction.

The most common affinity methods are receptor binding and direct detection of ligand binding. It also includes the interesting ability to use the protein target molecule itself to select the highest affinity compounds out of a large mixture of compounds that are combined with the protein target, called "affinity selection." Many reviews have been written that cover the various assays that have been developed to screen different kinds of targets or even disease conditions (Hill, 1998). It is beyond the scope of this chapter to review this wealth of information.

The number and quality of hits obtained is also highly dependent on the nature of the compound library that is available for screening. Clearly, the critical property for obtaining hits is the molecular diversity of the compounds. The broader the diversity the better. Evaluating diversity and selecting compounds for acquisition in assembling a compound collection has become a significant area of investigation in its own right today (Gordon and Kerwin, 1998). Most pharmaceutical company compound libraries are the accumulated result of efforts on specific projects over several years and thus contain a defined number of clusters of quite similar compounds. Thus, acquisition of compounds from outside commercial or academic sources is an excellent way to broaden the chemical diversity of the internal compound library.

4.3. Structure-Based Lead Discovery Methods

A novel method has been reported recently for discovering lead compounds for medicinal chemistry follow-up. This method, called "SAR by NMR" (Shuker et al., 1996) uses a distinctive nuclear magnetic resonance (NMR) fingerprint spectrum for the target protein that allows one to identify when a compound binds and basic binding site information, that is, near which residues of the protein the compound is binding. It permits detection of very weak binding compounds, as weak as 1–10 mM. In the strategy proposed for SAR by NMR, one tests ~10,000 very small compounds that are effectively fragments of typical drug compounds. If we consider a case where subsets of these fragments can potentially bind to each of three subsites in the active site of the molecular target, then the SAR by NMR screen is equivalent to screening a virtual library of all possible combinations of the 10,000 fragments at each of these three sites (A, B, and C in Fig. 7). This is formally equivalent to a traditional screen of a composite library (where A, B, and C fragments are chemically attached, by links such L1 and L2, in Fig. 7). That composite library would equal 10,000 × 10,000 × 10,000 or 1000 trillion compounds! The power of the method comes from the fact that these 10,000 fragment compounds can be tested by current NMR techniques in mixtures of 10–50 compounds at a time in several weeks. Thus, this vast virtual library can be examined in a relatively short period of time for the amount of information

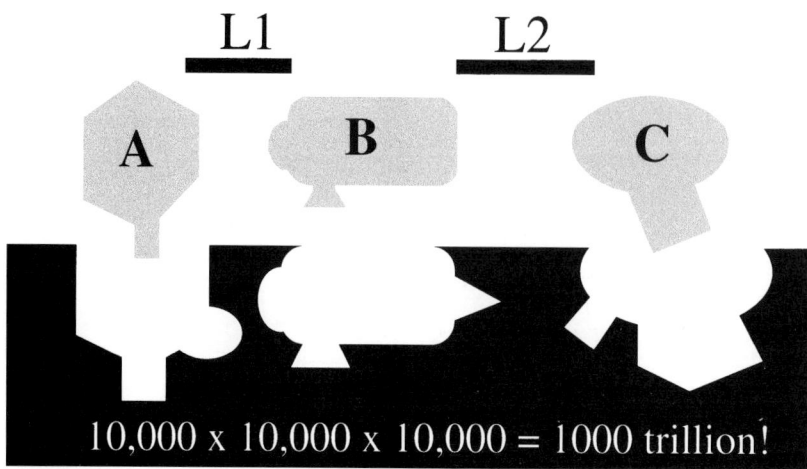

Fig. 7. The power of SAR by NMR comes from its screening of a virtual library. If one screens 10,000 fragments that can potentially bind at any of three sites, then the effective compound library is 10,000 cubed or 1000 trillion compounds if they were tested as linked compounds.

obtained. Furthermore, as noted earlier, the NMR spectra not only identify which fragments bind but also show near which residues these fragments lie. The SAR by NMR method has been used to find novel ligands for several proteins already including FK binding protein, papilloma virus E2, and stromelysin.

SAR by NMR is useful not only for determining which particular fragments bind to a site of interest, such as at the active site, but also because it provides structural information about where the fragments bind in the structure of the protein target that can be used by the chemists to link the fragments together rationally. Once two fragments or more are identified and are found to bind adjacent to each other, they can be linked chemically. The resulting cooperative binding of the two linked fragments gives a dramatic boost in binding affinity. One example from stromelysin will demonstrate this result (Fig. 8). Hydroxamic acid, a very rudimentary inhibitor of stromelysin at the active site, with a potency of 17,000 μM, was bound to stromelysin and the fragment library was screened by SAR by NMR for a second site binder. Such a compound was found at 280 μM. Simple analogs of this second site hit were examined and an analog was quickly found that exhibited a 10-fold improvement in potency at 20 μM. The two fragments, the hydroxamic acid and the biphenyl analog, were then linked by the chemists resulting in a linked compound with 0.025 μM potency, a huge increase in potency over the starting fragments. This very powerful method can be used on many proteins to help find lead compounds for medicinal chemistry and help direct the chemists, using the structural information, as to what to make to improve potency.

4.4. A Peptide or a Protein Lead— Peptidomimetics

When the natural or otherwise derived lead ligand is a peptide, it is usually desirable to modify the compound to remove the peptidic properties. The reason for this is that peptides are typically metabolically very labile in vivo and generally have very poor oral bioavailability. From that perspective, one would much rather start with a small molecule. Transformation of a lead peptide to a small molecule requires considerable work and is called peptidomimetics (Ripka and Rich, 1998). This involves relatively long and laborious modification of the peptide frequently taking years of work. Classic studies have been performed on the decapeptide peptide gonadotropin releasing hormone (GnRH) and the 14-residue peptide somatostatin, with only moderate success after many years of effort. One of the best success stories has been in the field of HIV protease inhibitors, in which a number of orally active compounds have been produced that are peptidic in character (Wlodawer and Vondrasak, 1998). In at least one case, using invaluable crystal structure information and structure-based drug design (see Section 5.2.), an ingenious peptidomimetic HIV protease inhibitor was designed and synthesized by a group at Dupont-Merck (Lam et al., 1994) as summarized in Fig. 9.

In many cases, because of the great difficulty in successfully creating a peptidomimetic in a reasonable amount of time and effort, the best strategy to

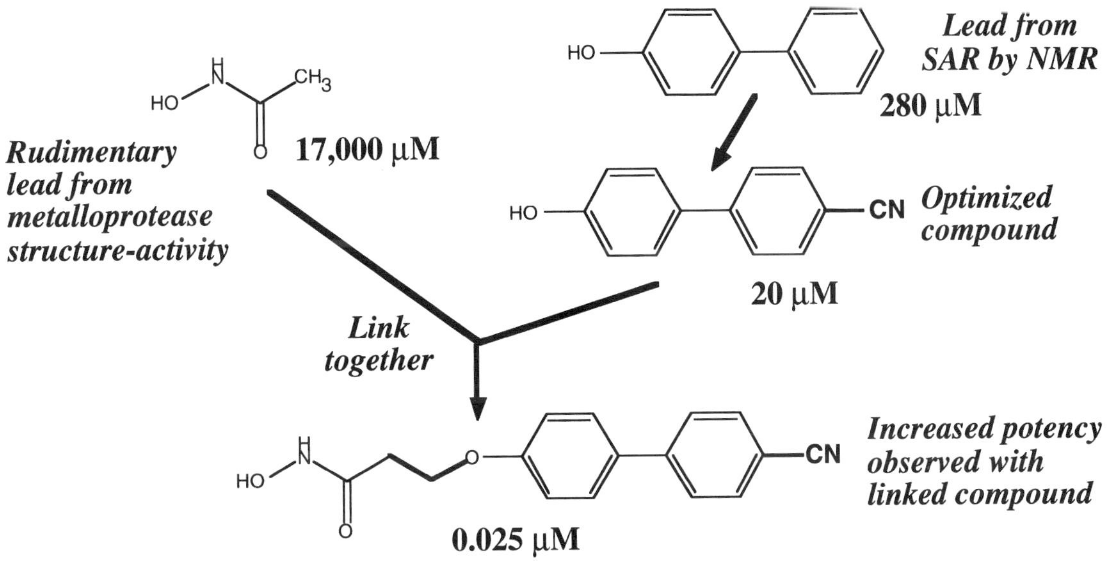

Fig. 8. Stromelysin inhibitor lead discovery. Starting with an extremely weak inhibitor of stromelysin, 17 m*M*, a fragment that binds to an adjacent site was identified by SAR by NMR, 280 µ*M*. This was then optimized using SAR by NMR to a 20 µ*M* compound. When the hydroxamic acid and optimized biphenyl fragments were linked by the chemists in a suitable fashion, the resulting linked compound was 25 n*M*, a 680,000-fold improvement in potency over the hydroxamic acid fragment and 8000-fold increase over the other fragment.

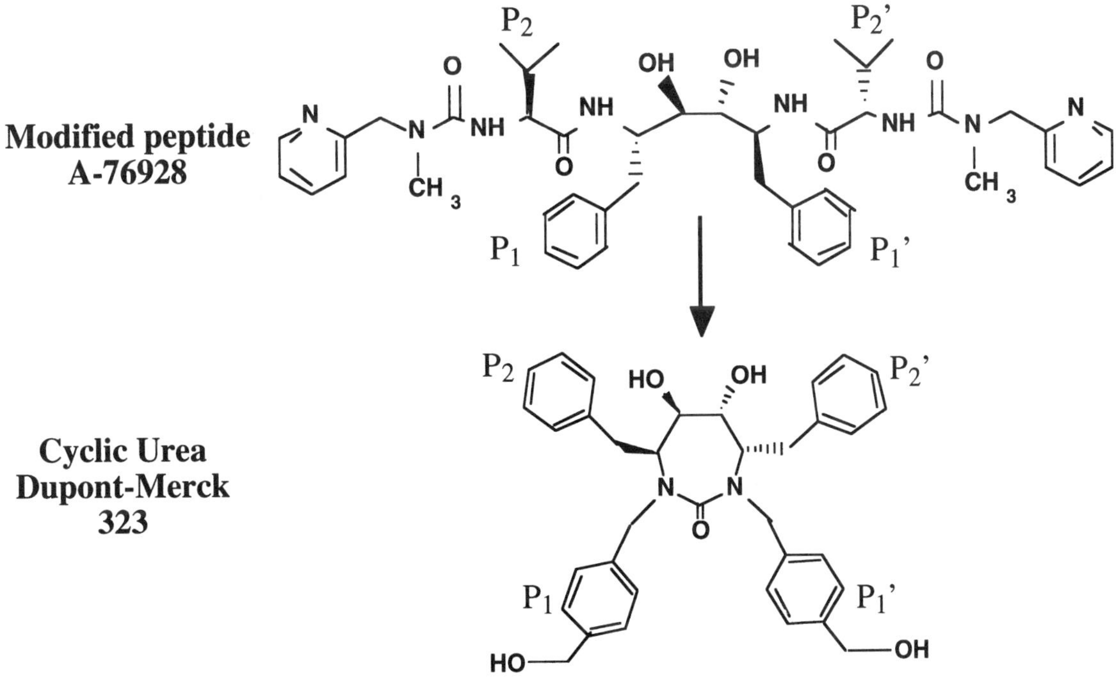

Fig. 9. Modification of a peptidic inhibitor, such as A-76928, to a peptidomimetic by forming the cyclic urea.

use with a molecular target that binds a peptide and/or a starting lead peptide is to screen one's small molecule compound library for hits using the molecular target and proceed with these hits to perform medicinal chemistry bypassing the peptide leads completely with this approach. Since the discovery of the chemical structure of GnRH in 1971 (Fig. 10A), tens of thousands of GnRH peptide analogs have been synthesized. Two relatively conservative changes in the structure produced a more potent and metabolically more stable "superagonist," leuprolide, that is currently the drug of choice in the field (Fig. 10B).

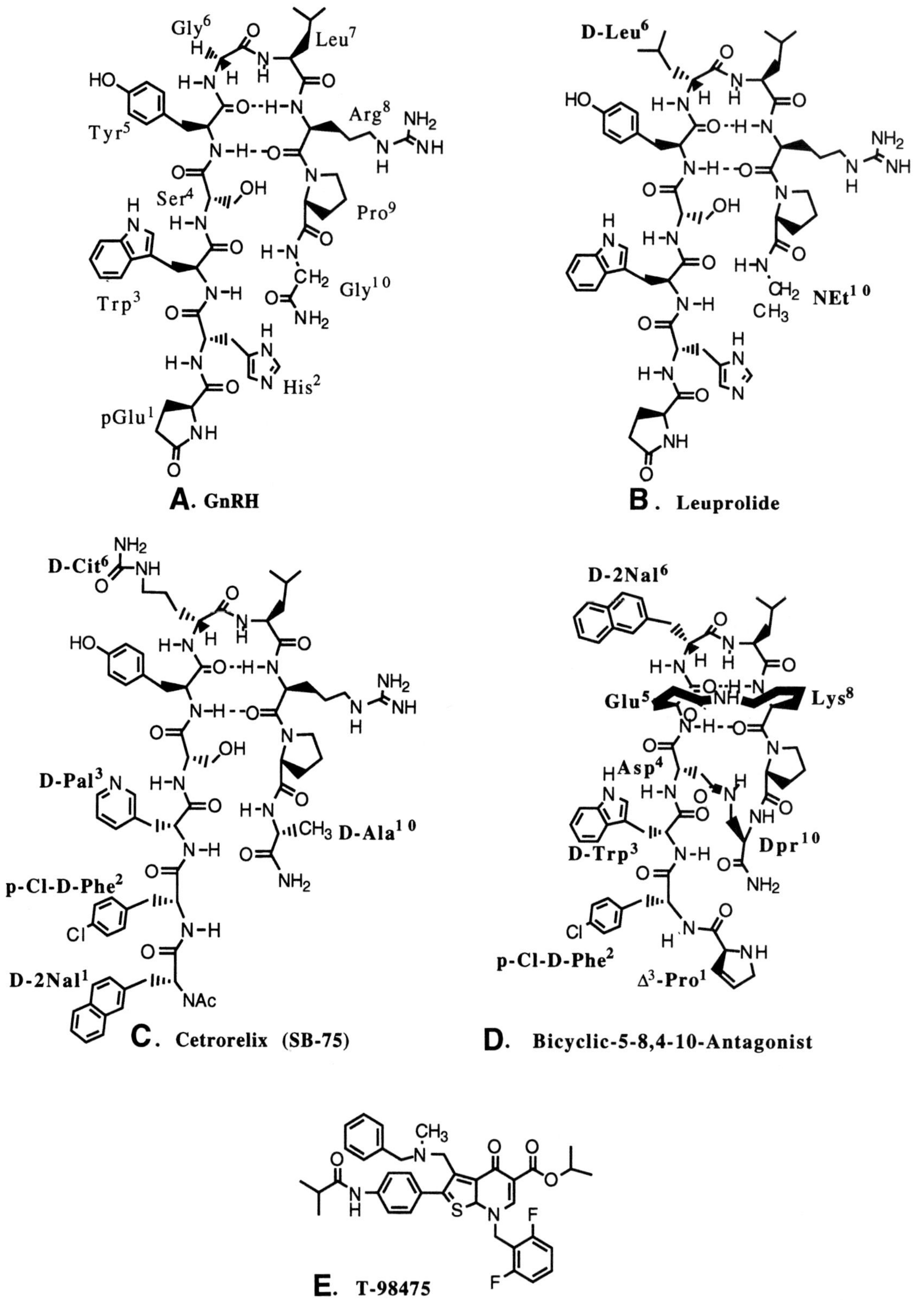

Fig 10. (A) The structure of the natural peptide ligand, GnRH or LHRH, a decapeptide. **(B)** Two changes were made: Gly^6 to D-Leu and Gly^{10} to NHEt forms a very potent agonist that is currently the drug of choice in the field. **(C)** An example of a peptide antagonist that is currently in the clinic. Note it has five D-amino acids and retains only three of the original natural amino acids of GnRH. **(D)** An illustration of a designed bicyclic peptide based upon molecular modeling studies. **(E)** A small antagonist with oral activity recently reported in the literature.

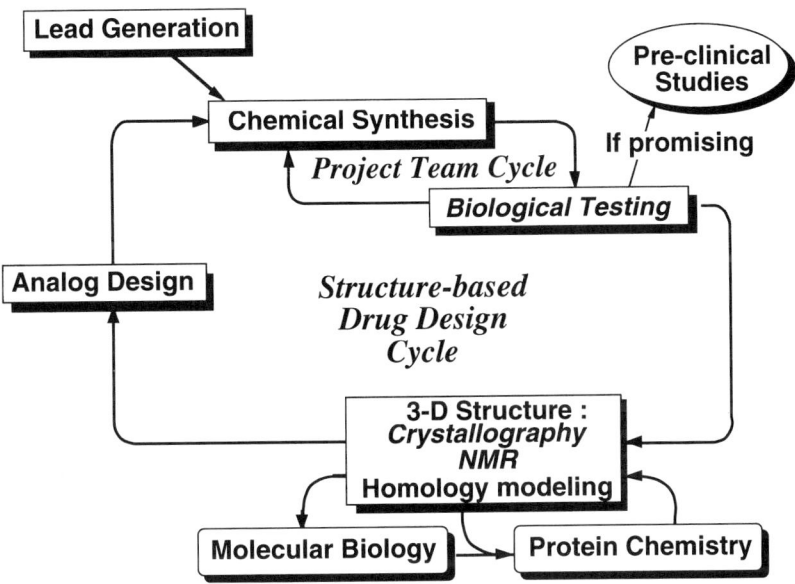

Fig. 11. Iterative structure-based drug design cycle. This is superimposed upon the medicinal chemistry cycle of Fig. 2. The three-dimensional structure of the target protein, as determined experimentally by crystallography, NMR, or homology modeling, is used to help design new analogs.

During the past decade and a half, the focus has been on producing antagonists. Compounds of this class typically have many changes from GnRH with four or five D-amino acids. An example of the one of the peptide antagonists currently in the clinic is shown in Fig. 10C. GnRH has also been the focus of many molecular modeling studies leading to the design and synthesis of a number of reasonably potent cyclic and bicyclic peptide antagonists (Fig. 10D). However, it was only recently, as a result of screening, that small molecule GnRH antagonists have begun to emerge that are orally active (Fig. 10E). Note that unlike the peptidomimetic example cited for HIV protease where the design was based and confirmed by structure-based drug design (Fig. 11), the relationship between the thienopyridine small molecule and the GnRH peptide analogs is not at all clear (Fig. 10).

4.5. Hits to Lead—Verification of a Hit

When a "hit" compound is found through screening the first step is to reconfirm its activity in primary and secondary in vitro assays. If the compound activity is reconfirmed, then the decision must be made whether to follow-up this hit with medicinal chemistry. This decision requires considering a variety of factors. The compound must be sufficiently potent (i.e., typically < 10 µM) to the project to be interesting. It must be medicinal chemistry "friendly." That is, it must be a stable small molecule with only one or at most two charges and a number of sites that can be easily replaced or substituted to permit efficient analoging. Ideally, it should have a molecular weight below 500 and few or no asymmetric centers. Once medicinal chemistry begins on the hit, the experience is that not all hits will produce productive leads that are worth pursuing. In the nonproductive scenario, typically 30–50 analogs will be prepared. Many inactive or weaker compounds will be found, but the original hit will continue to be the most potent of the series. In such cases, when virtually any change in the molecule, no matter how subtle, leads to loss in activity, the series should be abandoned and the chemists should move on to another hit. It is important to recognize the nonproductive hits as quickly as possible and terminate that series and move on to a more useful one. A "good" series will usually produce a 10-fold enhancement of potency/activity within the first 30–50 compounds prepared demonstrating that this series represents a true "lead," has potential, and should therefore be further pursued with a full program of medicinal chemistry optimization.

5. LEAD OPTIMIZATION

5.1. Medicinal Chemistry

Having found a suitable lead, the next step in the basic drug design pathway (Fig. 1) is to develop the series and optimize the compounds. Typically, thousands of compounds must be synthesized over 2–4 yr in order to produce one or a small number of compounds that have all the desired properties (Fig. 3) to be taken into the clinic and become a drug. Traditionally, this has been done using medicinal chemistry skills and experience (Fig. 2). The medicinal chemist looks carefully at the chemical structures that have been produced and the biological activity data generated in the critical primary and secondary assays, and uses the deduced structure–activity relationships to help decide what compounds to synthesize next. It soon became apparent that more rigorous methods were needed for analyzing these important structure–activity relationships. Quantitative structure–activity relationships methods (QSAR) were introduced to improve the analysis and the ability to predict what to make next (Martin, 1978). The major limitation of QSAR is that it is interpolative and is usually restricted to working within a single chemical series. It will suggest what the best member of the current series is likely to be, but it does not help the chemist move out of the current series to new chemical structures or series with significantly improved properties.

One of the most important challenges the pharmaceutical industry faces today is to expedite the compound optimization phase of drug discovery. How can this process be made faster and more effective?

5.2. Structure-Based Drug Design

It is intuitively obvious that if one knew the detailed interactions and spatial relationships of the lead compounds with the protein target active site, this would be valuable information to help in the design of novel and improved compounds. It has, in fact, taken perhaps a decade or more for this obvious fact to be reduced to practice and result in new, improved drugs. Structure-based drug design (Fig. 11) is now used extensively in the pharmaceutical industry to help accelerate the drug design process (Greer et al., 1994). It works as follows: The three-dimensional structure of the target protein is determined, either experimentally by X-ray crystallography or by NMR or theoretically by homology modeling. The lead compounds are examined in the active site to see the details of how they bind. The nature of the groups forming the active site permit the design of new compounds which are then synthesized and tested to determine their biological activity. When experimental structures are available, the new compounds can be examined in the crystal or the NMR tube to see if the compounds bind as predicted. Based upon these structures, a new set of compounds is designed, synthesized, tested, examined structurally, and the process iterated until suitable compounds are produced. Because detailed structural information is produced repeatedly, mistakes can be corrected and the true structure–activity relationship of the compounds followed and optimized more rapidly. In addition, the structural information can permit the design of quite novel compounds based upon the geometry and nature of the interactions that can be made to critical groups in the ligand binding site.

A couple of illustrative examples will demonstrate the power of the structural method. There is an ever increasing number of examples that continue to emerge of the use of structure-based drug design; however, the most extensive use has been in the field of inhibitors of HIV protease, a member of the aspartic proteinase family (Wlodawer and Vondrasek, 1998). The early determination of the crystal structure of this protein showed that a significant structural change occurred when an inhibitor was bound in the active site (Fig. 12). It also demonstrated clearly that the protein was a dimer with C2 symmetry. While many other companies were screening their libraries of renin inhibitors for aspartic proteinase inhibitors that would inhibit HIV protease, at Abbott it was decided to follow up on the structural motif by designing and synthesizing C2 symmetric compounds that would fit into the active site (Fig. 12). One of the first compounds prepared, A-74704, shown in Fig. 13, was highly potent at inhibiting the enzymatic activity with an IC_{50} of 3 nM. The crystal structure of the A-74704 enzyme complex showed that the compound bound at the active site exactly as the model design had proposed. Furthermore, the C2 symmetric inhibitors were not active against mammalian aspartic proteinases and thus were highly specific for HIV protease.

Another outstanding example of structure-based drug design in the HIV protease field was alluded to earlier in the peptidomimetics section (Section 4.4. and Fig. 9). The design of the cyclic urea analogs of the peptidic HIV protease inhibitors was a triumph of structure-based drug design. The cyclic urea imitates the core of the C2 symmetric diols (Figs. 9 and 13) perfectly and the oxygen of the urea carbonyl replaces a water molecule that is found trapped between the peptidic inhibitors and the enzyme (Fig. 14).

Thus, structure-based drug design can be used to discover new lead compound series that can bind at the active site. It permits the design of new series that are modifications of or significant departures from the existing series. It can also function effectively to expedite compound optimization within a series. These are just some of the ways in which structure-

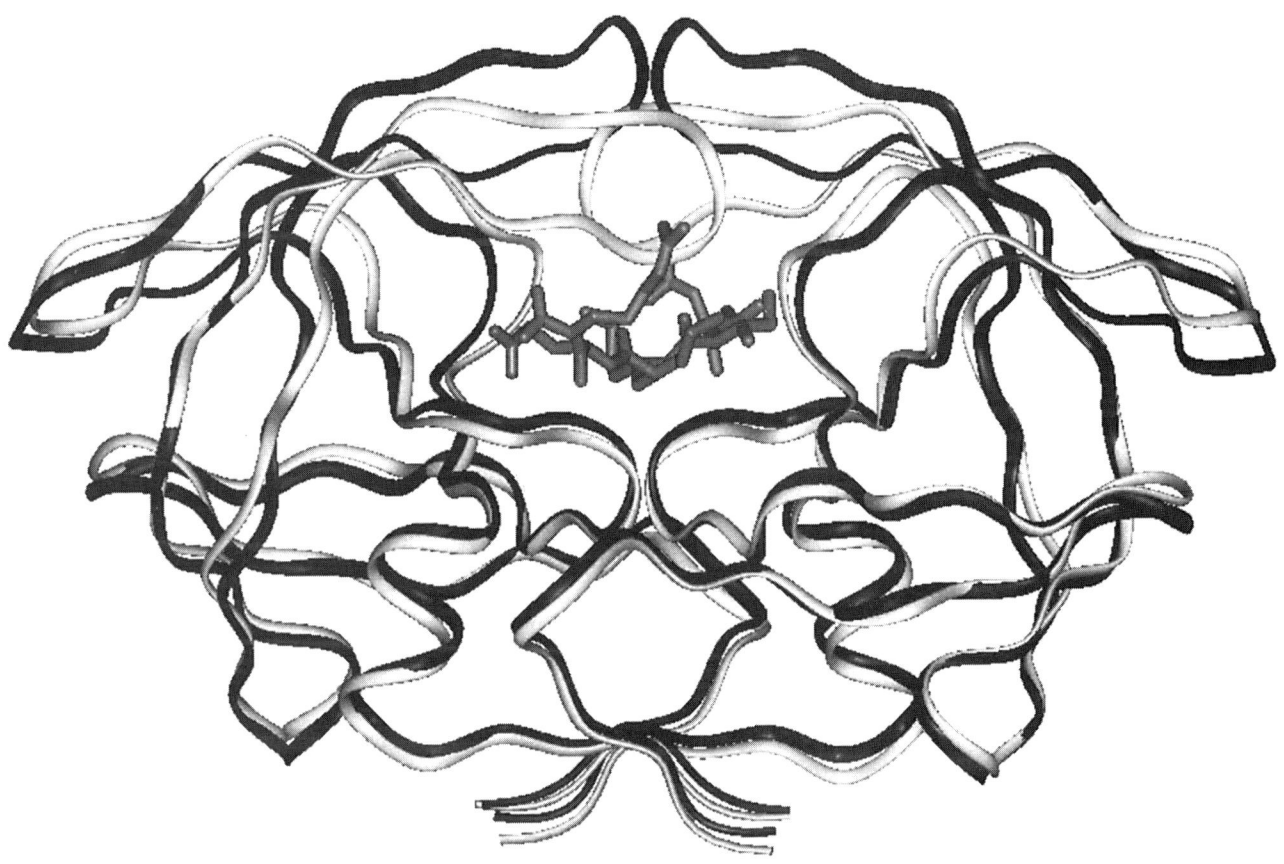

Fig. 12. Structure of HIV protease with *(white)* and without *(black)* ligand bound. The ligand, MVT-101, a peptide inhibitor, is shown in *gray*. Note the protein is a dimer, composed of two monomeric units that are related by a C2 or 180° rotation symmetry axis that lies vertically in the plane of the figure at the center of the protein structure. The inhibitor binds at the interface between the monomers.

based drug design is currently employed to add new ideas to the compound design process and help accelerate drug discovery.

5.3. Combinatorial/Parallel Chemistry

One of the newer technologies to emerge in the drug design process is combinatorial and parallel chemistry (Gordon and Kerwin, 1998). It is useful to define the terminology at the outset. Combinatorial synthesis refers to the synthesis of mixtures of compounds, typically hundreds to thousands in one vial. When many compounds are made simultaneously, but one per vial, it is called parallel synthesis.

Combinatorial synthesis is illustrated in Fig. 15. In general, combinatorial synthesis must be performed on solid resins. In the figure, three reagents, A, B, and C, are reacted with resin beads in three separate vials. These beads are mixed together and then divided into three aliquots. The three are each reacted with three further reagents, D, E, and F, respectively. Once again the beads are mixed and then divided into three and reacted with G, H, and I. The final three vials are left separate, each now containing nine compounds, for a total of 27. The virtue of this so-called "mix and split" process is that with 9 reactions, 27 compounds are produced. If one were to use 50 reagents at each step then 150 reactions would result in $50 \times 50 \times 50$, or 125,000 compounds, which is very efficient. If one chooses diverse substituents at each step, the resulting compounds will, in principle, be quite diverse.

In this way, very large numbers of compounds can be prepared. However, there are several problems with this approach. First of all, screening mixtures of up to thousands of compounds in one vial presents a rather formidable challenge. New methodologies have had to be developed to properly screen these mixtures. Furthermore, because of limitations in the kinds of chemistry that are amenable to combinatorial synthesis, the restrictions of all the compounds in a particular library having a common core, and the need to put three substituents on to get the efficiency boost of

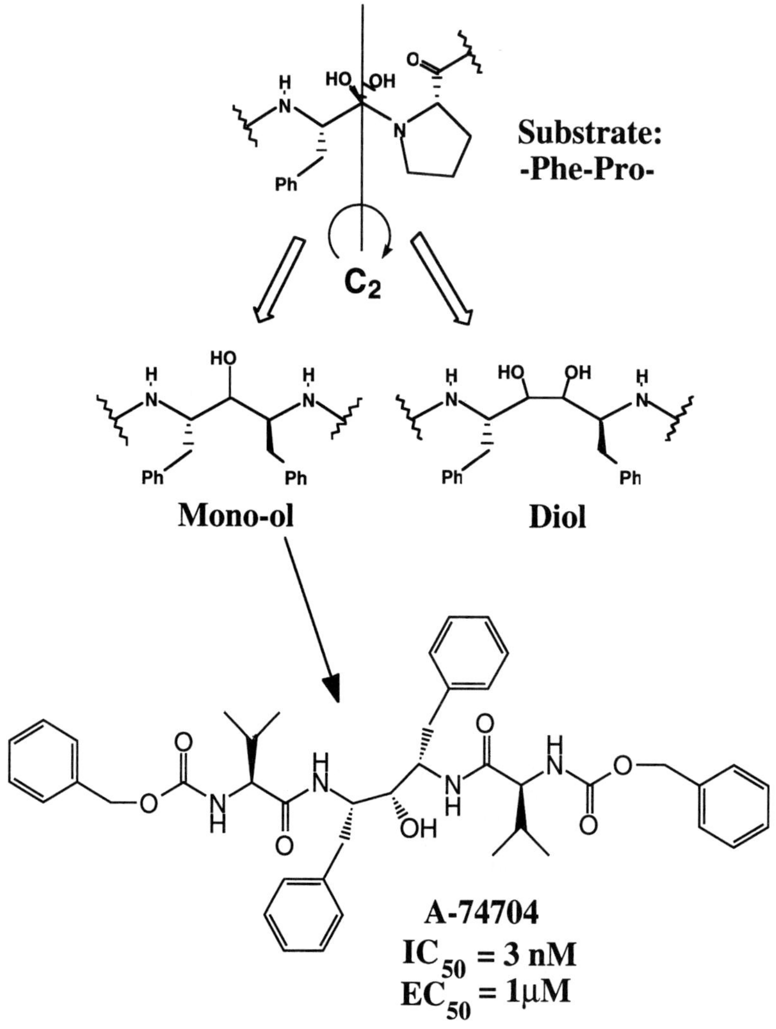

Fig. 13. The substrate transition state (**top**) is shown relative to the C2 axis of the HIV protease crystal structure. If one deletes the right hand side of the substrate and generates the second half of the inhibitor by the C2 operation one obtains the mono-ol shown on the **middle left.** If the substrate is positioned so the C2 axis lies halfway along the C–N bond that is cleaved by the enzyme and the C2 operation is performed, the diol on the **middle right** is produced. One of the first compounds prepared, A-74704, shown at the **bottom,** was highly potent at 3 nM inhibition of the enzyme. Very potent compounds were synthesized in both series by the chemists.

mix and split in the number produced, the resulting compounds are not as diverse or as properly behaved as one would like. Consequently, the true benefit and value of combinatorial libraries currently remains unclear.

Parallel synthesis, on the other hand, is expanding and becoming more and more widespread in the industry. It is apparent that considerable benefit and increase in efficiency can be obtained by preparing 10–20 compounds in the same time needed to synthesize one in the past. This can be accomplished using robotics and automation with relatively little decrease in chemical versatility. Parallel synthesis has typically been performed by a dedicated group of combinatorial chemists. However, improvements in robotics and automation are permitting these devices to be deployed more and more in the regular medicinal chemistry laboratory. It is expected that within several years, parallel synthesis will be integrated as a routine part of the repertoire of the practicing medicinal chemist.

The use of either combinatorial or parallel synthetic methods introduces new demands and new requirements upon the chemist that never existed before. If a chemist is to synthesize 20–100 analogs at one time routinely, computational tools need to be introduced

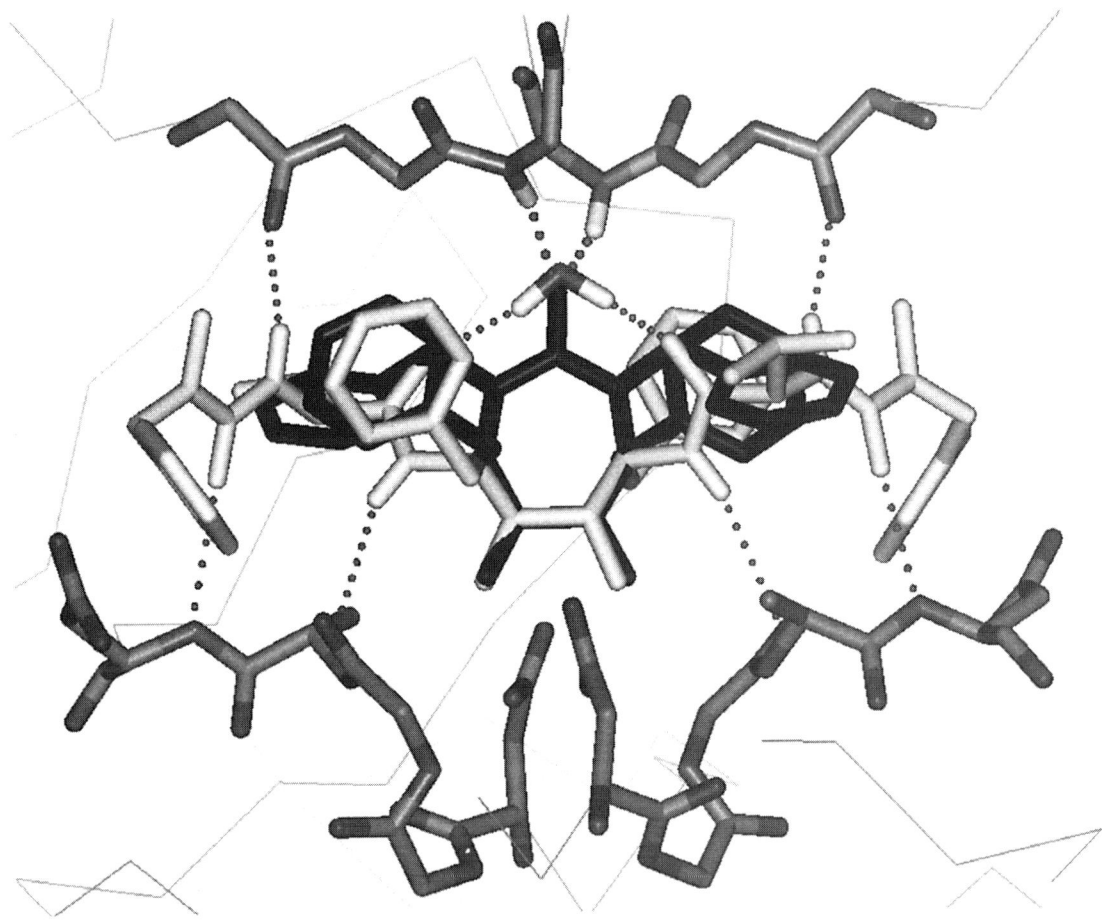

Fig. 14. Crystal structures of the active site of HIV protease *(gray)* is shown with the peptidic diol *(white)* and the cyclic urea *(black)* superimposed as they bind relative to the protein active site. Note how closely they overlap and form the same interactions with the active site residues. Note also the displacement of the water molecule by the urea carbonyl.

to permit the selection of suitable substituents to achieve the appropriate diversity and other chemical, physical, or pharmacophoric properties that are desired. Just the sheer effort to identify, order, receive, prepare, and purify the hundreds of starting materials for combinatorial or parallel synthesis becomes a major task. Computational tools to perform these functions become not a luxury but essential. Also necessary are the tools to record each synthesis, capture the purity and characterization data for each compound, and register the compounds automatically in the compound library. A number of groups and companies are working on developing software to perform these functions.

The capabilities of combinatorial and parallel synthesis relative to traditional medicinal chemistry are summarized in Table 1. Medicinal chemistry remains the mainstay of drug discovery, relevant at all stages and most versatile chemistry-wise, yet producing the smallest number of compounds. Parallel synthesis lies in the middle with the ability to do the routine, but not the most challenging chemistry, and therefore applicable at all but perhaps the last stages of drug discovery when final tuning of the compounds often requires more sophisticated chemistry. It offers hope for a significant increase in the number of compounds that can be prepared and hence in efficiency. The largest number of compounds can be made by combinatorial chemistry, but the chemistry must be simple, general, and capable of being performed on a solid resin and this perforce seriously limits its use in drug discovery.

5.4. Ease of Chemical Synthesis

With the economic pressures on the health care industry that have emerged over the past several years, the cost of manufacturing a compound can be an critical deciding factor in whether to pursue clinical development. It is important therefore to consider the synthetic difficulty of chemical series being pursued

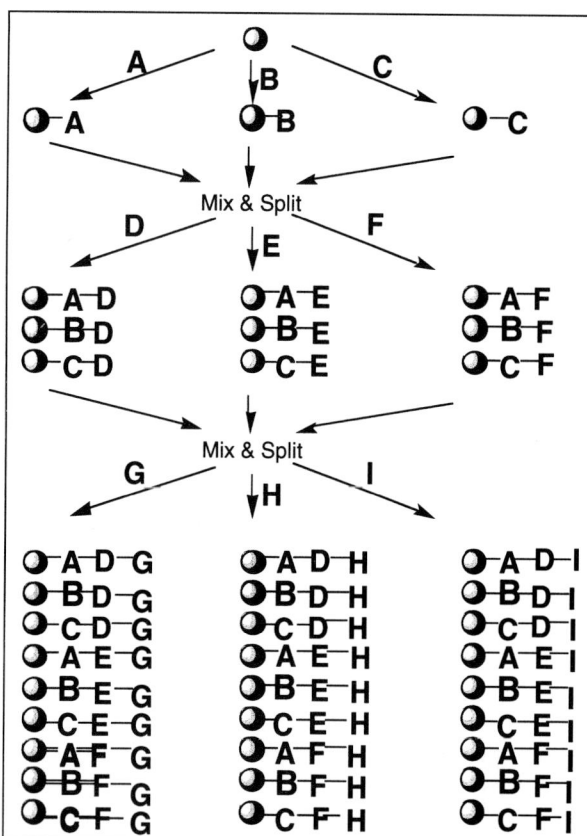

Fig. 15. Mix and split strategy. By reacting the beads at each step separately, then mixing them and reacting the mixtures with multiple reagents at the next step and continuing this process, one obtains a combinatorial mixture with a large number of compounds in each vial. For the example given where three reagents are used at each step one obtains three vials each containing nine compounds.

early in the series selection and optimization process. Indeed, this question should play a role as early as in deciding which screening hits or other leads to be pursued by the project team chemists. As the compounds become more potent and a drug candidate is close to being selected, it is valuable to enlist the advice and services of the process chemistry group to be sure that synthetic manufacturing issues are properly dealt with in choosing an appropriate drug candidate for clinical development.

6. THE END GAME

Compound optimization begins with improving the potency based upon in vitro assays as described earlier, typically enzyme activity measurements or receptor binding or the like. However, as pointed out in Fig. 3, it takes many other factors to transform an active compound into a drug. First and foremost is in vivo activity. Biological activity in an animal model depends not only on the intrinsic potency of the compound but also critically on the bioavailability by the selected route of administration and on the pharmacokinetics of the compound. As the desired route of administration for most drugs is oral, the barrier of good oral absorption must be overcome for a chemical series to succeed. Therefore, it is essential to begin testing pharmacokinetics and bioavailability early in the lifetime of a chemical series, even if the degree of intrinsic potency desired has not as yet been achieved. Many drugs are terminated early in the clinic because of formulation problems or inadequate pharmacokinetics or bioavailability costing much unnecessary wasted time and resources. The trend today is to aim for compounds that can be administered once or at most twice a day. More frequent dosing in most disease conditions, other than life threatening where the need is not otherwise satisfied, such as AIDS, is no longer acceptable from a marketing perspective. Indeed, because of these rather stringent requirements for pharmacokinetics and bioavailability, it may be useful for the chemists to devote some effort to optimize a second series that may be less potent but give better pharmacokinetics and bioavailability because of the critical importance of these properties to the identification of a successful drug.

Testing compounds in vivo is time consuming from several perspectives. Relative to in vitro experiments, large amounts of compound, tens to hundreds of milligrams, have to be prepared for each study. This is because at least three, and often much larger numbers of animals, have to be used at each dose and often dose ranging studies must be performed to determine minimum effective dose as well as optimal dose for efficacy. Some studies can be performed acutely, but

Table 1
Types of Chemistry and Their Capabilities for Drug Discovery

Method	Medicinal	Parallel	Combinatorial
Number of compounds:	1–10	100s–1000s	1000s–1,000,000s
Chemistry:	Challenging	Routine	Facile
Stage of project:	All	Early to late	Early to middle

for other disease conditions or for safety testing, the studies can require days to weeks of dosing. For in vivo studies, the number of animals that can be dosed and studied at one time is usually quite limited. Because of all these requirements, the studies are usually also very time consuming and relatively expensive. There is also a strong sentiment over the past number of years to decrease, as much as possible, the number of animals used in research and experiment. Consequently, there is a premium to discover methods of minimizing the number of in vivo studies or number of animals needed to identify a compound suitable for clinical development.

6.1. In Vivo Functional Activity

It is necessary to set up relevant functional animal models for the target disease. For some diseases, a single animal model may serve to provide sufficient information; in other cases, several models may need to be tested sequentially or in parallel to provide the information necessary to properly test the compounds and develop suitable in vivo structure–activity relationships. These relationships can be of crucial importance for the chemists to produce a compound that can become a drug candidate. Sometimes, multiple animal models may be necessary when no single animal model is considered a good test of the disease condition or when multiple indications are being considered for the compounds and the efficacy in different models may differ from compound to compound. For some disease conditions, acute dosing may be sufficient, while for others, chronic dosing may be needed for days or even weeks. Clearly, a dosing regimen that requires a very long time (months) is not useful or feasible for effective compound evaluation.

Sometimes there is no adequate or usable disease model; however, a suitable surrogate marker may be available to give an approximate readout of potential in vivo efficacy. Surrogate markers also serve to simplify and speed-up compound evaluation because they avoid the need to use a full-blown disease model. An example of this is the GnRH field described previously. Here, rather than use models of prostate cancer or endometriosis or some other indication for GnRH agonists or antagonists, one need only monitor testosterone levels in male animals or estrogen levels in females to see whether the compounds are effective or not.

A whole field may suffer in its drug discovery efforts from the paucity of suitable animal models. Thus, HIV drug candidates go to the clinic at risk because no suitable animal model exists to test them in vivo. The decision to go to the clinic is based on in vitro cellular antiviral assays for efficacy, and in vivo assessment is limited to pharmacokinetics and bioavailability. In the cancer field, the problems are even more complex. In humans, tumors are relatively slow growing, typically taking months for the cells to double and years for tumors to reach a significant size. An equivalent animal model for compound evaluation would be impossible to use because of the extremely long time factor. Furthermore, it is impossible to grow human tumors in normal animals because of foreign tissue rejection. Therefore, it is necessary to use specially selected, highly aggressive human tumor cell lines to get growth in the weeks to a couple of months time frame and to perform the experiments in the highly abnormal nude mouse model where human tumors will grow. Consequently, it is not at all surprising that many compounds that appear to work well in these artificial animal models are found to be ineffective in the clinic. This problem is exacerbated by the fact that the above animal models are even less meaningful for cytostatic cancer targets because such drugs may be truly effective only for the slow growing tumors found in humans.

Animal models serve not only as important indicators of the validity of the target and compound activity. They also permit estimating clinical dosage by considering what blood levels of compound are required in the animal models to achieve efficacy. By comparing such studies with the results of toxicity and side effect assessments, one can begin to estimate a therapeutic index (minimum nontoxic dose/efficacious dose) which is a crucial indicator of whether a compound is worth taking to the clinic.

6.2. Oral Bioavailability

Oral bioavailability is an essential characteristic for most drugs that are currently taken into the clinic. The oral route, is by far, the most desired route for drug administration in the United States and in many countries around the world. The percent oral bioavailability is typically calculated by measuring the area under the curve of blood levels of drug after oral administration vs blood levels of drug after intravenous administration. Thus, two in vivo studies have to be performed to determine this important property.

Oral bioavailability is the result of many different effects including absorption across the intestinal wall, metabolism in the liver from the hepatic circulation or the so-called "first pass effect," and distribution into tissue. It would be valuable to have a faster method for estimating oral bioavailability to permit more rapid SAR studies during compound optimization.

Recently, an in vitro method has been introduced that may permit pretesting compounds for their oral bioavailability potential, at least with respect to the intestinal absorption barrier. Caco-2 cells, a human colon adenocarcinoma cell line, form monolayers of differentiated epithelial cells that mimic the intestinal wall in their absorptive characteristics. These cells have proved useful for modeling compound transport. While in vivo studies would be performed to provide detailed oral bioavailability information, this in vitro test, together with in vitro metabolism studies described in the next subsection, could serve as a valuable, rapid prescreen of the project team compounds.

Introducing oral activity into a compound series can often be much more difficult than improving potency. Therefore, it is important to select early in a project the right kinds of lead compounds to pursue that are more likely to be orally bioavailable. This observation has spawned a detailed comparison of the properties of druglike compounds that are orally bioavailable vs those that show no blood levels after oral dosing. One such study led to the proposal of the so-called "rule of 5" in considering the kinds of compounds that would make suitable orally active drugs. This rule indicates that one should avoid compounds with molecular weight >500, with log P (the partition constant of the compound between water and octanol) >5, with more than 5 hydrogen donors, with more than 10 nitrogen and oxygen atoms. Rules, such as these, can easily be incorporated in computer programs that assess high-throughput screening hits and leads. They can evaluate compound suitability and can be used as guides to compound design during the medicinal chemistry optimization phase.

6.3. Pharmacokinetics and Metabolism

A critical property of a drug candidate is its pharmacokinetics (PK). The PK of a compound depends on a variety of factors including absorption, distribution, metabolism, and elimination (ADME) of the compound. In addition, the route of metabolism of a compound and the chemical nature and properties of any metabolites that are formed is important information that must be ascertained during the preclinical studies of a drug candidate. Most of these factors can be properly assessed only by in vivo experiments. However, progress is being made in obtaining some of the required information in vitro or improving the efficiency of the in vivo studies.

Significant strides have been made in determining the route of metabolism and nature of some of the metabolites using in vitro methods. These studies involve the use of liver homogenates, such as an S9 fraction or a preparation of hepatocytes, or even liver slices. Such experiments are typically done with rat, dog, and monkey and even human liver samples. This cross-species comparison can be quite important, as it may allow relating metabolism and pharmacokinetic results determined in animal models to that expected in the clinical setting. The in vitro studies also permit isolation of sufficient amounts of metabolites to allow identification using sophisticated mass spectrometry methods available today. The results found in such in vitro studies are often in excellent agreement with metabolism rates and metabolite patterns observed in vivo.

PK measurements, however, depend on so many effects involving organs and processes throughout the body that it is unlikely this will be completely modeled in a suitable in vitro system. To speed up the process of determining PK values for compounds during optimization, a number of groups have experimented with dosing animals with more than one compound at a time, called "combinatorial PK" (Tarbit and Berman, 1998). If five or ten compounds can be dosed per animal, then the process will be much more efficient, using less animals and taking significantly less time. Care must be taken to ensure that spurious results are not obtained because of interactions between the compounds and their respective metabolic pathways and the like. Therefore, it is important to run standard compounds with each mixture to ensure the reliability of the results obtained by this new method.

6.4. Formulation

A recent study reported that 41% of the compounds brought into the clinic were terminated because of inability to formulate the compound adequately. This is an astonishing statistic. When a compound cannot be properly formulated, it will fail in the clinic due to poor oral bioavailability, poor PK, and insufficient efficacy. Thus, a compound that has all the other properties cited in Fig. 3, arrived at with a great deal of hard work and years of effort, has a >40% chance of failing just because its physicochemical properties are not commensurate with a suitable formulation for humans.

One common property that may limit the ability to formulate is very high hydrophobicity. Many compounds become more potent as their hydrophobicity, as usually measured by log P, increases and thus many drug candidates have very high log P values, that is, above 5. Such compounds violate the so-called "rule of 5," described previously. Early attention to the overall physicochemical properties of the series is essential to avoid compounds that are likely to have poorer oral bioavailability, inferior PK, and the inability to be effectively formulated.

6.5. Toxicity

There are a number of ways that technological improvements in the area of toxicity would accelerate the drug discovery process. These include better prediction of toxicity either by computer program analysis of chemical structure or by high-throughput in vitro assays that are reliably indicators of toxicity, methods to permit more rapid determination of the mechanistic basis of an observed toxicity, and the use of these methods in the discovery phase of compound selection and optimization to help steer active chemical program away from toxic compounds to begin with (Todd and Ulrich, 1999).

A number of programs exist for predicting potential toxicities by analyzing chemical substructures. While these programs have some value, they have not reached the stage at which they can predict toxicity accurately enough to have a significant influence on the course of a discovery or development program.

In vitro toxicity assays date back more than two decades to the introduction of Ames testing for mutagenicity and carcinogenicity. More recently, an in vitro test for clastinogenicity has become accepted as a reliable predictor of in vivo chromosome breakage. In vitro tests described above for metabolism studies, such as microsome fractions or hepatocyte cultures, can prove useful for identifying metabolic pathways, such as cytochrome P450s, and indicate the possibility of potential drug–drug interactions. Of course, all such in vitro tests should be performed with human proteins, cells, or tissue samples, as species differences can play an important role in toxicity. Despite all these new tests, adequate in vitro tests for such common critical toxicity issues as hepatotoxicity, do not exist as yet and would be of great value in helping to eliminate unsuitable compounds early in preclinical development.

Ultimately in vivo animal testing, using traditional toxicity studies, remains the only reliable, accepted method for pretesting compounds for clinical studies.

7. CONCLUSION

Drug discovery is currently in the midst of a particularly challenging yet exciting phase. The discoveries of basic science highlighted by the Human Genome Project are revolutionizing the ability to determine the molecular bases of disease. However, by their very nature the pathways of the body are turning out in many cases to be remarkably complex and the result of the interplay of a large of number of proteins and biological factors. To assist in the utilization of this vast knowledge base of human biology are many new advanced technologies that impact upon every phase of drug discovery. These include the influence of genomics and functional genomics on target identification and validation, high throughput screening for hit discovery, structure-based drug design and combinatorial chemistry for lead optimization, and a host of new in vitro and in vivo strategies for speeding up the final optimization and preclinical testing of drug candidates. These innovative techniques are fueling the accelerated discovery of current drug candidates and will continue to foster the next generation of novel pharmaceutical agents. These methods have already shepherded a revolution in the treatment of AIDS and will be expected to have a major impact on our treatment of other worldwide major health care problems such as cancer, diabetes, and Alzheimer's disease in an efficacious and cost-effective way in the near future.

SELECTED READINGS

Gordon EM, Kerwin JF. Combinatorial Chemistry and Molecular Diversity in Drug Discovery. New York: John Wiley & Sons, 1998, p. 516.

Greer J, Erickson JW, Baldwin JJ, Varney MD. Application of the three-dimensional structures of protein target molecules to structure-based drug design. *J Med Chem* 1994; 37:1035–1054.

Hill DC. Trends in development of high-throughput screening technologies for rapid discovery of novel drugs. *Curr Opin Drug Disc Dev* 1998; 1:92–97.

Hudson TJ. The Human Genome Project: tools for the identification of disease genes. *Clin Invest Med* 1998; 21:267–276.

Lam PYS, Jadhav PK, Eyermann CJ, et al. Rational design of potent, bioavailable, nonpeptide cyclic ureas as HIV protease inhibitors. *Science* 1994; 263:380–384.

Martin YC. Quantitative Drug Design. New York: Marcel Dekker, 1978.

Ripka AS, Rich DH. Peptidomimetic design. *Curr Opin Chem Biol* 1998; 2:441–452.

Schmid MB. Novel approaches to the discovery of antimicrobial agents. *Curr Opin Chem Biol* 1998; 2:529–534.

Shuker SB, Hajduk PJ, Meadows RP, Fesik SW. Discovering high-affinity ligands for proteins: SAR by NMR. *Science* 1996; 274:1531–1534.

Tarbit MH, Berman J. High-throughput approaches for evaluating absorption, distribution, metabolism and excretion properties of lead compounds. *Curr Opin Chem Biol* 1998; 2:411–416.

Todd MD, Ulrich RG. Emerging technologies for accelerated toxicity evaluation of potential drug candidates. *Curr Opin Drug Discovery Dev* 1999; 2:58–64.

Wlodawer A, Vondrasek J. Inhibitors of HIV-1 protease: a major success of structure-assisted drug design. *Annu Rev Biophys Biomol Struct* 1998; 27:249–284.

INDEX

A

ABP-50, 107
ABP-120, 107, 110
ACAID, *see* Anterior chamber associated immune deviation
Acetylcholine receptor, *see* Nicotoninc acetylcholine receptor
Acetyl-coenzyme A carboxylase, phosphorylative regulation, 302
Actin,
   binding proteins,
      ABP-50, 107
      ABP-120, 107, 110
      α-actinin, 108
      aginactin, 107
      Cap32/34, 107
      cofilin, 106, 107
      coronin, 107, 108
      ponticulin, 108
      profilin, 108
      SCAR, 108
      severin, 108
      talin, 107
      WASp, 108, 109
   cytoskeletal regulation in *Dictyostelium* chemoattraction, 106, 110
α-Actinin, 108
Action potential-driven calcium influx,
   function, 161, 162
   modulation of influx,
      inwardly rectifying potassium channels, 165
      ligand-gated channels, 165, 166
      overview, 163
      sodium channels, 164, 165
      voltage-gated calcium channels, 161, 162
   spontaneous pacemaking and calcium transients, 162, 163
AD, *see* Alzheimer's disease
Adenylyl cyclase,
   calcium effects, 252, 253, 256, 257, 259
   catalytic reaction, 250
   chemotaxis role in *Dictyostelium*, activation and regulation, 102
   G protein regulation,
      activation, 250, 251
      Gbg, 253
      $G_i\alpha/G_o\alpha$, 253, 254, 260
   GTPase-activating activity, 259, 260
   history of study, 250
   molecular diversity, 251
   overview of signaling function, 249, 250
   phosphorylation, 257
   structure, 250, 251
   type-dependent regulation, 250, 251
ADP-ribosylation factor (ARF), phospholipase D1 regulation, 242, 243
β-Adrenergic receptor, 3, 4, 6–10, 12, 13, 24, 35
Aginactin, 107
AIF, 423
AKAP, 318, 319
AKAP 79, 215
Akt, 103, 104
ALS, *see* Amyotrophic lateral sclerosis
Alzheimer's disease (AD), nitric oxide role, 224, 225
α-Amino-3-hydroxy-5-methyl-4-isoxazole propionic acid (AMPA) receptor, 144, 145
   calcium signaling, 157
   subunits, 156
AMPA receptor, *see* α-Amino-3-hydroxy-5-methyl-4-isoxazole propionic acid receptor
Amyotrophic lateral sclerosis (ALS), nitric oxide role, 224
Androgen receptor, isoforms, 356
Animal testing, rational drug discovery, 456, 457
Anterior chamber associated immune deviation (ACAID), 434
Antihypertension drugs, rational drug discovery, 442
Antiviral agents, rational drug discovery, 444, 448, 452
AP-1, 222
Apaf-1, 423, 425
Apoptosis,
   Caenorhabditis elegans model, 416
   caspase,
      substrates, 416, 417
      types, 416
   definition, 415
   disassembly and phagocytosis of cell, 426, 427
   DNA degradation, 426
   energy metabolism, 422
   induction, 415, 416
   pathology, 427
   role,
      Bcl-2, 424–426
      calcium, 182, 424
      death receptors, 418–420
      mitochondria, 422–424
      mitogen-activated protein kinase, 417, 418
      nuclear receptors, 420
      physical damage, 422
ARF, *see* ADP-ribosylation factor
Arrestin, rhodopsin deactivation, 91
ATP,
   calcium wave propagation, 182
   receptors, *see* P2X channels
ATP-gated receptors, *see* P2X channels

B

Bacteriorhodopsin, structure, 6
Bcl-2, apoptosis role, 424–426, 433
Benign prostate hyperplasia (BPH), rational drug
    discovery, 442, 443
BMP-7, 402
BPH, *see* Benign prostate hyperplasia
BUB1, 410

C

Calcium,
    adenylyl cyclase regulation, 254, 255, 258, 259
    apoptosis role, 424
    ATPases, 152
    channel regulation, 181
    enzyme regulation, 180, 181
    homeostasis regulation, 149, 152, 187, 188
    intracellular and extracellular concentrations, 150
    patterns of signaling and cellular functions, 151, 152
    pools, 150
    processes under control,
        apoptosis, 182
        brain function, 184
        development, 182
        exocytosis, 182, 183
        fertilization, 182
        mitochondrial functions, 183, 184
        nuclear functions, 183, 188, 189
Calcium flux, *see also* Action potential-driven calcium
        influx; Ion channels; *specific channels*,
    blips, 174
    capacitative calcium entry,
        CRAC channels, 169, 170
        nonselective cationic channels, 170, 171
        overview, 169
        transient receptor potential protein channels, 171
    chemotaxis role in *Dictyostelium*, 104
    energetics, 137
    growth hormone receptor signal transduction, 76
    intercellular waves,
        extracellular messenger-mediated propagation,
            179, 180
        gap junctons and wave propagation, 178, 179
        pathways in mediation, 178
    intraorganelle signaling,
        endosome, 178
        Golgi apparatus, 178
        lysosome, 178
        mitochondria, 177, 178
        nucleus, 176, 177
        overview, 176
        secretory granule, 178
    ligand-gated channels, *see also* Nicotoninc
            acetylcholine receptor; *specific glutamate
            receptors,*
        agonists common with G protein-coupled
            receptors, 153, 154
        classification, 152, 153
        ligand delivery and regulation, 154
    localization of signals, 151 174, 176
    mobilizing receptors and V_m pathway,
        calcium-activated potassium channels, 172
        gonadotrophin-releasing hormone regulation, 171, 176
        M-channels, 173
        nonselective calcium activated cationic channels, 172
        store-operated calcium channels, 173, 174
    oscillators, 152
    pathways, overview, 150, 151, 175
    phototransduction in *Drosophila* regulation, 92–95
    puffs, 174
    sparks, 174
    specificity of signaling, 175, 176
    waves, 151, 174
Calcium release activated channel (CRAC), 102, 104,
        169, 170
Calmodulin,
    adenylyl cyclase regulation, 251, 255, 256, 257, 259
    avoidance behavior role, 202, 203
    binding to target proteins, 190–192
    calcium binding, 189, 190
    cameleon fluorescent indicators, 194
    cell proliferation regulation,
        Aspergillus nidulans, 202
        vertebrate cells, 201, 202
    enzyme regulation,
        calmodulin kinase I, 192, 193, 305, 306
        calmodulin kinase II, 192, 193
        calmodulin kinase IV, 193, 194
        myosin light chain kinase, 191, 192
        overview, 180, 190, 191
    inhibitors,
        antisense RNA, 196
        smooth muscle myosin light chain kinase peptide,
            development studies, 195, 196
            frog oocyte activation, 195
            sequence, 195
            transfection, 195
        W compounds, 194, 195
    learning and memory role, 203
    metaphase/anaphase transition role, 197–199
    mutant protein studies,
        enzyme-specific antagonists, 197
        insulin secretion, 196
    overview of significance, 203, 204

phototransduction in *Drosophila* regulation, 94
T lymphocyte activation role, 199, 201
cAMP, *see* Cyclic AMP
Cancer cytotoxic agents, rational drug discovery, 443
Cap32/34, 107
CAR, *see* Constitutive androstane receptor
Caspase,
    substrates, 416, 417
    types, 416
CBP, *see* CREB binding protein
CD45,
    cell adhesion regulation, 344, 345
    lymphocyte activation,
        B cells, 336, 337
        T cells, 332, 334–337
    structure, 334
Cdc2, 411–413
Cdc14, cell cycle regulation, 346, 347
Cdc25, 324–327, 329, 346, 411, 412
Cdc28, 413
Cell cycle,
    checkpoints,
        DNA replication checkpoint, 413
        functions, 408, 413
        G_1/S DNA damage checkpoint, 412, 413
        G_2/M DNA damage checkpoint, 411, 412
        morphogenesis checkpoint, 413
        spindle assembly checkpoint,
            evidence, 409
            overview, 408, 409
            receipt of signal to stop metaphase/anaphase transition, 410, 411
            signal transduction, 410
    clock, 407
    dual specificity phosphatase control,
        Cdc14, 346, 347
        Cdc25, 346, 411, 412
        overview, 345, 346
    ordering, 407, 409
c-fos, growth hormone and transcriptional activation, 77
CFTR, *see* Cystic fibrosis transmembrane regulator
cGMP, *see* Cyclic GMP
Chemotaxis, *Dictyostelium*,
    advantages of system, 99
    cyclic AMP as chemoattractant, 100
    cytoskeletal regulation,
        actin, 106, 110
        actin-binding proteins,
            ABP-50, 107
            ABP-120, 107, 110
            α-actinin, 108
            aginactin, 107
            Cap32/34, 107
            cofilin, 106, 107
            coronin, 107, 108
            ponticulin, 108
            profilin, 108
            SCAR, 108
            severin, 108
            talin, 107
            WASp, 108, 109
        myosin Is, 110
        myosin II, 109–113
    functions, 99, 100
    signaling components,
        adenylyl cyclase, activation and regulation, 102
        calcium flux, 104
        cyclic AMP receptors, 101, 102, 112, 113
        cyclic GMP, 103, 113
        G proteins, 102, 105, 106
        mitogen-activated protein kinase cascade, 105
        overview, 101
        PAKa, 106, 109
        phosphatidylinositol 3-kinase, 102, 104, 110
        phospholipase C, 104
        protein kinase B, 103, 104
        protein kinase C, 105
        Rac proteins, 105, 106
        Ras, 103
Chk1, 411
Chk2, 413
Cholera toxin, 21
Chromatin, modification by coactivators, 388
Coactivator, *see also* specific coactivators,
    chromatin modification, 388
    evidence for existence, 385, 386
    examples, 386–388
    modeling of nuclear receptor coactivators, 389
    pathology, 389
    transcriptional mediation, 388, 389
Cofilin, 106, 107
Combinatorial chemistry, rational drug discovery, 453–455
Constitutive androstane receptor (CAR),
    CARb signaling pathways, 381, 382
    features, 367, 368
Corepressor, *see also* specific corepressors,
    evidence for existence, 390, 391
    examples, 391–393
    mechanisms of action,
        covalent modification, 393, 394
        interference, 393
        sequestration, 393
    transcriptional repression, 389, 390
Coronin, 107, 108
COUP-TF, 372, 373
CRAC, *see* Calcium release activated channel
CREB, *see* Cyclic AMP responsive element-binding protein
CREB binding protein (CBP),
    differentiation and development role, 401, 402
    functional overview, 395, 396
    nuclear signaling integration, 402, 403
    recruitment,

CREB, 396–398
  nuclear receptors, 398, 399
  sequence-specific transcription factors, 396, 397
  transcriptional activation mechanisms, 399–401
CSF, *see* Cytostatic factor
Csw, mitogenic signaling, 339–341
CTLA-4, 432
Cyclic AMP (cAMP),
  accumulation pattern in cell, 262
  Dictyostelium chemoattractant, 100
  discovery, 277
  energy expenditure in nucleotide conversion, 262, 264
  overview of signaling, 249
Cyclic AMP-dependent protein kinase, *see* Protein kinase A
Cyclic AMP receptors,
  chemotaxis role in *Dictyostelium*, 101, 102, 112, 113
  overview, 277, 278
Cyclic AMP responsive element-binding protein (CREB),
  activation, 199, 201
  binding protein, *see* CREB binding protein
  long-term potentiation role, 203
Cyclic GMP (cGMP),
  chemotaxis role in *Dictyostelium*, 101, 111
  receptors, 277, 278
Cyclic GMP-dependent protein kinase, *see* Protein kinase G
Cyclic nucleotide gated channels, sensory transduction, 143, 144
Cyclic nucleotide phosphodiesterase (PDE),
  catalytic reaction, 261
  genes, 265, 266
  homeostatic regulation of cyclic nucleotide signaling and cell adaptation, 270–272
  inhibitors,
    cardiovascular disorder therapy, 273
    central nervous system regulation, 274
    development, 273
    historical perspective, 272, 273
    inflammatory disorder therapy, 273
  PDE6 photoactivation, 266, 267
  phosphorylation, 265, 269, 271
  signal compartmentalization, 272
  signaling pathway integration,
    overview, 267
    PDE1 and calcium signaling, 267, 268
    PDE2 and integration of cGMP and cAMP pathways, 268
    PDE3 role in insulin and growth factor action, 269, 269
    PDE7 regulation in T lymphocytes, 269, 270
  structure,
    domains, 264, 265
    sequence alignment of isozymes, 264
  superfamily of isoenzymes,
    classification and nomenclature, 264
    table, 263
Cystic fibrosis transmembrane regulator (CFTR), 144
Cytokines, *see also* specific cytokines,
  class I receptor superfamily, 116
  four a-helix bundle cytokines, 115, 116
  gp130-related cytokine receptors,
    cytokine functions,
      hematopoietic and immune effects, 125
      hypothalamic–pituitary–adrenal axis regulation, 125, 126
      overlapping function, 120, 126
    interleukin-6, 117, 118, 120
    interleukin-11, 118
    LIF receptor, 118
    OSM receptor, 119
    soluble receptors, 120
  immunosuppression mediation,
    immune privileged tissues, 434, 435
    oral tolerance, 433, 434
    regulation of T helper cell development, 433
  nitric oxide regulation, 222, 223
  receptor signaling pathways,
    inhibitors of JAK–STAT pathway,
      overview, 121
      PIAS, 124
      SHP-2, 122, 123
      SOCS, 123, 124
    Janus kinase, 118
    mitogen-activated protein kinase cascade, 119, 121
    STAT, 118, 119, 121
    tyrosine kinases, 121
  T helper cell expression, 430, 431
Cytostatic factor (CSF), 198

D

DAG, *see* Diacylglycerol
DAX-1, 378, 379
Diacylglycerol (DAG), protein kinase C activation, 238–240
Dictyostelium chemotaxis, *see* Chemotaxis, *Dictyostelium*
Dos, 341
DR1, 393
DRIP, 388
Drosophila phototransduction, *see* Phototransduction, *Drosophila*
Drug discovery, *see* Rational drug discovery

E

E6-AP, 387
EAI, *see* Endosomal acidic insulinase
EF-Tu, 27

Endosomal acidic insulinase (EAI), 43
Endosome, calcium signaling, 178
Epinephrine, regulation of response, 34, 35
ERK2, 102, 105
ERR, see Estrogen-related receptor
Estrogen receptor,
    coactivators, 358, 360
    corepressors, 360, 361
    ligand role in signal transduction, 357, 358
    selective estrogen receptor modulator, 354, 357, 358
    subtypes, 356
Estrogen-related receptor (ERR), 374
Exocytosis, calcium control, 182, 183

F

FADD, 418, 419
Farnesoid X receptor (FXR),
    features, 368, 371
    signaling pathways, 379, 380
Fas, 419, 423, 432, 433
F-box proteins, phosphorylation, 304
Fertilization, calcium control, 182
Follicle-stimulating hormone receptor, 3, 4, 8–12
14-3-3, phosphorylative regulation, 302, 303
Formulation, drugs, 458, 459
FTZ-F1, 374 375
FXR, see Farnesoid X receptor

G

G protein, see also Phototransduction, Drosophila;
    specific proteins,
    activation of receptors, 7, 8, 18, 19, 147
    adenylyl cyclase regulation,
        activation, 250, 251
        Gbg, 255
        $G_i\alpha/G_o\alpha$, 253, 254, 260
    chemotaxis role in Dictyostelium, 102, 105, 106
    classes, 17, 19
    crosstalk on small protein signaling pathways, 32
    crystal structures, 28, 30
    diseases and mutation, 30
    Ga activation specificity,
        G_{12}, 37–39
        G_i, 36
        G_o, 36, 37
        G_q, 35, 36
        G_s,
    guanine nucleotide binding and hydrolysis,
        overview, 18, 19
        regulators, 28, 29, 31–34, 39
    membrane association, 16, 17
    palmitoylation, 17
    phosphorylation, 17, 18
    receptor coupling domain, 11–14
    regulatory mechanisms, 23–25

    structure–function relationships of subunits,
        Ga, 19–21, 35
        Gbg, 21–23
    subunit features, 15, 16, 32
G protein-coupled receptors, see also specific receptors,
    family members, 3–5
    glycosylation, 9
    mRNA splicing variants, 9, 10
    palmitoylation, 8, 9
    regulation,
        desensitization, 14, 15
        G protein specificity, 24, 25, 34–39
        receptor conformation and signaling, 14
    structural overview, 3, 5
    structure–function relationships,
        G protein coupling domain, 11–14
        general features, 6–10
        ligand binding domain, 10, 11
Gab2, 341
Gap junction, calcium wave propagation, 178, 179
Gastrulation, G protein signaling, 37–40
GCN5, 400
GCNF, see Germ cell nuclear factor
Genomic resources, rational drug discovery, 444, 445
Germ cell nuclear factor (GCNF), 373
GH, see Growth hormone
GHBP, see Growth hormone binding protein
GHR, see Growth hormone receptor
Glucocorticoid receptor,
    apoptosis role, 420, 421
    subtypes, 356, 357
Glutamate receptors, see α-Amino-3-hydroxy-5-methyl-
    4-isoxazole propionic acid; Kainate receptor; N-
    Methyl-D-aspartate receptor
G_M, 317, 318
Golgi apparatus, calcium signaling, 148
Gonadotrophin-releasing hormone, analogs, 449, 451
GRIP-1, 387
Groucho, 393
Growth hormone (GH),
    clinical disorders, 58
    hypothalamo–pituitary axis regulation, 56, 57
    insulin-like growth factor-1 gene induction, 78
    metabolic actions, 56
    overview of research, 55, 56
    resistance and Laron syndrome, 58, 59
    sex differences in plasma profiles, 57
    somatogenic mechanisms, 56
    structure, 59
Growth hormone binding protein (GHBP),
    formation mechanisms, 63
    stoichiometry of growth hormone complexes, 63
Growth hormone receptor (GHR),
    dimerization, 61–63
    down-regulation,
        internalization, 81

nonproteolytic processing and internalization, 82, 83
ubiquitination and proteolysis, 82
homology between species, 59, 60
posttranslational modification, 60
signal transduction,
    calcium flux signaling, 76
    c-fos gene transcription activation, 77
    ERK, 76
    insulin receptor substrates, 77, 78
    JAK2 kinase,
        activation, 64, 65
        phosphorylation sites, 66
        receptor interactions, 65, 66
        STAT activation specificity, 69–71
    mitogen-activated protein kinase activation, 75, 76
    modulators and attenuators of signaling,
        protein tyrosine phosphatases, 78, 79
        SHP-1, 79
        SHP-2, 79, 80
        SOCS proteins, 80, 81
    overview, 63–65
    protein kinase C activation, 76, 77
    STAT,
        dimerization, 71, 72
        gene activation, 67–69
        specificity of STAT type activation, 69–71
        STAT5 functions, 69
        STAT5b in sexually dimorphic responses, 72–74
        STAT5b responses to growth hormone pulsatility, 74, 75, 78
        structures, 67
        types in activation, 66, 67
structure, 60, 61
variants, 61
GTPase, *see* G protein

H

HAT, *see* Histone acetyltransferase
Hepatocyte nuclear factor-4 (HNF-4), 373, 374
High-throughput screening, rational drug discovery, 447
HIPK, *see* Homeodomain interacting protein kinase
Histone acetyltransferase (HAT), 400, 401
HNF-4, *see* Hepatocyte nuclear factor-4
Homeodomain interacting protein kinase (HIPK), 393
Hypertension,
    nitric oxide role, 225, 226
    rational drug discovery, 442

I

IGF-1, *see* Insulin-like growth factor-1
IL-2, *see* Interleukin-2
IL-6, *see* Interleukin-6
IL-11, *see* Interleukin-11

Immune privileged tissues, 434, 435
INAD, phototransduction signalplex in *Drosophila*, 94–96
Inhibitor-1, 315, 316
Inhibitor-2, 316, 317
Inositol phosphates,
    calcium wave propagation, 179
    phototransduction in *Drosophila*, 91
Inositol 1,4,5-trisphosphate receptor channels,
    endoplasmic reticulum calcium excitability, 167
    overview of calcium signaling pathway, 166
    pharmacology, 167
    regulation, 166, 167
    structure, 166
Insulin,
    calmodulin mutant effects on secretion, 196
    endosomal degradation, 42, 43
Insulin-like growth factor-1 (IGF-1),
    induction by growth hormone, 78
    somatogenic mechanisms, 56
Insulin receptor, tissue distribution, 41
Insulin receptor kinase (IRK),
    activity, 42
    dephosphorylation by phosphotyrosine phosphatase,
        insulin action effects, 46, 47
        pervanadate effects, 44–46
        therapeutic targeting, 49
        types and characterization of phosphatases, 47–49
    diabetes role, 42, 49, 50
    endosomal acidification and deactivation, 49, 50
    internalization, 42
    structure, 41
Insulin receptor substrate (IRS),
    growth hormone receptor signal transduction, 77, 78
    interleukin-6 signal transduction, 123
Interleukin-2 (IL-2), gene regulation, 199
Interleukin-6 (IL-6),
    functions,
        hematopoietic and immune effects, 125
        hypothalamic–pituitary–adrenal axis regulation, 125, 126
        overlapping function with other gp130-related cytokines, 121, 126
    receptor, 117, 118, 120
    signaling pathways,
        inhibitors of JAK–STAT pathway,
            overview, 123
            PIAS, 125
            SHP-2, 123, 124
            SOCS, 123, 124
        Janus kinase, 120
        mitogen-activated protein kinase cascade, 122, 123
        STAT, 120, 122, 123
        tyrosine kinases, 123
Interleukin-11 (IL-11),
    functions,
        hematopoietic and immune effects, 125

Index 467

   hypothalamic–pituitary–adrenal axis regulation, 125, 126
   overlapping function with other gp130-related cytokines, 122, 126
  receptor, 118
  signaling pathways,
   inhibitors of JAK–STAT pathway,
    overview, 123
    PIAS, 125
    SHP-2, 123
    SOCS, 123, 124
   Janus kinase, 120
   mitogen-activated protein kinase cascade, 120, 122
   STAT, 120, 122, 123
   tyrosine kinases, 123
Ion channels, *see also specific channels*,
  action potential-driven calcium influx modulation,
   inwardly rectifying potassium channels, 165
   ligand-gated channels, 165, 166
   overview, 163
   sodium channels, 164, 165
   voltage-gated calcium channels, 163, 164
  calcium regulation, 181, 182
  cell growth and proliferation role, 143
  evolution,
   eukaryotic five-subunit channels, 147
   eukaryotic four-subunit channels, 147
   glutamate-gated channels, 147
   miscellaneous channels, 147, 148
   overview, 146
   prokaryotes, 146, 147
  gating, 137, 138
  intercellular communication role, 146
  ligand sensing, 140, 141
  metabolic sensing, 142, 143, 145, 146
  modulation,
   clustering with regulatory elements, 145
   G protein activation, 145
   phospholipid interactions, 145, 146
   phosphorylation, 144, 145
  overview of function, 135, 136
  permeation,
   energy for transfer, 137
   pore size, 136
   selectivity mechanisms, 136, 137
  resting membrane potential maintenance by potassium channels, 142
  secretion regulation by calcium channels, 142
  sensory transduction, 143, 144
  stretch sensing, 142
  structure,
   membrane topology, 138
   pore, 139, 140
   subunits, 138, 139
  voltage sensing, 141

 volume regulation, 143
IRK, *see* Insulin receptor kinase
IRS, *see* Insulin receptor substrate

J

Jak, *see* Janus kinase
JAK2 kinase, growth hormone receptor signal transduction,
  activation, 64, 65
  phosphorylation sites, 66
  receptor interactions, 65, 66
  STAT activation specificity, 69–71
Janus kinase (Jak),
  gp130-related cytokine signaling, 120
  inhibitors of JAK–STAT pathway,
   overview, 123
   PIAS, 125
   SHP-2, 123, 124
   SOCS, 124, 125
JIP1, regulation of kinases, 309
JNK kinase, 23

K

Kainate receptor,
  calcium signaling, 157
  subunits, 156
KAP-1, 392, 393
KRAB corepressors, 390, 391
KRIP-1, 393

L

Laron syndrome, 58, 59, 74
Leuprolide, 447
LIF,
  functions,
   hematopoietic and immune effects, 125
   hypothalamic–pituitary–adrenal axis regulation, 125, 126
   overlapping function with other gp130-related cytokines, 122, 126
  receptor, 118
  signaling pathways,
   inhibitors of JAK–STAT pathway,
    overview, 123
    PIAS, 125
    SHP-2, 123, 124
    SOCS, 124, 125
   Janus kinase, 120
   mitogen-activated protein kinase cascade, 122, 123
   STAT, 120, 122, 123
   tyrosine kinases, 123
Liver X receptor (LXR),

features, 371, 372
   signaling pathways, 380, 381
Long-term potentiation (LTP),
   calmodulin role, 203
   nitric oxide role, 224
LTP, see Long-term potentiation
Luteinizing hormone receptor, 3, 4, 8–12
LXR, see Liver X receptor
Lysosome, calcium signaling, 178

M

MAD1, 410
MAD2, 410
Magnesium, ion channel blocking, 138
MAPK, see Mitogen-activated protein kinase
MARCKS, phosphorylation, 216
Maturation promoting factor (MPF), 198
M-channels, 173
Medicinal chemistry, rational drug discovery, 451, 452
MEK kinase, 23
Metabotropic glutamate receptors, 12
N-Methyl-D-aspartate (NMDA) receptor,
   calcium signaling, 159
   nitric oxide regulation, 223, 224
   pathology, 157, 158
   subunits, 156, 157
Mitochondria,
   apoptosis role, 422–424
   calcium control of functions, 183, 184
   calcium signaling, 177, 178
Mitogen-activated protein kinase (MAPK),
   apoptosis role, 417, 418
   chemotaxis role in *Dictyostelium*, 105
   dephosphorylation, 341, 342
   gp130-related cytokine signaling, 119, 121
   growth hormone receptor signal transduction, 75, 76
   phospholipase A_2 activation, 232
   Ras activation, 23
   scaffold protein regulation, 311, 312
MLCK, see Myosin light chain kinase
MLCKA, 110
MPF, see Maturation promoting factor
Muscarinic receptors, 12, 13
Myosin Is, cytoskeletal regulation in *Dictyostelium* chemoattraction, 110
Myosin II,
   cytoskeletal regulation in *Dictyostelium* chemoattraction, 109–113
   phosphorylative regulation, 303, 304
Myosin light chain kinase (MLCK),
   calmodulin regulation, 193, 194
   phosphorylative regulation, 304
   smooth muscle myosin light chain kinase peptide as calmodulin antagonist,

   development studies, 197, 198
   frog oocyte activation, 197
   sequence, 197
   transfection, 197

N

NAB corepressors, 393
nAChR, see Nicotoninc acetylcholine receptor
NCoA-1, 386, 387
N-CoR, 391
Nerve growth factor induced gene B (NGFI-B), 375, 376
NF-kB, see Nuclear factor-kB
NGFI-B, see Nerve growth factor induced gene B
Nicotoninc acetylcholine receptor (nAChR),
   functional diversity, 156
   gating, 138
   physiology and pathology, 156
   pore, 139
   structure, 153, 154, 156
   subunits, 154, 156
NINAC, 94
NIPP-1, 317
Nitric oxide (NO),
   background of study, 219
   functions,
      cardiovascular system, 225, 226
      immune system, 225
      nervous system, 223–225
   reactivity, 219, 220
   targets,
      cytokines, 222, 223
      DNA, 222
      enzymes, 222
      receptors, 223
      structural proteins, 223
      transcription factors, 222
   therapeutics, 226, 227
Nitric oxide synthase (NOS),
   catalytic reaction, 219
   domains, 221
   regulation,
      dimerization, 222
      isoforms, 220, 221
      phosphorylation, 221, 300, 301
      regulatory proteins, 221
      transcript stability, 221, 222
NMDA receptor, see *N*-Methyl-D-aspartate receptor
NMR, see Nuclear magnetic resonance
NO, see Nitric oxide
NOS, see Nitric oxide synthase
NSD1, 392
Nuclear factor of activated T cells (NFATc), 199
Nuclear factor-kB (NF-kB), 222, 420
Nuclear magnetic resonance (NMR), rational drug

discovery, 447, 448, 452
Nucleus,
  calcium control of functions, 183, 188, 189
  calcium signaling, 176, 177
  phosphorylation of proteins and translocation, 303, 304
Nur77, 421, 422

O

Oral bioavailability, rational drug discovery, 457, 458
Oral tolerance, 433, 434
Orphan nuclear receptors,
  atypical receptors,
    DAX-1, 376, 377
    small heterodimer partner, 377
  classification, 367
  domains,
    amino terminal domain, 365, 366
    carboxy terminal domain, 365
    dimerization interface, 365, 366
    DNA binding domain, 365
    ligand binding domain, 365, 366
  drug targeting,
    overview, 382
    peroxisome proliferator activated receptor, 383, 384
    retinoid X receptor, 382
  homodimers and functions,
    COUP-TF, 372, 373
    germ cell nuclear factor, 373
    hepatocyte nuclear factor-4, 373, 374
  intracellular receptor superfamily, 363, 364
  ligand discovery techniques, 377–379
  monomers and functions,
    estrogen-related receptors, 374
    FTZ-F1, 374, 375
    nerve growth factor induced gene B, 375, 376
    retinoic acid-related orphan receptor, 376
  mutant phenotypes, 371, 384
  prospects for research, 384
  retinoid X receptor heterodimerization,
    constitutive androstane receptor, 367, 368
    farnesoid X receptor, 368, 371
    ligand response, 366, 367
    liver X receptor, 371, 372
    pregnane X receptor, 372
    RIP14, 371
  signaling pathways,
    CARb, 381, 382
    farnesoid X receptor, 379, 380
    liver X receptor, 380, 381
    pregnane X receptor, 381
    RIP14, 379, 380
    SF-1, 380, 381
  types in vertebrates, overview, 367, 369, 370
OSM,
  receptor, 119
  functions,
    hematopoietic and immune effects, 125
    hypothalamic–pituitary–adrenal axis regulation, 125, 126
    overlapping function with other gp130-related cytokines, 122, 126
  signaling pathways,
    inhibitors of JAK–STAT pathway,
      overview, 123
      PIAS, 125
      SHP-2, 124, 125
      SOCS, 124, 125
    Janus kinase, 120
    mitogen-activated protein kinase cascade, 122, 123
    STAT, 120, 122, 123
    tyrosine kinases, 123

P

P2X channels,
  inactivation properties, 158, 159
  overview, 153, 158
  physiology, 159
  structural and functional specificity, 158
p21, 223, 412, 413
p53
  acetylation, 401
  cell cycle control, 412, 413
  nitric oxide regulation, 221
  phosphorylation, 303
$p70^{s6k}$, phosphorylative regulation, 306–308
p300
  differentiation and development role, 401, 402
  functional overview, 395, 396
  nuclear signaling integration, 402, 403
  recruitment,
    CREB, 396–398
    nuclear receptors, 398, 399
    sequence-specific transcription factors, 396, 397
  transcriptional activation mechanisms, 399–401
P loop, ion channels, 139, 140, 146, 147
PAKa, 106, 109
Parallel chemistry, rational drug discovery, 454, 455
Parathyroid hormone receptor, 187
Parkinson's disease (PD), nitric oxide role, 225
PARS, *see* Poly(ADP-ribose) synthetase
p/CAF, 400, 401
PD, *see* Parkinson's disease
PDE, *see* Cyclic nucleotide phosphodiesterase
PDZ module, 95, 96, 145, 303
Peptidomimetics, rational drug discovery, 448, 449, 451
Peroxisome proliferator activated receptor (PPAR),
  apoptosis role, 422
  drug targeting, 383, 384

Pertussis toxin, 21
Pervanadate, insulin receptor kinase activation, 44–46
Pharmacokinetics, rational drug discovery, 458
Phorbol 12-myristate 13-acetate (PMA), discovery and structure, 205, 206
Phosducin, 24, 25
Phosphatidyl inositol, ion channel regulation, 145, 146
Phosphatidylinositol 3-kinase, chemotaxis role in *Dictyostelium*, 102, 104, 110
Phosphodiesterase, *see* Cyclic nucleotide phosphodiesterase
6-Phosphofructo-2-kinase, phosphorylative regulation, 300
Phospholipase A_1 (PLA_1),
    function, 232
    structure, 232
    substrate cleavage site, 227
Phospholipase A_2 (PLA_2),
    cytoplasmic isoforms,
        calcium-independent enzyme, 232
        catalysis, 230
        function, 232
        lysophospholipase activity, 231, 232
        regulation, 230, 231
        transcriptional regulation, 232
        types, 230
    low molecular mass isoforms, 229, 230
    substrate cleavage site, 229
Phospholipase C (PLC),
    calcium regulation, 188
    chemotaxis role in *Dictyostelium*, 104
    domains, 233
    functions,
        calcium mobilization, 237, 238
        protein kinase C activation, 240–242
    G protein regulation of PLCb,
        activation, 36, 37, 86
        G_i/G_o, 235, 236
        G_q, 233–235
    isozymes, 232, 233
    PLCd properties, 237
    substrate cleavage site, 229
    substrate specificity, 232, 237
    tyrosine kinase regulation of PLCg, 236, 237
Phospholipase D (PLD),
    cloning, 240
    functions, 243, 244
    isozyme properties, 240, 241
    phosphorylation, 243
    PLD1 regulation,
        ADP-ribosylation factor, 242, 243
        protein kinase C regulation, 241, 242
        Rho, 240
    PLD2 regulation, 243
    substrate cleavage site, 229
    substrate specificity, 241

Phototransduction, *Drosophila*,
    calcium in termination, 93, 94
    comparison with vertebrates, 86, 97
    electroretinograms, 86
    eye anatomy, 86
    genes, 86, 89
    genetic approaches to study, 86, 88
    INAD signalplex, 94–96
    inositol phosphate signaling, 91
    kinetics of cascade, 85, 86
    phospholipase C-dependent ion channels, 91–93, 97
    prolonged depolarization, 90, 91
    rhodopsin,
        activation, 89–91
        deactivation, 91, 94
        expression by cell type, 88, 89
        types, 88, 89
PiaA, *see* Pianissimo
Pianissimo (PiaA), 102
PIAS, inhibition of JAK–STAT pathway, 125
PKA, *see* Protein kinase A
PKB, *see* Protein kinase B
PKC, *see* Protein kinase C
PKG, *see* Protein kinase G
PLA_1, *see* Phospholipase A_1
PLA_2, *see* Phospholipase A_2
PLC, *see* Phospholipase C
PLD, *see* Phospholipase D
PMA, *see* Phorbol 12-myristate 13-acetate
Poly(ADP-ribose) synthetase (PARS), nitric oxide regulation, 224
Ponticulin, 108
Potassium channel, *see* Ion channels
PPAR, *see* Peroxisome proliferator activated receptor
Pregnane X receptor (PXR),
    features, 372
    signaling pathways, 381
Profilin, 108
Progesterone receptor, isoforms, 354, 356
Protein kinase A (PKA),
    activation, 291
    active site structure, 288, 289
    autoinhibition, 282, 283
    autophosphorylation, 283, 284
    catalytic reaction, 278, 289, 290
    catalytic subunits, 280
    cyclic nucleotide binding sites,
        fast and slow binding sites, 285, 286
        locations, 284, 285
        specificity of binding,
            analogs, 287, 288
            cyclic nucleotides, 287
        structural features, 286, 287
    dimerization, 281, 282
    domains, 281

feedback and feedforward control, 293, 294
inhibitors,
　cyclic nucleotide analogs, 291, 292
　　protein kinase inhibitor and peptides, 293
knockout mouse phenotypes, 294, 295
phosphorylative regulation, 278, 279, 282–284
regulatory subunits, 279–282
structure, 279
subcellular localization,
　anchoring mechanisms, 292
　translocation, 290, 291
　　substrate specificity, 278, 292, 291
Protein kinase B (PKB),
　chemotaxis role in *Dictyostelium*, 103, 104
　phosphorylative regulation, 306, 307, 309
Protein kinase C (PKC),
　chemotaxis role in *Dictyostelium*, 105
　discovery, 206, 207
　growth hormone receptor signal transduction, 76, 77
　inhibitors, 216, 217
　isozyme function examples, 216
　overview of regulation, 217
　phospholipase activation,
　　PLA$_2$, 230, 231
　　PLC, 238–240
　PLD1 regulation, 241, 242
　regulation,
　　anchoring proteins, 214, 215
　　calcium, 180, 181, 212
　　diacylglycerol, 210–212
　　membrane structure, 212
　　miscellaneous lipids and lipid products, 212
　　phorbol esters, 210–212
　　phosphatidylserine, 210
　　phosphorylation,
　　　activation loop switch, 213, 214
　　　autophosphorylation, 213, 304, 305
　　　C-terminal switch, 214
　regulatory moiety,
　　kinase domain, 209, 210
　　membrane-targeting modules,
　　　C1 domain, 208
　　　C2 domain, 208, 209
　　pseudosubstrate motif, 207, 208
　subclasses, 207
　substrates, 215, 216
　translocation, 207
Protein kinase G (PKG),
　activation, 291
　active site structure, 288, 289
　autoinhibition, 282, 283
　autophosphorylation, 283, 284
　catalytic reaction, 278, 289, 290
　cyclic nucleotide binding sites,
　　fast and slow binding sites, 285, 286

locations, 284, 285
specificity of binding,
　analogs, 287, 288
　cyclic nucleotides, 287
structural features, 286, 287
dimerization, 282
domains, 281
feedback and feedforward control, 293, 294
inhibitors, 291, 292
isoform structures and functions, 280, 281
knockout mouse phenotypes, 294, 295
phosphorylative regulation, 278, 279, 282–284
regulatory subunits, 282
subcellular localization,
　anchoring mechanisms, 292
　translocation, 292, 293
substrate specificity, 278, 290, 291
Protein serine/threonine kinase, *see also specific kinases*,
　catalytic activity regulation of targets,
　　cell motility and contractility regulation, 301, 302
　　metabolic enzymes, 299–301
　catalytic reaction, 297–299
　distribution in nature, 297
　evolution, 310
　overview of protein target regulation, 299
　protein–protein interaction regulation, 302, 303
　protein stability regulation, 304
　regulation,
　　intrasteric regulation, 304–306
　　overview, 302
　　phosphorylative regulation, 306–309
　　scaffold proteins, 309, 310
　subcellular localization regulation of proteins, 303, 308
Protein serine/threonine phosphatase,
　comparison of types,
　　catalytic subunits, 313, 314
　　regulatory subunits, 314
　　sequence homology, 313
　　structure, 312
　　substrate specificity, 312
　covalent modification of subunits, 319, 320
　history of study, 311, 312
　inhibitors and coordination of hormone signals, 315–317
　pathophysiology, 320
　prospects for study, 320, 321
　regulation, overview, 312, 314, 315
　subcellular targeting and function, 317–319
Protein tyrosine phosphatase (PTP), *see also specific phosphatases*,
　catalytic mechanism, 324, 328, 329
　classification,
　　Cdc25 phosphatases, 324–326
　　dual specificity phosphatases, 324
　　low molecular weight protein tyrosine phosphatases, 325

nonreceptor protein tyrosine phosphatases, 324
PTEN, 326
receptor protein tyrosine phosphatases, 324
STYX, 326
functions,
   CD45 in lymphocyte activation, 332, 334–337
   cell adhesion regulation,
      CD45, 344, 345
      low molecular weight protein tyrosine phosphatases, 345
      nonreceptor protein tyrosine phosphatases, 345
      receptor protein tyrosine phosphatases, 342–344
   cell cycle control by dual specificity phosphatases,
      Cdc14, 346, 347
      Cdc25, 346
      overview, 345, 346
   Csw in mitogenic signaling, 339–341
   mitogen-activated protein kinase dephosphorylation, 341, 342
   receptor tyrosine kinase regulation,
      low molecular weight protein tyrosine phosphatases, 342
      nonreceptor protein tyrosine phosphatases, 342
   SHP-1 and hematopoietic cell signaling, 337–339
   SHP-2 in mitogenic signaling, 339–341
growth hormone signaling modulation, 78, 79
history of study, 323, 324
regulation,
   autoinhibition, 331, 332
   overview, 329, 330
   phosphorylation, 332
   protein inhibitors, 332
   receptor protein tyrosine phosphatases, 332–334
   subcellular targeting, 330, 331
structure,
   Cdc25, 329
   crystal structures, 328–330
   low molecular weight protein tyrosine phosphatase, 329
   structural organization, 326, 327
substrate specificity, 331
therapeutic targeting, 349
PTEN, 328
PTP, *see* Protein tyrosine phosphatase
Purinoreceptors, *see* P2X channels
PXR, *see* Pregnane X receptor
Pyruvate kinase, phosphorylative regulation, 302

Q

QSAR, *see* Quantitative structure–activity relationship
Quantitative structure–activity relationship (QSAR), 454

R

Rac proteins, 105, 106
RACKS, 215, 239, 240
Rad proteins, phosphorylative regulation, 303
Raf-1, 23
Ras,
   chemotaxis role in *Dictyostelium*, 103
   mutation, 29
   regulators, 28, 29, 31
   signal transduction mediation,
      lipid signaling pathway, 30
      mitogen-activated protein kinase, 23, 30, 122
      vesicle transport pathway, 30
   superfamily of GTPases, 27, 28
Rational drug discovery,
   animal testing, 456, 457
   criteria for drugs, 441
   examples,
      antihypertension drugs, 442
      antiviral agents, 444, 448, 452
      benign prostate hyperplasia, 442, 443
      cancer cytotoxic agents, 443
   formulation, 458, 459
   genomic resources, 444, 445
   historical perspective, 439, 441
   lead identification,
      assay development, 445
      high-throughput screening, 447
      ligands, 445, 446
      literature or competitor's compounds, 446, 447
      peptidomimetics, 448, 449, 451
      SAR by NMR, 447, 448
      verification of hits, 451
   lead optimization,
      combinatorial chemistry, 453–455
      cost of synthesis, 455, 456
      medicinal chemistry, 451, 452
      parallel chemistry, 454, 455
      structure-based drug design, 452, 453
   multidisciplinary nature, 441
   oral bioavailability, 457, 458
   pharmacokinetics, 458
   target selection,
      overview, 441, 442
      subtype specificity, 443
   toxicology, 459
RegA, 102
Retinal, rhodopsin regulation, 89, 90
Retinoic acid-related orphan receptor (ROR), 376
Retinoid X receptor (RXR),
   drug targeting, 382, 383
   heterodimerization,
      constitutive androstane receptor, 367, 368
      farnesoid X receptor, 368, 371
      ligand response, 366, 367
      liver X receptor, 371, 372
      pregnane X receptor, 372
      RIP14, 371

subtypes, 382
RGS proteins, 19, 20, 33, 34, 36, 39
Rho, 38, 39, 242
Rhodopsin,
   Drosophila,
      activation, 89–91
      deactivation, 91, 94
      expression by cell type, 88, 89
      types, 88, 89
   structure and features, 3, 4, 6–8, 10, 12, 13
RICKs, 240
RIP13, 391
RIP14
   features, 371
   signaling pathways, 379, 380
ROCK, 318
ROR, *see* Retinoic acid-related orphan receptor
RPD-1, 392
RPF-1, 387
RXR, *see* Retinoid X receptor
Ryanodine receptor (RyR),
   global signals, 174, 175
   overview of calcium signaling pathway, 167
   pathology, 169
   pharmacology, 168, 169
   regulation, 167, 168
   structure, 167
RyR, *see* Ryanodine receptor

S

SCAR, 108
Secretin receptor, 4–6, 11
Serine/threonine kinase, *see* Protein serine/threonine kinase
Serine/threonine phosphatase, *see* Protein serine/threonine phosphatase
Serotonin receptors, 12
Severin, 108
SF-1, 382, 383
SHP, *see* Small heterodimer partner
SHP-1,
   growth hormone signaling modulation, 79
   hematopoietic cell signaling, 339–341
   subcellular targeting, 332
SHP-2,
   growth hormone signaling modulation, 79, 80
   inhibition of JAK–STAT pathway, 124, 125
   mitogenic signaling, 341–343
   subcellular targeting, 332, 333
SIN3, 394
Small heterodimer partner (SHP), 379
SMRT, 393, 394
SOCCs, *see* Store-operated calcium channels
SOCS proteins,
   growth hormone signaling modulation, 80, 81

   inhibition of JAK–STAT pathway, 125, 126
Sodium channel, *see* Ion channels
SRC, *see* Steroid receptor coactivator
SRG3, 422
SSN6, 394
STAT,
   gp130-related cytokine signaling, 120, 121, 123
   growth hormone receptor signal transduction,
      dimerization, 71, 72
      gene activation, 67–69
      specificity of STAT type activation, 69–71
      STAT5 functions, 69
      STAT5b in sexually dimorphic responses, 72–74
      STAT5b responses to growth hormone pulsatility, 74, 75, 78
      structures, 67
      types in activation, 66, 67
   inhibitors of JAK–STAT pathway,
      overview, 123
      PIAS, 126
      SHP-2, 124, 125
      SOCS, 125, 126
Ste5p, regulation of kinases, 311
Steroid hormone receptor, *see also specific receptors*,
   coactivators, 360, 362
   corepressors, 362, 363
   domains,
      amino terminal region, 355
      DNA-binding domain, 353–355
      hinge regions, 353
      ligand-binding domain, 351
      organization, 351, 352
   ligand role in signal transduction, 357, 358
   models of action, 353, 354, 361
Steroid receptor coactivator (SRC),
   SRC-1, 386–389
   SRC-2, 387–389
   SRC-3, 387
STICKs, 215
Store-operated calcium channels (SOCCs), 173, 174
STYX, 326
Sug1, 387
Sulfonylurea receptor, 142, 143, 145
SUN-CoR, 392
SWI/SNF, 388
SXR, 381

T

T cell,
   activation mechanisms, 429, 430
   adaptive immunity, 429
   antigen-presenting cell interactions, 429, 430
   autoimmunity mechanisms,
      innate immunity, 435, 436

lymphokine production loss, 435
regulating factor activity loss, 435
CD45 role in activation, 332, 334–337
helper cell types and cytokine expression, 430, 431
immunosuppression,
    active cell death, 432, 433
    anergy, 432
    cytokine mediated immunosuppression,
        immune privileged tissues, 434, 435
        oral tolerance, 433, 434
        regulation of T helper cell development, 433
    need for termination of activity, 431
    overview of mechanisms, 431, 432
    passive cell death, 433
    therapies using mechanisms, 436
PDE7 regulation, 269, 270
Talin, 107
Tamoxifen, 354, 357, 358
Theophylline, 273
TIF1, 388, 392
TIF2, 387
Toxicology, rational drug discovery, 459
TRAC2, 391, 392
TRADD, 419–422
TRAIL, 418
TRAMP, 418
Transcription factors, overview, 385

Transducin, 7, 86
TRAP, 390
Trip-1, 387
TRP, 91–93, 97, 171
TRPL, 91–93, 97, 171, 188
Tumor necrosis factor receptors, 418, 419
TUP1, 394

V

Vasointestinal polypeptide receptor, 4–6, 11
VGCCs, see Voltage-gated calcium channels
Visual transduction, see Phototransduction, *Drosophila*
Voltage-gated calcium channels (VGCCs),
    action potential-driven calcium influx,
        function, 159, 162
        modulation of influx, 163, 164
        spontaneous pacemaking and calcium transients, 162, 163
    cell distribution, 159
    classification, 159, 158
    subunits, 158, 159

W

WASp, 108, 109
WW domain, 303